Advances in Intelligent Systems and Computing

Volume 1334

The series “Advances in Intelligent Systems and Computing” contains publications on theory, applications, and design methods of Intelligent Systems and Intelligent Computing. Virtually all disciplines such as engineering, natural sciences, computer and information science, ICT, economics, business, e-commerce, environment, healthcare, life science are covered. The list of topics spans all the areas of modern intelligent systems and computing such as: computational intelligence, soft computing including neural networks, fuzzy systems, evolutionary computing and the fusion of these paradigms, social intelligence, ambient intelligence, computational neuroscience, artificial life, virtual worlds and society, cognitive science and systems, Perception and Vision, DNA and immune based systems, self-organizing and adaptive systems, e-Learning and teaching, human-centered and human-centric computing, recommender systems, intelligent control, robotics and mechatronics including human-machine teaming, knowledge-based paradigms, learning paradigms, machine ethics, intelligent data analysis, knowledge management, intelligent agents, intelligent decision making and support, intelligent network security, trust management, interactive entertainment, Web intelligence and multimedia.

The publications within “Advances in Intelligent Systems and Computing” are primarily proceedings of important conferences, symposia and congresses. They cover significant recent developments in the field, both of a foundational and applicable character. An important characteristic feature of the series is the short publication time and world-wide distribution. This permits a rapid and broad dissemination of research results.

Indexed by DBLP, EI Compendex, INSPEC, WTI Frankfurt eG, zbMATH, Japanese Science and Technology Agency (JST).

All books published in the series are submitted for consideration in Web of Science.

More information about this series at http://www.springer.com/series/11156

Harish Sharma · Mukesh Saraswat ·
Anupam Yadav · Joong Hoon Kim ·
Jagdish Chand Bansal
Editors

Congress on Intelligent Systems

Proceedings of CIS 2020, Volume 1

Springer

Editors
Harish Sharma
Department of Computer Science
and Engineering
Rajasthan Technical University
Kota, Rajasthan, India

Mukesh Saraswat
Department of Computer Science
and Engineering
Jaypee Institute of Information Technology
Noida, Uttar Pradesh, India

Anupam Yadav
National Institute of Technology
Jalandhar, Punjab, India

Joong Hoon Kim
Korea University
Seoul, Korea (Republic of)

Jagdish Chand Bansal
South Asian University
New Delhi, Delhi, India

ISSN 2194-5357 ISSN 2194-5365 (electronic)
Advances in Intelligent Systems and Computing
ISBN 978-981-33-6980-1 ISBN 978-981-33-6981-8 (eBook)
https://doi.org/10.1007/978-981-33-6981-8

This Springer imprint is published by the registered company Springer Nature Singapore Pte Ltd.
The registered company address is: 152 Beach Road, #21-01/04 Gateway East, Singapore 189721, Singapore

Preface

Congress on Intelligent Systems (CIS 2020) is a maiden attempt to bring the researchers, academicians, industries, and government personnel together to share and discuss the various aspects of intelligent systems. It was organized virtually during September 05–06, 2020. The congress is a brainchild of Soft Computing Research Society which is a non-profitable society. The theme of the congress was intelligent systems, machine vision, robotics, and computational intelligence. The conference witnessed multiple eminent keynote speakers from academia and industry from all over the world along with the presentation of accepted peer-reviewed articles.

This volume is a curated collection of the articles which are presented during the conference. This book focuses on the current and recent developments in intelligent systems on web-based evaluation models, decision-making process, anomaly detection, neural networks, classifiers, power system control, support vector machines, machine learning techniques, IoT applications, data protection techniques, error detection, Twitter data analysis, big data, data mining, text detection schemes, crop management system, intelligent control system, encryption in data communication, air quality monitoring system, language recognitions system, clustering techniques, modular robotics, and fuzzy classifiers.

In conclusion, the edited book comprises papers on diverse aspects of intelligent systems with many real-life applications.

New Delhi, India
October/November 2020

Harish Sharma
Mukesh Saraswat
Anupam Yadav
Joong Hoon Kim
Jagdish Chand Bansal

Contents

About the Editors

Harish Sharma is Associate Professor at Rajasthan Technical University, Kota, in the Department of Computer Science & Engineering. He has worked at Vardhaman Mahaveer Open University Kota, and Government Engineering College Jhalawar. He received his B.Tech. and M.Tech. degree in Computer Engineering from Government Engineering College, Kota, and Rajasthan Technical University, Kota, in 2003 and 2009, respectively. He obtained his Ph.D. from ABV-Indian Institute of Information Technology and Management, Gwalior, India.

He is the secretary and one of the founder members of Soft Computing Research Society of India. He is a lifetime member of Cryptology Research Society of India, ISI, Kolkata. He is Associate Editor of "International Journal of Swarm Intelligence (IJSI)" published by Inderscience. He has also edited special issues of the many reputed journals like "Memetic Computing", "Journal of Experimental and Theoretical Artificial Intelligence", and "Evolutionary Intelligence". His primary area of interest is nature-inspired optimization techniques. He has contributed to more than 65 papers published in various international journals and conferences.

Dr. Mukesh Saraswat is Associate Professor at Jaypee Institute of Information Technology, Noida, India. Dr. Saraswat has obtained his Ph.D. in Computer Science and Engineering from ABV-IIITM Gwalior, India. He has more than 18 years of teaching and research experience. He has guided 2 Ph.D. students, more than 50 M.Tech. and B.Tech. dissertations, and presently guiding 5 Ph.D. students. He has published more than 40 journal and conference papers in the area of image processing, pattern recognition, data mining, and soft computing. He was the part of successfully completed DRDE funded project on image analysis and currently running two projects funded by SERB-DST (New Delhi) on Histopathological Image Analysis and Collaborative Research Scheme (CRS), Under TEQIP III (RTU-ATU) on Smile. He has been an active member of many organizing committees of various conferences and workshops. He was also Guest Editor of the Journal of Swarm Intelligence. He is an active member of IEEE, ACM, and CSI Professional Bodies. His research areas include image processing, pattern recognition, mining, and soft computing.

Dr. Anupam Yadav is Assistant Professor, Department of Mathematics, Dr. B. R. Ambedkar National Institute of Technology Jalandhar, India. His research area includes numerical optimization, soft computing, and artificial intelligence; he has more than ten years of research experience in the areas of soft computing and optimization. Dr. Yadav has done Ph.D. in soft computing from Indian Institute of Technology Roorkee, and he worked as Research Professor at Korea University. He has published more than twenty-five research articles in journals of international repute and has published more than fifteen research articles in conference proceedings. Dr. Yadav has authored a textbook entitled "An introduction to neural network methods for differential equations". He has edited three books which are published by AISC, Springer Series. Dr. Yadav was General Chair, Convener, and a member of steering committee of several international conferences. He is Associate Editor in the journal of the experimental and theoretical artificial intelligence. Dr. Yadav is a member of various research societies.

Prof. Joong Hoon Kim Dean of Engineering College of Korea University, obtained his Ph.D. degree from the University of Texas at Austin in 1992 with the thesis title "Optimal replacement/rehabilitation model for water distribution systems". Prof. Kim's major areas of interest include optimal design and management of water distribution systems, application of optimization techniques to various engineering problems, and development and application of evolutionary algorithms. His paper which introduced the Harmony Search algorithm has been cited for more than 5000 times according to Google Scholar. He has been the faculty of School of Civil, Environmental and Architectural Engineering at Korea University since 1993. He has hosted international conferences including APHW 2013, ICHSA 2014 & 2015, and HIC 2016 and has given keynote speeches at many international conferences including 2013, GCIS 2013, SocPros 2014 & 2015, SWGIC 2017, and RTORS 2017. He is a member of National Academy of Engineering of Korea since 2017.

Dr. Jagdish Chand Bansal is Associate Professor at South Asian University New Delhi and Visiting Faculty at Maths and Computer Science, Liverpool Hope University UK. Dr. Bansal has obtained his Ph.D. in Mathematics from IIT Roorkee. Before joining SAU New Delhi, he has worked as Assistant Professor at ABV-Indian Institute of Information Technology and Management Gwalior and BITS Pilani. He is Series Editor of the book series Algorithms for Intelligent Systems (AIS) published by Springer. He is Editor-in-Chief of International Journal of Swarm Intelligence (IJSI) published by Inderscience. He is also Associate Editor of IEEE ACCESSS published by IEEE. He is the steering committee member and General Chair of the annual conference series SocProS. He is the general secretary of Soft Computing Research Society (SCRS). His primary area of interest is swarm intelligence and nature-inspired optimization techniques. Recently, he proposed a fission–fusion social structure-based optimization algorithm, Spider Monkey Optimization (SMO), which is being applied to various problems from engineering domain. He has published more than 70 research papers in various international journals/conferences.

He has supervised Ph.D. theses from ABV-IIITM Gwalior and SAU New Delhi. He has also received Gold Medal at UG and PG levels.

Modelling Subjective Happiness with a Survey Poisson Model and XGBoost Using an Economic Security Approach

Jessica Pesantez-Narvaez, Montserrat Guillen, and Manuela Alcañiz

Abstract The Living Conditions Survey of Ecuador contains a count variable measuring the subjective happiness of respondents. Two machine learning models are implemented to predict the level of happiness as a function of economic security among other factors. Even if the predictive performance is low, due to the fact that individuals tend to polarize extreme levels of happiness (either very low or very high), economic security is one of the most relevant determinants of a higher level of expected happiness, when we control for basic socio-demographic characteristics. Additionally, the analysis of missingness patterns in the target variable reveals some respondents' characteristics at the time of self-reporting satisfaction.

Keywords Machine learning · Living conditions · Count data · Occupation · Missingness

1 Introduction

The wealth of countries has been considered as a measurement of national economic progress for boosting the purchasing power and reducing poverty. However, there is evidence that it does not always lead social welfare because it may bring together income inequality. As a result, fields of economics happiness or sentiment analysis have emerged intensively due to their contributions in revealing drivers and factors of subjective satisfaction as well as their consequences. This entails a better understanding of individuals' behaviour and allows the design of public policy from a new perspective.

J. Pesantez-Narvaez · M. Guillen (✉) · M. Alcañiz
Department of Econometrics, Riskcenter-IREA, Universitat de Barcelona, 08034 Barcelona, Spain
e-mail: mguillen@ub.edu

J. Pesantez-Narvaez
e-mail: jessica.pesantez@ub.edu

M. Alcañiz
e-mail: malcaniz@ub.edu

H. Sharma et al. (eds.), *Congress on Intelligent Systems*, Advances in Intelligent Systems and Computing 1334, https://doi.org/10.1007/978-981-33-6981-8_1

In order to re-establish social welfare alternatively, the International Labour Organization (ILO) has devised the concept of economic security as a solution to the uncertainties, risks and exploitation that workers might face daily in the labour markets [12]. In fact, this notion carries an integral care of the worker since it provides a stable state of financial security to maintain dignified living conditions solvently. Considering [8], we induce that this approach could contribute to the objective happiness (i.e. good health status or low level of unemployment). In contrast, the query whether economic security has the same impact on subjective happiness (i.e. satisfaction and wellbeing) or not has not been directly answered.

For instance, there is a vast empirical economic literature devoted to understanding the link between individual's happiness and economic conditions. To summarize some of them, [7] demonstrated that informal sector workers have negative effects on their self-reported financial satisfaction. Chyi and Mao [5] studied the effect of household income and the number of generations under one roof on elderly's happiness. Flynn and MacLeod [9] found a significant relationship between financial security, schooling and happiness. Galletta [10] proved significant differences in the predicted levels of happiness given age class, branch of activity and disposable income. And Ferrer-i-Carbonell [8] found a relationship between job security and satisfaction.

These previous researches leave aside the impact analysis of economic security on worker's subjective happiness. Therefore, we aim to contribute to this strand of the modelling in happiness economic literature by predicting Ecuadorian workers' subjective happiness given certain levels of economic security (LES) and socio-demographic factors as control variables.

We use a survey Poisson model and XGBoost to predict the expected value of workers' subjective happiness and model it as a count variable with a complex survey designed data taken form the Living Conditions Survey (LCS) of 2014 to discover the effect that economic security has on subjective happiness.

This paper is divided into four parts after the introduction as follows: Sect. 2 contains data and descriptive statistics. Section 3 explains the methodology. Section 4 contains the discussion of results, and finally, Sect. 5 presents the conclusions of this research.

2 Data and Descriptive Statistics

We use the Living Conditions Survey (LCS) of Ecuador in 2014 performed by the Instituto Nacional de Estadísticas y Censos[1] (INEC). This poll was conducted by personal interview to 109,685 individuals through a two-stage stratified probabilistic sampling design, where the first unit of analysis is sectors and the second unit is dwellings. Questions particularly related to subjective wellbeing and emotional traits

[1] Instituto Nacional de Estadísticas y Censos is a public institution that provides and manages official statistics of Ecuador.

were only asked to workers older than 15 years old and younger than 64 years old, which corresponds to 47,132 observations.

The variables used in this research contained the following information for each respondent: living area (*rural/urban*), gender (*man/woman*), age (*age*), family order status under one roof (*head of the family, wife/husband, son/daughter, relatives, non-relatives who live in the same house*), respondent's ethnicity (*indigene, black/afroecuadorian, mulatto, mestizo, white/others*), respondent's years of schooling (*schooling*), respondent's civil status (*married/common-law marriage, divorced/widow, single*), respondent' work occupation group (*employer, day labourer, employer/freelance, non-paid worker, domestic worker*), respondent's economic security index (*economic security status*) and, finally, respondent's subjective happiness which is a proxy measured by the following question: *in the last 7 days, how many did you enjoy your life?*,[2] only answered by 46,002 individuals. Subjective happiness is the target variable we aim to predict given the aforementioned covariates.

The levels of economic security are measured with the Pesantez-Narvaez's economic security index [19], which is based on the dual cut-off methodology of Alkire and Foster's multidimensional poverty index [2]. It consists of five fundamental pillars or dimensions stated by ILO. Each achieved dimension is equal to 1, otherwise equal to 0. The sum of all the achieved dimensions by the worker will be the level of economic security, with a minimum of 0 and a maximum of 5. This index was built based on the guidelines of [13], Ecuadorian labour regulation and the availability of information of the LCS (2014).

- Labour market and income security are achieved if the worker has social security affiliation.
- Employment security is achieved if the worker has a written contract and receive 13°, 14° wage and paid vacation.
- Work security is achieved if the worker has a security insurance against harmful working environments.
- Skill reproduction security is achieved if the worker receives or gets enrolled in training courses.
- Representation security is achieved if the worker belongs to a union or has some collective protection.

Table 1 shows the descriptive statistics of the variables from the LCS 2014. Regarding the work occupation group, it shows that domestic workers, non-paid workers and day labourers reported fewer levels of subjective happiness than employers and employees. Men seem to achieve higher levels of subjective happiness than women. About the family order under one roof, family members report higher happiness than non-relatives. Respondents who are divorced or widowed have less happiness than married, common-married or single respondents. Individual who

[2]In our research, the subjective happiness is considered a cardinal measurement, as it ranges from 0 to 7. Ferrer-i-Carbonell and Frijters [6] stated cardinal and ordinal measures for subjective wellbeing. The first ones include [15] scales, and the second ones ask respondents transform numerical levels into verbal labels such as bad, good and very good.

Table 1 Mean subjective happiness for the variables in the living condition data set

Work occupation group	*Mean subjective happiness*	*Gender*	*Mean subjective happiness*	*Family order status*	*Mean subjective happiness*
Day labourer	4.773	Men	5.126	Head of the family	5.078
Domestic worker	4.858	Women	5.018	Wife/husband	5.086
Employee	5.299	*Area*	*Mean subjective happiness*	Son/daughter	5.105
Employer	5.001	Rural	4.674	Relatives	5.084
Non-paid worker	4.708	Urban	5.255	Non-relatives	4.357
Civil status	*Mean subjective happiness*	*Ethnicity*	*Mean subjective happiness*	*Economic security status*	*Mean subjective happiness*
Married/common-law marriage	5.109	Indigene	4.174	0	4.862
Divorced/widow	4.963	Mestizo	5.175	1	5.011
Single	5.064	Mulatto	5.196	2	5.152
		Black descendant/afroecuadorian	4.903	3	5.472
		White/others	5.003	4	5.701
				5	6.072

Data was obtained from the LCS 2014 with complete cases. Subjective happiness oscillates between 0 and 7. Sample weights from the survey are incorporated in the analysis, weighted estimates and frequencies which are presented. The correlation between respondent's age and subjective happiness is −0.047 and 0.133 between respondent's years of schooling and subjective happiness

considered themselves as indigenes as well as black descendant or afroecuadorian reports less happiness than the other ethnicity groups. Respondents' age shows a negative relationship with the subjective happiness while controlling the other covariates. However, their effect might be better seen with square of this variable since the literature found that if follows a U-inverted behaviour. There is a positive correlation between schooling and happiness, which means that education is not only linked to economic growth. Last but not least, the higher the economic security level, the happier the individual is.

3 Methodology

Let us assume that X_{ij} is the data matrix (set of covariates) of i observations $i = 1, \ldots, n$ and j covariates $j = 1, \ldots, k$. Let Y_i be the target or independent variable we aim to predict that corresponds to "subjective happiness". In this case, Y_i is the count of the number of days per week that the respondent feels satisfied.

Survey Poisson model as well as the XGBoost with Poisson objective seems appropriate to predict our data and will be briefly described.

3.1 Survey Poisson Model

The Poisson regression model also known as log-linear model assumes that Y_i follows a Poisson distribution:

$$Y_i \sim \text{Poisson}\left(e^{\beta_o + \sum\limits_{j=1}^{k} X_{ij}\beta_k}\right) \tag{1}$$

where β_o, β_1, ..., β_p are model parameters. Y_i is conditional on parameter λ equal to $e^{\beta_o + \sum_{j=1}^{k} X_{ij}\beta_k}$. Thus, the expected value of Y_i is equal to:

$$E(Y_i|X_{ij}) = e^{\beta_o + \sum\limits_{j=1}^{k} X_{ij}\beta_k} \tag{2}$$

Applying logarithms to both sides of the equation, we obtain:

$$\log(E(Y_i|X_{ij})) = \beta_o + \sum_{j=1}^{k} X_{ij}\beta_k \tag{3}$$

A weighted maximum likelihood method is used to estimate the Poisson model, where a weighted likelihood function L (likelihood function with a vector of sampling weights W_i) is maximized through an iteratively reweighted least squares algorithm, also known as the Fisher scoring method (further details in [11]).

$$L = \sum_{i=1}^{n} W_i\left(Y_i\left(\beta_o + \sum_{j=1}^{k} X_{ij}\beta_k\right) - e^{\beta_o + \sum\limits_{j=1}^{k} X_{ij}\beta_k} - \log(Y_i!)\right) \tag{4}$$

In addition to this, stratified probabilistic design-based standard errors are incorporated, obtaining a survey Poisson regression model proposed by [18]. A simple Poisson model would assume that observations are independent, which is not our case.

3.2 XGBoost

The XGBoost proposed by [4] is one of the most accurate predictive methods among machine learning algorithms. It builds D classification and regression trees (CART) within D iterations (each iteration denoted by d, $d = 1, \ldots, D$) subsequently in reweighted versions. Thus, each model (tree) $f_d(X_i)$ is trained and corrected using the residuals of the previous tree in order to predict the target variable. The XGBoost borrows the boosting ensemble method which is the sum of the D trained models:

$$\hat{Y}_i = \sum_{d=1}^{D} f_d(X_i) \quad i = 1, \ldots, n \tag{5}$$

The penalization used to correct newer versions of the models is adjusted with an optimal weighting procedure once a Poisson loss function plus a regularization term is minimized:

$$\mathcal{L} = \sum_{i=1}^{n} \left(-Y_i \left(\beta_o + \sum_{j=1}^{k} X_{ij}\beta_k \right) + e^{\beta_o + \sum_{j=1}^{k} X_{ij}\beta_k} + \log(Y_i!) \right) + \sum_{d=1}^{D} \acute{\eta}(f_d) \tag{6}$$

The regularization term $\acute{\eta}$ is also known as the shrinkage penalty which avoids the problem of overfitting by penalizing the complexity of the model. For further details of the methodology, see [20].

All in all, monitoring the differences in the predictive capacity of these algorithms is relevant for this research. Survey Poisson model can incorporate the sampling design to the estimation procedure, while the XGBoost is not able to do it. Conversely, the XGBoost has shown to obtain more accurate predictions, due to its tree-based designed, than in certain cases with some generalized linear models like Poisson [1].

3.3 Missing Data

The analysis is focused on a univariate missing data pattern case since the missing values are only detected in the target variable. Little and Rubin [17] stated that missingness can occur due to three mechanisms: missing completely at random (MCAR), missing at random (MAR) and missing not at random (MNAR).

Missing completely at random (MCAR) occurs when the non-response of data does not depend on observed either unobserved outcomes. Missing at random (MAR) occurs when the non-response of data depends on observed, but not on unobserved outcomes. An example is that a certain category of a variable is less likely to fill in a satisfaction survey, but this has nothing to do with their level of satisfaction.

Missing not at random (MNAR) occurs when the data is neither MAR nor MCAR. This non-response is also known as non-ignorable. Under a pattern mixture[3] approach, the sensitivity analysis is highly recommended. The idea is to create some scenarios with changes in the offset parameter δ representing an average difference between the missing value Y_{miss} and the observed value Y_{obs} in the following way: $Y_{\text{miss}} = Y_{\text{obs}} + \delta$ [14, 21]. Note that the greater the value of δ is, the great adjustment will receive the missing values. If $\delta = 0$, the effect is nule or too small, so one might suppose to be in the MCAR case. The most desired scenario would be the one where the observed values are the most similar to the imputed missing values.

[3]Little [16] represented pattern mixture models when the overall distribution of the variable of interest can be seen as a mixture of the distribution of the observed as well as the missing values.

4 Results

This section presents, firstly, the results of the missingness analysis; secondly, the predictions of the subjective happiness with the survey Poisson model and XGBoost; and thirdly, the interpretation of the determinants in the achievement of subjective happiness.

4.1 *Results of the Sensitivity Analysis of Missing Data*

The first step in order to uncover the type of missingness distribution is to evaluate if the outcome per individual is observed or not. To do so, let us denote R as the missing indicator variable that turns into 1 for missing values and 0 for observed values. R will be regressed by the covariates explained in Sect. 2 with a logistic regression.

Table 2 presents the results of a logistic regression model to evaluate the missingness distribution. A priori we may discard the MCAR case since almost all the coefficient estimates are significant. This fact reveals that non-response depends at least on observed values. Now, having a closer look in Table 2, one may say that individuals who are female, divorced/widow or live in urban areas are less likely to refuse to answer (or more likely to answer), but at the same time, they reported the lowest levels of subjective happiness with respect to other individuals of their corresponding group. Regarding the family order status, there is not a clear behaviour since wife/husband of the head of the family are more likely to answer,

Table 2 Results of a logistic regression model to evaluate the missingness distribution

Variables	Coefficient	Variables	Coefficient
Intercept	−4.221***	Mestizo	−0.944*
Economic security	−0.102*	Mulatto	−1.169***
Urban	−0.379**	Black descendant/afroecuadorian	−0.302*
Female	−0.917***	White/others	0.242
Age	0.096**	Schooling	−0.122***
Age^2	−0.001·	$Schooling^2$	0.003**
Wife/husband	−0.376*	Day labourer	−0.429*
Son/daughter	0.786***	Employer	−1.043***
Relatives	1.141***	Non-paid worker	−0.254*
Non-relatives	0.688·	Domestic worker	−1.289**
Divorced/widow	−1.149***		
Single	0.158		

The significance of the coefficients is given in the following way: ***, **, *, ·, to the 0%, 0.1%, 1%, 5% of significance level, respectively. The base categories are rural, male, head of the family, married/common-law marriage, indigene and employee

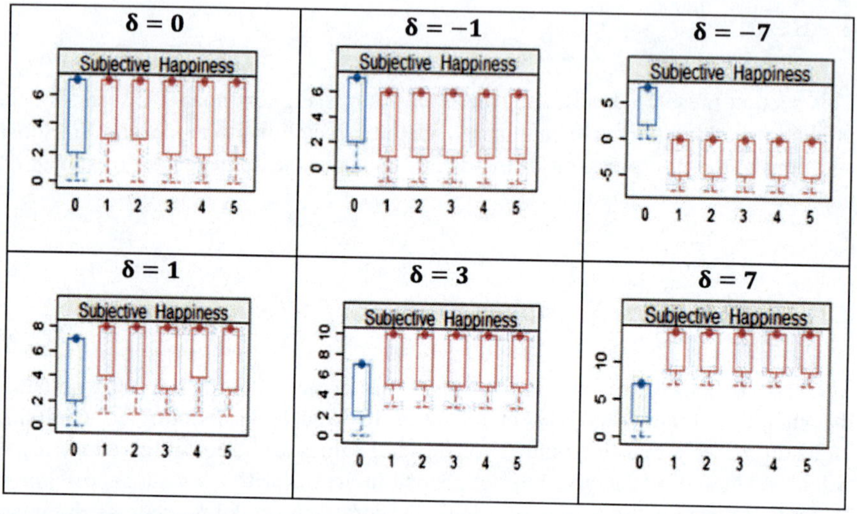

Fig. 1 Scenarios of the offset parameter δ under NMAR case

while son/daughters (ones who reported more happiness) are more likely to avoid an answer. Regarding the occupation group, day labourer, employer, non-paid workers and domestic workers are more likely to answer the subjective happiness question with respect to the employees. Finally, people with fewer years of schooling are more likely to answer until some point when they have more years of schooling and decide not to answer.

Overall, it seems that individuals with less subjective happiness tend to avoid refusals to answer, whereas individuals with more happiness are more likely to refuse to answer. This pattern would probably place into the MAR case; however, we will work under a NMAR case with a sensitivity analysis to determine possible adjustments for missing values in case they are nonignorable.

Figure 1 shows different scenarios of the offset parameter δ under NMAR case. Since the missing values are found in the target variable (discrete variable that oscillates from 0 to 7), we pose scenarios of δ with values above and below the bounds. The boxplots drawn in blue represent the observed data, and the ones in red represented the imputed data within five iterations by multivariate imputation by chain equations (MICE) [3]. Here, we expect the red boxplot to have almost the same shape as blue boxplots. When $\delta = 0$, the imputed iterations tend to follow the pattern; however, the distribution of the imputed data changes substantially when δ is increased or decreased with respect to the observed data.

It seems so far that the imputed data with $\delta = 0$ is the most favourable and accurate because their pattern is quite similar to the observed data in comparison with the other scenarios. In addition to this, $\delta = 0$ means that no adjustments are needed, so we can consider a MCAR case; however, since there is dependence on observed data, we consider to be in MAR case, and then, MICE method is appropriate to impute missing data.

4.2 Predicted Subjective Happiness

Table 3 shows the results of the predictions in the testing and training data sets for both data sets, without missingness and completed with imputation by chained equations. Findings show that the proportion of cases in each level of subjective happiness remains quite constant between both data sets.

The largest accumulation of cases is seen in the lowest and highest levels of subjective happiness, revealing that great majority of individuals of the sample tend to feel very happy, a second largest subsample of individuals who feel exactly the opposite, and the rest of interviewees who are at intermediate levels. The XGBoost seems to slightly outperform the survey Poisson due to the proximity of accumulated cases by each predicted level of happiness with the real ones. Nevertheless, both

Table 3 Results of the predictions in the testing and training data sets before and after the multiple imputation by chained equations

Observed subjective happiness (levels)	Observed cases		Survey Poisson model		XGBoost—Poisson objective	
	Case (a) (without missingness)	Case (b) (completed data)	Case (a) (without missingness)	Case (b) (completed data)	Case (a) (without missingness)	Case (b) (completed data)
Testing data set						
0	1645	1701	101	103	128	131
1	1044	1011	481	492	550	566
2	979	873	1163	1193	1247	1288
3	691	680	1898	1953	1949	2012
4	399	416	2354	2422	2345	2416
5	598	555	2359	2426	2307	2369
6	643	557	1991	2043	1925	1970
7	8140	8006	1455	1487	1401	1427
Training data set						
0	3871	4051	236	235	306	307
1	2300	2351	1136	1119	1309	1294
2	2197	2158	2767	2708	2970	2907
3	1536	1587	4539	4426	4652	4530
4	880	923	5641	5488	5607	5449
5	1425	1434	5661	5505	5521	5366
6	1369	1335	4777	4649	4613	4489
7	18,625	19,154	3484	3398	3556	3275

Case (a) shows the predictions with the data set without missing values corresponding to 46,002 observations, while case (b) shows the predictions with the data set where imputed values were included corresponding to 47,132 observations. Both data sets were randomly divided into training data set which contains the 70% of the total samples (32,203 for case (a) and 32,993 for case (b)) and testing data set which contains the 30% of the total samples (13,799 for case (a) and 14,139 for case (b)) methods tend to overestimate the intermediate cases and underestimate the extreme ones.

4.3 Regression Analysis

The interpretation of the factors associated of the subjective happiness will be through the parameter estimates of the survey Poisson model.

Table 4 shows the parameter estimates of the survey Poisson model. These results show that higher levels of economic security are positively associated to the expected levels of subjective happiness. Individuals who live in urban areas have more expected value of happiness than individuals who live in rural areas. Women have less expected value of happiness than men. The older the individual, the less happy they are. The coefficient estimate of age^2 is not significant in the model. Therefore, here, an individual's age does not have a U-inverted behaviour at the time of explaining the subjective happiness. Moreover, the wife/husband of the head of the family seems to decrease the subjective happiness with respect to the head of the family, while the effect of son/daughter, relatives and non-relatives seems not to be significant in this analysis. Ethnicity such as mestizo, mulatto or white/others have a positive effect on the expected value of self-reported subjective happiness with respect to indigene individuals. Apart from this, more years of schooling associate positively on the

Table 4 Results of parameter estimates of the survey Poisson model

Variables	Coefficient	Variables	Coefficient
Intercept	1.519 (2e−16)***	Mestizo	0.065 (0.009)**
Economic Security	0.015 (6.58e−08)***	Mulatto	0.150 (9.16e−11)***
Urban	0.052 (4.39e−06)***	Black descendant/afroecuadorian	0.249 (1.36e−07)***
Female	−0.058 (0.000)***	White/others	0.122 (0.010)*
Age	−0.004 (0.028)*	Schooling	0.007 (0.029)*
Age^2	0.000 (0.488)	$Schooling^2$	0.000 (0.604)
Wife/husband	0.035 (0.009)**	Day labourer	−0.026 (0.036)*
Son/daughter	−0.007 (0.472)	Employer	−0.0257 (0.036)*
Relatives	0.004 (0.659)	Non-paid worker	−0.025 (0.027)
Non-relatives	−0.192 (0.137)	Domestic worker	−0.042 (0.689)
Divorced/widow	0.016 (0.162)		
Single	−0.004 (0.741)		

The significance of the coefficients is given in the following way: ***, **, *, ·, to the 0%, 0.1%, 1%, 5% of significance level, respectively. The base categories are rural, male, head of the family, married/common-law marriage, indigene and employee. The data used for this model corresponds to the completed testing data set modelled with a survey Poisson regression expected value of subjective happiness. Finally, day labourers and employers have a negative effect on self-reported subjective happiness.

5 Conclusions

This analysis reveals that economic security can be considered as a contributor of individuals' subjective happiness given socio-demographic factors as control variables. Unlike other economic measures such as GDP, economic security might influence positively on both individuals' satisfaction and wealth.

The subjective happiness is a count variable, modelled with a survey Poisson model with a stratified probabilistic sampling design, and also, with a XGBoost with Poisson objective. From a statistical view, the predictive capacity of both methods underestimates the predicted cases that correspond to the extreme values of subjective happiness. Alternative proposed modelling techniques are suggested to capture more accurately the distribution of this dependent variable.

Last but not least, even when the sensitivity analysis of missing values discloses a MCAR pattern, certain type of individuals who tend to omit their self-reported levels of happiness corresponds to groups with low levels of self-satisfaction.

Acknowledgements We thank the Spanish Ministry of Economy, FEDER grant ECO2016-76203-C2-2-P.

References

1. Adam-Bourdarios, C., Cowan, G., Germain-Renaud, C., Guyon, I., Kégl, B., Rousseau, D.: The Higgs machine learning challenge. J. Phys. Conf. Ser. **664** (2015)
2. Alkire, S., Foster, J.: Counting and multidimensional poverty measurement. J. Public Econ. 476–487 (2011)
3. Azur, M.J., Stuart, E.A., Frangakis, C., Leaf, P.J.: Multiple imputation by chained equations: what is it and how does it work? Int. J. Methods Psychiatr. Res. **20**(1), 40–49 (2011)
4. Chen, T., Guestrin, C.: XGBoost: a scalable tree boosting system. In: 22nd ACM SIGKDD International Conference on Knowledge Discovery and Data Mining, pp. 785–794 (2016)
5. Chyi, H., Mao, S.: The determinants of happiness of China's elderly population. J. Happiness Stud. **13**(1), 167–185 (2012)
6. Ferrer-i-Carbonell, A., Frijters, P.: How important is methodology for the estimates of the determinants of happiness? Econ. J. **114**(497), 641–659 (2004)
7. Ferrer-i-Carbonell, A., Gërxhani, K.: Financial satisfaction and (in) formal sector in a transition country. Soc. Indic. Res. **102**(2), 315–331 (2011)
8. Ferrer-i-Carbonell, A.: Happiness economics. SERIEs **4**(1), 35–60 (2013)
9. Flynn, D.M., MacLeod, S.: Determinants of happiness in undergraduate university students. Coll. Stud. J. **49**(3), 452–460 (2015)
10. Galletta, S.: On the determinants of happiness: a classification and regression tree (CART) approach. Appl. Econ. Lett. **23**(2), 121–125 (2016)

11. Green, P.J.: Iteratively reweighted least squares for maximum likelihood estimation, and some robust and resistant alternatives. J. Roy. Stat. Soc.: Ser. B (Methodol.) **46**(2), 149–170 (1984)
12. ILO (International Labour Office): Economic Security for a Better World. International Labour Office (2004)
13. ILO (International Labour Office): The Rules of the Game. A Brief Introduction to International Labour Standards. Review Edition 2009 (2009)
14. Leurent, B., Gomes, M., Faria, R., Morris, S., Grieve, R., Carpenter, J.R.: Sensitivity analysis for not-at-random missing data in trial-based cost-effectiveness analysis: a tutorial. Pharmacoeconomics **36**(8), 889–901 (2018)
15. Likert, R.: A technique for the measurement of attitudes. Arch. Psychol. **140**, 55 (1932)
16. Little, R.J.: Pattern-mixture models for multivariate incomplete data. J. Am. Stat. Assoc. **88**(421), 125–134 (1993)
17. Little, R.J., Rubin, D.B.: Statistical Analysis with Missing Data. Wiley, New York (2002)
18. Lumley, T.: Analysis of complex survey samples. J. Stat. Softw. **9**(1), 1–19 (2004)
19. Pesantez-Narvaez, J.: Limited dependent variable modelling: an econometric analysis of migration and economic security in the Ecuadorian labour market. Master's thesis, Universitat Politècnica de Catalunya (2018)
20. Pesantez-Narvaez, J., Guillen, M., Alcañiz, M.: Predicting motor insurance claims using telematics data—XGBoost versus logistic regression. Risks **7**(2), 70 (2019)
21. Van Buuren, S.: Flexible Imputation of Missing Data. CRC Press (2018)

Monitoring Web-Based Evaluation of Online Reputation in Barcelona

Jessica Pesantez-Narvaez, Francisco-Javier Arroyo-Cañada, Ana-María Argila-Irurita, Maria-Lluïsa Solé-Moro, and Montserrat Guillen

Abstract In the hotel sector, online reputation and customer satisfaction help measure the quality of service based on the opinions of the survey participants. This research takes information provided by TripAdvisor from a sample of 247 hotels in Barcelona in order to obtain users' reaction to each establishment. A robust compositional regression is modelled to diagnose the score given to each hotel as a result of the customers' profiles. Additionally, a principal component analysis is proposed to visualize customers' behavioural patterns of scoring. The results let us detect a particular behaviour for the business travellers' scale seen alongside other groups of tourists who also rated hotels with the top marks at this time. Furthermore, the score given by travellers who arrived during the summer is shown as not being significant to achieving a final high score. The proposed model can also be used to track the stability of score over time and to identify suspicious deviations from benchmark levels.

Keywords Profile · Online reputation · Compositional regression · Hospitality

J. Pesantez-Narvaez (✉) · M. Guillen
Department of Econometrics, Riskcenter-IREA, Universitat de Barcelona, 08034 Barcelona, Spain
e-mail: jessica.pesantez@ub.edu

M. Guillen
e-mail: mguillen@ub.edu

F.-J. Arroyo-Cañada · A.-M. Argila-Irurita · M.-L. Solé-Moro
Department of Business, University of Barcelona, Barcelona, Spain
e-mail: fjarroyo@ub.edu

A.-M. Argila-Irurita
e-mail: aargila@ub.edu

M.-L. Solé-Moro
e-mail: mlsolesole@ub.edu

H. Sharma et al. (eds.), *Congress on Intelligent Systems*, Advances in Intelligent Systems and Computing 1334, https://doi.org/10.1007/978-981-33-6981-8_2

1 Introduction

Online reputation has become a popular tool to support purchase decision making in the tourism sector. Customers are therefore putting their trust in the opinions provided by hotel comparison sites due to their useful information about travel experiences and recommendations, accommodation alternatives, as well as the possibility of making actual hotel bookings. Thus, understanding customers' behaviour at the time of scoring a hotel seems relevant for managerial actions. Here, we provide a method to present web-based evaluations based on travellers' characteristics to obtain a comparable measure for hotels that have different characteristics.

Measuring hotel reputation has led to several studies being conducted to date. To name some, Wang et al. [15] analysed certain hotel comparison in order to evaluate their trustworthiness and detect how much online information could be trusted. Swamynathan et al. [13] designed reliable reputation systems. Koskinen et al. [8] profiled the reviewers who post reliable reviews. And still other studies looked at reputation profiles to detect malicious sellers in electronic marketplaces, such as [14]. Also, Radojevic et al. [12] analyzed the impact of business travelers on the final evaluation of hotel services.

Previous findings tend to analyse online reputation in hotels focusing mainly on customer's personal characteristics or the service itself, leaving aside alternative characteristics such as the type of traveller (alone or accompanied), or the season of travelling or even both simultaneously. In fact, there is evidence about the influence of the type of accompaniment of a person as well as the time of the year on perceptions, subjective wellbeing and decisions.

On the one hand, the perceived service satisfaction might differ whether the customers come accompanied or not. A research supported by [9] in which individuals who tend to socialize, or form part of a group is more likely to demonstrate higher life satisfaction. Faulmüller et al. [6] realized that members of a group suffer from information bias because they mainly use the information that is available to all of them and ignore what is known to only few. Moreover, group thinking theory states that an external attack or negative impact on the group is even worse than one coming from a single member of the group.

On the other hand, the perceived service satisfaction and emotions might be influenced by the time of year. For example, an empirical research from [5] as well as [3] found that seasons influence on emotions. Denissen et al. [4] found that, even when a person was in a low mood, his mood could be raised on days when the temperature was higher. In contrast, Hsiang et al. [7] proved that extreme temperatures (heat and high levels of rain) may stimulate aggression in humans. And [8] tested that a lower level of activation may find it more difficult to perform tasks during winter months.

The purpose of this research is to identify travellers' behavioral patterns through a principal component analysis. We aim to study the influence either positively or negatively of the type of traveler and seasons on the final score of a hotel through a robust compositional regression model. This exploratory analysis is done in the hotel sector of Barcelona, one top travel destination in Europe.

This paper is divided in the following four parts. In Sect. 2, we present the methodology description of a robust compositional regression model and principal component analysis for compositional data. Section 3 describes the compositional data used in this research. Section 4 presents the results and their discussions. And Sect. 5 contains the final conclusions and some managerial implications.

2 Methodology Description

This research has two methodological research contributions. The first one is to implement a principal component analysis to detect similar traveller satisfaction patterns at the time of evaluating a hotel. The objective is to start from highly correlated variables (behaviour of customers' evaluations) and transform them into a smaller number of uncorrelated variables known as principal components in order to ease the interpretation. The second contribution consists of a robust compositional regression to measure the effect that some compositional factors have on the hotel's final score given some additional explanatory variables. We present a brief statistical description of the aforementioned techniques as follows:

Let Y_i, $i = 1, \ldots, n$ be our response variable and X_{ik}, $k = 1, \ldots, K$ be the corresponding set of covariates. In this article, Y_i is a numeric variable and X_{ik} is compositional data. Each covariate X_k is a vector of D strictly positive components, $d = 1, \ldots, D$, and the sum of all its parts is a constant c. It can be defined in a simplex sample space S^D as:

$$S^D = \left\{ \left(X^1, X^2, \ldots, X^D\right)' : X^d > 0; \sum_{d=1}^{D} X^d = c \right\} \tag{1}$$

2.1 Principal Component Analysis

The principal component analysis (PCA) is a mathematical procedure that allows to represent n observations of a K-dimensional space into a smaller dimension space optimally. The method builds uncorrelated synthetic variables known as *principal components* that contains the maximum variability of information. The purpose is to represent the n observations of a K-dimensional space into a reduced dimensional space by projecting the observations into new axes having the least deformation as possible.

Let $a_{ik} = (a_{ik}, \ldots, a_{iK})'$ be the unit normal vector for each observation i. The projection of each observation i towards the unit normal vector direction is:

$$Z_i = \sum_{k=1}^{K} X_{ik} a_{ik}, \tag{2}$$

giving a resulting projected vector $Z_i a_i$. We are interested in minimizing a distance r_i between an observation i and its projected vector $Z_i a_i$:

$$\text{Minimize} \sum_{i=1}^{n} r_i^2 = \sum_{i=1}^{n} |X_i - Z_i a_i|^2. \tag{3}$$

Let's consider that minimizing $\sum_{i=1}^{n} r_i^2$ is equivalent to maximize $\sum_{i=1}^{n} Z_i^2$. And the projections Z_i are zero mean variables, so maximizing the sum squares is equivalent to maximize the variance.

All in all, the first principal component is the lineal combination of the original variables that has a maximum variance found in Z_i. For further details see in [11].

In the case of principal component analysis with compositional data, the vector Z_i is computed:

$$Z_i = \sum_{k=1}^{K} \sum_{d=1}^{D} X_{ik}^{d} a_{ik}. \tag{4}$$

2.2 *Robust Compositional Regression Model*

The compositional analysis was first introduced by [1]. After [2] explained that compositional regression is quite analogous to multiple linear regression, but with more complex considerations. An isometric log-ratio transformation is applied before modelling to obtain a singular transformed covariance matrix. It allows that conclusions can be drawn for compositions since they only provide a relative magnitude information, and the total sum of its parts is irrelevant.

The isometric log-ratio "ilr" of a covariate X_k is defined as:

$$\text{ilr}(X_k) = \left(\langle X^1, e_1\rangle, \langle X^2, e_2\rangle, \ldots, \langle X^{D-1}, e_{D-1}\rangle\right) \tag{5}$$

where $e_1, e_2, e_3, \ldots, e_{D-1}$ is the simplex orthogonal base. A standard specification model is:

$$Y_i = a + \sum_{d=1}^{D-1} \beta_d \text{ilr}(X_i) + \varepsilon_i, \tag{6}$$

where Y_i is the response or dependent variable, a is the constant coefficient, and β_k are the estimated coefficients of the independent variables. Note that if residuals do not hold the normality assumption, it can be solved by regressing a robust error compositional regression, mostly known as robust compositional regression.

3 Data Description

We use a publicly available compositional data set [10] collected from the official TripAdvisor website. It has a sample of two hundred and forty-seven hotels located in Barcelona, Spain. It contains information related to the type of traveller (reviewer) and the hotel's final score.

Dependent and independent variables are used in the robust compositional regression model. The dependent variable is the "hotel's final score" given by TripAdvisor based on customers' scores. The ranking goes from 1 (the lowest) to 5 (the highest). The independent variables are: type of traveller (business, family, lone traveller, couple, and friends), time of year (March–May, June–August, September–November, and December–February), location (represents the neighbourhood where the hotel is located[1]), and finally the hotel's number of stars (determines the hotel category). Each category of the variables is rated to a five-scale score (excellent, very good, average, poor and terrible).

To explain it deeper, let's take an example with the variable: type of traveller with the category of business travellers. They have rated a hotel as excellent, very good, average, poor and terrible, so one has five compositional parts instead of only one (business traveller—excellent, business traveller—very good, business traveller—average, business traveller—poor, business traveller—terrible). Consequently, the type of traveller has twenty-five parts or subcategories, and the time of the year has twenty parts or subcategories. Let's consider them as transformed variables onwards.

The specification of our model is:

$$\begin{aligned} Y_i = a &+ \sum_{d=1}^{19} \beta_d \text{ilr}_{(i)}(\text{score given by time of year})_i \\ &+ \sum_{d=1}^{24} \beta_d \text{ilr}_{(i)}(\text{score given by the type of traveller})_i \\ &+ \beta_{25}\text{Location} + \beta_{26}\text{Num.Stars} + \varepsilon_i, \end{aligned} \tag{7}$$

where Y_i is the response variable, in this case hotel final's score, a is the constant coefficient, and β_k are the estimated coefficients related to the independent variables.

Table 1 shows the average scores given to hotels in Barcelona according to the type of traveller by TripAdvisor. The descriptive statistics show that overall all the

[1] Neighbourhoods registered in Barcelona are: Barceloneta, Barrio Gótico, Ciudat Vella, El Born, El Prat de Llobregat, El Raval, L'eixample, Gava, Gràcia, Horta Guinardó, La Vila Olímpica de Poblenou, Les Corts, L'Hospitalet de Llobregat, Nou Barris, Poble Sec, Poblenou, San Marti, Sant Andreu, Sant Cugat del Vallès, Sants Montjuic, Sarrià, Sant Gervasi.

Table 1 Average scores given to hotels in Barcelona according to the type of traveller by TripAdvisor

	Friends (%)	Business (%)	Lone (%)	Couple (%)	Family (%)
Excellent	44.210	36.077	42.898	47.355	46.015
Very good	39.949	38.892	37.095	36.681	37.778
Average	11.059	16.081	12.923	10.537	10.673
Poor	3.381	5.520	4.339	3.393	3.234
Terrible	2.401	3.431	2.475	2.035	2.3

Own source calculations based on Pesantez [10] data set. Average scores are calculated as column percentages, that is as the proportion of travellers in each category (Friends, Business, Lone, Couple and Family) that voted each opinion ranked from terrible to excellent

Table 2 Average scores given to hotels in Barcelona according to seasons by TripAdvisor

	Spring (%)	Summer (%)	Autumn (%)	Winter (%)
Excellent	45.232	44.517	42.885	47.237
Very good	37.432	37.408	37.184	37.110
Average	11.453	11.610	12.013	10.415
Poor	3.620	3.825	3.959	3.177
Terrible	2.263	2.640	3.959	2.062

Own source calculations based on Pesantez [10] data set. Average scores are calculated as column percentages, that is the proportion of travellers that voted each opinion ranked from terrible to excellent in each season (time of the year when the traveller stayed in the hotel)

different customer profile groups gave mainly high ratings (excellent and very good) to the hotels in Barcelona and there was a low percentage of lower ratings (less than 7% for poor and terrible). A slightly lower score within the high ratings (excellent and very good) is given by business travellers. Table 2 shows the average scores given to hotels in Barcelona according to seasons by TripAdvisor. When it comes to seasonality, on average all the scores given to the hotels are mainly in high ratings, then and the level of satisfaction seems to be same during the whole year.

4 Discussion of Results

This section presents the results obtained by the principal component analysis and the compositional regression model.

The principal component analysis of hotel scores given by the type of traveller shows that the first two components accumulate the 61% of the variability of data, and the first four components accumulate the 77% of data. We take the first two components for the analysis because main patterns can be already detected.

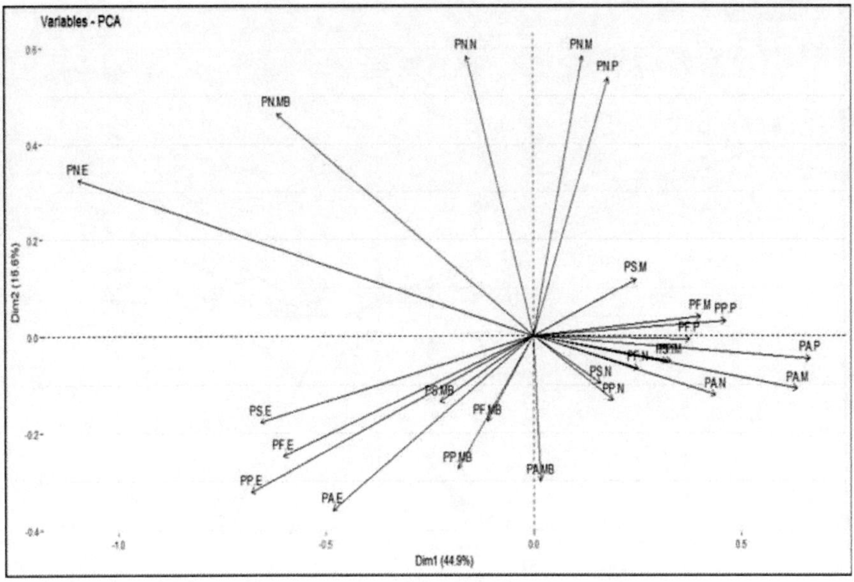

Fig. 1 Biplot of hotels' scores given by type of traveller (abbreviations in Appendix 1)

Figure 1 shows the biplot of hotels' scores given by type of traveller. High scores (excellent, very good) given by lone travellers, couples, family and friends are highly correlated, which means that overall, these groups seem to be satisfied by the types of services, infrastructures and accommodation provided by hotels in Barcelona. Business travellers, however, rate as excellent or very good a hotel based on other criteria. The way in which friends, couples and families rate hotels positively is uncorrelated to the way in which business travellers rate them.

The principal component analysis of hotel scores given by time of year shows that the first two components accumulate the 71% of the variability of data, and the first four components accumulate the 80.6% of data. The same as in the previous analysis, we focus on the first two components.

Figure 2 shows the biplot of hotels' scores given by time of year. Scores from all the season which are rated as excellent are highly correlated, which means that people will have the same criteria for being satisfied, no matter what time of year it is. Hotels in Barcelona are also adapting their services efficiently according to climate changes and tourist reception management. Even though it is safe to say that the reasons for rating a hotel with high scores are closely related, the reasons for rating a hotel as poor and terrible might not be that straight. For instance, negative external shocks outside the group of customers caused a greater disconformity than what it was real, also specific service details for business customers should be considered since they reported slightly lower levels of satisfaction compared to other customers.

Table 3 presents the results obtained by the estimated compositional regression model from (7). It shows that when one evaluation rating a hotel as poor and coming

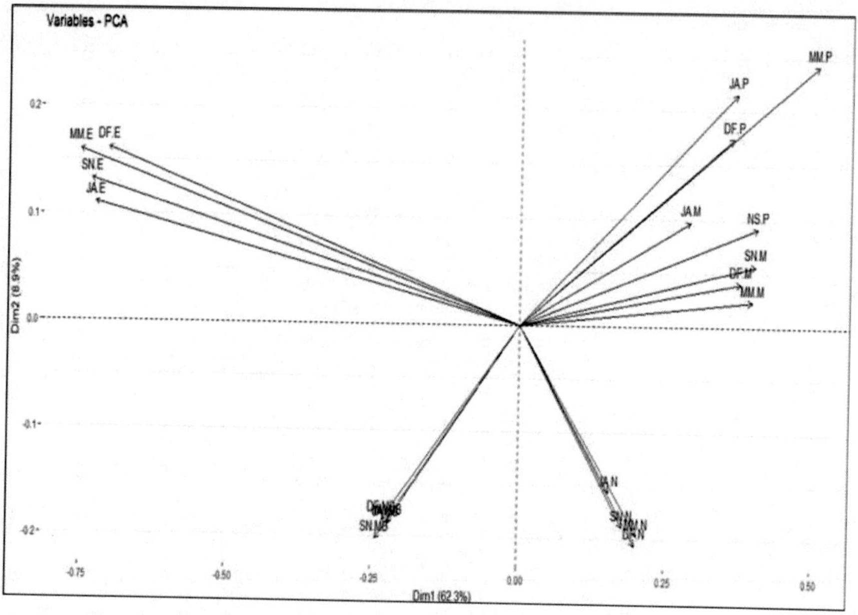

Fig. 2 Biplot of hotel scores as given by time of year (abbreviations in Appendix 2)

from a group of friends occurs, then the hotels' final score decreases substantially. In contrast, an evaluation rating a hotel as very good coming from business travellers may increase the hotel's final score by 0.053, while one score as excellent will decrease it by 0.109. This result is not surprising given that business travellers seem to appreciate features that differ from those that are highly valuated by other types of visitors. Therefore, if business visitors rate the hotel as excellent, then others may underrate this same score causing the final average to decrease.

Business travellers' satisfaction might have different ways of being interpreted, one of these is that their level of exigence is higher than that of other types of travellers, while at the same time, needing a specific infrastructure.

The results of the location of the hotel seem to indicate that all areas have on average a lower score than Barceloneta which is located near the beach, and other neighbourhoods that do not have a significant coefficient (Sant Andreu, Gràcia, etc.).

The interaction between a bad evaluation (poor, average or very good as opposed to excellent) given between March and May reduces a lower final average score. This means that hotels should try to improve satisfaction in spring and possibly also autumn, thus making a "satisfactory" score move to "excellent", while scores given in summer are not significant within the model.

The category of the hotel measured in stars does not have a significant impact.

Table 3 Results of the robust compositional regression (only significant coefficients)

Variables	Coefficient	Variables	Coefficient
Intercept	3.637 (0.124)***	Ciutat Vela	−0.307 (0.073)***
Friends' score as poor	−0.039 (0.019)*	El Born	−0.171 (0.093)·
Business score as excellent	−0.109 (0.034)**	El Ensanche	−0.122 (0.055)*
Business score as very good	0.053 (0.030)·	El Prat de Llobregat	−0.384 (0.123)**
Business score as poor	−0.015 (0.023)*	El Raval	−0.155 (0.062)*
Score as very good when the traveller arrived between March and May	−0.104 (0.053)·	L'hospitalet de Llobregat	−0.208 (0.069)**
Score as average when the traveller arrived between March and May	−0.060 (0.033)·	La Vila Olímpica de Poblenou	−0.026 (0.053)***
Score as poor when the traveller arrived between March and May	−0.037 (0.020)·	Les Corts	−0.123 (0.068)·
Score as excellent when the traveller arrived between September and November	−0.173 (0.095)·	Poblenou	−0.111 (0.062)·
Score as average when the traveller arrived between September and November	−0.079 (0.029)**	Sant Cugat del Valles	−0.200 (0.082)*
Score as average when the traveller arrived between December and February	−0.030 (0.017)·	Sants Montjuic	−0.218 (0.064)***
Barrio Gótico	−0.128 (0.058)*		

The significance of the variables is given in the following way: ***, **, *, ·, to the 0%, 0.1%, 1%, 5% and 10% of significance level respectively. The values in parenthesis represent the standard error of each variable. The base category for local is Barceloneta, so all significant locations will be compared with this one. The model is globally significant at 5% ($F_{\text{statistic}} = 0.0068$, $F_{4,243,0.05} = 2.839$). A robust error compositional regression is built in order to correct the normality assumption violation since Shapiro test rejects the null hypothesis of normality in the base model ($P_{\text{value}} = 4.895\text{e}{-06}$)

5 Conclusions and Managerial Implications

The results show that business travellers and leisure visitors have different opinions; therefore, carefully differentiating the service provided mainly to these two types of customers could further improve customer satisfaction and thus the online reputation provided. Following on, hotel managers could satisfy the needs of all these travellers without losing efficiency (providing unnecessary resources).

A good result is that scores are stable over the different seasons. However, hotels should improve satisfaction levels in spring. Moreover, summer does not appear to be significant within the model, which means that vacation time and longer sunnier hours may relax customers' emotions with neither extreme positive nor negative effects on the final hotels' score.

The non-significance of the category of the hotel (measured in stars) demonstrates that online reputation is gaining more importance, where previously the number of stars of a hotel had been used as a mark of reputation.

As a result, patterns of customers' behaviour and their relationship with time of the year when evaluating the hotel service are detected, which could be used to let hotel managers establish more personalized client service strategies without losing efficiency in resources endowment.

This methodological approach allows scenarios to be constructed about the predicted score of a hotel. It uses the main characteristics of its visitors and a composition of evaluations (from terrible to excellent) given by these visitors, in order to visualize effectively changing behavioural. Therefore, when hotels receive an average score when they had been predicted to achieve a higher value, we can suspect that unfair negative opinions have been given in order to manipulate the poll. The proposed model can also be used to track the stability of score over time and to identify deviations from benchmark levels, which would be an indication of deterioration in the service.

Acknowledgements J. P-N and M. G thank the Spanish Ministry of Economy, FEDER grant ECO2016-76203-C2-2-P.

Appendix 1: Abbreviation and Corresponding Meaning of Fig. 1

Abbreviation	Meaning	Abbreviation	Meaning
PA.E	Friends' score as excellent	PS.N	Lone traveller's score as average
PA.MB	Friends' score as very good	PS.M	Lone traveller's score as poor
PA.N	Friends' score as average	PS.P	Lone traveller's score as terrible
PA.M	Friends' score as poor	PP.E	Couple's score as excellent
PA.P	Friends' score as terrible	PP.MB	Couple's score as very good
PN.E	Business traveller's score as excellent	PP.N	Couple's score as average
PN.MB	Business traveller's score as very good	PP.M	Couple's score as poor
PN.N	Business traveller's score as average	PP.P	Couple's score as terrible

(continued)

(continued)

Abbreviation	Meaning	Abbreviation	Meaning
PN.M	Business traveller's score as poor	PF.E	Families' score as excellent
PN.P	Business traveller's score as terrible	PF.MB	Families' score as very good
PS.E	Lone traveller's score as excellent	PF.N	Families' score as average
PS.MB	Lone traveller's score as very good	PF.M	Families' score as poor
		PF.P	Families' score as terrible

Appendix 2: Abbreviation and Corresponding Meaning of Fig. 2

Abbreviation	Meaning	Abbreviation	Meaning
MM.E	Score as excellent when the traveller arrived between March and May	SN.E	Score as excellent when the traveller arrived between September and November
MM.MB	Score as very good when the traveller arrived between March and May	SN.MB	Score as very good when the traveller arrived between September and November
MM.N	Score as average when the traveller arrived between March and May	SN.N	Score as average when the traveller arrived between September and November
MM.M	Score as poor when the traveller arrived between March and May	SN.M	Score as poor when the traveller arrived between September and November
MM.P	Score as terrible when the traveller arrived between March and May	SN.P	Score as terrible when the traveller arrived between September and November
JA.E	Score as excellent when the traveller arrived between June and August	DF.E	Score as excellent when the traveller arrived between December and February
JA.MB	Score as very good when the traveller arrived between June and August	DF.MB	Score as very good when the traveller arrived between December and February
JA.N	Score as average when the traveller arrived between June and August	DF.N	Score as average when the traveller arrived between December and February

(continued)

(continued)

Abbreviation	Meaning	Abbreviation	Meaning
JA.M	Score as poor when the traveller arrived between June and August	DF.M	Score as poor when the traveller arrived between December and February
JA.P	Score as terrible when the traveller arrived between June and August	DF.P	Score as terrible when the traveller arrived between December and February

References

1. Aitchison, J.: The statistical analysis of compositional data. J. R. Stat. Soc. 139–177 (1982)
2. Van den Boogart, K.G., Tolosana-Delgado, R.: Analysing Compositional Data with R. Springer-Verlag, Berlin (2013)
3. Connolly, M.J.: Some like it mild and not too wet: the influence of weather on subjective well being. J. Happiness Stud. **14**, 457 (2013)
4. Denissen, J., Butalid, L., Penke, L., van Aken, M.: The effects of weather on daily mood: a multilevel approach. Emotion 662–667 (2008)
5. Ettema, D., Friman, M., Olsson, L.E., Gärking, T.: Season and weather effects on travel-related mood and travel satisfaction. Front. Psychol. **8**, 140 (2017)
6. Faulmüller, N., Kerschreiter, R., Mojzisch, A., Schulz-Hardt, S.: Beyond group-level explanations for the failure of groups to solve hidden profiles: the individual preference effect revisited. Group Process. Intergr. Relat. 653–671 (2010)
7. Hsiang, S.M., Burke, M., Edward, M.: Quantifying the influence of climate on human conflict. Am. Assoc. Adv. Sci. **341**, 6151 (2013)
8. Koskinen, O., Pukkila, K., Hakko, H., Tiihonen, J., Väisänen, E., Särkioja, T., Räsänen, P.: Is occupation relevant in suicide? J. Affect. Disord. 197–203 (2002)
9. Li, N., Kanazawa, S.: Country roads, take me home… to my friends: how intelligence, population density, and friendship affect modern happiness. Br. J. Psychol. 675–697 (2016)
10. Pesantez, J.: Visualizing online—hotel reputation in Barcelona through a robust compositional regression model and a principal component analysis. Mendeley Data **2** (2018)
11. Peña, D.: Análisis de Datos Multivariantes. McGraw Hill (2002)
12. Radojevic, T., Stanisic, N., Stanic, N., Davidson, R.: The effects of traveling for business on customer satisfaction with hotel services. Tour. Manag. **67**, 326–341 (2018)
13. Swamynathan, G., Almeroth, K., Zhao, B.: The design of a reliable reputation system. Electron. Commer. Res. 239–270 (2010)
14. Sänger, J., Pernul, G.: TRIVIA: visualizing reputation profiles to detect malicious sellers in electronic. J. Trust Manag. 3–5 (2016). https://doi.org/10.1186/s40493-016-0026-8
15. Wang, Y., Chan, F., Leong, H.V., Chi, S., Ngai, G., Au, N.: Multi-dimension reviewer credibility quantification across diverse travel communities. Knowl. Inf. Syst. **49** (2016)

Information Technology for the Synthesis of Optimal Spatial Configurations with Visualization of the Decision-Making Process

Sergiy Yakovlev, Oleksii Kartashov, Kyryl Korobchynskyi, Oksana Pichugina, and Iryna Yakovleva

Abstract The paper describes models and methods for solving the problem of optimal synthesis of spatial configurations of geometric objects. Special attention is paid to the issue of using modern information technologies, which allow automatically designing models of optimization problems and their subsequent solving by specialized software packages. It proposed tools for visualizing geometric information throughout the solution process. To a decision-maker, the presented technology allows changing the current configuration interactively. This instrument is of particular importance since most of the problems under consideration are NP-complete, and there are no methods for treating them guaranteed finding the optimal global solution.

Keywords Geometric information · Spatial configuration · Information technologies · Visualization

1 Introduction

Nowadays, interest in the development and implementation of information technologies for decision-support and decision-making, that require conducting a synthesis of complex technical systems and taking into account the spatial shape of their elements, is growing dramatically. This task is inextricably connected with geometric design problems, where a mapping of geometric information on material objects into Euclidean space is performed. Among such practical areas of applications are transportation and storage services logistics, aerospace objects' layout, the security of fuel and energy complexes, developing and improving 3D printing techniques, etc. They are all required considering the shape of the corresponding prototypes and

---

S. Yakovlev (✉) · O. Kartashov · K. Korobchynskyi · O. Pichugina
National Aerospace University "Kharkiv Aviation Institute", 17 Chkalova Street, Kharkiv 61070, Ukraine

I. Yakovleva
National University of Urban Economy, Kharkiv, Ukraine

H. Sharma et al. (eds.), *Congress on Intelligent Systems*, Advances in Intelligent Systems and Computing 1334, https://doi.org/10.1007/978-981-33-6981-8_3

imposing constraints on their relative positions. The synthesis of such spatial configurations of material objects includes their processing, transformation, and storage of geometric information about the objects. It is based on constructing and exploiting information-analytical models of the corresponding problems, developing and implementing modern information technologies. An integral feature of such technologies is a constant visualization of design decisions allowing their adjusting in an automated mode under decision-maker.

This research field is developing intensively worldwide. An example is a series "Springer Optimization and Its Applications" (https://www.springer.com/series/7393) focusing on models, methods, and information technologies of layout synthesis in aerospace engineering. Another illustration is a series "Lecture notes in logistics" (https://link.springer.com/bookseries/11220) dedicated to studying problems of Logistics Transportation, Warehousing and Distribution Services. Important results on mathematical modelling and software developing in geometric design problem are presented in [1–14].

Today, developing information systems for solving problems of synthesis of optimal configurations of complex spatial objects requires using mathematical models' designed in the automatic mode. For instance, a computer simulation assumes a geometric information transformation and constant visualization of the changes and current solutions. It is a stage of high importance and responsibility that requires human creativity. We pose the following problem, which effective solution can significantly facilitate this stage of decision making.

Problem statement: It is required to develop relevant information technology that

- forms an information-analytical model for describing data structures when creating complex spatial configurations;
- makes consolidation of data structures of geometric information used in optimization and visualization of spatial configurations;
- performs a dynamic transformation of the data structures when the synthesis and visualization of complex objects' spatial configurations are carried out.

2 Basic Concepts and Definitions

A class of problems of spatial configurations' synthesis associated with mapping of geometric information about a collection of material objects subject to certain constraints is singled out. Geometric information $g = (\{s\}, \{m\}, \{p\})$ about such an object includes a spatial shape $\{s\}$, metric parameters $\{m\} = (m_1, \ldots, m_k)$ that specify its sizes, and placement parameters $\{p\} = (p_1, \ldots, p_l)$ about the object position in space [15]. The class of problems associated with such a mapping of geometric information are solved in the scope of geometric design field. It is known that these optimization problems and extremely hard because of their multiextremality and high dimensionality. Typically, objects of the following shape are considered: a sphere, a cone, a parallelepiped, a polygon, as well as complex objects composed from them.

Their common feature is fixed metric characteristics of involved geometric objects. Problems of spatial configurations' synthesis for objects of arbitrary shape possess their own characteristics, thus requiring developing new mathematical models and special methods for their solution. Moreover, the necessity to make a synthesis of spatial configurations also requires further development and improvement of methods for processing, converting, and storing geometric information about material objects.

To describe an external structure and a type of a collection of material objects or their components, configuration space (CS) of geometric objects is formed [15–17] based on a formalization of a concept of geometric information $g = (\{s\}, \{m\}, \{p\})$. For an object S, an equation of its boundary

$$f(\xi, m) = 0,$$

where $\xi = (x, y)$, if $S \subset R^2$, and $\xi = (x, y, z)$ if $S \subset R^3$ underlies determining the components $\{s\}$ and $\{m\}$ of geometric information g. It is proposed using the boundary equation of 2D and 3D objects, whose basic classes are presented in papers [9, 10].

Let us consider a coordinate system $Oxyz(Oxy)$ in space $R^3(R^2)$, respectively, further referred to it as a fixed one. With an object S, we associate their own coordinate system, which origin is called a pole. A relative position of these coordinate systems characterizes the placement parameters $p = (p_1, \ldots, p_\beta) = (v, \theta)$ of the object. Here, v is a vector of the pole's coordinates of the object S in the fixed coordinate system, while v is a vector of angular parameters determining a relative position of the axes of their own and the fixed coordinate systems. The equation of its general position gives a position of a geometric object (GO) relative to the fixed coordinate system

$$F(\xi, m, p) = f[A(\xi - u), p] = 0$$

Here, A is an orthogonal operator defined through the angular parameters v. This equation underlies the formation of CS of GO, which determines a set of values of geometric variables called the generalized coordinates. They yield a position of GO or its components in space, relative to each other as well as to the fixed coordinate system. In the scope of this paper, we assume that $\Xi(S)$—the configuration space of GO S—is induced by the generalized variables $g = (m, p)$.

3 Configuration Space of Geometric Objects Based on Geometric Information

Let us consider a collection of objects $\Omega = \{S_1, \ldots, S_n\}$. By $\Xi(S_i)$, we denote a configuration space of an object S_i, which the generalized variables are $g^i = (m^i, p^i)$, $i \in J_n$, where $J_n = \{1, \ldots, n\}$. To each point $g^i \in \Xi(S_i)$, a parameterized object

$S_i(g^i) \subset R^3(R^2)$ will be associated. Let us form CS

$$\Xi(\Omega) = \Xi(S_1) \times \cdots \times \Xi(S_n)$$

of a set of basic GOs given by the generalized variables $g = (g^1, \ldots, g^n)$.

A mapping $\xi : \Omega \to \Xi(\Omega)$ satisfying a given set of constraints Λ determines a spatial configuration of GOs S_i, $i \in J_n$. In turn, this spatial configuration defines a collection $S_i(g^i)$, $i \in J_n$ of parameterized GOs. By set-theoretic operations, the last ones form a complex of GOs $S_B = B(S_1, \ldots, S_n)$, where B is an operator determining a structure of the collection S_i, $i \in J_n$. In the configuration space $\Xi(\Omega)$, the object S_B corresponds to a parameterized GO

$$S_B(g^1, \ldots, g^n) = B(S_1(g^1), \ldots, S_n(g^n)).$$

When the generalized variables $g^i = \hat{g}^i$, $i \in J_n$ are fixed, a point $\hat{g} = (\hat{g}^1, \ldots, \hat{g}^n) \in \Xi(\Omega)$ determines an image of a complex object

$$S_B(\hat{g}^1, \ldots, \hat{g}^n) = B(S_1(\hat{g}^1), \ldots, S_n(\hat{g}^n)).$$

Presence of the constraints Λ allows offering a typology of spatial configurations depending on relations GOs. Note that approaches of formalizing the relations are completely determined by choice of the generalized variables of CS, by constraints on a mutual location of GOs, and by their physicomechanical characteristics (see Fig. 1).

For the objects S_i, $i \in J_n$, let us introduce a binary relation $\{*\}$ of a pairwise non-intersection. We will use a notation $S' * S''$ if $\operatorname{int} S' \cap \operatorname{int} S'' = \emptyset$, i.e. if objects S' and S'' have no common internal points. If $S_i(g^i) * S_j(g^j)$ for all $i, j \in J_n$, $i \neq j$, then the spatial configuration is called a packing configuration. In most of the packing problems, one more object S_0 called a container is involved. In this case, all objects S_i, $i \in J_n$ must be placed into the container S_0.

Now, on the set of GOs, let us introduce one more binary relation—an inclusion relation $\{\circ\}$, where a denotation $S' \circ S''$ is used to reflect that $\operatorname{int} S' \subset S''$.

Assume that an object S_0 has the generalized variables g^0 in configuration space $\Xi(S_0)$. Let us form a new configuration space $\Xi(S_0, \Omega) = \Xi(S_0) \times \Xi(\Omega)$. Then a set of the generalized variables will define a layout configuration, if $S_j(g^j) \circ S_0(g^0)$ and $S_i(g^i) * S_j(g^j)$ for any $i, j \in J_n$, $i < j$.

The generalized variables $g^0, g^1, \ldots, g^n$ of configuration space $\Xi(S_0, \Omega)$ may also be subject to additional constraints. This induces special classes of packing configurations, such as a layout configuration [16, 18] when constraints on the minimum and maximum distances between objects are given. If the objects S_i, $i \in J_n$ are solid bodies with given masses, then a balanced system of such bodies yields a balanced packing configuration [16, 19].

Let a functional $\xi : \Xi(S_0, \Omega) \to R^1$ be given in configuration space $\Xi(S_0, \Omega)$. Now, we can formulate a problem of finding an optimal configuration of GOs S_i, $i \in$

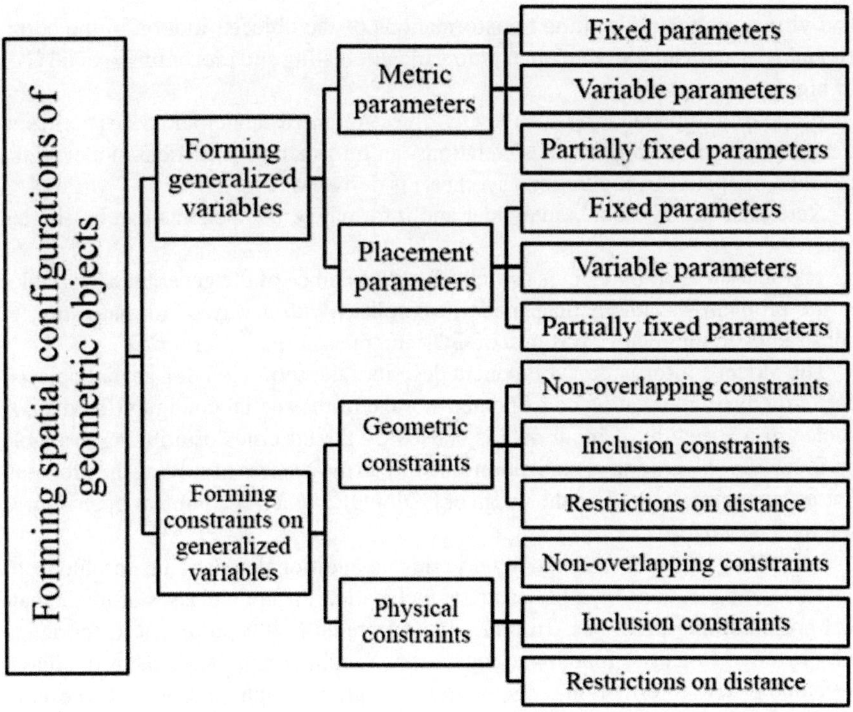

Fig. 1 Process of forming spatial configurations of geometrical objects

J_n. So, it is required to find

$$g^* = \arg \min_{g \in W \subseteq \Xi(S_0, \Omega)} \xi(g), \tag{1}$$

where W is a set of admissible configurations satisfying certain constraints Λ.

The generalized variables $g = \left(g^0, g^1, \ldots, g^n\right)$ of configuration space $\Xi(S_0, \ \Omega)$ take real values and allow equivalent formulating the problem (1) as a mathematical programming problem. To formalize constraints on relative positions of GOs, Φ-functions [20] and ω-functions [21] may be used. In this study, existing methods for constructing Φ-functions of basic 2D and 3D objects were summarized and generalized. They concern placement and packing problems with variable metric parameters of GOs and are utilized when the BaseObjects database was formed.

We propose an object-oriented model of a GO consisting of an abstract class GeometryObjectBase, which implements a collection of virtual methods with common operations for all geometric objects (GOs). Examples of such operations are reading information from a file or database, saving data, visualizing configurations, etc. A GeometryObjectBase class' descendants have two implementations depending on the GO-dimension—2D or 3D. Each of these classes contains fields and virtual methods for affine transformations of the objects' motion in the corresponding space. Such an implementation allows creating and processing spacial GOs of highly complex shapes.

Based on the peculiarities of the GO-object-oriented model and the specifics of mathematical modelling of GOs' relations, an information-analytical model of the problem of spatial configurations synthesis is designed (Fig. 2).

Note that, in the model, analytical and informative components can be singled out.

The analytical component is associated with a choice of the generalized variables in the problem's mathematical model, as well as with a way of formalization of constraints on the relative position of GOs and present quality criteria.

The model's information component describes the formation of a spatial objects' data structure and creating consolidated storage of the spatial configuration data. A choice of a structure of input data is caused by the specifics of utilizing available optimization solvers and visualization software in the process of solving the problem. For instance, we focused on the usage of COIN-OR, which is a popular open-source software package.

Depending on a chose of quality criteria and additional constraints, problems of optimal configurations' synthesis can be included in different classes of mathematical programming problems. To find a local extremum, it is sufficient to formalize the objective function, functional constraints, Jacobian and Hessian. In this case, developing new information technologies for optimal configurations' synthesis in complex systems requires constructing their information-analytical models in automatic mode. With this regard, the advantage of our approach is the possibility of

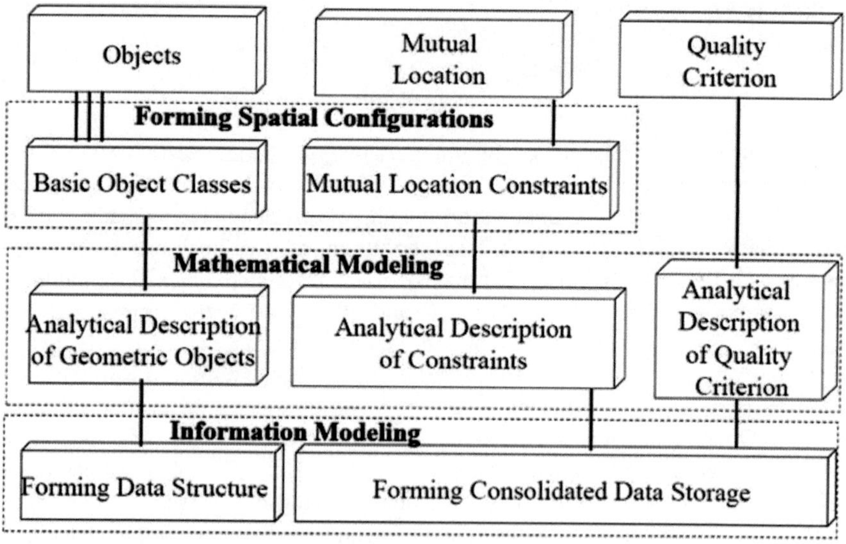

Fig. 2 Process of forming information-analytical model

implementing methods for solving such problems disregarding a research domain. Also, note that the approach is applicable for solving NP-hard problems ignoring their high dimensionality and multiextremality.

4 Information Technology Model for the Synthesis of Spatial Configurations

The information technology of spatial configuration synthesis consists of six interconnected blocks and is depicted in Fig. 3 in the form of an IFEF0 chart. In the figure, it is shown how the system functional blocks interact with each other. Also, the required steps to obtain a new configuration are shown.

Block 1 depicts a process of forming the model's analytical component with respect to a quality criterion and additional constraints.

In Block 2, depending on initial data, a spatial configuration data structure is formed using the generalized variables and configuration space.

Block 3 is responsible for the synthesis of the locally optimal spatial configurations with respect to the chosen quality criterion and generalized variables. Then, the designed mathematical model is analysed, and a choice of the software COIN-OR is justified. After that, the geometric information data structure is adapted for utilizing the solver. Finally, limitations on computing resources (memory size and runtime) are established, and recommendations on a relevant solution accuracy are developed. An output of the block is a spatial configuration meeting all the constraints.

In Block 5, processing and storage of results of spatial configurations' synthesis in a data warehouse "Configuration Repository" using DBMS are carried out.

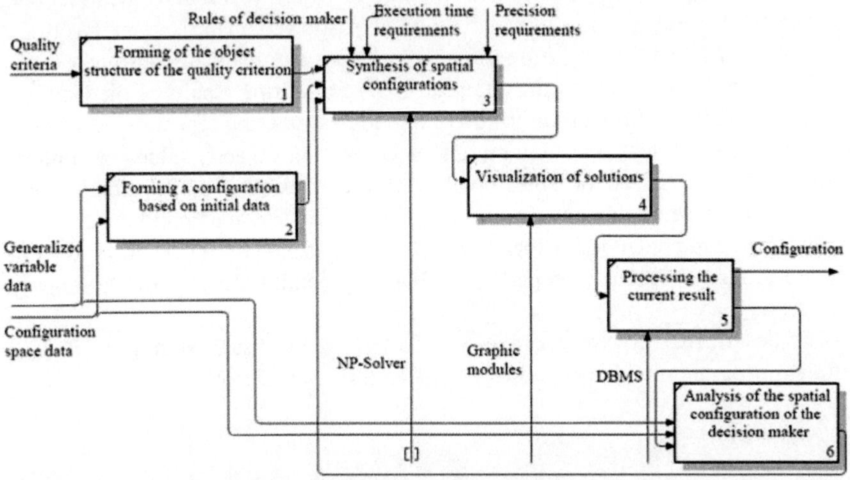

Fig. 3 Information technology for the synthesis of spatial configurations

Finally, Block 6 describes the possibilities of involving a decision-maker (DM) in an analysis of the found spatial configuration. Depending on the quality of a decision found so far, DM can either leave the decision unchangeable or proceed in order to improve it. In the latter, DM sets the new spatial configuration's generalized variables and repeat the above process again. Thus, parameters of the optimization model may be changed at the following iterations. Note the initial spatial configuration may be infeasible, but, according to the technology proposed, a locally optimal configuration are automatically synthesized, which satisfy all available constraints.

5 The Software Package Description

In the scope of models and methods of information technologies of layout synthesis of spatial configurations, a software application has been developed, including [22]:

- software application—a solver for implementation of methods of spatial configurations' optimization;
- repository program—a consolidated repository for storing information on the process of solving the problems;
- software applications for interactive visualization of the solutions obtained.

The software complex has a hierarchical structure and consists of three functional parts. The first component is the software application (SA) GeneralSolver that integrates a chosen Solver with necessary computational software components. GeneralSolver is used to dynamically transform the data structure of geometric information in the process of synthesizing spatial configurations of complex objects. In turn, this *SA* is based on the .NET Framework. The program validates the data entered by a user and verifies their correctness and compliance with the format. If incorrect data are entered, an error warning with clarification is displayed on the screen. If a critical error occurs while executing a program, the program automatically terminates.

The second component of the complex is SA for storing results of the GeneralSolver calculations. In addition to the evaluation results, the repository stores full information about spatial configurations of underlying objects, values of configuration space generalized variables involved in the synthesis, and information on a mathematical formulation of problems (quality criteria and constraints).

The third component is SA for 2D or 3D visualization. For 3D visualization, one can use different visualization packages: 3Ds Max Studio, Revit, VRED, Stingray, etc.

The described software package was developed in Visual Studio 2017 using different programming languages (C#, C++, and ECMA Script).

6 An Example of Implementing the Results

As an example of implementing the results, consider the following packing problem for spherical objects. It is required to place a set of spheres of a given radius into a sphere of minimum radius. Generalized variables in this problem are the coordinates of the centers of spheres (placement parameters) and radii (metric parameters). The mathematical model as a nonlinear optimization problem is given in [23–26]. NLP-Solver IPOPT (https://tomopt.com/tomlab/) was used to find local extrema. The interactive environment 3Ds Max Studio was used to display the results and interact with DM.

First, the format of the initial data of the mathematical model is converted into the form of the Solver representation. Then the initial values of the generalized variables are set, and the optimization process is carried out using IPOPT. The result (local extremum) is stored in a special format in the consolidated database and converted to the 3Ds Max Studio data format. As a result of the analysis of solution, the decision-maker can make any changes using the interactive graphical environment and (or) the associated numeric parameter values. Taking into account the changes made, we get a new starting point, which is again fed to the Solver. We have an iterative process that is repeated until a solution is obtained that satisfies the DM.

Below we will illustrate the visualization of the described process. Let's randomly generate the initial configuration of spherical objects. Note that the initial configuration may not satisfy the constraints of the problem, that is, the spheres may intersect, be outside the placement area and at large distances from each other. However, using IPOPT ensures that all the constraints are met. The DM can modify the obtained solution. As a result, using the export script, a new spatial configuration is formed, which is used as the starting point (Fig. 4). The improved locally optimal configuration is shown in Fig. 5. Other results of numerical experiments are described in [27]. The results are saved in text files, which allows you to return to any stage of the calculation at any stage. If the spatial configuration satisfies the DM, then it is saved in the form of graphic data, files with initial, intermediate and resulting data (Fig. 6).

7 Conclusion

The paper proposes models and methods for solving the optimization problem of the synthesis of spatial configurations of geometric objects. They utilize the concepts of geometric information and configuration space. It made it possible to create information technology for interactive solving such problems. This technology involves specially developed methods as well as modified already known methods that take into account specifics of problems under consideration. The presented approach allows for solving a wide class of geometric design problems. In particular, it was applied for layout, covering, partition problems and for monitoring and control systems.

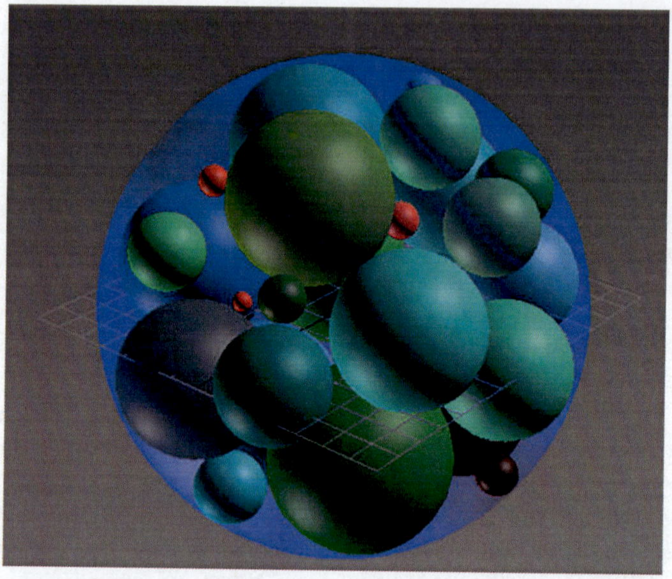

Fig. 4 The locally optimal solution

Fig. 5 The decision, modified by the decision-maker

Fig. 6 The improved configuration

References

1. Fasano, G.: A modeling-based approach for non-standard packing problems. In: Optimized Packings with Applications, vol. 105, pp. 67–85 (2015). https://doi.org/10.1007/978-3-319-18899-7_4
2. Bortfeldt, A., Wascher, G.: Constraints in container loading: a state-of-the-art review. Eur. J. Oper. Res. **229**(1), 1–20 (2013). https://doi.org/10.1016/j.ejor.2012.12.006
3. Waescher, G., Haussner, H., Schumann, H.: An improved typology of cutting and packing problems. Eur. J. Oper. Res. **183**, 1109–1130 (2007). https://doi.org/10.1016/j.ejor.2005.12.047
4. Fadel, G.M., Wiecek, M.M.: Packing optimization of free-form objects in engineering design. In: Fasano, G., Pintér, J.D. (eds.) Optimized Packings with Applications, pp. 37–66. Springer International Publishing, Cham (2015). https://doi.org/10.1007/978-3-319-18899-7_3
5. Sun, Z.-G., Teng, H.-F.: Optimal layout design of a satellite module. Eng. Optim. **35**(5), 513–529 (2003). https://doi.org/10.1080/03052150310001602335
6. Coggan, J., Shimada, K., Yin, S.: A survey of computational approaches to three-dimensional layout problems. CAD Comput. Aided Des. **34**(8), 597–611 (2002). https://doi.org/10.1016/S0010-4485(01)00109-9
7. Tian, T., Zhu, W., Lim, A., Wei, L.: The multiple container loading problem with preference. Eur. J. Oper. Res. **248**(1), 84–94 (2016). https://doi.org/10.1016/j.ejor.2015.07.002
8. Drira, A., Pierreval, H., Hajri-Gabouj, S.: Facility layout problems: a survey. Annu. Rev. Control. **31**(2), 255–267 (2007). https://doi.org/10.1016/j.arcontrol.2007.04.001
9. Stoyan, Y., Romanova, T.: Mathematical models of placement optimization: two- and three-dimensional problems and applications. In: Fasano, G., Pintér, J. (eds.) Modeling and Optimization in Space Engineering, vol. 73, pp. 363–388. Springer, New York (2013). https://doi.org/10.1007/978-1-4614-4469-5
10. Bennell, J., Scheithauer, G., Stoyan, Y.G., Romanova, T.: Tools of mathematical modelling of arbitrary object packing problems. J. Ann. Oper. Res. **179**(1), 343–368 (2010). https://doi.org/10.1007/s10479-008-0456-5
11. Stoyan, Yu., Pankratov, A., Romanova, T.: Placement problems for irregular objects: mathematical modeling, optimization and applications. In: Butenko, S., et al. (eds.) Optimization Methods and Applications, pp. 521–558. Springer, New York (2017). https://doi.org/10.1007/978-3-319-68640-0_25
12. Kiseleva, E.M., Koriashkina, L.S.: Theory of continuous optimal set partitioning problems as a universal mathematical formalism for constructing Voronoi diagrams and their generalizations. Cybern. Syst. Anal. **51**(3), 325–335 (2015). https://doi.org/10.1007/s10559-015-9725-x
13. Yakovlev, S.V.: On a class of problems on covering of a bounded set. Acta Math. Hung. **53**(3), 253–262 (1999). https://doi.org/10.1007/BF01953365
14. Shekhovtsov, S.B., Yakovlev, S.V.: Formalization and solution of one class of covering problem in design of control and monitoring systems. Avtom. Telemekh. **5**, 160–168 (1989). https://mi.mathnet.ru/eng/at6296
15. Stoyan, Y.G., Yakovlev, S.V.: Configuration space of geometric objects. Cybern. Syst. Anal. **54**(5), 716–726 (2018). https://doi.org/10.1007/s10559-018-0073-5
16. Yakovlev, S.V.: Configuration spaces of geometric objects with their applications in packing, layout and covering problems. In: Advances in Intelligent Systems and Computing, vol. 1020, pp. 122–132 (2019). https://doi.org/10.1007/978-3-030-26474-1_9
17. Yakovlev, S., Kartashov, O., Pichugina, O., Yakovleva, I.: Geometric information and its mapping in monitoring and control systems. In: 2019 IEEE 2nd Ukraine Conference on Electrical and Computer Engineering (UKRCON), pp. 1003–1006 (2019). https://doi.org/10.1109/UKRCON.2019.8879998
18. Stoyan, Yu.G., Semkin, V.V., Chugay, A.M.: Optimization of 3D objects layout into a multiply connected domain with account for shortest distances. Cybern. Syst. Anal. **50**(3), 374–385 (2014). https://doi.org/10.1007/s10559-014-9626-4

19. Stoyan, Yu., Romanova, T., Pankratov, A., Kovalenko, A., Stetsyuk, P.: Balance layout problems: mathematical modeling and nonlinear optimization. In: Space Engineering. Modeling and Optimization with Case Studies, vol. 114, pp. 369–400 (2016). https://doi.org/10.1007/s10559-015-9746-5
20. Scheithauer, G., Stoyan, Yu., Romanova, T.: Mathematical modeling of interactions of primary geometric 3D objects. Cybern. Syst. Anal. **41**(3), 332–342 (2005). https://doi.org/10.1007/s10559-005-0067-y
21. Yakovlev, S.V.: Formalizing spatial configuration optimization problems with the use of a special function class. Cybern. Syst. Anal. **55**(4), 581–589 (2019). https://doi.org/10.1007/s10559-019-00167-y
22. Yakovlev, S.V., Kartashov, O., Korobchynskyi, K.: The informational analytical technologies of synthesis of optimal spatial configuration. In: IEEE 13th International Scientific and Technical Conference on Computer Sciences and Information Technologies, vol. 1, pp. 140–143 (2018)
23. Hifi, S.M., M'Hallah, R.: A literature review on circle and sphere packing problems: models and methodologies. Adv. Oper. Res. **7**, 1–22 (2009). https://doi.org/10.1155/2009/150624
24. Stoyan, Yu.G., Scheithauer, G., Yaskov, G.N.: Packing unequal spheres into various containers. Cybern. Syst. Anal. **52**(3), 419–426 (2016). https://doi.org/10.1007/s10559-016-9842-1
25. Sutou, A., Day, Y.: Global optimization approach to unequal sphere packing problems in 3D. J. Optim. Theory Appl. **114**, 671–694 (2002). https://doi.org/10.1023/A:1016083231326
26. Yaskov, G.: Methodology to solve multi-dimensional sphere packing problems. J. Mech. Eng. **22**(1), 67–75 (2019). https://doi.org/10.15407/pmach2019.01.067
27. Yakovlev, S., Kartashov, O., Korobchynskyi, K., Skripka, B.: Numerical results of variable radii method in the unequal circles packing problem. In: 2019 15th International Conference on the Experience of Designing and Application of CAD Systems, CADSM 2019—Proceedings, pp. 1–4 (2019). https://doi.org/10.1109/CADSM.2019.8779288

Hybrid Approach to Combinatorial and Logic Graph Problems

Vladimir Kureichik, Daria Zaruba, and Vladimir Kureichik Jr.

Abstract The article is concerned with one of the possible hybrid approaches to combinatorial and logic graph problems. They belong to the class of NP-hard optimizations problems. The authors have suggested a hybrid approach to solve these issues more effectively. The search process is divided into several levels, which constitutes a distinctive feature of this approach. At the first level, the graph model is compressed by the method of fractal aggregation. Then, at the second level, the genetic algorithm is performed, and the graph model is decompressed at the third level. In order to realize this approach in practice, it has been developed the three-level hybrid algorithm which can obtain quasi-optimal solutions in polynomial time and avoid local optimum areas. In this article, it is considered a particular combinatorial and logic problem—a placement of graph vertices in the lattice. A software application has been developed to carry out a series of computational experiments on the basis of PECO test circuits. Conducted tests have shown the advantage of the suggested hybrid approach in comparison with already known optimization methods.

Keywords Hybrid approach · Combinatorial and logic problems · Graph · Fractal sets · Evolutionary modeling

1 Introduction

The use of a graph theory in state-of-the-art computer science and information technologies is becoming increasingly common. The development of effective methods to solve combinatorial and logic graph problems is essential in terms of modern artificial intelligence challenges [1–3]. Graph models are applied in various branches of science and have a high degree of formalization and universality.

V. Kureichik · D. Zaruba (✉) · V. Kureichik Jr.
Southern Federal University, Rostov-on-Don, Russian Federation
e-mail: dvzaruba@sfedu.ru

V. Kureichik
e-mail: vkur@sfedu.ru

H. Sharma et al. (eds.), *Congress on Intelligent Systems*, Advances in Intelligent Systems and Computing 1334, https://doi.org/10.1007/978-981-33-6981-8_4

New effective methods and algorithms to solve graph problems have been taking place through many years and remain highly relevant. This is due to the fact that combinatorial and logic problems are NP-complete and NP-hard [3]. Therefore, the development of universal methods and algorithms which can find precise optimal solutions in acceptable time is very difficult. One of the approaches to solve this issue is the use of bioinspired methods and its hybridization. It allows to reduce the search space and obtain quasi-optimal solutions in polynomial time.

2 The Problem Description

There are many combinatorial and logic graph problems but let us consider the graph partitioning and its placing on the plane. The partitioning (placement) problem is to find such a partition (variant of placement) from a set of possible alternative solutions to which the objective function is minimized (maximized) and taking into account all restrictions of the problem [4, 5].

Partitioning of a graph (hypergraph) into parts (or placement of graph vertices) belongs to the class of discrete optimization problems, because the objective function is discrete and there are a lot of restrictions on variables. Therefore, these problems belong to a special class of combinatorial tasks. In partitioning, placement, finding the maximum clique and independent sets, etc., the total number of solutions is equal to the number of permutations of n vertices, i.e.,

$$C_n = n!. \tag{1}$$

Taking into account restrictions on the constitution of subsets (m is the number of subsets), the overall number of alternative solutions is equal to

$$C_n^m = \frac{n!}{(n-m)!} \tag{2}$$

3 The Hybrid Approach

To date, there are several approaches to addressing NP-hard problems. The first approach is to simplify the algorithm, i.e., to reduce its time complexity using heuristic search procedures. The second one is to simplify the problem be reducing its dimensionality or its decomposition [6–11].

The authors suggest hybridizing these approaches and develop the multilevel method which implements various algorithms at different search levels. The key idea is that, at the first level, it is reduced dimensionality of the problem, but, at the second level, it is performed the optimization algorithm based on heuristic search

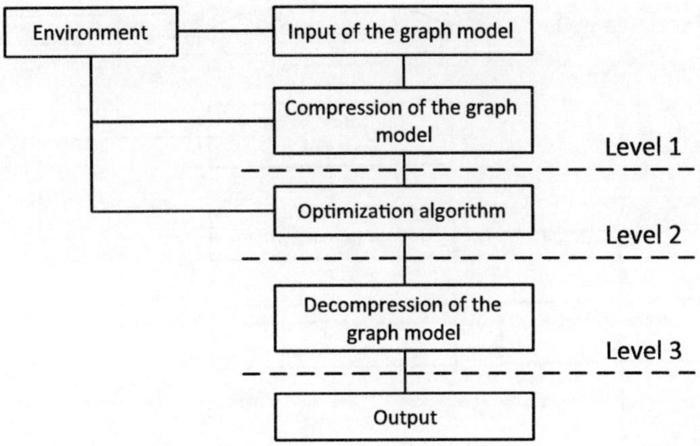

Fig. 1 Three-level hybrid search architecture

procedures. Lastly, at the third level, the reverse decomposition is implemented. The three-level hybrid search architecture is shown in Fig. 1.

Let us describe this architecture in more detail. At first, it is initialized the graph model of the combinatorial and logic problem. Then, it is compressed to reduce the problem dimensionality using well-known fractal aggregation mechanisms [12]. Compression is carried out in two steps: compression of the graph model horizontally and implementation of optimization algorithms based on evolutionary modeling methods [13, 14].

These methods can process large amounts of data and find quasi-optimal solutions to the problem in polynomial time.

After optimization, the graph is decompressed. Figure 2 shows the search architecture in more detail.

The block "Environmental" manages input data, graph compression, and implementation of evolutionary and genetic algorithms.

4 Hybrid Three-Level Algorithm

The authors have developed the hybrid three-level algorithm based on the suggested architecture which is represented in Fig. 3.

The hybrid three-level algorithm involves three levels: compression of the graph model, optimization procedures on the basis of the genetic algorithm and decompression of the graph model.

The hybrid three-level algorithm involves the following levels. At the first level, the graph model is compressed by the fractal aggregation to reduce the dimensionality of the problem. At the second level, there are implemented optimization procedures on the basis of the genetic algorithm. Here, a decision-maker chooses the evolutionary

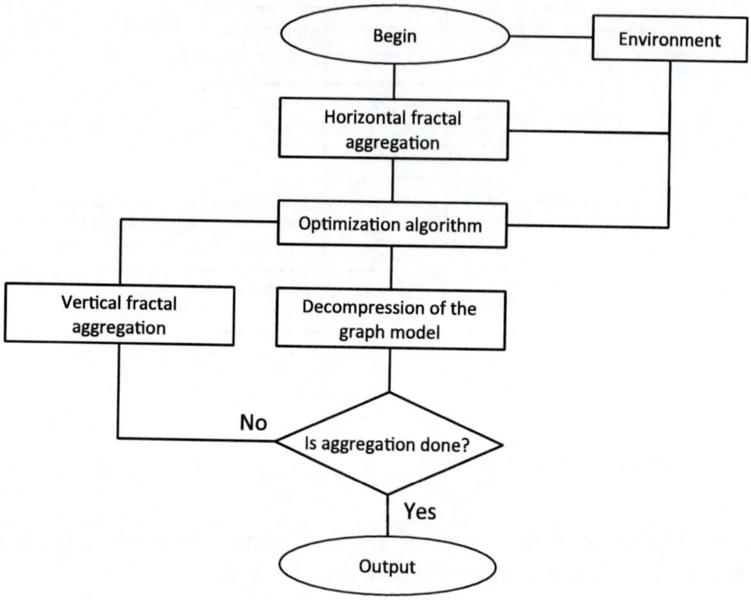

Fig. 2 Detailed hybrid search architecture

model, the selection method, and specified search operators [15]. At the third level, the graph model is decompressed.

Let us describe each block in the hybrid three-level algorithm in more detail.

1. Input of the graph model.
2. Compression of the graph model horizontally.
3. Generation of the initial population of alternative solutions taking into account external effects (the environment).
4. Calculation of minimum, maximum, and average values of the objective function for each chromosome.
5. Calculation of the average value of the objective function for population.
6. Reproduction, i.e., selection of parent chromosomes for further crossing.
7. Reproduction of descendants (the implementation of crossover, mutation, and inversion operators) with "the best" features.
8. Calculation of minimum, maximum, and average values of the objective function for each chromosome.
9. Selection of prospective alternative solutions on the basis of a combination of Darwin's and Lamarck's evolutionary theories [14].
10. Checks the condition "Is stop criterion reached?". If "yes", go to item 13. If "no", go to item 11.
11. The evolutionary adaptation.
12. Generation of the new population of alternative solutions and iterative implementation of the genetic algorithm until the stop criterion will be reached.

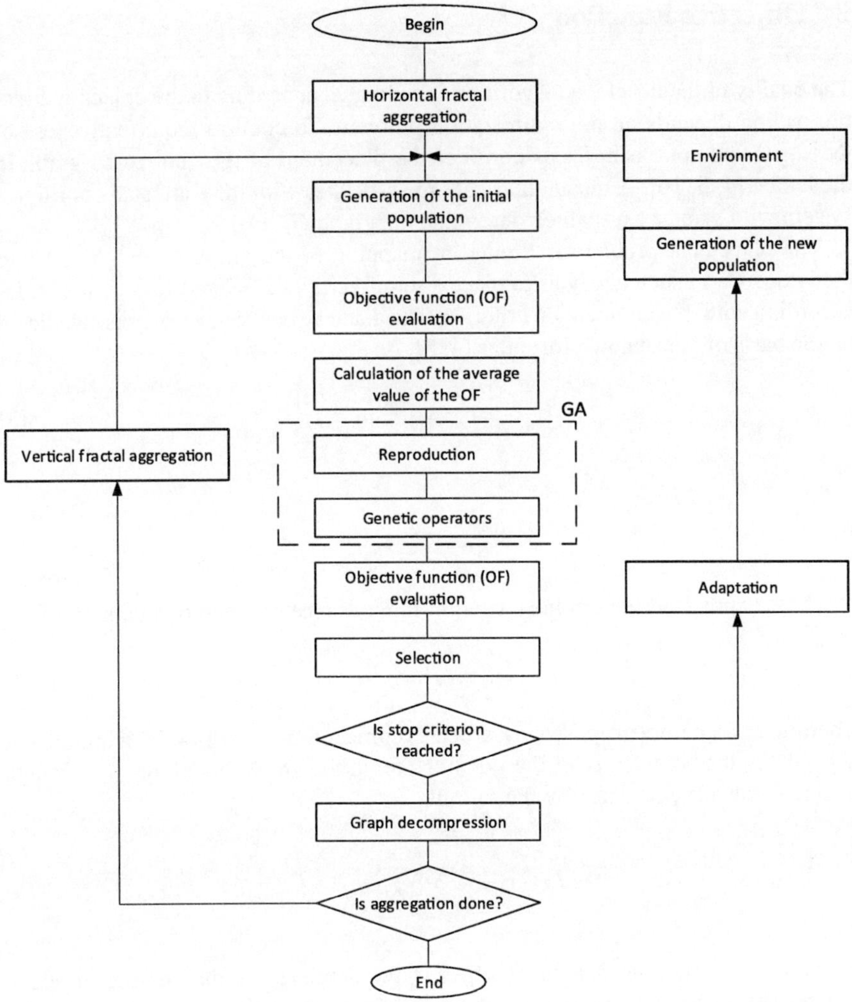

Fig. 3 Hybrid three-level algorithm

13. The influence of the environment.
14. Decompression of the graph model.
15. Check the condition of fractal aggregation.
16. Compression of the graph model vertically and implementation of the genetic algorithm until the stop criterion will be reached.

The suggested approach can reduce the dimensionality of the problem and obtain quasi-optimal solutions in polynomial time.

5 Objective Function

The quality of the developed algorithm is estimated according to the objective function which depends on the optimization problem. To confirm the effectiveness of the suggested approach, let us consider the placement of the undirected graph in the lattice [4, 5, 16]. A rectangular pattern associates with the Cartesian coordinate system with s and t axes which determine the graph G_r.

The placement problem is limited to mapping of the graph $G = (X, U)$ into the lattice G_r in such a way as to place a set of vertices $X = \{x_i\}$ into lattice nodes according with the optimization criterion. The distance between vertices is calculated on the basis of well-known formulas [4, 5]

$$d_{ij} = |s_i - s_j| + |t_i - t_j| \tag{3}$$

or

$$d_{ij} = \sqrt{|s_i - s_j|^2 + |t_i - t_j|^2}. \tag{4}$$

These expressions determine a straight-line distance between two points:

$$d_{ij} = (s_i - s_j)^a + (t_i - t_j)^a \tag{5}$$

where a is a coefficient posed by a decision-maker, $a = 1 \div 4$. This parameter depends on the complexity of the graph. Note, in the graph model, the total length of connections is calculated by the formula:

$$L(G) = \frac{1}{2}\sum_{i=1}^{n}\sum_{j=1}^{n} d_{ij} * c_{ij} \tag{6}$$

where $L(G)$ is the total length of edges in the graph, c_{ij} is the number of edges which connect vertices x_i and x_j. The aim of optimization is to minimize $L(G)$ ($L(G) \to \min$).

6 Experiments

The authors have developed a software application to demonstrate the effectiveness of the proposed three-level algorithm and determine its characteristics. The experiments have been conducted on the PECO benchmarks [17–19] with the use of well-known algorithms KraftWerk [20] and Capo 8.6 [21] and the hybrid three-level algorithm.

The total length of connections has been considered as the objective function. The main characteristics of PECO test circuits are presented in Table 1.

Table 1 PECO test circuits

№	Lattice size	The number of elements	The number of connections
1	10 × 10	100	184
2	95 × 95	9025	17,864
3	100 × 100	10,000	19,804
4	190 × 190	36,100	71,824
5	200 × 200	40,000	79,604

The dependencies of the quality of these algorithms on test circuits are shown in Table 2 and Fig. 4.

Having analyzed the obtained results, it can be concluded that the developed three-level algorithm exceeds the KraftWerk by 5% and the Capo 8.6 by 8% with comparable implementation time.

Table 2 Comparison of the placement results obtained by the KraftWerk, the Capo 8.6, and the three-level algorithm according to the objective function value

№	Optimal	KraftWerk	Capo 8.6	Hybrid algorithm
	HPWL	HPWL	HPWL	HPWL
1	184	202	184	184
2	17,884	18,302	22,764	18,012
3	19,804	20,519	21,314	20,135
4	71,864	75,384	89,814	73,386
5	79,604	82,335	98,041	81,674

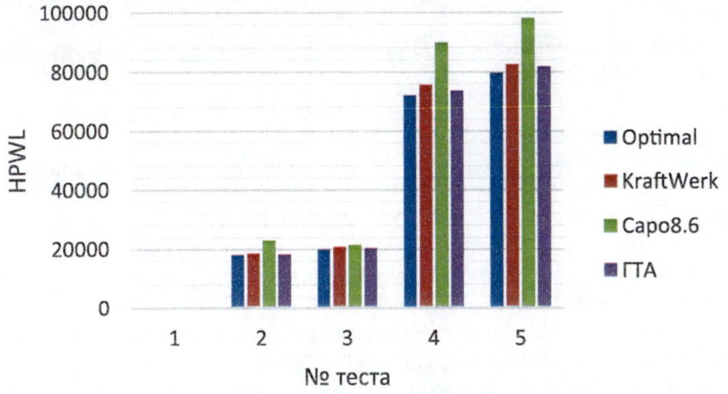

Fig. 4 Comparison of objective function values obtained by placement algorithms

7 Conclusion

The article provides the hybrid approach to solve combinatorial and logic problems on graphs. The distinctive feature of this approach is to divide the search process into several levels. At the first level, the graph model is compressed on the basis of fractal aggregation. At the second level, the optimization is implemented by the genetic algorithm, and, at the third level, the graph model is decompressed. To realize this approach, the authors developed the three-level algorithm which can obtain quasi-optimal solutions in polynomial time. Such combinatorial and logic problem as the placement of circuit elements in the lattice is considered to confirm the suggested approach. On the basis of the developed software application, it is conducted a set of experiments on the PECO benchmarks. The experimental results show advantage of the developed hybrid approach for combinatorial and logic problems in comparison with well-known methods.

The quality of placement in the lattice, obtained by the hybrid three-level algorithm, exceeded the KraftWerk and the Capo 8.6, on average, by 5–8% with comparable implementation time.

Acknowledgements This research is supported by a grant of the Russian Foundation for Basic Research (RFBR), project #19-01-00059.

References

1. Haggarty, R.: Discrete Mathematics for Computing. Pearson Education, UK (2002)
2. Ore, O.: The Theory of Graphs. Moskow (2008)
3. Cormen, H., Leiserson, C., Rivest, R., Stein, C.: Introduction to Algorithms, 3rd edn. MIT Press (2009)
4. Sherwani, N.A.: Algorithms for VLSI Physical Design Automation, 3rd edn. Kluwer Academic Publisher, USA (2013)
5. Alpert, C.J.: Handbook of Algorithms for Physical Design Automation. Auerbach Publications Taylor & Francis Group, USA (2009)
6. Hendrickson, B.: A multilevel algorithm for partitioning graphs. In: Proceedings of the 1995 ACM/IEEE Conference on Super Computing, pp. 626–657
7. Schloegel, K., Karypis, G., Kumar, V.: Multilevel Diffusion Schemes for Repartitioning of Adaptive Meshes. University of Minnesota, Department of Computer Science (1997)
8. Karypis, G., Kumar, V.: Analysis of Multilevel Graph Partitioning. Technical Report TR 95-037, Department of Computer Science, University of Minnesota (1995)
9. Hauck, S., Borriello, G.: An evaluation of bipartitioning techniques. In: Proceedings of Chapel Hill Conference on Advance Research in VLSI (1995)
10. Kureichik, V., Jr., Bova, V., Kureichik, V.: Hybrid approach for computer-aided design problems. In: International Seminar on Electron Devices Design and Production (SED). Proceedings. Prague (2019)
11. Kureichik, V., Zaruba, D., Kureichik, V.: Hybrid approach for graph partitioning. In: Advances in Intelligent Systems and Computing, vol. 573, pp. 64–73. Prague (2017)
12. Kureichik, V.V., Kureichik, V.M.: A fractal algorithm for graph partitioning. J. Comput. Syst. Sci. Int. **41**(4), 568–578 (2002)

13. Kacprzyk, J., Kureichik, V.M., Malioukov, S.P., Kureichik, V.V., Malioukov, A.S.: Evolutionary models of decision making. Stud. Comput. Intell. **212**, 23–114 (2009)
14. Kureichik, V.V., Kureichik, V.M., Sorokoletov, P.V.: Analysis and a survey of evolutionary models. J. Comput. Syst. Sci. Int. **46**(5), 779–791 (2007)
15. Kureichik, V.V., Kureichik, V.M.: Genetic search-based control. Autom. Remote Control **62**(10), 1698–1710 (2007)
16. Kureichik, L., Kureichik, V., Kureichik, V., Leschanov, D., Zaruba, D.: Hybrid approach for VLSI fragments placement. In: Advances in Intelligent Systems and Computing, vol. 679, pp. 349–358 (2018)
17. Kacprzyk, J., Kureichik, V.M., Malioukov, S.P., Kureichik, V.V., Malioukov, A.S.: Experimental investigation of algorithms developed. In: Studies in Computational Intelligence, vol. 212, pp. 211–223+227–236 (2009)
18. Adya, S.N., Markov, I.L.: Consistent placement of macro-blocks using floor planning and standard-cell placement. In: Proceedings of International Symposium on Physical Design, pp. 12–17 (2002)
19. Wang, M., Yang, X., Sarrafzadeh, M.: Dragon 2000: standard-cell placement tool for large industry circuits. In: ICCAD, pp. 260–263 (2000)
20. Caldwell, A.E., Kahng, A.B., Markov, I.L.: Can recursive bisection alone produce routable placements. In: DAC, pp. 477–482 (2000)
21. Roy, J.A., Papa, D.A., Markov, I.L.: Congestion-driven placement for standard-cell and RTL netlists with incremental capability. In: Nam, G.J., Cong, J. (eds.) Modern Circuit Placement. Series on Integrated Circuits and Systems, pp. 123–146. Springer Science + Business Media, LLC, Boston, MA (2007)

Energy-Efficient Algorithms Used in Datacenters: A Survey

M. Juliot Sophia and P. Mohamed Fathimal

Abstract The count of datacenters is drastically increasing as the need for cloud computing is increasing day by day. Datacenters are the major electricity consumers in the world (Khan et al. in IEEE Trans Cloud Comput, 2019 [1]). Successful cloud service providers (CSP) such as Google, Amazon, and Microsoft are having a huge number of datacenters worldwide with hundreds of servers (PMs) to satisfy their customers' requirements (Wang et al. in Comput Netw 161, 2019 [2]; Cheng et al. in Heterogeneity aware workload management in distributed sustainable datacenters. In: IEEE 28th International Parallel and Distributed Processing Symposium, 2014 [3]). Servers along with their network and cooling equipment require a huge amount of energy to operate. A study has shown that data centers will use around 3–13% of global electricity in 2030 compared to 1% in 2010 (Anders et al. in MDPI Challenges, 2015 [4]). Data centers not only consume a lot of energy but also release a lot of greenhouse gas (GHG). Energy consumption and carbon dioxide emission of datacenters are the major concerns of cloud service providers (CSPs). The benefits of cloud datacenters are given to us by the cost of global warming since the servers used in datacenters dissipate a lot of heat (Xu et al. in IEEE Trans Sustain Comput 4(1), 2019 [5]). Implementation of holistic approaches such as advanced frameworks, power-efficient electronic devices, and well-designed cooling and heating equipment are not sufficient to address the problem of energy consumption. Effective use of servers in datacenters will help the CSPs to face the energy efficiency challenge to a greater extent. Allocating services to power-efficient PMs and reducing the count of running PMs are the best practices followed in datacenters. The purpose of this survey is to study different kinds of algorithms used in datacenters in recent years to achieve energy efficiency targets and to minimize carbon dioxide emissions.

Keywords Consolidation · Virtualization · Migration · Containers · Virtual machines · Metrics

M. Juliot Sophia (✉) · P. Mohamed Fathimal
Department of Computer Science and Engineering, SRM Institute of Science and Technology, Chennai, India
e-mail: jm1071@srmist.edu.in

P. Mohamed Fathimal
e-mail: fathimap@srmist.edu.in

H. Sharma et al. (eds.), *Congress on Intelligent Systems*, Advances in Intelligent Systems and Computing 1334, https://doi.org/10.1007/978-981-33-6981-8_5

1 Introduction

Majority of electricity consumption in datacenters is because of non-technical equipments such as UPS, cooling infrastructure, and networking devices [6]. As shown in Fig. 1, only 26% of power consumption is only by servers and storage devices and the rest of the power consumption are by other devices [4]. Cloud service providers (CSPs) are taking all necessary measures to minimize the energy consumption and in turn to minimize carbon dioxide emission. Power management in datacenters can be acquired by using both hardware and software methods. Hardware methods include installation of energy-efficient servers, hot/cold aisle arrangement of servers, proper power distribution methods, and efficient cooling methods such as water control systems and air control systems. Software method requires policies and algorithms which can make the servers more efficient. Data centers also implement energy efficiency techniques through various layers such as application layer, datacenter layer, cluster layer, RAID layer, node layer, virtualization layer, OS layer, processor layer, disk layer, memory layer, and network layer [7]. Though datacenters comprise of thousands of servers, it is not true that all the servers are providing services all the time. Most of the time datacenter servers are either idle or underutilized [8]. Idle servers consuming unnecessary energy and without contributing in the execution of services are called 'comatose' or 'Zombie' [9]. Following power efficient methods to allocate the services and identifying zombie servers and putting them in sleep mode or turning OFF [10] are the best practices followed in datacenters. When the servers are turned OFF or in sleep mode, and if the number of client requests suddenly increase, then CSP cannot provide the service on time. Turning ON the server may take a while, so the requested service cannot be provided on time and with expected performance [11] which leads to SLA violation which is another greatest challenge encountered by CSPs. In order to accomplish customers' satisfaction with minimal amount of energy consumption, CSPs have started using different types of techniques in datacenters. Usage of bare metal alone cannot meet customer's growing demand so concepts such as virtual machines (VMs) and containers are introduced.

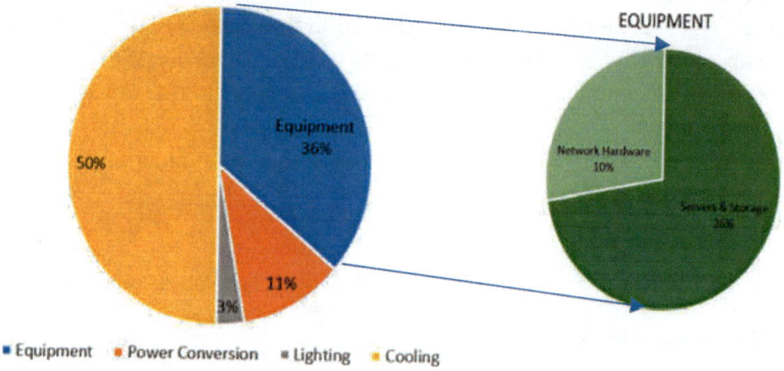

Fig. 1 Energy consumption in data centers

In this survey paper, we have discussed different techniques and energy-efficient metrics used in datacenters. We have conducted a detailed study on different types of algorithms available to minimize the energy consumption of data centers with both containers and VMs. Unlike other surveys instead of concentrating on the same type of algorithms, we have covered various algorithms based on different power efficiency metrics.

2 Techniques Used in Datacenters to Maintain Energy Efficiency

Cloud service providers such as Google, Amazon, and Microsoft are taking preventive measures to reduce the impact of datacenters they own in the environment. Google cloud use outside air for cooling and also building custom servers [12]. Amazon acquires energy efficiency by using a more energy-efficient server population and higher server utilization [13]. Microsoft concentrates on IT operational efficiency, IT equipment efficiency, datacenter infrastructure efficiency, and also use of renewable electricity to minimize energy consumption [14]. Along with the aforementioned techniques, cloud service providers are also following four main techniques for maintaining energy efficiency in their datacenters. Those techniques are as follows,

Virtualization [15]: Each server (PM) in a datacenter is capable of hosting more than one virtual machine (VM). VM is nothing but the emulation of PM. VM exhibits the behavior of a separate computer by abstracting the physical structure of numerous technologies, including hardware, operating system (OS), storage, and other networking resources. So by the means of virtualization, one physical server is capable of running more than one virtual system each with different operating systems. This in turn reduces the need for the number of physical servers up in the datacenters and so reduces the energy consumption and carbon dioxide emission.

Containerization [16]: Unlike virtualization instead of creating different virtual systems with different operating systems, single physical machine can host a number of containers without creating separate operating systems for the containers. Containers encapsulate applications with their required libraries and binaries in order to provide the application as a service [17]. It is proven that container-based data centers use less energy than virtualized data centers since containers would not require their own operating system and can use only the resources required for the application upon container start [18].

Live Migration of VMs/Containers: VMs and containers from underutilized/overutilized PMs can be migrated to energy-efficient PMs and idle PMs can be switched OFF or can be put into sleep mode. Live migration of VMs/containers between the PMs will reduce the number of running PMs in datacenters [8] (Fig. 2).

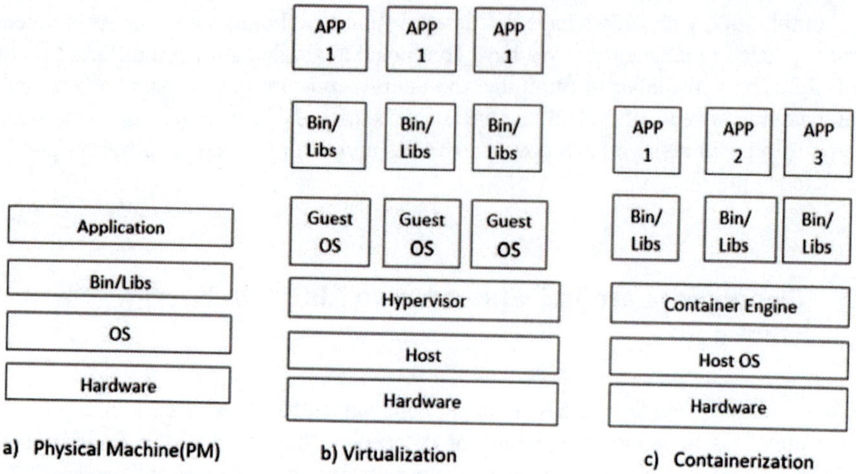

Fig. 2 Physical machine, virtualization, and containerization

Consolidation: Reducing the number of working servers in datacenters is called consolidation [19]. Energy consumption can be significantly reduced by consolidating applications on less number of servers and making idle servers sleep or power OFF [20]. Consolidating workload on an optimized number of servers and switching OFF idle ones is one of the most common solutions. Shutdown techniques can save at least 84% of the energy that would be otherwise lost to power idle nodes [21]. Resizing the number of servers in data centers can be implemented using various consolidation techniques [19]. Consolidation is the method of moving a VM/container from one PM to another [22]. This can be done by following the steps below,

1. Maintaining the list of servers according to the load
2. Set threshold (upper threshold, lower threshold)
3. Identify PMs with VMs/containers, not under the threshold
4. Identify the efficient PMs that can accommodate these VMs/containers
5. Move the VMs/containers to efficient PMs
6. Inefficient PMs should be power OFF/change to sleep mode.

After consolidation, the number of working PMs in datacenters will be less. The main drawback of consolidation technique is the possibility to violate service level agreement (SLA). A client's expectation from a CSP is maintained as a SLA agreement between them. It mainly involves,

- Completion of job on time,
- Low budget,
- Security, and
- Fulfilling the requirements.

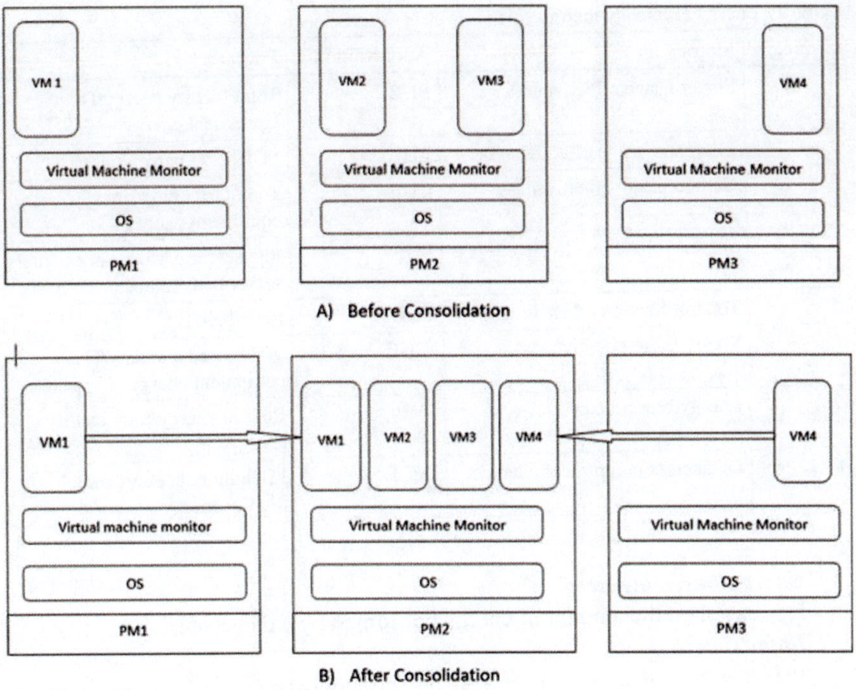

Fig. 3 Consolidation

There should not be a compromise in quality of service. As per [23] scalability, availability, security, portability, interconnectability, and cost are used to ensure quality of service in datacenters (Fig. 3) [20].

3 Energy Efficiency Metrics Used in Datacenters

CSPs follow different energy efficiency metrics to analyze energy consumption in datacenters. Power efficiency can be achieved by implementing modern state-of-the-art techniques at the cooling, power delivery, and management levels of the datacenters [24] but that is alone not enough to minimize energy consumption. IT workload should also be handled effectively. Power efficiency metrics in datacenters can be classified as follows [25],

a. ***Resource usage metrics***:
Utilization of resources like CPU, memory, bandwidth, and storage capacity by a single server or group of servers in a datacenter [4].

b. ***Heat aware metrics***:
Efficiency can be determined based on heat generated by a datacenter.

Table 1 Energy efficiency metrics

S. No.	Metrics	Abbreviation	Formula
1	Power usage effectiveness	PUE	Total facility energy/IT equipment energy
2	Energetic usage effectiveness	EUE	PUE/CPU percentage
3	Carbon usage effectiveness	CUE	Total GHG emissions/IT equipment energy
4	Server utilization effectiveness	SUE	Idle server performance/actual server performance
5	Total utilization effectiveness	TUE	PUE * SUE
6	Water usage effectiveness	WUE	Water used annually/IT equipment energy
7	Energy reuse factor	ERF	Reused energy/total facility energy
8	Datacenter compute efficiency	DcCE	IT equipment energy/total facility energy

c. ***Energy-based metrics***:
It is based on the amount of energy consumed by a datacenter.
d. ***Impact metrics***:
Performance of a datacenter can be measured based on its impact on the environment and economics.

Refer Table 1 for energy efficiency metrics used in datacenters [6, 26–28].

4 Different Types of Algorithms Used in Datacenters

Migration of VM/container is considered to be an optimization problem [29]. There are many optimization algorithms available to support VM migration. But choosing an efficient algorithm plays a major role in getting the best solution. Traditional optimization algorithms are proved as less effective when compared with heuristic/metaheuristic algorithms. Optimization problems can be effectively handled by three types of algorithms as follows,

Heuristic algorithms: Used to find solutions among all possible solutions but the best result is not guaranteed. E.g.: TSP, Tabu search, etc.

Metaheuristic algorithms [30]: The best result is guaranteed for the given problem by executing the available solutions iteratively. E.g.: simulated annealing, harmony search, etc.

Bio-inspired algorithms [31]: It is a type of metaheuristic algorithm with the ability to resolve complex problems with very simple initial conditions and without any search space knowledge. It is inspired by biological evolution. Solution for a given problem is identified by the way in which nature works. E.g.: genetic algorithm, ACO, firefly algorithm, etc.

5 Related Work

Ali et al. proposed a virtual machine replacement algorithm (VMR) [21] which ensures QoS and energy awareness in datacenters by fixing a threshold of the number of VMs that can run in a physical machine. All the PMs are arranged in increasing order of number of VMs accommodated by them $(1 - N)$. If a PM is under the threshold (first PM in the list), VMs from those PMs are allocated to PM (N) in last, if that PM is full and does not have resource provision to accept the new VM, then the VM can be placed in next PM $(N - 1)$.

Beloglazova et al. [32] state that though IT power consumption of data centers depends upon CPU, memory, disk storage, and network interfaces of PMs, the majority of power consumption is based on CPU utilization of the PMs. CPU utilization is proportional to the overall system load. CPU utilization will change over time so it is represented as a function of time. This paper proposes a method in which the initial VM placement is based on modified best fit (MBFD) algorithm, VMs are placed in decreasing order of their CPU utilization and placed in a PM which shows only minimal increase in power consumption as the result of new VM's placement. VMs for migration are selected based on double threshold, an upper and a lower threshold limit based on CPU utilization. If a PM falls under lower threshold, all the VMs from that PM should be migrated. If it falls above upper threshold, a few of the VMs should be migrated from that PM.

Usman et al. [33] used an interior search algorithm to allocate the VM's energy efficiently. ISA algorithm works on the basis of two groups "composition group and mirror group" [34]. Composition group consists of a set of fittest PMs identified for VM migration. In mirror group, fittest PMs and the best PM identified so far for migration is compared to provide efficient migration. A tuning parameter α (ranges from 0 to 1) is used to test PMs fitness.

Li et al. [35] suggested a work for both initial VM placement and dynamic migration of VMs by considering energy consumed by the servers and cooling device computer room air conditioner (CRAC). Total energy consumption is the sum of computing energy and cooling energy. GRANITE algorithm proposed in this paper works by the assumption that CPU capacity, memory size of VM, and task utilization are known or can be predicted in advance by CSP. GRANITE algorithm works in two phases: initial placement and dynamic migration.

Heba et al. [36] proposed a model based on locust-inspired scheduling algorithm. Unlike other models, VM allocation and migration are based on individual PM instead of a centralized server. As there is no centralized server, each PM will transfer itself based on processing capability as powerful server and weak server and based on load as heavy server and light server. It is similar to the behavior of locust which can change itself to be in a solitary phase or in a gregarious phase.

Liu et al. [37] proposed a model that works on the basis of prediction with a help of a linear regression algorithm container consolidation with CPU usage prediction (CNCUP). This paper is about providing container as a service (CaaS). CaaS is provided with the help of virtualized datacenter. For container consolidation, the requirement of the container and the VM, it has been placed is taken into account. The method used in this paper predicts the CPU utilization of PMs in the future through two phases-overutilized PM detection and underutilized PM detection. CNCUP algorithm is based on dynamic characterization. It optimizes the allocation of containers by means of both current and future requirements. Before placing a container in a PM, the changes in the PMs resource utilization are calculated and only if it falls under acceptable threshold the container will be placed in that corresponding PM.

Wang et al. [38] proposed thermal-aware datacenter workload placement algorithms TASA and TASA-B. The scheduling works in three steps: (1) Increase in temperature is predicted, (2) workload is distributed, and (3) Thermal information is updated. The proposed model defines datacenter in terms of total number of nodes and a thermal map. Thermal map represents ambient temperature in a datacenter which is a variable of three-dimensional space and time. Model proposed in this paper contains datacenter model and workload model, and a thermal-aware scheduler is also used to place the workload in an appropriate energy efficient PM. Periodic information of datacenter is used to backfill the PMs at correct time.

Yavari et al. [39] proposed a model for VM migration based on CPU utilization, memory consumption and temperature of servers. Model proposed by them consist of local and global manager. Each PM acts as local manager and records information such as CPU utilization, memory usage, and temperature and pass the information to global manager. One particular node will be used as global manager where the information of all local managers is stored. The execution of algorithm also takes place in global manager. VM placement is done with the help of heuristic energy and temperature-aware virtual machine consolidation (HET-VC) and firefly energy and temperature-aware virtual machine consolidation (FET-VC).

Choi [40] proposed a virtual machine placement algorithm with temperature and reliability of servers as the main constraints of VM placement. Based on the resource, availability servers are considered for VM placement but only the reliable servers with less power consumption and which avoid heat islands are considered. When a server is identified as capable of accommodating migrated VM, it is verified for its outlet temperature and if it is identified as over the maximum allowable temperature, VM would not be placed in that server in order to avoid heat island. Heat island occurs when heat generated during operation of the server will get accumulated as hot air behind the rack of that particular server. Heat islands occurrence in server racks affect the reliability of servers. This algorithm is formulated to minimize the power consumption of servers and ensure the reliability of the servers by maintaining availability.

Li et al. [41] proposed a model suitable for VM load balancing between both inter-host machines (inter-HMs) and intra-host machines (intra-HMs). MOVMrB is the algorithm proposed to optimize load balancing of multiple resources across host machines and within the individual host machine. Each resource is considered as different dimensions, whereas, in many other multiple resource algorithms, all the resources are considered as single dimension. They proposed a hybrid live migration algorithm to support both serial and parallel migration with the help of different transfer routes. This model also proposes an invalid migration filter which avoid migration of VM with same resource demands among HMs with same capacity.

Mishra et al. [42] proposed modified FAT tree architecture for building network topology in cost-effective datacenters. They considered, energy optimization problem as a MILP—mixed integer linear programming problem and proposed a MILP algorithm APFFR: active path first finding routing algorithm which uses an optimization solver software package CPLEX. The proposed model is used to minimize energy consumption of active servers, links, and switches. Energy minimization goal can be achieved by the controller which acts as an intermediate device between user and data center network (DCN). An objective function calculates the minimum energy consumed by active devices to transfer data from source IP (controller) to destination IP (server). User's data, controller's IP address, destination server IP address and objective functions are used as input. Minimized energy required to find the optimal route for data transfer from source to destination is the output.

Moon et al. [43] proposed a slave ants based ant colony optimization algorithm (SACO) for task scheduling in cloud datacenters. Global optimization problem is solved using slave ants to avoid long paths which are wrongly accumulated by pheromones of leading ants. The algorithm works with task manager, task scheduler, and resource manager of a datacenter. SACO algorithm does scheduling by the means of following three steps: (i) Cloud tasks and resource information is collected from resource manager, (ii) check whether the resource of VM meet the requirement of a task, (iii) if resources required is available, then the task is allocated to the VM by the scheduler.

Adhikary et al. [44] proposed a model of cloud infrastructure with clusters of servers. VM migration can be done within a cluster or among all the clusters within the whole datacenter. This paper is about the usage of two different algorithms inter-cluster and intra-cluster VM scheduling algorithms. Clusters in datacenters can be in active mode or in sleep mode. In active mode, the cluster is fully functional where all its servers and resource managers are fully functional and will be serving users' request, whereas, in sleep mode, whole cluster can shut its entire components. Each cluster with its information record database (IRDB) stores information of resource requirement of previous tasks. For each new request, IRDB checks the resource information and update the status to resource predictor (RP). RP with the help of resource manager (RM) and resource allocator (RA) assigns VMs for the requests. Each cluster calculates its occupancy ratio and gets occupancy ratio of its neighboring clusters and maintains neighborhood occupancy matrix (OMat).

Eswaran et al. [45, 46] proposed an optimization model with two basic properties (i) implementing solution representation with CPU and memory utilization as parameters. (ii) Finding optimal solution in an iterative manner with an algorithm, which works in the same way as intelligent water drops [7]. IWD algorithm mimics the way river flows toward the sea. River flow relies on the velocity of water drops, soil carried by water drops and soil deposited in the edge.

Yang et al. [47] used a hypergraph coverage model to provide energy-efficient storage in cloud data centers. Instead of using distributed storage system like Hadoop distributed file systems [HDFS] which follows same replication and storage strategy for all type of data. $\tilde{\kappa}$-covering algorithm is used so the replication and storage is based on user's data requirement only and not common for all the data. The coverage method proposed shows many, too many relationships among files, data blocks, racks, and nodes. Based on the data requirement of users, required data nodes will be functioning. Other data nodes which are not required will be identified and switched OFF (Table 2).

6 Conclusion

In this survey paper, we have discussed about different metrics and algorithms used by datacenters to identify zombie servers in order to minimize energy consumption without violating SLA. Instead of concentrating on only algorithms based on resource usage metrics, this survey covers algorithms of other metrics such as heat, energy, and impact also. From this study, it is understood that metaheuristic and bio-inspired algorithms can work more efficiently than heuristic algorithms. Most of the algorithms discussed here are suitable for virtualized as well as containerized datacenters with homogenous servers. Power-efficient algorithms suitable for both homogenous and heterogeneous datacenters are required. Most of the algorithms on resource usage metrics are considering only CPU utilization other resources such as memory and network should also be taken in to account. From this survey, we can conclude that the number of algorithms based on heat and impact metrics is very less while comparing to resource usage and energy-based metrics. So more research work should be devised to create models based on heat and impact metrics which can be used in heterogeneous datacenters by implementing bio-inspired algorithms to attain energy efficiency.

Table 2 Comparison of algorithms

S. No.	Algorithm	Description	Factors considered for migration	Advantages	Disadvantages
1	Virtual machine replacement algorithm (VMR) [23]	Servers are always rearranged based on the VMs run by them. If the count of VMs falls below the threshold, then those VMs will be moved to PM from the last of the list	Number of VMs	Very simple to execute since only number of VMs are considered for migration	Can work efficiently only on homogeneous workload
2	Modified best fit decreasing algorithm (MBFD) [25]	VMs are placed in decreasing order of CPU utilization. PMs which are not under double threshold will be taken into account for migration	Upper and lower CPU utilization threshold	Can work efficiently on heterogeneous workload	Migration is purely based on CPU utilization and other resources such as memory and network are not considered
3	Interior search algorithm (ISA) [27]	Best suitable PMs are placed in composition group based on their fitness value. Each PM in composition group is compared with global fitness PM identified so far with the help of tune factor. If the tune value is less the value of best PM identified so far, then it will be placed in mirror group and can be used for migration	Uses a tuning parameter α which ranges from 0 to 1	Very simple to execute since it requires only one tuning parameter. It is best suitable for both global as well as local search	Setting up the tuning factor is difficult

(continued)

Table 2 (continued)

S. No.	Algorithm	Description	Factors considered for migration	Advantages	Disadvantages
4	A greedy based scheduling algorithm minimizing total energy (GRANITE) [31]	The algorithm follows greedy method for initial VM placement. PM with least increase in power consumption is selected for VM placement. Dynamic migration is done in regular interval by verifying the threshold of CPU temperature. PM with more than 10% CPU temperature is selected for migration	CRAC temperature, CPU utilization	It works by considering the operations and thermal characteristic of datacenters	Performance degradation
5	A locust-inspired scheduling algorithm (LACE) [29]	Selection of phase and acceptance of VMs are the duty of individual PMs. LACE works with the help of global migration rule (GMR) and local migration rule (LMR)	Total number of free resources	Since the selection is based on individual PMs by themselves overloading can be avoided to a greater extent	Number of PMs will be ready to accept VMs during initial allocation and consolidation competition between the PMs is not addressed properly
6	Container consolidation with CPU usage prediction (CNCUP) [30]	Uses a prediction method to choose a PM for container placement by following two conditions. Condition 1: check whether the PM contains all necessary resources required to run the container. Condition 2: check whether the placement of container will not make the PM overutilized in near future	Power consumption, SLA violation rate, total number of container migration, average number of active PMs	Prediction of all the resources such as CPU, memory, and network takes place	It is only a prediction algorithm and for migration this algorithm should be implemented with some other PM selection algorithms

(continued)

Table 2 (continued)

S. No.	Algorithm	Description	Factors considered for migration	Advantages	Disadvantages
7	Thermal-aware scheduling algorithm (TASA) and thermal-aware scheduling algorithm with backfilling (TASA-B) [32]	TASA: jobs are scheduled periodically and a thermal map is maintained to record current temperature of all the nodes in datacenter. Jobs are arranged as hottest to coolest based on the temperature required to complete the job and nodes are arranged as coolest to hottest. Job with high temperature requirement can be placed in the node which coolest among all the nodes available. TASA-B: works the same as TASA but with a backfilling of nodes only if there is no violation in the starting time of the job and raise in maximum temperature	Node temperature, datacenter temperature, response time	Used to reduce the temperature generated in datacenters	Increase in response time of jobs

(continued)

Table 2 (continued)

S. No.	Algorithm	Description	Factors considered for migration	Advantages	Disadvantages
8	Heuristic energy and temperature-aware virtual machine consolidation (HET-VC) [33]	An objective function based on CPU utilization, memory usage and temperature of server is calculated for all non-under loaded and overloaded servers. Server with least objective function is selected for VM placement. Priority is given to server with high performance then the medium performance and last the low performance servers	Total energy consumption, number of migrations, number of VM replacements	Decreased number of VM migrations and minimized energy consumption	Trapping at local optimum is possible
9	Firefly energy and temperature-aware virtual machine consolidation (FET-VC) [33]	Intensity of servers is calculated using an objective function and if there are servers with lower intensity than the server in which the VM currently runs then the firefly algorithm is used to update new parameters in terms of CPU, memory, and temperature for all candidate servers. Finally, the algorithm selects the server with minimum objective function, CPU, memory, and temperature utilization is selected for VM placement	Total energy consumption, number of migrations, number of VM replacements, ESV-energy and SLA violations	Convergence in local optimum is avoided. Minimizes number of migrations	Increase in time complexity

(continued)

Table 2 (continued)

S. No.	Algorithm	Description	Factors considered for migration	Advantages	Disadvantages
10	Virtual machine placement algorithm (energy and reliability) [34]	Redundancy ratio is calculated to satisfy target reliability of servers. VM demand is sorted in descending order. PM with enough capacity for VM placement is selected, then power consumption of server and server rack is calculated. VM is placed in that particular VM only if the outlet temperature is within the maximum allowable temperature. Backup servers should also be arranged to ensure reliability	Outlet temperature, redundancy ratio, server capacity	Reliability of servers are maintained by saving the power consumed by servers	Higher failure rate of VM placements
11	Multi-objective VM rebalancing algorithm (MOVMrB) [35]	It is a biogeography-based optimization algorithm which divides the complex system into two subsystems Subs-H for inter-HM and Subs-V for intra-HM. The algorithm includes three major operations initialization, subsystem optimization, and elitists selection	Number of subsystems, number of islands per subsystems	Each resource is treated as separate dimension and load balance of each resource is maximized simultaneously in both inter and intra-host machines	Algorithm is tested only in tree network architecture with access switches and aggregate switches and not in other network architecture such as SDN

(continued)

Table 2 (continued)

S. No.	Algorithm	Description	Factors considered for migration	Advantages	Disadvantages
12	Active path first finding routing algorithm (APFFR) [36]	The algorithm finds all the path from source to destination and store it in an array path. First, path from array is taken and all the devices in the path are verified to check whether they are ON or not. If the devices are OFF they will be turned ON. Once all the devices are turned ON, an echo packet is sent in that path to check the traffic in that route. If high traffic is identified in that route, then that path will be deleted from array, and next path will be verified, if all devices are turned, and if there is no bottle neck identified, then the path will be selected, and data transfer will be completed. Once data transfer completed, all the devices in that path will be turned OFF	Number of servers, links, switches, and network traffic	Linear programming mathematical technique (MILP) is used to minimize or maximize a linear function such as cost or output consisting of several variables	CPLEX tool is required to formulate objective function

(continued)

Table 2 (continued)

S. No.	Algorithm	Description	Factors considered for migration	Advantages	Disadvantages
13	Slave ants based ant colony optimization algorithm (SACO) [37]	A slave ant with best make span among the group is selected as normal ant. Among all normal ants one ant with best mapping information between tasks and resources will be selected as queen ant. Global updating procedure will be done and if the queen ant's mapping is best then it will be assigned to the map	Number of cloud tasks, file size of tasks, number of CPUs, VMs. and hosts, CPU capacity, RAM, bandwidth	Long paths are avoided using diversification and reinforcement strategies. Minimal preprocessing overhead while comparing with other ACO algorithms	Produce efficient results only in homogenous cluster of servers
14	Inter-cluster VM scheduling algorithm [38]	When occupancy ratio of a cluster gets down of lower threshold (LTH). It checks its neighboring cluster which has minimum occupancy ratio. The occupancy ratio of selected neighboring cluster should be greater than the upper threshold (UTH). If it is going to be minimum than UTH, then all the VMs of PS cluster are moved to NS cluster selected and PS cluster will be put to sleep mode	Occupancy ratio, workload ratio, performance ratio	Putting whole cluster of servers in sleep mode will save huge amount of power consumption	Turning ON the cluster of servers during peak load all at a time is an additional overhead. Tested only in homogenous cluster of servers

(continued)

Table 2 (continued)

S. No.	Algorithm	Description	Factors considered for migration	Advantages	Disadvantages
15	Intelligent water drops algorithm (IWD) [41]	Like water drops always prefer the passageway with lesser soil, IWD algorithm finds out the best solution for efficient VM placement among all possible solutions available	It is based on the distance between the PMs. Only the PMs within the shortest distance will be selected for migration	Can handle energy efficiency metrics in heterogeneous datacenters	Not suitable for homogenous datacenters
16	$\tilde{\kappa}$-covering algorithm [44]	Data blocks are replicated number of times only based on dynamic user requirement. $\tilde{\kappa}$-covering is a complex problem in which K is not fixed. $\tilde{\kappa}$ traverse of hypergraph is solved using implicit enumeration	Data migration is not appreciated since there is a possibility of data collision during transmission of data among data nodes	$\tilde{\kappa}$-covering algorithm is best suited for dynamic requirement of users. It solved availability problem of various types of data by enabling only data nodes which are required to be up and running	Suitable for small-scale data centers only. Scalability and overhead issues are to be addressed in order to implement the model in large scale data centers

References

1. Khan, A.A., Zakarya, M., Buyya, R., Khan, R., Khan, M., Rana, O.: An energy and performance aware consolidation technique for containerized datacenters. IEEE Trans. Cloud Comput. (2019). https://doi.org/10.1109/TCC.2019.2920914
2. Wang, J.H., Wang, J., An, C., Zhang, Q.: A survey on resource scheduling for data transfers in inter-datacenter WANs. Comput. Netw. **161** (2019). https://doi.org/10.1016/j.comnet.2019.06.011
3. Cheng, D., Zhou, X., Ding, Z., Wang, Y., Ji, M.: Heterogeneity aware workload management in distributed sustainable datacenters. IEEE Trans. Parallel Distrib. Syst. In: IEEE 28th International Parallel and Distributed Processing Symposium (2014). https://doi.org/10.1109/IPDPS.2014.41
4. Anders, S., Andrae, G., Edler, T.: On global electricity usage of communication technology: trends to 2030. MDPI Challenges (2015). https://doi.org/10.3390/challe6010117
5. Xu, M., Toosi, A.N., Buyya, R.: iBrownout: an integrated approach for managing energy and brownout in container based clouds. IEEE Trans. Sustain. Comput. **4**(1) (2019). https://doi.org/10.1109/TSUSC.2018.2808493s
6. Uddin, M., Darabidarabkhani, Y., Shah, A., Memon, J.: Evaluating power efficient algorithms for efficiency and carbon emissions in cloud data centers: a review. Renew. Sustain. Energy Rev. **15** (2015). https://doi.org/10.1016/j.rser.2015.07.061
7. You, X., Lv, X., Zhao, Z., Han, J., Ren, X.: A survey and taxonomy on energy-aware data management strategies in cloud environment. IEEE Access **8** (2020). https://doi.org/10.1109/ACCESS.2020.2992748
8. Zhang, Q., Liu, L., Pu, C., Dou, Q., Wu, L., Zhou, W.: A comparative study of containers and virtual machines in big data environment. In: 2018 IEEE 11th International Conference on Cloud Computing (CLOUD), San Francisco, CA, USA, pp. 178–185, (2018) https://doi.org/10.1109/CLOUD.2018.00030
9. Whitney, J., Delforge, P.: Scaling up energy efficiency across the data center industry: evaluating key drivers and barriers. In: NRDC, Apr 2014
10. Khan, A.A., Zakarya, M., Khan, R.: Energy aware dynamic resource management in elastic cloud datacenters. Simul. Modell. Pract. Theory **92** (2019). https://doi.org/10.1016/J.SIMPAT.2018.12.001
11. Singh, P., Gupta, P., Jyoti, K.: Energy aware VM consolidation using dynamic threshold in cloud computing. In: IEEE International Conference on Intelligent Computing and Control Systems (2019). https://doi.org/10.1109/ICCS45141.2019.9065427
12. https://www.google.co.in/about/datacenters/efficiency/. Last accessed 10/08/2020
13. https://sustainability.aboutamazon.com/environment/the-cloud#section-nav-id-2. Last accessed 10/08/2020
14. https://blogs.microsoft.com/on-the-issues/2018/05/17/microsoft-cloud-delivers-when-it-comes-to-energy-efficiency-and-carbon-emission-reductions-study-finds/. Last accessed 10/08/2020
15. Wang, H., Tianfield, H.: Energy-aware dynamic virtual machine consolidation for cloud datacenters. IEEE Access **6** (2018). https://doi.org/10.1109/ACCESS.2018.2813541
16. Saleh, N., Mashaly, M.: A dynamic simulation environment for container-based cloud data centers using container CloudSim. In: Ninth International Conference on Intelligent Computing and Information Systems (2019)
17. Preeth, E.N., Mulerickal, F.J.P., Paul, B., Sastri, Y.: Evaluation of Docker containers based on hardware utilization. In: IEEE International Conference on Control Communication and Computing, India (2015). https://doi.org/10.1109/ICCC.2015.7432984
18. Mohallel, A., Bass, J.M., Dehghantaha, A.: Experimenting with Docker: Linux container and base OS attack surfaces. In: International Conference on Information Society (2016). https://doi.org/10.1109/I-SOCIETY.2016.7854163
19. Varasteh, A., Goudarzi, M.: Server consolidation techniques in virtualized data centers: a survey. IEEE Syst. J. **11**(2) (2017). https://doi.org/10.1109/JSYST.2015.2458273

20. Bermejo, B., Juiz, C.: Virtual machine consolidation: a systematic review of its overhead influencing factors. J Supercomput **76**, 324–361 (2020). https://doi.org/10.1007/s11227-019-03025-y
21. Rais, I., Orgerie, A.-C., Quinson, M., Lefèvre, L.: Quantifying the impact of shutdown techniques for energy-efficient data centers. In: Concurrency and Computation Practice and Experience (2018)
22. Ali, R., Shen, Y., Huang, X., Zhang, J., Ali, A.: VMR: virtual machine replacement algorithm for QoS and energy-awareness in cloud data centers. In: IEEE International Conference-2017. https://doi.org/10.1109/GREENCOMP.2010.5598295
23. Sawhney, J., Raisinghani, M.S., Idemudia, E.: Quality management in a data center: a critical perspective. In: 49th Annual Meeting of the Decision Sciences, Chicago (2018)
24. Mavus, Z., Angın, P.: A secure model for efficient live migration of containers. J. Wirel. Mob. Netw. Ubiquitous Comput. Depend. Appl. (JoWUA) **10**(3), 21–44 (2019). https://doi.org/10.22667/JOWUA.2019.09.30.021
25. Cupertino, L., Costa, G.D., Oleksiak, A., Piatek, W., Pierson, J., Salom, J., Sisó, L., Stolf, P., Sun, H., Zilio, T.: Energy-efficient, thermal-aware modeling. Ad Hoc Netw. **25**, 535–553 (2015). https://doi.org/10.1016/j.adhoc.2014.11.002
26. https://itpeernetwork.intel.com/the-elephant-in-your-data-center-inefficient-servers/#gs.cg3o3x. Last accessed 10/08/2020
27. https://www.missioncriticalmagazine.com/articles/87294-the-pursuit-of-low-pue. Last accessed 10/08/2020
28. Mentzelioti, G.L.D., Gritzalis, D.: A new methodology toward effectively assessing data center sustainability. Comput. Secur. **76** (2017). https://doi.org/10.1016/j.cose.2017.12.008
29. Ali, H.M., Lee, D.C.: Optimizing the energy efficient VM placement by IEFWA and hybrid IEFWA/BBO algorithms. In: International Symposium on Performance Evaluation of Computer and Telecommunication Systems (SPECTS), Montreal (2016), pp. 1–8. https://doi.org/10.1109/SPECTS.2016.757051
30. Fong, S.: Meta-zoo-heuristic algorithms. In: Seventh International Conference on Innovative Computing Technology (2017). https://doi.org/10.1109/INTECH.2017.8102456
31. Kumar, A., Bawa, S.: Generalized ant colony optimizer: swarm based meta-heuristic algorithm for cloud services execution. Computing **101**(11), 1609–1632 (2019). https://doi.org/10.1007/s00607-018-0674-x
32. Beloglazov, A., Abawajy, J., Buyya, R.: Energy-aware resource allocation heuristics for efficient management of data centers for cloud computing. Future Gener. Comput. Syst. **28**(5) (2010). https://doi.org/10.1016/j.future.2011.04.017
33. Usman, M.J., Ismail, A.S., Chizari, H., Aliyu, A.: Energy-efficient virtual machine allocation technique using interior search algorithm for cloud datacenter. In: 6th ICT International Student Conference (2017). https://doi.org/10.1109/ICT-ISPC.2017.8075327
34. Gandomi, A.H.: Interior search algorithm (ISA): a novel approach for global optimization. ISA Trans. **53**(4), 1168–1183 (2014). https://doi.org/10.1016/j.isatra.2014.03.018
35. Li, X., Garraghan, P., Jiang, X., Wu, Z., Xu, J.: Holistic virtual machine scheduling in cloud datacenters towards minimizing total energy. IEEE Trans. Parallel Distrib. Syst. **29**, 1 (2017). https://doi.org/10.1109/TPDS.2017.2688445
36. Kurdi, H.A., Alismail, S.M., Hassan, M.M.: LACE: a locust-inspired scheduling algorithm to reduce energy consumption in cloud datacenters. IEEE Access **6**, 1 (2018)
37. Liu, J., Wang, S., Zhou, A., Xu, J., Yang, F.: SLA-driven container consolidation with usage prediction for green cloud computing. Front. Comput. Sci. **14**(1) (2020). https://doi.org/10.1007/s11704-018-7172-3
38. Wang, L., Khan, S.U., Dayal, J.: Thermal aware workload placement with task-temperature profiles in a data center. J. Supercomput. **61**, 780–803 (2012). https://doi.org/10.1007/s11227-011-0635-z
39. Yavari, M., Rahbar, A.G., Fathi, M.H.: Temperature and energy-aware consolidation algorithms in cloud computing. J. Cloud Comput. **8**, 13 (2019). https://doi.org/10.1186/s13677-019-0136-9

40. Choi, J.: Virtual machine placement algorithm for energy saving and reliability of servers in cloud data centers. J. Netw. Syst. Manag. **27**, 149–165 (2019). https://doi.org/10.1007/s10922-018-9462-3
41. Li, R., Zheng, Q., Li, X., Wu, J.: A novel multi-objective optimization scheme for rebalancing virtual machine placement. In: IEEE 9th International Conference on Cloud Computing (2016). https://doi.org/10.1109/CLOUD.2016.0099
42. Mishra, J., Sheetlani, J., Reddy, K.H.K.: Data center network energy consumption minimization: a hierarchical FAT-tree approach. Int. J. Inf. Technol. (2018). https://doi.org/10.1007/s41870-018-0258-1
43. Moon, Y., Yu, H., Gil, J.-M., Lim, J.B.: A slave ants based ant colony optimization algorithm for task scheduling in cloud computing environments. Hum. Cent. Comput. Inf. Sci. **7**, 28 (2017). https://doi.org/10.1186/s13673-017-0109-2
44. Adhikary, T., Das, A.K., Razzaque, M.A.: Energy efficient scheduling algorithms for data center resources in cloud computing. In: 2013 IEEE International Conference on High Performance Computing and Communications (HPCC) & 2013 IEEE International Conference on Embedded and Ubiquitous Computing (EUC). https://doi.org/10.1109/HPCC.and.EUC.2013.244
45. Eswaran, S., Daniel, D., Jayapandian, N.: An augmented intelligent water drops optimization model for virtual machine placement in cloud environment. IET Netw. (2020). https://doi.org/10.1049/iet-net.2019.0165
46. Shah-Hosseini, H.: The intelligent water drops algorithm: a nature-inspired swarm-based optimization algorithm. Int. J. Bio-Inspir. Comput. **1** (2009). https://doi.org/10.1504/IJBIC.2009.022775
47. Yang, T., Pen, H., Li, W., Yuan, D., Zomaya, A.Y.: An energy-efficient storage strategy for cloud datacenters based on variable K-coverage of a hypergraph. IEEE Trans. Parallel Distrib. Syst. **28**(12), 3344–3355. https://doi.org/10.1109/TPDS.2017.2723004

Software Log Anomaly Detection Method Using HTM Algorithm

Rin Hirakawa, Keitaro Tominaga, and Yoshihisa Nakatoh

Abstract In software development, log data plays an important role in understanding the behavior of the system at runtime and also serves as a clue to identify the cause of the defects. In the log messages generated by systems such as Android, a wide variety of component outputs are mixed and makes it very difficult to identify the problem from the vast amount of log data. The ultimate goal of this research is to develop a GUI tool to visualize anomalies in log messages and reduce the burden on the log analyst. In this paper, we propose a method for learning time-series patterns of logs based on features created from structured log messages. The proposed model is evaluated using the open dataset HDFS, and confirmed its effectiveness through detection score comparison with the baseline methods. As a result, it was shown that proposed method can detect anomalies in the accuracy comparable to baseline methods customized for log data.

Keywords Anomaly detection · Log data analysis · Hierarchical temporal memory · Time-series learning · Online prediction · Unsupervised learning · Anomaly score

1 Introduction

Log data is important for understanding the behavior of software systems and analyzing the causes of problems. However, in recent years, as systems have become larger and faster, manual log data analysis is becoming less and less practical [1]. Especially in the system like Android generates log data where the output from many components is interleaved. The procedures for analyzing such data are redundant and

R. Hirakawa · Y. Nakatoh (✉)
Kyushu Institute of Technology, 1-1 Sensuicho, Tobata Ward, Kitakyushu, Fukuoka, Japan
e-mail: nakatoh@ecs.kyutech.ac.jp

K. Tominaga
Panasonic System Design Co., Ltd., 3-19, Shinyokohama, Kohoku-ku, Yokohama 222-0033, Japan
e-mail: tominaga.keitaro@jp.panasonic.com

H. Sharma et al. (eds.), *Congress on Intelligent Systems*, Advances in Intelligent Systems and Computing 1334, https://doi.org/10.1007/978-981-33-6981-8_6

slow down the project. As a result, tools to visualize log statistics and changes over time steps are increasingly becoming essential in log analysis.

The goal of our study is to achieve a visualization tool that helps analysts to discover the cause of the anomaly, by automatically assigning temporal anomaly scores to the log data. In this paper, we propose a method for calculating anomaly score based on the internal state of the model that learns the temporal patterns of structured log data. In general, it is difficult to assign an anomaly label to log data that exceeds tens of thousands of lines, so our method uses unsupervised time-series learning model Hierarchical Temporal Memory (HTM) [2, 3] for anomaly detection.

2 Background

In general, anomaly detection methods based on log data can be divided into two categories: supervised and unsupervised learning. If the configuration of the system evolves over time, the nature of the anomalies in the log data may change as well, so that regular model updates are required. Unsupervised learning methods do not require labels for learning, so constant learning online can prevent degradation quality of predictions. Unsupervised learning is also suitable for cases where the amount of data in the log is too large to be labeled.

Related Works. There are several methods for time-series learning of log data, such as DeepLog [4], LogAnomaly [5] and LogRobust [6], and these methods use LSTM. These methods are capable of detecting anomalies in large log datasets with high accuracy. It has been shown that anomaly detection (online prediction) is also possible in parallel with learning.

There is also the method [7] for distance-based anomaly detection by clustering some features extracted from log data. However, such a method requires expert knowledge of the software domain to select the appropriate parameters in advance. A more generalized log-based anomaly detection method should ideally be able to automatically extract event templates and parameters from log data online.

Previous Study. In our previous studies, we have treated log messages as natural language information and attempted to train some models using word IDs, word distributed representation, and document distributed representation [8].

When inputting information such as log template IDs, these methods must allocate embedding to respective IDs and consumes a lot of memory. This requires a lot of time and resources to training the model. In our proposed method, we encode the log's template IDs into a sparse binary array and conduct one-shot learning for the time series patterns of the log data. This makes the learning of time series patterns more efficient than the methods of related and previous studies.

3 Proposed Method

3.1 Log Parser

Since the proposed method uses IDs of the log template as input, structuring of the log messages in advance is necessary. Using the toolkit published in Logparser [9], log event templates can be extracted. The log parser learns event templates from unstructured logs and convert raw log messages into a sequence of structured events. Automatic log analysis can also extract parameters, which are strings embedded in event templates. The proposed method uses only the information of templates.

3.2 Method Overview

The proposed method uses a log parser to convert the raw log messages to the event templates category sequence $K = \{k_1, k_2, \ldots, k_n\}$, and then encodes each element of K into a binary array $S = \{s_1, s_2, \ldots, s_n\}$ representing each category (Fig. 1). s_i is a sparse matrix of fixed-length l whose element values from $\frac{l}{c}j$th to $\frac{l}{c}(j+1)$th are 1 and all other are 0, when the total number of template categories is c, and the category ID is j. The HTM layer receives the ith element s_i of S in time step i, and learns the time series pattern of S. The layer predicts the next time input s_{i+1} based on past input information up to $s_1, s_2, \ldots, s_i$. Since the predictions are represented as firing patterns of cells (neurons) in the HTM layer, the time series anomaly score

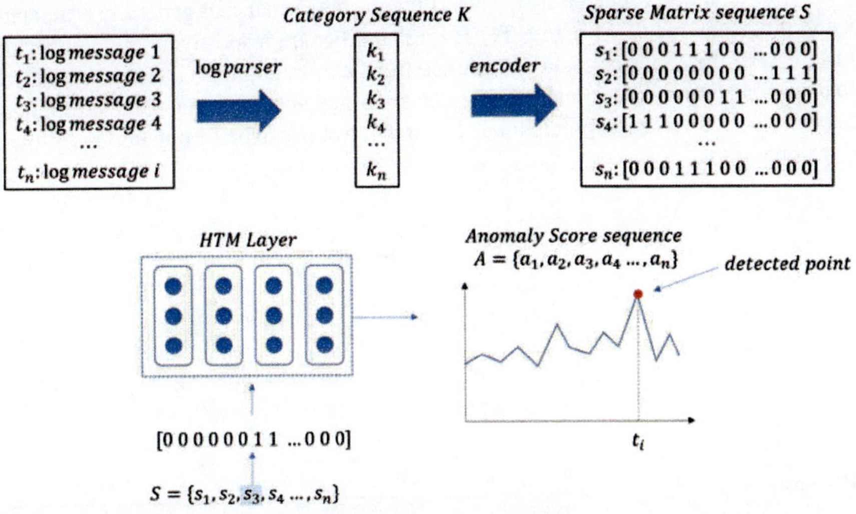

Fig. 1 Overview of the proposed log anomaly detection method

at the next time $i + 1$ can be calculated by assessing the rate of overlap with the firing pattern when s_{i+1} actually enters. Anomaly scores $A = \{a_1, a_2, \ldots, a_n\}$ is output corresponding to the time of each log message. The definition of anomaly score is similar to our previous study [8]. At the anomaly detection stage, the anomaly score sequence A is output similar to training stage, with the internal parameter update stopped. Then, the time t_i at which the abnormal log message is input is determined by performing threshold processing based on the empirically set threshold.

4 Evaluation

4.1 Experiment Setting

Dataset. As a dataset for evaluation experiments, we use the demo version HDFS dataset (100 k) in the Loglizer [1] toolkit. HDFS [10] is log data collected on a Hadoop distributed file system, which has been manually labeled as anomalous by Hadoop domain experts. Labels are given on a block-by-block basis, with 313 of the total 7940 blocks being anomalous blocks (Table 1).

Baselines. We compare the proposed method to the seven baseline methods implemented in the Loglizer benchmark toolkit. Five of these are unsupervised learning methods (PCA [11], Invariants Miner [12], Log Clustering [13], Isolation Forest [14], LR [15]); the other two are supervised learning methods (Decision Tree [16], SVM [17]). All method parameters have not been changed from the initial state of the tool.

Metrics. For anomaly detection, the classification performance is generally evaluated using precision, recall, $f1$-score. TP, TN, FP and FN are respectively the number of instances of the abnormal data determined to be abnormal, normal data determined to be normal, normal data determined to be abnormal and abnormal data determined to be normal. In this case, precision, recall and $f1$-score can be formulated as follows (Formulas 1–3).

$$\text{precision} = \frac{\text{TP}}{\text{TP} + \text{FP}} \tag{1}$$

$$\text{recall} = \frac{\text{TP}}{\text{TP} + \text{FN}} \tag{2}$$

Table 1 Breakdown of HDFS datasets

# of block	Total	Anomaly	Normal
Train	588	37	1551
Test	6352	276	6076

Table 2 Condition of the HTM layer

Parameters	Value
boostStrength	3.5158429366572728
columnShape	(2048)
localAreaDensity	0.04884751668248205
potentialPct	0.9076699930632756
potentialRadius	2363
synPermActiveInc	0.032426261405406734
synPermConnected	0.14503910955639598
synPermInactiveDec	0.009950307527727686
activationThreshold	11
cellsPerColumn	8
initialPerm	0.22083745592443638
maxSegmentsPerCell	161
maxSynapsesPerSegment	48
minThreshold	8
newSynapseCount	18
permanenceDec	0.0806042024027492
permanenceInc	0.08121052484196703

$$f1\text{-score} = \frac{2 \times \text{precision} \times \text{recall}}{\text{precision} + \text{recall}} \tag{3}$$

Also, in a model that outputs a decision score for each instance as the proposed method, the analysis is needed to set up the appropriate threshold. This is usually determined using visualizations such as precision–recall (PR) curve and ROC curve. We used these visualizations to investigate the variability characteristics of the classification performance of the proposed method with respect to threshold variations. In comparison to each baseline, the values of precision and recall are used at which the $f1$ score of the proposed method is best.

Implementation. The HTM layer in the proposed method is implemented based on htm.core [18] library which is the community fork of the nupic.core [19] (official implementation of HTM algorithm). The ScalarEncoder included in the library is used in the conversion of the categorical sequence into the sparse matrix sequence.

Model Condition. The conditions of the HTM layer to be used in the proposed method and the list of encoder parameters of the sparse matrix are shown in Tables 2 and 3. The number of classes of encoder parameters is automatically determined by the number of unique log templates in the training data.

Table 3 Sparse matrix encoder parameters

Parameter	Value
Num class	14 + 2 (OOV, PAD)
activeBits	21
Total length	21 * 16 = 336

4.2 *Input Format*

Preprocessing. In the HDFS dataset, anomaly labels are assigned on a block-by-block basis, so the regular expression is used for to separate the log data into blocks. The first 20% of the total dataset was used to train the models for each method, and the accuracy of anomaly detection is compared using the remaining 80% of them. The training data is randomly shuffled.

Streaming Format. In the proposed method, the log template ID sequence for each block is entered only once, starting from the beginning. The breakdown of the instances to be entered is the same as in Table 1. The proposed method also requires assigning a single representative anomaly score for each instance. Therefore, the average value of the anomaly score calculated for each element of the template ID sequence in each instance is used as an anomaly score for each block. Also, the internal state of the model is reset when the input reached the end of the template ID of each instance.

Online Prediction. Assuming that log anomality changes over time, we evaluate the accuracy when the model is trained in parallel with calculation of anomaly scores for the entire log data. The input method is a streaming format, and the accuracy evaluation is performed as the other methods.

4.3 *Results*

Accuracy. A comparison of the precision, recall and $f1$-score of each model is shown in Table 4. Each baseline model shows an overall high precision, but there is considerable variation in the value of the recall. The learning model in the streaming format of the proposed method is particularly good for recall values.

Discussion. The learning curve of the proposed method in streaming format is shown in Fig. 2. The anomaly scores at each training step (instance) converge early in the training data, and we can see that the model is able to learn stably the time series pattern. There is a part where the anomaly score is swept away like an outlier on the way, and we confirmed that the value corresponds to an abnormal instance included in the training data. The behavior of the model during training is a subject for future investigation.

Table 4 Accuracy comparison on HDFS datasets

Model	Precision	Recall	$F1$
PCA	0.976	0.304	0.464
Invariants miner	0.979	0.518	0.678
Log clustering	0.979	0.518	0.678
Isolation forest	0.979	0.518	0.678
LR	1.000	0.232	0.376
SVM	1.000	0.232	0.376
Decision tree	1.000	0.004	0.007
HTM-streaming	0.974	**0.533**	**0.689**
HTM-online prediction	0.902	**0.533**	0.67

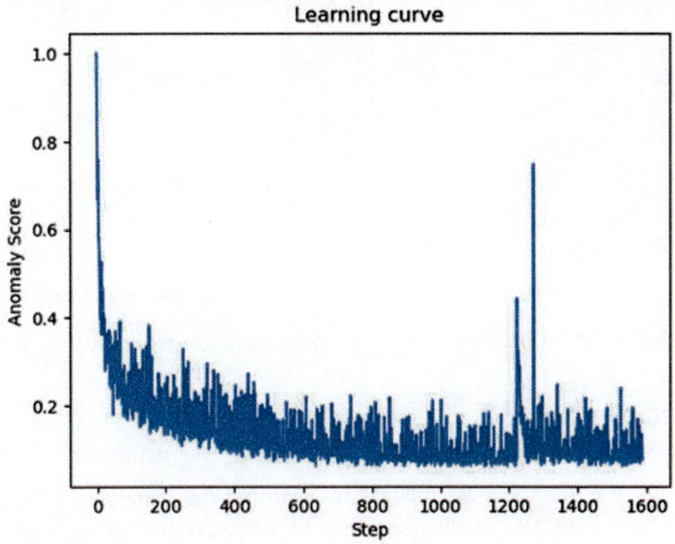

Fig. 2 Learning curve of HTM model in streaming format

Online Prediction. Figure 3 shows the precision–recall curves and the ROC curves for streaming learning and online prediction methods. It is shown in Table 4 that the precision and recall for the best $f1$ score in streaming learning are better than online prediction. The visualization in Fig. 3 shows that the accuracy variation of anomaly detection by online prediction is smooth with respect to threshold variations. It is possible that this method of learning is superior from a practical point of view.

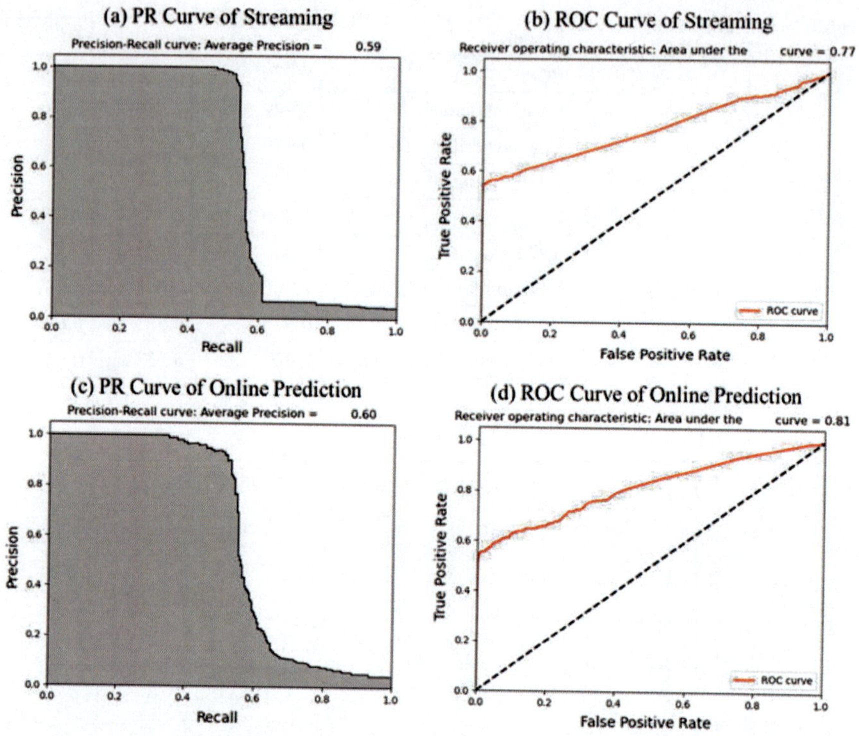

Fig. 3 Comparison of PR and ROC curves in streaming and online prediction formats

5 Conclusion

In this paper, we proposed the method for learning the pattern of structured log data using only template information in order to determine the time of anomaly input. In the accuracy evaluation using the anomaly score for each instance calculated by the proposed method, it was shown that anomalies can be detected with an accuracy comparable to that of the log data-specific baseline methods. Future challenges include using the full data of HDFS for comparison of accuracy to DeepLog and LogAnomaly in online prediction format. Methods for estimating the optimal hyperparameters of the model should also be considered.

Acknowledgements We thank Hideki Itai and his section members for their helpful feedback on the paper. This work is supported by a grant from Panasonic System Design.

References

1. Shilin, H., Jieming, Z., Pinjia, H., Lyu, M.R.: Experience report: system log analysis for anomaly detection. In: IEEE International Symposium on Software Reliability Engineering (ISSRE) (2016)
2. Subutai, A., Alexander, L., Scott, P., Zuha, A.: Unsupervised real-time anomaly detection for streaming data. Neurocomputing **262**, 134–147 (2017)
3. Cui, Y., Ahmad, S., Hawkins, J.: Continuous online sequence learning with an unsupervised neural network model. Neural Comput. **28**, 2474–2504 (2016)
4. Du, M., Li, F., Zheng, G., Srikumar, V.: Deep log: anomaly detection and diagnosis from system logs through deep learning. In: ACM SIGSAC Conference on Computer and Communications Security, pp. 1285–1298 (2017)
5. Meng, W., Liu, Y., Zhu, Y., Zhang, S., Pei, D., Liu, Y., et al.: Loganomaly: unsupervised detection of sequential and quantitative anomalies in unstructured logs. In: Proceedings of 28th International Joint Conference on Artificial Intelligence (IJCAI), vol. 7, pp. 4739–4745 (2019)
6. Zhang, X., Xu, Y., Lin, Q., Qiao, B., Zhang, H., Dang, Y., et al.: Robust log-based anomaly detection on unstable log data. In: Proceedings of the ACM Joint Meeting on European Software Engineering Conference and Symposium on the Foundations of Software Engineering, pp. 807–817 (2019)
7. Zhaoli, L., Tao, Q., Xiaohong, G., Hezhi, J., Chenxu, W.: An integrated method for anomaly detection from massive system logs. IEEE Access **6**, 30602–30611 (2018)
8. Hirakawa, R., Tominaga, K., Nakatoh, Y.: Study on software log anomaly detection system with unsupervised learning algorithm. In: Ahram, T., Taiar, R., Gremeaux-Bader, V., Aminian, K. (eds.) Human Interaction, Emerging Technologies and Future Applications II. IHIET 2020. Advances in Intelligent Systems and Computing, vol. 1152, pp. 122–128. Springer, Cham (2020)
9. Jieming, Z., Shilin, H., Jinyang, L., Pinjia, H., Qi, X., Zibin, Z., Lyu, M.R.: Tools and benchmarks for automated log parsing. In: International Conference on Software Engineering (ICSE) (2019)
10. Xu, W., Huang, L., Fox, A., Patterson, D., Jordan, M.I.: Detecting large-scale system problems by mining console logs. In: Proceedings of the ACM SIGOPS 22nd Symposium on Operating Systems Principles (SOSP'09). ACM, New York (2009)
11. Xu, W., Huang, L., Fox, A., Patterson, D., Jordan, M.: Large-scale system problems detection by mining console logs. In: Proceedings of SOSP'09 (2009)
12. Jian-Guang, L., Qiang, F., Shengqi, Y., Ye, X., Jiang, L.: Mining invariants from console logs for system problem detection. In: Proceedings of USENIX Annual Technical Conference (ATC), pp. 231–244 (2010)
13. Qingwei, L., Hongyu, Z., Jian-Guang, L., Yu, Z., Xuewei, C.: Log clustering based problem identification for online service systems. In: Proceedings of International Conference on Software Engineering (ICSE), pp. 102–111 (2016)
14. Liu, F.T., Ting, K.M., Zhou, Z.-H.: Isolation forest. In: Proceedings of the 8th IEEE International Conference on Data Mining (ICDM'08), pp. 413–422. IEEE Computer Society Press (2008)
15. Bodik, P., Goldszmidt, M., Fox, A., Woodard, D., Andersen, H.: Fingerprinting the datacenter: automated classification of performance crises. In: Eurosys (2010)
16. Chen, M., Zheng, A.X., Lloyd, J., Jordan, M.I., Brewer, E.: Failure diagnosis using decision trees. In; ICAC'04: Proceedings of the 1st International Conference on Autonomic Computing, pp. 36–43 (2004)
17. Liang, Y., Zhang, Y., Xiong, H., Sahoo, R.: Failure prediction in ibm bluegene/l event logs. In: ICDM'07: 7th International Conference on Data Mining (2007)
18. htm.core repository. https://github.com/htm-community/htm.core. Accessed 24 June 2020
19. nupic.core repository. https://github.com/numenta/nupic.core. Accessed 25 June 2020

Logical Inference in Predicate Calculus with the Definition of Previous Statements

Vasily Meltsov, Nataly Zhukova, and Dmitry Strabykin

Abstract At the present stage of computer engineering and information technology development, one of the most actual problems is the creation of high-performance artificial intelligence systems. Along with the successful use of neural networks and various machine learning algorithms, the theory and methods of logical inference play an important role. These mechanisms are indispensable in the design of such knowledge processing systems as expert systems, decision support systems, enterprise management systems, software verification systems, medical and technical diagnostics systems, battle control systems, etc. One of the promising applications of inference is logical prediction. The authors propose an original method for the inference of conclusions in predicate calculus with the definition of previous statements. This approach allows us to reduce the problem of forecasting the situation development, including the transition of the situation to a desired phase, to the task of deductive inference. The method is based on an iterative repetition of the partial and complete "disjunct division" procedure. The article gives a formal and meaningful definition of the problem of the logical inference of conclusions in predicate calculus with the definition of previous statements. Features of all method's stages implementation are illustrated by examples. The main advantage of the proposed method is the expansion of the intellectual system functionality and the possibility of "disjunct division" operations parallel execution in the inference procedure.

Keywords Knowledge processing · Deductive inference · First-order predicate calculus · Previous statements · Logical prediction

V. Meltsov (✉) · D. Strabykin
St. Petersburg State Electrotechnical University "LETI", Vyatka State University, Kirov, Russia

N. Zhukova
St. Petersburg State Electrotechnical University "LETI", SPIIRAS, St. Petersburg, Russia

H. Sharma et al. (eds.), *Congress on Intelligent Systems*, Advances in Intelligent Systems and Computing 1334, https://doi.org/10.1007/978-981-33-6981-8_7

1 Introduction

Today, due to the rapid development of information and computer technology, it becomes possible to create high-performance intelligent systems [1, 2]. The emergence of such systems in various spheres of human activity contributes to the development and modification of methods for reasoning modeling [3, 4]. The use of such artificial intelligence methods is especially effective in solving problems in such areas as enterprise management [5], transport logistics [6], image processing, logical forecasting [7], software verification [8], semantic analysis of speech [9], intellectual control of knowledge [10], medical and technical diagnostics [11, 12], etc.

One of the most interesting and promising areas of science that studies both theory (methods) and applied reasoning modeling algorithms is mathematical logic. The use of inference as a mathematical basis for creating intelligent systems is becoming increasingly relevant. Here, probably, the starting method should be considered the resolution method developed by Robinson [13]. There are other equally powerful logical inference methods: Maslov's method, inference based on Beta's interpretations [14], pattern matching [15], etc. [16].

Today one of the fastest methods is the parallel inference method of disjunct division (IMDD) [17]. However, the experience gained by the authors in the process of solving applied problems of logical forecasting showed that when designing intelligent systems for predicting the development of a situation in a given phase [18], there is a need to modify the existing inference method, namely to develop a specialized IMDD with the definition of previous statements. In this case, the given (including the final) phase can be described using the inferred conclusion, and the phases preceding it can be described by the statements preceding the conclusion.

2 Definition of the Problem

2.1 Meaningful Definition of the Problem

We will use the logical model of knowledge representation, in which situations are described by first-order predicate calculus formulas. The previously known simple one-liberal statements about certain events or phenomena (facts) are represented by predicate-facts that contain constants instead of variables. The relations between one-letter statements are defined by rules (formulas) that contain several predicates and special propositional connectives. Since any formula in the logic of first-order predicates (including sequent) can be represented in a disjunctive form, further description of problem area will contain a special kind of disjuncts—clause [17]. Thus, the facts and rules form the set of initial premises for the logical inference of the conclusion, which, in the simplest case, is a predicate that describes the predicted phase of the development of the situation [7].

The determination of the phase preceding the predicted one is possible using the inverse deductive inference. In this case, for the inferred conclusion, there are many previous statements (atomic formulas) from which it directly follows. The set of statements obtained also describes the phase of the situation preceding the predicted one. One part of these atomic formulas can be pre-known facts, and the other, large part is the desired statements, that is, statements obtained in the process of each step of the logical conclusion. The latter, in turn, can also be determined from the set of statements preceding them. The process of inference ends when at the next step all the previous statements obtained turn out to be facts.

Thus, the inference task is to determine the (inverse) sequence of sets of statements, the first of which contains a conclusion, the last only facts.

2.2 Formal Definition of the Problem

The problem of the logical inference of conclusions with the definition of previous statements in the logic of first-order predicates can be formulated as follows. There are initial consistent premises (set of premises) defined in the form of the set of clauses $M = \{D1, D2, \ldots, DI\}$, and each clause contains one positive (without inversion) literal. The set M includes a subset MF of one-left clause facts and a subset MP of the original rules. There is also a conclusion represented by the derivable clause d, which consists of positive literals (usually one). The problem is formulated as follows:

1. Establish the derivability of the conclusion d from the set of initial premises given by the set of clauses M;
2. For the inferred conclusion d, define a family of sets of statements $G^H = \{G_1, \ldots, G_h, \ldots, G_H\}$, in which each set G_h ($h = 1, \ldots, H$) consists of statements (represented by positive literals) that provide using the set of premises M^h ($M^h \subseteq M$) is the inference of the statements of the set G_{h-1}: $G_h, M^h \Rightarrow G_{h-1}$ and $G_0 = \{d\}$, and each literal e of the set G_h is either a consequence of the literals of the sets $G_{h+1}, \ldots, G_H$ or coincides with one of the facts f of the subset M^F after applying the unifying substitution $\lambda(\lambda e = \lambda f)$ to them. In particular, the substitution may be empty.

The problem of inference is that the preceding statements in predicate calculus can be solved using the disjoint division method after modifying the procedures used in it. The method is built on the basis of partial and complete disjuncts division, as well as a special inference procedure, which are obtained by modifying similar known procedures from work [14].

3 Basic Processing Procedures

3.1 *Partial "Disjuncts Division"*

Partial disjuncts division is performed using a special procedure for generating residues, which is one of the main procedures used in the method of disjuncts division for predicate logic. The procedure assumes that the premises and the conclusion are presented in the form of clauses (disjuncts). The initial expressions of premises and conclusions in the predicate calculus are reduced to the required form using special algorithms [19, 20]. The following notation is used to describe the procedure for the formation of residues.

$\omega = <b, d, r, q, n>$—procedure for generating remainders, in which:

- *b*—divisible-remainder (disjunct of premise) used to obtain the residues;
- *d*—divisor-remainder (disjunct of conclusion) involved in the formation of residues;
- *r*—quotient—a set of literals excluded from the clause acting as a dividend in the formation of the divisible-remainder *b* (for the clause, the reference $r = 0$);
- *q*—A particular feature of the solution, which has three meanings: "0"—at least one zero remainder is received; "1"—all remainders obtained are equal to one; "*g*"—received more than one remainder which is unequal to unit in the absence of zero remainders;
- $n = \{<b_t, d_t, r_t>, t = 1, \ldots, T\}$—many triples consisting of a new divisible-remainder b_t and the corresponding divisor-remainder d_t, as well as set of literals r_t, used in the formation of the remainder b_t as a result of applying the procedure ω.

Determination of «derivative» clause $b[L]$, containing literal *L*, with literal L' clause *d* is the following:

- $\frac{\partial b[L]}{\partial L'} = 1$, if literals *L* and L' are not unified;
- $\frac{\partial b[L]}{\partial L'} = 0$, if literals *L* and L' are unified and clause $b[L]$ contains only literal *L*;
- $\frac{\partial b[L]}{\partial L'} = b_a$, if literals *L* and L' are unified and clause $b[L]$ contains more than one literal. Here $b_a = \lambda b[L] \div \lambda L$, i.e., represents the remainder obtained from clause $b[L]$ after applying unifying substitution λ to it and exception of literal *L*.

Matrix of «derivative» clause *b* with clause d' is determined by the following way:

$$\mu[b, d] = \left| \frac{\partial b[L_j]}{\partial L'_k} \right| = |\Delta_{kj}|$$

where $j = 1, \ldots, J$ and $k = 1, \ldots, K$, moreover *J*—number of literals in clause *b*, and *K*—number of literals in clause *d*.

Also denote the set of new divisible-remainder as $n\hat{} = \{b_t, t = 1, \ldots, T\}$.

In the procedure for the formation of residues, the following actions are performed.

1. Calculation of «derivative» matrix $M[b, d]$.
2. The condition for the formation of reminders is checked. If all derivatives in matrix $\mu[b, d]$ equal to unit, then $q = 1, n = \{<1, 1, \emptyset>\}, n\hat{} = \{1\}$ is accepted and point 6 is performed, otherwise the next point.
3. The set of $n\hat{}$ new ones not equal to unity of divisible-remainder ones is determined. This set includes only various remainders of the matrix $\mu[b, d]$, as well as remainders equal to zero, which differ from each other by the sets of literals removed from them:

$$n^\wedge = \{b_t | t = 1, \ldots, T\} = \overset{A}{\underset{a=1}{U}} \{b_a\},$$

where A—total number not equal to unit remainders,

T—the number of different remainders (including zero remainders that differ in the sets of literals removed from them).

4. The condition for a successful solution (answer) is checked. A solution is considered successful if the remainders b_1, …, b_v, …, b_V, $V \geq 1$ of set $n\hat{}$, formed by excluding from the divisible-remainder b after applying the corresponding unifying substitutions λ_1, …, λ_v, …, λ_V, of literals, placed in the corresponding sets (r'_1, …, r'_v, …, r'_V), equal to zero. In this case $q = 0, n = \{<0, 0, r^*>\}$ is accepted, where $r^* = (\lambda_1 r \cup r'_1) \cup \ldots \cup (\lambda_1 r \cup r'_v) \cup \ldots (\lambda_1 r \cup r'_V), n\hat{} = \{0\}$ and point 6 is performed, otherwise the next point.
5. The set $n = \{<b_t, d_t, r_t>, t = 1, \ldots, T\}$ of triples is determined, consisting of new divisible-remainder b_t and remainders-dividers d_t, and also sets of literals r_t, used in the formation of the remainder b_t. Set r_t calculated as follows: $r_t = (\lambda_{t1} r \cup r_{t1}) \cup \ldots \cup (\lambda_{tw} r \cup r_{tw}) \cup \ldots \cup (\lambda_{tW} r \cup r_{tW})$, where W—the number of identical remainders in the matrix of derivatives obtained using substitutions $\lambda_{t1}, \ldots, \lambda_{tW}$.

For every residue-dividend b_t using substitution λ_t, used in its formation, the corresponding divisor-remainder is calculated: $d_t = \lambda_t(d \div w)$, where w—auxiliary clause containing literals excluded from the remainder d. When calculating the remainder d_t from remainder d literals, L'_h are excluded, for which in matrix $\mu[b, d]$, one of the following two conditions is true:

- $\left|\frac{\partial b[L_j]}{\partial L'_h}\right| = 1$ for all $j = 1, \ldots, J$; i.e., the row of units in matrix corresponding to the literal L'_h;
- $\left|\frac{\partial b[L_j]}{\partial L'_h}\right| = 1$ for all j, except $j = u$, such, that $\left|\frac{\partial b[L_u]}{\partial L'_h}\right| = b_t$, i.e., in the row of the matrix corresponding to the literal L'_h all «derivatives» equal to unit, except the only, corresponding to the current remainder b_t.

If at least one divisor-remainder d_t is not equal to zero, then $q = g$ is accepted, and else—$q = 1$. Next point is performed.

6. The procedure calculation results are fixed ω: q and $n = \{\langle b_t, d_t, r_t \rangle, t = 1, \ldots, T\}$.

An example of constructing a matrix of "derivatives" and the formation of a pair of residues is considered in detail in [7].

To determine the statements preceding the conclusion, the calculation of quotients is included into the well-known procedure for partial disjuncts division. The quotient is a set of literals excluded from the disjunction that acted as a dividend in the formation of the residue-dividend.

3.2 Complete "Disjunct Division"

The complete disjuncts division is aimed at obtaining the so-called final remainders (non-reducible remainders) and the formation of literals set used in the process of inference. It can be said that that the divisible-remainder b is final for the divisor-remainder d, if applying procedure $\omega = \langle b, d, n, q, r \rangle$ to it does not generate new remainders, differ from unit.

The complete disjuncts division is performed taking into account the facts (single-litre initial clauses) using a special procedure.

$\Omega = \langle D, d, R, Q, N \rangle$—the procedure for the complete division of the clause of premise D by clause d, in which:

- R—the set of literals used to infer the clause d (for the clause of the inferred conclusion $R = \emptyset$);
- Q—sign of solution having three meanings: «0»—solution is found, «1»—clause D does not have remainders, differ from unit, with clause d, «G»—set of final remainders is received;
- $N = \{\langle Bj, Rj \rangle, j = 1, \ldots, J\}$—set of couples, consisting of the final remainder Bj and set of literals Rj, used in its formation. With $Q = 0$ $N = \{\langle 0, R' \rangle\}$, where R' consists of literals, used for clause d inference, and literals, used for zero remainders formation. With $Q = 1$ $N = \{\langle 1, \emptyset \rangle\}$.

The formation of remainders set is carried out by repeated use of ω-procedures and consists of a series of steps. On every step, ω-procedure is applied to existing divisible-remainder and divisible-remainder, forming new divisible-remainder and new divisor-remainder, which are used as input in the next step. The process ends when on the next step ω-procedure is detected, in which zero remainders are received ($q = 0$), or in every ω-procedures of current step, signs will be generated indicating the receipt of the final residues ($q = 1$).

Consider the complete disjuncts division by the example 1.

Example 1. Initial data are the following sentences [21].

(1) like(Jill, vine): $P(a, c)$;
(2) like(John, food): $P(b, e)$;
(3) like(John, vine): $P(b, c)$;

(4) man(John): $P_1(b)$;
(5) like(y, vine) and man (y) → escort(x, y):

$$P(y, c)P_1(y) \to R(x, y) \text{ or } \neg P(y, c) \vee \neg P_1(y) \vee R(x, y);$$

(6) escort(Mary, z) ("Does Mary have escort, who is he ?"): $R(m, z)$.

It should also be noted that the literal conclusion (6) is not unified with any of the literal facts (1–4).

In procedure $\Omega = <D, d, R, Q, N>$ for the formation of final remainder, clause (5) will be as a remainder disjuncts-divisible D, and as a disjuncts-divider, a clause (6), wherein $R = \emptyset$, because d—clause of inferred conclusion.

1. *Preparatory step.* Remainder formation procedure in progress $\Omega = <D, d, q, n, r>$. The "derivatives" matrix is calculated:

	$\neg P(y, c)$	$\neg P_1(y)$	$R(x, y)$
$R(m, z)$	1	1	Δ_{13}

In matrix "derivatives" $\Delta_{11} = \Delta_{12} = 1$, and "derivative" Δ_{13} defined using unification substitution $\lambda 13 = \{m/x, z/y\}$: $\Delta 13 = \lambda 13D \div \lambda 13R(m, z) = \neg P(z, c) \vee \neg P1(z) = b1$. Set of literals to exclude form a literal $\lambda 13R(x, y) = R(m, z)$. Since the remainder $b1$ nonzero and not equal to unity is received, then $q = g$, $r1 = \lambda 11R \cup rt1 = \{R(m, z)\}$ is accepted, and the source data for the main step is formed: $n = \{<b1, d1, r1>\}$. Clause $d1$ formed by supplementing the clause $d1^*$ with inversions of fact literals. In this case, clause $d1^*$ is obtained from clause d by eliminating the literal, which corresponds to a row in the matrix that contains only one "derivative" that is not equal to unity, which represents the remainder $b1$. Thus, we obtain: $d1^* = 0$, and $d1 = \neg P(a, c) \vee \neg P(b, e) \vee \neg P(b, c) \vee \neg P1(b)$.

2. *Main step (h = 0).* For the triple $\{<b1, d1, r1>\}$, the ω-procedure is performed. The procedure generates three triples $n1 = \{<b1.t, d1.t, r1.t>, t = 1, 2, 3\}$:

	$\neg P(z, c)$	$\neg P_1(z)$	
$\neg P(a, c)$	Δ_{11}	1	$\lambda_{11} = \{a/z\}$, $\Delta_{11} = \neg P_1(a) = b_{1.1}$; $r_{1.1} = \lambda_{11}r_1 \cup \{\neg P(a, c)\}$
$\neg P(b, e)$	1	1	$\Delta_{12} = \Delta_{21} = \Delta_{22} = \Delta_{32} = \Delta_{41} = 1$
$\neg P(b, c)$	Δ_{31}	1	$\lambda_{31} = \{b/z\}$, $\Delta_{31} = \neg P_1(b) = b_{1.2}$; $r_{1.2} = \lambda_{31}r_1 \cup \{\neg P(b, c)\}$
$\neg P_1(b)$	1	Δ_{42}	$\lambda_{42} = \{b/z\}$, $\Delta_{42} = \neg P(b, c) = b_{1.3}$; $r_{1.3} = \lambda_{42}r_1 \cup \{\neg P_1(b)\}$

$n1 = \{<\neg P1(a), \neg P(b, c) \vee \neg P1(b), \{R(m, a), \neg P(a, c)\}>,$
$<\neg P1(b), \neg P(a, c) \vee \neg P1(b), \{R(m, b), \neg P(b, c)\}>,$
$<\neg P(b, c), \neg P(a, c) \vee \neg P(b, c), \{R(m, b), \neg P1(b)>\}\}.$

Sign $q1 = g$, therefore $h = 1$ is accepted, and the main step is again performed for the new initial data.

3. *Main step (h = 1)*. With this implementation of the main step, for each triple of clauses $<b_{1.t}, d_{1.t}, r_{1.t}>$ of set n_1 ω-procedure is preformed: $\omega_{1.t} = <b_{1.t}, d_{1.t}, q_{1.t}, n_{1.t}>$, $t = 1, 2, 3$:

$\omega_{1.1}$		$\omega_{1.2}$		$\omega_{1.3}$	
	$\neg P_1(a)$		$\neg P_1(b)$		$\neg P(b, c)$
$\neg P(b, c)$	1	$\neg P(a, c)$	1	$\neg P(a, c)$	1
$\neg P_1(b)$	1	$\neg P_1(b)$	0	$\neg P(b, c)$	0

In procedures $\omega 1.2$ and $\omega 1.3$ signs $q1.2 = q1.3 = 0$, since remainders equal to zero are obtained. In procedure $\omega 1.2$, a zero remainder is obtained after exclusion from a disjunct-divisible literal $\neg P1(b)$ and in procedure $\omega 1.3$—literal $\neg P(b, c)$. In both cases, empty unifying substitutions $\lambda = \theta$ were used. $Q1 = 0, N1 = \{<0, R'>\}$ is accepted and the set of literals $R' = R \cup r^*1 \cup r^*2$ is calculated, where $R = \emptyset$, $r^*1 = \theta r1.2 \cup \{\neg P1(b)\} = \{R(m, b), \neg P(b, c), P1(b)\}$, $r^*1 = \theta r1.3 \cup \{\neg P(b, c)\} = \{R(m, b), \neg P1(b), \neg P(b, c)\}$. Then, $R\square = \{R(m, b), \square P(b, c), P1(b)\}$. There is a transition to the final step of Ω-procedure.

4. Final step. The results of the procedure are recorded. Since there are signs $q1.2 = q1.3 = 0$, then $Q = 0, N = \{<0, R'>\}$ is accepted, there $R' = \{R(m, b), \neg P1(b), P(b, c)\}$. Logical inference completed successfully. The desired value of z is determined by the substitution $\lambda 31 = \lambda 42 = \{b/z\}$. Thus, a solution is obtained: "escort(Mary, John)," i.e., Mary have an escort and he is John. Moreover, a set of the statements made by literals $R\square$ preceding the conclusion are formed, on which inference of the conclusion is based: escort(Mary, John)—$R(m, b)$, man(John)—$P1(b)$ and like(John, vine)—$P(b, c)$.

To determine the statements preceding the conclusion, the calculation of the set of literals used to derive the clause and the sets of literals used in the formation of the final remainders is included clause into the well-known procedure for the complete disjuncts division.

4 Procedure of Inference

The procedure allows you to make a step of inference, transforming the inferred clause into set of new clauses necessary to continue the inference in the next step.

$W = <M, d, f, o, p, m>$—inference procedure, in which:

- $M = \{D_1, D_2, \ldots, D_i, \ldots, D_I\}$—set of initial clauses;
- $Di = L_1^i \vee L_2^i \vee ... \vee L_j^i \vee ... \vee L_{Ji}^i$—*i*th initial clause consists of literals L_j^i;
- $d = L_1 \vee L_2 \vee \ldots \vee L_k \vee \ldots \vee L_K$—inferred clause consists of literals L_k;
- $o = <c, C>$—a pair of sets of current residues, consisting of remainders sets formed before (c) and after (C) the procedure;

- p—sign of the end of the inference, having three values: «0»—inference completed successfully, «1»—inference failed, «G»—continue of inference is required for the resulting output sequences;
- f—set of literals, used for clause d inference from the clauses of the set M (for the clause of the inferred conclusion, this is an empty set);
- $m = \{<d_g, R_g>, g = 1, \ldots, G\}$—set of pairs: new inferred clause d_g and corresponding to it set of literals R_g, used for its inference. With $p = 0$ $m = \{<1, f'>\}$, where clause f' consists of literals, used for inference of clause d, and literals, used for zero remainder formation. With $p = 1$ $m = \{<0, \emptyset>\}$.

The inference procedure uses the previously considered Ω-procedure of remainders formation. If the procedure is applicable, then at the first application of the procedure $f = \emptyset$. The input set of current remainders c is supplemented with inversions of the literals of the inferred clause d: $c = \{\neg Lk, k = 1, \ldots, K\}$.

The inference procedure differs from the well-known procedure [19] in defining the set of literals used to infer the clause d from the set of clauses M, as well as in forming a set of pairs: a new inferred clause and the corresponding set of literals used in its inference.

5 Method of Inference

Inference by the method of disjuncts division is reduced to the repeated use of W-procedures and consists of a number of steps. At each step of the inference, the W-procedures are applied to the existing inferred and initial clauses, forming new inferred clauses used in the next step. The sets of literals used to infer the clauses of the step are combined into the set of literals of the previous statements (G_h) and form a family of statements sets $G^h = \{G_1, \ldots, G_h\}$. The inference process ends when at the next step a clause is detected for which inference is not possible ($p = 1$), or the sign of successful inference ($p = 0$) will be formed for every inferred clause on current.

For a more complete description of the method, the index function $i(h)$ will be used, in which h will denote the number of the inference step. Denote the general sign of the end of the inference by P. Then, the description of the method can be represented as follows.

Before starting the inference, it is checked: Are the inferred clauses not the direct consequences of the one-literal initial clauses? In the set of initial clauses M, one-literal clauses are distinguished, and for each clause d_k ($k = 1, \ldots, K$) being inferred, it is checked whether it contains at least one literal that matches (after applying the necessary substitution) with at least one one-literal clause (fact). Check is carried out using partial clause division procedures: $\omega_k = <D^*, d_k, r_k, q_k, n_k>$, in which the auxiliary clause D^* acts as a dividend, composed of one-literal initial clauses. If during the partial division procedure for the inferred clause d_e ($e = 1, \ldots, E$), the remainder is equal to zero, and then clause d_e is excluded from the set of inferred clauses, and the facts r_e used in the formation of the remainder are included in the set $G_0 = r_1 \cup \ldots \cup r_E$. With the exception of all inferred clauses $P = 0$, $G^0 = \{G_0\}$ is accepted and inference does not performs. Otherwise, the remaining clauses are included in the set of inferred clauses $m = \{d_t, t = 1,2, \ldots, T\}$ (moreover, if all inferred clauses are included in the set m, then $G_0 = \emptyset$), $h = 1$ is accepted and the first step of the inference is performed.

Inference step h—For every inferred clause $d_{i(h)}$ of all sets $m_{i(h-1)}$ w-procedures applied:

$$W_{i(h)} = <M, d_{i(h)}, f_{i(h)}, o_{i(h)}, p_{i(h)}, m_{i(h)}>, \; i(h) = i(h-1) \cdot t_{i(h-1)},$$
$$t_{i(h-1)} = 1, \ldots, T_{i(h-1)}.$$

Set of literals $f_{i(h)}$, used for current step inference, are combined into the set of literals of previous statements (G_h), and a family of sets of statements $G^h = \{G_1, \ldots G_h\}$ is formed, and also the general sign of the end of the output at the hth step is calculated.

$$G_h = \bigcup_{A=1}^{B} f_{i(h)}, \; G^h = G^{h-1} \cup \{G_h\}, \; P_h = \bigvee_{A=1}^{B} p_{i(h)}, \text{ где } A = t_{i(h-1)}, B = T_{i(h-1)}.$$

If Ph = G, then h increases by one, and process continues (($h + 1$)th step of inference is performed), else inference finishes. Moreover, with Ph = 1 inference finishes unsuccessful, and with Ph = 0—successful, and a family of statements sets *Gh* consists of literals sets of the preceding statements, which are necessary for the successful inference of a given conclusion.

A feature of the method is the formation of a family of sets of literals of the preceding statements.

Example 2. The application of the inference of conclusions with the definition of previous statements will be considered as an example (Table 1).

Inference consists of three steps and succeeds. At the first step, a family of previous statements sets is formed $G1 = \{G1\}$, where $G1 = \{P(c, b), O(e, b)\}$.

Table 1 The set of initial premises and conclusion

Knowledge	Premises/conclusion	Clauses (disjuncts)
Facts	(1) direct (Serge, Boris)	$D1 = P(c, b)$
	(2) direct (Boris, Anna)	$D2 = P(b, a)$
	(3) direct (Anna, Alex)	$D3 = P(a, e)$
Rules	(4) direct $(x, y) \rightarrow$ report (y, x)	$D4 = \neg P(x, y) \vee O(y, x)$
	(5) direct (z, u) and report$(s, u) \rightarrow$ report (s, z)	$D5 = \neg P(z, u) \vee \neg O(s, u) \vee O(s, z)$
Conclusion	(6) "Does Alex report to Serge?"	$d = O(e, c)$?

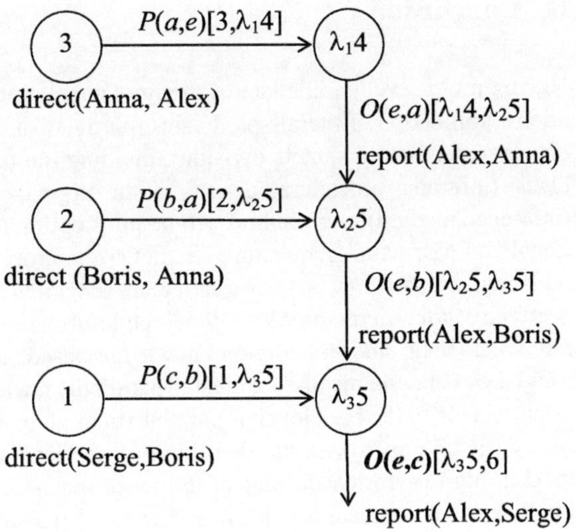

Fig. 1 Conclusion inference scheme «reports (Alex, Sergey)»

At the second step, the family is replenished with the set $G2 = \{P(b, a), O(e, a)\}$: $G2 = \{G1, G2\}$. In the third step, the family is replenished with set $G3 = \{P(a, e)\}$: $G3 = \{G1, G2, G3\}$.

The result is the following family of sets preceding the conclusion $O(e, c)$ of statements $G3 = \{P(c, b), O(e, b)\}, \{P(b, a), O(e, a)\}, \{P(a, e)\}$. An inference circuit (in figure) can represent inference in predicate calculus. The inference schemes in the predicate calculus have two features: The graph arcs are marked with literals, and the original premises modified by unifying substitutions are mapped to the top of the graphs. The inference diagram for the considered example is shown in Fig. 1, where $\lambda 1 = \{c/z1, b/u1, e/s1\}, \lambda 2 = \{b/z2, a/u2, e/s2\}, \lambda 3 = \{e/x3, a/y3\}$—unifying substitutions.

The considered method for conclusions inference in predicate calculus with the definition of previous statements, initially oriented toward the logical forecasting of the development of situations, can also be applied in other intelligent systems. An important advantage of the method is the high degree of parallelism of the performed calculations. Parallelism of calculations is shown on four levels: unification of literals-predicates (partial disjuncts division), division pairs of clauses (complete disjuncts division), dividing the original clauses by the inferred clause (inference procedure), and division of the original clauses with all inferred clauses (inference method). Given the possibility of performing a parallel algorithm for unifying all terms in comparable predicates, this method allows its effective application in the implementation of software intelligent systems on modern high-performance multi-core computing platforms [2].

6 Conclusion

Having multi-level parallelism of the implementation of the main processing operations: unification of literals-predicates (partial disjuncts division), division pairs of clauses (complete disjuncts division), dividing the original clauses by the inferred clause (inference procedure), division of the original clauses with all inferred clauses (inference method), this method will be most optimally implemented not on a traditional von Neumann architecture for modern microprocessors, but on the basis of a non-standard data flow control architecture (dataflow architecture). This architecture is effective for solving problems in which information is represented by knowledge rather than data; since for solving knowledge-based tasks, it is necessary to maintain a high level of dynamically manifested parallelism, which the considered method can provide. Due to the high level of parallelism implemented in the considered method of logical inference with the definition of previous conclusions, it allows it to be used in high-performance logical inference machines, which are the basis of any intelligent devices, and provides the necessary performance, in contrast to sequential methods (for example, SLD-resolutions method). Determining the statements needed to reach a final conclusion is an important element of any forecasting. This approach allows you to identify the chain of consequences required for the occurrence of a necessary event or to achieve a certain result. For example, in the field of enterprise management, this method will form a chain of optimal step-by-step decisions that must be taken to achieve the required financial profit from the enterprise. Also now, the authors are negotiating with the Federal Antimonopoly Service in the Kirov region on the introduction of such an intelligent system to identify the behavior of companies that have entered into anticompetitive agreements (cartel).

Acknowledgements The paper was prepared in Saint-Petersburg Electrotechnical University (LETI), and is supported by the Agreement № 075-11-2019-053 dated 20.11.2019 (Ministry of Science and Higher Education of the Russian Federation, in accordance with the Decree of the Government of the Russian Federation of April 9, 2010 No. 218), project «Creation of a domestic high-tech production of vehicle security systems based on a control mechanism and intelligent sensors, including millimeter radars in the 76–77 GHz range».

References

1. Czarnowski, I., Howlett, R., Jain, L.: Intelligent Decision Technologies 2019. Springer Singapore (2019). https://doi.org/10.1007/978-981-13-8303-8
2. Meltsov, V.Yu.: High-Performance Systems of Deductive Inference: Monograph. Science Book Publishing House, Yelm (2014)
3. Kulik, B., Fridman, A., Zuenco, A.: Logical Inference and Defeasible Reasoning in N-Tuple Algebra. Book News Inc., Portland (2013)
4. Khemani, D.: Artificial Intelligence: Knowledge Representation and Reasoning. IIT Madras. https://swayam.gov.in/nd1_noc20_cs30/preview
5. Szczerbicki, E., Sanin, C.: Knowledge Management and Engineering with Decisional DNA. Springer International Publishing (2020). https://doi.org/10.1007/978-3-030-39601-5

6. Chen, Y., Duong, T.Q.: Industrial networks and intelligent systems. In: Proceeding of 3rd International Conference, INISCOM 2017, Ho Chi Minh City (2017)
7. Strabykin, D.A.: Logical method for predicting situation development based on abductive inference. J. Comput. Syst. Sci. Int. **52**(5), 759–763 (2013)
8. Strabykin, D., Meltsov, V., Dolzhenkova, M., Chistyakov, G., Kuvaev, A.: Formal verification and accelerated inference. In: Artificial Intelligence Perspectives in Intelligent Systems, pp. 203–211. Springer International Publishing, Switzerland (2016)
9. Filchenkov, A., Pivovarova, L., Žižka, J.: Artificial intelligence and natural language. In: Proceeding of 6th Conference AINL-2017, pp. 3–14. St. Petersburg (2017)
10. Meltsov, V.Yu., Lesnikov, V.A., Dolzhenkova, M.L.: Intelligent system of knowledge control with the natural language user interface. In: Quality Management, Transport and Information Security, Information Technologies (IT&QM&IS), pp. 671–675. St. Petersburg (2017)
11. Sambukova, N.: Machine learning in studying the organism's functional state of clinically healthy individuals depending on their immune reactivity. In: Diagnostic Test Approaches to Machine Learning and Commonsense Reasoning Systems (2013). https://doi.org/10.4018/978-1-4666-1900-5.ch010
12. Al-Emran, M., Shaalan, H., Hassanien, A.: Recent Advances in Intelligent Systems and Smart Applications. SSDC, vol. 295, pp. 46–54. Springer, Cham (2020). https://doi.org/10.1007/978-3-030-47411-9
13. Resolution Method in AI: https://www.tutorialandexample.com/resolution-method-in-ai/
14. Russell, S., Norvig, P.: Artificial Intelligence: A Modern Approach, 3rd edn. Prentice Hall, Upper Saddle River (2010)
15. Munoz, D., Grubb, A., Bagnell, J., Hebert, M.: Using Inference Machines for Perception Tasks. The Robotics Institute, Carnegie Mellon University, Pittsburgh (2013)
16. Putzky, P., Welling, M.: Recurrent Inference Machines for Solving Inverse Problems: Under Review as a Conference Paper at ICLR 2017. University of Amsterdam (2017)
17. Strabykin, D.A.: Parallel computation method for abductive knowledge-based inference. J. Comput. Syst. Sci. Int. **39**(5), 766–771 (2000)
18. Meltsov, V., Kuvaev, A., Zhukova, N.: Knowledge processing method with calculated functors. In: Arseniev, D., Overmeyer, L., Kälviäinen, H., Katalinić, B. (eds.) Cyber-Physical Systems and Control. CPS&C 2019. Lecture Notes in Networks and Systems, vol. 95. Springer, Cham (2020). https://doi.org/10.1007/978-3-030-34983-7_19
19. Dolzhenkova, M., Meltsov, V., Strabykin, D.: Method of consequences inference from new facts in case of an incomplete knowledge base. Indian J. Sci. Technol. **9**(39), 100413 (2017). https://doi.org/10.17485/ijst/2016/v9i39/100413
20. Meltsov, V.Y., Strabykin, D.A., Kuvaev, A.S.: Model of the operating device with a tunable structure for the implementation of the accelerated deductive inference method. Adv. Intell. Syst. Comput. **1156**, 155–164 (2020). https://doi.org/10.1007/978-3-030-50097-9_16
21. Dolgenkova, M.L., Chistyakov, G.A.: Sequences Inference Method out of New Facts by Disjuncts Division in Predicate Calculus (Example). https://zenodo.org/record/57859

Congestion Management Considering Demand Response Programs Using Multi-objective Grasshopper Optimization Algorithm

Jimmy Lalmuanpuia, Ksh Robert Singh, and Sadhan Gope

Abstract The increase in consumption of electrical energy has become one of the major challenges in power system operation and control. In the meantime, transmission congestion is one of the major concerns for power system operator and planner. Thus, this paper presents a multi-objective grasshopper optimization technique for congestion line management considering the demand response programs. The objective of this congestion management problem is to reduce the emission and also the total operation cost of the system. The above-mentioned optimization technique is tested on IEEE 30-bus test system with varying load conditions. It is found that demand response programs can solve the congestion management problem significantly. Results also show that multi-objective grasshopper optimization technique performs better as compare to other optimization techniques discussed in this work.

Keywords Demand response programs · Congestion management · Multi-objective grasshopper optimization algorithm

1 Introduction

Restructuring power industry results in increasing demand of electricity and thus utilization of full capacity of transmission lines becomes very important. The increased demand leads to outages of transmission lines and generators, damage of equipment's, contraction in proper maintenance and violates the thermal limits, voltage limits, and stability limits [1]. Violation of the above-mentioned limits results in congestion in power system. This congestion must be managed so that the system capacity is optionally utilized for the benefits of the market participants. Since all the market participants are trying to maximize their benefits, huge competition is seen in electricity market sector [2]. However, the transmission line flows were determined by the system operator in power market within their allowable range at lower cost. Thus, the management of congestion in a restructured power system is a bit complex, and

J. Lalmuanpuia (✉) · K. R. Singh · S. Gope
Department of Electrical Engineering, Mizoram University, Aizawl 796004, India

H. Sharma et al. (eds.), *Congress on Intelligent Systems*, Advances in Intelligent Systems and Computing 1334, https://doi.org/10.1007/978-981-33-6981-8_8

construction of new transmission lines is impractical solution because of the restriction in the environmental issues, transmission rights, and economic issues [3]. Many solution for this congestion problems have come up, one being the nodal pricing method which enable the independent system operator to generate electricity from a specific place and receive at another specific place [4]. A new control technique using nodal pricing is proposed which identify steadily the limits of the transmission line and enable the power system to operate at stable economic which solves the transmission line congestion [5]. A market-based locational marginal pricing on the sub-network has been used in determining the congestion in the power system [6]. Also, bidding strategy is widely used for the improvement of the problems of congestion management with the implementation of two-level optimization problem [7]. With the advance and increase in knowledge and technology, researchers also carry out using artificial intelligent optimization algorithm such as bat swarm optimization, particle swarm optimization, firefly optimization algorithm, and many other evolutionary algorithm in order to mitigate and solve the congestion problem [8–10].

FACTS devices allow us to control the transmission line flows in addition within its capability of supplying the additional reactive power in the network. It also reduces the unnecessary use of generation rescheduling, which reduce the overall generation cost and system losses along with an increase in the stability limits of the network by reducing the transmission congestion [11, 12]. Generation rescheduling techniques are generally used for mitigating transmission congestion in power system [13, 14]. Generation rescheduling along with load shedding was employed for management of congestion in [15]. Application of pump hydroelectric storage system is another efficient technique for congestion management [16].

Demand side management (DSM) is a very new approach for minimization of the operation cost in peak hours. It also minimizes the total congestion cost and also save fuels for the operation [17]. DSM can be categorized as demand response (DR), spinning reserve (SR), load shifting program (LSP), and energy efficiency (EE). With the advancement in power industry, the congestion management problem deals with newly useful methods DR's programs [18–20]. In this approach, the consumers were given an opportunity to change their energy consumptions with some incentives in accordance with the fluctuation in energy price in electricity market, which indirectly helps to mitigate the congestion.

In this paper, we are going to implement demand response to mitigate congestion. A multi-objective grasshopper optimization algorithm (MOGOA) will be used in response to solve this congestion problem. The main focus of this paper is to mitigate the congestion along with minimizing emission and generation cost with demand response cost.

2 Materials and Methods

2.1 Demand Response Programs Cost

In electricity market, the participation of the consumers with response to market prices can be defined as demand response programs (DRPs). Two factors can be considered in acceptance of their response. In Factor I, all the prices of electricity changes at the retail level which reflect changes in the real price of electricity and in Factor II, the introduction of incentive program which, in order to satisfy the power consumption at the critical time by making the customer reduces their consumption with paid incentives [21]. The demand response capacity can be estimated based on the incentive along with the penalty and customer benefits. The change in load at the m_{th} bus is obtained after the implementation of DR and is presented in Eq. (1).

$$\Delta d_m = do_m - d_m \tag{1}$$

where do_m and d_m are the amount of load before and after DRs are implemented at the selected m_{th} bus, respectively. An incentive is paid to the customer for each load reduction in each unit, and the total incentive paid for their involvement in DRPs can be expressed as in Eq. (2).

$$inc_m = inc \times [do_m - d_m] \tag{2}$$

Those customers who are participating in the demand response program but do not response to the minimum load reduction requirement as per the contract were given penalty and the total penalty, which are given to the customer who does not response for the responsive load bus m_{th} as requested by the ISO, can be calculated as written as in Eq. (3).

$$pen_m = pen \times [\text{LR}_m - \Delta d_m] \tag{3}$$

In this paper, the values of *pen* factor are assumed to be 0, and the *inc* is 0.1 times the price of electricity ($\rho 0$) to 10 times the price of electricity ($\rho 0$) before the implementation of DRP's. So, in order to express the customer revenue in terms of load, the load model used can be written as in Eq. (4).

$$d_m = do_m \times \left[1 + E \times \frac{\rho - \rho_0 + inc - pen}{\rho_0}\right] \tag{4}$$

Here, $\rho 0$ represent the price of electricity before implementing the DRPs and ρ represents the electricity price after implementing the DRPs, E is the load elasticity [21].

The main objectives of this work are as follows

Cost minimization

The cost of generation in addition to the amount of incentive paid to the customer in case of response to the load can be calculated using Eq. (5).

$$\text{Minimize} \sum_{p=1}^{NG} a_p + b_p P_{Gp} + c_p P_{Gp}^2 + \sum_{m=1}^{N_{\mathrm{DR}}} inc_m \tag{5}$$

where a_p, b_p, and c_p indicate the coefficient of generation cost of the p_{th} generating unit.

Emissions minimization

The production of electricity power plant need to burn fuel and this causes emission problem, thus the amount of emission produced need to be reduced and the same is presented in Eq. (6).

$$\text{Minimize} \sum_{p=1}^{NG} \alpha_p + \beta_p P_{Gp} + \gamma_p P_{Gp}^2 \tag{6}$$

where α_p, β_p, and γ_p indicate emission coefficient of the p_{th} generating unit.

Equality constraints

The equality constraint of the optimization problem is defined for the power flow equation in order to balance the active and reactive power in the system and the same are presented in Eqs. (7) and (8).

$$P_{G_q} - \left(P_{D_q} - d_q\right) - V_q \sum_{a=1}^{N_B} V_a \left(G_{qa} \cos\theta_{qa} + B_{qa} \sin\theta_{qa}\right) \quad q = 1, \ldots, N_B \tag{7}$$

$$Q_{G_q} - Q_{D_q} - V_q \sum_{a=1}^{N_B} V_a \left(G_{qa} \sin\theta_{qa} - B_{qa} \cos\theta_{qa}\right) \quad q = 1, \ldots, N_B \tag{8}$$

Inequality constraints

Generator constraints

At each generator units, the amount of allowable ranges and limits for active, reactive, and voltage are determined within minimum and maximum limit. Equations (9)–(11) represent the allowable minimum and maximum limits for voltage, active power, and reactive power at each generation units.

$$V_{G_p}^{\min} \leq V_{G_p} \leq V_{G_p}^{\max};\ p = 1, \ldots, N_G \tag{9}$$

$$P_{G_p}^{\min} \leq P_{G_p} \leq P_{G_p}^{\max};\ p = 1, \ldots, N_G \tag{10}$$

$$Q_{G_p}^{\min} \leq Q_{G_p} \leq Q_{G_p}^{\max};\ p = 1, \ldots, N_G \tag{11}$$

Incentive paid to the responsive loads constraints

The customer incentive in accordance with their participation in DRPs is determined within allowable minimum and maximum range as in Eq. (12).

$$inc_m^{\min} \leq inc_m \leq inc_m^{\max};\ m = 1, \ldots, N_{\mathrm{DR}} \tag{12}$$

Security constraints

The voltages as well as the thermal limits of the power system have its own limits within an allowable range and should not exceed this range, and the same are presented in Eqs. (13) and (14).

$$V_{L_r}^{\min} \leq V_{L_r} \leq V_{L_r}^{\max};\ r = 1, \ldots, N_L \tag{13}$$

$$S_{L_n}^{\min} \leq S_{L_n} \leq S_{L_n}^{\max};\ n = 1, \ldots, N_{\mathrm{BR}} \tag{14}$$

3 Multi-objective Grasshopper Optimization Algorithm

Nature-inspired grasshopper life cycle consists of larva with the characteristic of slow and steady movement, followed by the swarm with fast react and long-range characteristics are an efficient way of solving optimization in real world. The other character posse by grasshopper is their ability to seek their food in two ways separated as exploration and exploitation [22]. In the exploitation, grasshoppers are moving locally while suddenly moving to the search agent. The mathematical model for representing the behavior is represented in Eq. (15).

$$X_i = S_i + G_i + A_i \tag{15}$$

here X_i is the ith grasshopper situation, S_i the interaction with social, G_i is the force acting on the ith grasshopper by the gravitational force, and A_i is the advection of wind direction. Thus, Eq. (15) can again be written with respect to a random no (r) in [0, 1], and the same is presented in Eq. (16).

$$X_i = r_1 S_i + r_2 G_i + r_3 A_i \tag{16}$$

$$S_i = \sum_{\substack{j=1 \\ j \neq i}}^{N} s(d_{ij})\hat{d}_{ij} \tag{17}$$

where d_{ij} grasshopper ith and jth distance as $d_{ij} = |x_j - x_i|$, and $\hat{d}_{ij} = \left|\frac{x_j - x_i}{d_{ij}}\right|$ is a unit vector from the ith grasshopper to the jth grasshopper. The interaction of grasshopper with social (s) can be defined as in Eq. (18).

$$s(r) = f e^{\frac{-r}{l}} - e^{-r} \tag{18}$$

The gravitational force acting on the grasshopper is represented in Eq. (19)

$$G_i = -g\hat{e}_g \tag{19}$$

g represents the constant gravitational force, and $\hat{e}_g$ represent vector toward earth. The parameter of A (i.e., advection of wind) is presented in Eq. (20).

$$A_i = u\hat{e}_w \tag{20}$$

where u is a constant, and $\hat{e}_w$ represent vector of wind direction. Substituting the value of Eqs. (17), (19) and (20) in Eq. (15), we have

$$X_i = \sum_{\substack{j=1 \\ j \neq i}}^{N} s(|x_j - x_i|)\frac{x_j - x_i}{d_{ij}} - g\hat{e}_g + u\hat{e}_w \tag{21}$$

where N is the number of population of grasshoppers. The later Eq. (22) is the modified form which is used for solving the optimization problem.

$$X_i^d = c\left(\sum_{\substack{j=1 \\ j \neq i}}^{N} c\frac{ub_d - lb_d}{2} s(|x_j - x_i|)\frac{x_j - x_i}{d_{ij}}\right) + \hat{T}_d \tag{22}$$

The coefficient c, which reduces the comfort zone, is given in Eq. (23).

$$c = c\max - l\frac{c\max - c\min}{L} \tag{23}$$

Here, l represents iteration at present, and L represents the maximum number of iteration. During MOGOA optimization, the target is assumed to be the best of the objective value or the fittest grasshopper. So, this will help each of the iteration in the MOGOA to save the most outstanding target in the spaces, and the grasshopper will approach it. Until and unless the stopping criterion is satisfied, the position of the best target is updated. After satisfaction, the position and fitness of the best target are finally returned as the best solution for the global optimum.

4 Results and Discussions

In this paper, we have discussed the management and evaluation of congestion with the help of IEEE 30-bus test system. The test network consists of six generators along with 20 loads and 41 transmission lines. The sets of network generators are located at 1, 2, 13, 22, 23, and 27 buses where 1 being the slack bus. The total load of the test system is 189.2 MW. The information about the transmission lines data along with the amount of active and reactive power demand, including power generation within its maximum and minimum generation limits, the emission and generation coefficient is given in [21]. In order to implement the demand response successfully, we have selected a load bus to participate in the demand response programs on the basis of their influence on network response [12]. The load buses nos. 7, 8, 12, 17, 19, 21, and 30 were selected, and we assumed that they can reduce their load up to 10% for each bus [12]. The elasticity coefficient of E is assumed to be equal to −0.1 [12], and the maximum price of DRP is assumed to be 50 $/MW.

In order to mitigate and relief the congestion in the transmission line, we have used the DRPs for +120% and 130% load condition. To verify the impact of DRPs, the problem had also been solved without DRPs. The multi-objective grasshopper optimization algorithm was used for minimizing the objective function of the total cost and emission of the test system. The parameters of the MOGOA are given in Appendix. To verify the effectiveness of the results, initially base case problem is solved and is reported in Table 1.

From Table 1, it is observed that the MOGOA is a suitable technique for lowering the cost of the applied system. From this table, it is also observed that with DRPs, the total cost is reduced, compared to the case without DRPs. For verifying the effectiveness of the presented approach, two cases of the system were investigated.

Table 1 Base case results comparison with different algorithm

S. No	Method applied	Total cost ($/h)
1 [21]	BF-NM	648.86
2 [21]	MOPSO (without DR)	588.33
3 [21]	MOPSO (with DR)	601.35
4	MOGOA (without DR)	586.795
5	MOGOA (with DR)	592.954

Case 1: 120% load demand, i.e., 227.04 MW load considering with and without DR.

Case 2: 130% load demand, i.e., 245.96 MW load considering with and without DR.

Case 1: In this case, total system load is 227.04 MW, which is 20% more than the base load of the system. With this load, load flow program is run and is found that line 1–2 and line 6–8 are congested. To mitigate this congestion, initially MOGOA was used, and generators are rescheduled. After rescheduling the generators, line flows are reduced in the above-said lines, and the congestion has been removed. To verify the impacts of DRPs in congestion, again MOGOA is used with considering DRPs and obtained the line flows of the same lines. It is also found that line congestion is removed, and the system is in safe condition. Table 2 shows the line flow before and after the congestion management with and without considering DRPs.

Observing the numerical results obtained from Table 3, it is noted that the total cost of generation reduces considering DRPs, which means that the implementation of DRPs reduces the generation of expensive generators in the system. The generation cost is reduced from 739.85 to 737.9064 $/h, which is 1.944 $/h reduction in cost after implementation of DRPs.

From the results, it is also observed that by using DRPs, emission is reduced in compared to the normal case since the conventional power plant generation reduces which contributed a lot in environmental pollution. The emission is reduced from 462.977 to 457.096 ton/h after implementation of DRPs which is the reduction of 5.881 (ton/h) in emission. The power loss is also decreased as expected (i.e., the power loss is higher without DRPs than with DRPs). From Table 3, it is seen that total power loss is 3.28266 MW in normal case and 3.24887 MW in DRPs case.

Table 2 Congested line MVA flow with 120% load case

S. No	Congested line No	Congested line flow limit	Actual line flow	Line flow after running MOGOA	Line flow after running MOGOA and DR
1	1–2	130	131.99	118.32	118.32
2	6–8	32	33.39	28.47	27.3198

Table 3 Optimal results with 120% load case

S. No	Name of the parameters	Values without DR	Values with DR
1	Generation cost ($/h)	739.85	737.906
2	Demand response cost ($/h)	0	71.8286
3	Total cost	739.85	809.735
4	Emission (ton/h)	462.977	457.096
5	Power generation (MW)	230.3227	230.2889
6	Power loss (MW)	3.28266	3.24887

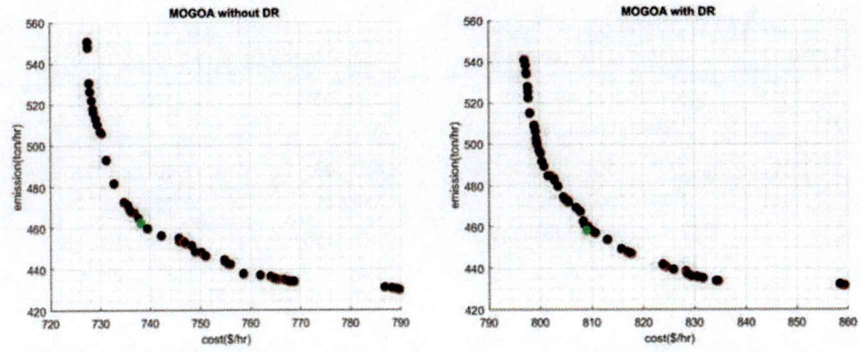

Fig. 1 Convergence characteristics for 120% load

Figure 1 shows the comparative convergence characteristics for 120% load case with and without DRPs using MOGOA.

Case 2: In this case, total system load is 245.96 MW, which is 30% more than the base load of the system. With this load, load flow program is run and is found that lines 1–2 and lines 6–8 are congested. To mitigate this congestion, initially MOGOA was used and generators are rescheduled. After rescheduling the generators, line flows are reduced in the above-said lines, and the congestion has been removed. To verify the impacts of DRPs in congestion, again MOGOA is used with considering DRPs and obtained the line flows for the same lines. It is also found that line congestion is removed, and the system is in safe condition. Table 4 shows the line flow before and after the congestion management with and without considering DRPs.

Observing the numerical results obtained from Table 5, it is noted that the total cost of generation reduces considering DRPs, which means that the implementation of DRPs reduces the operation of expensive generators in the system. The generation cost is reduced from 820.763 to 818.396 $/h, which is 2.367 ($/h) reduction in cost after implementation of DRPs.

It is also seen from Table 5 that emission can be reduced by using DRPs. It means that the conventional power generation share and contribute a lot in environmental pollution. The emission is reduced from 559.541 to 557.258 ton/h, which is 2.283 ton/h reduction in emission after DRPs.

Table 4 Congested line MVA flow with 130% load case

S. No	Congested line No	Congested line flow limit	Actual line flow	Line flow after running MOGOA	Line flow after running MOGOA and DR
1	1–2	130	146.41	123.94	123.94
2	6–8	32	36.44	29.66	29.66

Table 5 Optimal results with 130% load case

S. No	Name of the parameters	Value without DR	Value with DR
1	Generation cost ($/h)	820.763	818.396
2	Demand response cost ($/h)	0	79.0056
3	Total cost	820.763	897.4016
4	Emission (ton/h)	559.541	557.258
5	Power generation (MW)	249.4186	249.403
6	Power loss (MW)	3.45858	3.44303

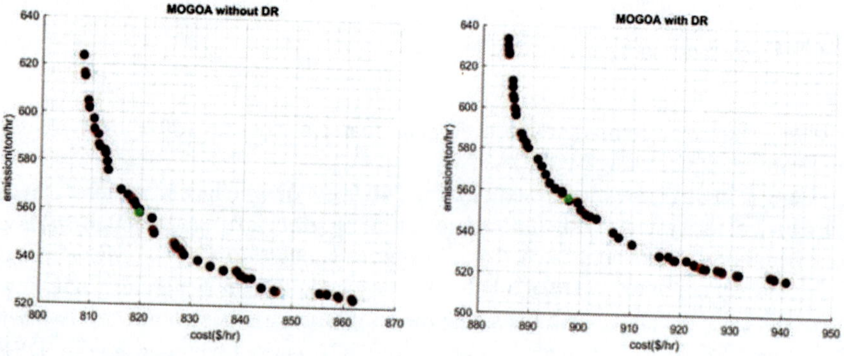

Fig. 2 Convergence characteristics for 130% load

Table 6 Decrease in load after implanting DR at selected bus in MW

Case No	Bus 7	Bus 8	Bus 12	Bus 17	Bus 19	Bus 23	Bus 30
1	0.86	1.41	0.42	0.34	0.36	0.12	0.40
2	0.91	1.20	0.44	0.36	0.38	0.12	0.42

Figure 2 shows the comparative convergence characteristics for 130% load case with and without DRPs using MOGOA.

Table 6 shows the DRPs share for mitigating the line congestion in the system. The numerical results also show the amount decrease in load at the selected bus for successful implementation of DRPs for both the cases.

5 Conclusion

The congestion management on IEEE 30-bus test system is studied, and the same is examined using demand response programs. Result shows that this work is able to reduce the congestion with minimum emission and generation cost. The performance of the optimization technique is also tested on this system with different transmission line loading. It is found that multi-objective grasshopper optimization technique is able to solve the congestion problem effectively as compared to other optimization techniques. It is also found that demand response programs is able to mitigate the congestion occurs in the transmission lines with reduced emission and cost. Thus, demand response programs became efficient tools for congestion management under extreme level in deregulated power systems. The work may be extended considering renewable energy sources and demand side management.

Appendix: Parameters of MOGOA

S. No	Name of the parameter	Value
1	Max no. of iteration	100
2	Population size	200
3	Cmax	1
4	Cmin	0.0004

References

1. Amjady, N., Hakimi, M.: Dynamic voltage stability constrained congestion management framework for deregulated electricity markets. Energy Convers. Manage. **58**, 66–75 (2012)
2. Romero-Ruiz, J., Pérez-Ruiz, J., Martin, S., Aguado, J., De la Torre, S.: Probabilistic congestion management using EVs in a smart grid with intermittent renewable generation. Electr. Power Syst. Res. **137**, 155–162 (2016)
3. Hosseini, S.A., Amjady, N., Shafie-Khah, M., Catalão, J.P.: A new multi-objective solution approach to solve transmission congestion management problem of energy markets. Appl. Energy **165**, 462–471 (2016)
4. Ding, F., Fuller, J.D.: Nodal, uniform, or zonal pricing: distribution of economic surplus. IEEE Trans. Power Syst. **20**(2), 875–882 (2005)
5. Han, J., Papavasiliou, A.: Congestion management through topological corrections: a case study of Central Western Europe. Energy Policy **86**, 470–482 (2015)
6. Kang, C., Chen, Q., Lin, W., Hong, Y., Xia, Q., Chen, Z., Wu, Y., Xin, J.: Zonal marginal pricing approach based on sequential network partition and congestion contribution identification. Int. J. Electr. Power Energy Syst. **51**, 321–328 (2013)
7. Jain, A.K., Srivastava, S.C., Singh, S.N., Srivastava, L.: Bacteria for aging optimization based bidding strategy under transmission congestion. IEEE Syst. J. **9**(1), 141–151 (2015)
8. Esfahani, M.M., Sheikh, A., Mohammed, O.: Adaptive real-time congestion management in smart power systems using a real-time hybrid optimization algorithm. Electr. Power Syst. Res. **150**, 118–128 (2017)
9. Verma, S., Mukherjee, V.: Firefly algorithm for congestion management in deregulated environment. Eng. Sci. Technol. Int. J. **19**, 1254–1265 (2016)

10. Chellam, S., Kalyani, S.: Power flow tracing based transmission congestion pricing in deregulated power markets. Int. J. Electr. Power Energy Syst. **83**, 570–584 (2016)
11. Hooshmand, R.-A., Morshed, M.J., Parastegari, M.: Congestion management by determining optimal location of series FACTS devices using hybrid bacterial for aging and Nelder-Mead algorithm. Appl. Soft Comput. **28**, 57–68 (2015)
12. Yousefi, A., Nguyen, T., Zareipour, H., Malik, O.: Congestion management using demand response and FACTS devices. Int. J. Electr. Power Energy Syst. **37**(1), 78–85 (2012)
13. Hemmati, R., Saboori, H., Ahmadi Jirdehi, M.: Stochastic planning and scheduling of energy storage systems for congestion management in electric power systems including renewable energy resources. Energy **133**, 380–387 (2017)
14. Verma, S., Mukherjee, V.: Optimal real power rescheduling of generators for congestion management using a novel ant lion optimizer. IET Gener. Transm. Distrib. **10**(10), 2548–2561 (2016)
15. Reddy, S.S.: Multi-objective based congestion management using generation rescheduling and load shedding. IEEE Trans. Power Syst. **32**(2), 852–863 (2017)
16. Gope, S., Goswami, A.K., Tiwari, P.K., Deb, S.: Rescheduling of real power for congestion management with integration of pumped storage hydro unit using firefly algorithm. Int. J. Electr. Power Energy Syst. **83**, 434–442 (2016)
17. Li, C., Yu, X., Yu, W., Chen, G., Wang, J.: Efficient computation for sparse load shifting in demand side management. IEEE Trans. Smart Grid **8**, 250–261 (2017)
18. Shayesteh, E., Moghaddam, M.P., Yousefi, A., Haghifam, M.R., Sheik-El-Eslami, M.: A demand side approach for congestion management in competitive environment. Int. Trans. Electr. Energy Syst. **20**, 470–490 (2010)
19. Tabandeh, A., Abdollahi, A., Rashidinejad, M.: Reliability constrained congestion management with uncertain negawatt demand response firms considering repairable advanced metering infrastructures. Energy **104**, 213–228 (2016)
20. Abdi, H., Dehnavi, E., Mohammadi, F.: Dynamic economic dispatch problem integrated with demand response considering non-linear responsive load models. IEEE Trans. Smart Grid **7**, 2586–2595 (2016)
21. Zaeim-Kohan, F., Razmi, H., Doagou-Mojarrad, H.: Multi-objective transmission congestion management considering demand response programs and generation rescheduling. Appl. Soft Comput. **70**, 169–181 (2018)
22. Mirjalili, S.Z., Mirjalili, S., Sarem, S., Faris, H., Aljarah, I.: Grasshopper optimization algorithm for multi-objective optimization problems. Appl. Intell. J. **80**, 805–820 (2018)

A Hardware Accelerator Implementation of Multilayer Perceptron

VS Thasnimol and Michael George

Abstract In recent years, there has been a surge in the use of machine learning models for various important applications in a wide variety of fields. Neural networks play a huge role in this area. The hardware acceleration of neural network models has been a recent area of active research. In hardware acceleration-based implementation of computing tasks, the throughput of the system is increased by decreasing latency. This work focuses on developing a digital system implementation of the most common structure for a feedforward neural network (FFNN) known as a multilayer perceptron (MLP). In pursuit of this aim, the proposed architecture is designed by considering the performance, accuracy, and resource usage and also the remarkable benefits of the design to accelerate the feedforward neural network computation.

Keywords Feedforward neural network · FPGA implementation · Multilayer perceptron · Neural network acceleration

1 Introduction

Feedforward neural networks are the most used machine learning algorithms, with numerous applications. In real case applications, the fast processing time is required, but the computer-based system is not able to maintain adequate throughput. The size of the feedforward neural network is growing due to big data applications and the complexity of problems, but the computational speed and power consumption are the major issues.

Based on the computational resources and internal architecture possibilities, FPGA is used as an independent device. By using synchronous computation and interconnections, FPGA devices allow the parallelization of neural networks. Based on the application, it can be rearranged with various weights and topologies. There are cases in which migrating a neural network from software to FPGA hardware,

V. Thasnimol (✉) · M. George
Department of Electronics and Communication Engineering,
Rajiv Gandhi Institute of Technology, Kottayam, Kerala, India
e-mail: michaelgeorge@rit.ac.in

H. Sharma et al. (eds.), *Congress on Intelligent Systems*, Advances in Intelligent Systems and Computing 1334, https://doi.org/10.1007/978-981-33-6981-8_9

i.e., FFNN can be designed, simulated, and programmed on FPGA hardware for implementation including the training process.

Regarding the hardware topologies, feedforward neural network architecture consists of different hardware entities to perform multiplication and addition. Such designs are called fine-grained architecture. But this type of architecture is impractical due to heavy hardware requirements, which leads to high resource usage, high power, and low speed. The number of neuron units and interconnection synapses in FFNN maintains an exponential relation. With every new neuron unit in the FFNN, the arrangement of connections becomes more complicated and consumes part of resources.

The coarse-grained architecture can be mainly designed for large-sized networks, where a small number of processing elements execute the time-multiplexed serial computations of the network. If the processing node is simpler, the working speed is faster, but it requires extra clock cycles. Here in this case, the hardware accelerator implementation avails from short point to point data lines and pipelined operations.

Systolic massive parallel architecture (SYMPA) [1] for feedforward neural networks mainly depends on the neural processing elements (NPE) having memory for weight storage, data inputs, address, and control lines. Instead of using a large number of activation function block (AFB), a single one is used for the entire FFNN architecture, thereby forming a mixture of fine-grained and coarse-grained architecture with parallel processing. Here, the proposed architecture is appropriate for any type of FFNN like multilayer perceptron, auto-encoder, or logistic regression. Also, it can extend to random size, only limited by available resources. This paper proposes a specialized hardware architecture and a design approach for the hardware acceleration of FFNN.

2 Literature Review

A great variety of researches related to the hardware implementation of artificial intelligence-based algorithms. The approach proposed by Ferreira et al. [2] in this architecture needs a lower transfer rate of input data, and another aspect is the possibility of tuning. The use of memory LUTs is significantly higher in this case. All networks in this unit achieved a maximum frequency of 300MHz, and it denotes that low effect in the fan-out complexity with the surge of the inputs from the network.

A customized and efficient topology for the artificial neural network was introduced by Nedjah et al. [3]. This design needs a reduced hardware area due to the reuse of circuitry by both weighted sum and activation function and also requires more clock cycles to complete the computation. So, the overall performance is improved. In the case of 34-unit, computation requires 356 clock cycles, and sigmoid is used as the activation function. Also, a finite state machine is needed to control the entire operation.

Vuk Vranjkovic et al. [4] proposed architecture for machine learning classifiers on the Virtex-7 FPGA platform. This system is considered as a one-dimensional array of reconfigurable processing nodes, proficient to implement classifiers such as decision trees, artificial neural networks, and SVMs. It provides an average resource occupation results and a maximum frequency of 113 MHz. This architecture uses more registers and DSPs and slightly lower value for memory.

The work accelerator implementation of deep neural networks based on FPGA developed by Huynh [5]. They used a similar approach to that of the proposed system. The system is trained with stacked sparse auto-encoder, and also here the feedforward phase was implemented using fixed-point arithmetic and the systolic array techniques. The benefits of LUT's to store the weight values corresponding to each neuron granted the reduction of area occupied by the device. This design obtains a much lower clock frequency, and it is slightly better in DSP and memory but uses more LUTs and registers.

A reconfigurable neural network implementation proposed by Oliveira et al. [6], Here, 20 neurons were implemented in FPGA cyclone, using 77.8 MHz. Also, this method provides flexibility and reconfiguration in execution time due to the invariability of the implemented area of the network.

3 FFNN Computation Algorithm

According to the application, the number of units in inputs, outputs, and hidden layers is varied. The arrangement of neural networks differs mainly due to their interconnections. All FFNN having the following properties such as there is no connection exist amid the units in the same layer, the same activation function can be shared by all units and computed independently, and the output of a particular layer is a function of the preceding layer input and a bias.

Define a vector $\mathbf{N} = (N_0,..N_i..N_L)$, associated among the number of neuron units in the input layer(N_0), i'th layer (N_i), and the number of outputs (N_L) where L is the total number of layers including the output layer. General case $\mathbf{Y_i}$ is the layer output from the preceding layer output $\mathbf{Y_{(i-1)}}$ and W^i, b_i is the weight and bias vector of ith layer, and F is represented as the activation function, respectively. The computation that corresponds to a single neuron unit is given as in Eq. (1):

$$Y_{ij} = F(\sum_{k=1}^{N_{i-1}}(W^i_{jk}) * Y_{(i-1)k} + b_{ij}) \tag{1}$$

For the algorithm computation, the bias value is renamed as $\mathbf{b_i} = \mathbf{w_i}$. Also, the layer matrix W'^i is formed from the concatenation of weight matrix and bias column vector.

$$W'^i = \left[\mathbf{w_{0i}}\ W^i\right]$$

Then, the sum of products $\mathbf{S_i} = W'^i \times Y'_{i-1}$ (Eq. 2.a). FFNN computation requires two operations such as vector by matrices operation and activation function (Eq. 2b).

$$a.\ \mathbf{S_i} = W'^{i} \times \begin{bmatrix} 1 \\ \mathbf{Y}_{(i-1)} \end{bmatrix} = W'^{i} \times Y'_{(i-1)}$$

$$b.\qquad \mathbf{Y}_i = (F(S_{i1}), ..F(s_{ik})..F(S_{iN_1})) \qquad (2)$$

The output values correspond to units in a layer are used as inputs of the next layer. The layer-wise structure and computation procedure are repeated for all layers, and Fig. 1. represents the layer-wise, feedforward computation procedure for $N_0 \times N_1 \times N_2 = 3 \times 3 \times 2$ FFNN ($L = 2$). Here, the inputs are serially processed, after computation, the resulting values $\mathbf{S_i}$ ($i = 1, \ldots, L$) are hoarded in memory, and the same structure is repeated for all layers. The sum of product values without the activation function is represented as S_i. If no matter the FFNN sizes, a single activation function can be used for whole architecture.

The computation procedure consists of a one-directional dataflow, and all units are processed in a parallel manner. So, every layer output value is reliant upon the output values of the preceding layer. The variation in the number of neuron units also affects the required computational blocks.

The proposed layer-wise parallel architecture is mainly based on neural processing elements having a local weight memory like BRAM, multiply-accumulate unit (MAC), input lines, command, and control lines. For the interlayer memory for storing the weighted sum values, scratchpad registers are also included in that unit, and these are arranged in a daisy chain model forming an SR-ring structure. These stored results are externally available after the computation of each layer.

The FFNN computation algorithm mainly consists of three parts: First is accepting the network input, then through the network feedforward propagation of the signal, and finally obtaining the output. In the first section, inputs are multiplied with weights of each unit in corresponding NPE. By MAC operation, these results are added to the accumulator register, and also the bias values are stored into the corresponding weight index places. After first layer computation, the resulting values in the accumulator are

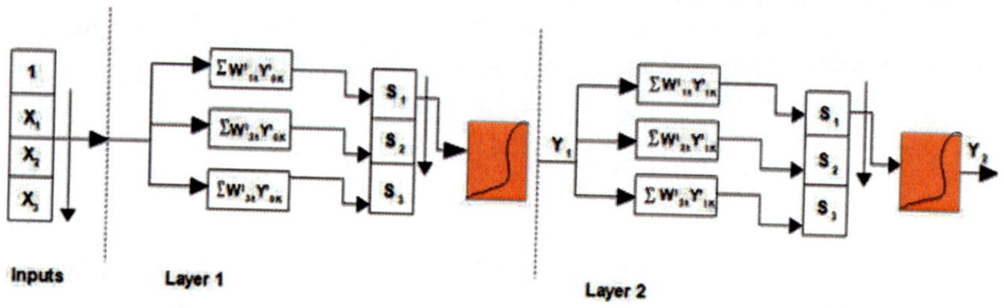

Fig. 1 Computation procedure of 3 ×3×2 MLP

hoarded to SR, and also NPE is ready for the next computation. The network output can be externally accessed using the AFB port, after the sum of product calculation of the network.

4 Hardware Architecture

From the FFNN computation algorithm-specific hardware blocks in FPGA like appropriated memories, arithmetic units and various inter-connections are considered. The main aim is to an optimal and efficient implementation of commercial hardware, and it favors portability to any FPGA. According to the availability of hardware resources, architecture can be enlarged in several layers and units per layer.

The partial sum of products of the prevailing unit is calculated and stored in the ALU accumulator of the NPE, and after that, stored value is transferred into the scratchpad register through the ring-like structure into the activation function block. The same process can be repeated for the next layers by feeding back the AFB output values. The entire operation is controlled by the finite state machine (FSM) block. The loading and addressing of weight values into memory are performed by using this control unit.

Modular structure and minimization of connection are important specifications of this architecture. The external connection in an NPE is addressing, write enable (we) into the memory, and a scratchpad ring structure. In arranging NPEs, the rightmost one contains the last unit of each layer, and the order is from right to left direction. Figure 2 shows 3×3×2 MLP with the output layer only contains two units. The data input to the system and weight updating is done using the DIN port. This peculiarity is effective when weight updating without using device reprogramming.

The memory size demanded weights in full design is given by Eq. 3, and the weights are distributed in different memories with the maximum size for NPE is given by Eq. 4. So, the required information in the case of FFNN is:

$$a.\quad Total\,Weights = (\sum_{i=0}^{L-1}(N_i + 1) * (N_i + 1) \tag{3}$$

$$b.\quad Max\,Weights\,in\,NPE = (\sum_{i=0}^{L-1}(N_i + 1) \tag{4}$$

For the proper generation of an FFNN structure, a number of neural processing elements, neuron units in all layers, and the bit size are required.

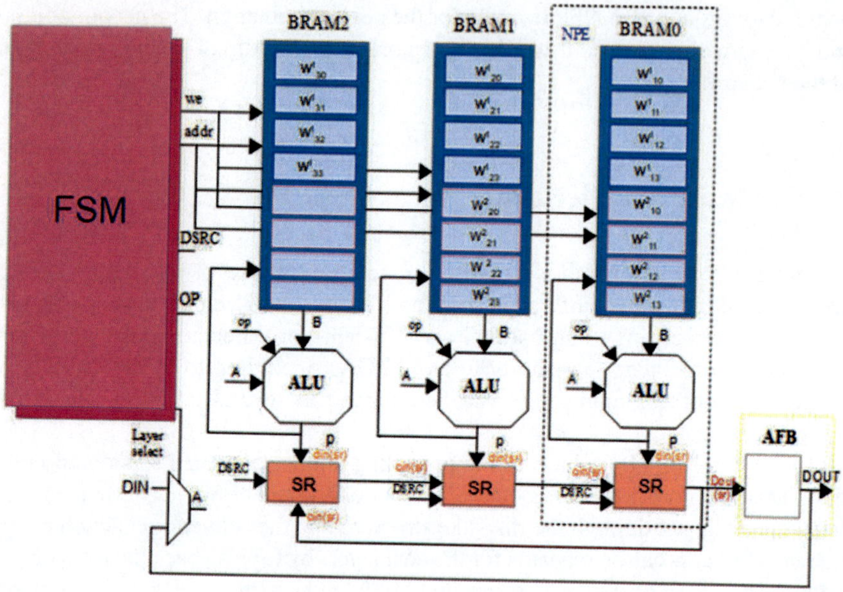

Fig. 2 Hardware architecture of 3 ×3×2 multilayer perceptron

4.1 *Neural Processing Element*

A single NPE consists of three blocks such as RAM block, a scratchpad registers SR, and an ALU. Concurrent NPE operation is allowed by RAM block with single-cycle read/ write access. For arithmetic computations, an ALU with two N-bit data inputs A, B, and the required output is hoarded as P (accumulator), it performs operations such as $P = 0$, $P = A * \mathrm{B}$, $P = A * \mathrm{B} + P$, $P = A$. The last one is the scratchpad register, and it is connected to nearby SR in a ring-like manner. The register contents and the source of input data are selected through the DSRC signal from the control unit. The SR contents latch through dout (SR) for every clock cycle. When the DSRC value is 1, it acts as a rotating register, and when it is zero SR value is updated from ALU. Then the values are shifted through AFB, also AFB output is feedback to the input for the next layer.

4.2 *Activation Function Block (AFB)*

The neural computation of the entire architecture is done by using a single activation function due to the systolic nature and hardware complexity of the proposed system. The ReLU activation function is faster and simpler defined in the range [0, ∞]. It is implemented in fixed-point arithmetic with a sign bit evaluation and conditional signal assignment (Eq. 5).

$$F(X) = \begin{cases} X, & \text{if } x \geq 0. \\ 0, & \text{if } x < 0. \end{cases} \quad (5)$$

4.3 *Network Control Sequences*

The control unit was designed to execute the neural computations according to the defined parameters such as the number of neuron units in inputs, hidden layer, and the output layer. Two counters for generating address and counting the cycle of operation are used for the execution of control operations.

4.4 *Number of Clock Cycles of Execution*

During layer computation, in the case of Ni inputs data, requires Ni + 1 clock cycles along with an interlayer delay including ALU (T_ALU), SR (T_SR), and AFB (T_AFB). All layers account for bias calculation time except the output layer by adding one clock cycle. So, the total clock cycles needed for the FFNN computation time can be described as (Eq. 6):

$$C = (\sum_{i=0}^{N_L-1} (N_i + 1) + N_L + (L * (T_{ALU} + T_{SR} + T_{AFB})) \quad (6)$$

5 Results and Discussion

The architecture coded in Verilog simulations is performed using Xilinx Vivado IDE for synthesis and implementation with Zybo-Z7-20 based on Xilinx all programmable system-on-chip architecture as a target device.

5.1 *Hardware Blocks*

The hardware architecture mainly consists of finite state machine (FSM), neural processing elements (NPE), and an activation function block (AFB). Here, the proposed system uses standard blocks for efficient and optimal implementation on the hardware platform. The NPE mainly consists of three parts BRAM, ALU, and scratchpad register.

The design of block RAM in IP integrator is done by using Vivado. The contents in BRAM can be loaded using a Xilinx coefficient file (COE) or by using the default data operation. The data ports (dina, dout) width is selected in the Vivado IDE. The address (addr) port width is determined by using memory depth respect for each port. Here, PortA operations are mainly used for accessing data from the COE file.

The entire order of operation of architecture is controlled by using finite state machine and the weight loading and updating in the BRAM, controlling the operand values in ALU, the DSRC signal in scratchpad register, etc. The state chart diagram in Fig. 3 shows the corresponding control signals and the entire flow of operation between different blocks. Here, mainly considering five states from BRAM to AFB including a reset state, i.e., s0 to s5. The control signals used in the finite state machine are DSRC, operand 'op', write enable (we) and address (addr) for BRAM, and each stage is designed as states.

Activation function block is necessary to achieve the neural computations of the architecture. By serial feeding of data into this part, calculate the output function for the sum of products generated in the network. Here, the faster and simplest activation function ReLU is implemented with fixed-point arithmetic.

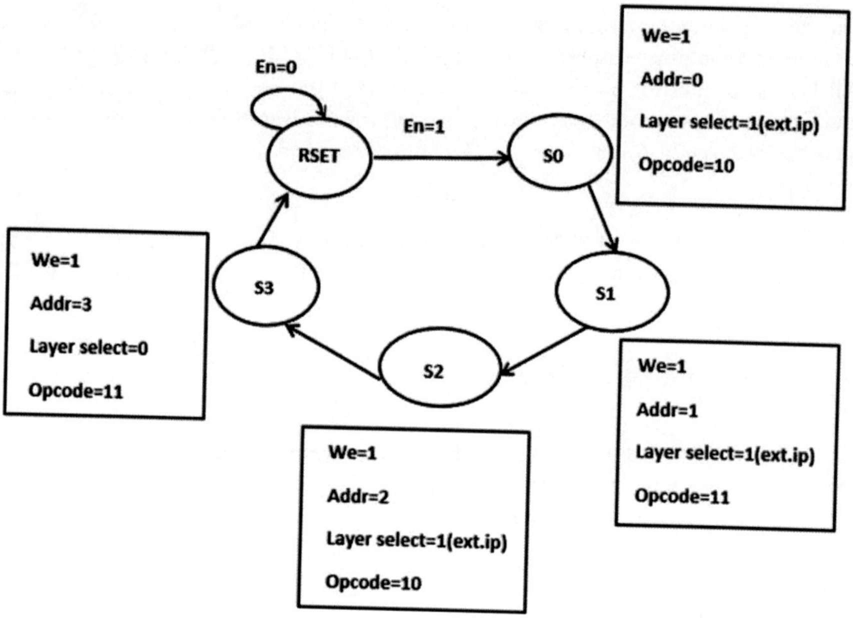

Fig. 3 State diagram for FSM

5.2 *Hardware Architecture of 3×3×2 MLP*

The obtained block-level designs will be combined to be used for the entire hardware architecture design and implementations in the proposed system. The simulation was done using 16-bit fixed-point arithmetic as shown in Fig. 4. FPGA devices require less power and a small board to work compared to other PC-based systems. The standardized format is used to store floating-point numbers in memory as IEEE 754 arithmetic. Here, the output obtained using IEEE 754 floating-point arithmetic with single precision of 32 bit is shown in Fig. 5.

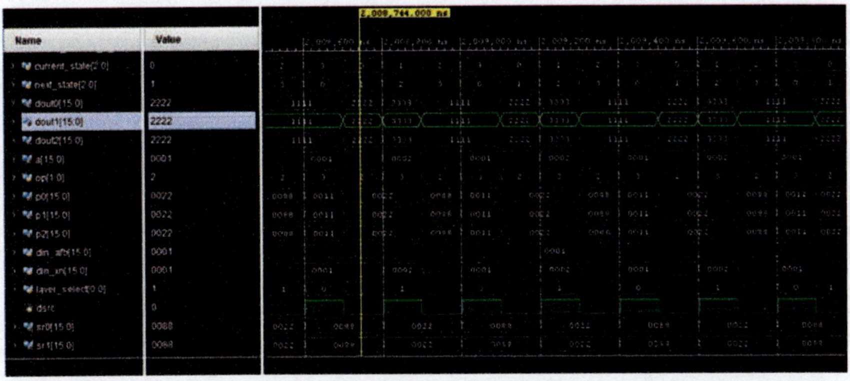

Fig. 4 Simulation of hardware architecture of 3×3×2 MLP using fixed-point operations

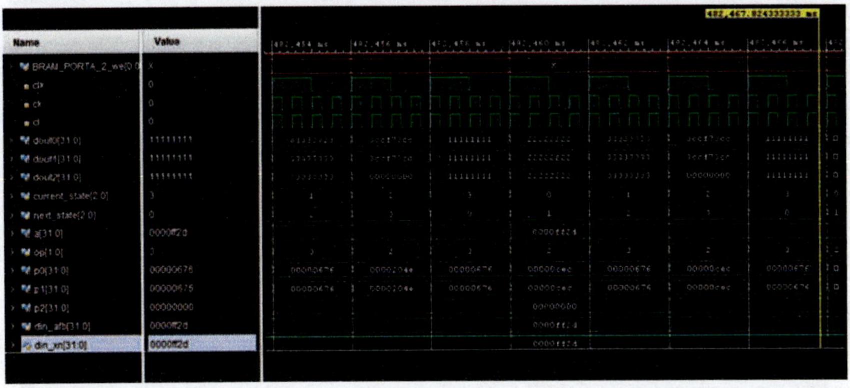

Fig. 5 Simulation of 3×3×2 MLP using IEEE754 floating-point arithmetic

Table 1 Resource usage and performance of 3 × 3 × 2 MLP using fixed-point operation

Type	Used (Util %)/Available)
Slice Registers	13(0.01)/106400
Register as flip flop	5(<0.01) / 106400
Register as latch	8(<0.01) / 106400
Slice LUTs	6(0.01) / 53200
Clocking (BUFGCTRL)	1(3.13)/ 32

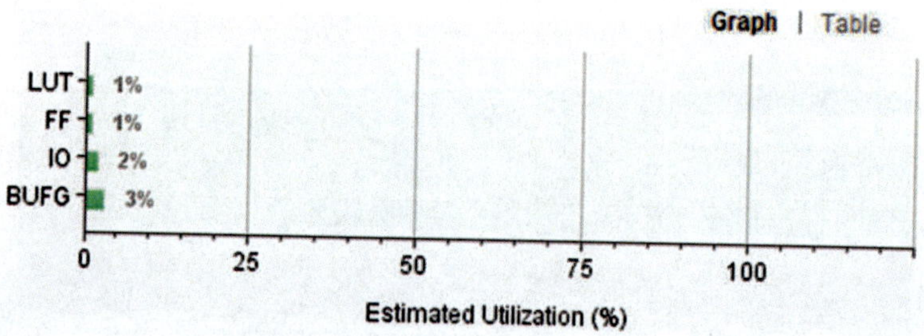

Fig. 6 Resource utilization graph of MLP

5.3 *Resource Utilization and Performance Evaluation*

This section includes performance evaluation from the synthesis report of 3×3×2 MLP using fixed-point operation including the resource utilization of different components. Table 1 shows the number of available and used slice registers and slice LUTs as in the case of both logic and memory. By considering the RAM usage both 16- and 32-bit size memory generators are used. The resource utilization is graphically represented as in Fig. 6. The power report shows the on-chip power estimation of resources after the synthesis of hardware architecture with 0.332 W (95%) power consumption of BRAM and the remaining part consisting of 5% including signals (0.005 W), I/O (0.008 W), and logic part (0.006 W).

6 Application

To test the performance of the hardware architecture an application is used here. This application aims to reconstruct the hardware architecture by using the weights and bias values, that can be extracted from the encoding and decoding section of an auto-encoder. Auto-encoder is a type of FFNN, which aims to extract features used for the reconstruction of the exact input signal from the features.

Here, an auto-encoder with ten inputs, a single hidden layer which consists of 5 neurons, and 10 neuron units in the output which is created from a 784 × 32 × 784

Fig. 7 784 × 32 × 784 Auto-encoder with original (Top) and reconstructed (Bottom) digits

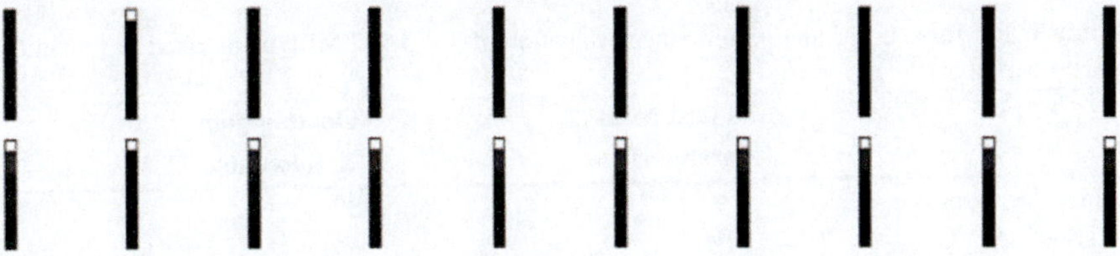

Fig. 8 10 × 5 × 10 AE in Keras with original (Top) and reconstructed (bottom) parts

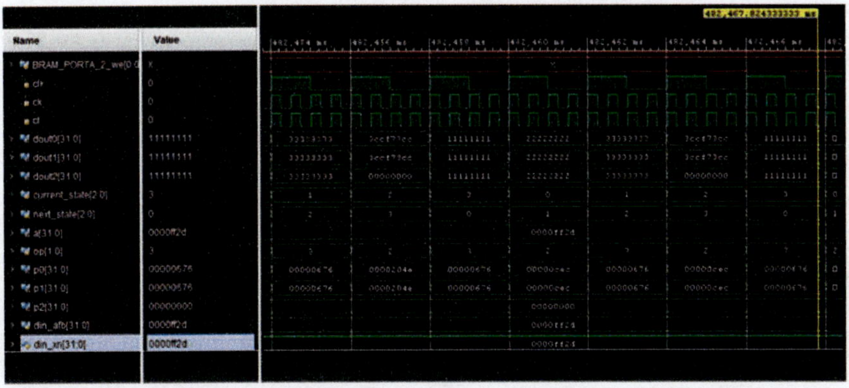

Fig. 9 10 × 5 × 10 AE in Keras with original (Top) and reconstructed (bottom) parts

AE with the MNIST dataset are shown in Fig. 7 and 8. Here, the encoder and decoder model values in decimal form are converted to standard 32-bit IEEE floating-point format.

The ReLU activation function was implemented in this application. The simulation of architecture including the AFB was configured using IEEE754 floating-point format as shown in Fig. 9. Table 2 shows resource utilization and overall performance.

By comparing, 3 × 3 × 2 MLP architecture using fixed-point and floating-point arithmetic is shown in Table 3. Here, the floating-point architecture shows higher resource utilization compared to the other with an extra usage of DSP (3%), and also the LUTs, IO, FF, and BUFG are increased and incorporating IEEE754 floating-point operation increases the accuracy of the system. The architecture turns into a good candidate for its use in practical applications.

Table 2 Performance analysis of 10 × 5 × 10 AE using floating-point operations

Type of FFNN	10 × 5 × 10 AE
Slice registers	6362 (5.98%)
Slice LUTs	7333 (13.78%)
DSP48E1	20 (9.09%)
Clocking (BUFGCTRL)	12.50%

Table 3 Resource usage and performance evaluation of 3 × 3 × 2 MLP using fixed and floating-point operation

	Fixed point architecture	Floating point architecture
DSP	-	3%
LUTs	1%	4%
IO	2%	6%
Clocking (BUFGCTRL)	3%	16%
Total on-chip power	0.461 W	36.64 W
Accuracy	16 bits	32 bits

7 Conclusion

The architecture shows remarkable benefits, to accelerate neural network computation mainly for feedforward neural networks. It is also used for the computation of large FFNN but limited by the available resources in hardware, which is mainly useful for deep multilayer neural networks. Also, the usage of single AFB and the external accessing of the intermediate layer outputs are the main peculiarities of the architecture. This feature can be used for debugging purposes. Also, weight updation during normal operation is possible. By incorporating, the IEEE754 floating-point operations increase the accuracy of the entire system. The architecture becomes a good candidate for its use in practical applications. In the future, the proposed design can easily be adapted for any deep learning applications.

Acknowledgements This work was supported by APJ Abdul Kalam Technological University—Center for Engineering Research & Development (APJAKTU-CERD) Research Seed Money Project (No. KTU/RESEARCH 2/4068/2019).

References

1. Medus, L.D., Iakymchuk, T., Frances Villora, J.V., Bataller Mompean, M., Rosado Munoz, A.: A novel systolic parallel hardware architecture for the FPGA acceleration of feedforward neural networks. In: IEEE Access 2920885, vol. 7 (2019)

2. Ferreira, A.P., Barros, E.N.: A high performance full pipelined arquitecture of MLP neural networks in FPGA. In: Proceedings 17th IEEE International Conference Electron, Circuits System, pp. 742–745 (2010)
3. Nedjah, N., Silva, R.M., Mourelle, L.M.: Compact yet efficient hardware implementation of artificial neural networks with customized topology. Expert Syst. Appl. **39**(10), 9191–9206 (2012)
4. Vranjkovic, V. , Struharik, R.: Coarse-grained reconfigurable hardware accelerator of machine learning classifiers. In: Proceedings of International Conference Systems, Signals Image Process (IWSSIP), pp. 1–5 (2016)
5. Huynh, T.V.: Deep neural network accelerator based on FPGA. In: Proceedings of 4th NAFOSTED Conference Inf. Computer Science, pp. 254–257 (2017)
6. Oliveira, J.G.M., Moreno, R.L., Oliveira Dutra, O., Pimenta, T.C.: Implementation of a reconfigurable neural network in FPGA. In: Proceedings of International Caribbean Conference Devices, Circuits System (ICCDCS), pp. 41–44 (2017)

Building a Classifier of Behavior Patterns for Semantic Search Based on Cuckoo Search Meta-heuristics

Victoria Bova, Elmar Kuliev, Ilona Kursitys, and Dmitry Leshchanov

Abstract The paper deals with the problems of extracting and classifying the thematic categories representing user preferences to increase the pertinence of semantic search in the knowledge-based information systems. The authors propose building the classifier of behavioral user profiles by analyzing the information requirements, implicit preferences, and knowledge resources interaction. The method is based on the probabilistic algorithm for Expectation-Maximization clustering of user characteristics and distributed knowledge resources to generate the structure of the classifier input parameters. Optimization is implemented by the mechanism of selecting informative features based on revealing user latent interests and the resource ability to satisfy the users. To reduce the search space, the paper proposes a bio-inspired algorithm based on the cuckoo search. The algorithm is compared with fuzzy c-means algorithm on Machine Learning Repository of the University of California, Irvine. The results show the effectiveness and prospects of the proposed method in the semantic search systems.

Keywords Semantic search · Behavioral user profile · The problem of classification · EM-algorithm · Bio-inspired methods · Cuckoo search meta-heuristics

V. Bova · E. Kuliev · I. Kursitys (✉) · D. Leshchanov
Southern Federal University, Rostov-on-Don, Russia
e-mail: kursitys@sfedu.ru

V. Bova
e-mail: vvbova@sfedu.ru

E. Kuliev
e-mail: ekuliev@sfedu.ru

D. Leshchanov
e-mail: leshanov@sfedu.ru

H. Sharma et al. (eds.), *Congress on Intelligent Systems*, Advances in Intelligent Systems and Computing 1334, https://doi.org/10.1007/978-981-33-6981-8_10

1 Introduction

The present days, the tasks of intelligent processing the large volumes of weakly structured information related to knowledge discovery in the open information space are considered as essential tasks in terms of the development of intelligent recommendation and analytical systems, business intelligence systems, and systems of semantic search and analysis of the content on the Internet [1–3]. When working with the large volumes of non-structured data in the intelligent systems (IS), it is necessary to search for hidden dependencies, extract, structure, and classify the thematic categories representing the users' interests and preferences [1, 3–6]. One of the measures allowing us to estimate the quality of the generated suggestions is their relevance to the hidden information needs of the user-pertinency. The problems that appear in the process of forming the user's request can be related to the ambiguity of their formulation, lack of the keywords which can clearly determine the semantics of the required resources or lack of traditional terminology in the interested domain. The mentioned facts determine the relevance and importance of the research aimed at solving the main task of the semantic search—to adequately reflect the users' information needs based on the analysis of their implicit behavior to increase the pertinence of the search [2–4].

The paper proposes to study the implicit profile of the users' behavior and formulate the rules of its processing to improve the accuracy and pertinence of the information search. We suggest a method for classification of users' behavioral patterns (profiles) to model the interaction between the main subjects and objects of the semantic search. To optimize the parameters of the profile classifier, we suggest a bio-inspired algorithm based on the cuckoo search.

The problem of effective meeting the users' information needs is one of the key problems in the information systems related to knowledge management [1, 4]. Most of the modern applications are focused on using and processing the knowledge on the domain that the user is interested in. This knowledge is embedded in the information system at the stage of its development and extracted dynamically from the available information resources. As a rule, access for information resources is provided through the Web, where we can use various formalization models, methods, and formats of their storage, access rules, and processing methods. Thus, the problem of knowledge extraction is related to identification of information objects associated with the tasks that the users need to solve [6–8].

Generally, the problem involves the following statement: to meet the users' information needs effectively, we must obtain the formalized description of these information needs based on the knowledge of what the user wants and in which form. To achieve the desired goal, we need to solve the following tasks [1]:

- clustering of information resources similar in several features;
- clustering of the users' profiles with similar interests and behavior characteristics;
- classification of behavior patterns based on the generalization of the aggregation of both explicit and implicit data on the users' information needs.

The problem of extraction of the useful knowledge on the user preferences and their processing becomes complex due to the large volumes and heterogeneity of the accumulated data and their fast changes (or updates) during the time [4, 7].

In terms of big data analysis and finding patterns in them, we can face the problem of selecting informative features for developing the classification model [2, 5]. Too many informative features can increase model complexity and require a lot of computational resources and time. The rejection of the features that seem to be insignificant or appearing at the noise level can lead to the loss of important information. Another problem is to determine the relationship between the noise features and the objective variable of the studied task. The need of reducing the size of the initial data and removing noisy and redundant features to interpret the data explains the importance of the research and use of the methods for data mining and machine learning and bio-inspired methods for finding the optimal solutions [5, 7–9].

2 Formulation of the Problem

Let us formulate the problem studied in this paper as follows. Let *Res* and *User* denote the set of resources and users, respectively. Assume that each user $u \in User$ has a set of preferences (interests) T which are called themes, each theme is $t \in T$, represented in resource $r \in Res$.

To compare the user profiles and resources, we introduce the concept of generalized thematic (search) profile of a resource and user which combines the features of the resources and user preferences [1]. The vector of the parameters is represented as: $\theta_j = (\alpha_{i1}, \alpha_{i2}, \ldots, \alpha_{in}, \beta_{j1}, \beta_{j2}, \ldots, \beta_{jm})$, where $\alpha_{i1}, \alpha_{i2}, \ldots, \alpha_{in}$ denotes the weight of the feature I for the resource $r_i \in Res$; $\beta_{j1}, \beta_{j2}, \ldots, \beta_{jm}$ denotes the weight of the feature j for the user $u_j \in User$.

Let X denote a set of objects with the assigned probabilistic measure P for $p(u, r)$—the probability of choosing the resource r by the user u. The problem of generation of the classifier structure includes the assessment of parameter vector $\Theta = (w_1, \ldots, w_k, \theta_1, \ldots, \theta_k)$ based on the object sampling $X_i = \{x_1, \ldots, x_z\} \subset X$ and the number of vector components k.

Assuming that the distribution density of information features on sampling X has a form of k distributions (classifiers):

$$p(x) = \sum_{j=1,k} w_j p_j(x_i), \ \sum_{j=1}^{k} w_j = 1, \tag{1}$$

where $p_j(x_i) = \max f(x_i, \theta_j)$ is a likelihood function of the classifier j, w_j is a priori probability, defined as $p(u, r)$.

Optimization of the parameters involves selecting the parameters of the likelihood function and resulting labels of classifiers in such a way that the accuracy criterion is maximized on the training sample:

$$E(\theta) = \frac{\sum_{i=1}^{z} \begin{cases} 1, \text{ if } p(x_i) = f(x_i, \Theta) \\ 0, \text{ otherwise} \end{cases}}{z} \to \max, \quad (2)$$

where $f(x_i, \Theta)$ is a function describing the classifier, Θ is a vector of the parameters of the behavior profile, and z is the number of training datasets.

3 Method for Building the Classifier of Behavior Pattern

The process of building the classifier is based on the consequent solving of the following tasks (Fig. 1):

(1) selection of informative features of the profile based on the idea of identifying latent interests and preferences of users;
(2) generation of a classifier structure (vector description of structural characteristics) of users' profiles and information resources to form patterns of users' behavior with similar thematic preferences based on the probabilistic EM algorithm [10];
(3) optimization of the classifier parameters using the cuckoo search meta-heuristics aimed at reducing the dimension of the feature space of implicit large-volume data about users' preferences [6, 11].

The first step in solving the task includes the process of feature construction (feature engineering) [4]. In the considered case, the objects represent the users, and the features denote the search categories of the information resources visited by the users. This feature description allows us to form the behavioral patterns of the users as a vector of their interests and thematic preferences. The features that are found among not more than one user are considered non-informative.

The initial information on the feature space is used to select the features at the pre-processing stage for the further construction of the classifier structure [9, 12].

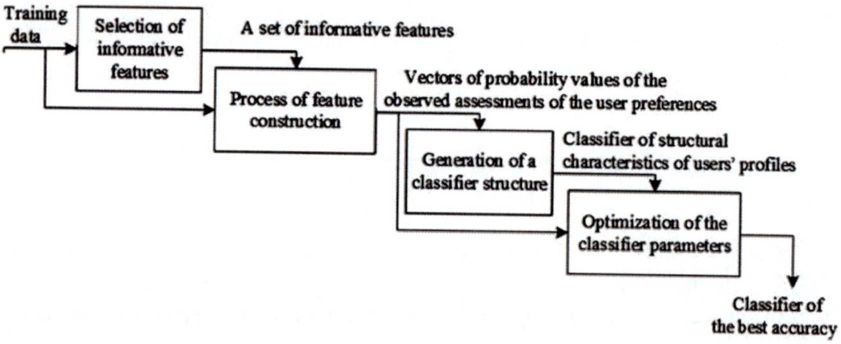

Fig. 1 Process of building the classifier

At the stage of feature selection on the basis of the training data, we form a set of informative features represented by the vectors of probability values of the observed assessments of the user preferences for choosing the information resources. The purpose of selecting the features is to avoid overtraining of the model, reduce the volume of the analyzed data, improve the effectiveness of classification, remove the irrelevant and noisy features, and increase the interpretability of the obtained results [5, 7].

Then, the profile classifier is clarified according to the users' interaction with the information resources (resource selection actions). After the first step is finished, we need to select a subset of the most informative and reliable feature to build a model.

Using the method of collaborative filtering [1, 13], the data are reduced to matrix form of the training sample in the matrix of preferences factorization $D = (u_i, r_i)_{i=1}^{l}$, where l denotes the size of the preferences features space.

Let us assume that the user u has an interest $t \in T$ with a probability $p(t|u)$. The profile u can be determined as a vector of probability values $p(t|u)$, $t = 1, \ldots, |T|$, where $\sum_{t \in T} p(t|u) = 1$. Each resource r corresponds to a certain set of themes $t \in T$ and satisfies them with a probability $p(t|r)$. Hence, a profile r denotes the vector of the probabilities $p(t|u)$, $t = 1, \ldots, |T|$, where $\sum_{t \in T} p(t|r) = 1$. Let us denote w_j as probability of prediction of choosing the information resource ч r, if we meet the conditions $p(t|r)$ and $p(t|u)$. Thus, $p(u|r)$ denotes the probability of choosing the resource r by the user u and can be represented as:

$$p(u, r) = \sum_{t \in T} p(t|r) \times p(t|u). \tag{3}$$

To restore the feature space of the profiles $p(t|u)$ and $p(t|r)$ in terms of the sparse data and exclusion of redundant features which do not affect the results of the classification in the training sample D, we suggest using the Expectation-Maximization (EM) algorithm for dividing the mixture of distributions [10]. The algorithm is based on the method of iterative calculations of the maximum likelihood assessment of the function describing the classifier.

At the stage of structure generation, let us introduce the intermediate vector Q representing hidden variables to obtain the initial representation of the structure of the processed data. The EM algorithm is based on the iterative repetition of the steps: at step E, the algorithm calculated the expected values of Q using the current approximation of the profile parameters vector Θ; at the stage M, the algorithm solves the task of the likelihood maximization and finds the next approximation of the vector Θ using the current values of Q and Θ.

At step E, let us define the combined density of the probability of obtaining the object x_i generated by the jth component of the composition $p(x_i$ and $\theta_j) = p(x_i)$ $P(\theta_j|x_i) = w_j\, p_j(x_i|\theta_j)$ using the values of $p(x_i$ and $\theta_j)$.

Let the vector $q_{ij} \equiv P(\theta_j|x_i)$ define a posteriori probability of generating the training object x_i by the jth component of the composition. These parameters are considered as hidden values. Let us denote $Q = (q_{ij})_{m \times k} = (q_1, \ldots, q_j)$, where q_j is a jth column of the matrix Q assuming that each object can be generated by one component. Using the parameters w_j and θ_j, we can calculate q_{ij} for each I and j:

$$q_{i,j} = \frac{w_j p_j(x_i, \theta_j)}{\sum_{s=1}^{k} w_s p_s(x_i, \theta_s)}. \quad (4)$$

At step M, we need to solve k independent tasks of maximizing the weighted likelihood to form the parameter vectors of the profile θ_j.

$$f(x) = F(X^z, \Theta) = \sum_{i=1}^{z} \ln \sum_{j=1}^{k} q_{i,j} p_j(x_i, \theta_j) \to \max_{\Theta}. \quad (5)$$

The objects of the training sampling are considered with their weights q_{ij} and their distribution is individual for each of k components.

4 Bio-inspired Algorithm Based on Cuckoo Search

To solve the task of building the classifier of the best accuracy which can save the good generalization ability in terms of big data, the paper proposes the populational cuckoo search algorithm (CSA). Considering the fact that the real feature space Θ is of high dimension and can be scattered, and the number of objects is high, the process of the algorithm parameters optimization (its training) is carried out on the training sampling—a certain finite subset $X_i \subseteq X$.

CSA creation was inspired by cuckoo behavior in the process of brood parasitism. The algorithm represents the simulation of the random walk and applies the cuckoo behavior strategy to the optimization of the search space. CSA is based on three following rules [12, 14]:

(1) each cuckoo specie can lay just one egg and put it in a randomly chosen nest. If the host notices strangers' eggs with a certain possibility ρ_a (0; 1), the host will throw these eggs away or leave the nest and build another nest somewhere else [11];
(2) the nests with high-quality eggs are inherited by the next generations;
(3) the number of the available nests is fixed.

In terms of CSA, each egg in the nest represents a solution, and the cuckoo's egg represents a new solution. The purpose is to replace the worst solutions in the nests with new and potentially better cuckoo's solutions with a high value of fitness function.

The general process of simulating the cuckoo search involves initializing several nests, calculating the fitness function of each nest, and updating the cuckoo's position according to the Levi flight principle unless the bird finds the global point of the potentially best solution [12]. The scheme of the cuckoo behavior base on the Levi flight is demonstrated in Fig. 2.

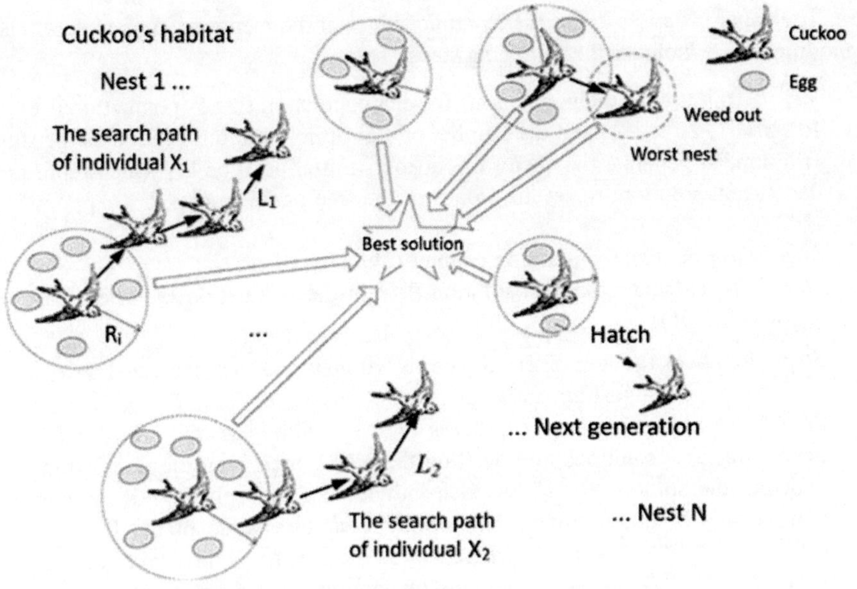

Fig. 2 Simulation model of a cuckoo's behavior

Let us consider the task of global unconstrained maximization. In general terms, the CSA can be described in the following way:

(1) to initialize the population S of the host nets and the cuckoo, i.e., define the initial values of the vector X_i;
(2) to perform a certain number of the random movements of the cuckoo in the search space based on the Levi flight according to the parameters describing the scale of the search area with high fitness value and to calculate the new position of the cuckoo X_n;
(3) to choose the nest s_i randomly and if $f(X_n) > f(X_i)$, replace the egg in the nest with the cuckoo egg assuming $X_i = X_n$;
(4) to delete a certain number of randomly chosen worst fitted nests with the possibility ρ_a (probably including the net s_i), then to build the same number of new nests according to the rules of step 1;
(5) if the stop criterion is not met, go to step 2.

In the classic CSA the probability ρ_a and Levy Flight parameters are constant [12], and the information on the high-quality solutions is not considered. To diverse the search at first iterations, we propose using high values of ρ_a—probability with which the nest can be left by the host (probability of deleting the solution vectors) and ν—the free parameter of the probability vector of cuckoo flight to nests with a high fitness value. At the final iterations, we use lower values of ρ_a and ν to improve the accuracy of the extremum localization (intensification of the search procedure).

The paper proposes to use the dynamic values of the mentioned parameters. The modified CSA includes the following steps.

Step 1. To initialize the population. To set a population size S (population is $\Theta = \{\theta^s; s = 1, 2, \ldots, S\}$) and the number of algorithm iterations N used as the stop criterion; to generate the vector of cuckoo's initial position θ^s, which represent the current solution; to set the values of the free parameters $\rho_a{}^{\min}$, $\rho_a{}^{\max}$, $\nu^{\min}$, $\nu^{\max}$.

Step 2. To initialize the iteration counter $Counter = 1$ до N.

Step 3. To calculate the best solution θ^{best} in the population Θ, where $best = \arg\max\limits_{0 \le s \le S}(E(\theta^s))$.

Step 4. To check the stop criterion. The algorithm works iteratively and if $Counter = N$, to stop the algorithm and take the solution as θ^{best} (*шаг* 9).

Step 5. To start the iterative process of the optimum search which includes generating new solutions and deleting the worst ones from the population. To generate the solution $\theta^{s,\,new}$ for each individual s by replacing all the current components of the solution vectors with a random value of the Levi flight: $\theta_l^{s,new} = \theta_l^{s,counter-1} + \text{Levi}_l$, where $l = 1, 2, \ldots, m$ (m is the size of the search vector θ^s); $\theta_l^{s,counter}$ is the value of the lth component of the population vector θ^s at the iteration *Counter*. The value of the Levi flight is calculated as follows:

$$\text{Levi}_l = \frac{\gamma_l \lambda_l}{|\nu_l|^{1/\beta}}, \tag{6}$$

where γ_l is the Levi flight coefficient, recommended value of which is 0.01 [11], $\beta = 1.5$; λ_l is a normally distributed value $\lambda_l \sim N(0; r_\lambda^2)$, $\nu_l = \nu^{\max}\exp(d^n)$, and

$$d\frac{1}{n}\ln\left(\frac{\nu^{\max}}{\nu^{\min}}\right) \cdot r_\lambda^2 = \left\{\frac{\Gamma(1+\beta)\sin\left(\pi * \frac{\beta}{2}\right)}{\Gamma\left(\frac{1+\beta}{2}\right) * 2^{(\beta-1)/2}}\right\}^{1/\beta}, \tag{7}$$

where $\Gamma(x)$—is gamma-function [11].

Step 6. To start the local search. To estimate the quality of a new solution after applying the new positions based on the random walk of the Levi distribution. If the new solution vector shows better accuracy than the corresponding population vector, replace it with the new solution, i.e., if $E(\theta^{s,\,new}) > E(\theta^{s,\,counter-1})$, then $\theta^{s,\,counter} = \theta^{s,\,new}$.

Step 7. To delete the worst solution from the population Θ after analyzing the solutions in the current search area. If $rnd \le \rho_a$, then we delete such a solution that $k = \arg\min\limits_{0 \le s \le S}(E(\theta^s))$ where rnd is a random probability value taken from $[\rho_a{}^{\max}, \rho_a{}^{\min}]$.

Step 8. To generate a new solution randomly instead of the deleted one and go to *Step 2*.

Step 9. To save the solution with the best classification accuracy.

The algorithm returns the solution vector that shows the best accuracy of the classifier.

5 Experimental Research

To study the productivity of the proposed method, we conducted a computational experiment on the datasets from the open storage Machine Learning Repository [15]. The UCI Machine Learning Repository is the largest open storage of the real and modeled task of the intelligent data analysis. To carry out the experiments, we chose the User Knowledge Modeling Data Set. The proposed method was compared with the machine learning method fuzzy c-means in terms of the classification accuracy on the test datasets on different training samplings, which were used to train the classifiers. The results of the comparative analysis for the task of training the classifiers by the CSA and fuzzy c-means algorithms according to the criteria of the accuracy are demonstrated in Table 1.

The calculated data show that the classification accuracy obtained with the developed method is 8–10% better on average than the fuzzy c-means. If we reduce the sampling size to minimal (from 140 to 60), the classification accuracy stays maximal. The analysis of the obtained data demonstrates that the CSA allows us to form a set of profile informative features on a little amount of the test data without reducing the classification accuracy. Based on the previous experiment, we can conclude that the best CSA accuracy is obtained in terms of the initial sampling divided in the percentage ratio of 70:30, and with the size of the test sample of 60. The sample is used for conducting the experiments to determine how accuracy and execution time

Table 1 Comparative analysis of the accuracy of CSA and fuzzy c-means

Method	Size of the training sample	Size of the test sample	Accuracy on the test data
CSA	160	40	0.941
Fuzzy c-means	160	40	0.879
CSA	120	60	0.964
Fuzzy c-means	120	60	0.867
CSA	100	80	0.935
Fuzzy c-means	100	80	0.844
CSA	90	100	0.917
Fuzzy c-means	90	100	0.833
CSA	80	120	0.914
Fuzzy c-means	80	120	0.796
CSA	60	140	0.893
Fuzzy c-means	60	140	0.770

depend on the number of iterations. The results of are demonstrated in Table 2 and Figs. 3 and 4.

According to Table 2 and figures, the proposed modified CSA demonstrates high accuracy and better results than fuzzy c-means by 15–20% on the average starting with the 50th iteration and achieves a stable high rate. This fact confirms the hypothesis that even little amount of the initial data allows the CSA to classify the profiles of user needs with high accuracy.

The visual representation of the comparison results shows that the execution time of both algorithms has linear complexity. CSA works much faster, and the increasing number of iterations does not significantly increase the execution time. Thus, the developed algorithm is effective for solving the classification task.

Table 2 Effect of the number of iterations on the accuracy and runtime of the algorithms

Comparison criterion	Number of iterations							
	25	50	75	100	150	200	250	500
CSA accuracy	0.75	0.86	0.91	0.95	0.96	0.95	0.98	0.98
CSA time (ms)	27	28	37	43	43	54	50	53
Fuzzy c-means accuracy	0.70	0.71	0.65	0.78	0.81	0.78	0.76	0.86
Fuzzy c-means time (ms)	36	40	44	53	59	60	64	71

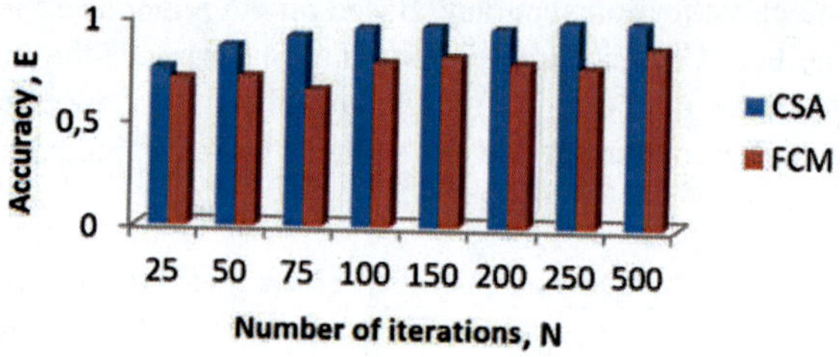

Fig. 3 Diagram comparing the classification accuracy by CSA and fuzzy c-means

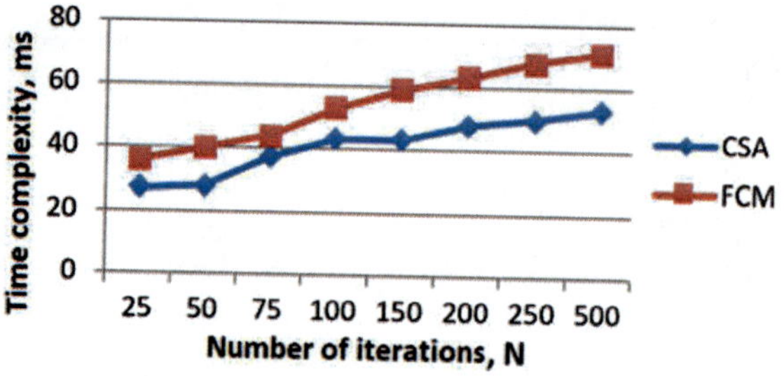

Fig. 4 Time complexity of the CSA and fuzzy c-means

6 Conclusion

The results include the method for classification of the behavioral profiles of the users based on a meta-heuristic approach to optimize the dimension of the feature space of the category data on user interests and needs. To increase the accuracy and speed of solving the task of classification of users' behavioral profiles, we developed a bio-inspired algorithm for optimization of the classification parameters based on the cuckoo search meta-heuristics. The benefits of the proposed modified cuckoo search algorithm (CSA) involve simple implementation, ability to identify noisy objects, good generalizing capability. The experimental results demonstrate that in terms of significant reduction of the test data, CSA shows high accuracy of classification, which is 15–20% higher than the fuzzy c-means algorithm. This proves the effectiveness of the suggested method for the classification of user behavioral profiles. The authors conclude that the method is useful for modeling the interaction between the main subjects and objects of the semantic search in the information systems related to knowledge management.

Acknowledgements The research was funded by RFBR according to the research project № 18-07-00050 and project № 19-07-00099.

References

1. Bova, V., Kravchenko, Yu., Rodzin, S., Kuliev, E.: Hybrid method for prediction of users' information behavior in the Internet based on bioinspired search. J. Phys. Conf. Ser. (2019). https://doi.org/10.1088/1742-6596/1333/3/032008
2. Kravchenko, Y.A., Kuliev, E.V., Kursitys, I.O.: Information's semantic search, classification, structuring and integration objectives in the knowledge management context problems. In: 10th IEEE International Conference on «Application of Information and Communication Technologies—AICT 2016», pp. 136–141. IEEE Press, Baku (2016)
3. Bova, V.V., Kureichik, V.V., Leshchanov, D.V.: The model of semantic similarity estimation for the problems of big data search and structuring. In: 11th IEEE International Conference on «Application of Information and Communication Technologies—AICT 2017», pp. 27–32 (2017)
4. Jalalirad, A., Tjalkens, T.: Using feature-based models with complexity penalization for selecting features. J. Signal Process. Syst. **90**(2), 201–210 (2018)
5. Bova, V.V., Kureichik, V.V., Zaruba, D.V.: Data and knowledge classification in intelligence informational systems by the evolutionary method. In: 6th International Conference «Cloud System and Big Data Engineering», Noida, pp. 6–11 (2016)
6. Chifu, V.R., Pop, C.B., Salomie, I., Niculici, A.N.: Optimizing the semantic web service composition process using cuckoo search. J. Intell. Distrib. Comput. **5**, 93–102 (2012)
7. Kravchenko, Yu.A., Markov, V.V., Kursitys, I.O.: Bioinspired algorithm for acquiring new knowledge based on information resource classification. In: Proceedings of IEEE International Russian Automation Conference (RusAutoCon) (2019). https://doi.org/10.1109/RUSAUTOCON.2019.8867663
8. Kravchenko, Y., Kursitys, I., Bova, V.: The development of genetic algorithm for semantic similarity estimation in terms of knowledge management problems. J. Adv. Intell. Syst. Comput. **573**, 84–93 (2017)

9. Hodashinsky, I.A., Anfilofiev, A.E., Bardamova, M.B.: Metaheuristics for parameters optimization of fuzzy classifiers. J. Inform. Math. Technol. Sci. Manag. **1**(27), 73–81 (2016)
10. Bova, V.V., Scheglov, S.N., Leshchanov, D.V.: Modified EM-clustering algorithm for integrated big data processing tasks. J. Izv. SFEDU Eng. Sci. **4**(198), 154–166 (2018)
11. Yang, X.S., Deb, S.: Multiobjective cuckoo search for design optimization. J. Comput. Oper. Res. **40**(6), 1616–1624 (2013)
12. Sarin, K., Hodashinsky, I.: Identification of fuzzy classifiers based on the mountain clustering and cuckoo search algorithms. In: Proceedings of International Siberian Conference on Control and Communications (SIBCON), pp. 1–6. IEEE Press, Astana (2017)
13. Guo, G., Zhang, J., Thalmann, D.: Merging trust in collaborative filtering to alleviate data sparsity and cold start. J. Knowl.-Based Syst. **57**, 57–68 (2014)
14. Coelho, L.S., Guerra, F.A., Batistela, N.J., Leite, J.V.: Multiobjective cuckoo search algorithm based on Duffings oscillator applied to Jiles-Atherton vector hysteresis parameters estimation. IEEE Trans. Magn. **49**(5), 1745 (2013)
15. Machine Learning Repository. https://www.ics.uci.edu/~mlearn/MLRepository.html

TETRA Enhancement Based on Adaptive Modulation

Ali R. Abood and Alharith A. Abdullah

Abstract Every day, millions of people place their trust in products based on Terrestrial Trunked Radio (TETRA). TETRA is a Private Mobile Radio (PMR) criteria purposed for meeting the special requirement for voice communication, which is reliable and secure communication links. The data rate's narrowband and on-demand applications required big data rate, fast communication, and security in different environments. So, new TETRA Release introduces in this work to support novel high data rate applications which require wideband such as video streaming and adapting with cases of congestion or density of communication when special cases of application occur which they effect on link cause bottlenecks and stop services like denial of service (DoS). An adaptive TETRA system has been suggested in the selection of the advanced network based on the input of the modulation objects. As well as, based on these modulations (π/4 DQPSK, 4 QAM, 16 QAM, 64 QAM), they give us more data rate from the single module modulation-based system. All parts and techniques of the proposed work are implemented in 'OMNET++' tool of simulation and programming language of visual C# with visual studio. Finally, we show that our proposed work gives us more adaptation way when we use multiple modulation comparing with single modulation.

Keywords TETRA modulation · MAC layer · Π/4 DQPSK · 4 QAM · 16 QAM · 64 QAM

1 Introduction

Terrestrial Trunked Radio (TETRA) is a digital wireless program designed mostly for work in government fields and their applications, which are time and security that are the most important parameters taken into consideration when working in this type of networks where it is possible to make an individual and collective call and transfer

A. R. Abood · A. A. Abdullah (✉)
College of Information Technology, University of Babylon, Babil, Iraq
e-mail: alharith@itnet.uobabylon.edu.iq

H. Sharma et al. (eds.), *Congress on Intelligent Systems*, Advances in Intelligent Systems and Computing 1334, https://doi.org/10.1007/978-981-33-6981-8_11

data in an unlimited amount, especially in applications that require streaming for video data within easy reach future [1, 2].

Achieving TETRA to solve all problems related to technological and security issues and services that meet customized needs, it is developed basically for voice services depending on the TETRA standards for the connection.

Private Mobile Radio (PMR) is a mobile service for users who want to be contacted over a short coverage area with a central unit which is achieved for a particular necessity in a suitable price. TETRA is considered as a principal concept in the PMR systems' future generations [3].

Radio network attack is one of the most common types of attacks of networks, for instance, overhearing sensitive conversation as one of the most classic attacks of this type. There are many modern cryptography techniques based on cipher algorithms to protect data from exposure to attacks which are divided into two types:

The first type does not modify the transferred data, but only listen, eavesdropping and monitoring to it, and thus takes important and confidential information.

The second type is to modify the exchanged data through modification, injection, jamming or theft, etc.

The TETRA standard supports two levels of security management:

- Air interface encryption level for anyone connected to the network or GSM.
- End-to-end encryption level for a specific user on demand which commonly involved paying the cost for key management. TETRA system is based on three layers of Open Systems Interconnection (OSI) model in which they structured as physical layer, data link layer and network layer [4].

There are many proposed developed systems for the applications of TETRA; for example, in [5], the author suggested a security system based on authentication protocol for the TETRA system verification tool called Scyther which is implemented with encryption and hash function and other supported functions. Scyther constructor is applied to analyze and verify the security properties of the TETRA system.

Intelligent intentional electromagnetic interference (IEMI) describes in [6] and how TETRA deals with and resilience, and TETRA's modulation scheme was also investigated such as $\pi/4$ differential quadrature phase-shift keying (QPSK) for interference signals and jamming.

In [7], the author suggests a novel technique based on modulation and coding according to channel behavior, and this technique is evaluated by taking into consideration channel behavior, concrete layer and mobility conditions for TETRA cellular systems.

In [8], the author describes the security concepts in TETRA systems in general which are included:

- Authentication: allowing legitimate user to login into the network with authentication mechanism.

- Security features: To ensure security, several measures can be taken from an operational point of view, and there is a privilege to get action and doing something within network.
- Protecting the traffic from outsiders: Countermeasures for this problem consist of three cases: The first one is the end-to-end encryption, in other words, security from one endpoint to the other in the system. The second is based on radio links encryption, while the third is to protect the links between the mobile switch and TETRA base station (TBS).
- Management: It is considered as one of the network security aspects; for instance, in TETRA systems, we can manage devices remotely using secured connections like Secure Shell (SSH) and Secure File Transfer Protocol (SFTP); thus, management can be touched with users' behaviors into an audit server.
- Denial of service (DoS): Although this attack does not directly affect data, for instance, stealing of data, it causes a disaster if it occurs. To counter-measure DoS, TETRA base stations are prepared with detection techniques such as jamming to avoid this problem.

In [9], the author shows the modulation schemes used in TETRA system with multi-sub-carrier approach and modulation symbol rate of each type 4-QAM, 16-QAM or 64 QAM; also, the author explains bit error rate (BER) performance evaluation and additive white Gaussian noise (AWGN).

We mentioned security concepts for TETRA system which is classified as encryption algorithms and modulation schemes in Fig. 1.

The study is totally presented as follows: In Sect. 2, we explain the proposed TETRA module layers and which layers implemented in the proposed system. In Sect. 3, we describe the proposed TETRA system structure. Section 4 is devoted for pointing out proposal frame of adaptive TETRA System, whereas Sect. 5 deals with simulation and results. Finally, we conclude this paper in Sect. 6.

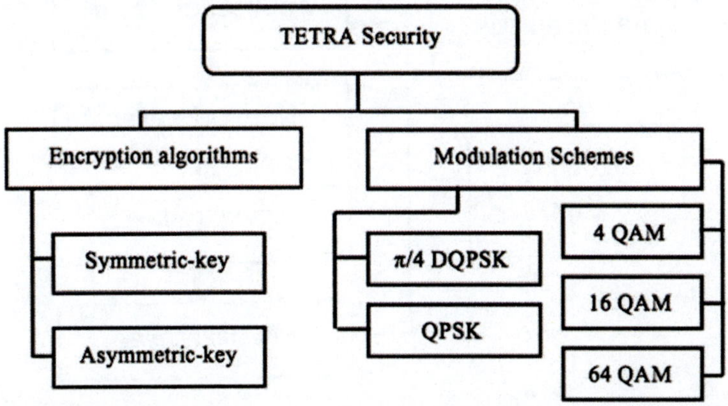

Fig. 1 TETRA security classifications

2 The Proposed TETRA Modules Layers

The proposed system contains four layers, and we sort them from the medium channel layer to the end user layer (MAC layer, physical layer, network layer, application layer). Furthermore, the proposed system involving other dependency layers we called (signaling and communication link, Data resources management module). We explain functions of each layer used for the proposed system.

- Physical Layer: It simulates physical Motorola device components.
- MAC Layer: It describes the proposed system's main components through frames and how the proposed system encapsulates different modulation types within.
- Network Layer: It explains routing features to pass packet routes between call and listener.
- Application Layer: It presents end user graphics to deal with call signaling.

Signaling and communication link functions are timing managements for different types of data signals and passing parameters between different layer modules. The module of management of data resources simulates database management in order to pass message calls objects as an xml within database of OMNET++. Above, we depict the proposed layers and their communication, as Fig. 2 shows.

In the simulation's run time in the first stage, we commence input call voices to pass over C# programming language as objects to OMNET++ straightway within the first setting state, which has a difference in time size of data. These calls go through the specific field of data within payload field of the frame. We explain call objects in Fig. 3.

Each node (mobile) can do one or more than one call as a group communication which is described as a point-to-point call.

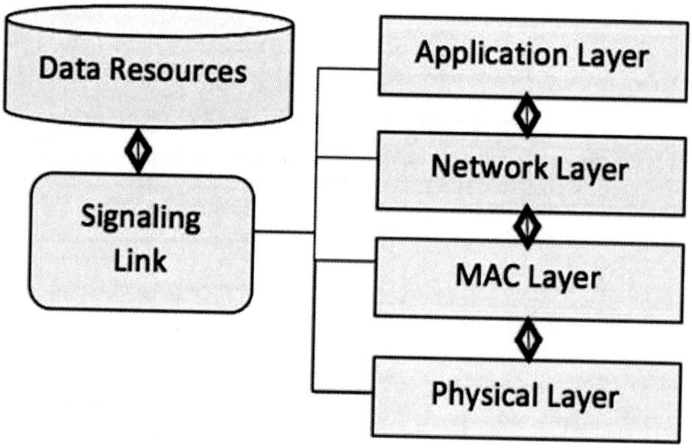

Fig. 2 TETRA layers proposed

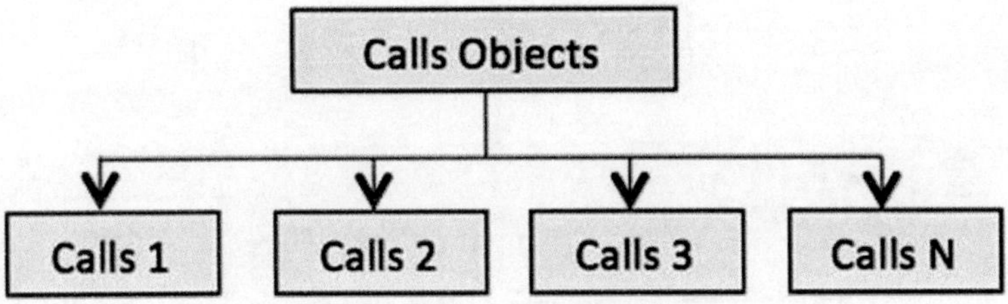

Fig. 3 Calls objects for the proposed system

3 The Proposed TETRA System Structure

In the second stage, after getting the voice calls, the modulation type is chosen from the four proposed types (π/4 DQPSK, 4 QAM, 16 QAM, 64 QAM) with specific features for each one of them to implement the application effects of each type within OMNET++ environment, which they are passed as an object from the physical layer (modulation layer) to MAC layer frame fields payload segment parameters within the part of encapsulating field. In addition, data (voice calls) from the first simulation stage builds calls.

All these processes will be entered to TETRA system structure. Subsequently, statistics and results (in many diagrams, log file, as well as analysis files) will be built by implementing proposal TETRA adaptive system. In Fig. 1, we initialize our proposed system with input modulation selection through 'OMNET++' simulator tool at running simulation time. Each of these modules and general steps for the simulation process are shown above in Fig. 4.

We show the block diagram as the main steps for simulating the proposed system with these steps:

1. Initial and updating setting state.
2. Passing voice calls.
3. Selection modulation type.
4. Recording events and statistics.
5. Building comparison parameters.
6. Results (log file, vectors, scalars, histogram and analysis files).

The proposed simulation exchanges messages between system components represented by the first component TETRA system servers (initial and update) messages. In contrast, initially describes the first time initial setting and timing managements and update describes any changes that occur for the network, while the second component mobile node messages (RTS, CTS, ACK, voice call object).

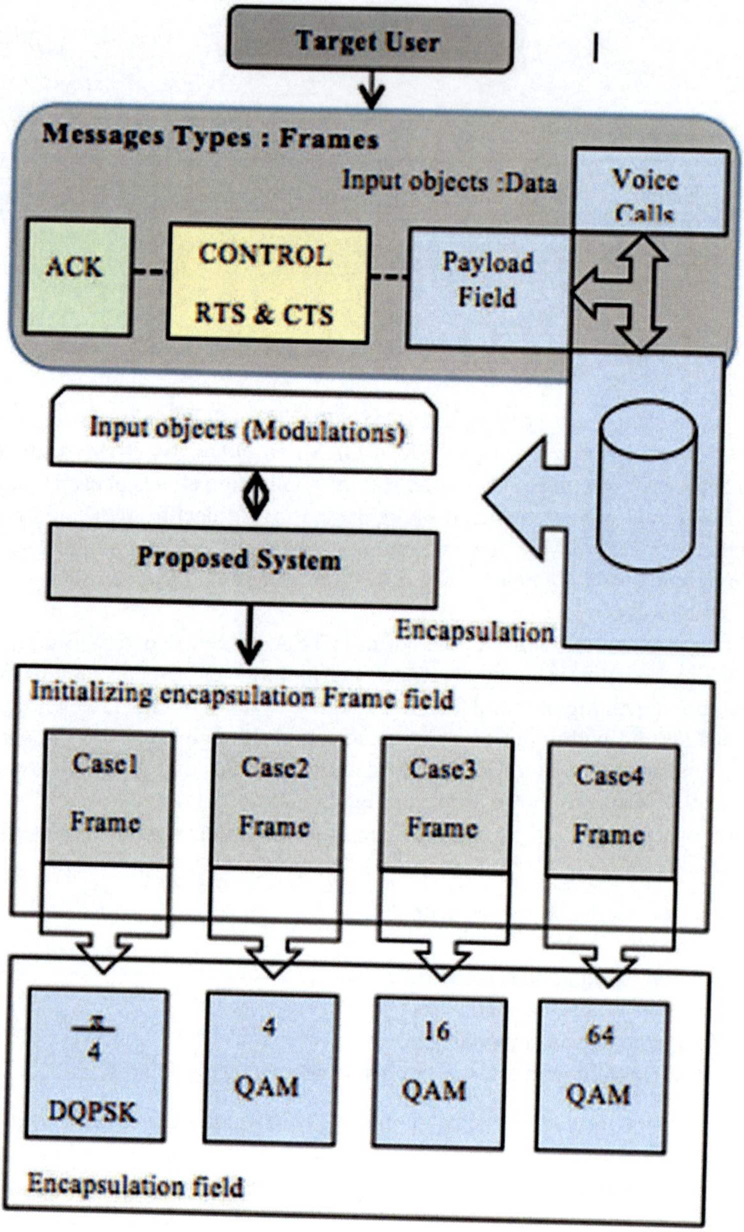

Fig. 4 Proposed TETRA structure block diagram

4 Proposal Frame of Adaptive TETRA System

The frame's basic feature is defined by the header segment bits, and the main data (voice calls) is contained in payload segment (0–254 bytes). It consists of the proposed components below, and Fig. 3 describes them:

Proposed channel: It represents the free channel that performs as an idle channel for transmitting messages.

Frame ID: It is purposed as a position of a slot. Frame ID denotes in which slot the frame should be transmitted. In a one communication course, a frame ID is used only once on each channel. Each frame has a particular specified frame ID matching with a particular slot. The frame ID ranges from 1 to 2047 (00000000001 to 11111111111), and the frame ID 0 is an invalid frame ID.

Data length: It is utilized to refer to the size of the encapsulation field. The size of encapsulation field is encoded in this field by means of setting it to the number of encapsulation data bytes divided into two (data length $\times$ 2 = number of encapsulation data bytes).

Source (Src): It describes source MAC address.

Destination (Des): destination MAC address.

Encapsulation: It contains security features for the proposed system (modulation adaptation).

Data (Voice Calls): It specifies input voice calls entered during initial simulation state. In Fig. 5, we describe frame data transmission field in TETRA message format.

We will explain the general algorithm for our proposed system through all simulation process. Surely, this algorithm is not all algorithm and code of implementation,

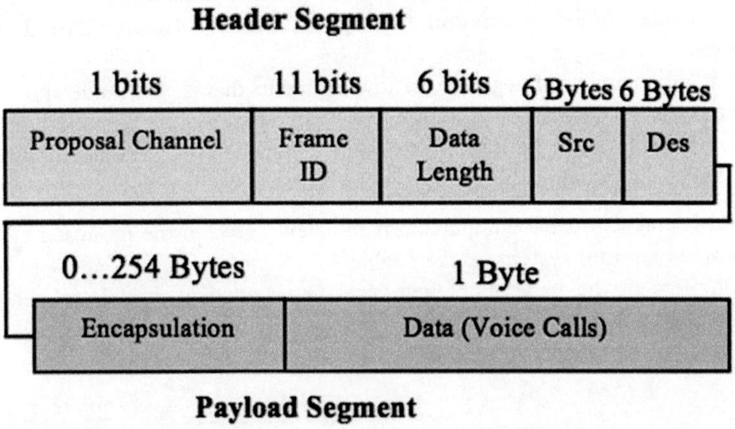

Fig. 5 Proposal frame format

because many steps and different code sections are there to simulate the entire TETRA structure (four layers and interconnection layers).

In Fig. 6, we demonstrate MAC layer algorithm as a main algorithm within our proposed adaptation TETRA system. This layer is concerned with data types incoming from the physical layer. It passes that data to the network layer via send method for calling messages there using different steps we describe in the proposed MAC layer algorithm of mobile nodes.

We also have the physical layer implemented with algorithm in our proposed system. This layer is in charge of inspecting about the taking charge of the variable that begins transmitting through the transmitter with the cognitive wireless node, and this variable is known (address) through getParentModule function. We can also mark how many available channels for users and idle one suggested from FHSS. These steps are illustrated in Fig. 8, algorithm for physical layer process in cognitive radio network.

Note from Table 1, we illustrate the simulation parameters setting for the entire proposed system in 'OMNET++' simulator.

In Fig. 6, we will explain MAC layer algorithm as a main algorithm within our proposal adaptation TETRA system. This layer deals with data type incoming from the physical layer and pass to network layer with the sent method for calling message there using different steps we describe in the proposed MAC layer algorithm of mobile nodes.

Our simulation results are given as follows:

- Figuring the number of active nodes of mobile that represent the network's lifetime and how this can be employed if cluster-based architecture is implemented for emergencies and high momentum of communication.
- Calculating the voice calls data signal that arrive to the receiver without error, for every reception node.
- Assessing our proposed system's quality by means of employing message statistic test for messages scheduled through overall time of simulation.
- Figuring the number of control message exchange between transmitter and receiver.
- Figuring the number of negative acknowledgments that is given indicates the not arrived or missed calls.
- Calculating the throughput of each system case which gives us an indication of how the proposed adaptation case is better than the single case.

We will represent these comparison parameters between the proposed systems which implement four system cases in Table 2.

We implement the general mathematical and statistical models to get these simulation parameters values for each proposed simulation case where:

Throughput is a rate of:

$$\text{Throughout} = \frac{\text{Number of max bytes}}{\text{Transmission time}} \tag{1}$$

Fig. 6 Proposal MAC layer algorithm of nodes

Proposal MAC Layer algorithm
Input/output : lower layer(Control objects, DATA(Voices objects)), upper layer (Control objects, DATA(Voices objects))
Phase 1 : Initialize case the number of total Frames, current Destination, current Data Channel, proposed Channel, Modulation changer.

Phase 2 : **start Handlers functions**
A-Handle Controller state:
Handle RTS : if (isTransmitting & isReceiving = true) then {delete msg }
else
{ I am idle, select the proposed channel }
Handle CTS : clear an RTS to get idle state by :
1- Get Proposed Channel.
2- Send Data.
B- Handle Node Data proposed:
1- Handle Data node: if(ackEnabled == true && currentDataChannel != 0)
{ pass values for source, destination, Number Of frames ,Proposed Channel
begin send(ack, dataLower) }
else { delete msg}
 	
2- Handle Ack node: clear timing and next sending attempt
if (currentDataChannel == 0) // channel is lost Don't send next packet
else { Send the next packet}
C- Handle TETRA Proposed : Initial and Setting TETRA components
if(isTransmitting == true) // I transmitted and be within
Case 1: if(ackEnabled == true)
1-Cancel the Acknowledgement timer .
 	
2-Cancel handover channel for TETRA .
3-Preparing to start RTS/CTS for a new channel.
else
Stop sending the acknowledgment
Case 2: if(ackEnabled == false)
received getIdle and handover channel.

*\ Timers RTS timer : RTS is sent. // Set RTS timer : clear RTS // idle state. set Ack Time Out : message arrive // Channel is free to prepare for next message during send Data function.
Phase 3 : **Send Functions** A-Send RTS : if (rtsAttempts >= 1) 1-Send RTS on a free channel with given parameters (source, Destination, ProposedChannel) Then if no response is received 2-Send another RTS if multiple RTS are enabled through rtsAttempts parameter} Else -Failed RTS Attempts , -Inform app layer by sending a nack -Re-initialize rtsAttempts parameter to the number of RTS attempts for next session. -Send nack through ctrlUpper to application layer . B-Send CTS : 1-Send CTS with current parameters(Source, Destination, ProposedChannel) 2-Make isReceiving = true to begin send data. C-Send Data : 1-Send data process with two cases : Case 1: if (currentDataChannel = 1, 2, 3 or 4) send Voice Calls. Case 2: if ackEnabled == true (Send the next packet only after receiving ACK) with 2 steps : step 1: if currentDataChannel !=0 then Frame sending step 2: else get proposal Idle function (getIdle) E-send Nack : transmitting and receiving unstable case for instance problem in data lower, ctrl upper or missed calls .
End Algorithm

Table 1 Simulation parameters setting

Parameters	Value
Simulation time	20 m
Total data channel	4
Voice length	Dynamic
Source format Proposal channel Link layer ack Modulation scheme Number of TETRA system Number of nodes Topology area	.Wave 1, 2, 3 or 4 Enable π/4 DQPSK, 4 QAM, 16 QAM or 64 QAM 4 TETRA, 4 MSO 8 787 × 447
Number of cluster	Four clusters
MAC layer	802.11b standard
Data type	Voice calls

Table 2 Performance comparison parameters

Comparison parameters	π/4 DQPSK	4 QAM	16 QAM	64 QAM
Number active	2	4	6	8
Number of voice call	23	59	108	258
Missed call	7	3	2	1
Throughput	11.50	29.50	80.25	168.66

The mean is: the usual average.

$$\text{Mean}\ (\mu) = \frac{\text{Summation of All data signals}}{N} \tag{2}$$

So to find mean, we will add summation of all data rate created in each node and then divide the number of active node.

In addition, we find standard deviation (SD) of the proposed system which is based on (x) as the data signal object of each node and then find mean divided on (N) number of active node in the network [10].

$$\text{SD} = \sqrt{\frac{\sum |x - \mu|^2}{N - 1}} \tag{3}$$

Time management which can provide a delay for data signals (voice call objects) depending on simulator interval through simulation. In other words, OMNET++ gives us a speed changer feature during running time, so we can increase and decrease speed for data signal which remains in the TETRA system from the transmitter to the receiver. In addition, there is a time synchronization counter for data signals in which they exchange between layers.

5 Simulation and Results

We used 'OMNET++' to implement our proposed system because of easy to use with graphical user interface (GUI). As well as the flexibility to learn where it depends on C++ programming language, another feature we exploit through it contains many frameworks, libraries, models, etc. That saves researchers' effort and time for carrying out research simulations as if in real environments. In addition, the lack of requirements for the installation state in Windows Operating System. 'OMNET++' is used as the simulation platform. According to the general network architecture of the proposed TETRA system, all the communication takes place on the TETRA sites act as a base station and Mobile Switch Office (MSO) acts as a setting server unit and target users (mobility nodes). In this research, we have designed a proposed TETRA system in 'OMNET++' which comprises four clusters. The general network topology within one area and each region cluster has two base components and two nodes. Users may be clarified with fire bridges, police forces, emergency health or other target user units.

Statistical analysis of end-to-end also explained which describes mean and standard deviation for the data signal of each system module through the same simulation time in Fig. 7.

Maximum data rates of the proposed system are explained in Fig. 8. They were summarized depending on how many data signal correctly acknowledged without errors. The proposed adaptive system has a better data rate which takes consideration as an efficient step to working within the stream data rate (video) system. We compare with the traditional system that works within just one system module like (π/4 DQPSK) against the proposed system explained above.

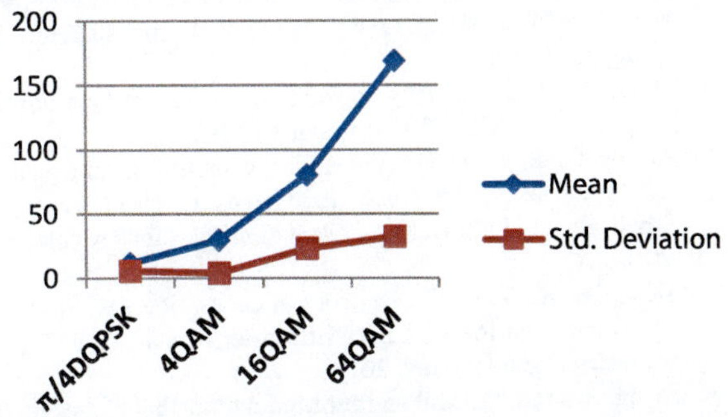

Fig. 7 Statistical analysis of the proposed adaptive system

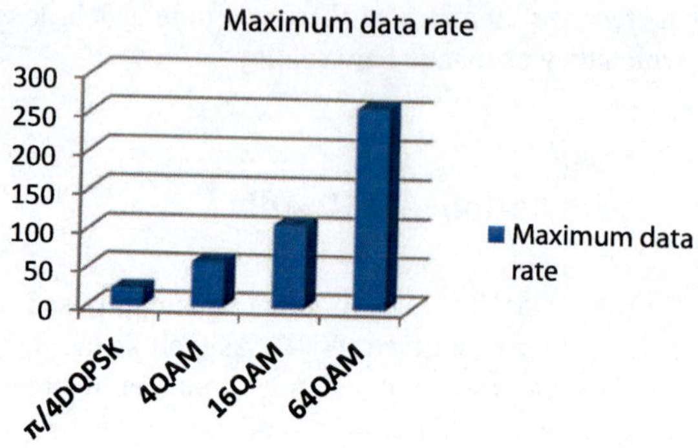

Fig. 8 Maximum data rates of the proposed system

6 Conclusion

The proposed systems described the improvement of TETRA system by using combination of modulation types instead of single modulation. We implement voice calls object in OMNET++ simulator and C# programing language. Our results show high data rate and acceptance of stream data flow allowed which gives us more directions to working within video data, and we evaluate the proposed system by computing bandwidth, uplink–downlink data, throughput, subcarriers and other standard features for each type of modulation. Finally, we are convinced that the proposed TETRA enhancement presented is practical and can provide significant opportunity for work on advanced TETRA networks efficiently.

References

1. Sodersrtrom, E.: TETRA—TErrestrial Trunked Radio. Linkping University (2016)
2. Stavroulakis, P.: Terrestrial Trunked Radio-TETRA: A Global Security Tool. Springer Science & Business Media, Berlin Heidelberg New York (2007)
3. ETSI European Telecommunications Standards Institute: Terrestrial Trunked Radio (TETRA); Voice Plus Data (V+D), Designers Guide. European Telecommunications Standards Institute, Valbonne (1997)
4. Duan, S., et al.: Security analysis of the terrestrial trunked radio (TETRA) authentication protocol. In: NISK-2013 Conference (2013)
5. Van de Beek, S., et al.: Vulnerability of terrestrial-trunked radio to intelligent intentional electromagnetic interference. IEEE Trans. Electromagn. Compat. **57**(3) (2015)
6. Tarchi, D., et al.: Novel link adaptation for TETRA cellular systems. Int. J. Commun. Syst. (2009)
7. Alejandro, R., Taleb, T.: Delivery of encryption keys in TETRA networks. Thesis submitted for examination for the degree of Master of Science in Technology, Aalto University School of Electrical Engineering, 2016
8. Zhang, X., Yao, D.: Multicarrier Modulation Techniques for TETRA. State Key Laboratory of Rail Traffic Control and Safety, Beijing Jiaotong University, Beijing. IEEE (2011)

9. Hoymann, C., et al.: Performance analysis of TETRA and TAPS and implications for future broadband public safety communication systems. In: International Workshop on Broadband Wireless Ad-Hoc Networks and Services, ETSI, Sophia Antipolis (2002)
10. Madushanka, L.: Statistics and Standard Deviation. General Sir John Kotelawala Defence University—Southern Campus, Sri Lanka, Mathematics Learning Centre

Dynamic Stability Enhancement of Grid Connected Wind System Using Grey Wolf Optimization Technique

Ashish Khandelwal, Nirmala Sharma, Ajay Sharma, and Harish Sharma

Abstract To maintain the dynamic stability of a system is a significant problem in the field of grid connected wind system. The dynamic stability may disturb due to any type of fault or disturbance incurred in a wind system. To overcome this dynamic stability problem, in this article a wind energy converting system is used. Here, super conducting magnetic energy storage (SMEs) unit is applied to the considered system. The SMES controller is designed by a voltage source converter (VSC) and a chopper. This SMEs unit supplies active and reactive powers when required, thus balances frequency and voltage of the system which ultimately results in enhancement of dynamic stability of the system. In this work, a recent grey wolf optimization technique (GWO) is applied to optimize the duty cycle of DC-DC chopper. According to the GWO optimized duty cycle signal of chopper charging and discharging of SMEs unit has been controlled. The validation of GWO based control scheme is done with the help of MATLAB simulation results.

Keywords Dynamic stability · GWO · SMEs · Chopper · Wind system

1 Introduction

As[1] a deficiency of conventional energy sources, the aim of researchers become to generate energy from non conventional energy resources like solar and wind. The main problem of continuous generation is facing in such systems. Presently, energy storage from these system is one of major challenge.

[1]This paper is output of sponsored project funded by RTU(ATU) under TEQIP-III.

A. Khandelwal (✉) · A. Sharma
Engineering College, Jhalawar, India

N. Sharma · H. Sharma
Rajasthan Technical University, Kota, India
e-mail: nsharma@rtu.ac.in

H. Sharma
e-mail: hsharma@rtu.ac.in

H. Sharma et al. (eds.), *Congress on Intelligent Systems*, Advances in Intelligent Systems and Computing 1334, https://doi.org/10.1007/978-981-33-6981-8_12

The energy storage has been performed by different methods. The some of popular methods are battery storage systems, flywheel storage, capacitor storage, and super conducting magnetic energy storage (SMEs) systems [1]. The SMEs has properties of high storage, fast response, and it's charging and discharging capacity. The SMEs works on the principle of electric energy storage in the magnetic field. The SMEs is capable to deliver and receive the apparent power as per need. To improve the system stability many works have been already carried out with the interface of SMEs. The literature shows the SMEs capability. The various control techniques are also available in literature like PI, fuzzy logic [2] and adaptive artificial neural network [3] etc. but still SMEs control techniques are limited. The new control methods can develop a more robust SMEs system.

Presently, The nature inspired algorithms (NIA) become popular for optimal control parameters for any control methods. These algorithms are become popular because of their simple structure, and avoidance of local optima [4–7]. Further, The NIA technique namely grey wolf optimization (GWO) has previously solved various real world problems like TNEP problem [8–11], frequency modulated sound waves parameter estimation problem [12], harmonic estimator design [13], automated face retrieval problem [14], etc. and the outcomes are found more accurate.

This paper introduces a GWO controlled SMEs for more stable wind energy conversion system. To control the SMEs unit a voltage source converter (VSC) and chopper are used. The GWO based controller controls the duty cycle of the chopper. The GWO based control scheme is validated with the help of MATLAB simulation results.

The remaining paper might be described in the following parts which are as follows: The mathematical modeling and system description of grid connected wind system are produced in Sect. 2. The SMEs unit modeling details are provided in Sect. 3 The basics of GWO is presented in Sect. 4. In Sect. 5, the control strategy of the wind system using GWO describes the design procedure for SMEs control model using GWO. In Sect. 6 the simulation results of the proposed model are given. In the end, the work has been concluded in Sect. 7.

2 Grid Connected Wind Turbine System

The conversion from wind power to mechanical torque is performed by the wind turbine. The calculation of the mechanical torque of the turbine has been done from wind power. The coefficient of power of the turbine C_p is considered. The mathematical expression of power can be represented as follows [15]:

$$P_m = 0.5C_p(\lambda, \beta)\rho S V_w^3 \tag{1}$$

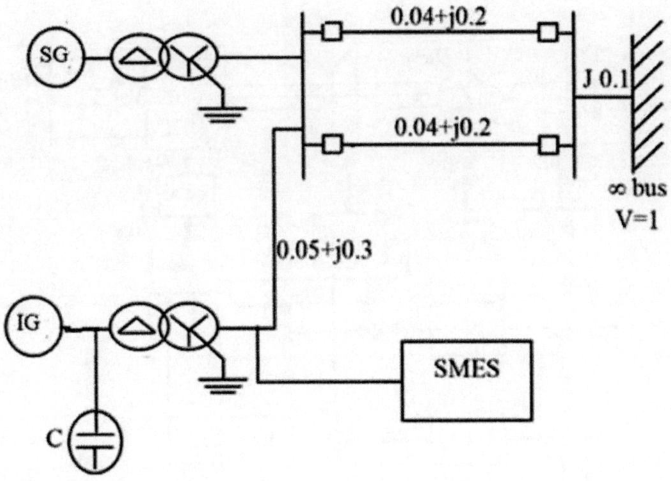

Fig. 1 Grid integrated wind system

where
P_m = turbine output mechanical power
C_p = turbine power coefficient
ρ = air density
S = turbine swept area
V_w = wind speed
λ = tip speed ratio
β = pitch angle

The grid integrated wind system with provision for improvement of stability is provided in Fig. 1. It connects synchronous generator (SG) and for wind system an induction generator (IG). The double circuit line is used to feed the energy to the infinite bus. The C is the capacitor which is added to the terminals of wind system for compensation of the reactive power. For the energy storage and stability purpose, SMEs system is added to the terminal end of wind system.

3 Modeling of SMEs System

The SMEs system modeling is described in Fig. 2. This system has mainly the following parts:

1. IGBT based voltage source converter (VSC)
2. 60 mF DC-link capacitor
3. Three-phase transformer
4. Super conducting coil
5. IGBTs based chopper

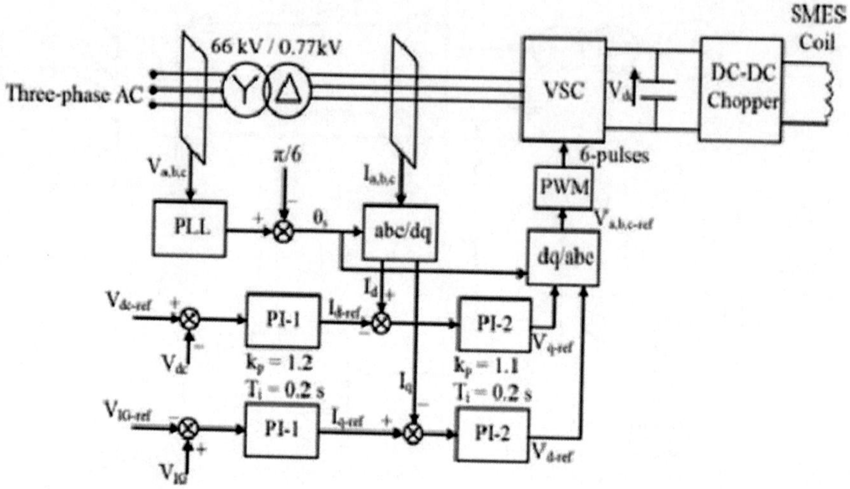

Fig. 2 Controller of SMEs unit of wind system

The SMEs stored energy (EN) and the rated power of SMEs (P) are mathematically written as per following equations:

$$EN = \frac{1}{2} L_{sme} I_{sme}{}^{2} \tag{2}$$

$$P = \frac{dEN}{dt} = L_{sme} I_{sme} \frac{dI_{sme}}{dt} = V_{sme} I_{sme} \tag{3}$$

where
L_{sme} = super conducting coil inductance
I_{sme} = current from the coil
V_{sme} = Instantaneous voltage

3.1 *Modeling of VSC*

The VSC is a medium to connect the AC bus and super conducting coil. Figure 2 shows the control scheme. The reference frame transformation is used to relate the *dq* and *abc* parameters. The work of phase-locked loop (PLL) is to identify the angle of transformation. The voltage error between V_{dc-ref} and V_{dc} is supplied to the PI-1 controller. The PI-1 produces the current signal I_{d-ref}. Also the error of IG voltage V_{IG-ref} and V_{IG} is supplied to the PI-1. The PI-1 produces the signal I_{q-ref}.

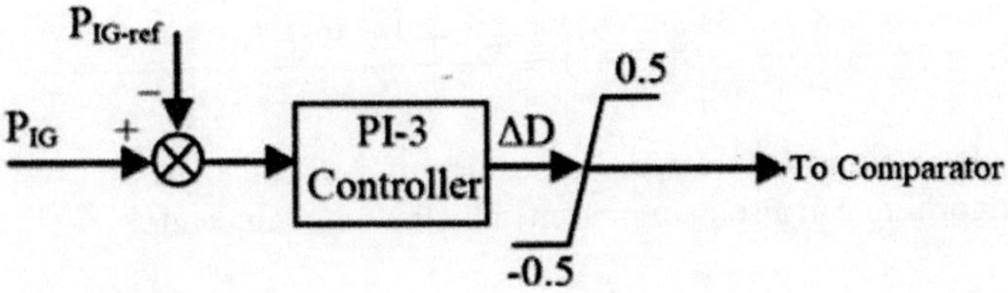

Fig. 3 DC-DC chopper controller

The subtraction signal between I_d and I_{d-ref} and I_q and I_{q-ref} gives as an input signal to PI-2. The PI-2 provides the V_{d-ref} and V_{q-ref}. The obtained signals provides $V_{abc-ref}$. This reference voltage used to generate the firing pulses.

3.2 Modeling of Chopper

The terminal voltage of the chopper is maintained by delivering or storing the energy in the coil. This voltage control of the chopper is performed by regulating the duty cycle of it. According to the duty cycle, the SMEs is charged or discharged. The 50 percent duty cycle is reference when no charging and discharging occurs. The duty signal controller of chopper is given in Fig. 3. The power error signal between Real(IG) and Real(IG-ref) is given as an input signal to PI-3. The PI-3 output is used duty cycle signal updation process.

4 Grey Wolf Optimization (GWO) Technique

The NIAs have been designed to solve complex optimization problems from the social behavior of natural species. The GWO algorithm has been inspired by grey wolves (Gws). The Gws have a peculiar behaviour of group hunting. This behavior was mathematically formed and presented in the form of GWO algorithm by Mirjalili et al. [16].

These Gws are living in groups known as packs. These Gws are divided into four parts according to their social behavior which is known as omega (ω), delta (δ), beta (β), and alpha (α). The α is superior one followed by β, δ and ω. In mathematical formulation α is the best solution. The inventor of GWO has represented the encircling, hunting and attacking behavior in the form of mathematical equations. For the solution of the problem, the optimal solution search process has been started by the generation of a random population of Gws. During the iterations α, β and δ Gws update their positions so their distance from prey is updated. The position update equations are as follows:

$$\vec{D}(t+1) = \frac{\vec{D_1} + \vec{D_2} + \vec{D_3}}{3} \tag{4}$$

where D_1, D_2 and D_3 are updated position of α, β and δ Gws.
This algorithm can also be represented by the following code.

Algorithm 1 GWO

```
Random generation of Gws population D_i
Initialize the algorithm parameters
Define the ζ= number of maximum runs
Calculate fitness of each Gws
D1 = α search agent
D2 = β search agent
D3 = δ search agent
while Time < ζ do
  for each Gws do
    Modify the position of Gws by equation 4
  end for
  Modify algorithm parameters
  calculate the fitness of all Gws
  Improve D1, D2 and D3
  Time = Time+1
end while
return D1
```

5 Control Strategy of Wind System Using GWO

In the control part, The duty cycle of the chopper is controlled using GWO based PI controller. For this purpose the PI-3 controller parameters are calculated with the help of the GWO algorithm.

The response surface method (RSM) is applied to design an empirical model by getting correlation of system response and its design variables [17]. For designing the PI-3 parameters three phase fault (3LG) is created for the system. The parameters of response for voltage of common bus selected are maximum overshoot percentage (MOP) and time of settling (ToS) These parameters are affected by the design variables of PI-3. The PI-3 controller parameter K_{p3} and T_3 are the design parameter. The RSM 13 simulation runs are created. The value of MOP and ToS for each run has been stored. Then the RSM model is formulated by the MATLAB. This obtained model can be expressed by following equations:

$$MOP = 3.4850 + 0.1564K_{p3} - 2.0084T_3 - 2.0292K_{p3}T_3 + 1.6924{K_{p3}}^2 + 0.1943{T_3}^2 \tag{5}$$

$$ToS = 0.9421 + 1.3234K_{p3} - 0.5492T_3 - 0.1192K_{p3}T_3 + 0.4499K_{p3}{}^2 + 0.8248T_3{}^2 \tag{6}$$

The MOP equation is the objective function and ToS equation is the nonlinear constant of the problem. The GWO is run for 30 Gws and 500 runs. The process has been repeated 30 times in search of the best result. The fitness function the difference between voltage and voltage maximum of the system. The GWO is applied to the minimization of the fitness function. The obtained value of design parameters K_{p3} and T_3 are found 0.62 and 1.25 respectively.

6 Simulation Results

The simulation of the model has been performed using the MATLAB environment. The speed of the wind is fixed at rated speed. The three phase fault is created on the line. The fault is created at 0.1 sec. The real and reactive power of the SMEs unit is changed during the disturbance. The IG terminal voltage remains almost constant when GWO based PI controlled SMEs unit is connected compare to without SMEs unit and only PI controlled SMEs unit is considered. The waveform of the IG terminal voltage is shown in Fig. 4. It can be concluded from the waveform that a better dynamic response is achieved when a GWO based PI controlled SMEs is considered.

Figure 5 and 6 shows the real and reactive power response of the wind system. This response proves that wind system becomes stable when GWO based PI controlled SMEs is considered.

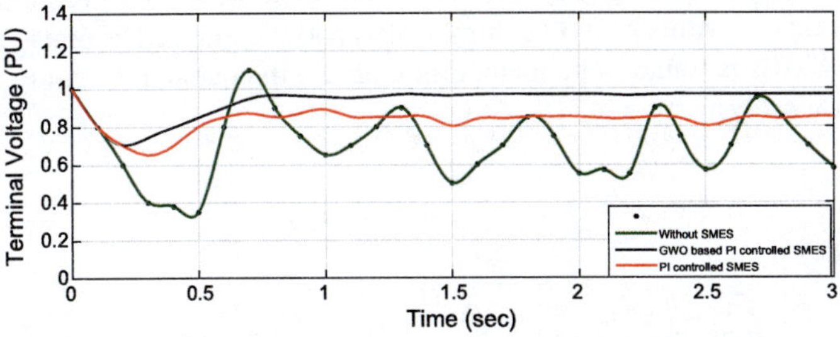

Fig. 4 Response of terminal voltage for various systems under 3LG fault

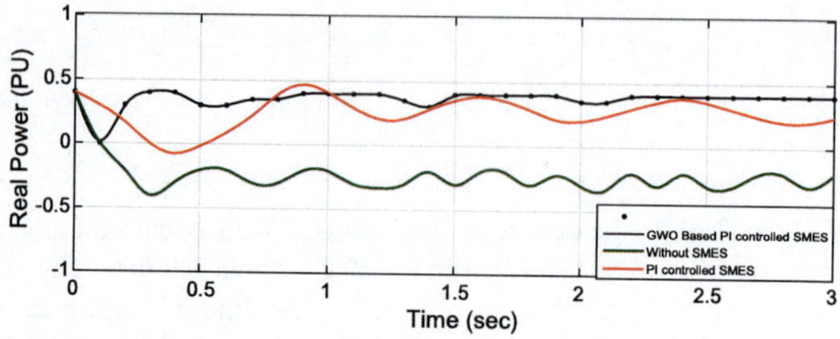

Fig. 5 Response of real power for various systems under 3LG fault

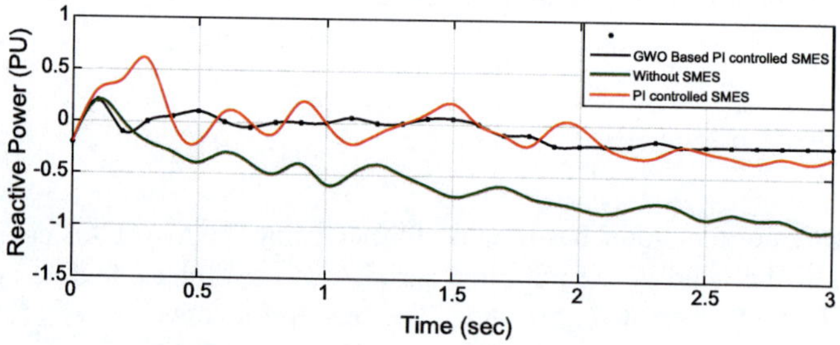

Fig. 6 Response of reactive power for various systems under 3LG fault

7 Conclusion

In this article, Grey wolf optimization technique (GWO) controlled super conducting magnetic energy storage (SMEs) has introduced which improve the dynamic stability of wind system. The GWO has been used to find the optimal parameters of PI controller which controls the duty cycle chopper. The simulation result shows that the dynamic response of wind system with GWO controlled SMEs has found more better than that of without GWO control SMEs and without SMEs unit. The GWO controlled SMEs is found to be method of enhance the dynamic stability of a grid connected wind system.

References

1. Buckles, W., Hassenzahl, W.V.: Superconducting magnetic energy storage. IEEE Power Eng. Rev. **20**(5), 16–20 (2000)
2. Ali, M.H., Murata, T., Tamura, J.: A fuzzy logic-controlled superconducting magnetic energy storage for transient stability augmentation. IEEE Trans. Control Syst. Technol. **15**(1), 144–150 (2006)
3. Hasanien, H.M., Abdelaziz, A.Y.: An adaptive-controlled superconducting magnetic energy storage unit for stabilizing a grid-connected wind generator. Electric Power Components Syst. **43**(8-10), 1072–1079 (2015)

4. Sharma, A., Sharma, H., Bhargava, A., Sharma, N.: Power law-based local search in spider monkey optimisation for lower order system modelling. Int. J. Syst. Sci. **48**(1), 150–160 (2017)
5. Sharma, A., Sharma, H., Bhargava, A., Sharma, N.: Fibonacci series based local search in spider monkey optimisation for transmission expansion planning. Inderscience (2017) (In press)
6. Sharma, A., Sharma, H., Bhargava, A., Sharma, N., Bansal, J.C.: Optimal power flow analysis using lévy flight spider monkey optimisation algorithm. Int. J. Artif. Intell. Soft Comput. **5**(4), 320–352 (2016)
7. Sharma, A., Sharma, H., Bhargava, A., Sharma, N., Bansal, J.C.: Optimal placement and sizing of capacitor using limaçon inspired spider monkey optimization algorithm. Memetic Comput., 1–21 (2016)
8. Khandelwal, A., Bhargava, A., Sharma, A., Sharma, H.: Modified grey wolf optimization algorithm for transmission network expansion planning problem. Arab. J. Sci. Eng. **43**(6), 2899–2908 (2018)
9. Khandelwal, A., Bhargava, A., Sharma, A., Sharma, H.: Acopf-based transmission network expansion planning using grey wolf optimization algorithm. In: Soft Computing for Problem Solving, pp. 177–184. Springer (2019)
10. Khandelwal, A., Bhargava, A., Sharma, A.: Voltage stability constrained transmission network expansion planning using fast convergent grey wolf optimization algorithm. Evol. Intell., 1–10 (2019)
11. Khandelwal, A., Bhargava, A., Sharma, A., Sharma, N.: Security constrained transmission network expansion planning using grey wolf optimization algorithm. J. Stat. Manage. Syst. **22**(7), 1239–1249 (2019)
12. Saxena, A., Kumar, R., Das, S.: β-chaotic map enabled grey wolf optimizer. Appl. Soft Comput. **75**, 84–105 (2019)
13. Saxena, A., Kumar, R., Mirjalili, S.: A harmonic estimator design with evolutionary operators equipped grey wolf optimizer. Expert Syst. Appl. **145**, 113125 (2020)
14. Shukla, A.K., Kanungo, S.: Automated face retrieval using bag-of-features and sigmoidal grey wolf optimization. Evol. Intell., 1–12 (2019)
15. Hasanien, H.M.: Shuffled frog leaping algorithm-based static synchronous compensator for transient stability improvement of a grid-connected wind farm. IET Renew. Power Generation **8**(6), 722–730 (2014)
16. Mirjalili, S., Mirjalili, S.M., Lewis, A.: Grey wolf optimizer. Adv. Eng. Soft. **69**, 46–61 (2014)
17. Khuri, A., Mukhopadhyay, S.: Response surface methodology. Wiley Interdisc. Rev. Comput. Stat. **2**(2), 128–149 (2010)

A Comparative Analysis on Wide-Area Power System Control with Mitigation the Effects of an Imperfect Medium

Mahendra Bhadu, K. G. Sharma, D. K. Pawalia, and Jeetendra Sharma

Abstract The increasing interest in wide-area damping control poses significant challenges for the consistent operation, control and stability of the complete power system. This study investigates various techniques for mitigation of the effect of imperfect communication medium in a wide-area damping control system. The Modelling of an imperfect communication network is done by considering the signal latency, packet drop-out, and random noises in the remote wide-area signal. The signal latency is modelled using the third order of the Pade approximation method, and the random white noise is Gaussian in nature. Different control strategies are applied in a wide-area power system for mitigating the effects of signal latency and noise along with overall stability enhancement of the complete grid. The performance of various controllers as a wide-area damping controller (WADC) is evaluated upon a standard test power system through suitable time-domain simulations using the nonlinear model on MATLAB/Simulink platform.

Keywords Power system control · Power system stabilizer (PSS) · Wide-area control (WAC) · Wide-area damping controller (WADC)

1 Introduction

The modern power system is very complex and distributed in nature, in which a disturbance leads to the oscillations in the complete system. As per the classification of the modes of oscillations, the local modes of oscillations have a frequency range from 0.8 to 2 Hz, and inter-area modes of oscillations have from 0.1 to 0.8 Hz. The

M. Bhadu (✉)
Engineering College Bikaner, Bikaner, India

K. G. Sharma
Engineering College Ajmer, Ajmer, India

D. K. Pawalia
Rajasthan Technical University, Kota, India

J. Sharma
The Imperial Electric Company, Kolkata, India

H. Sharma et al. (eds.), *Congress on Intelligent Systems*, Advances in Intelligent Systems and Computing 1334, https://doi.org/10.1007/978-981-33-6981-8_13

local mode of oscillations is damped out locally by local PSS. In contrast, the inter-area mode of oscillations needs the WADCs since the inter-area modes of oscillations may or may not be controllable and observable from the same area [1, 2].

In the WAC system, the selected remote signal is routed through the communication medium to the WADC venue. The selected remote signal may be rotor speed deviation, phase angle, tie-line power, etc. The modal residue-based method is usually adopted for the signal selection as well as for the selection of the location of WADC [3]. The different communication medium has different issues in terms of constant latency, randomly varying latency, random signal drop-out, and random process noise. The total signal latency may vary from 2 ms^{-2} s in a WAC system [4]. The imperfectness of the communication network in the WAC system leads to a reduction in system stability [5, 6]. In the recent past, some work has been carried out regarding the communication aspects of wide-area control [7, 8]. However, in a comparative analysis point of view, their work seems to be inadequate. Recently, the LQG-based WADC has also been proposed from a continuous and discrete mode of operation point of view [9], and in [10, 11], the uncertainties had used from system perspective rather than the transmission medium.

The scope of this work is to perform a comparative analysis in a WAC system. The ultimate goal is to handle the issues of imperfect communication medium via using different controllers like lead–lag, multi-band PSS, and LQG-based WADC. The imperfect communication medium is considered by including the random process and measurement noise, random packet drop-out, and constant as well as varying signal latency in the wide-area remote signal. The MB-PSS, in particular, 4B-PSS-based WADC, is compared with conventional lead–lag-based WADC for different scenarios, from a signal latency point of view. The variation in time delay is kept in between 2 ms and 2 s. However, the constant time delay is also modelled using Pade order approximation method, as it gives the representation of latency in the frequency domain. Further, the performance of lead–lag PSS and MB-PSS-based WADC is evaluated with random signal drop-out and random Gaussian noise. To mitigate the effects of imperfectness as well as to meet the damping objectives, the robust LQG-based WADC is placed. The performance of designed controllers is assessed on the nonlinear model of the two-area four-machine test power system in the MATLAB platform. A pictorial representation of WAC is depicted in Fig. 1.

The remaining part of the paper is structured, as mentioned below. Section 2 describes the modelling of the test power system. Various controllers to be used as WADC are mentioned in Sect. 3. Section 4 gives a brief description of the random Gaussian white measurement and process noise used in the test power system. The results and discussions are mentioned in Sect. 5, followed by conclusions drawn from the results in Sect. 6. Finally, the conclusion section is followed by references.

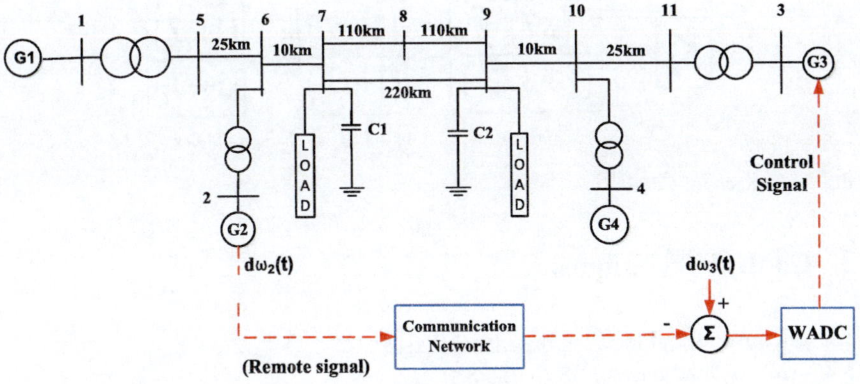

Fig. 1 Conceptual representation of WADC-based two-area four-machine power system. Here, dotted line represents the continuous communication link

2 Modelling of the Test Power System

2.1 Kundur's Two-Area Four-Machine Test System

Kundur's two-area four-machine system is used in the current study, which contains a two-symmetrical area connected via tie-lines of 230 kV, as shown in Fig. 1. At the given operating point, area-1 transports nearly 413 MW power to another area. A three-phase auto-clearing fault of 0.133 s duration is considered for the perturbation of the system, near bus 8. Each generator in this system is equipped with its local PSS. The configurations of local PSS, as well as time delay representations, are considered as per [5, 12–15]. Further, a supplementary lead–lag controller is designed and attached to the system.

2.2 Selection of Control Location and Wide-Area Remote Signal

The suitable location of the WADC and the corresponding remote signal is identified using the residue method. For the given test power system, WADC is placed at machine 3, and the corresponding remote signal is adopted as the speed deviation of generator 2 (M2).

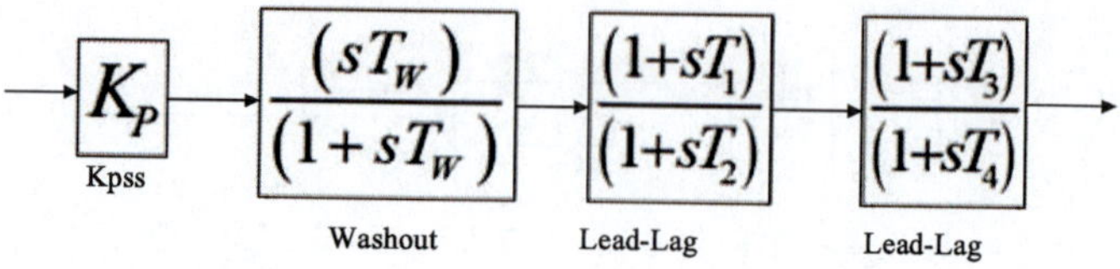

Fig. 2 Structure of CPSS

3 Control Techniques

3.1. Conventional lead-lag based stabilizer
3.2. Multi-band based PSS (MB-PSS)
3.3. Linear quadratic Gaussian (LQG) based controller.

3.1 Conventional PSS (CPSS)

The framework of CPSS is as per appeared in Fig. 2. The CPSS includes the gain block, two lead–lag compensating blocks, and one washout block [16, 17].

3.2 Multi-band-based PSS (MB-PSS4B)

The multi-band power system stabilizer is taken as per the IEEE St. 421.5 PSS 4B sort model. The structure of MB-PSS4B consists of various working frequency bands. The MB-PSS4B of the MB-PSS family consists of three different bands, namely as low-frequency, intermediate-frequency, and high-frequency bands. Further, these low-, intermediate-, and high-frequency bands are concerned with global mode, inter-area mode, and local mode of oscillations, respectively [18].

3.3 Linear Quadratic Gaussian (LQG) Controller

The LQG-based WADC is designed using the linearized state-space model of the continuous-time system that contains the random noise. The communication and measurement noise are taken into consideration during the state-space modelling of the power system [9, 19, 20].

4 Noise in System

The power spectral density of Gaussian random noise is constant over the whole frequency spectrum. As there is no correlation between any of two samples of a Gaussian random noise process at various time instances, hence the auto-correlation or the auto-covariance of white noise is zero. So, the auto-covariance of the white noise process will be an impulse function at origin lag. The measurement white noise considered in this manuscript is normally distributed pseudorandom in nature, having covariance as $6 \times 10^{-6}\, I$. Here, I is the unity matrix of the appropriate dimension [9, 21].

5 Results and Discussions

For the analysis purpose, different kinds of WADC are armed with the test power system at the selected location. The time-domain responses of rotor angle deviation of machine 1 (M1) concerning machine 5 (M4) are plotted for various cases, specifically:

Case (1): Comparison without and with lead–lag-based WADC,
Case (2): Response of lead–lag WADC with different time delays,
Case (3): Comparison between lead–lag WADC and MB-PSS WADC,
Case (4): Impact of random time delay and random packet drop-out,
Case (5): Impact of measurement and process noise.

In Case (1), a comparison of rotor angle deviation is represented to show the efficacy of WADC in the given system. As shown in Fig. 3, the damping is improved when lead–lag WADC is used along with the local power system stabilizers. The

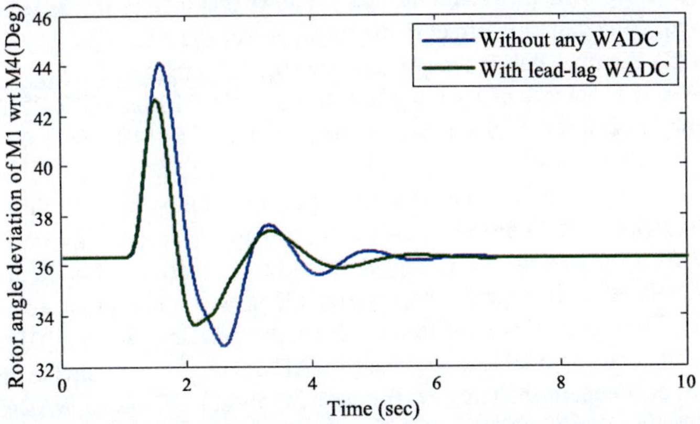

Fig. 3 Rotor angle deviation of M1 wrt M4, with and without conventional lead–lag-based WADC

Table 1 Eigenvalue analysis corresponding to case 1

Mode	Without WADC		With WADC	
	Freq. (Hz)	Damp. factor	Freq. (Hz)	Damp. factor
Local mode-1	1.54	0.29	0.97	0.46
Local mode-2	1.62	0.28	1.67	0.42
Inter-area mode	0.63	0.26	0.57	0.35

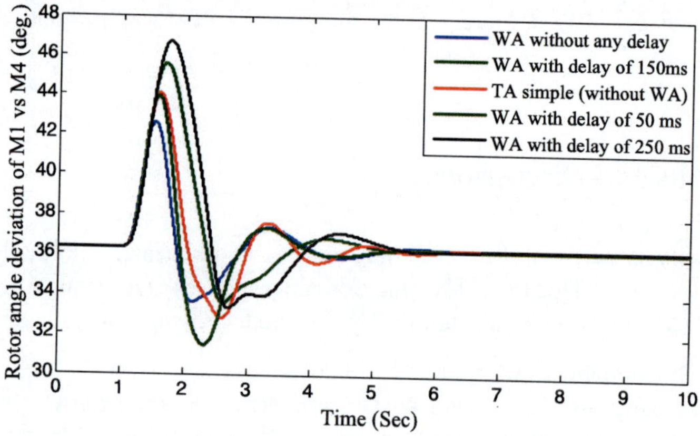

Fig. 4 Response of lead–lag-based WADC for different values of time delay in remote WA signal

corresponding eigenvalue analysis is also done to strengthen the outcome derived from the time-domain response, as mentioned in Table 1.

In Case (2), the performance of lead–lag-based WADC is evaluated with three different values of delay, like 50, 150, and 250 ms, as shown in Fig. 4. The response of lead–lag WADC with different time delays shows that increasing the value of time delay degraded the damping effect in the given power system.

In Case (3), the time-domain response is compared, as depicted in Fig. 5, with two different WADCs, namely of lead–lag WADC and MB-PSS-based WADC having a signal latency of 250 ms in the wide-area remote signal. The MB-PSS-based WADC shows a better damping effect as corresponded to lead–lag WADC.

Case (4) is considered to show the impact of a randomly varying time delay in addition to random packet drop-out also. Figures 6 and 7 represent the distant wide-area signal with random drop-out of packets and a comparison of wide-area signal with delay only with the signal having delay along with random packet drop-out, respectively. The phenomenon mentioned above deteriorates the overall damping of the system, and conventional lead–lag-based WADC is unable to retain the prescribed damping effect. Henceforth, it leads to the use of a robust LQG-based WADC in place of the conventional one, which mitigates the effects of various delays, as shown in Fig. 8.

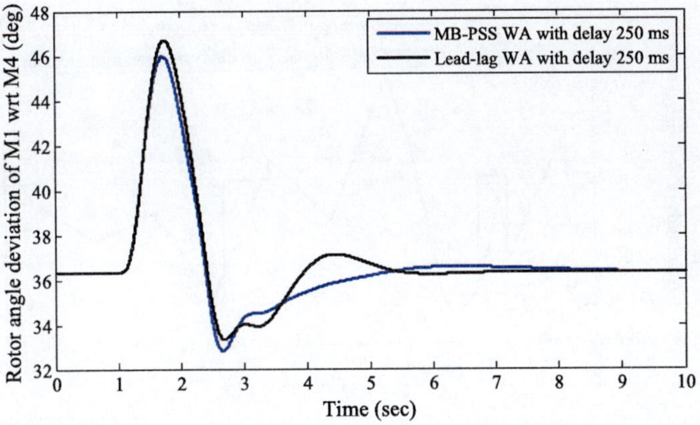

Fig. 5 Comparison of rotor angle deviation between lead–lag and MB-PSS-based WADC, having delay of 250 ms in remote signal

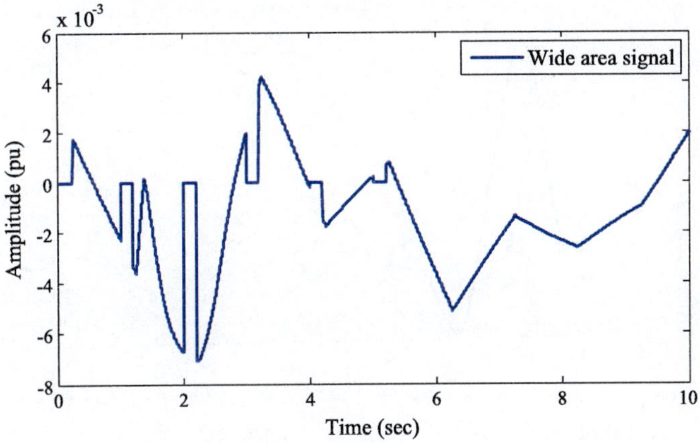

Fig. 6 Wide-area signal with random packet drop-out

Further, to reflect the imperfectness of the communication line, process and measurement noise is taken into consideration along with the signal latency, as taken in Case (5). The conventional lead–lag and MB-PSS-based WADC is not capable of maintaining the prescribed damping, as depicted in Fig. 9. However, the LQG-based controller is robust adequate to deal with the problems of imperfectness, and in addition to this, it stabilizes the system.

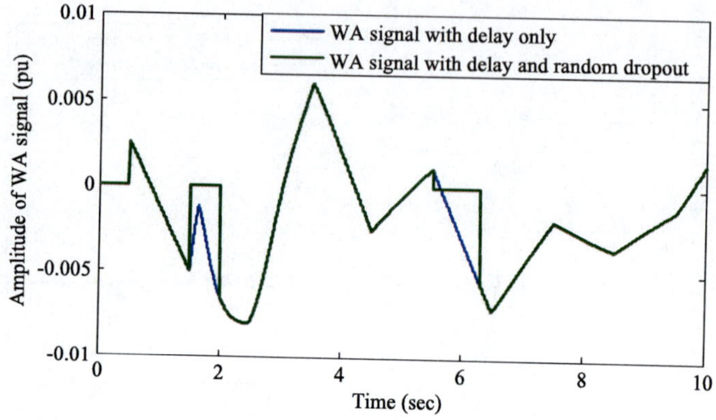

Fig. 7 A comparison between wide-area signal with delay only and the signal having delay along with random packet drop-out

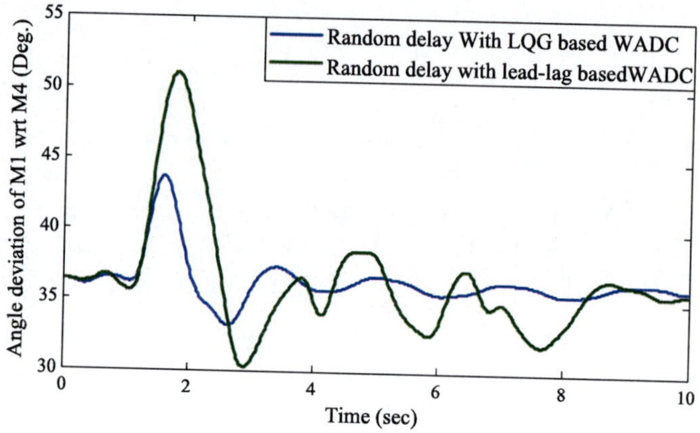

Fig. 8 A comparison of rotor angle deviation having a random delay in WA signal with lead–lag and LQG-based WADC

6 Conclusions

This work presents the significant role of a WADC in the modern power system, along with the comparative performance analysis of different controllers used as the WADC. In a wide-area control, various issues arise due to the imperfectness of the communication medium. The imperfectness includes the fixed signal latency, randomly varying latency, random signal drop-out, random measurement, and transmission noise. The conventional lead–lag-based and MB-PSS-based WADC is good enough to improve the damping of inter-area oscillatory modes only when there is a perfect transmission medium. Still, these controllers are incapable of handling the effects arising due to the imperfectness of the transmission medium. However, the

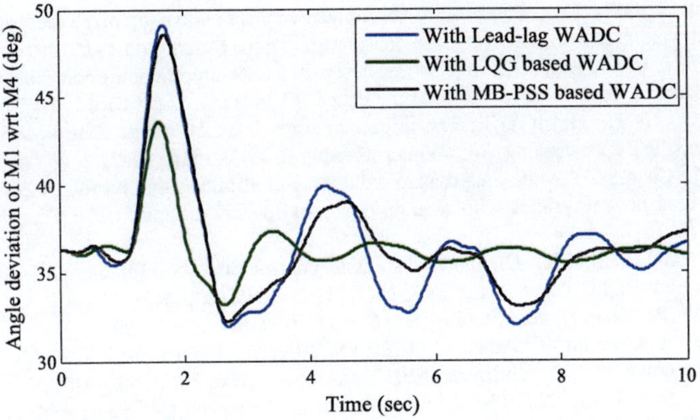

Fig. 9 A comparison of rotor angle deviation having signal latency, measurement noise in WA signal with lead–lag, MB-PSS and LQG-based WADC

effects of an imperfect medium are expertly handled by using the LQG-based WADC while maintaining the damping objectives.

References

1. Klein, M., Rogers, G.J., Kundur, P.: A fundamental study on inter-area oscillation in power systems. IEEE Trans. Power Syst. **6**(3), 914–921 (1991)
2. Zhang, Y., Bose, A.: Design of wide-area damping controllers for inter-area oscillations. IEEE Trans. Power Syst. **23**(3), 1136–1143 (2008)
3. Power System Relaying Committee of the IEEE PES: IEEE Standard for Synchrophasor Data Transfer for Power Systems. IEEE Std C37.118.2-2011 (Revision of IEEE Std C37.118-2005), 28 Dec 2011, pp.1–53
4. Bhadu, M., Senroy, N., Janardhanan, S.: Discrete wide area power system damping controller using periodic output feedback. Electr. Power Compon. Syst. J. **44**(17), 1892–1903 (2016)
5. Wu, H., Tsakalis, K.S., Heydt, G.T.: Evaluation of time-delay effects to wide-area power system stabilizer design. IEEE Trans. Power Syst. **19**(4), 1935–1941 (2004)
6. Chaudhuri, B., Majumder, R., Pal, B.C.: Wide-area measurement-based stabilizing control of power system considering signal transmission delay. IEEE Trans. Power Syst **19**(4), 1971–1979 (2004)
7. Zhang, S., Vittal, V.: Design of wide-area power system damping controllers resilient to communication failures. IEEE Trans. Power Syst. **28**(4), 4292–4300 (2013)
8. Zhang, S., Vittal, V.: Wide-area control resiliency using redundant communication paths. IEEE Trans. Power Syst. **29**(5), 2189–2199 (2014)
9. Bhadu, M., Senroy, N., Kar, I.N., NairS, G.: Robust linear quadratic Gaussian based discrete mode wide area power system damping controller. IET Gener. Trans. Distrib. **10**(6), 1470–1478 (2016)
10. Roy, N.K., Pota, H.R., Mahmud, M.A., et al.: Voltage control of emerging distribution systems with induction motor loads using robust LQG approach. Int. Trans. Electr. Energy Syst. **24**(7), 927–943 (2014)
11. Zenelis, I., Wang, X., Kamwa, I.: Online PMU-based wide-area damping control for multiple inter-area modes. IEEE Trans. Smart Grid 1–11 (2020)

12. Ghosh, S., Folly, K.A., Patel, A.: Synchronized versus non-synchronized feedback for speed-based wide-area PSS: effect of time-delay. IEEE Trans. Smart Grid 1–10 (2016)
13. Roy, S., Pstel, A., Kar, I.N.: Analysis and design of a wide-area damping controller for inter-area oscillation with artificially induced time delay. IEEE Trans. Smart Grid 1–10 (2018)
14. Wu, H., Ni, H., Heydt, G.T.: The impact of time delay on robust control design in power systems. IEEE Power Eng. Soc. Winter Meeting **2**, 1511–1516 (2002)
15. Li, M., Chen, Y.: A wide-area dynamic damping controller based on robust $H\infty$ control for wide-area power systems with random delay and packet dropout. IEEE Trans. Power Syst. 1–10 (2017)
16. Son, K.M., Park, J.K.: On the robust LQG control of TCSC for damping power system oscillations. IEEE Trans. Power Syst. **15**(4), 1306–1312 (2000)
17. Kundur, P.: Power System Stability and Control. McGraw-Hill (1994)
18. Kamwa, I., Grondin, R., Trudel, G.: IEEE PSS2B versus PSS4B: the limits of performance of modern power system stabilizers. IEEE Trans. Power Syst. **20**(2), 903–915 (2005)
19. Bhadu, M., Senroy, N.: Real-time simulation of a robust LQG based wide-area damping controller in power system. In: Innovative Smart Grid Technologies Conference Europe (ISGT-Europe), IEEE PES 2014, 12–15 Oct 2014, pp. 1–6
20. Zolotas, A.C., Chaudhuri, B., Jaimoukha, I.M., et al.: A study on LQG/LTR control for damping inter-area oscillations in power systems. IEEE Trans. Control Syst. Technol. **15**(1), 151–160 (2007)
21. Rathore, B., Bhadu, M., Bishnoi, S.K.: Modern controller techniques of improve stability of AC microgrid. In: Fifth International Conference on Signal Processing & Integrated Networks, SPIN 2018, 22–23 Feb 2018

Adware Attack Detection on IoT Devices Using Deep Logistic Regression SVM (DL-SVM-IoT)

E. Arul and A. Punidha

Abstract Malware represents an imminent threat to businesses and users every day. It is a fact that protection systems cannot compete today whether it is phishing emails or backdoor, exploits spread across the Internet, along with various evasion methods and other vulnerabilities to security. The availability and usefulness of systems like veil, shelter and others are considered to be used by pen testing professionals. By training a deep logistic regression and support vector machine (DLR–SVM) with multiple input clusters with a malicious or benign API, one single output unit is called malicious or benign. It was then trained to discover a malicious pattern in deep LR's unknown IoT firmware. The result shows that the rating for the truth positive is 98.11%, and that the attack on the adware is 0.07% positive.

Keywords IoT · Backdoors · Malware · API calls · Deep learning · SVM · Logistic regression learning · Adware

1 Introduction

Adware is a harmful platform developed to view ads on your computer, most commonly in an Internet browser. Several security professionals see it as the precursor to present-day potentially unacceptable program (PUP). It usually uses a deceitful technique to either mask itself as genuine, or put a spin to another software to manipulate them into downloading it on the PC, laptop, or smartphone. For each modern organization, malware is a nightmare. Attackers and cybercriminal always have new malicious software available to attack their targets. Security vendors do their best to protect themselves from malware attacks, but, unfortunately, they cannot do that by discovering millions of malware every month. There is therefore a need for new

E. Arul (✉)
Department of Information Technology, Coimbatore Institute of Technology, Coimbatore, Tamil Nadu, India

A. Punidha
Department of Computer Science and Engineering, Coimbatore Institute of Technology, Coimbatore, Tamil Nadu, India

H. Sharma et al. (eds.), *Congress on Intelligent Systems*, Advances in Intelligent Systems and Computing 1334, https://doi.org/10.1007/978-981-33-6981-8_14

approaches like deep learning. It is necessary to learn and learn the other different architectures of artificial neural networks before diving into technical details and steps for the practical implementation of the DL method [1, 2].

Active opponents (malware writers) work constantly to avoid detection and release new malware file versions which are significantly different from those seen during training. Thousands of software companies produce new kinds of benign executable that are substantially different. The training set lacked data on these types, but the model must however be seen as good [3].

Attacks on IoT devices have increased substantially, and it does not seem to slow down in the near future. Things like the thermostat nest and the doorbell ring are very media attention, but the IoT industry, utilities and other industries are also very widespread, and the compromise of an IoT device can have a much more serious effect than anyone else who has to watch your front porch or mess at home [4].

A botnet is a malware control and distribution network of systems [5]. Hackers use them to a huge extent to get private information, target DDoS and use data online. Botnets, nowadays, consist of a wide range of computer, laptop and so-called 'smart devices' connected devices. Such interconnected stuff has two unique features, namely the Internet is enabled, and a large amount of data is transmitted via a network. Because IoT is mostly about connected devices, the network is connected to a large number of devices, and therefore the number of botnets and cyber-attacks is that [6].

Machine learning is an application of artificial intelligence (AI) that enables systems to learn and improve experience automatically without being explicitly scheduled. Machine study focuses on software that can access and use data for its very own purpose [3]. The process starts with observations or information such as examples, direct experience or training in order to find patterns in the data and make more decisions in future in the form of the examples we provide [7]. The main objective is to allow computers to learn and change behavior automatically without the need for human intervention or support.

Controlled learning machines can use labeled examples to predict future events to apply what has been learned in the past to new data. The learning algorithm generates an inferred function to predict the output values based on an analysis of a specified training dataset [8]. After sufficient training, the program can provide goals for any new input. The study algorithm can also compare its output with the correct, desired output and detect errors to change the model in line with it.

Machine learning allows massive data analysis. This usually provides quicker and more accurate results to identify cost-effective opportunities or risky threats, but may also require extra time and resources to properly train this. Machine learning with IA and cognitive technology can make the processing of large quantities of information even more efficient.

2 Related Work

Di Xue, the identification of malware plays a major role in tracing computer security attack origins [9]. The current methods of static analysis are quickly categorized, but in some malware, using labeling and obfuscation techniques they are unsuccessful. The dynamic methods of analyses are more standard for packaging and obscuring which cause excessive classification costs [10, 11]. We suggest a classification method Malscore based on the calculation of probability and machine learning, setting the likelihood threshold for static analytics (called phase 1) and dynamic analysis (called phase 2) to resolve these shortcomings [10, 12]. In phase 1 gray scale images, variable ngrams and machine learning were analyzed by the convolutionary neural networks with spatial pyramid pooling, and in phase 2, the native API calling (dynamic features) were analyzed. The researcher performed 174,607 samples of malware from 63 malware families in experiments [13, 14]. The results show that Malscore achieves 98.82% malware accuracy. In addition, Malscore was contrasted with the static and dynamic analysis process [15]. The time required for preprocessing and testing decreased by 59.58 and 61.70%. In addition to accelerating the static analysis process using CNN in image recognition, Malscore combined static analysis with dynamic analysis also seemed more resistant to dynamic analysis [16].

Mahmoud Kalash et al., a profound learning system for malware detection. The volume of malware has risen enormously in the last few years, putting financial institutions, companies and individuals seriously at risk for security [17, 18]. New strategies are needed so that malware samples are easily identified and recognized to determine their actions to avoid malware spread [19, 20]. Machine learning approaches are common for malware detection, but most current malware learning methods use low-level learning algorithms (e.g., SVM). Recently, the CNN, a profound learning approach, has demonstrated superior performance in comparison with conventional learning algorithms, particularly in tasks such as image classification [8]. Motivated by this success, we are proposing a malware samples identification CNN-based architecture [21]. Then, convert malware binaries to gray images and then classify a CNN. Experiments of Malimg and Microsoft, two demanding malware classification datasets, show that our approach is better than modern results [22]. The method proposed achieves accuracy in Malimg and Microsoft, respectively, of 98.52% and 99.97% [23].

Ajit Kumar, smartphones in particular are particularly popular on the mobile android platform, making them at the top of the target list of attackers and growing their 80% share in the global smartphone operating systems market [24]. More private data and poor security allow the attacker to write a variety of malware for smartphones differently, and the ability to blur the malware detection applications using different coding techniques gives the attacker more energy. Many ways of detecting malware have been suggested by software review, which now seriously tackle the code obfuscation issue and high calculation demand [25, 26]. They propose a machine learning approach to detect malware in android by examining the visual representation in grayscale, RGB, CMYK and HSL of binary formatted APK files. GIST functionality has been extracted from the malware and the friendly dataset and used for training computer algorithms. Initial research results are promising and computer efficient. Random forest has achieved 91% highest accuracy for gray scale image among machine learning algorithms, which can be improved further through the harmonization of different parameters [4, 27].

3 Theoretical Background

3.1 *Delineation of IoT Adware Attacks Classification and Identification Using (DL-SVM-IoT)*

The aim of the supporting vector machine algorithm is to find a hyperplane in a N-dimensional space (N—the number of features) of IoT executables [2]. There are numerous possible hyperplanes to choose to differentiate the two groups of data points that is malicious or benign. We aim to find a plane with the maximum margin of the data points of both classes, i.e., the maximum distance between them [8, 28]. Maximizing the limit distance provides some strengthening to enhance the confidence of future data points, i.e., various features used. Hyperplanes are boundaries of decision which allow for the classification of data points used in IoT executable service call classification [29]. There are different classes of data points falling on either side of the hyperplane in malicious and benign executables. The hyperplane size also depends on the IoT executables and its number of characteristics [21]. If the input number of characteristics is 2, the hyperplane is only a single line. The hyperplane becomes a two-dimensional plane if the number of input features is three that are taken on each benign and malware IoT executables. When the number of features is over 3, it is hard to imagine [19, 20].

Support vectors are data points closer to the hyperplane and affect the hyperplane's direction and orientation of IoT executables. We maximize the margin of the classifier with these support vectors. The removal of vectors of support will alter the hyperplane position on each clusters of malicious and benign [30]. These are the things that help us build our SVM. In a logistic regression, we use the sigmoid function to squash the value within the range of each service call attribute's [0, 1]. If the squashed value reaches a threshold value (0.5) for each classes of results in Table 1, we will give it a label class 0 to class 9 in Table 1, otherwise, we will give it a cluster classes [2]. In SVM, we take the output for the logistic regression function and identify it with one class if the output is greater than 1, and if the output is −1, it can identify it as a different class. Since in SVM, we adjust the threshold values to 1 and −1, we get this margin of reinforcement value ([−1, 1]) for each clusters of IoT executables [1, 31].

We strive to maximize the gap from the data to the hyperplane in the SVM algorithm given by Eq. (1).

Table 1 In-house DL-SVM-IoT-based Internet network call groups classification report (x_test, y_pred) are established

Various classes of API calls	Precision	Recall	$f1$-score support	$f2$-support
Class 0	1.00	1.00	1.00	1
Class 1	1.00	1.00	1.00	1
Class 2	1.00	1.00	1.00	1
Class 3	1.00	1.00	1.00	1
Class 4	1.00	1.00	1.00	1
Class 5	1.00	1.00	1.00	1
Class 6	1.00	1.00	1.00	1
Class 7	1.00	1.00	1.00	1
Class 8	1.00	1.00	1.00	1
Class 9	1.00	1.00	1.00	1
Micro avg.	1.00	1.00	1.00	10
Macro avg.	1.00	1.00	1.00	10
Weighted avg.	1.00	1.00	1.00	10

$$c(x, y, f(x)) = (1 - y * f(x))_+ \tag{1}$$

The cost is 0 when the value and the value of the expected value are the same. If it is not, the loss cost is determined. In addition, we add a cost function regulatory parameter for every IoT malicious and benign cluster calls [32]. The goal of the regulatory parameter is to balance the maximization and loss of the margin each cluster benign and malicious cluster. The cost functions look like below, after adding the regularization parameter [33].

$$\min w\, \lambda||w||2 + \sum_{i=1}^{n}(1 - y_{i\langle x_i, w\rangle})_+ \tag{2}$$

With the loss function now, we take partial derivatives Eqs. (3) and (4) to evaluate the gradients in relation to the weights of IoT executable clusters. We can change our weights with the gradients sample in Figs. 1 and 2.

$$\frac{\delta}{\delta W_k}\lambda||W||2 = 2\lambda W_k \tag{3}$$

$$\frac{\delta}{\delta W_k}(1 - y_i\langle x_i, w\rangle)_+ = \begin{cases} 0, & \text{if } y_{i\langle X_i, w\rangle \geq 1} \\ -y_i x_{ik}, & \text{else} \end{cases} \tag{4}$$

When no error occurs, i.e., our model correctly forecasts the class of our data point, we only need the gradient from the regularization parameter to be updated by Eq. (5) [34].

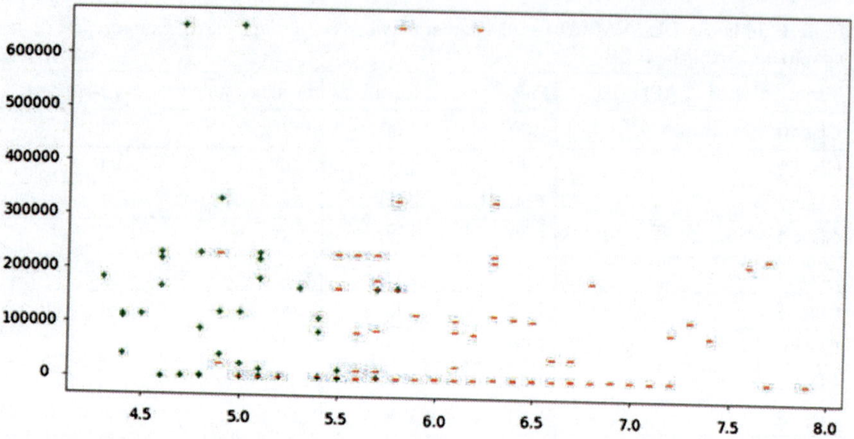

Fig. 1 Results of various feature (like *X*-axis IATRVA, *Y*-axis ResSize values) IoT malware and benign set, data points visualization

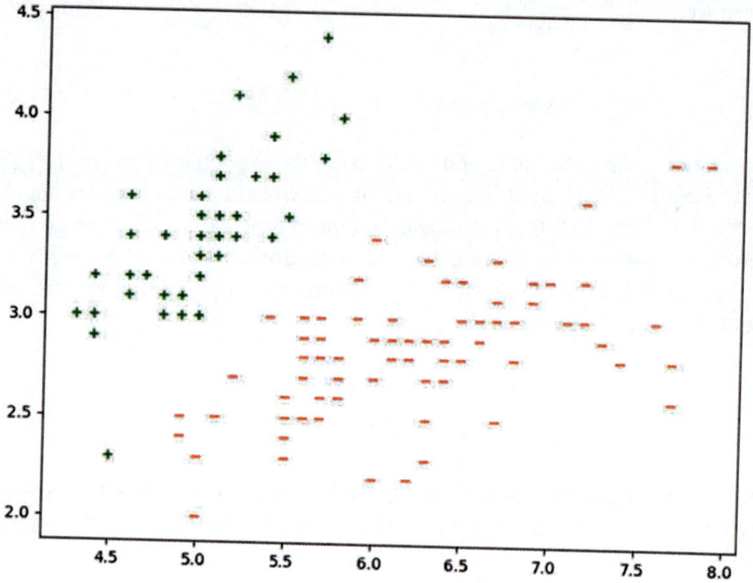

Fig. 2 IoT *X*-axis malware API and *Y*-axis malware benign API set, data points visualization

$$\omega = \omega + \alpha \cdot (2\lambda\omega) \tag{5}$$

If there is an incorrect classification, i.e., our model errs in predicting our data point class, the loss is included together with the regularization parameter to update the gradient in Eq. (6) [33].

$$\omega = \omega + \alpha \cdot (y_i \cdot x_i - 2\lambda\omega) \quad (6)$$

4 Experimental Results and Comparison

The IoT runtime database [35, 36] is the dataset we will use to implement our SVM algorithm. As we have three classes of IoT malware, we will delete one of the classes. This leaves us with a problem with the binary classification as shown in Fig. 2.

Many features are also available to us. We only use two features, i.e., IATRVA and ExportSize. We are taking these two characteristics and designing them. From the graph, you can conclude that the data points can be separated by a linear line [31].

We extracted API calls from the adware files and divide into training and testing data the features needed. For training, 90% of the data are utilized and the remainder 10% to validate the adware dataset. Let us now use the numpy library to construct our SVM template [30].

$\alpha(0.0001)$ is learning rate and μ is set to 1/epochs. $\alpha(0.0001)$ the regularization value, therefore, reduces the number of epochs. We now divide the weights, since there are only 10 data points in the test data. We remove the properties and predict values from the test data. We collect the projections to compare the current values and print out the accuracy of our model. Accuracy of our SVM model 1.0.

SVM has applied on IoT firmware's to identify the more relative internal communication around malware sample API calls, concentrating on default IoT system API calls, and then finding more internal proximity to interrelated malicious network connection API calls. The logistic regression cluster result applied to the cluster algorithm of the dynamic model based on SVM. Further specific network connection API calls identified are part of an IoT-related internal malware API [30] call in Table 1. The findings in Table 2 were compared to the related work with the proposed SVM-IoT.

Table 2 Compared with existing malware methods of the proposed DL-SVM-IoT

Methods	Number of malware are detected	TP ratio (%)	FP detected	FP ratio (%)
Di Xue	1023	91.99	149	0.12
Mahmoud Kalash	990	89.02	193	0.15
Proposed SVM-IoT	1091	98.11	095	0.07

Total number of malware file taken for analysis: 1112
Total number of normal file taken for analysis: 1211

5 Conclusion and Future Work

Most of the malware inserted into the target machine uses IoT-related API-related calls to perform untested malicious activities. Malware robs the user of personal information and sends it to the hacker server, spreads malicious spam mail and makes full use of the bandwidth of the IoT device network. Deep LR SVM training—IoT clustering algorithm used for malicious executable and clustered firmware API calls conducting adware operation explicitly in this proposed work. Finally, to look for further similarities to any malicious activity of any executable, a DLR-SVM learning algorithm was used. The outcome is a true positivity score of 98.11% and a false positive ratings of the various IoT devices firmware assault of 0.07%. This work will be done in future with other IoT APIs that allow the execution of malicious network activities.

References

1. https://towardsdatascience.com/support-vector-machine-introduction-to-machine-learning-algorithms-934a444fca47
2. https://expertsystem.com/machine-learning-definition/
3. https://dzone.com/articles/malware-detection-with-convolutional-neural-networ
4. Makandar, A., Patrot, A.: Malware analysis and classification using artificial neural network. In: Proceedings of IEEE International Conference on Trends in Automation, Communications and Computing Technology, Dec 2016, pp. 1–6
5. Kilgallon, S., De La Rosa, L., Cavazos, J.: Improving the effectiveness and efficiency of dynamic malware analysis with machine learning. In: Proceedings of IEEE Resilience Week, Sept 2017, pp. 30–36
6. https://resources.infosecinstitute.com/machine-learning-malware-detection/#gref
7. Cai, H., Meng, N., Ryder, B.G., Yao, D.: DroidCat: effective android malware detection and categorization via app-level profiling. IEEE Trans. Inf. Forensics Secur. **14**(6), 1455–1470 (2019)
8. Microsoft malware classification challenge (big 2015) first place team: say no to overfitting. https://blog.kaggle.com/2015/05/26/microsoft-malware-winners-interview-1st-place-no-to-overfitting/ (2017). Accessed 22 Apr 2017
9. Xue, D., Li, J., Lv, T., Wu, W., Wang, J.: Malware classification using probability scoring and machine learning. IEEE Access **7**, 91641–91656 (2019). https://doi.org/10.1109/ACCESS.2019.2927552
10. Lyda, R., Hamrock, J.: Using entropy analysis to find encrypted and packed malware. IEEE Secur. Priv. **5**(2), 40–45 (2007)
11. Yan, J., Qi, Y., Rao, Q.: Detecting malware with an ensemble method based on deep neural network. Secur. Commun. Netw. **2018**, Art. No. 7247095 (2018)
12. Han, K., Kang, B., Im, E.G.: Malware analysis using visualized image matrices. Sci. World J. **2014**, Art. No. 132713 (2014)
13. Tobiyama, S., Yamaguchi, Y., Shimada, H., Ikuse, T., Yagi, T.: Malware detection with deep neural network using process behavior. In: Proceedings of IEEE 40th Annual Computer Software and Applications Conference, June 2016, pp. 577–582

14. Kalash, M., Rochan, M., Mohammed, N., Bruce, N.D.B., Wang, Y., Iqbal, F.: Malware classification with deep convolutional neural networks. In: 9th IFIP International Conference on New Technologies, Mobility and Security (NTMS). IEEE (2018). https://doi.org/10.1109/NTMS.2018.8328749
15. Dash, S.K., Suarez-Tangil, G., Khan, S., Tam, K., Ahmadi, M., Kinder, J.: DroidScribe: classifying android malware based on runtime behavior. In: Proceedings of IEEE Security and Privacy Workshops (SPW), pp. 252–261 (2016)
16. Chen, S., Xue, M., Tang, Z., Xu, L., Zhu, H.: StormDroid: a streaminglized machine learning-based system for detecting android malware. In: Proceedings of the 11th ACM on Asia Conference on Computer and Communications Security, pp. 377–388 (2016)
17. Gibert Llauradó, D.: Convolutional neural networks for malware classification. Master's thesis, Universitat Politècnica de Catalunya (2016)
18. Simonyan, K., Zisserman, A.: Very deep convolutional networks for large-scale image recognition. arXiv preprint arXiv:1409.1556 (2014)
19. Sung, A.H., Xu, J., Chavez, P., Mukkamala, S.: Static analyzer of vicious executables (save). In: Annual Computer Security Applications Conference, pp. 326–334. IEEE (2004)
20. Rieck, K., Holz, T., Willems, C., Düssel, P., Laskov, P.: Learning and classification of malware behavior. In: International Conference on Detection of Intrusions and Malware, and Vulnerability Assessment, pp. 108–125. Springer (2008)
21. Kumar, A., Pramod Sagar, K., Kuppusamy, K.S., Aghila, G.: Machine learning based malware classification for android applications using multimodal image representations. In: 2016 10th International Conference on Intelligent Systems and Control (ISCO). IEEE. https://doi.org/10.1109/ISCO.2016.7726949
22. Kolter, J.Z., Maloof, M.A.: Learning to detect malicious executables in the wild. In: Proceedings of the Tenth ACM SIGKDD International Conference on Knowledge Discovery and Data Mining, pp. 470–478. ACM (2004)
23. Siddiqui, M., Wang, M.C., Lee, J.: A survey of data mining techniques for malware detection using file features. In: Proceedings of the 46th Annual Southeast Regional Conference on XX, pp. 509–510. ACM (2008)
24. Suarez-Tangil, G., Tapiador, J.E., Peris-Lopez, P., Ribagorda, A.: Evolution, detection and analysis of malware for smart devices. IEEE Commun. Surv. Tutor. **16**(2), 961–987 (2014)
25. Cai, H., Meng, N., Ryder, B., Yao, D.: DroidCat: unified dynamic detection of android malware. Ph.D. dissertation, Department of Computer Science, Virginia Polytechnic Institute and State University, Blacksburg (2016)
26. Schultz, M.G., Eskin, E., Zadok, F., Stolfo, S.J.: Data mining methods for detection of new malicious executables. In: IEEE Symposium on Security and Privacy, pp. 38–49. IEEE (2001)
27. Kalash, M., Rochan, M., Mohammed, N., Bruce, N.D.B., Wang, Y., Iqbal, F.: Malware classification with deep convolutional neural networks. In: Proceedings of IEEE IFIP International Conference on New Technologies, Mobility and Security, Feb 2018, p. 15
28. Idika, N., Mathur, A.P.: A Survey of Malware Detection Techniques, vol. 48. Purdue University (2007)
29. You, I., Yim, K.: Malware obfuscation techniques: a brief survey. In: International Conference on Broadband, Wireless Computing, Communication and Applications, pp. 297–300. IEEE (2010)
30. Arp, D., Spreitzenbarth, M., Malte, H., Gascon, H., Rieck, K.: Drebin: Effective and explainable detection of android malware in your pocket. In: Network and Distributed System Security Symposium, pp. 23–26 (2014)
31. https://media.kaspersky.com/en/enterprise-security/Kaspersky-Lab-Whitepaper-Machine-Learning.pdf
32. Gascon, H., Yamaguchi, F., Arp, D., Rieck, K.: Structural detection of android malware using embedded call graphs. In: Proceedings of 2013 ACM Workshop on Artificial Intelligence and Security—AISec'13, pp. 45–54 (2013)
33. Ham, Y.J., Lee, H.-W.: Detection of malicious android mobile applications based on aggregated system call events. Int. J. Comput. Commun. Eng. **3**(2), 149–154 (2014)

34. Suarez-Tangil, G., Tapiador, J.E., Peris-Lopez, P., Blasco, J.: Dendroid: a text mining approach to analyzing and classifying code structures in android malware families. Expert Syst. Appl. **41**(4 PART 1), 1104–1117 (2014)
35. https://contagiodump.blogspot.in/2010/11/links-and-resources-for-malware-samples.html
36. https://tuts4you.com/download.php?list.89

Intrusion Detection System for Securing Computer Networks Using Machine Learning: A Literature Review

Mayank Chauhan, Ankush Joon, Akshat Agrawal, Shivangi Kaushal, and Rajani Kumari

Abstract Network security is becoming very important for the networking society in recent years due to increasingly evolving technology and Internet infrastructure. Intrusion detection system is primarily any security software, capable of identifying as well as immediately warning administrators in case somebody or something tries to access the network system by performing malicious practices. So, intrusion detection system (IDS) is extremely important in order to provide security to network systems. It is a tool that attempts to defend the networks against a hacker. IDS is helpful not only to predict successful intrusions but also to monitor activities that attempts to breach security. This literature review aims to find out the importance of detecting intrusions using machine learning methods. This paper presents answers to questions like what machine learning techniques have been used so far for IDS, how effective these methods are for detecting intrusions in network systems, i.e., how much predictive accuracy the models provided, what are the demerits of previous studies and what areas are still open for research in this field.

Keywords Network security · Machine learning · Intrusion detection systems · Predictive performance · Accuracy

1 Introduction

Any unauthorized behavior on a computer network is an intrusion to a network. Most often such unauthorized actions endanger network security. Network intrusions involve the stealing of important system resources and often seriously harm network security and/or its data. Intrusion can be defined as any type of unauthorized activity that leaves confidentiality, availability, or data integrity in an information system compromised. Cyberthreats that can potentially harm the network are of 4 main types [1].

M. Chauhan (✉) · A. Joon · A. Agrawal · S. Kaushal · R. Kumari
Amity University, Gurugram, Haryana 122413, India

H. Sharma et al. (eds.), *Congress on Intelligent Systems*, Advances in Intelligent Systems and Computing 1334, https://doi.org/10.1007/978-981-33-6981-8_15

- *DOS (Denial of Service)*: DoS attacks exhaust targeted system's resources and disable them from providing any services, for example, back, neptune, and teardrop.
- *Root to local attack (R2L) (Illegal remote access)*: R2L threat allows unauthorized access to remote intruders, for example, ftp-write, spy, and warezmaster.
- *User to root attack (U2R) (untrusted local super-user privileges access)*: U2R assault attempts to obtain privileges of the superusers, for example, buffer-overflow, and rootkit.
- *Probing*: Probe intrusion checks for target system's vulnerabilities, for example, ipsweep, nmap, portsweep, and satan.

Intrusion detection systems are applications capable of detecting some form of disruptive threats, malicious content or encroachment that may cause harm to our network. Network security has become a significant concern of the networking community in recent years due to increasingly evolving technology and Internet infrastructure. Several techniques and frameworks have been designed to make processes, networks, and software more secure.

Traditionally, security professionals chose login credentials security mechanism, encryption algorithms, and access controls in addition to firewalls as a measure to safeguard the network. However, these mechanisms are not sufficient to protect the platform [2]. Network systems will become more susceptible as the percentage of devices in the network continues to grow. This makes it easy for hackers to hack records, privacy of users, and business secrets. Even though people have done their hardest to secure their sensitive information, cyber-attacks occur frequently due to the nature of the network infrastructure and the abundance of security threats. In these conditions, cyber-attack detection techniques should be stronger and even more effective than ever before to identify and avoid the hacking attacks [3]. Consequently, many administrators intend to use intrusion detection systems (IDSs) to detect security threats by monitoring network traffic.

The goal for the machine learning classification algorithm, based on the dataset of a computer network, is to train a predictive model, i.e., a classifier, able to differentiate between authorized and unauthorized connections within a network.

1.1 Machine Learning for IDS

The intrusion detection system is a protection against security breaches which try to steal data stored on different systems, such as servers and personal computers. For the case of common threats, identifying and processing it quickly is easier for the administrator, but evaluating unknown suspicious data is difficult, and the cost of repair grows as the dealing is postponed. Machine learning approaches are commonly used for IDS due to its capability to identify regular/threat network traffic by learning trends in the data previously gathered [4].

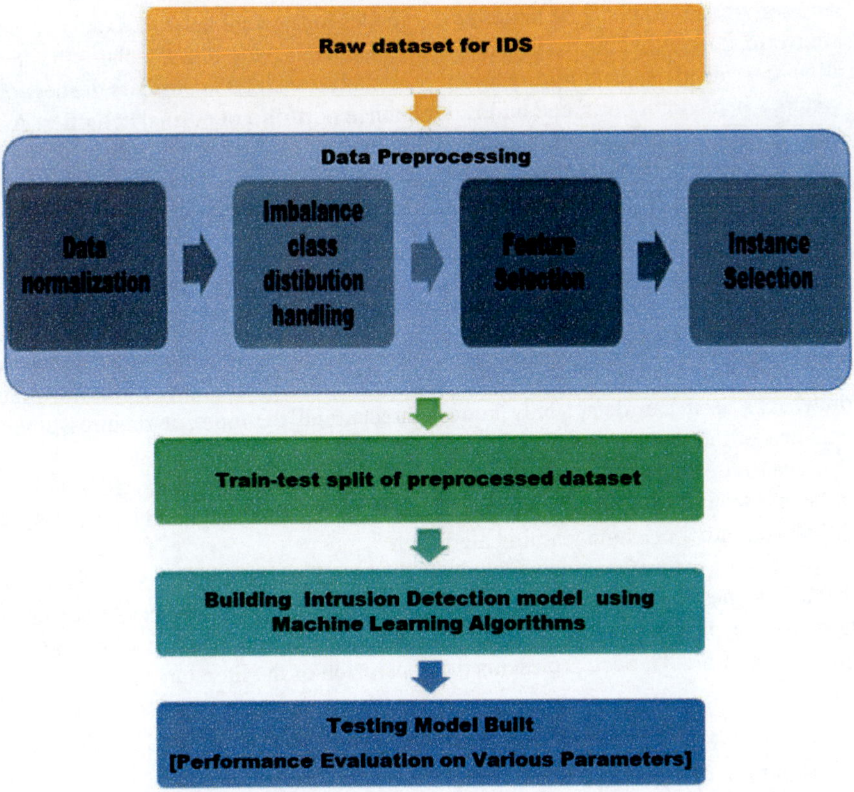

Fig. 1 Machine learning framework for IDS

The trained classifier safeguards against cyber-attacks and intruders on a network of computers. Figure 1 shows the basic machine learning framework for intrusion detection system. In this system, raw dataset of IDS will move for further preprocessing steps then go for standard train and test split for so that mode training will be done and then training model will ready for work.

1.2 Types of IDS

IDSs can be categorized in various types, the most familiar being signature-based, misuse-based, and anomaly-based IDSs.

Misuse-based IDS. They can accomplish skillfully the task of detecting known attacks like snort. There is less false alert rate to this sort of IDS. It is unable to detect new attacks that do not match any previous intrusion [5].

Anomaly-based IDS. They first create a model for normal network behavior, and afterward, it identifies any critical differences from this pattern and declares the deviation to be intrusion. In simple words, the anomaly-based IDS considers the network system's behavioral norm and builds a reference portfolio of normal activities. Any action that does not fit the behavioral norm of the network will be considered as intrusion. This form of IDS can detect both known and new threats, but it does experience increased false alert rates. To lower the false alert rates, different machine learning strategies are implemented [6].

Signature-based IDS. SIDSs archive the signatures of the deceitful operations in a information base and begin applying pattern matching techniques to detect intrusions [2]. In the signature-based framework, patterns of threats or intruder's actions are simulated, and a warning is raised if a match is found. But this framework has drawbacks as it can detect only known threats, and therefore, it requires timely updation of the threat signatures.

Because of the benefits of AIDS, most current IDSs either use an AIDS directly or benefit from it in a hybrid approach. Such IDSs need to be trained by processing the dataset into a machine learning algorithm.

The rest of the review paper is organized into following section: Sect. 2 defines the review question that forms the base of conducting this literature survey. Section 3 presents the papers in review. Section 4 illustrates the results to various review question, and finally, Sect. 5 presents the conclusion of the literature survey.

2 Review Questions

To keep the review focused, review questions were defined before studying the various works in the field of intrusion detection system. Table 1 shows the research questions that represents the objectives of conducting this review.

Table 1 Review questions for the literature survey

S. No.	Research question
1.	Why IDS is important?
2.	What machine learning techniques have been used so far in the field of IDS?
3.	What datasets are available for building intrusion detection models?
4.	What parameters are suitable for evaluating performance of IDS?
5.	What are the limitations of existing IDS?

3 Papers in Review

Divyatmika et al. (2016). In this paper, they proposed a two-tiered framework that uses neural networks and machine learning to detect intrusions at network level. Since their research relies on the behavior of the network, they regarded the TCP/IP traffic packets as their input data. They applied parameter filtering technique to preprocess the dataset and thereafter employed agglomerative hierarchical clustering algorithm to construct an automatic framework on training dataset. They used KNN algorithm to classify the data as normal traffic or intrusion packets. They applied MLP algorithm for misuse detection and reinforcement learning for detecting anomalies. Their architecture's TP score was 0.99 and 0.01 for false positives. Their framework, therefore, offers a higher level of protection by offering higher TP and lower false positive rates. The Kappa statistical and MSE values are 0.9988 and 0.014, which indicate high detection accuracy and lower error rates [7].

Halimaa et al. (2019). This research applied SVM and Naïve Bayes models on the NSL-KDD dataset. This dataset contains 19,000 observations. Rate of accuracy and rate of misclassification are used as measures of performance. The SVM model obtained accuracy value above 0.90, whereas Naïve Bayes model obtained very low accuracy rate of 0.71. Also, SVM offered a low misclassification rate of around 0.02 and Naïve Bayes provided a bad misclassification rate of around 0.4. The model can detect the attacks that it had learned while training phase. It cannot detect new unknown attacks [5].

Ali et al. (2018). Supervised IDS is a software capable to predict new threats using the information of past threats. They have used ANN, the fast learning network (FLN) based on particle swarm optimization (PSO) for intrusion detection was introduced in this study and is termed PSO-FLN. The model was evaluated on popular KDD99 dataset for intrusion detection. PSO-FLN almost surpassed other learning models regardless of the number of neurons in the hidden layer. It gave accuracy value of 0.98 and above. The model is not able to handle class imbalance problem, and that is why obtained less accurate results for some classes of attacks like R2L [8].

Karatas et al. (2020). They used KNN, random forest, gradient boosting, Adaboost, decision tree, and LDA techniques to implement machine learning-driven IDSs. The dataset, CSE-CIC-IDS2018, was chosen to develop a more practical IDS, rather than outdated and often used datasets. The dataset selected was also imbalanced. Then the imbalance ratio is minimized by using an oversampling technique called SMOTE. The model performance was evaluated on different parameters like recall, precision, *f*-score, and accuracy. All the six algorithms when used with SMOTE technique gave high-performance rates on all parameter (average value 0.99). The oversampling technique improves the accuracy for minority class detection but most of the algorithms used provide high accuracy value with original dataset for majority classes. So, there is not much improvement in accuracy value for majority class [2].

Riaz Khan et al. (2019). This paper used convolution neural network for building intrusion detection system. They tested the model on KDD99 dataset for accuracy performance evaluation parameter. Using Adam optimisation technique, the pooling method implements downsampling to minimize the loss function. Their model detected the intrusions with an accuracy value of 99.23%. Only accuracy was considered while validating the method, whereas in IDS false positive rate in very crucial parameter. FPR of IDS has not been taken care of [3].

Gautam et al. (2018). They have applied ensemble machine learning model for setting up IDS. The KDD99 dataset was used to train the model. Accuracy, precision, and recall rates were considered as performance validation parameters. They applied information gain attribute selection technique to find out most important features of dataset for building the model and then used bagging ensemble technique to detect the intrusions. The result showed that bagging provide good results (accuracy = 99.97, precision = 99.99, recall = 99.98) as compared to Naïve Bayes, PART, and Adaboost algorithms [9].

Lin et al. (2018). This research was centered on network intrusion detection using LeNet-5-based convolution neural networks (CNNs) to identify network attacks. Experimental results indicated that the IDS predictive accuracy goes up to 99.65% with observations greater than 10,000. The aggregate accuracy value is 97.53%. The experiment was conducted on dataset KDD99. The adaptive delta optimization technique was utilized to fine-tune the parameters of the model and mitigate the prediction errors by using back-propagation error derivatives and to provide quick intrusion detection mechanism [10].

Ali H. Mirza (2018). They applied ensemble machine learning technique (on LR, NN, and DT) by using weighted majority voting rule to predict intrusions in network. PCA algorithm was used as feature selection method. The model was built using 10% of the KDD99 dataset. They got an accuracy value of 97.53%. They focused only on binary classification and cannot differentiate between different kind of threats. FPR was not considered while evaluating model's predictive ability [1].

Park et al. (2018). Kyoto 2006+ dataset was used in this study. Dataset Kyoto 2006+ includes data on traffic flow obtained from Nov 2006 through Dec 2015. They applied random forest algorithm to build ID model. The performance was measured on accuracy, recall, precision, $f1$-score, and $f2$-score parameters. The overall results were good (all values = 0.99). But as the framework did not handle class imbalance problem, the prediction accuracy was very poor for certain classes of attack (like for shellcode attack). The advantage of this study is that is worked on one of the recent datasets available for IDS and able to detect different categories of attacks [4].

Taher et al. (2019). They used supervised machine learning techniques ANN and SVM along with wrapper filter selection method as intrusion detection model. The model was trained on NSL-KDD dataset. It is capable of detecting four types of attacks, i.e., DOS, probe, spy, and worm. The SVM model gave an accuracy of

82.34% and ANN provided an accuracy value of 94.02%. The model did not consider false positive rate parameter and class imbalance problem in the dataset [11].

Rahmani et al. (2015). They used a hybrid machine learning technique that is a combination of genetic algorithm as feature selection method and SVM as the classification algorithm to detect intrusions. They used KDD99 dataset for their experiment. They got good results in terms of TPR (= 0.973) and FPR (= 0.017) which is desirable for any intrusion detection model [6].

Yaseen et al. (2016). They proposed a real-time multi-agent system for intrusion detection system. It used two machine learning algorithms SMV and extreme learning machine (ELM) on the KDD99 dataset. They got an accuracy of 95%, *f*-score of 97%, and FPR approx. 2.13%. It is capable of predicting unknown attacks in real-time efficiently in less processing time [12].

The literature review Table 2 shows the machine learning methods and datasets used, performance obtained and disadvantages of the studies taken up for this literature survey.

4 Discussions

The result section provides the answers to the review questions (RQ) defined in Sect. 2 of this review paper.

RQ1 (Importance of IDS). Network security has become a significant concern of the networking community in recent years due to increasingly evolving technology and Internet infrastructure. Network vulnerabilities allow hackers to invade into computers using not only proven but also new forms of intrusions are highly difficult to detect. To secure the networks from them, one of the most common security methods is the intrusion detection system (IDS) [2].

Even though people have done their hardest to secure their sensitive information, cyber-attacks occur frequently due to the nature of the network infrastructure and the abundance of security threats. Here comes the importance of IDS, as it is primarily a security application that is capable of identifying and immediately warning administrators in case somebody or something tries to access the network system by performing malicious practices or breaching security policies [6].

IDS is a tool that attempts to defend the networks against a hacker. It is helpful not only to predict actual intrusions but also to monitor activities that try to breach security, that enables it for supplying valuable information for timely defensive-measures. IDS is used to decide whether the user behavior or traffic patterns being tracked was nefarious. If a suspicious event is identified, it would trigger a warning [8].

RQ2 (Machine learning methods used by previous studies for IDS). Various machine learning techniques have been used to implement IDS. Bayesian networks, decision tree, genetic algorithms, K-means clustering, NN, and SVM are some of

Table 2 Summary of papers in review

Year	Authors	Machine learning methods used	Dataset used	Predictive performance	Disadvantages
2019	Halimaa et al.	SVM, Naïve Bayes	NSL-KDD	*SVM model* Accuracy = 0.90 Misclassification rate = 0.02 *Naïve Bayes* Accuracy = 0.71 Misclassification rate = 0.4	It cannot detect new unknown attacks efficiently
2018	Ali et al.	ANN based on particle swarm optimization (PSO)	KDD99	Accuracy value = 0.98	The model is not able to handle class imbalance problem It obtained less accurate results for some classes of attacks like R2L
2020	Karatas et al.	KNN, Random Forest, Gradient Boosting, Adaboost, Decision Tree, and LDA	CSE-CIC-IDS2018	Recall, precision, *f*-score, and accuracy (average value 0.99, for all parameters)	There is not much improvement in accuracy value for majority class
2019	Riaz Khan et al.	Convolution neural network	KDD99	Accuracy value = 99.23%	In IDS false positive rate is very crucial parameter. FPR of IDS has not been taken care of
2018	Gautam et al.	Ensemble machine learning model	KDD99	Accuracy = 99.97, precision = 99.9, recall = 99.98	It cannot handle multiclass classification (i.e., it cannot detect all types of attacks)

(continued)

Table 2 (continued)

Year	Authors	Machine learning methods used	Dataset used	Predictive performance	Disadvantages
2018	Ali H. Mirza	Ensemble machine learning technique (on LR, NN, and DT)	KDD99	Accuracy = 97.53%	They focused only on binary classification and cannot differentiate between different kind of threats. FPR was not considered while evaluating model's predictive ability
2018	Park et al.	Random forest	Kyoto 2006+	Accuracy, recall, precision, $f1$-score, and $f2$-score (all values = 0.99)	The framework did not handle class imbalance problem
2019	Taher et al.	ANN, SVM	NSL-KDD	ANN accuracy = 94.02% SVM accuracy = 84.32%	The model did not consider false positive rate parameter and class imbalance problem in the dataset
2016	Ashfaq et al. [13]	RWNN algo	KDD99	Accuracy value = 95.32%	Do not handle class imbalance problem. Does only binary classification. Do not consider detecting types of attacks
2016	Kim et al. [14]	LSTM-based recurrent neural network deep learning classifier	KDD99	Accuracy = 0.96 and FAR = 0.10	Do not handle class imbalance problem, so obtained poor results for certain classes (like for probe attack class detection rate was merely 54%)

Table 3 Datasets utilized for developing intrusion detection systems

S. No.	Name of dataset	Description
1.	KDD99	This is a standard dataset for IDS. It contains a diverse set of simulated attacks data in military network environment
2.	NSL-KDD	This dataset is an improved version of KDD99 dataset
3.	Kyoto dataset	Kyoto 2006 + dataset was created using the three-year actual network traffic (Nov. 2006 to Aug. 2009) that is collected from different forms of honeypots
4.	CSE-CIC-IDS2018	It is a recent dataset for IDS. It includes a complete description of threats for systems, protocols, or lower level network entities. The dataset contains seven different types of attacks, including brute force, Heartbleed, Botnet, DoS, DDoS, Web attacks, and inward network intrusion

the commonly used algorithms for IDS [15]. Other techniques most widely used are KNN, Naive Bayes, ANN, random forest, gradient boosting, LDA, Adaboost, and ensemble machine learning methods.

Addition to machine learning methods, some researches also applied deep learning algorithms for detecting intrusions. The deep learning techniques used for IDS are self-taught learning [16], convolution neural networks [17], RWNN algorithm [13], and LSTM-based recurrent neural network [14].

RQ3 (Datasets available in the field of IDS). From the literature survey, we found that four datasets are available for intrusion detection in networks via machine learning. The datasets are KDD99 dataset, NSL-KDD dataset, Kyoto dataset, and CSE-CIC-IDS2018 dataset. Table 3 shows the brief description of these datasets.

RQ4 (Performance evaluation parameters used for IDS). An intrusion detection system's efficiency is primarily based on accuracy. To decrease false alerts and improve the identification rate, accuracy for intrusion detection needs to be strengthened [5]. As datasets available for ID are imbalanced in nature, so accuracy alone is not good enough to measure the performance of the system. That is why other performance evaluation parameters are also considered by some researchers in the field of IDS [4].

Two forms of alerts in IDS are false positives and false negatives. The false positive alert corresponds to regular activities which are mistakenly classified as an intrusion. Whereas, in false negative alert actual threats are reported as regular actions or events. The aggregate of these two alerts is defined as the false alert rate, and it is used for performance assessment to determine the efficacy of a specific IDS technique [6].

The frequently used parameters for validating the predictive capability of intrusion detection models are accuracy, false alert rates, recall, *f*-score, precision, TPR, and FPR.

RQ5 (Limitation of existing studies done in the area of IDS). Although a lot of machine learning frameworks are already implemented for detecting intrusions in the network systems, most of the existing studies suffer from one or more of the below given drawbacks:

- Because of the enormous amounts of data, the false warning rate of network intrusion system is high, and the accuracy of detection is decreased. This is one of the main challenges when the network is up against unexpected/unknown attacks [5]. Models developed mostly for IDS can detect the attacks that it had learned while training phase, but they cannot detect new unknown attacks efficiently. Present intrusion detection systems can detect mainly known threats. Because of the large misclassification rates of the existing methods, predicting unknown threats or zero-day threats still is still an open area for research [11].
- The datasets were gathered for IDS and are from some specific networks over a specific period and generally do not include up-to-date information. So, they do not guarantee to detect intrusions in real-time system.
- The datasets are also imbalanced and therefore cannot contain enough data for all forms of attacks. Most of the machine learning methods studied in literature did not tackled the class imbalance distribution problems.
- In majority of work, only accuracy was considered while validating the method. Whereas in IDS false positive rate in very crucial parameter. FPR of IDS has not been taken care of.
- In some of the research works, although the framework used gave good results, but it can detect only two classes, i.e., attack or not attack. It cannot handle multiclass classification (i.e., it cannot detect all classes of attacks separately).
- The standard IDS should be capable of detecting new threats in real time and respond appropriately to such attacks with the minimum response time to avoid any impact on the network. Most of the studies done in the field of IDS lack in satisfying the above requirement [12].

5 Conclusion

The intrusion detection model is used to analyze suspicious activities that happen within a network. Intrusion Detection is application or tool that checks for distrustful behavior of a network. The standard IDS should be capable of detecting new threats in real time and respond appropriately to such attacks with the minimum response time to avoid any impact on the network. IDS is helpful not only to predict successful intrusions but also to monitor attempts to breach security, that enables it for supplying valuable information for timely defensive-measures.

From literature survey, we analyzed several machine learning as well as deep learning techniques such as SVM, naïve Bayes, ANN, random forests, KNN, convolution neural networks, and more were used to implement IDS. Although the methods used provide satisfactory results in terms of predictive accuracy, there are certain issues that are not handled properly.

Because of the enormous amounts of data, the false warning rate of network intrusion system is high, and the accuracy of detection is decreased. This is one of the main challenges when the network is up against unexpected attacks [5]. The studies focused mostly on binary classification and cannot differentiate between different kind of threats. FPR was not considered while evaluating model's predictive ability [1]. The datasets available for ID are imbalanced in nature, so accuracy alone is not good enough to measure the predictive capability of the system. That is why it is important to consider other parameters such as precision, f-score and recall, and misclassification rates [4]. Present intrusion detection systems can detect mainly known threats. Because of the high misclassification rate of the existing methods, predicting unknown threats or zero-day threats still is still an open area for research [11]. The standard IDS should be capable of detecting new threats in real time and respond appropriately to such attacks with the minimum response time to avoid any impact on the network. Most of the studies done in the field of IDS lack in satisfying the above requirement [12].

The results obtained from literature survey indicate that the IDS based on machine learning frameworks have capability to efficiently detect intrusions of both known and unknown types. If certain issues mentioned above are tackled carefully, IDS using machine learning methods will provide very effective and strong techniques for securing network infrastructures against malicious attacks and will help in timely detection of security threats as well as will help administrators to take appropriate corrective measures to prevent any type of damage to the system.

References

1. Mirza, A.H.: Computer network intrusion detection using various classifiers and ensemble learning. In: 2018 26th Signal Processing and Communications Applications Conference (SIU), pp. 1–4 (2018)
2. Karatas, G.: Increasing the performance of machine learning-based IDSs on an imbalanced and up-to-date dataset. IEEE Access **8**, 32150–32162 (2020)
3. Khan, R.U., Zhang, X., Alazab, M., Kumar, R.: An improved convolutional neural network model for intrusion detection in networks. In: 2019 Cybersecurity and Cyberforensics Conference (CCC), pp. 74–77 (2019)
4. Park, K.: Classification of attack types for intrusion detection systems using a machine learning algorithm. In: 2018 IEEE Fourth International Conference on Big Data Computing Service and Applications (BigDataService), pp. 282–286 (2018). https://doi.org/10.1109/BigDataService.2018.00050
5. Halimaa, A.A., Sundarakantham, K.: Machine learning based intrusion detection system. In: 2019 3rd International Conference on Trends in Electronics and Informatics (ICOEI), Tirunelveli, pp. 916–920 (2019). https://doi.org/10.1109/ICOEI.2019.8862784
6. Rahmani, R., Chizari, M., Maralani, A., Eslami, M., Golkar, M.J., Ebrahimi, A.: A hybrid method consisting of GA and SVM for intrusion detection system. Neural Comput. Appl. (2017). https://doi.org/10.1007/s00521-015-1964-2(2015)
7. Divyatmika, Sreekesh, M.: A two-tier network based intrusion detection system architecture using machine learning approach. In: 2016 International Conference on Electrical, Electronics, and Optimization Techniques (ICEEOT), Chennai, pp. 42–47 (2016). https://doi.org/10.1109/ICEEOT.2016.7755404

8. Ali, M.H., Al Mohammed, B.A.D., Ismail, A., Zolkipli, M.F.: A new intrusion detection system based on fast learning network and particle swarm optimization. IEEE Access **6**, 20255–20261 (2018). https://doi.org/10.1109/ACCESS.2018.2820092
9. Kumar, R., Gautam, S.: An ensemble approach for intrusion detection system using machine learning algorithms. In: 2018 8th International Conference on Cloud Computing, Data Science & Engineering (Confluence), pp. 14–15 (2018)
10. Lin, W., Lin, H., Wang, P., Wu, B., Tsai, J.: Using convolutional neural networks to network intrusion detection for cyber threats. In: 2018 IEEE International Conference on Applied System Invention (ICASI), pp. 1107–1110 (2018)
11. Taher, K.A.: Network intrusion detection using supervised machine learning technique with feature selection. In: 2019 International Conference on Robotics, Electrical and Signal Processing Techniques (ICREST), pp. 643–646 (2019)
12. Al-Yaseen, W.L., Othman, Z.A., Zakree, M., Nazri, A.: PT US CR. Pattern Recognit. Lett. (2016). https://doi.org/10.1016/j.patrec.2016.11.018
13. Aamir, R., Ashfaq, R., Chen, Y.H.D., Chen, D.: Toward an efficient fuzziness based instance selection methodology for intrusion detection system. Int. J. Mach. Learn. Cybern. (2016). https://doi.org/10.1007/s13042-016-0557-4
14. Kim, J., Kim, J., Le, H., Thu, T., Kim, H.: Long short term memory recurrent neural network classifier for intrusion detection, Sept 2017, p. 5 (2016). https://doi.org/10.1109/PlatCon.2016.7456805
15. Das, S., Nene, M.J.: A survey on types of machine learning techniques in intrusion prevention systems. In: 2017 International Conference on Wireless Communications, Signal Processing and Networking (WiSPNET), Chennai, pp. 2296–2299 (2017). https://doi.org/10.1109/WiSPNET.2017.8300169
16. Niyaz, Q., Sun, W., Javaid, A.Y., Alam, M.: A deep learning approach for network intrusion detection system (2016). https://doi.org/10.4108/eai.3-12-2015.2262516
17. Xiao, Y., Xing, C., Zhang, T., Zhao, Z.: An intrusion detection model based on feature reduction and convolutional neural networks. IEEE Access **7**, 42210–42219 (2019). https://doi.org/10.1109/ACCESS.2019.2904620

An Exploration of Entropy Techniques for Envisioning Announcement Period of Open Source Software

Anjali Munde

Abstract Through the rising intricacies of the software, the quantity of probable bugs is furthermore growing promptly. These bugs hamper the prompt software improvement series. Bugs, if deferred unanswered, may initiate complications in the elongated track. Moreover, with no former information around the position and the quantity of bugs, administrators might not be competent to assign supplies in a beneficial way. In order to affect this trouble, investigators have formulated abundant bug estimation methods till now. These source encryptions practice periodic variations in order to encounter the novel characteristic introduction, characteristic improvement, and faults fix. A significant part of concern for OSS is when to announce a latest edition. In this paper, a method by assuming the quantity of faults documented in numerous announcements of Bugzilla software has been established and distinctive degrees of entropy specifically, Shannon entropy and Kapur entropy aimed at variations in several software revisions during interval periods have been computed. A simple linear regression is employed initially to forecast the faults that are still impending. By means of these anticipated faults and entropy degrees in multiple linear regression, the announcement period of the software has been forecasted. Data visualization using Python has been elucidated. The outcomes are significantly effective for the software administrators to announce the edition on that interval. The outcomes of projected versions through the prevailing in the texts are evaluated and discovered that the projected simulations are beneficial fault forecaster since they have exhibited substantial enhancement in their operations.

Keywords Entropy · Envisaging bugs · Software release · Shannon entropy · Kapur entropy · Open source software · Statistical model · Regression · Software bug repository

A. Munde (✉)
Amity University, Noida, Uttar Pradesh, India

H. Sharma et al. (eds.), *Congress on Intelligent Systems*, Advances in Intelligent Systems and Computing 1334, https://doi.org/10.1007/978-981-33-6981-8_16

1 Introduction

Software is essential in contemporary era. The advancement of software monitors distinct development simulations. Throughout the previous eras, because of the increasing progression in software functions, burden develops, escalating on the software businesses creating consistent software in a small frame of period. Lately, an exemplary shift has been taking place in software advancement because of progressions in the transmission tools that have risen to the development of open source software.

An imperative feature of open source software is when to announce a new version. The main aim is announcing the software at appropriate interval and within confined budget constraint. The strategy behind releasing software is a difficult task. Announcement scheduling performs a very vital role in handling popular announcements to the consumers. It is essentially a group of latest characteristics that are counted to a prevailing production atmosphere. The latest announcement is just after the growth procedure and additional succeeding announcements are established upon the preceding announcements, number of supplies and faults fixed. Each announcement comprises a set of characteristics fulfilling particular restrictions of the organization as needed by the consumers. Judgment concerning what characteristic to be comprised in which announcement version is a boring task. A few open source software systems announce their software version regularly by means of milestone wise, fixed time, important patch wise, etc. An extensive technique for bug estimation is employing entropy of modifications as suggested by Hassan [5–8]. Kapur et al. [10, 11] considered an ideal resource allocation project to lessen the outflow of software throughout the examination stage below the vigorous setting and retained Optimal Control Theoretic system to examine the conduct of active models in software analysis stage and operational phase. Kapur et al. [12] predicted an accurate method for determining when to end software inspection anticipating halt attention and cost as two elements simultaneously. The recommended model is originated on multi trait efficacy investigation which profits companies to originate a cautious outcome on the best system of the software. Ambros and Robbes [1, 2] suggested a standard for prediction of bugs and showed an extensive evaluation of approved bug prediction approaches. Singh and Chaturvedi [17] founded the complexity of code change and used it for predicting bugs. Singh et al. [18] determined the theoretical bugs staying stationary in the software by proposing three methods precisely software reliability growth models, hypothetical complexity of code changes established models and complexity of code changes constructed models. Chaturvedi et al. [3] developed a model to predict the theoretical modification complexity in the software and experimentally approved the model using historic code changes facts of open source software Mozilla's components. Chaturvedi et al. [4] indicated a complexity of code change model to direct the NRP utilizing multiple linear regression model and calculated the functioning using several residual statistics, goodness of fit curve and R2.

Kaur et al. [13] recommended the metrics attained by course of entropy of alterations to calculate five machine learning systems directed at predicting faults. Kaur and Chopra [14] proposes Entropy Churn Metrics established on History Complexity Metrics and Churn of Source Code Metric. The exploration assessed the software subsystems on three constraints to determine which properties of software techniques create HCM or ECM additionally chosen above others. Kumari et al. [15] proposed fault dependency grounded arithmetical simulations by examining the instantaneous description of faults and remarks projected through operators in conditions of the entropy-grounded degrees.

In this paper, the faults existing in foremost announcements of bugzilla project are taken into attention with certain information preprocessing. Then distinct measures of entropy specifically, Shannon entropy [16] and Kapur entropy [9] are computed for the variations in numerous software renews from 2011 to 2019. A technique is suggested with the support of linear regression to conclude the announcement time of the bugzilla open source software by means of entropy measures and faults detected. Regression analysis has been employed in two stages. In the first phase, simple linear regression has been employed among the entropy computed and experimental faults for each announcement through every interval to attain the regression coefficients. The regression coefficients are consumed verifying the forecasted faults for each announcement in each time period. In the second stage, multiple linear regression is utilized to forecast subsequent announcement time of the software product for each time period using the entropy measure and the faults envisaged in stage one.

1.1 Information Measures

Shannon [16] initiated the notion of entropy in communication theory and originated the topic of information theory. The stochastic technique has a significant property identified as entropy which is extensively tapped in numerous areas.

The Shannon entropy, H_{n} is stated as

$$H(P) = -\sum_{s=1}^{n} p_s \log p_s \tag{1}$$

where $p_s \geq 0$ and $\sum_{s=1}^{n} p_s = 1$.

Consistent algebraic concepts from information theory as Kapur entropy (α–β entropy) and Shannon entropy have been developed. These entropies are recognized as basic entropies and because of the presence of limitations, they are used for association functions in miscellaneous subjects for instance coding theory, statistical decision theory, etc. There are several measures of entropy though in this study, and only Shannon entropy and Kapur entropy have been assumed to study conclusions initiated on these entropies. For advance enquiry, supplementary measures of entropy may be developed into observation and the review and comparative assessment is new topic of investigation.

Kapur articulated a systematic line to attain a simplification of Shannon entropy. He represented entropy of order α and degree β described as follows:

$$H_{\alpha,\beta}(P) = \frac{1}{1-\alpha}\log\left(\sum_{i=1}^{n} p_i^{\alpha+\beta-1} \Big/ \sum_{i=1}^{n} p_i^{\beta}\right),$$

$$\alpha \neq 1, \alpha > 0, \beta > 0, \alpha + \beta - 1 > 0 \quad (2)$$

2 Predicting Bugs and Estimating Release Time of Software

A project is selected and then sub-systems are chosen. The CVS logs are browsed and the bug reports of all sub-systems are accumulated. The bugs are obtained through the descriptions and organized on annual base aimed at every sub-system. Additionally, Shannon entropy and Kapur entropy aimed at every interval utilizing these bugs registered in favor of every sub-system is computed. The bugs are envisaged by applying simple linear regression for upcoming year grounded on the entropy analyzed for every existing interval.

This statistics facts has been employed to compute the probability of every constituent for the 9 time periods from 2011 to August 2019. By means of these probabilities, Shannon entropy is estimated correspondingly for every interval. Likewise, Kapur entropy can be determined. Table 1 exhibited under represents the Shannon entropy.

The Kapur entropy for quantities of $\alpha = 0.1$ and $\beta = 1.0$ has been reviewed in Table 2 beneath.

As of this investigation, it has been examined that Shannon entropy is highest in the year 2017 and lowest in the year 2012. Kapur entropy is highest in the year 2017 and lowest in the year 2012.

Table 1 Shannon entropy aimed at every time period

Year	Shannon entropy
2011	0.2441
2012	0.1955
2013	0.3684
2014	0.4320
2015	0.4579
2016	0.3727
2017	0.6041
2018	0.5533
2019	0.5717

Table 2 Kapur entropy aimed at every time period

Year	Kapur entropy
2011	0.2947
2012	0.2873
2013	0.4602
2014	0.4675
2015	0.4638
2016	0.4637
2017	0.6875
2018	0.5961
2019	0.5986

After Shannon entropy and Kapur entropy have been computed for numerous announcements of bugzilla software in every time period and then by utilizing linear regression procedures, the announcement time of the software is computed. Regression analysis is an arithmetical procedure which aids in examining the associations between the variables. Simple linear regression model comprises two variables in which one variable is forecasted by the other variable. It is the utmost fundamental prototype which is normally employed for extrapolation. Multiple linear regression is a frequently utilized arithmetical data investigation method, in which two or more independent variables are utilized to envisage the value of the dependent variable. In this paper, regression analysis has been implemented in two phases. In first phase, simple linear regression method has been used among the entropy measure and experimental faults existing in every announcement in every time period to envisage faults which are still to arrive in forthcoming announcement in every time period. In second phase, multiple linear regression method has been utilized among entropy measure, experimental announcement time, and the faults forecasted in first stage to envisage the announcement time of the software for every announcement.

Therefore, the subsequent linear regression model is suggested:

$$Y = a + bX_1 \tag{3}$$

$$Y_1 = c + dX_1 + eY \tag{4}$$

where *a*, *b*, *c*, *d*, and *e* are regression coefficients.

In the first phase, the simple linear regression has been employed for the exogeneous variable X_1 and endogeneous variable Y with Eq. (3) to acquire the regression coefficients a and b. These regression coefficients are then employed to attain the forecasted number of faults, i.e., Y for every announcement in every time period. In the second phase, we consider X_1 and Y as independent variables and Y_1 as dependent variable. Therefore, multiple linear regression in this phase has been utilized for the independent variables X_1 and Y and dependent variable Y_1 utilized Eq. (4) to attain the regression coefficients *c*, *d*, and *e*. Once the assessment of these regression coefficients are anticipated, we can forecast the time of every announcement. Here,

X_1, i.e., entropy is unique in every situation as two entropy measures specifically, Shannon entropy and Kapur entropy are derived to forecast the announcement time of every announcement for every time period.

The subsequent record represents the envisaged bugs and the predicted release time of Software using the Shannon entropy measure and data visualization using Python (Fig. 1 and Table 3).

In Fig. 2, we have elucidated the relationship between the observed and predicted release time over a period of time from 2011 to 2019 using Shannon entropy.

The subsequent record represents the envisaged bugs and the predicted release time of software using Kapur entropy measure by way of discrete estimates of α and β restriction and data visualization using Python (Fig. 3 and Table 4).

In Fig. 4, we have elucidated the relationship between the observed and predicted release time over a period of time from 2011 to 2019 using Shannon entropy.

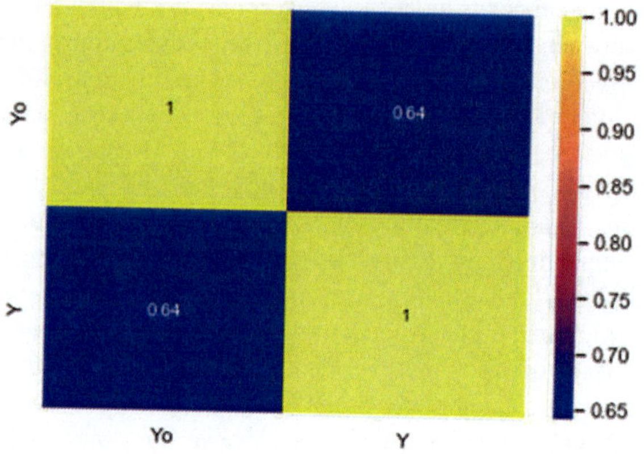

Fig. 1 Relationship between observed and predicted bugs over a period of time using Shannon entropy

Table 3 Shannon entropy aimed at every time period

Year	X_1	Y_o	Y	Observed release time	Y_1
2011	0.2441	4	7.0866	7	10.8923
2012	0.1955	12	5.5639	13	10.8405
2013	0.3684	9	10.9812	11	11.0247
2014	0.432	9	12.9740	11	11.0925
2015	0.4579	5	13.7855	12	11.1201
2016	0.3727	13	11.1160	14	11.0293
2017	0.6041	16	18.3663	5	11.2758
2018	0.5533	22	16.7746	21	11.2217
2019	0.5717	24	17.3511	15	11.2413

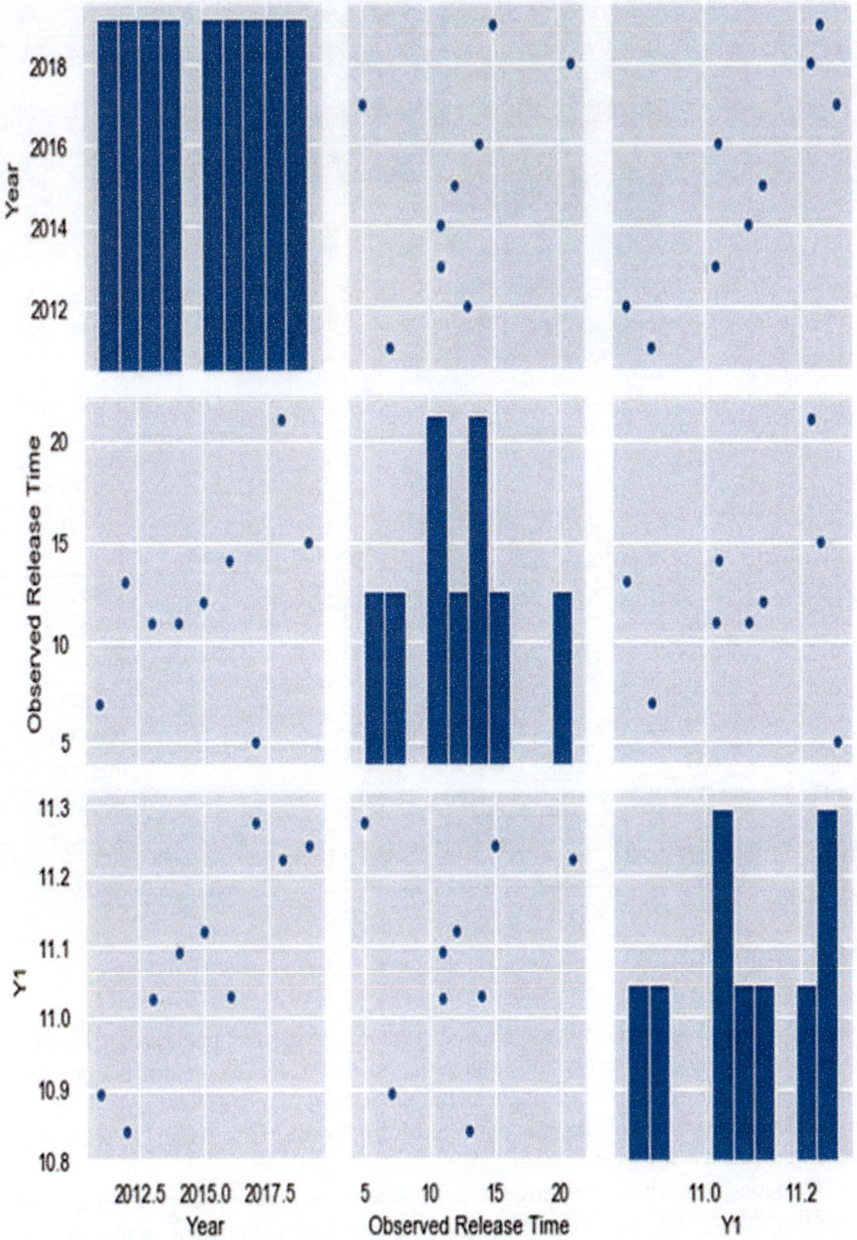

Fig. 2 Relationship between year, observed and predicted release time over a period of time using Shannon entropy

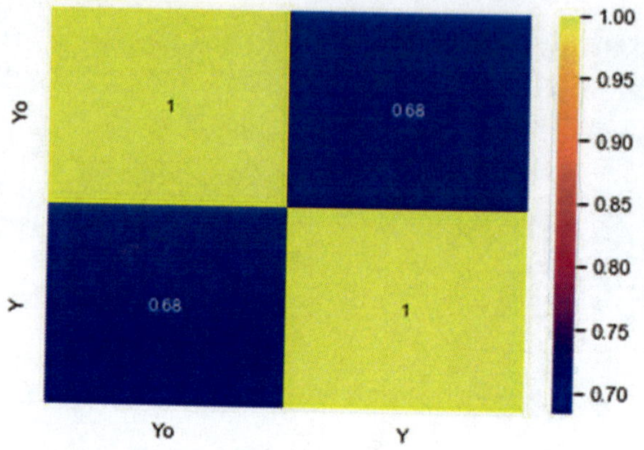

Fig. 3 Relationship between observed and predicted bugs over a period of time using Kapur entropy

Table 4 Kapur entropy aimed at every time period

Year	X_1	Y_o	Y	Observed release time	Y_1
2011	0.2947	4	6.0789	7	11.9290
2012	0.2873	12	5.8157	13	11.9909
2013	0.4602	9	11.9648	11	10.5434
2014	0.4675	9	12.2244	11	10.4823
2015	0.4638	5	12.0929	12	10.5133
2016	0.4637	13	12.0893	14	10.5141
2017	0.6875	16	20.0487	5	8.6405
2018	0.5961	22	16.7981	21	9.4057
2019	0.5986	24	16.8870	15	9.3847

3 Conclusion

In this paper, a methodology is established to determine the announcement time of the software established on faults static and variations in the data. A method is suggested to establish the forecasted time of every announcement of the OSS-bugzilla utilizing regression analysis. The data was collected from the Bugzilla Web site, www.Bugzilla.org, for each software release.

The paper comprised selecting the bugs present in foremost announcements of bugzilla project with certain data preprocessing. Various measures of entropy specifically, Shannon entropy, and Kapur entropy for the alterations in numerous software updates have been analyzed. In the first stage, simple linear regression is related amongst the entropy computed and detected bugs for all announcement through each period to acquire the regression coefficients. Then, these regression coefficients are operated to ascertain the anticipated bugs for each announcement in each

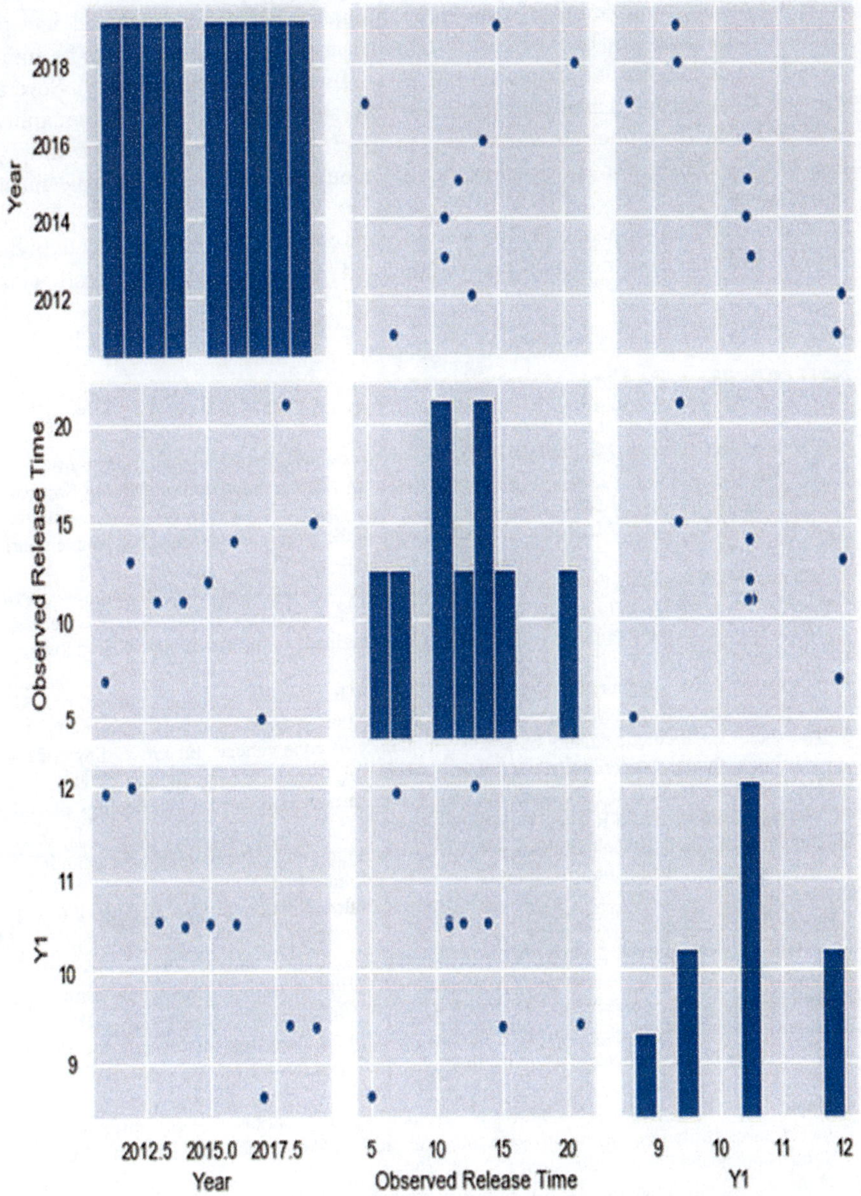

Fig. 4 Relationship between year, observed and predicted release time over a period of time using Kapur entropy

time period. In multiple linear regression, amount of faults envisaged and entropy computed are measured as exogeneous variables and duration in months to be anticipated as a endogeneous variable for bugzilla software (www.bugzilla.org). Several measures of entropy, namely Shannon entropy and Kapur entropy are computed independently for numerous announcements of Bugzilla software and one entropy at a time is chosen for anticipation of the announcement time in multiple linear regressions.

In this study only Shannon entropy and Kapur entropy have been engaged. In the future, another entropy measures may be pondered with separate constraint estimates.

References

1. Ambros, M.D., Lanza, M., Robbes, R.: An extensive comparison of bug prediction approaches. In: MSR'10: Proceedings of the 7th International Working Conference on Mining Software Repositories, pp. 31–41 (2010)
2. Ambros, M.D., Lanza, M., Robbes, R.: Evaluating defect prediction approaches: a benchmark and an extensive comparison. Empir. Softw. Eng. **17**(4–5), 571–737 (2012)
3. Chaturvedi, K.K., Bedi, P., Mishra, S., Singh, V.B.: An empirical validation of the complexity of code changes and bugs in predicting the release time of open source software. In: Proceedings of the IEEE 16th International Conference on Computational Science and Engineering. Sydney, pp. 1201–1206 (2013)
4. Chaturvedi, K.K., Kapur, P.K., Anand, S., Singh, V.B.: Predicting the complexity of code changes using entropy based measures. Int. J. Syst. Assur. Eng. Manag. **5**(2), 155–164 (2014)
5. Hassan, A.E.: Predicting faults based on complexity of code change. In: Proceedings of the 31st International Conference on Software Engineering, Vancouver, pp. 78–88 (2009)
6. Hassan, A.E., Holt, R.C.: Studying the chaos in code development. In: Proceedings of 10th Working Conference on Reverse Engineering (2003)
7. Hassan, A.E., Holt, R.C.: The chaos of software development. In: Proceedings of the 6th IEEE International Workshop on Principles of Software Evolution (2003)
8. Hassan, A.E., Holt, R.C.: The top ten list: dynamic fault prediction. In: Proceedings of ICSM, pp. 263–272 (2005)
9. Kapur, J.K.: Generalized Entropy of Order α and β, The Maths Semi, pp. 79–84 (1967)
10. Kapur, P.K., Chanda, U., Kumar, V.: Dynamic allocation of testing effort when testing and debugging are done concurrently. Commun. Depend. Qual. Manag. **13**(3), 14–28 (2010)
11. Kapur, P.K., Pham, H., Chanda, U., Kumar, V.: Optimal allocation of testing effort during testing and debugging phases: a control theoretic approach. Int. J. Syst. Sci. **44**(9), 1639–1650 (2013)
12. Kapur, P.K., Singh, J.N.P., Sachdeva, N., Kumar, V.: Application of multi attribute utility theory in multiple releases of software. In: International Conference on Reliability, Infocom Technologies and Optimization, pp. 123–132 (2013)
13. Kaur, A., Kaur, K., Chopra, D.: An empirical study of software entropy based bug prediction using machine learning. Int. J. Syst. Assur. Eng. Manag. 599–616 (2017)
14. Kaur, A., Chopra, D.: Entropy churn metrics for fault prediction in software systems. Entropy (2018)
15. Kumari, M., Misra, A., Misra, S., Sanz, L., Damasevicius, R., Singh, V.: Quantity quality evaluation of software products by considering summary and comments entropy of a reported bug. Entropy (2019)
16. Shannon, C.E.: A mathematical theory of communication. Bell Syst. Tech. J. **27**(379–423), 623–656 (1948)

17. Singh, V.B., Chaturvedi, K.K.: Improving the quality of software by quantifying the code change metric and predicting the bugs. In: Murgante, B., et al. (eds.) ICCSA 2013, Part II. LNCS, vol. 7972, pp. 408–426. Springer-Verlag, Berlin, Heidelberg (2013)
18. Singh, V.B., Chaturvedi, K.K., Khatri, S.K., Kumar, V.: Bug prediction modelling using complexity of code changes. Int. J. Syst. Assur. Eng. Manag. **6**(1), 44–60 (2014)

Occurrence Prediction of Pests and Diseases in Rice of Weather Factors Using Machine Learning

Sachit Dubey, Raju Barskar, Anjna Jayant Deen, Nepal Barskar, and Gulfishan Firdose Ahmed

Abstract Rice is one of the major cash crops in India and has been eaten in every part of the Indian subcontinent in every shape or form. In terms of production, India is one of the top producers of rice in the world along with China. Every year farmers lose a large amount of there produce to pest and disease infection. Often the change in climatic conditions favors the development of pest and disease. In this research paper, we discuss the possibility of using machine learning techniques to identify the climatic conditions which are favorable to pest and disease associated with rice. We also propose to develop a model that will identify whether a given weather condition will support the occurrence of pest and disease.

Keywords Rice · Pest · Machine learning · K-NN · Decision tree · Random forest · SVM · Logistic regression

1 Introduction

Rice is the most popular part of Indian cuisine and is consumed at least once in a day every region. In terms of rice production, India is the largest global rice producer in terms of area harvested. While some of the rice produced is exported globally most of it is used to meet the local demand. In the year 2019, the rice yield was estimated to be around 2.6 thousand kilograms per hector; over the years, there has been an increase

S. Dubey (✉) · R. Barskar · A. J. Deen
Department of Computer Science and Engineering, University Institute of Technology, RGPV, Bhopal, M.P., India

R. Barskar
e-mail: rajubaraskar@rgtu.net

N. Barskar
Department of Computer Science and Engineering, Institution of Engineering and Technology, DAVV, Indore, M.P., India

G. F. Ahmed
College of Agriculture, JNKVV, Powarkheda, Hoshangabad, M.P., India
e-mail: dr.gfahmed17@jnkvv.org

H. Sharma et al. (eds.), *Congress on Intelligent Systems*, Advances in Intelligent Systems and Computing 1334, https://doi.org/10.1007/978-981-33-6981-8_17

in overall yield figures to compete with China [1]. With the increase in population, the annual rice consumption has also increased which is reflected in the county's production and consumption volumes. Every year a significant amount of rice yield is lost to various natural causes one of them being the infestation of pest and disease. Every year farmers may lose an estimated 30% of there produce to pest attacks and disease infestation [2]. Like other plant life, the occurrence of pest and disease in rice largely depends upon the changing weather parameters like temperature, humidity, rainfall, and wind speed. Over the years, various attempts have been made to exploit the relationship between the pest occurrence and climatic pattern to forecast the probability of there occurrence in advance. A well-developed forecasting model for pest infestation not only helps the farmers to take the necessary measure to safeguard their crops but also helps them to choose appropriate pesticides in the right amount to reduce the pollution level caused by uncontrolled use of non-organic pesticide.

The forecasting system makes use of various parameters to predict the outbreak of disease or pest [3]. One of the first attempts to forecast and predict the occurrence of disease was done for Stewart wilt in wheat using the winter temperature index. Netherlands developed a multi-disease/pest for winter wheat known as EPIdemiology, PREdiction, and PREvention (EPIPRE) which used the relationship between host and pathogen to predict the outbreak of disease [4]. Various statistical models have been used to forecast the pest occurrence most popular being the multiple linear regression model. In this research article, we make use of an emerging machine learning technique to predict whether a given weather condition supports the occurrence of pest and disease or not.

Machine learning has been an emerging area in both statistical and computational research various machine learning algorithms like K-NN, decision tree, and random forest provided better efficiency than traditional regression model. Performance of various machine learning algorithms against past weather and pest data is taken from the crop DSS system crated by ICAR is reviewed. Machine learning has not been traditionally used in the past for developing the forecasting models but in recent years, various machine learning technique has been used in this field which has been discussed below.

2 Related Work

In 2006, Jain et al. [5] used machine learning technique like logistic regression and decision tree to predict the presence of powdery mildew in mango they made use of past weather and disease incidence data to develop and test the model in which RJP algorithm gave the best performance by providing an accuracy of 84%.

Ramesh and Vydeki [6] used a machine learning technique to find the presence of blast disease in rice leaves with the help of image segmentation and processing they made use artificial neural network (ANN) and K-NN for identifying the presence of blast disease in rice leaves using image segmentation and processing. They compared the performance of both K-NN and ANN to find the best suiting algorithm for identification. ANN achieved a maximum accuracy of 99–100%.

Sahith and Vijaya [7] used the machine learning technique to identify the presence of three rice diseases such as bacterial blight, brown spot, and leaf smut. They collected 120 images of diseased plants and used image processing and machine learning technique to build a classification model. They compared the performance of various decision tree algorithms like random forest, REP tree, and J48. The result showed that random forest proved to be superior providing maximum accuracy of 76.19%.

Skawsang et al. [8] used a machine learning technique for predicting the population density of rice pests. They used past weather data from field trials carried out for brown planthopper using a light trap experiment. The authors made use of an artificial neural network to build a regression model for predicting the population density of brown planthopper during the cropping season. The result showed that the artificial neural network performed better as compared to old traditional algorithms by providing $R2 = 0.77$ and RMSI = 1.68.

Kaundal et al. [9] used artificial neural network for forecasting the occurrence of leaf blast in rice using past weather data. For forecasting, the occurrence of leaf blast in rice six significant weather variables was identified, namely temperature, humidity, rainfall, and wind speed. The result showed that the support vector machine or SVM generated minimum mean absolute error (%MAE) of 44.12% compared to other alternatives.

Jia-You and Hsieh used artificial neural network (ANN) for predicting the presence of rice blast disease using historic weather data from 2014 to 2018 of several districts in China. The result provided an accuracy of 89% in identifying the presence of rice blast in a given district.

Studied the performance of long short-term memory (LSTM) recurrent neural network [10] to provide early forecasting of blast disease in rice. Historic weather data from three locations in China, namely Chelwon, Icheon, and Milyang, were used. The model showed a maximum accuracy of 62% for Cheown 61.5 for Iceolwon and 46.9% for Milyang.

3 Proposed Work

The workflow of the whole project has been described below. First, an appropriate dataset of past weather and pest is selected for predictive modeling. The data is then preprocessed before being fed to the development the preprocessing involves various steps like data cleaning, normalization, and standardization of data followed by resampling of data to balance out the imbalance dataset, and finally, the data is restructured and fed in for the analysis. The preprocessed dataset is then divided into training and testing data 27% of data is taken for testing while another part is used for training various machine learning models. The result then is evaluated using various criteria like accuracy, $f1$-score, and precision to chose the best possible algorithm.

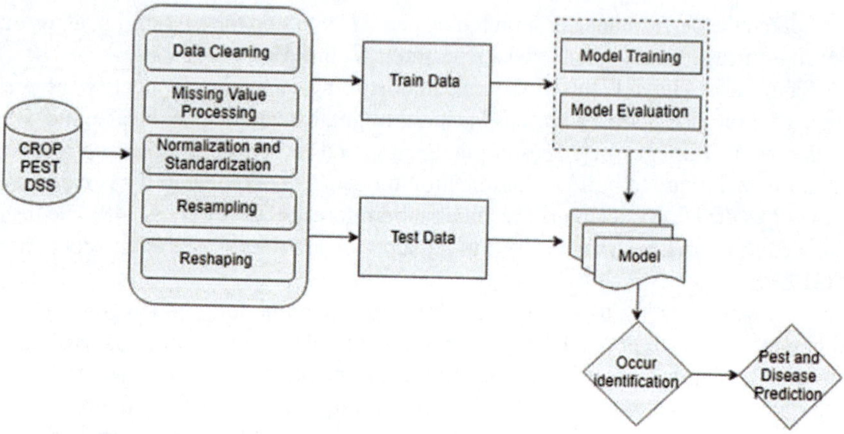

Fig. 1 Overview of analysis and prediction of rice pests and diseases

The detailed overview of each step is shown in Fig. 1 and is elaborated in detail below.

3.1 Data Collection

The past weather and pest data have been taken from the CROP PEST-DSS database (https://www.crida.in:8080/naip/AccessData.jsp) which has been established by ICAR and CIDR. It contains 34,472 weekly pest records for 11 insects and disease in rice and 13 insects and disease in cotton surveyed in 12 important research locations across India. The oldest record dates back to 1975 for yellowsteamborer in rice.

3.2 Data Preprocessing

The raw data taken from the given database is transformed along various stages of preprocessing. A detailed description of all the steps involved in transforming data before feeding it to the machine learning algorithms for analysis.

i. *Data Cleaning*: Data cleaning is the process of getting rid of unwanted data observation which is not relevant to the concerned study. This involves dealing with missing values removing outlier observation. It is an essential step in preprocessing of data as an uncleaned dataset often reduces the performance as well as generate various unwanted anomalies.
ii. *Missing Value Processing*: The dataset contains a significant amount of missing value that should be dealt with properly. Ether the missing is assigned significant

values or is completely ignored from the scope of Consideration. Often this depends on the dataset some time other values are used to predict or estimate real value.

iii. *Normalization and Standardization*: Normalization and standardization processes are necessary to put all the concerned parameters within the same scale of measurement. Normalization and standardization improve the efficiency of algorithms as well as put different values in different columns within the same scale. Normalization is not always necessary when all the feature values are bounded by the same scale of measurement.

iv. *Resampling*: Resampling is a process of balancing the dataset. Sometimes the dataset is unbalanced; i.e., there is a mismatch of record frequency of different classes which may lead to the development of a biased classification model that may favor the identification of the classes with maximum frequency. There are various methods of resampling data some of them are described below:

  (a) *Upsampling*: The process of increasing the frequency of the minority class to that of the majority class to balance out the distribution of different classes within a dataset is known as upsampling.
  (b) *Downsampling*: The process of decreasing the frequency of the majority class to that of the majority class to balance out the distribution of different classes within a dataset is known as downsampling.

v. *Restructuring*: The dataset is then compiled after resampling to completer the last step of the preprocessing step before feeding the dataset for predictive modeling.

3.3 Modeling of Data for Testing and Training Purpose

The dataset is then split into training and testing set for predictive modeling. 30% of data is used for testing while the other 70% of it is used for training. The dataset is then evaluated on various parameters that have been described below.

3.4 Machine Learning

Machine learning is the ability of the system to integrate knowledge using large scale observation to improve upon and extend systems knowledge rather than hard programming it to perform a specific task. Machine learning makes use of computational methods to transform empirical data into a usable model. Machine learning can be defined as "A study of making machines acquire new knowledge, new skills, and recognize current knowledge" [11]. Machine learning techniques that have been discussed in this study which are described below.

i. *K-Nearest Neighbors (K-NN)*
K-nearest neighbors are supervised techniques that make use of a concept that similar things tend to be near one another. It makes use of proximity of a data from an unlabeled dataset to categorize and classify a given data point [12]. The K-nearest neighbors make use of various mathematical techniques to evaluate the proximity between data point s like Euclidean distance and Gaussian distance [13]. This technique assigns each unlabeled data to the class containing the data point having the closest proximity to that point [14].

ii. *Decision Tree*
Decision tree is a supervised learning algorithm that makes use of a flowchart like structure in which the internal nodes represent the test on attribute each branch represents the outcome of the test and each leaf represents the label associated with the classification class the topmost node of a decision tree is called as root. Decision tree algorithm makes use of a cost function to perform recursive partition of a feature set until further splitting may not add to the value of prediction [15].

iii. *Random Forest*
It is a supervised learning algorithm which makes use of various decision tree to reach on a common conclusion random forest makes use of a process known as voting to classify an unlabeled data item. It takes into account various labels generated by individual trees within a forest and select the label with maximum frequency. Random forest provides more accuracy than a decision tree and is more effective in a case where a small change in training data can lead to a large variation in results in a decision tree [16].

iv. *Support Vector Machine (SVM)*
Support vector machine or SVM is a supervised learning algorithm that makes use of boundary points that separate class know as support vectors for classification of unlabeled data. The main aim of SVM is to find an optimal data plane that divides two classes of data points. SVM provides better accuracy in cases where the dimensional space is high. It also is memory efficient as it makes use of a subset of training points for classification [17].

v. *Logistic Regression*
It is one of the important statistical methods which is used machine learning directly. It makes use of the equation to develop using past data values to predict an outcome for a binary variable, y for one or more response variable x. However, in the case of logistic regression, the response variable can be categorical or continuous. The logistic regression equation calculates group membership for a particular class and then assigns unlabeled data accordingly [18].

4 Results

There are about 16,584 records of about 10 pests and diseases associated with rice. Out of all these records, top size dataset is chosen. These dataset consists of various time series records which are associated with yellowsteamborer, Greenleafhopper, and Gallmidge. Figure 2 shows the composition of records associated with various pests and diseases provided by the crop Pest-DSS.

After selecting the dataset for analysis, each dataset is divided into the training set and test set. Where first their quarter of the dataset is considered for training and rest is used as a test set. After dividing the dataset into training and test subset, performance of each dataset is evaluated against various machine learning algorithm, namely K-nearest neighbors (K-NN), logistic regression, decision tree, random forest, and support vector machine (SVM). Against selected criteria, namely accuracy, $F1$-score, and area under the curve (AUC). All the generated results are provided in Table 1.

As seen from results generated from the table, performance of random forest proved to be superior for all three datasets.

In the case of yellowsteamborer, random forest provides an overall accuracy of 92% with an $F1$-score of about 0.92 and an AUC of about 0.97 which proved to be superior when compared with those of SVM ({Accuracy: 86%, $F1$-score: 0.86, AUC: 0.92}), Decision Tree ({Accuracy: 88%, $F1$-score: 0.88, AUC: 0.88}), K-NN ({Accuracy: 86%, $F1$-score: 0.86, AUC: 0.92}), and Logistic Regression ({Accuracy: 68%, $F1$-score: 0.68, AUC: 0.75}).

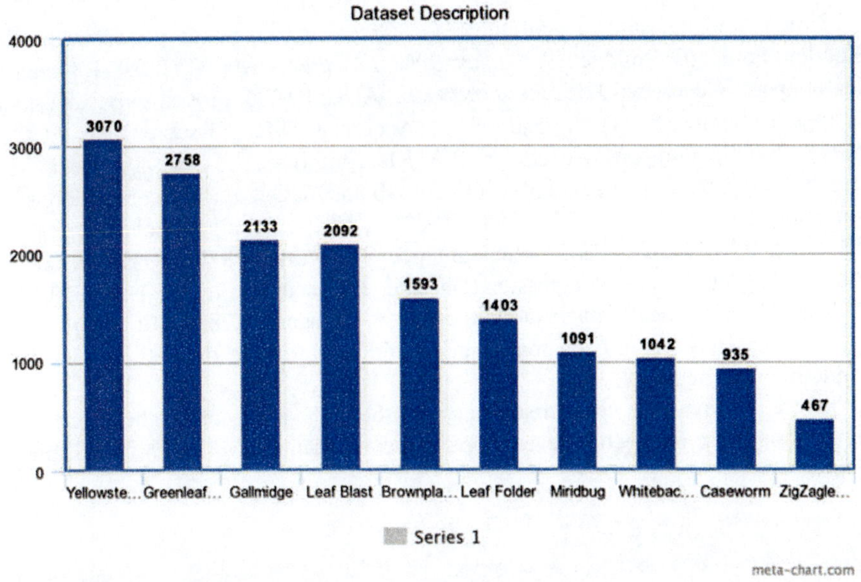

Fig. 2 Dataset description

Table 1 Result obtained after implementing various machine learning algorithms against each dataset

Algorithm	Pest	Accuracy	$F1$-score	Area under curve (AUC)
K-nearest neighbor (K-NN)	Yellowsteamborer	0.86	0.86	0.92
	Gallmidge	0.87	0.87	0.92
	Greenleafhopper	0.87	0.87	0.93
Logistic Regression	Yellowsteamborer	0.68	0.68	0.75
	Gallmidge	0.72	0.72	0.77
	Greenleafhopper	0.78	0.78	0.85
Decision Tree	Yellowsteamborer	0.88	0.88	0.88
	Gallmidge	0.85	0.85	0.85
	Greenleafhopper	0.91	0.91	0.91
Random Forest	**Yellowsteamborer**	**0.92**	**0.92**	**0.97**
	Gallmidge	**0.90**	**0.90**	**0.97**
	Greenleafhopper	**0.92**	**0.92**	**0.97**
Support Vector Machine (SVM)	Yellowsteamborer	0.86	0.86	0.92
	Gallmidge	0.89	0.89	0.94
	Greenleafhopper	0.88	0.88	0.95

Source: Bold highlights the results generated by our experiment in which performance of Random Forest Classifier came out to superior among all the algorithms

Similarly in the case of Gallmidge and Greenleafhopper, the performance of Random Forest (Gallmidge: {Accuracy: 90%, $F1$-score: 0.90, AUC: 0.97}, Greenleafhopper: {Accuracy: 92%, $F1$-score: 0.92, AUC: 0.97}), proved to be superior to those generated by SVM (Gallmidge: {Accuracy: 89%, $F1$-score: 0.89, AUC: 0.94}, Greenleafhopper: {Accuracy: 88%, $F1$-score: 0.88, AUC: 0.95}), Decision Tree (Gallmidge: {Accuracy: 85%, $F1$-score: 0.85, AUC: 0.85}, Greenleafhopper: {Accuracy: 91%, $F1$-score: 0.91, AUC: 0.91}), K-NN (Gallmidge: {Accuracy: 87%, $F1$-score: 0.87, AUC: 0.92}, Greenleafhopper: {Accuracy: 87%, $F1$-score: 0.87, AUC: 0.93}) and Logistic Regression (Gallmidge: {Accuracy: 72%, $F1$-score: 0.72, AUC: 0.77}, Greenleafhopper: {Accuracy: 78%, $F1$-score: 0.78, AUC: 0.85}).

Figure 3 shows confusion matrix generated for all three datasets against the random forest method.

Figure 4 provides a comparison between the AUC value of all the algorithm by plotting there Respective reverse operations characteristics (ROC) for all three dataset.

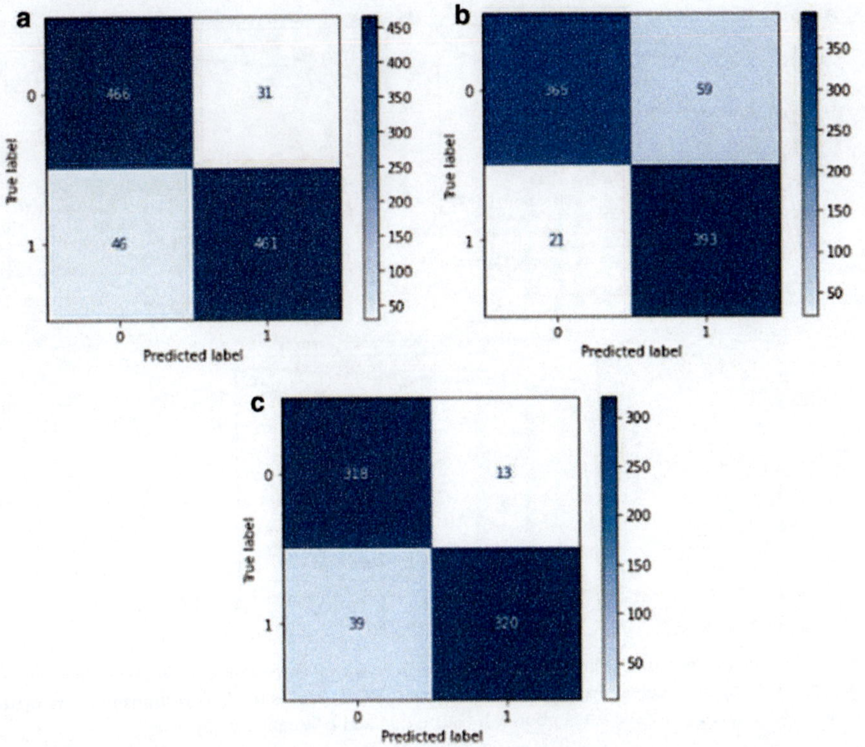

Fig. 3 Confusion matrix on the three kinds of the dataset with random forest. Subfigures **a**, **b**, and **c** show the confusion matrix of rice pests and disease occurrence of Yellowsteamborer, Gallmidge, and Greenleafhopper, respectively. Here, the blue bar representative model predicts the correct number of samples. **a** Yellowsteamborer, **b** Gallmidge, and **c** Greenleafhopper

5 Conclusion

In this paper, we discussed the analyzed performance of various machine learning algorithms in predicting the occurrence of major pests and diseases of rice. The scope of this study covered the implementation of various supervised learning algorithms like K-NN, logistic regression, decision tree, random forest, and SVM. All the algorithms were tested against datasets of three major rice pest, namely Yellowsteamborer Gallmidge and Greenleafhopper. The result showed that the performance of the random forest algorithm proved to be superior as compared to other supervised learning algorithms for all three datasets moderate results were obtained from K-NN, SVM, and decision tree while the results generated by logistic regression turn out to be least favorable.

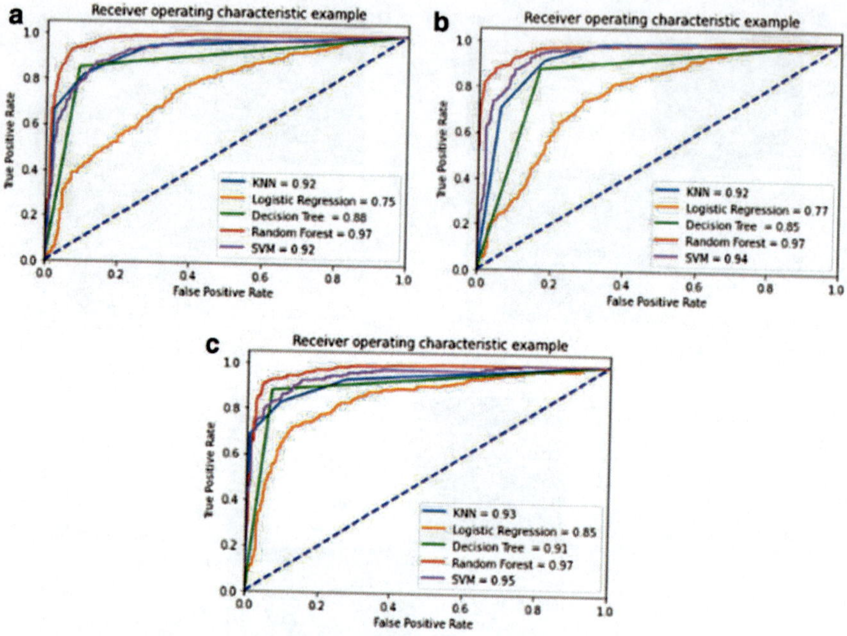

Fig. 4 Receiver operation characteristics (ROC) curve for all three kinds of datasets. Subfigures **a**, **b**, and **c** show the receiver operation characteristics (ROC) curves of all algorithms implemented against each dataset. **a** Yellowsteamborer, **b** Gallmidge and **c** Greenleafhopper

6 Future Scope

In this study, only one major set of factors, namely weather parameters were studied. More research work can be done in building a more cohesive dataset that takes into account the life cycle of pests and pathogens associated with diseases as well as genetic markup of the host involved. This will provide a better outlook and framework for predicting the hazard level of pests and diseases. A pest intruder detection system can also be developed using cloud computing [19].

References

1. Jaganmohan, M.: Annual yield of rice India FY 1991–2019 (2020). Retrieved from https://www.statista.com/statistics/764299/india-yield-of-rice/
2. Dhaliwal, G.S., Jindal, V., Mohindru, B.: Crop losses due to insect pests: global and Indian scenario. Indian J. Entomol. **77**(2), 165–168 (2015)
3. Munir, M. (ed.): Plant disease epidemiology: disease triangle and forecasting mechanisms in highlights. Hosts Virus. **5**(1), 7–11 (2018)
4. Reinink, K.: Experimental verification and development of EPIPRE, a supervised disease, and pest management system for wheat. Neth. J. Plant Pathol. **92**(1), 3–14 (1986)

5. Jain, R., Minz, S., Ramasubramanian, V.: Machine learning for forewarning crop diseases. J. Ind. Soc. Agric. Stat. **63**(1), 97–107 (2009)
6. Ramesh, S., Vydeki, D.: Application of machine learning in detection of blast disease in South Indian rice crops. J. Phytol. 31–37 (2019)
7. Sahith, R., Vijaya Pal Reddy, P., Nimmala, S.: Decision tree-based machine learning algorithms to classify rice plant diseases. Int. J. Innov. Technol. Explor. Eng. (IJITEE) **9**(1), 5365–5368 (2019)
8. Skawsang, S., Nagai, M., Tripathi, N.K., Soni, P.: Predicting rice pest population occurrence with satellite-derived crop phenology, ground meteorological observation, and machine learning: a case study for the Central Plain of Thailand. Appl. Sci. **9**(22), 4846 (2019)
9. Kaundal, R., Kapoor, A.S., Raghava, G.P.: Machine learning techniques in disease forecasting: a case study on rice blast prediction. BMC Bioinform. **7**(1), 485 (2006)
10. Kim, Y., Roh, J.H., Kim, H.: Early forecasting of rice blast disease using long short-term memory recurrent neural networks. Sustainability **10**, 34 (2017)
11. Woolf, B.P.: Machine learning. In: Building Intelligent Interactive Tutors: Student-Centered Strategies for Revolutionizing E-Learning, pp. 221–297. Morgan Kaufmann/Elsevier, Burlington (2009). https://doi.org/10.1016/B978-0-12-373594-2.00007-1
12. Talabis, M.R.: Information Security Analytics, pp. 1–12. Syngress is an Imprint of Elsevier, Waltham (2015)
13. McCue, C.: Identification, characterization, and modeling, Chap. 7. In: Data Mining and Predictive Analysis, 2nd edn., pp. 137–155. Butterworth-Heinemann, Amsterdam (2015)
14. Keller, J.M., Gray, M.R., Givens, J.A.: A fuzzy k-nearest neighbor algorithm. IEEE Trans. Syst. Man Cybern. **4**, 580–585 (1985)
15. Wright, R.E.: Logistic regression (1995)
16. Friedl, M.A., Brodley, C.E.: Decision tree classification of land cover from remotely sensed data. Remote Sens. Environ. **61**(3), 399–409 (1997)
17. Breiman, L.: Random forests. Mach. Learn. **45**(1), 5–32 (2001)
18. Suykens, J.A., Vandewalle, J.: Least squares support vector machine classifiers. Neural Process. Lett. **9**(3), 293–300 (1999)
19. Jaiswal, S., Ahirwar, M., Baraskar, R.: Intruder notification system & security in cloud computing: a review (2017)

A Literature Review on Generative Adversarial Networks with Its Applications in Healthcare

Viraat Saaran, Vaishali Kushwaha, Sachi Gupta, and Gaurav Agarwal

Abstract Since the discovery of neural networks, there is a trend of training the data (structured or unstructured) and predicting the outcome based on these training sets. Whereas these new neural networks, Generative Adversarial Networks or GANs are achieving popularity because of its capability to produce new data instead of classifying them. It operates on two neural networks fighting against one another, that would be able to co-train through plain old backpropagation. Its adaptive learning allows it to generate new data without replicating the previous outcome. Due to these advantages, GANs have opened a wide range of applications in the medical and healthcare sector. Essential applications like image segmentation, image-to-image translation, style transfer, and classification are getting more recognized. Given the growing trend of GANs in the medical community, we present an overview of Generative Adversarial Networks with its potential applications in the medical field. The main objective of this briefing is to investigate and provide a descriptive view of GANs and its applications in the healthcare sector. This paper also makes an effort in identifying GANs' advantages and disadvantages. Finally, we conclude the paper with future scope and conclusion.

Keywords Generative Adversarial Networks (GANs) · Neural network (NN) · Data augmentation (DA)

1 Introduction

> Generative Adversarial Networks is the most interesting idea in the last ten years of Machine Learning.—Facebook AI Director, Yann LeCun

Since the discovery of neural networks, there is the reinvigoration of deep learning in computer vision [1], the endorsement of artificial intelligence models (especially deep learning branch of AI) in healthcare community has increased dramatically. According to the report of Zion market research, in 2018 the global market of medical

---

V. Saaran · V. Kushwaha · S. Gupta (✉) · G. Agarwal
Raj Kumar Goel Institute of Technology, Ghaziabad, India

H. Sharma et al. (eds.), *Congress on Intelligent Systems*, Advances in Intelligent Systems and Computing 1334, https://doi.org/10.1007/978-981-33-6981-8_18

imaging was almost USD 34 billion, and by the year 2025, it is expected to increase over USD 48.6 billion at a compound annual growth rate of 5.2%. Hence, the wide endorsement of artificial intelligence models in healthcare fields is because of its ability to recognize a pattern, detect, and analyze the trend and also to augment the representation of images along with image classification. Here in our study, we concentrated on one of the amazing researches done in the history of artificial intelligence GANs or Generative Adversarial Networks along with its implementation in the healthcare and medical field.

GANs are newly discovered neural networks that instead of following the trend of training the data (structured or unstructured) and predicting the outcome, generate random data (both audio and video) by adding the noise [2]. It performs two main steps—one is the generation of the data (generated or fake data) and another one is focused on distinguishing discrimination. For any machine learning model, the most promising approach is to use generative models that learn to discover the essence of data and find the best distribution to represent it. Therefore, GANs are more preferable than any other NNs because they are able to generate the data (images or audios) while having the same distribution of the features.

Contrary to deep learning which has its root tracked back to the early 90s, GANs are a relatively new twenty-first-century adversarial model which is discovered by Goodfellow in 2014 [3]. As a result, there are only a few instances where this technology is applied. It is still an evolving technology in the field of medical healthcare, like the work done by Han et al. [4] where he showed how the MR image dataset generated from GANs can yield a better result than classic data augmentation in the detection of brain tumor. Hence, by the implementation of GANs we can get viable solutions to the problems persisting in the current healthcare system.

The purpose of this research is to present a review of the generative adversarial networks or GANs and their current or future practical applications in the field of the medical healthcare industry.

2 What are Generative Adversarial Networks or GANs

GANs or Generative Adversarial Networks are a class of AI structure for assessing generative models. In this network, there are two NN which are prepared at the same time by the adversarial method: a generative model, which is used for producing data, and a discriminative model which is liable for evaluating the likelihood that a data was drawn from the training data (real data) or was produced by the generative model (generated data). The generative model can be thought of as comparable to a group of forgers, attempting to create counterfeit money and use it without recognition, while the discriminative model is closely resembling the police, attempting to distinguish the fake cash [3]. The generative model alone will not create any data it will simply produce some irregular clamor or random noise. Theoretically, the discriminative model in GANs gives direction to the generative model on what kind of data (or images) is to be created. During training, the generator turns out to be continuously

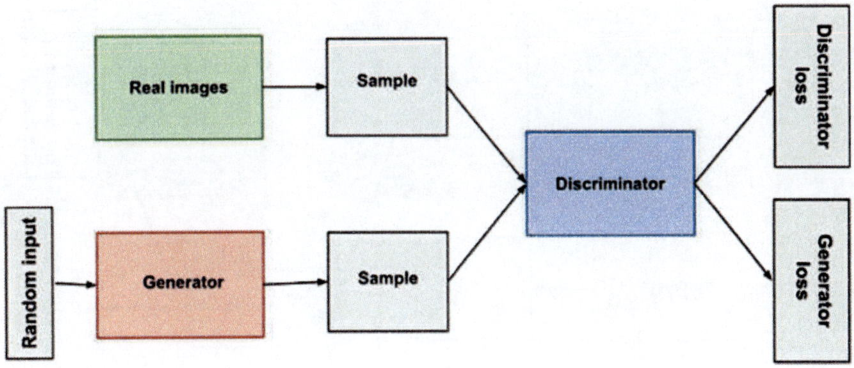

Fig. 1 GAN architecture [6]

better at creating pictures (or any other data), until the discriminator is not, at this point ready to recognize genuine pictures (data) from counterfeit. GANs framework is inspired by a minmax two-player game. In-game theory, the GANs model converges when the discriminative model and the generative attain Nash equilibrium [5]. This is the optimal point for the minmax equation (Fig. 1).

Different parts of GANs are:

- Generative Model—I create fake data
- Discriminative Model—I detect what is fake and what is true.

2.1 Generative Model

The generative model or the generator of GAN is also a NN. The generator part learns to generate/produce fake data or false data by getting regular feedback from the discriminative model. It eventually learns how to fool the discriminative model and makes the discriminator believe that the output is real. As any other NNs need some form of input, this generative model also takes input as a noise. By adding noise in the model, we can use GANs to generate very diverse data. We can use uniform distribution while adding noise as it is easy to sample and the noise distribution does not matter much. For comfort, the space of the noise sample is usually of lower dimensionality of the output (generated data) space (Fig. 2).

To begin with, we test some noise (clamor) z taking a normal (or uniform) distribution. Accepting input as z, a generator G will produce a picture (or any other data) x where $x = G(z)$, and G is a differentiable capacity spoken to by a multilayer (numerous) perceptron. Theoretically, z is used to represent the equivalently hidden features of the images (or any other data) produced by a generator, like sharpness, color, and shape. As in ML classification, we only tune the hyperparameter and do not supervise the features of the learning model. In the same way, in Generative Adversarial Networks, we do not supervise the semantic importance of z, we left

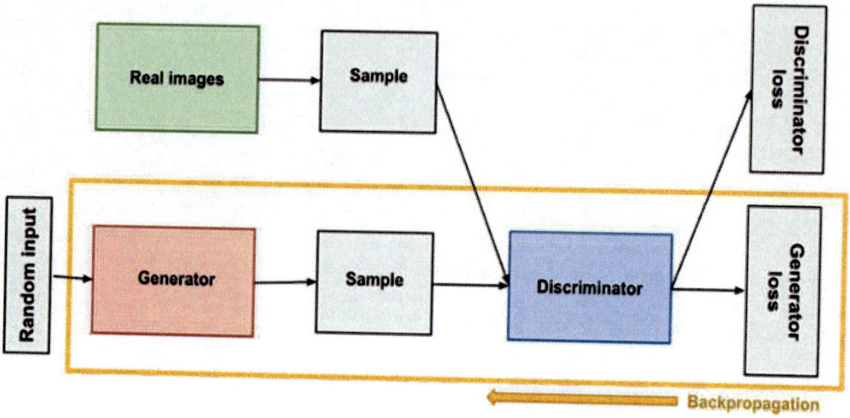

Fig. 2 Generative model with backpropagation [6]

learning on the training of the model. After the output is generated by the model, it is passed to the discriminative model for the evaluation of output as fake or real by the discriminative classifier.

Here we detect the generative loss to be minimum, so that the model can generate more real looking output. Generative Loss function is given below:

$$L(G) = \min\left[\log(D(x)) + \log(1 - D(G(Z)))\right] \tag{1}$$

This loss function should be minimum in the case of the generative model. Where $G(z)$ is the output of the generative model when noise z is applied. $D(G(z))$ is the odds of the discriminative model that the generate/fake data is real. $D(x)$ is the odds of the discriminative model that real data is genuine.

2.2 *Discriminative Model*

Discriminative model or the discriminator in a Generative Adversarial Networks or GANs is just a classification model which attempts to recognize the genuine information from the phony information made by the generator. In this model, any NN architecture can be used appropriately to what data it is classifying. The discriminator or the discriminative model treats both the real data (training data) and fake data (generated data) differently. It differentiates the probability of the output to be fake or real by a function $D(x)$.

The training samples of the discriminator come from two different origins: Real data instances and fake data instances. Instances of real data, for example, the Original image of a painting. The discriminator or the discriminative model uses these real instances as positive examples during the process of training. On the other hand, fake

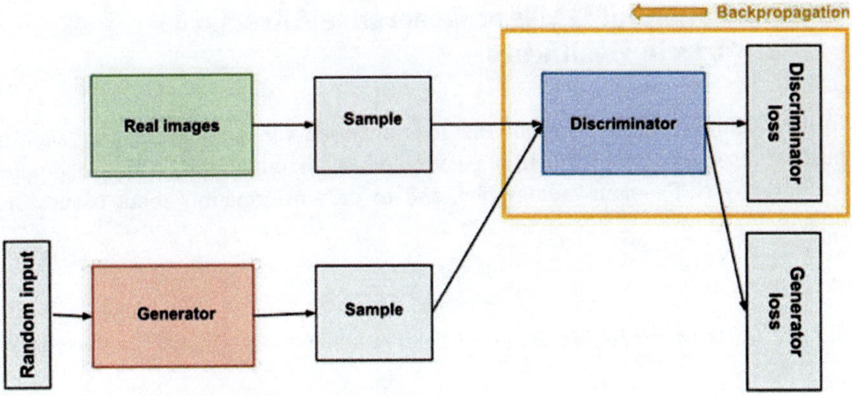

Fig. 3 Discriminative model with backpropagation [6]

instances generated by the generative model or the generator. The discriminative model treats these instances as negative instances or examples during the process of training.

In Fig. 3 given, we have two different sample boxes that illustrate the two data sources (Real data and fake data instances) channeling into the discriminative model or discriminator. We train the discriminative model just as we train any deep NN classifier. If $D(x) = 1$ then it means input is real and if the input is generated, then the value of $D(x)$ is 0. By getting the values of $D(x)$ we can identify the contribution of the real data over the fake data.

After classifying both fake and real data from the generative model or generator, we calculate loss (Discriminator or Generator loss) that misclassifies a fake example as a real or real example as fake. If data is real (i.e., 1) then loss function will be:

$$L(D(x), 1) = \log(D(x)) \tag{2}$$

And for data coming from generator, if the data is fake (i.e., 0) then loss function is:

$$L(D(G(z)), 0) = \log(1 - D(G(x))) \tag{3}$$

Now, for a classifier to classify the fake and real dataset Eqs. (2) and (3) should be maximized and final loss function for the discriminator is:

$$L(D) = \max\left[\log(D(x)) + \log(1 - D(G(x)))\right] \tag{4}$$

After finding the loss using backpropagation, we train our generative model or generator. The generator will then receive these targeted values; i.e., we train the generative model to create any fake data and this data will fool the discriminator, and it will classify it to be real instead of fake data.

3 Applications of GANs or Generative Adversarial Networks in Healthcare

In the healthcare sector, various methods are used for medical imaging such as medical resonance imaging (MRI), positron emission tomography (PET), computed tomography (CT), plain radiography, and to gain information about tissues and organs.

3.1 *Electronic Health Record*

An electronic health record (EHR) is an ordered collection of patient's health information stored in a digital format. EHRs comprise of a mixture of data such as vital signs, laboratory test results, demographics, radiology images, medication and allergies, medical history, immunization status, billing information, and personal statistics like age and weight.

EHRs are increasingly employed as digital inpatient information systems because of the growing importance of various technical factors considered in hospitals. These systems ensure the safety, efficiency, and quality of healthcare provision. But, to gain these advantages it is highly dependent on system optimization.

The adoption of electronic health records (EHR) with large quantity and quality of data by healthcare organizations (HCOs) has led to an eruption in computational health. However, EHR data are known to be complex due to their multi-modality, mixing categorical, and continuous data with semi-structured and free-text medical notes.

EHR generated by using GANs has recently shown impressive performance [7]. In GANs, the generator acquires a random prior $z \in R^r$ and produces synthetic output $G(z) \in R^d$, whereas the discriminative model tries to discriminate the sample whether it is genuine or fake.

3.2 *Retinal Image Synthesis*

Image-to-image translation gives training mapping amidst an output and an input picture and comes under the category of graphics and vision. Till now, many different deep NN models have been presented and do not achieve much to give more realistic pictures that have used to acknowledge the issue of image-to-image translation.

To overcome this limitation, a generative adversarial networks (GANs) approach came. It is considered as a boost in the field of medical image analysis, with numerous applications. The approach focuses on the nature of the target translation picture instead of *t* focusing on pixel-to-pixel dissimilarity. For upgradation in detail and quality of synthetic images, a pipeline is illustrated with three key aspects: the resolution of paired images, the input mask, and, the architecture of Generative Adversarial Networks.

Image-to-image translation using GANs is used "**Towards Adversarial Retinal Image Synthesis**" [8]. Synthesizing pictures of the eye fundus approached earlier is done by contriving multiplex models of the diagnosis of the eye but it is a stimulating piece of work. By examining suitable parameter space, new images can be generated such that synthesizing of eye fundus images will become proficient directly from data.

3.3 Skin Lesion Analysis

A skin lesion is a chunk of the skin having an anomalous growth or emergence as compared to the skin throughout it. It may occur on any part of the body. Skin lesions comprise hardening, discolorations, rash, blisters, cysts, swelling, pus-filled sacs, or any other change in or on the skin which may result in serious issues. Skin cancer is a very ordinary cancer globally, with melanoma being the deadliest form. It causes most deaths. The accurate recognition of melanoma is tremendously stimulating because of several reasons like visual resemblance or low contrast in melanoma and non-melanoma. Precisely recognizing melanoma in a premature stage can crucially reduce the risk and shoot up the endurance rate of patients.

A skin lesion style-based GAN model is initiated which is based on the architecture of styleGAN [9]. It is worthwhile for producing skin lesion images with rich diversity and high resolution. GANs forces synthesizing samples on the real images distribution model such that they are indistinguishable from real images. To distinguish benign from a malignant skin lesion, the images must have high resolution for better classification of skin cancer.

3.4 Medical Image Segmentation

Medical image segmentation is the procedure of semiautomatic or automatic recognition of boundaries within a picture. The continuous advancement of different methods like positron emission tomography, computed tomography (CT), X-ray, and magnetic resonance imaging (MRI) are used to create medical images.

Magnetic resonance (MR) and computer tomography (CT) imaging are primarily used for the segmentation. MR is a flexible and dynamic technology which by altering the imaging parameters and using different pulse sequences achieves variable image contrast. While a CT scan is an imaging method that requires the human body's functional and structural information by using X-rays.

For the above modalities, medical imaging is the essential process as it finds the ROI or region of interest. The ROI can be achieved through an automatic or semiautomatic process. The division of areas on the image is based on the specific description, like segmenting body organs or tissues for detection of border, tumor, and mass.

As inspired by the GANs, the new architecture called SegAN can be used for semantic segmentation [10]. SegAN is having a multi-scale loss function in which a critic network (C) and a segmentor network (S) will be trained to maximize and minimize an objective function. Due to the multi-scale loss function in SegAN architecture, it enforces the learning of hierarchical features more straightforwardly and efficiently.

4 Discussion

This section gives advantages and drawbacks of adopting GANs. The future of using GANs in medical and healthcare is also discussed in this section.

4.1 Advantages

Generative Adversarial Networks have huge implications for the evolution of generative models. One of the great significances of Generative Adversarial Networks is that no annotation of the shape is required for the probability distribution of the generative model. The scope of generated data samples is broadened by adopting a neural network structure for generating high-dimensional data which does not limit the generation dimension. In comparison with other generative methods especially using predefined probability density [11], GANs have these following advantages:

- We can parallelize the produced data during sampling. Thus, samples can be produced in parallel and bring about a significant accelerate for sampling, and due to this advantage, it opens new doors for GANs to be utilized in different applications [12].
- Generally, for the training criteria, two adversarial neural networks are used to train GANs, and we can also use backpropagation for training. In GANs training, we are not dependent on the Markov chain method; thus, many diverse functions can be integrated into the GANs model.
- It does not need to proximate a probability by introducing a minimum boundary that improves the training efficiency and decreases the difficulty during training. The tedious sampling sequence is avoided in the GANs generation process where we review a few sample data and then are able to produce newly generated data. For increasing the diversity of the generated samples, the training in GANs avoids simply coping the features of original data instead take some amount of probability distribution.

- In conduct, the data produced by Generative Adversarial Networks are easy to follow by humans, as GANs can represent very fine, very real looking data, even degenerate distributions, and if we consider Markov chain then this model will produce blurry data which will be difficult for humans to understand. Hence, GANs contribute a promising result in creative data that is relevant to humans.

4.2 Disadvantages

Generative Adversarial Networks have untangled a lot of problems in the area of generative models and are being adopted in many other fields than the healthcare sector, but they still have some limitations. Instability of GANs training and unable to converge still persist in GAN model. Points like equilibrium existence and model convergence are yet to be proved by GANs. Cost functions of the generative model and discriminative model are maintained at the equilibrium (Nash equilibrium) by the GANs, i.e., their parameters should remain at a minimum. But, in practice sometimes the increase in cost function in the discriminative model may lead to a decrease in the generative model's cost and vice-versa. Therefore, a convergence of the Generative Adversarial Networks may regularly fall flat and eventually becomes unstable.

Another essential concern with GANs is the problem of mode collapse. Mode collapse refers to a scheme in which the generative model generates the data (fake data) that contain some same features which are enough for the model to distinguish the data but a little challenging for human understanding. This concern is very destructive for Generative Adversarial Networks that are tested in real applications since GANs lack diversity due to mode collapse restriction. Different researches tried to introduce different object functions to reduce or minimize the mode collapse or summing up new factors or components [13, 14]. Whereas, GANs can solve the mode collapse for a very complex and multi-dimensional distribution of genuine data.

4.3 Future Scope

GANs were formally introduced to generate conceivable counterfeit images and have shown exciting achievement in the area of computer vision. In recent years, GANs have gained a significant increase in a research area related to medical imaging. It is showing some exciting applications in the medical area as discussed in the paper. Along with some applications like MR image to CT, brain MR image augmentation for tumor detection is also gaining attention. But several drawbacks like mode collapse, training instability saddle points, and non-convergence of a model still is a challenge. Other research areas can be the model's inability to learn the different concepts of living objects rather than a static object from sample data.

The specialty of GANs is that it can be modified to the different frameworks according to the need of the application. Several popular GANs framework is cGAN, cycleGAN, styleGAN, medGAN, stackGAN, pix2pix, IcGAN, deblurGAN. The success of GANS in unsupervised and semi-supervised inconsistency detection has the potential to be farther continued for disclosure of embedded devices, such as pacemaker, tubes, staples, and wires on X-rays.

Finally, we would like to point out, although we have discussed several applications of GANs in this paper, the endorsement of Generative Adversarial Networks in the healthcare sector is still in early stage and there are no such breakthrough applications that are being adopted by the world, it is still used in limited areas of the medical sector.

5 Conclusion

Since the discovery of GANs in 2014 by I.Goodfellow and its colleagues, it has been gaining its attention in the AI community. We studied that the core idea of GANs is based on minmax game in game theory. GANs comprises of two DL model generative model and discriminative mode both execute iteratively, where the generative model produces the fake data and discriminative model predicts the fake data to be fake or real. The primary goal is to produce fake data and to fool the discriminator making it believe that fake data is genuine data. The capability of GANs to produce "infinite" new data from probability distributions makes it significantly used in many medical applications.

References

1. Krizhevsky, S., Hinton, G.E.: Imagenet classification with deep convolutional neural networks. In: Advances in Neural Information Processing Systems, pp. 1097–1105 (2012)
2. Wang, D., Qin, X., Song, F., Cheng, L.: Stabilizing training of generative adversarial nets via Langevin Stein variational gradient descent. arXiv:2004.10495v1 (2020)
3. Goodfellow, I., Pouget-Abadie, J., Mirza, M., Xu, B., Warde Farley, D., Ozair, S., Courville, A., Bengio, Y.: Generative adversarial nets. In: Advances in Neural Information Processing Systems, pp. 2672–2680 (2014)
4. Han, C., et al.: Combining noise-to-image and image-to-image GANs: brain MR image augmentation for tumor detection. IEEE Access **7**, 156966–156977 (2019)
5. Kreps, D.M.: Nash equilibrium. In: Eatwell, J., Milgate, M., Newman, P. (eds.) Game Theory. The New Palgrave, pp. 167–177. Palgrave Macmillan, London (1989)
6. Generative Adversarial Networks Homepage: https://developers.google.com/machine-learning/gan/. Accessed 15 June 2020
7. Choi, E., Biswal, S., Malin, B., Duke, J., Stewart, W.F., Sun, J.: Generating multi-label discrete patient records using generative adversarial networks. arXiv:1703.06490v3 (2017)
8. Costa, P., Galdran, A., Meyer, M.I., Abràmoff, M.D., Niemeijer, M., Mendonça, A.M., Campilho, A.: Towards adversarial retinal image synthesis. arXiv:1701.08974v1 (2017)
9. Bissoto, A., Rerez, F., Valle, E.: Skin lesion synthesis with generative adversarial networks. arXiv:1902.03253v1 (2019)
10. Han, C., Murao, K., Satoh, S., Nakayama, H.: Learning more with less: GAN-based medical image augmentation. arXiv:1904.00838v3 (2019)

11. Salimans, T., Goodfellow, I., Zaremba, W., Cheung, V., Radford, A., Chen, X.: Improved techniques for training GANs. In: Advances in Neural Information Processing Systems, pp. 2234–2242 (2016)
12. Alqahtani, H., Kavakli-Thorne, M., Kumar, G.: Applications of generative adversarial networks: an updated review. Arch. Comput. Methods Eng. (2019). https://doi.org/10.1007/s11831-019-09388-y
13. Che, T., Li, Y., Jacob, A.P., Bengio, Y., Li, W.: Mode regularized generative adversarial networks. arXiv:1612.02136v5 (2017)
14. Ghosh, A., Kulharia, V., Namboodiri, V.P., Torr, P.H., Dokania, P.K.: Multi-agent diverse generative adversarial networks. In: Proceedings of the IEEE Conference on Computer Vision and Pattern Recognition, pp. 8513–8521 (2018)

A Federated Search System for Online Property Listings Based on SIFT Algorithm

Yasser Chuttur and Yashi Arya

Abstract Facilitating access to accurate real estates information can contribute positively in the social and economic development of developing countries. Recently, in Mauritius, it has been observed that despite the increasing number of real estate agents marketing their products on the Internet, potential buyers are still facing difficulties in obtaining relevant information due to several duplications with varying information observed across multiple real estate platforms. Such a situation can leave buyers confused and reluctant to take a decision, thereby preventing progress. In an attempt to help buyers obtain adequate information about a property advertised online, a federated search system, based on the SIFT algorithm, that can aggregate and display similar real estate property listings from multiple websites into one location is proposed. Buyers are able to then view all the details for the same property in a single location to decide which online platform to visit and pursue with their quest for a property acquisition. In this paper, the architecture, implementation and evaluation results of the proposed system are presented.

Keywords E-service architecture · Federated search · SIFT algorithm

1 Introduction

Over the past ten years, Mauritius has rapidly transitioned into an e-service society where several companies have invested in Information and Communication Technologies (ICT) to provide their products and services online. In the real estate business, a similar trend has been observed with a growing number of agencies adopting the Internet to post their advertisements on the Internet. In contrast to traditional approach, where potential buyers would rely on property agents to search for properties on their behalf, contemporary approaches involving online property search

Y. Chuttur (✉) · Y. Arya
University of Mauritius, Reduit, Mauritius
e-mail: y.chuttur@uom.ac.mu

Y. Arya
e-mail: yashi.arya@umail.uom.ac.mu

H. Sharma et al. (eds.), *Congress on Intelligent Systems*, Advances in Intelligent Systems and Computing 1334, https://doi.org/10.1007/978-981-33-6981-8_19

promote disintermediation, thus reducing cost of search [1]. Along with property agencies, owners can also post their properties for sale online. In fact, the ease of online real estate advertisement has given way to a new approach for marketing a large number of properties to a large number of audiences. Potential buyers are thus expected to visit a number of websites to obtain information on different properties advertised. An example of a property advertisement on propertycloud.mu is shown in Fig. 1.

In Mauritius, however, it is a normal tendency for property owners to contact multiple property agencies to have their properties listed for sale. Consequently, the same property can be posted multiple times on different websites, and the related advertisements often differ when it comes to pricing and related details given.

An example is shown in Fig. 2, where differences in the two posts observed are: image orientation, description of property on sale, and size indicated for the property.

Fig. 1 Sample property information available online

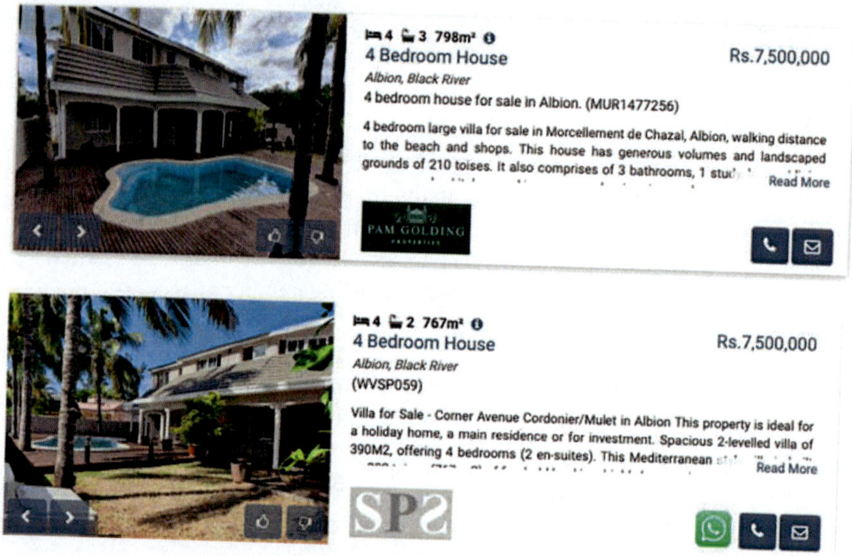

Fig. 2 Dissimilar information for same property posted by two real estate agencies

Consequently, buyers are provided with different information when viewing the same property on different sites, with the net result that buyers are confused as to whether the information displayed is accurate and up to date.

One way to address the problem of finding accurate information about properties posted online would be to manually visit multiple websites and compare the results individually. Ideally, two or more websites will have the same information listed for the same property enabling the buyer to be somewhat confident on the information displayed. In the worst case, when no websites will have identical information, the buyer will have to contact one or more real estate agents to obtain further information on a property of interest. Unfortunately, given the large number of posts and websites that have properties posted online, manual scanning would be a very tedious process and cost ineffective with effects such as (1) discouraging the adoption of the Internet for property search and (2) hindering purchase decisions of potential buyers with an overall negative impact on social and economic development.

To address the problem of manually scanning multiple websites and comparing property information individually, it is proposed to have an automated way that can visit multiple real estate websites to display properties (unique listing and duplicate listing) in one location for easy comparison. In other words, the focus is on the design and implementation of a federated search system that will enable buyers to obtain information on properties collected from different real estate websites in one place.

Federated search is not new. Multiple implementations of federated search systems already exist in several domains [2] but to our knowledge, none exist for real estate properties. In our case, the focus will be to determine whether two or more properties listed on different websites are for the same property based on the images posted.

For example, as shown in Fig. 2, the images of the properties shown are taken from different angles but yet, it is clear that the two properties are the same. The focus is on images because, in general, the textual description of properties listed can vary greatly based on the style of marketing used by the real estate agent and other information such as property size may not be correctly listed. Consequently, detecting similar properties using textual description may not yield accurate results. In contrast, detecting properties based on the images posted seems more relevant in our case. This is because property images usually remain invariant, except if taken from different angles as shown in Fig. 2. In this proposal, the SIFT algorithm which is a popular image processing algorithm used to detect similarity in images is used.

The rest of this paper is structured as follows: Sect. 2 provides an overview of federated search systems including a description of three popular image matching techniques. Section 3 describes the architecture and implementation details of our proposed real estate property federated search system. In Sect. 4, the performance evaluation of the implemented system is presented and discussed. Finally, the paper is concluded in Sect. 5.

2 Online Federated Search System

Federated search, also known as parallel search or 'meta' search, is a technique used to search multiple sources simultaneously [2]. Users submit queries via a single interface and multiple sources are searched in parallel and returned results are combined into a single list to be displayed in one location. Federated search is usually more useful than individual search platforms in situations where the returned results must be compared and evaluated across different sites. For example, federated search is common in implementations for online hotel bookings, airline reservations or car rental where several providers offer their services on different websites. In this case, a federated search system will pull relevant information from multiple service providers' websites and aggregate all the results in one location for users to compare and evaluate the offers available. Popular federated services include hotels.com, Expedia,[1] Trip Advisor,[2] Trivago[3] and Kayak.[4] In all those websites, users have the ability to make a single search to obtain various results that can help them decide about which service to book. By using a federated search system, users do not have to search multiple sources individually and are thus able to save time when searching for information online [3].

Techniques and implementations for federated search have been extensively conducted since 1994 [4] with variations in application domains and optimization techniques notably in areas of information extraction and results ranking [5, 6]. But, in general, it can be said that federated search systems have in common the following main components: *source crawling*, *collection creation*, and a *user interface* for query formulation and results presentation [7].

2.1 *Source Crawling*

Unlike conventional search engines like Google, federated search systems do not obtain their information from the general web [8]. Rather federated search systems have specific sources or websites that they collect information from to serve a specific purpose. The approach used to obtain information from specific sources differs depending on the type of access made available to the federated system. For instance, some federated systems have direct access to the databases of the sources [9–11], and this is typically true for online airline booking sites such as expedia.com. The use of wrappers is common in other cases, where federated search systems rely on returned results from other search engines [12–14]. The wrapper converts a user query into a format used by the particular web search engine, submits the query, extracts the search results, and then translates the extracted information back into a standard

[1]https://www.expedia.com/.

[2]https://www.tripadvisor.com/.

[3]https://www.trivago.com/.

[4]https://www.kayak.com/.

format for the federated system to display the results [4]. Another approach is to use web scraper APIs [15, 16]. Several web scraper APIs are available as open-source software, and those allow a developer to easily train a crawler to locate and download specific contents from selected source websites.

2.2 Collection Creation

Unless real-time information is required, federated search systems, with the exception of meta search engines (where multiple search engines are searched simultaneously), maintain their own databases populated with information collected from source websites [17, 18]. Crawlers collect selected information from source websites, and information retrieved is fed back to the federated search back-end system for further processing before storage [19]. The type of processing will depend on the goal of the federated search system. Different types of common processing are reformatting collected data for storage optimization, filtering out duplicate information, truncating textual contents, etc. [20, 21]. In the present case, the goal is to identify real estate properties that are deemed similar from different sources as a result of the images posted online. To this end, an appropriate image matching algorithm is required. Three most popular image matching techniques are discussed here, namely SIFT, SURF, and ORB [22].

SIFT

Scale invariant feature transform (SIFT) was developed by Lowe [23]. The algorithm works by extracting unique features of image areas, which remain unchanged to image scale, rotation and are robust to variations in noise, illumination, distortion and viewpoint. Four main steps are necessary in the SIFT algorithm: (1) scale–space extrema detection; (2) keypoint localization; (3) orientation assignment; and (4) keypoint descriptor.

In the first step, SIFT feature algorithm finds keypoints that can be reliably extracted within the scale space of an image. Extrema over all scales and image locations are searched, and keypoints are identified as local minima or maxima based on the difference-of-Gaussian (DoG) images across scales generated in this step. From the first step, local feature keypoints are generated and those represent a specific image pattern, which contrasts to neighboring pixels. Each keypoint can be extracted and described using keypoint descriptors, enabling image matching or object recognition. Since scale–space extrema detection generates too many keypoint candidates, some of which, are unstable, some keypoints must be removed. In keypoint localization, the keypoints are filtered so that only stable keypoints are retained. Following which each keypoint is assigned an orientation that makes the keypoint descriptor invariant to rotation. An orientation histogram of local gradients from the closest smoothed image is used in this case. Eventually, each keypoint is assigned as descriptors an image location, scale, and orientation, which ensure the invariance to image rotation, location, and scale. Euclidean distance between each invariant feature descriptor can then be used to evaluate image matching, object recognition, etc. The implementation of the SIFT algorithm is well documented in [24, 25].

Due to its robustness toward image scaling, intensity, orientations, and other image variations, SIFT is the most widely image matching algorithm used [26] and has been extensively researched since 2004. As reported in [27, 28], SIFT is considered to be the most accurate feature-detector descriptor for scale, rotation, and affine variations compared to SURF, ORB, and other image matching algorithms.

SURF

Speed-up robust feature (SURF) proposed by Bay [29] is an approximation of SIFT and is a blob detector. Similar to SIFT, SURF also employs Gaussian scale–space analysis of images but with the addition of detectors based on *determinant of Hessian Matrix*, which exploits integral images to improve feature detection speed. Rather than using Gaussian functions to average an image, squares are used for approximation, as convolution with square is much faster when the integral image is used. SURF makes use of a 64 bin descriptor, which describes each detected feature of an image with a distribution of *Haar wavelet responses* within a given area or neighborhood. Selected neighborhood around a key point is divided into sub-regions and for each sub-region, the *Haar wavelet* responses are taken and represented to obtain SURF feature descriptor. Features are then compared only if they have same type of contrast (based on Laplacian sign already calculated) for faster image matching. The main benefit of SURF over SIFT has been found to be its low computational cost and speed [27]. However, although SURF features are invariant to rotation and scale, they suffer from little affine invariance and lower performance compared to SIFT [28].

ORB

SIFT is mainly criticized for its dependency on heavy computational resources, which makes it unsuitable for real-time systems and low-power devices. To address the need for faster image matching needs especially within environment wit no GPU acceleration, Oriented FAST and Rotated BRIEF (ORB) was suggested by Rublee et al. [30]. At its core, ORB makes use of FAST keypoint detector [31] and BRIEF descriptor [32]. FAST is used to determine key points to obtain the top N points following application of a Haris corner measure. Harris corner detector is based on Moravec algorithm [33]. Following design of a detecting window in the image, the average intensity variation is computed by shifting the window in small increment in different directions. Each window center point is then marked as a corner point.

In contrast to SIFT, FAST is orientation and rotation variant. Around each located corner, an intensity-weighted centroid is computed. The directions of generated vectors from the corner point are used to obtain the image orientation while moments are used to improve rotation invariance and create a rotation matrix. Since the performance of BRIEF descriptors is affected by orientation, BRIEF descriptors in ORB must be configured according to the orientation computed. A feature store of ORB descriptors paired with coordinates from a given key frame is then built. The store can be searched using ORB descriptors to find the best match for input descriptors [34].

Performance comparison study conducted by [28] indicates that ORB is faster than SIFT and SURF. In most situations, ORB is found to have similar or slightly better performance than SIFT, but in situations where images are rotated at certain degrees and have varying intensity, ORB is found to exhibit lower performance than SIFT.

2.3 User Interface

Federated search systems apply data abstraction to provide a uniform user interface to allow users enter a single query to search and retrieve data from multiple databases [35]. The method used to let users formulate their query and visualize results will depend on the underlying architecture and goal of the federated system. For federated systems that retrieve data from other search engines, queries are formulated and results are returned and displayed in native format with the same ranking or order as if one has entered the query in the source website [36]. Results can be merged in rows or kept as separate columns in the same window.

For pre-harvested data, whereby federated search systems crawl source websites to create their own collections and maintain their own databases, the user interface allows users to select specific filters to sort returned results accordingly [37, 38]. Some federated search systems may also include advanced filters to refine the query formulated by users, such as applying boolean operators like AND, OR, NOT, specifying specific search attributes such as author name, title, subjects, etc. [39].

Regardless of the approach used, the user interface of any federated search system acts as a single point of query formulation and results display for search conducted across multiple sources [40].

3 Implementing the Property Federated Search System

Following lessons learnt from past studies, the four main components required for our federated search system are constructed, i.e., a web scraper to retrieve data from multiple local real estate websites; a data cleansing function to reformat collected data; a property matching algorithm, to detect similar properties and a website interface to allow users to formulate their search query and view returned results in one single interface. Figure 3 shows a general architecture of our federated search system. Implementation details are given further.

The system shown in Fig. 3 was developed using Python 3.8.2, PyCharm IDE, and MySQL. Software libraries used are: *Fnmatch2*, for Unix filename pattern matching; *spaCy*, for natural language processing/text processing; *Numpy*, for arrays and scientific computing, *Pandas*, for data analysis and manipulation, *OpenCV*, for image

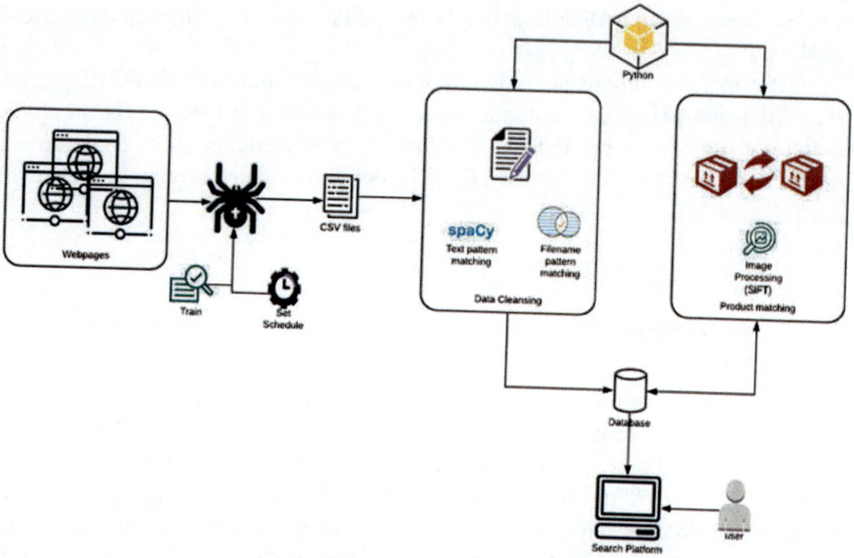

Fig. 3 General architecture of implemented real estate federated search system

processing; *Mysqlclient Package*, to create and manage link between our Python application and the MySQL database; and finally *Flask* for creating a web interface.

Four local real estate property websites as source websites were used for this study. The websites are: *lexpressproperty.mu, propertycloud.mu, molakaz.mu*, and *seef.mu*. Those websites contain sufficient properties to crawl and are popular source of information on real estate properties on sale in Mauritius. An existing web scraping API, namely *import.io*, to collect desired property data from the selected source websites is used. Multiple extractors were trained to retrieve specific information such as property descriptions (links, details, etc.) and associated images links for each property found by the crawler. A schedule could also be set for the crawler to visit the source real estate properties websites regularly at set intervals.

All crawled data was downloaded as *csv* files and then processed by our data cleansing algorithm mainly to remove special or meaningless characters prior to storage in a database. To detect similar properties on different websites, SIFT algorithm is chosen for its reported accuracy and performance in processing images of varying colour intensity, orientation and scale.

For each image match detected for different property posted on different source websites, our system aggregates all the data for the matched properties in a separate table of our database such that when a user searches for a given property, the search is conducted on the matched property table with all merged results returned to the user. An example of returned results is shown in Fig. 4. All the information for similar properties posted on multiple sites is available in one single location allowing for easy comparison. In Fig. 4, it is noted that the same properties have different prices listed on three different source websites. In this way, a user is presented with clear

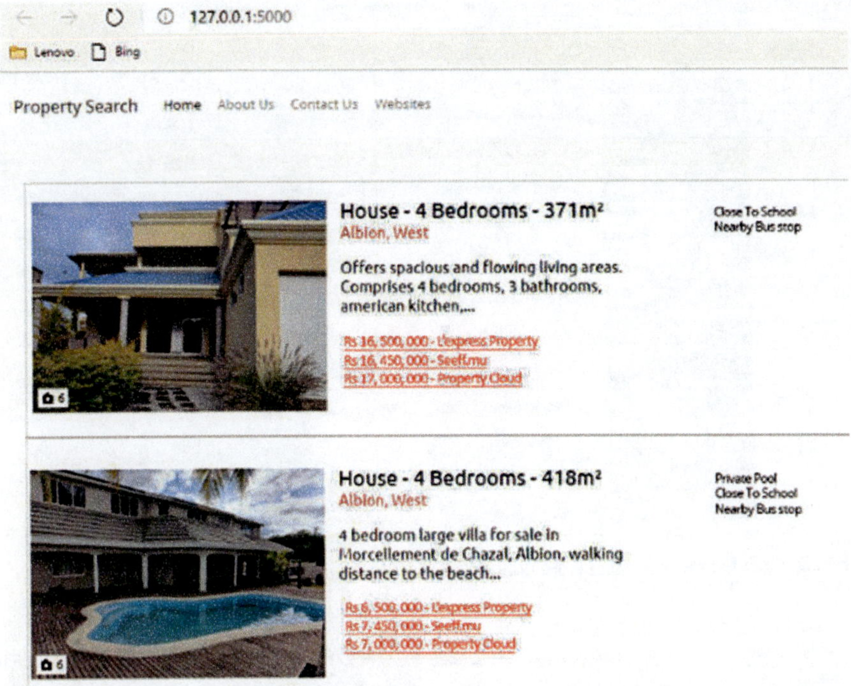

Fig. 4 Aggregated real estate property data from different sources

information that he/she may use to decide on the next plan of actions. Our system allows a user to further click on any of the source websites in order to obtain further details of the property listed and proceed with the buying decision. To aid in creating advanced queries, refinements were included in the user interface to allow users to filter our displayed results based on desired attributes such as *close to school, close to shop, close to bus stop, fenced, garden*, etc.

4 Evaluation and Discussions

A total of 3000 property records with a total of 14,343 images were crawled and downloaded from the four source websites used in this study. The trained crawler *import.io* successfully identified desired data to be collected from all source websites and saved the records accordingly on our processing server accordingly. Figure 5 shows a snapshot of the collected data.

For property matching, SIFT algorithm performance was evaluated with a sample of 300 known properties manually identified as identical from two or more of the source websites. Figure 6 shows the key points identified by the SIFT algorithm for two image pairs (a) at different scale (b) with different orientation.

Fig. 5 Crawled property data from 3000 URLs by *import.io*

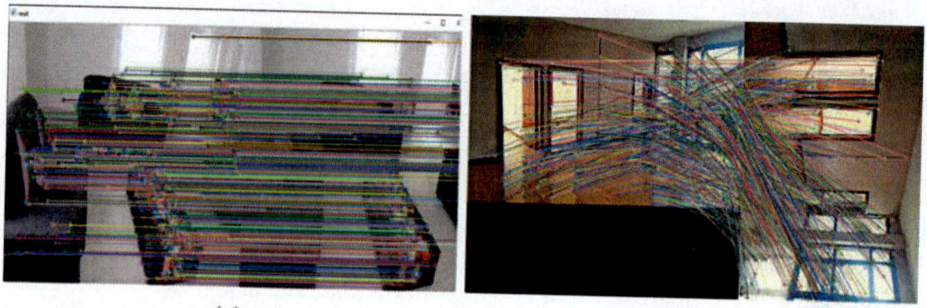

(a) (b)

Fig. 6 SIFT key points matching for image pairs **a** at different scale, **b** with different orientations

Features that are extracted and matched by the SIFT algorithm are indicated by lines drawn between key points located for the two images. The larger the number of lines linking two images, the closer a match is determined. In the example shown, it is considered that the SIFT algorithm could correctly find matches between the images used. For the sample of 300 properties known to be identical, however, a fair accuracy of 86.7% was obtained, i.e., 260 out of the 300 properties were correctly matched. Our results corroborate with previous findings reported in [28], thus demonstrating the effectiveness of using SIFT algorithm for online property matching. As far as the properties that could not be matched, manual analysis reveals that when images are flipped for the same properties, the SIFT algorithm does not perform as expected failing to find a corresponding match.

5 Conclusions

Property search online is a tedious process especially when varying information is posted on multiple websites for the same properties. In this paper, the implementation of a federated system specifically designed to operate on property images was presented. Given that properties may be described in different ways by real estate agents, the assumption that same properties can be easily identified if the images posted are similar was tested. To address the issue of images having different orientation, intensity, scale, etc., the SIFT algorithm was chosen for similar image detection. Results obtained were fairly high (over 80%), but improvement in the image matching process is further warranted for better user satisfaction. As future works, matching other attributes such as property description and location data may be considered for better performance. Since the field of image analysis is also subject to further development, more performing algorithm over SIFT may also be studied.

References

1. Srinivasan, R.: Organising the unorganised: role of platform intermediaries in the Indian real estate market. IIMB Manage. Rev. **29**(1), 58–68 (2017)
2. Shokouhi, M., Si, L.: Federated search. Found. Trends Inf. Retr. **5**(1), 1–102 (2011)
3. Randall, S.: Federated searching and usability testing: building the perfect beast. Ser. Rev. **32**(3), 181–182 (2006)
4. Avrahami, T.T., Yau, L., Si, L., Callan, J.: The FedLemur project: federated search in the real world. J. Am. Soc. Inform. Sci. Technol. **57**(3), 347–358 (2006)
5. White, M.: One search to search them all? eLucidate 16(2) (2020)
6. Baeza-Yates, R., Cuzzocrea, A., Crea, D., Bianco, G.L.: An effective and efficient algorithm for ranking web documents via genetic programming. In: Proceedings of the 34th ACM/SIGAPP Symposium on Applied Computing, 8 Apr 2019, pp. 1065–1072
7. Sholtis, S., Tachibana, R.G., Auga, T., Henderson, K.J., Bhamidipati, V.S., Cadence Design Systems Inc.: Federated system and methods and mechanisms of implementing and using such a system. U.S. Patent 7,392,255 (2008)
8. Olston, C., Najork, M.: Web crawling. Found. Trends Inf. Retr. **4**(3), 175–246 (2010)
9. Baer, W.: Federated searching: friend or foe? Coll. Res. Libr. News **65**(9), 518–519 (2004)
10. Boyd, J., Hampton, M., Morrison, P., Pugh, P., Cervone, F.: The one-box challenge: providing a federated search that benefits the research process. Ser. Rev. **32**(4), 247–254 (2006)
11. Georgas, H.: Google vs. the library (part II): student search patterns and behaviors when using Google and a federated search tool. Portal Libr. Acad. **14**(4), 503–532 (2014)
12. Collarana, D., Lange, C., Auer, S.: FuhSen: a platform for federated, RDF-based hybrid search. In: Proceedings of the 25th International Conference Companion on World Wide Web, 2016 Apr 11, pp. 171–174
13. Graupmann, J., Biwer, M., Zimmer, P.: Towards federated search based on web services. In: BTW 2003–Datenbanksysteme für Business, Technologie und Web, Tagungsband der 10. BTW Konferenz 2003. Gesellschaft für Informatik eV
14. Salampasis, M., Hanbury, A.: PerFedPat: an integrated federated system for patent search. World Patent Inf. **1**(38), 4–11 (2014)
15. Alba, A., Bhagwan, V., Grandison, T.: Accessing the deep web: when good ideas go bad. In: Companion to the 23rd ACM SIGPLAN Conference on Object-Oriented Programming Systems Languages and Applications, 19 Oct 2008, pp. 815–818

16. Singh, A., Tucker, C.S.: A machine learning approach to product review disambiguation based on function, form and behavior classification. Decis. Support Syst. **1**(97), 81–91 (2017)
17. Wegrzyn-Wolska, K.: Statistical classification of search engines interrogated by the meta-search system. In: AMT, 19 May 2006, pp. 317–322
18. Yoshii, A., Yamada, T., Shimizu, Y.: Development of federated search system for sharing learning objects between NIME-glad and overseas gateways. Educ. Technol. Res. **31**(1–2), 125–132 (2008)
19. Salampasis, M.: Federated patent search. In: Current Challenges in Patent Information Retrieval 2017. Springer, Berlin, Heidelberg, pp. 213–240
20. Coiera, E., Walther, M., Nguyen, K., Lovell, N.H.: Architecture for knowledge-based and federated search of online clinical evidence. J. Med. Internet Res. **7**(5), e52 (2005)
21. Wang, Y., Mi, J.: Searchability and discoverability of library resources: federated search and beyond. Coll. Undergrad. Libr. **19**(2–4), 229–245 (2012)
22. Chien, H.J., Chuang, C.C., Chen, C.Y., Klette, R.: When to use what feature? SIFT, SURF, ORB, or A-KAZE features for monocular visual odometry. In: 2016 International Conference on Image and Vision Computing New Zealand (IVCNZ), 21 Nov 2016. IEEE, pp. 1–6
23. Lowe, D.G.: Distinctive image features from scale-invariant keypoints. Int. J. Comput. Vis. **50**(2), 91–110 (2004)
24. Wang, G., Rister, B., Cavallaro, J.R.: Workload analysis and efficient OpenCL-based implementation of SIFT algorithm on a smartphone. In: 2013 IEEE Global Conference on Signal and Information Processing, Dec 2013, pp. 759–762. IEEE
25. Huang, H., Guo, W., Zhang, Y.: Detection of copy-move forgery in digital images using SIFT algorithm. In: 2008 IEEE Pacific-Asia Workshop on Computational Intelligence and Industrial Application, Dec 2008, vol. 2, pp. 272–276. IEEE
26. Grabner, M., Grabner, H., Bischof, H.: Fast approximated SIFT. In: Asian Conference on Computer Vision, Jan 2006, pp. 918–927. Springer, Berlin, Heidelberg
27. Tareen, S.A.K., Saleem, Z.: A comparative analysis of sift, surf, kaze, akaze, orb, and brisk. In: 2018 International Conference on Computing, Mathematics and Engineering Technologies (iCoMET), Mar 2018, pp. 1–10. IEEE
28. Karami, E., Prasad, S., Shehata, M.: Image matching using SIFT, SURF, BRIEF and ORB: performance comparison for distorted images, 7 Oct 2017. arXiv preprint arXiv:1710.02726
29. Bay, H., Ess, A., Tuytelaars, T., Van Gool, L.: Speeded-up robust features (SURF). In: Computer Vision ECCV 2006. Lecture Notes in Computer Science, vol. 3951, pp. 404–417 (2006)
30. Rublee, E., Rabaud, V., Konolige, K., Bradski, G.: ORB: an efficient alternative to SIFT or SURF. In: 2011 International Conference on Computer Vision, 6 Nov 2011, pp. 2564–2571. IEEE
31. Rosten, E., Drummond, T.: Machine learning for highspeed corner detection. In: European Conference on Computer Vision, vol. 1 (2006)
32. Calonder, M., Lepetit, V., Strecha, C., Fua, P.: Brief: binary robust independent elementary features. In: European Conference on Computer Vision (2010)
33. Adel, E., Elmogy, M., Elbakry, H.: Image stitching system based on ORB feature-based technique and compensation blending. Int. J. Adv. Comput. Sci. Appl. **6**(9) (2015)
34. Weberruss, J., Kleeman, L., Drummond, T.: ORB feature extraction and matching in hardware. In: Australasian Conference on Robotics and Automation, Dec 2015, pp. 2–4
35. Nguyen, D., Demeester, T., Trieschnigg, D., Hiemstra, D.: Federated search in the wild: the combined power of over a hundred search engines. In: Proceedings of the 21st ACM International Conference on Information and Knowledge Management, Oct 2012. pp. 1874–1878
36. Boss, S.C., Nelson, M.L.: Federated search tools: the next step in the quest for one-stop-shopping. Ref. Libr. **44**(91–92), 139–160 (2005)
37. Burke, J.: Discovery versus disintermediation: the new reality driven by today's end-user. In: VALA 2010 Conference, Melbourne, Feb 2010, p. 382
38. Boyer, G.M., Besaw, M.: A study of librarians' perceptions and use of the summon discovery tool. J. Electron. Resour. Med. Libr. **9**(3), 173–183 (2012)

39. Muir, S.P., Leggott, M., Mah, C., Stranack, K.: dbWiz: open source federated searching for academic libraries. Library Hi Tech (2005)
40. Mukherjee, R., Jaffe, H., Verity Inc.: System and method for dynamic context-sensitive federated search of multiple information repositories. U.S. Patent Application 10/743,196 (2005)

Digital Brain Building a Key to Improve Cognitive Functions by an EEG–Controlled Videogames as Interactive Learning Platform

P. K. Parthasarathy, Archana Mantri, Amit Mittal, and Praveen Kumar

Abstract This investigation provides a methodical review of electroencephalography (EEG)-oriented brain–computer interfaces (BCI) integrated learning platform through videogames, a vast field of research that gives a path through for all questions concerning the future direction of BCI—Games. The vision is to develop a basis of hypothesis on upgrading the mental skill through digital brain-building therapy. Everyone overcomes problems of any current situation through various debates, but at the same time, the difficulty faced tends us to fail. It is all that you do not have the specific skill, but our brain is not been trained to solve the problem that you face on the day to day life. Digital brain-building therapy will be as easy as taking a medicine for which gaming would be the indirect approach for training. The performance of the training through gaming is based on the psychological state of the person assigning the situation during the course of training. The paper tests the process of BCI research integrating with video games and shows that the video game platform offers plenty of benefits. One game application is been used to examine future directions.

Keywords Brain–computer interface · Computer games · Game design · EEG · Human–computer interaction · NeuroSky · Neurogaming

---

P. K. Parthasarathy
Chitkara Design School, Chitkara University, Rajpura, Punjab, India
e-mail: Parthasarathy@chitkara.edu.in

A. Mantri
Chitkara University Institute of Engineering and Technology, Chitkara University, Rajpura, Punjab, India
e-mail: archana.mantri@chitkara.edu.in

A. Mittal · P. Kumar (✉)
Chitkara Business School, Chitkara University, Rajpura, Punjab, India
e-mail: praveen.kumar@chitkara.edu.in

A. Mittal
e-mail: amit.mittal@chitkara.edu.in

H. Sharma et al. (eds.), *Congress on Intelligent Systems*, Advances in Intelligent Systems and Computing 1334, https://doi.org/10.1007/978-981-33-6981-8_20

1 Introduction

Video games have become a widespread leisure activity, pervasive form of entertainment and a substantial field of research. Video games use has increased steadily over time [1]. The immersive gaming technology enhances training in industry, academia and military. In a variety of tasks, video game players (VGPs) perform better than non-videogame players (NVGPs). The present study hypothesizes that VGPs are able to shift attention faster than NVGPs of one of the video game genres like action video game (AVG) have become popular among the young generation. The researchers and neuroscientists around the world have been trying to explore the impact created by technologies like AVG on the human brain and its cognitive functions. Also playing AVGs results in a wide range of behavioral benefits, including enhancements in low-level vision, visual attention, speed of processing and statistical inference, and these skills are not just gaming skills, but real-world skills [2].

Why do not we play a game giving more exercise to the brain with the help of doing all the functional activities without using any controller? In the current era, we are all totally dependent on technology which has led us to human–computer interaction. Digital brain building as therapy the future based on EEG measurements, the user's mind is trained to bridge new connection and to either increase or decrease the use of specific brain function which is achieved by interconnecting with gaming appliances [3]. This interconnecting is done through the brain–computer interface which is based on recognition of subject-specific EEG.

We contribute to the literature by recorded patterns EEG signals from sensorimotor areas during mental imaginations of specific movements are classified into controls [4]. Neuroergonomics theory suggests that knowledge of brain-behavior relationships can be applied to optimize the design of environments to accelerate learning. The convergence of recent advances in ultra-low power consumer electronics, ubiquitous computing and wearable sensor technologies enables real-time monitoring of cognitive and emotional states providing objective, timely and ecologically valid assessments. These technologies allow access to the psychophysiological states associated with learning. Increasing evidence suggests that physiological correlates of attention, alertness, cognitive workload and arousal [5].

1.1 User State Monitoring

Playability is defined as a set of properties that describe the player experience using a specific game system whose main objective is to provide enjoyment and entertainment, by being credible and satisfying, when the player plays alone or in the company. Playability is characterized by different attributes and properties to measure the video game playing experience like satisfaction, learning, efficiency, immersion, motivation, emotion and socialization [6]. Individuals will invariably have different reactions to a given game, and without an assessment tool that can be employed online, researchers will experience difficulties in identifying the cause of these differences. Video games are a natural fit for brain–computer interface (BCI) and electroencephalography (EEG) equipment is widely used to record brain signals in BCI systems. Early BCI applications have targeted disabled users who have mobility or speaking issues. Their aim was to provide an alternative communication channel for those users. But later on, BCI enters the world of healthy people as well. It works as a physiological measuring tool that retrieves and uses information about an individual's emotional, cognitive or effectiveness state. The target of brain signals utilization has been extended beyond controlling some object or offering a substitution for specific functions in what is called passive BCI [7]. This creates an objective measure of the gamer's state.

1.2 Neurogaming Using NeuroSky

EEG games are the future of gaming, and to play, you are going to need an EEG headset. Luckily for you, the world's most affordable EEG headsets—NeuroSky. Gamers of all ages are always looking for the next exciting thing, and with EEG games, the way you play is going to change forever. EEG headsets have completely revolutionized gaming—and that is because the only controller you need to play with the EEG headset is your mind. Gaming hardware companies who wish to broaden their stable of interface device solutions should strongly consider the benefits of designing and manufacturing BCI headsets targets user audiences. NeuroSky enables its OEM partners to create custom BCI headsets that are affordable, consumer-friendly and able to extend the gaming experience into new realms. We hope BCI and EEG biosensor technology will act as a springboard for creating different genres of games and extending the way existing genres are played. As a potential BCI gaming hardware OEM/ODM, it might interest you to know how the developer community you rely on to drive sales of your peripheral products via their content perceives opportunities in the BCI space. Obviously, the more compelling the use case is for BCI, the greater the demand for your piece of hardware.

1.3 EEG Acquisition

Pioneering video game developers leveraged BCI as a secondary (supplement) controller to immerse players into telekinetic experiences. From knocking over cylindrical towers, bending state control empowered users and simultaneously gave them a basic understanding of how brain activity could be harnessed.

With the benefit of being non-invasive to the wearer, it is the tool that is practical of use by game developers [8] were able to find significant differences in the beta and gamma bands among various stimulus modalities. They also found an increase in

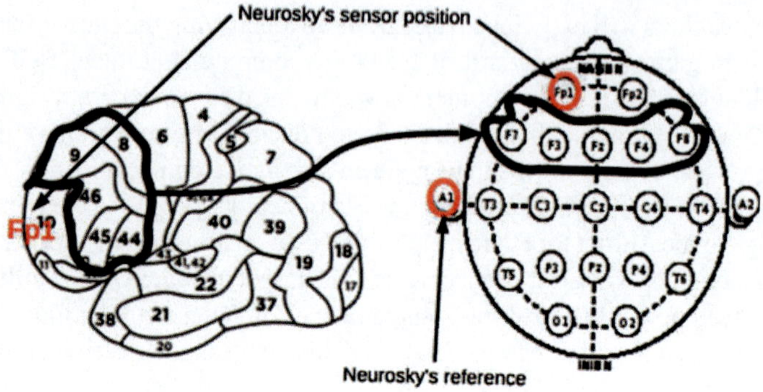

Fig. 1 NeuroSky's sensor positing

power estimates during high-intensity gameplay. The authors conclude that their findings suggest that the EEG can be used to assess differences in frequency bands when persons are experiencing various stimulus modalities using off-the-shelf EEG-based gaming technology [9]. There is potential for EEG data to proffer valuable information about the participants felt cognitive and affective processing. Although there have been growing efforts in the neurogaming literature to recognize a user's cognitive and affective states in real-time using EEG. Establishing the optimal relation among frequency bands. Task engagement and arousal states are one of the main goals of neurogaming, a standardized method has to be established. The ideal research situation would test classifiers with in the same context, users, feature extraction methods and protocol [10] (Fig. 1).

1.4 Neuroergonomics Application

Digital brain-building therapy feedback to be integrated into society, there is a need for the feedback to be accepted. People may find the problem in acceptance. This problem is not just based on the well-being of the mental state but also on the well-being of the physical state. Is this a possible way for therapy is still an open topic to debate? So, there is a need for implementation on the training procedures that provides research, development and applications.

This paper would create a platform which integrates BCI in evaluating the mental well-being with the destination to create an interactive learning platform (ILP) in increasing the skill set of an individual. These ILP games will help the user play for the benefits and get fun as a bonus. The game mechanics and psychology that are involved are meant to motivate and create arousal for the interactive learning platform which will be designed to enhance the cognitive skill and provide proper evidence with solid training in multiple tasks.

2 Methodology

2.1 Mathematical Model

The statistical model for discovering how the attention span was generated is proven to be a challenging task. The unfiltered and raw data from the sensors is automatically transferred and extracted from the chip located in the headset. The method of filtration uses an algorithm called the (CooleyTukey Fast Fourier Transform (FFT) algorithm by J. W. Cooley and John Tukey in 1965) (Eq. 3). A divide and conquer algorithm is the basis for the FFT algorithm where the workload of the system is divided to make calculations more effective.

$$X(k) = \sum_{n=0}^{N-1} x(n) \cdot e^{-j[\frac{2\pi}{N}]nk} \tag{1}$$

where $k = 0, 1, \ldots N - 1$.

If,

$$W_N = e^{-j[\frac{2\pi}{N}]} \tag{2}$$

From Eqs. 1 and 2

$$X(k) = \sum_{n=0}^{N-1} x(n) \cdot W_N^{nk} \tag{3}$$

where $k = 0, 1, \ldots N - 1$.

The formula for the FFT algorithm using $O[N\log N]$.

FFT is further based on the discrete Fourier transform (DFT) to isolate different EEG waves. DFT uses an $O[N\hat{}2]$ computational complexity to calculate, whereas a more powerful FFT uses an $O[N\log N]$ computer complexity. This implies that when calculating DFT $x(n)$ the FFT algorithm is more efficient. (O is a command).

$$X_k = \sum_{n=0}^{N-1} x_n \cdot e^{-i2k\pi\frac{n}{N}} \tag{4}$$

where $k = 0, 1, \ldots N - 1$.

The formula for the DFT algorithm using $O[N\hat{}2]$.

Fig. 2 Brain Runner game

2.2 *Neurogame and Educational Concept*

The ideation behind this development is to make the process understandable and easily set up a function to coordinate with the EEG headset. The game was developed by the authors with existing game development software (Unity) and the OpenBCI communication system. Both applications are open source and a detailed coding will be found in the source code (Fig. 2).

2.3 *Brain Runner*

The perfectly named Brain Runner (Fig. 1) was coded on Unity. Its uses two OEM partners to compete with each other. At the first stage of testing and development, the user was made to compete with the AI with that the cognitive skill, which is the concentration level of the user was tested through a continuous process of monitoring and converted into graphical format. In between the game, the user goes through the obstacle which was a challenge and by the gesture the obstacle needs to be crossed. In the second stage of testing, the user compete with the other user with the help of two EEG device. The user was given a choice of selecting the level of the game in three modes like easy, intermediate and difficult. The more the user focus and turns his attention in the racing at the difficulty level, the easier for him to win the race. In the second stage of testing and development, user 1 competed challenging the user 2. Now, the two human brains compete with each other. Here, the motor cortex function by automatically giving movement to the body and which makes the user giving more and more activeness to the cognitive function of the brain.

Brain Runner application divided into 2 modules, the first module is connectivity and the second one main game. The connectivity module is responsible for establishing a connection between the NeuroSky headset and Unity3d application. Once connection established successfully it activates the main game module. In the main game module, the Unity3d application continuously receives attention value from the NeuroSky module in the main loop. According to received attention, it updates player character speed and updates the graph. In the main loop program continuously update enemy speed and check if either of them completed the race, accordingly show win or lose screen.

1. The brain while playing these games is kept in constant entertainment, so it will make it stronger.
2. Depending on how the player develops and react while playing, different brain conditions can be diagnosed.
3. Hopefully, this game can be used to treat different mental illnesses.

2.4 Hybrid BCI

The research examination into BCI and games so far has been directed on a single modality of interaction, in other words, that BCI was utilized as the main channel for communication. Considering that the performance of such implementations depends on the amount of noise in the environment as well as the degree of users movement, this aversion is understandable. In any case, if BCI turns into a prime gaming interface, then these approaches need to be examined, considering the most end-user. Hybrid BCI is defined as pure hybrid BCIs that combine two or more BCI paradigms, physiological hybrid BCIs, which combine a BCI with another physiological signal, or mixed hybrid BCIs, which combine a BCI with non-physiological input (see Figs. 3, 4 and 5) [11].

Another opportunity that has not been researched is using present BCI modalities with touch [12]—or auditory-based stimuli. While every BCI devices hold merit for gaming applications, there are various preferences of utilizing different ideal models in a solitary implementation. At the point when the contribution from different ideal models is incorporated in the design, this cooperation framework is viewed hybrid system. In a survey of 55 journal articles on the topic of hybrid BCIs [13], it was found that in most cases hybrid BCIs were used to improve accuracy and to provide additional control signals. Multimodal control was also established through an eye tracker and an SSVEP BCI [14]. The authors concluded this to be a well-rounded control modality since it solves the problem with eye-trackers where unintentionally attended items are mistakenly selected. In our game, there was only one example of pure hybrid approaches implemented for BCI games.

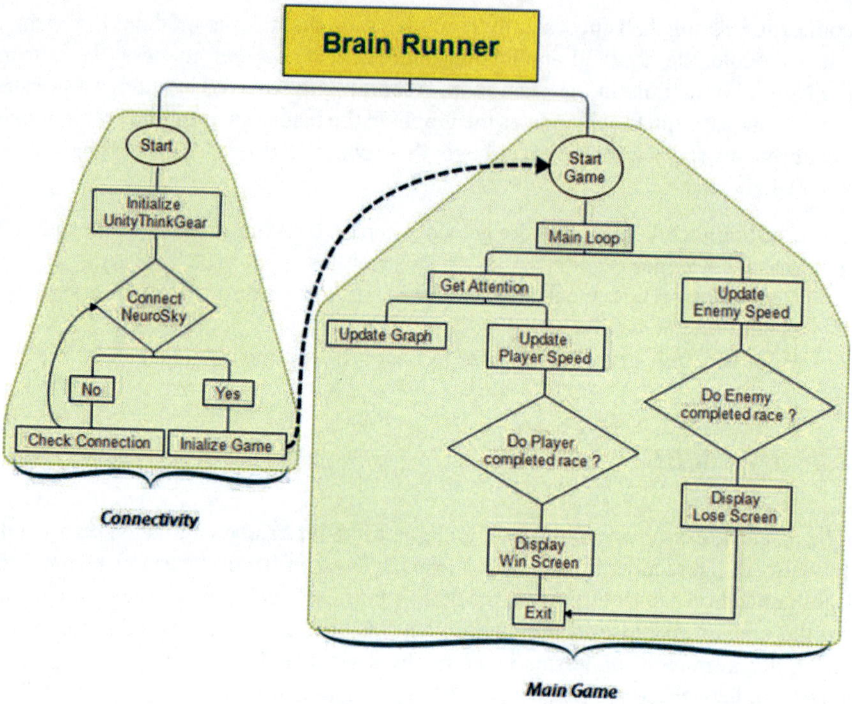

Fig. 3 Brain Runner Flowchart

Fig. 4 Hybrid BCI

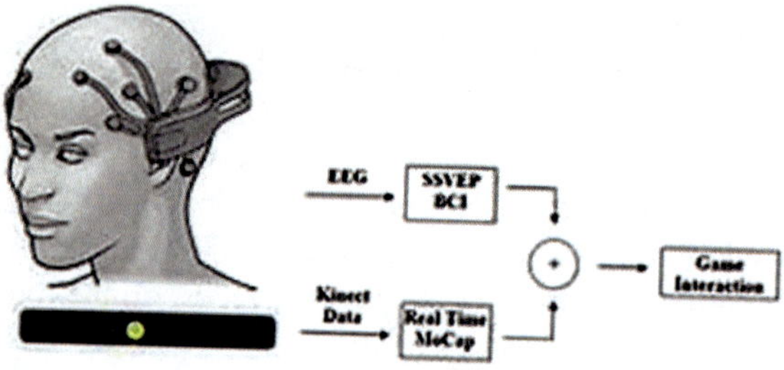

2.5 Interactive Session

Brain Runner game was performed at leading university of India. Staffs of different age group on both genders participated in the testing activity. The session was characterized into three phases: Introduction, Setup and Gameplay. A session started with a brief lecture: an introduction on neurogaming, the activity, expectations and questionnaire session. During the setup, the designer would explain the concept of the EEG recording and the electrical impulses that the neurons generate through which the signals are converted into brain–computer interaction. In the Gameplay,

Fig. 5 Hybrid BCI Architect

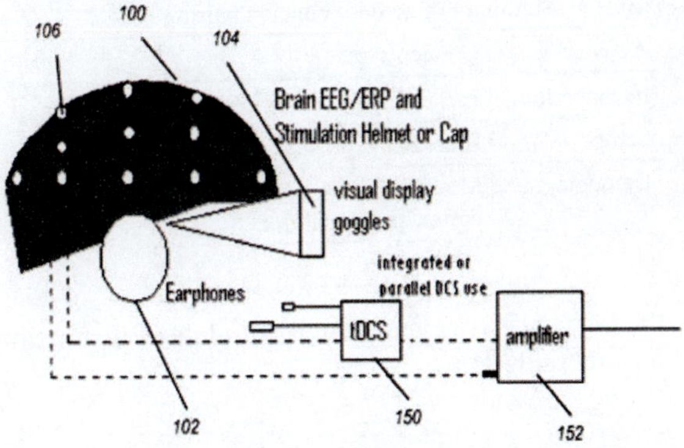

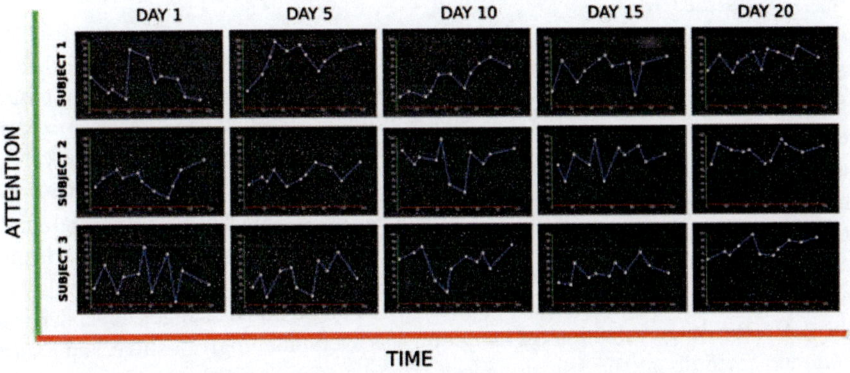

Fig. 6 Time for processing

the staffs where been taken through the User Interface (UI) of the game and then the answers for the questions. All together for one staff, the testing lasted for 15–20 min where their dopamine levels raised high and made the user play the game twice and trice. After a staff played finishing the game, they were been given feedback in the means of graphical form in which it represented their concentration level switched every second on various attempts and then once when they continuously played it also showed the improvement of their concentration and attention level (Fig. 6).

3 Results

Polling showed that the staffs of CURIN and Dept of CSA&D as a whole had an overwhelmingly positive effect on interest in giving more attention in playing without any controllers rather the entire action was through thinking. More interestingly those who played the game, everyone wanted to play again and again so that they can win

Table 1 Summary of activity concept pairing

Activity	Concepts
Introduction	Increasing the attention level of the brain
Setup	Electroencephalography device
Gameplay	Motor imagery, emotion representation, the role of The frontal lobe, prefrontal cortex and dopamine

the race and also even though with lots of distraction their concentration kept them trying (Table 1).

4 Conclusion

Gaming implementation of BCIs tackles the biggest challenges of the BCI of today. For an interface to be effective, the interaction must be seamless, and user experience must be of prime concern. Since all the BCI device depends on user motivation and mood, games are a natural research tool for a possible positive area of development. In this paper, we demonstrated a skeleton that could be utilized for upgrading neurogaming education. While the proof unequivocally recommends a positive inspirational part in the action, progressively a survey is required to determine the utility of neurogaming as an educational apparatus. We likewise noted changes dependent on the staffs. While the valuation for the action was overwhelmingly positive, some staff communicated dissatisfaction at the lack of completion of the game. Since sessions allowed relatively short training periods in shortcoming by using it as a point of conversation for illustrating the problems of neuroprosthetics and the complex nature of the human motor system which was excepted.

With further work, portions of the encouraging experience could be incorporated with the games themselves, automating the teaching and requiring less tutor guidance. This world requires heaps of arranging and discussion with increasingly experienced educators so as to structure an ideal game/instructing balance. We likewise accept that extra ideas can be given distinctive game designs. Our own design depended on evident properties: concepts of motor movement with motor imagery and concepts of emotional structures with emotional game balance. We would like to see a conversation between game designers and neuro specialists and neuro physiotherapist on extra, blending among design and education. While more work is required primarily instructive testing on which the game has more noteworthy potential as an educational tool.

References

1. Rideout, V.J., Foehr, U.G., Roberts, D.F.: Generations M2: Media in the Lives of 8- to 18-Years Olds (2010). Retrieved from https://www.kff.org/entmedia/entmedia012010nr.cfm
2. Bavelier, D., Green, C.S., Pouget, A., Schrater, P.: Brain plasticity through the life span: learning to learn and action video games. Ann. Rev. Neurosci. **35**, 391–416 (2012)
3. Van Aart, J., Klaver, E., Bartneck, C., Feijs, L., Peters, P.: Neurofeedback gaming for well-being. In: Brain-Computer Interfaces and Games Workshop at the International Conference on Advances in Computer Entertainment Technology, pp. 3–5 (2007)
4. Mak, J.N., Wolpaw, J.R.: Clinical applications of brain-computer interfaces: current state and future prospects. IEEE Rev. Biomed. Eng. **2**, 187–199 (2009)
5. Berka, C., Pojman, N., Trejo, J., Coyne, J., Cole, A., Fidopiastis, C., Nicholson, D.: Merging Cognitive Neuroscience & Virtual Simulation in an Interactive Training Platform (2012). Retrieved from https://www.researchgate.net/publication/236610554
6. Francisco-Aparicio, A., Gutiérrez-Vela, F.L., Isla-Montes, J.L., Sanchez, J.L.G.: Gamification: analysis and application. In: New Trends in Interaction, Virtual Reality and Modeling, pp. 113–126. Springer, London (2013). Retrieved from https://link.springer.com/chapter/10.1007%2F978-1-4471-5445-7_9
7. Brouwer, A.M., Van Erp, J., Heylen, D., Jensen, O., Poel, M.: Effortless passive BCIs for healthy users. In: International Conference on Universal Access in Human-Computer Interaction, July 2013, pp. 615–622. Springer, Berlin, Heidelberg. Retrieved from https://doi.org/10.1007/978-3-642-39188-0_66
8. Parsons, T.D., McMahan, T., Parberry, I.: Neurogaming-based classification of player experience using consumer-grade electroencephalography. IEEE Trans. Affect. Comput. (in press)
9. Liu, Y., Sourina, O., Nguyen, M.K.: Real-Time EEG-Based Emotion Recognition and its Applications. Retrieved from https://link.springer.com/chapter/10.1007%2F978-3-642-22336-5_13
10. Lotte, F., Congedo, M., Lécuyer, A., Lamarche, F., Arnaldi, B.: A Review of Classification Algorithms for EEG-Based Brain-Computer Interfaces. Retrieved from https://pubmed.ncbi.nlm.nih.gov/17409472/
11. Allison, B.Z.: Toward a hybrid brain-computer interface based on imagined movement and visual attention. J. Neural Eng. **7**(2) (2010a)
12. Allison, B.: BCI demographics: how many (and what kinds of) people can use an SSVEP BCI? IEEE Trans. Neural Syst. Rehabil. (2010b)
13. Banville, H., Falk, T.H.: Recent advances and open challenges in hybrid brain-computer interfacing: a technological review of non-invasive human research. Brain Comput. Interfaces (2016)
14. Kos'Myna, N., Tarpin-Bernard, F.: Evaluation and comparison of a multimodal combination of BCI paradigms and eye tracking with affordable consumer-grade hardware in a gaming context. IEEE Trans. Comput. Intell. AI Games (2013)

P. K. Parthasarathy is a professor at Chitkara Designs School, Chitkara University, Punjab, India. He has a decade of domestic and international experience in academic and industry leadership, teaching, consulting, and mentorship. His current mandate is to manage the Postgraduate and Undergraduate (Art and Design) program specially designed for young creatives minds and for working professionals at Chitkara University and also has additional charge as Dean, Chitkara Design School with a mandate to manage M.Des (UI/UX), B.Des (UI/UX), B.Des (Game Design), BFA (Applied Art), BA (Photography and Visual Art), B.Tech (Specialization in Game Programming). He is also active with various activities in AR/VR Laboratory under Chitkara University Research and Innovation (CURIN). His other exposure of experiences are as he has worked on Regional, National and International movies, designed and executed games for school in US according to US standard core syllabus, designed and executed AR/VR projects for L&TTafeTCS, Smart Stone Technology (Australia), Growlib (China), designed course curriculum and mentored for University of Wales and Birmingham City University—UK affiliated programs in India.

Archana Mantri is Ph.D. in Electronics and Communication Engineering with 30 years of experience in Research, Development, Training, Academics and Administration of Institutes of Higher Technical Education. Her areas of expertise are Project Management, Problem and Project Based Learning, Curriculum design and development, Pedagogical Innovation and Management. Areas of interest include Change Management, Education Technology, Cognitive Sciences, Predictive Analysis, Technical Writing, Assessment Technologies, Augmented Reality and Electronics and Communication Engineering. She has handled numerous International and National research projects and has guided several Ph.D. and ME scholars.

Amit Mittal is a professor at Chitkara Business School, Chitkara University, Punjab, India. He has over 19 years of domestic and international experience in academic leadership, teaching, research, consulting, training and mentorship. His current mandate is to manage the Ph.D. (Management) program specially designed for working executives at Chitkara University and also has additional charge as Dean, Chitkara College of Sales and Marketing with a mandate to manage MBA (Sales and Retail Marketing), MBA (BFSI) and MBA (Pharmaceutical Management). He is also active with various activities under the Chitkara University Office of Strategic Initiatives. He served the Oshwal Education and Relief Board (Nairobi, Kenya) for 3 years before joining Chitkara University in 2014. His areas of research and consulting expertise are International Marketing, Consumer Behavior, Brand Management, Shopping Behavior, Business Research Methods, Higher Education Curriculum Development and New Institution Building. He is an active resource person for FDPs, MDPs and Corporate Trainings. He is a known Afrophile and loves exploring different cultures. In his free time, he listens to western music and seeks (and tweets) information on everyday global issues.

Praveen Kumar is presently working as an assistant professor at Chitkara Business School, Chitkara University, Punjab, India. He earned his Ph.D. in finance from the Department of Business Administration, National Institute of Technology, Kurukshetra, India. He completed his MBA in Finance from the Department of Business Administration, Chaudhary Devi Lal University SirsaHaryana in 2014. He graduated in commerce from Manohar Memorial P.G College, Fatehabad (Haryana), affiliated to Kurukshetra University, Kurukshetra in 2012. He has published more than ten research papers in various international journals of repute including *Meditari Accountancy Research (Emerald Publishing Ltd), Managerial Finance (Emerald Publishing Ltd), Journal of Environmental Accounting & Management, SCMS Journal of Indian Management* etc. and presented eight research papers in different national and international conferences. He is also three times UGC-NET/JRF qualified in Management. At present, his research interests include ESG reporting and carbon accounting. Praveen Kumar is the corresponding author and can be contacted at praveen.kumar@chitkara.edu.in.

Intelligent Car Cabin Safety System Through IoT Application

Rohit Tripathi, Nitin, Honey Pratap, and Manoj K. Shukla

Abstract In the present study, focus has been made on car cabin or driver safety and monitoring. Safety is the main concern of many industries and countries while designing a car. This study aims to develop a smart driving system to attain a safe and done up traffic society for aged drivers. By analyzing the data of total number of road accident 35–40% and it is due to drowsiness or dizziness so to mind this issue, an attempt has been taken to solve this problem. Here, one hardware has been designed with help of sensors: alcohol sensor (MQ3), smoke sensor (MQ135), vibration sensor (SW-420) and gyroscope (ADXL335) and IP camera as well as GPS module (GY-GPS6MV2). Raspberry Pi has been used as microcontroller. It comes up with a solution of monitoring, alarming and minimizing entirely the respective percentage of accident due to mentioned reason. It is seen that the maximum drowsiness has been found in driver in evening time period, whereas it gives alcohol drunk level with accuracy of ±0.05 mg/L. Every driver can be monitored and diagnosed with the help of data using sensors, GSM and GPS and shared over the Internet using cloud to stop situation any mishappening while on road.

Keywords Face recognition · Raspberry pi · Arduino · GSM · GPS · Safety features

1 Introduction

In India, more than thousand road accidents occur every day around, and 20 children under the age of 15 die every day due to such accidents every in the country. The mentioned and many more are the reasons behind and unfortunate accident taking place out there in the world frequently. Altered for 181.36 million vehicles and

R. Tripathi (✉) · Nitin · H. Pratap
Department of Electronics and Communication Engineering, Galgotias University, Greater Noida, U.P., India

M. K. Shukla
Department of Electronics Engineering, Harcourt Butler Technical University (H.B.T.U.), Kanpur, U.P., India

H. Sharma et al. (eds.), *Congress on Intelligent Systems*, Advances in Intelligent Systems and Computing 1334, https://doi.org/10.1007/978-981-33-6981-8_21

its approximately 1.29 billion populations, India reported a traffic collision rate of about 0.8 per 1000 vehicles in 2015 compared to 0.9 per 1000 vehicles in 2012 and a 10.79 fatality rate per 100,000 people in 2015. The thought process behind this project occurred at this point of time when we thought of solving this issue which is immense in nature and is causing misfortunate to those who are affected by it. Though we cannot design something which considered every aspects or reason behind an accident, we tried to come up with a solution for a few. We narrowed the reasons down to which we can find a solution and design a prototype which when actively included in the system deduct and significant amount of mishappening on the road.

In literature review, many findings are present like the intelligent safety system has capably designed with the higher accuracy. By analyzing the system, it is found that such system extensively removed the entirely happened accident and to also pledge there were succeeding some respective rules like a fasten seat belt also as being come round in driving [1]. Inattentive driving is the cause of vehicle collisions. A person image was stored in a database as an entrance for sack out calculations in real time and the camera captured the image of the driver. This has given information that through image processing, software captured whether the driver was sack out or not by comparing the value of real-time database [2]. The system attained above 90% accuracy by analyzing the data including all major operation and the false alarm rate in on the road area almost below 5%, we practically showed that the head texture estimation algorithm was robust to extreme image deformation wireless system which provides embolden results, and it was also predicted that the image feature detection was in tackle situation and also improved the system performance and lastly further enhanced the gaze estimation [3]. It is also considered to use in other driver assistance system although the disturbance warning system might not have the substantial effect in teaching the driver to keep the eyes on road every time, and it might still be useful in critical situation [4]. In this system, it removed the number of accident and system monitored driver's image cures, it delivered alert of driver on critical condition, and also it gave driver voice to text facilities for some collision emergencies and traffic if system will detect the collision and instantly send the call to help nearby police station and hospital by using the data with the help of cloud [5]. In this work, the smart phone-based car accident detection system was not easy to handle it. It was nearly with the many obstacle that keep from the researchers from achieving higher accurate detection system. The object was determining the occupant which was inner or outer the vehicle. When the vehicle was moving at a low speed, then it minimized the impact of this obstacle which was proved in particle result controlled in this work [6]. The idea of accident detection, alert system with SMS to user-defined mobile number the GPS tracking and GSM alert-based algorithm was designed and implemented in embedded domain. This detection system was more accurate and with no loss of time, but there was one delay caused because of message; it was queue-based technique which could be reduced but given more priority to the message communicated to the controller [7]. In this project, the study presented accident detection and alert system with the SMS to define mobile number. The GPS tracking and GSM alert-based algorithm was designed and implemented within embedded system domain. It can also track a geographical information automatically and also send an alert message, it showed higher accuracy and sensitivity, and also this project is made more user-friendly. This method was verified to be highly beneficial for the industry [8]. It also saved the time and the accident location automatic detection which helped to provide security to the vehicle and lives of the people, and these studies provided a workable solution to traffic hazard [9]. In the detection system, the output of the sensor was analog in nature firstly which was converted in digital form with the help of analog to digital convertor of the microcontroller unit, and this unit controller was considered the entire circuit. When the output of system reached the threshold, the controller switches got ON and then the relay got cut off automatically and buzzer created sound [10]. The idea of accident prevention system used eye blink sensor which is based on infrared. It is noted if eye closed, this means the output was high, and similarly if the eye opened, it meant, the output was low. If the driver closes eye more than 3 s, then engine will be stop automatically and message is sent to owner [11]. The aim of this study was all about controlling theft of vehicle which made vehicles more secure by using GPS. It would be more convenient for the user to trace the target by using camera and also developing a mobile-based application to get the real-time view of the vehicles [12].

A non-intrusive prototype computer vision system for real-time monitoring system was developed. Firstly, the validation of the measure of the accuracy for vision technology is involved, and second, the validity of the fatigue parameter by analyzing the practical results which show fatigue monitor system, reasonable reliable, accurate in characterizing human fatigue, and it exhibited the state of the art in real time nonintrusive fatigue monitoring [13]. An efficient solution was provided to develop the smart system for the vehicle which gave various parameters of vehicle in between the constant time period. A system contained Arduino, alcohol sensor, GPS and GSM module which also improved the safety feature consequently proving to be an effective development in the automobile industry [14]. This system aimed successfully to detect and consequently identify a concentration of a gas, butane gas. It adapted with the given request and acted upon it in terms of controlling ignition of the two-wheeler car and indicated the exceeding of specific prohibited gas threshold, however it could. This system also improved by removing the parts of two-wheeler car with more reasonable component like as relay [15].

Few habits are drowsiness or dizziness, drunk drivers, sudden mishappening or as overheating of parts leading to system failure for drowsiness or dizziness. An attempt has been taken with the help of advanced image processing where, in which a procedure of blinking eyes pattern flame of the driver has been studied and analyzed, if he/she is sleepy or not in addition to this, there is a buzzer which will be active if the image study results in dizziness. For drunk driving, alcohol detection has been measured through sensor which gives an alert to nearby police station and hence will help in halting the driver from driving any further. Smoke detectors or detecting overheating of a part in the car resulting in system failure is also sensed, and an alert has been sent to driver and owner both at the same time. Using modern technology and approaches, the present study has been extended to high-end level where the exact location of the accident has been identified through GPS, and the output data of all sensors is going to save on cloud IoT application. Here, think peak has been used for cloud due to its free source.

2 IoT (Internet of Things)

Internet of Things (IoT) is an advanced automation and analytics system which exploits networking, sensing, big data and artificial intelligence technology to deliver complete systems for a product or service. These systems allow greater transparency, control and performance when applied to any system. The most important features of IoT include connectivity, sensors, active engagement and small device use. The main components of IoT-based system are microcontroller like Arduino or Raspberry Pi, sensors and actuators which help the designed system to operate wirelessly with the availability of Internet only. The application of IoT exits across all smart lifestyle and businesses or industries. IoT involvement offers improved customer engagement, technology optimization, reduced waste, enhanced data collection, remote operation and real-time monitoring.

3 System Description and Principal

In this study, a multipurpose and multi-level verification-based access control system has been designed. Before designing the present system, the detailed review has also been carried out. It is found that Arduino is the best microcontroller for IoT application, and it is widely available and having low cost. After that, Arduino has been taken. Here, three-level verification protocol has been adopted where the first verification stage is fingerprint through fingerprint scanner, the second one is password through keypad, and if access is allowed after both verification stages, SMS has been delivered to owner or authorized person as alert in both access on as well as access off stage. The block diagram of the system has been shown in Fig. 1.

The locking system setup mainly consists of Arduino UNO. It also contains of 4 × 4 keypad, fingerprint sensor (R305), GSM module (900 A), solenoid lock, jumper wires, etc. The Arduino IDE has been used as software and uses code in C++ language. Interfacing between the hardware and the software has been done from Arduino UNO to PC by connecting the Arduino cable. The circuit diagram of the designed system has been shown in Fig. 2. The specifications of all sensors have been given in Table 1. Once the hardware and software are connected, the "enroll" code is uploaded and then the fingerprints for the input purpose are taken from human finger with commands on the serial monitor and stored for the further verification purpose.

After the installation of all the components, the enroll code is entered and uploaded to the Arduino UNO board, and then the fingerprints have been taken and enrolled from multiple users and then the final code is uploaded to the board as a password/key code to be entered and function code for the process. Now, the hardware is ready

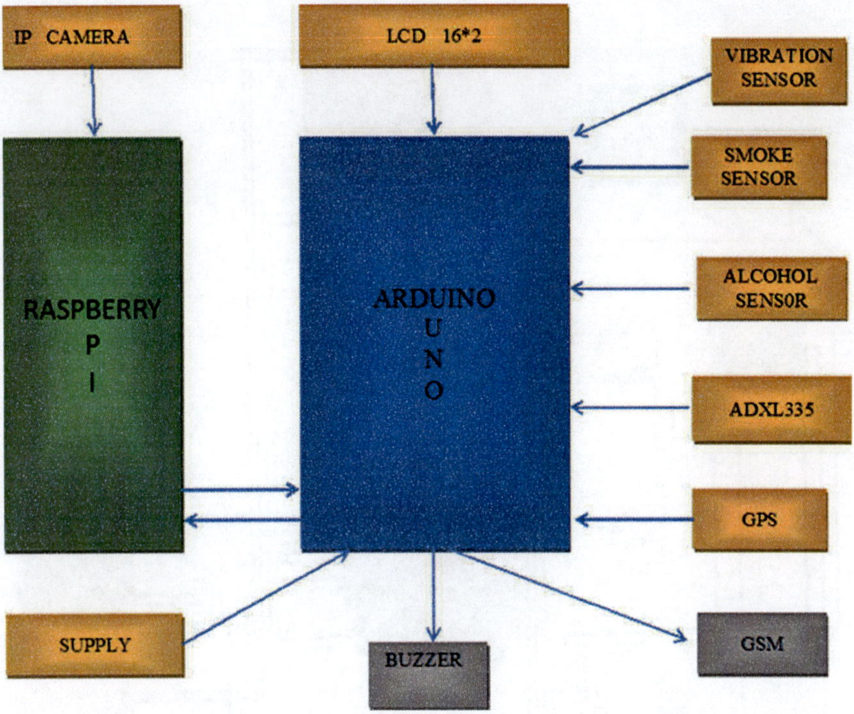

Fig. 1 Block diagram of smart car cabin system

for working in a proper manner with accessible only for authorized persons who can get access in particular location or area. Such systems can be implemented where restricted areas are unsafe for general public like bank lockers, confidential cupboards or rooms, VIP areas or any highly restricted areas. The actual hardware pictures of system have been given in Fig. 3.

The drowsiness of driver is detected by eye aspect ratio. Shape predictor with 68 facial landmarks is used. For finding the eye aspect ratio, eye landmark is used. Sleepiness is measured with the help of EAR, eye landmark ($X1$, $X2$, ... $X6$ and $Y1$, $Y2$, ... $Y6$) as shown in Fig. 4. Eye aspect ratio is measured by calculating the Euclidean distance between the eyes. The eye aspect ratio can vary person to person. The system is accurate by measuring eye closing rate in every 0.5 s. The average open eye aspect ratio of Indian people is 0.35, and the threshold value is 0.2 for closed eyes as shown in Figs. 5 and 6.

$$\text{EAR} = \frac{|X2 - X6| + |X3 - X5|}{2|X1 - X4|}, \frac{|Y2 - Y6| + |Y3 - Y5|}{2|Y1 - Y4|} \tag{1}$$

where $X1, \ldots, X6$ and $Y1, \ldots, Y6$ are the eye landmark for both eyes, respectively. The eye aspect ratio is stable around 0.35 when eyes were open and starts approaching to

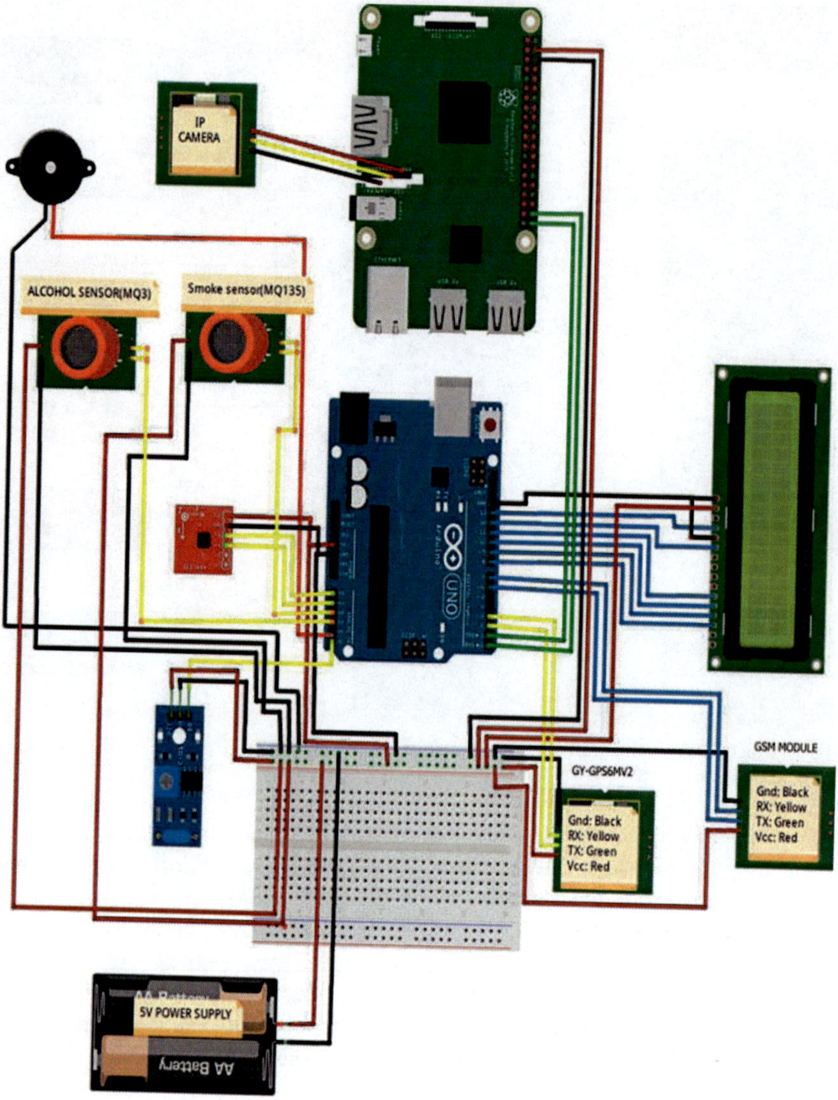

Fig. 2 Circuit diagram of proposed smart car cabin system

threshold value as eyes were closing. When it crosses, the threshold value drowsiness is detected.

Then the predefine program starts the buzzer in car and sends an alert text message to saved mobile number of owner or any family member as well as saved in web server.

Table 1 Name and specifications of sensor/input device

S. No.	Sensor/input device name	Work	Range
1.	Alcohol sensor (MQ3)	To sense the alcohol level of driver	0.04–4 mg/L
2.	Smoke sensor (MQ135)	To sense the smoke in vehicle	10.6 m × 10.6 m (coverage)
3.	Vibration sensor (SW-420)	To sense the vibration of vehicle during accident	
4.	Gyroscope (ADXL335)	Detect the inclined angle of vehicle with respect to horizontal surface	0 to ±180°
5.	IP camera	Streams live video of driver's face	
6.	GPS module (GY-GPS6MV2)	Collect the live location of vehicle	

Fig. 3 Actual picture of prototype of proposed smart car cabin system

4 Methodology

See Fig. 7.

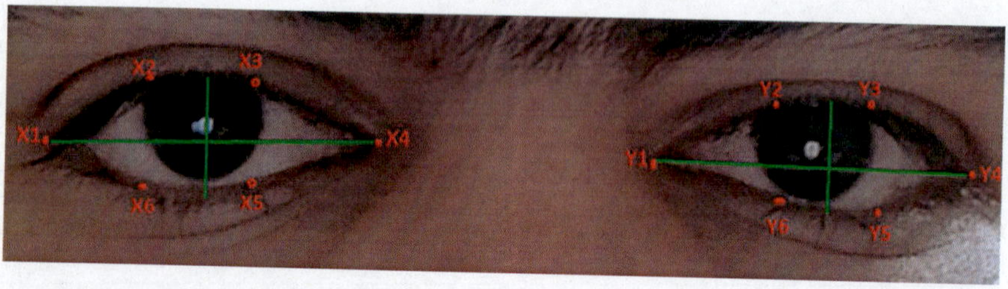

Fig. 4 Eye landmark (*Xi*, *Yi*) for open eyes

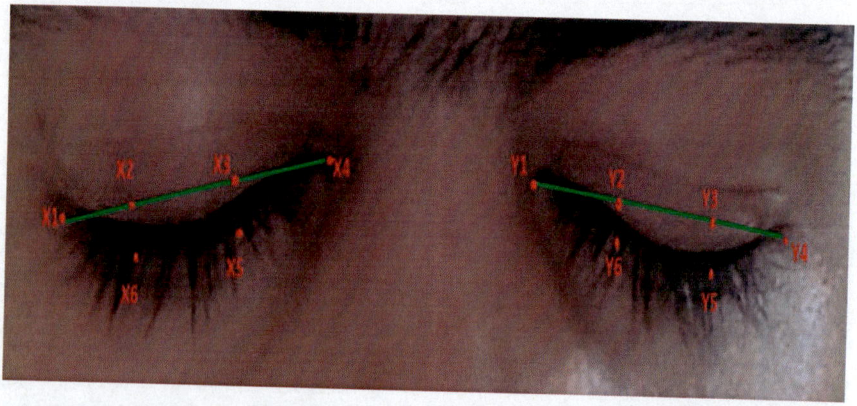

Fig. 5 Eye landmark (*Xi*, *Yi*) for closed eyes

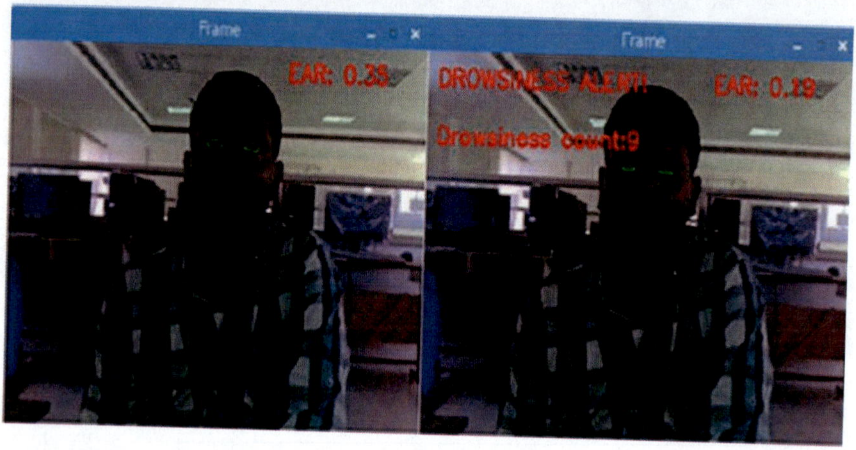

Fig. 6 EAR of open and closed eyes for drowsiness detection

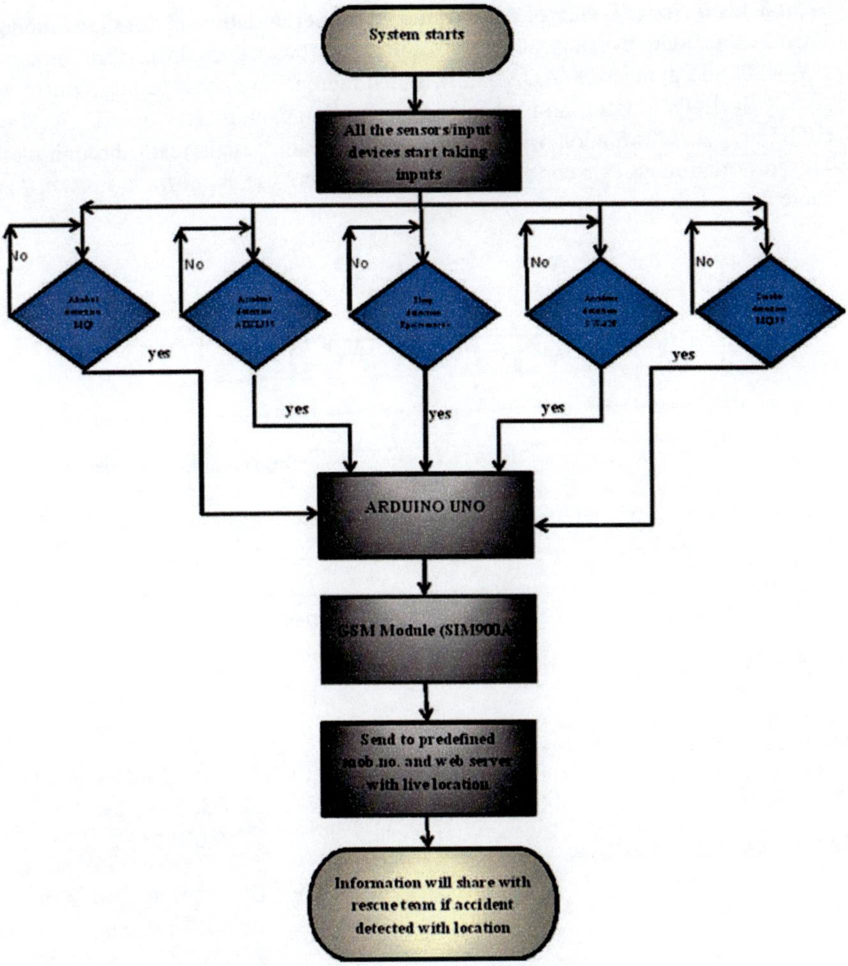

Fig. 7 Flowchart of system working through two-stage verifications

5 Results and Discussions

The drowsiness of driver and alcohol level of driver can be measured by our system. The system will calculate the eye aspect ratio (EAR) with the help of Raspberry Pi and Raspbian camera. The eye aspect ratio can vary person to person. The system is accurate with measuring eye closing rate in every 0.5 s, the average eye aspect ratio is 0.35 in India, and if the value crosses the existed threshold value that is 0.2, then we receive an alert signal from buzzer connected to the Arduino present in our system. Similarly, the range of alcohol sensor is 0.5–10 mg/L, and the threshold value is 1.0 mg/L in our system. Alcohol module will capture the alcohol level of driver, and if the value crosses predefine threshold value, then an alert message goes to predefined No. of owner of car or family member with the help of GSM module. Then if an accident happens, our system will detect it with the help of vibration sensor (SW-420) and gyroscope (ADXL335). When there is no mishappening, sensor will give logic LOW output, and when there is mishappening, sensor will give logic HIGH output. Information will be sent to family and rescue team through cloud. This information will also be saved in cloud for future use as shown in Figs. 8, 9 and Table 2.

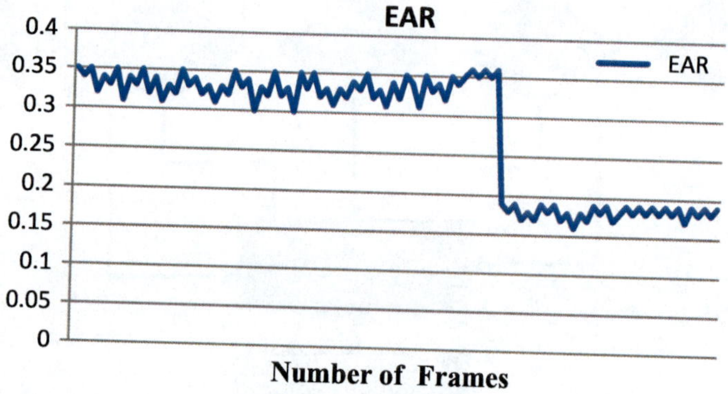

Fig. 8 Graph of eye aspect ratio for drowsiness detection

Fig. 9 Screenshot of alert messages

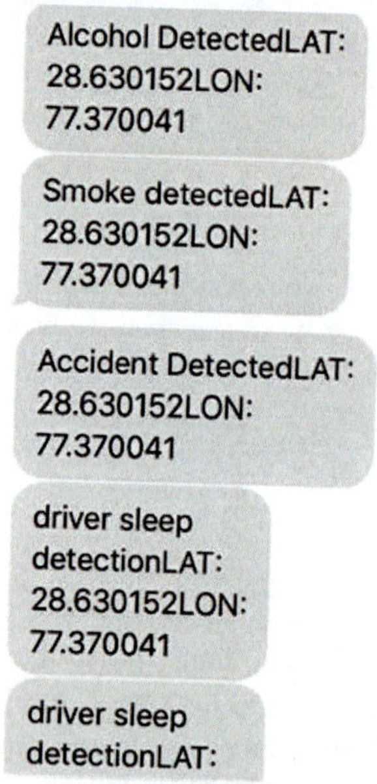

Table 2 Final output values of system

Driver		Drowsiness (EAR)		Drunk driven (mg/L)		Mishappening alert	
Age (years)	Time (h)	Affected	Unaffected	Affected	Unaffected	Affected	Unaffected
18–28	6:00–18:00	–	0.35	5.1	–	HIGH	–
	18:00–6:00	0.19	–	–	0.4	–	LOW
29–39	6:00–18:00	0.17	–	4.3	–	HIGH	–
	18:00–6:00	–	0.34	5.7	–	–	LOW
40–50	6:00–18:00	–	0.35	–	0.1	–	LOW
	18:00–6:00	0.18	–	4.9	–	HIGH	–

6 Conclusions

To reduce the no. of fatalities due to sleepy driver and drunk driver, a drowsiness detection and alcohol detection system is designed. The system will alert driver from sleepy state. The Raspberry Pi along with the Raspbian camera is used to detect and calculate the EAR of driver for drowsiness detection. Even if there was an accident, it will be captured by vibration and gyroscope sensors present in our system, and the information will share through GSM with live location for immediate rescue. This system can be implemented in vehicle of India to avoid the large number of accidents happening every year and to provide better results.

References

1. Hammadi, M., Smaeel, M.: Intelligent car safety system. In: International Conference of Institute of Electrical and Electronics Engineers, pp. 319–322 (2016)
2. Pingale, P., Gote, P., Rai, R., Madankar, S., Suresh, S.: DIP based monitoring and control of drowsy driving for prevention of fatal traffic accidents. Int. J. Innov. Eng. Sci. **2**(6), 89–92 (2017)
3. Vicente, F., Huang, Z., Xiong, X., Torre, F., Zhang, W., Levi, D.: Driver gaze tracking and eyes off the road detection system. In: International Conference of Institute of Electrical and Electronics Engineers, pp. 1–14 (2015)
4. Ahlstrom, C., Kircher, K., Kircher, A.: A gaze-based driver distraction warning system and its effect on visual behavior. Int. Conf. Inst. Electr. Electron. Eng. **14**(2), 965–973 (2013)
5. Charde, S., Bobade, N., Dandekar, R.: A methodology: IoT based drowsy driving warning warning and traffic collision information system. Int. Res. J. Eng. Technol. **5**(2), 3355–3357 (2018)
6. Gupta, T., Tripathi, R., Shukla, M.K., Mishra, S.: Design and development of IoT based smart library using line follower robot. Int. J. Emerg. Technol. **11**(2), 1105–1109 (2020)
7. Murkut, H., Patil, F., Yadav, V., Deshpande, M.: Automatic accident detection and rescue with ambulance. SSRG Int. J. Electron. Commun. Eng. **2**(6), 24–29 (2015)
8. Prabha, C., Sunitha, R., Anitha, R.: Automatic vehicle accident detection and messaging system using GSM and GPS modem. Int. J. Adv. Res. Electr. Electron. Instrum. Eng. **3**(7), 10723–10727 (2014)

9. Goud, V., Padmaja, V.: Vehicle accident automatic detection and remote alarm device. IJRES **1**(2), 1–6 (2012)
10. James, N., Aparna, C., John, T.: Alcohol detection system. Int. J. Res. Comput. Commun. Technol. **3**(1), 59–64 (2014)
11. Katkar, S., Kumbhar, M.M., Kadam, P.: Accident prevention system using eye blink sensor. Int. Res. J. Eng. Technol. **3**(5), 1588–1590 (2016)
12. Verma, P., Bhatia, J.: Design and development of GPS-GSM based tracking system with google map-based monitoring. Int. J. Comput. Sci. Eng. Appl. **3**(3), 33–40 (2013)
13. Ji, Q., Zhu, Z., Lan, P.: Real-time nonintrusive monitoring and prediction of driver fatigue. Int. Conf. Inst. Electr. Electron. Eng. **53**(4), 1052–1068 (2004)
14. Bhuta, P., Desai, K., Keni, A.: Alcohol detection and vehicle controlling. Int. J. Eng. Trends Appl. **2**(2), 92–97 (2015)
15. Al-Youif, S., Ali, M., Mohammed, M.: Alcohol detection for car locking system. In: International Conference of Institute of Electrical and Electronics Engineers, pp. 230–233 (2018)

Utilization of Delmia Software for Saving Cycle Time in Robotics Spot Welding

Harish Kumar Banga, Parveen Kalra, and Krishna Koli

Abstract In automobile manufacturing industry, resistance spot welding is widely used. Car's body is built by welding sheets of metal. In industry, common application where robots are being used is spot welding. In this paper, robot movement between two welding points, path followed while spotting, gripping, and pay-load-carrying activities, number of holds, moves, and possibility to enhance interaction between four robots were analyzed on offline robot simulation software 'Delmia V5'. Body shop assembly line has four Fanuc robots that perform about 209 welding spots in 532 s. After modification and proper sequencing, 12.7% reduction in cycle time was observed.

Keywords Assembly line balancing · Robotic welding · Offline programming

1 Introduction

Offline programming (OLP) is an attractive option for many manufacturers. Using simulation software, it is possible to digitally recreate robots, tools, fixtures, and the entire cell, and then define a program complete with motions, tool commands, and logic. That program can be processed and downloaded to the robot, similar to the process of programming any CNC machine with CAM software. Even if line stoppage is not a problem, however, there are still cases where designing a robot program in simulation is preferable to manual programming. For example, in a palletizing application involving dozens or even hundreds of boxes, a typical program could include hundreds of points. Teaching all these points manually would be tedious and time consuming. In these cases, OLP comes into play [1]. Automotive robotic cell is shown in Fig. 1.

After survey, it is found that with some sequencing and elimination of few activities, cycle time could be further reduced. But manual implementation of observed

H. K. Banga (✉)
Mechanical Engineering Department, GNDEC, Ludhiana 141006, India

P. Kalra · K. Koli
Production and Industrial Engineering Department, PEC, Chandigarh 160012, India

H. Sharma et al. (eds.), *Congress on Intelligent Systems*, Advances in Intelligent Systems and Computing 1334, https://doi.org/10.1007/978-981-33-6981-8_22

Fig. 1 Automotive robotic cell

points would be dangerous or hazardous. There might be chance of interference and clashing of two robots while working. Thus, we opt for offline working method for simulation and analysis. This research paper focuses on reduction of work cell cycle time and elimination of unnecessary cost by simulating the model on offline robotic software 'Delmia V5' [2].

2 Problem Statement

In SML, automotive industry is having robotic cell in which four FANUC robots have been used. They are having load capacity of 280 kg, which means each can bear their end effector load up to 280 kg. In robotic cell, there are two C gun and two X gun which are handed over to each robot diagonally. Truck cabin is having 216 spots which are supposed to be welded [3–5].

Three years before when JBM got contract for designing of robotic cell for ISUZU, they suggested to perform 160 spotting on respective truck cabin in 'robotic body shop cell' while remaining 56 spots to be welded in 're-spotting cell'. They simulated their suggested model in offline simulating software namely 'Delmia V5'. Cycle time generated was approximately 7 min 27 s (447 s). When all these things started implementing here in Ropar plant, ISUZU people asked them to add few more spots in order to reduce manual work in re-spotting section, thus enhancing productivity.

So activities like floor re-spotting and roof header re-spotting were added in robotic tasks. Number of spots to be welded in robotic cell then increased to 200. Remaining 16 spots at door-floor sections have been shifted to 're-spot section' due to geometrical constraints. But here problem got arrived. Incremented spotting has not been simulated for optimum time management. Similarly, in initial tasks, gapping and spacing between two activities were too much because of inter-dependence. Hence, cycle time is increased to 8 min 52 s (532 s) for 200 spots. By thorough observation standing in front of robotic cell and simulating same cycle in offline programming software, we ensured that there is a great chance of cycle time reduction by doing proper sequencing in PERT chart [6–10].

3 Evolutionary Approach

In order to develop well-planned, clash-less, and optimized model, we created the same in offline simulation software that is in 'Delmia V5'. By doing sequencing in PERT chart, reducing holding time and its time of occurrence, by adding or eliminating few tasks, we developed one cycle. Overall cycle time reduced by 68 s to 7 min 44 s (464 s) [11].

Points that we considered while optimizing cycle time virtually:

- Reach and Feasibility—We tried to check whether particular group of spots could be reached and performed by certain robot or not. It helped to decrease a workload from a robot which is working little excessively [12]. After putting roof by robot 2, it goes to home, picks gun, performs geo-spotting of roof with back panel, drops gun, and then puts off the gripper to home position. We tried to check whether same geo-spotting was in reach as well as feasible by robot 1 and 3 or not. We found that it was feasible. So, it reduced work load as well as performance time of robot 2 (Figs. 2 and 3).
- Task Sequencing—We noted down time taken for each task by each robot. Also, we noted down hold time by each robot at both time that is before and after lifting. We tried to interfere and re-sequenced tasks in this empty period by fulfilling required environmental conditions for each task [13].
- After Lifting Spot Adaption—Activities which left working alone in before lifting conditions were tried to adapted in after lifting conditions by checking time of both activity and empty spaces. Front panel re-spot was the activity performed by robot 4 and was adapted in after lifting condition. It was performed once side re-spotting of snake member was done [14–16].
- Motion Followed—Joint motion is to be applied while non-spotting activities such as gripping or putting off of any object, free movement non-carrying activities, and pay-load-carrying tasks. It helps to get smooth and flawless motion. While 'linear motion' is applied for spotting actions where path is linear and well directed, it gives us very disciplined motion. Time taken by joint motion is

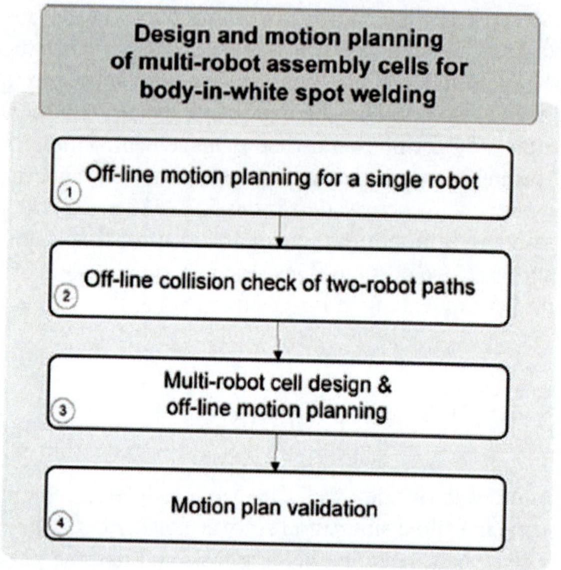

Fig. 2 Motion planning algorithm

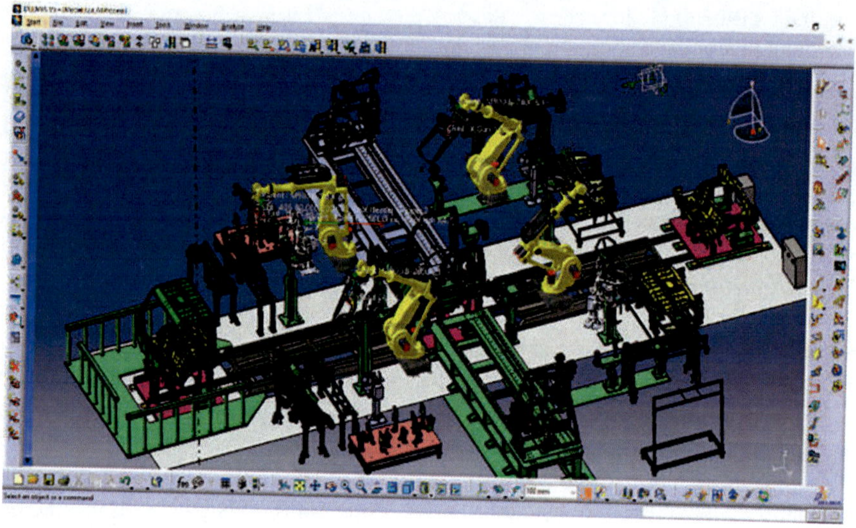

Fig. 3 Delmia V5 user interface

more comparatively linear motion. Thus, whole simulation is combination of both motions [17, 18].

- Limiting Conditions—Each robot is having six degrees of freedom. External seventh axis is added in the form of end effector may be C gun, X gun, or any kind of gripper. We took care that rotation of joint 5 would not exceed more than

85% and rotation of remaining axis not more than 88% as per norms. These are not critical limiting values, but here we consider factor of safety.

4 Simulation Experiments

There are a total of 10 activities in the current working cycle, whose overall working time is 204 s out of 532 s which means robot 1 is standing idle for the remaining 328 s. Through thorough observation, we found that robot 1 among all four robots is having maximum idle time. Also, some time robot 2 and robot 4 were waiting for robot 1 to complete its task. This thing we eliminated and made those two robots to finish their task and commanded robot 1 to perform afterward. We shared a few tasks from other robots with robot 1 as shown in Fig. 4. After doing geo-spotting of side panel, robot 1 remains idle. We inserted its X gun through the side panel and also made geo-spotting of a roof with a back panel. Similarly, at the end of the front panel re-spotting, it becomes free. Thus, we inserted it through the side panel to perform roof–roof header re-spotting.

There are total 13 activities in the current working cycle, whose overall working time is 400 s out of 532 s which means robot 1 is standing idle for remaining 132 s. After thorough observation, we found that robot 2 is busy enough, but its sequence is such that at some moments, robot 1 and robot 2 were waiting for robot 2 as shown in Fig. 5 to place roof at the dock. So, before performing side panel re-spot, we commanded it to dock roof before. As well as due to change in sequence of robot 3, collision was taking place in side panel region. Thus, we upgraded side-spot activity just after door grippers starts moving off. After docking of the roof, robot 3 used picks the gun used to perform geo-spotting, and then, gripper was removed. We shared this geo-spotting task by robot 1 and 3. Thus, overall working time reduced.

There are total 13 activities in the current working cycle, whose overall working time is 304 s out of 532 s which means robot 1 is standing idle for remaining 228 s. After thorough observation, we found that robot 3 as shown in Fig. 5 is busy enough, but re-sequencing is needed in order to make whole cycle compact. Thus, we placed front gripper first before doing side panel geo-spotting as robot 4 was idle in that time. Similarly, as discussed before, it shared geo-spotting of back panel and re-spotting of roof with robot 1. It was getting tackled while performing side re-spotting activity, thus its sequence changed from roof header re-spot to back panel re-spotting (Fig. 6).

There are total 10 activities in the current working cycle, whose overall working time is 367 s out of 532 s which means robot 1 is standing idle for remaining 165 s. After thorough observation, we found that robot 4 as shown in Fig. 4 also busy enough. We reduced its holding time due to cumulative response and support from other three. Front panel brought by robot 3 and roof by robot 2 early as compared to working one helped robot 4 to perform its task restlessly. Roof re-spot is the task which got shared by robot 1 and 3 made its work easier (Fig. 7).

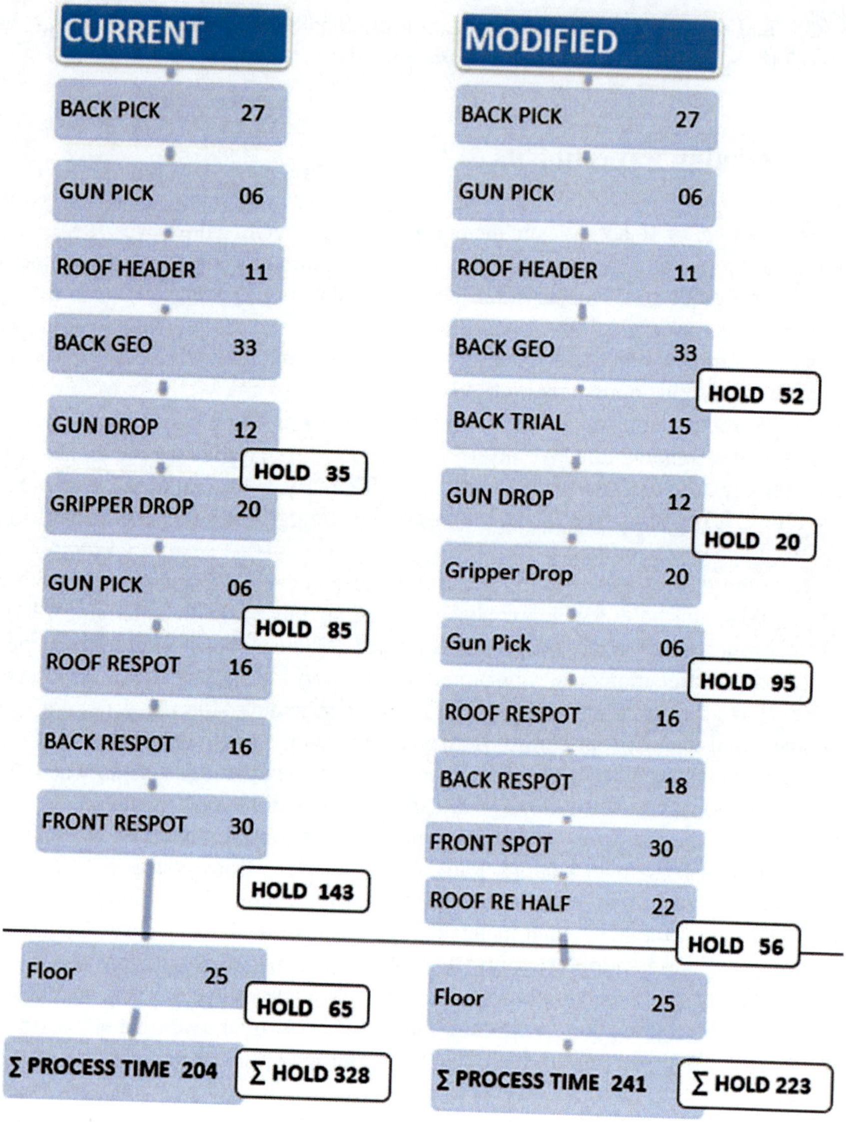

Fig. 4 Robot 1

- Modified sequence for each robot is '1–2–3–4–*–5–6–7–8–9 10–*', '1–2–4–5–#–#–#–9–10–3–13–11–12', '1–2–4–5–6–3–*–8–11–12–13–*–9–8–7', and '1–2–3–4–7–5–6–8–9–10–#', respectively [17], where '*' represents an added activity and '#' represents removed activity.
- Holding time for each robot has been reduced due to proper sequencing. Process time of robot 2 is reduced by 59 s (Fig. 8) due to removal of back geo activity

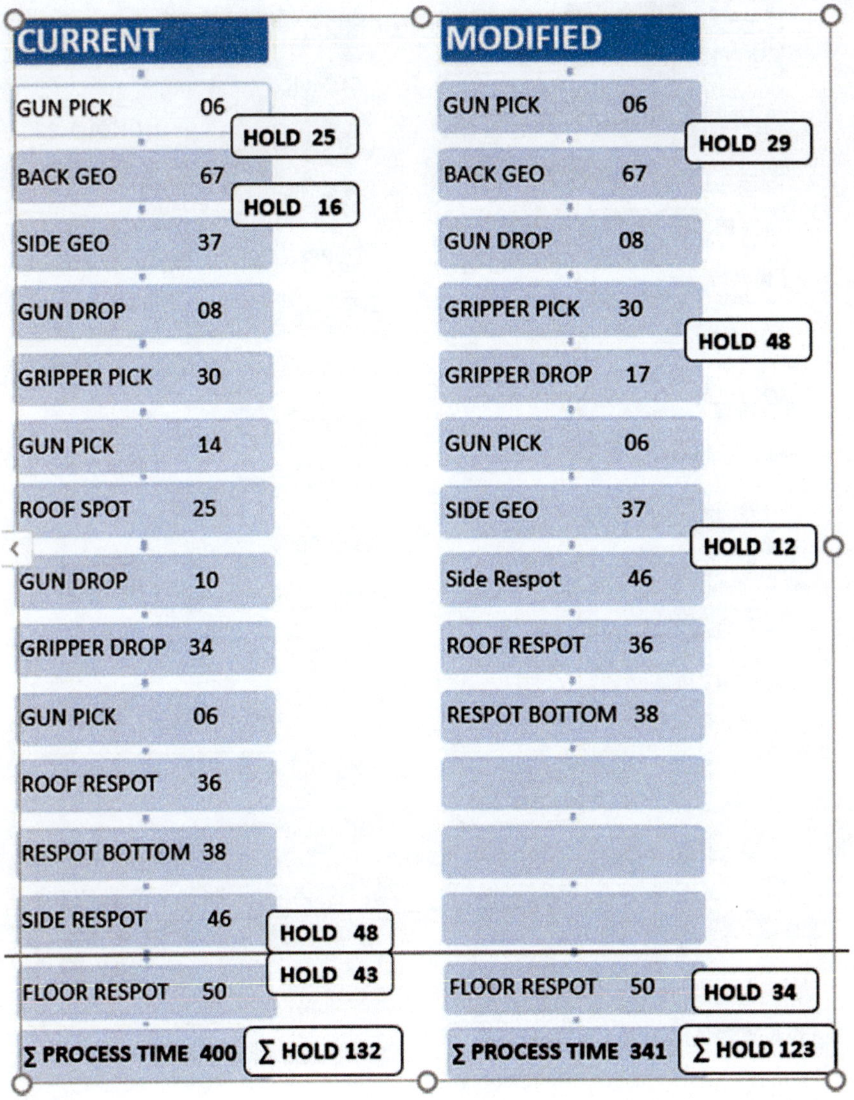

Fig. 5 Robot 2

while process time of robot 4 is reduced by 40 s due to removal of front re-spot activity. But due to addition of both activities as mentioned above has been shared by robot 1 and robot 3, their operation time increased by 37 s (Fig. 9) and 34 s, respectively.

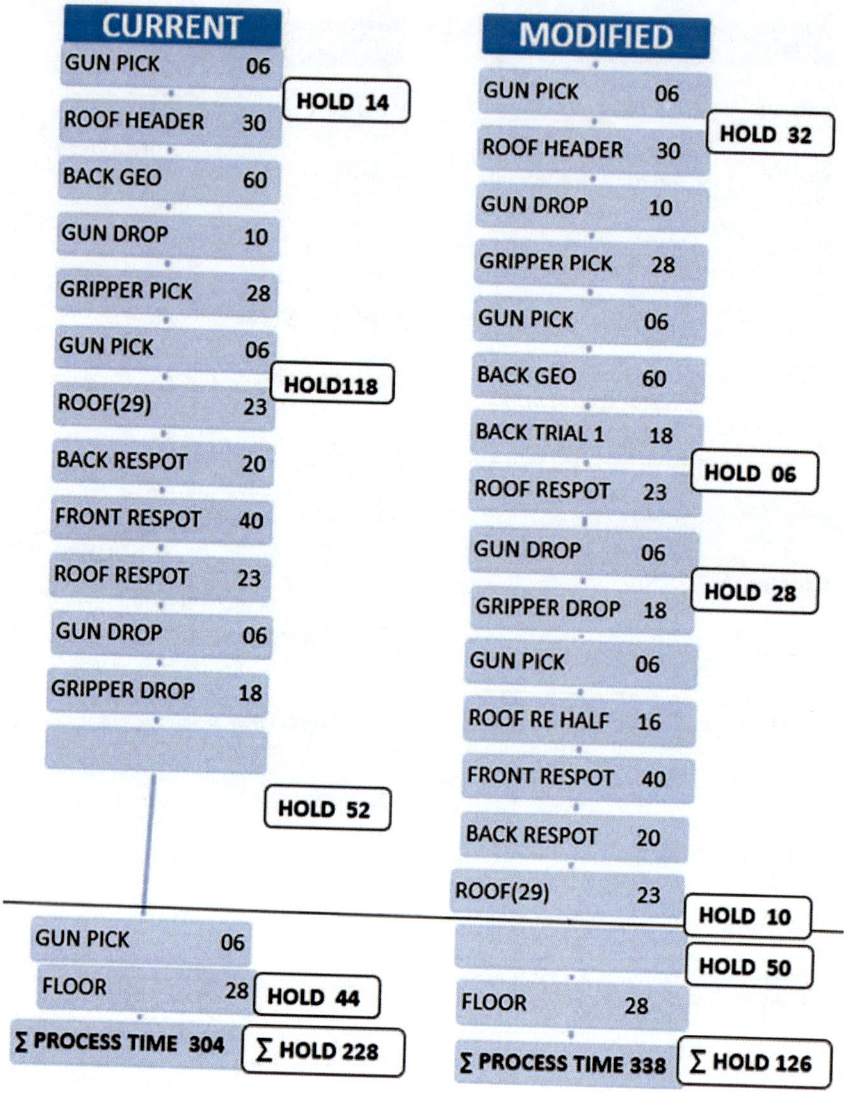

Fig. 6 Robot 3

5 Conclusion

By creating virtual model of body shop spot welding assembly line with help of simulating and controlling program, overall efficiency of a project significantly enhanced. Detailed leveling and application of simulating environment result in achieving accurate minimization of work cell cycle time. We successfully reduced overall cycle time by 68 s. Current cycle time is 532 s which is reduced to 464 s using Delmia V5.

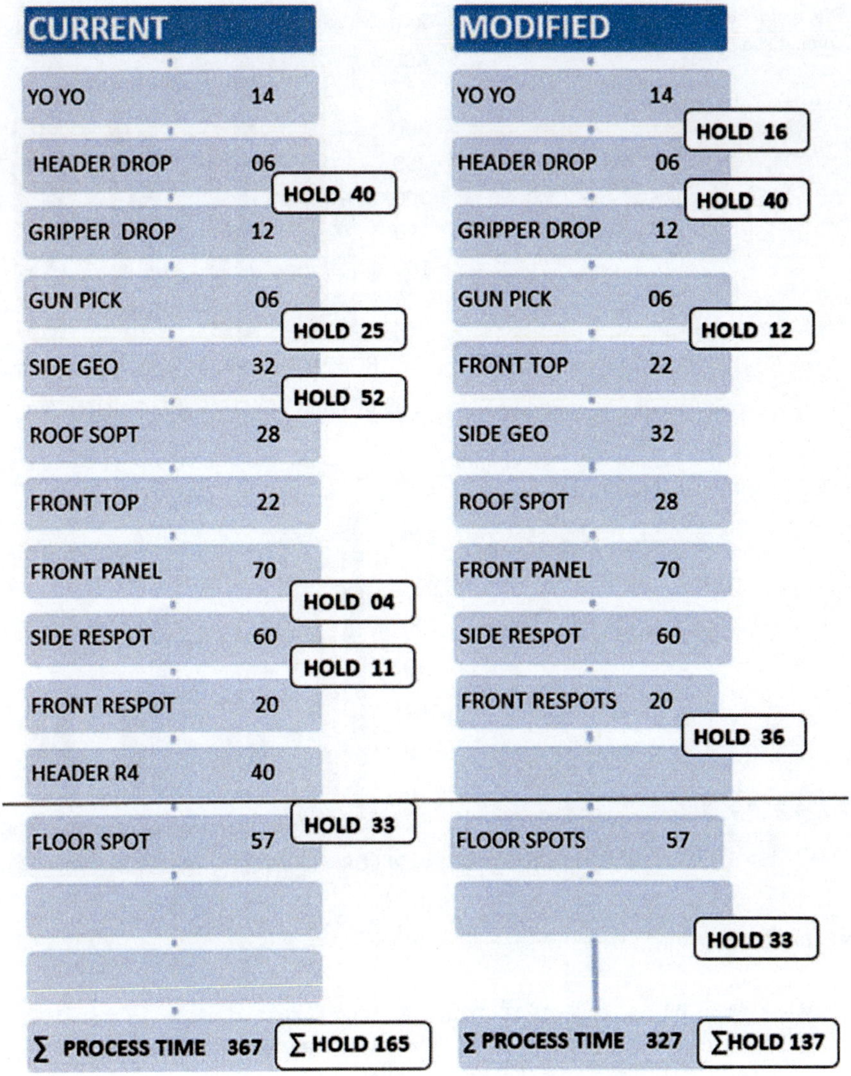
CURRENT
YO YO 14
HEADER DROP 06
HOLD 40
GRIPPER DROP 12
GUN PICK 06
HOLD 25
SIDE GEO 32
HOLD 52
ROOF SOPT 28
FRONT TOP 22
FRONT PANEL 70
HOLD 04
SIDE RESPOT 60
HOLD 11
FRONT RESPOT 20
HEADER R4 40
FLOOR SPOT 57
HOLD 33
∑ PROCESS TIME 367
∑ HOLD 165
MODIFIED
YO YO 14
HOLD 16
HEADER DROP 06
HOLD 40
GRIPPER DROP 12
GUN PICK 06
HOLD 12
FRONT TOP 22
SIDE GEO 32
ROOF SPOT 28
FRONT PANEL 70
SIDE RESPOT 60
FRONT RESPOTS 20
HOLD 36
FLOOR SPOTS 57
HOLD 33
∑ PROCESS TIME 327
∑HOLD 137

Fig. 7 Robot 4

Fig. 8 Process time comparison

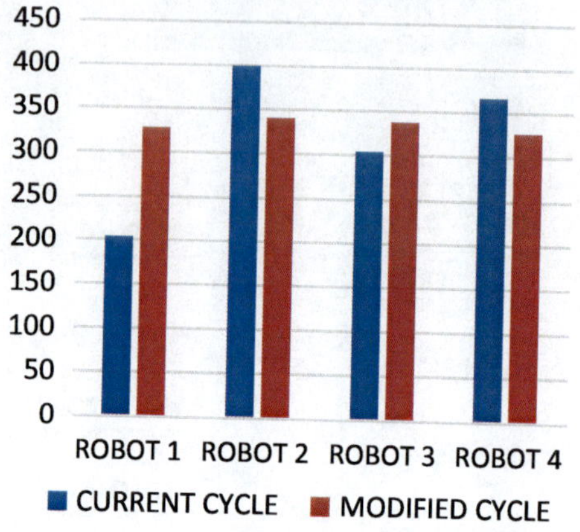

Fig. 9 Idle time comparison

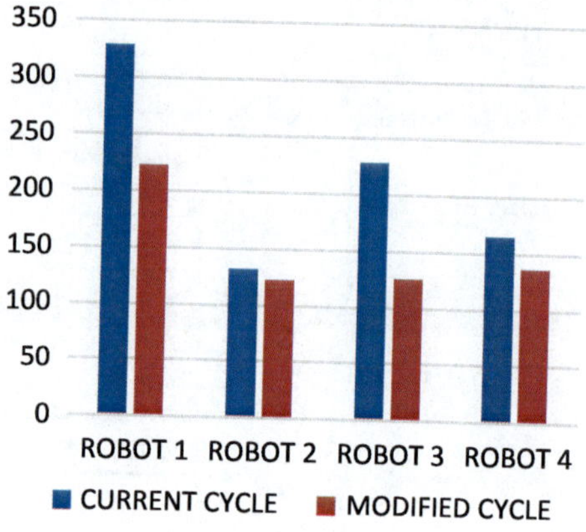

References

1. Akturk, M.S., Tula, A., Gultekin, H.: Design of a fully automated robotic spot-welding line. In: ICINCO 2011—Proceedings of the 8th International Conference on Informatics in Control, Automation and Robotics, vol. 2, pp. 387–392 (2011)
2. Aniwaa: 3D Scanning Process. Available: https://www.aniwaa.com/3D-scanning-technologies-and-the-3Dscanning-process. Accessed Feb 2016
3. Aniwaa: Artec Eva. Available: https://www.aniwaa.com/product/3D-scanners/artec-eva. Accessed Feb 2016
4. Annabi, N., Tamayol, A., Uquillas, J.A., Akbari, M., Bertassoni, L.E., Cha, C., Khademhosseini, A.: 25th anniversary article: rational design and applications of hydrogels in regenerative medicine. Adv. Mater. **26**(1), 85–124 (2014)
5. Apeagyei, P.R.: Application of 3D body scanning technology to human measurement for clothing fit. Change **4**(7) (2010)
6. Arbace, L., Sonnino, E., Callieri, M., Dellepiane, M., Fabbri, M., Idelson, A.I., Scopigno, R.: Innovative uses of 3D digital technologies to assist the restoration of a fragmented terracotta statue. J. Cult. Herit. **14**(4), 332–345 (2013)

7. Banga, H.K., Parveen, K., Belokar, R.M., Kumar, R.: Rapid prototyping applications in medical sciences. Int. J. Emerg. Technol. Comput. Appl. Sci. (IJETCAS) **5**(8), 416–420 (2014)
8. Banga, H.K., Belokar, R.M., Madan, R., Dhole, S.: Three dimensional gait assessments during walking of healthy people and drop foot patients. Def. Life Sci. J. (2017)
9. Banga, H.K., Kumar, R., Kumar, P., Purohit, A., Kumar, H., Singh, K.: Productivity improvement in manufacturing industry by lean tool. Mater. Today Proc. (2020)
10. Banga, H.K., Belokar, R.M., Kumar, R.: A novel approach for ankle foot orthosis developed by three dimensional technologies. In: 3rd International Conference on Mechanical Engineering and Automation Science (ICMEAS 2017), vol. 8, no. 10, pp. 141–145. University of Birmingham, UK (2017)
11. Banga, H.K., Belokar, R.M., Kalra, P., Madan, R.: Fabrication and stress analysis of kid's ankle foot orthosis with additive manufacturing. J. Mech. Eng. I-Manager J. **7**(1) (2017)
12. Chae, M.P., Rozen, W.M., McMenamin, P.G., Findlay, M.W., Spychal, R.T., Hunter-Smith, D.J.: Emerging applications of bedside 3D printing in plastic surgery. Front. Surg. **2** (2015)
13. Chang, J.W., Park, S.A., Park, J.K., Choi, J.W., Kim, Y.S., Shin, Y.S., Kim, C.H.: Tissue-engineered tracheal reconstruction using three-dimensionally printed artificial tracheal graft: preliminary report. Artif. Organs **38**(6), E95–E105 (2014)
14. Ciocca, L., Scotti, R.: CAD-CAM generated ear cast by means of a laser scanner and rapid prototyping machine. J. Prosthet. Dent. **92**(6), 591–595 (2004)
15. Dimitriadis, S.G.: Assembly line balancing and group working: a heuristic procedure for workers' groups operating on the same product and workstation. Comput. Oper. Res. **33**, 2757–2774 (2006). https://doi.org/10.1016/j.cor.2005.02.027
16. Dubravčik, M., Kender, Š: Application of reverse engineering techniques in mechanics system services. Procedia Eng. **48**, 96–104 (1992)
17. Nilakantan, J.M., Huang, G.Q., Ponnambalam, S.G.: An investigation on minimizing cycle time and total energy consumption in robotic assembly line systems. J. Clean. Prod. **90**, 311–325 (2015). https://doi.org/10.1016/j.jclepro.2014.11.041
18. Segeborn, J., Segerdahl, D., Ekstedt, F., Carlson, J.S., Andersson, M., Carlsson, A., Söderberg, R.: An industrially validated method for weld load balancing in multi station sheet metal assembly lines. J. Manuf. Sci. Eng. **136**, 1–7 (2013). https://doi.org/10.1115/1.4025393

Data Protection Techniques Over Multi-cloud Environment—A Review

Rajkumar Chalse and Jay Dave

Abstract Data protection is an essential and important process in a multi-cloud environment. Researchers proposed various techniques for data protection. This paper highlights the recent literature review on data protection. The paper showcases various techniques for data protection, i.e., data protection as a service (DPaaS), secure data storage (SCS), live migration, intrusion detection systems (IDS), secret sharing made short, and ESCUDO-CLOUD. It is explored that the machine learning also could be used with existing cloud services like DPaaS which is a dynamic service for protecting data in the cloud. The hybrid or combined approach using machine learning, physical parameter retrieval, and client-based validation is recommended for improvement of various frameworks leveraged across the platforms. In this paper, we analyze various data protection techniques to perform comparative analysis which lead to proposing of a secure client system with code hidden feature in context of multi-cloud environments.

Keywords Data protection · Secure data storage · Live migration · Multi-cloud

1 Introduction

The cloud technology from its inception has focused majorly on providing various utilities as a service like an infrastructure, storage, and software to its wide consumer base which is geographically separated. The nature of the storage utility has always been the prime concern because of the probability of conduction of various attacks by intruders during the migration and post-migration. Especially the multi-cloud environment [3] is more vulnerable to security-relevant threats, and thereby, it enables the robust requirement for leveraging cutting edge cloud platforms like data protection as a service (DPaaS) based SCS framework [1] IKaaS platform [2], protected fragmentation and distribution of data encrypted [4], storage as a Service (STaaS)-based secure cloud storage framework [5] and elliptic curve cryptography-based framework [6].

R. Chalse (✉) · J. Dave
Department of Computer Engineering, Indus University, Ahmedabad, Gujarat, India

H. Sharma et al. (eds.), *Congress on Intelligent Systems*, Advances in Intelligent Systems and Computing 1334, https://doi.org/10.1007/978-981-33-6981-8_23

Various cloud service providers have been using data encryption as a service (DEaaS) [1] to deal with security-related issues. However, data encryption as a service suffers when it comes to multi-cloud environments because of the inter-cloud exchange of data along with the dependency parameters and policy pertaining problems. However, DPaaS could play vital role while providing the solution as provided by the SCS structure. The proposed DPaaS model is not only capable of providing multiple encryption and decryption [6] options, but also it is capable of providing a multi-cloud access mechanism and policy preservation based on the segregation of managing and controlling the data from encryption and decryption of the data. The mentioned model operates at depth levels like at cloud consumer account level, virtual machine level, and the physical hardware level of each virtual machine.

The IKaas platform [2] supports secure storage as well as secure inter-cloud migration with the help of designated local clouds. The local cloud acquires and makes the data available for further reference post privacy certification approval. The PCA legitimates the data as per the policy of the government where it would be stored on the cloud virtual machine. The PCA would be followed by a security gateway mechanism placed on the verge of the local cloud where each such local cloud meets the global cloud. Each cloudlet has to pass through the gateway in the form of a query form. The response of such a query form usually follows the same channel until it meets the intended user.

There is variety of attacks from DOS to distributed disk operating system [3] which could be left unidentified by some of the security mechanisms like firewalls. Hence, the intrusion detection model may result in huge success if used with adequate training datasets with machine learning algorithms. However, in the multi-cloud environment, it becomes quite cumbersome to identify the category of the attack as per the little research work [3] which has been carried out. The supervised learning mechanism could be utilized for the purpose of detection as well as the categorization of random attack or vulnerability. The categorization is possible with the help of a couple of methods like single-type attack categorization [3] and stepwise attack categorization [3].

The fragmenting data is no longer a novel approach, but some research has shown that the retrieved sub-blocks of data could be compressed and considered as a plain text [4] [JD] in continuation to which the cipher text will be achieved after enforcing encryption algorithms. The final step will be taken before the dispersal which is the composition of cipher text blocks [4].

Furthermore, the research has indicated that elliptic curve cryptography-based novel approaches taken for vehicular networks [7] in cloud system have opened up possibilities for leveraging said approach for real-world applications.

There exist a prime need of a robust system which can get rid of all the probabilities of data exploitation and eliminate the chance of various attack conduction in both states of transaction and rest. The paper discusses the system which effectively enforces various encryption techniques at client side. Furthermore, the users have the privilege to decide the occurrence of enforcing the encryption algorithm on the retrieved segmentation of the semi-structured data.

2 Literature Survey

Providing data security in unsecured or untrusted third parties in general and cloud environment is often provided through control policies and encryption methods [7]. As example, Yu et al. introduced a secured access control for data in untrusted cloud storage system [8] based on combination of different techniques like attribute based encryption, proxy re-encryption, and lazy re-encryption.

In addition of the above research work, to address the issue of key block loss, Huang et al. show a key recovery scheme in cloud [9], a framework provided data privacy built on top of sector [10]. The basic research of this solution is to maintain secret key shares at the partially trusted parties for the key pair recovery.

Another related works to our research are [11] and [12]. On the one hand to [11], both their work and our target multi-cloud system provide several options for user to manage the services according to their greed's. The difference, however, is that their work focuses on protecting the virtual machine with firewall and security tools while focus is specifically on the encrypted data. On the other hand, our research work can be considered as an related of [12] because both works share the same idea of providing DEaaS. However, while [12] presented primitive ideas shown a high-level hard coded framework, our work introduced ideas with clear, proposed design framework. Besides, the working ideologies between this work and ours are also little different.

3 Comparative Analysis or Research Scope

Past more researches has been conducted with respect to data protection concerns into multi-clouds environment which we have given in Table 1, and this table shows the details of the literature survey in different researcher paper which includes the author name with published year of the paper, framework/Techniques used, issues addressed, security applied on, and benefits in the given researcher paper.

The table shows comparison of data protection technique in existing papers for providing data protection in cloud computing environment. It also describes various frameworks or techniques used, issue addressed, security applied on, and benefits.

4 Visibility of Comparative Study

Protecting data while transferring over cloud is still open research and found challenging task for researcher in cloud computing area. It required detail and real-time research for finding solution for protecting data while transferring over the cloud. The visibility of comparative study is to improve the performance of different frameworks which are based on various algorithm or techniques leverage by various frameworks.

The paper has clearly shown that each technique has its own issues and benefits over each other. Further, comparison of various research papers in the term of frameworks or techniques used, its issues addressed, data security applied on, and benefits have been given in the comparison Table 1. These techniques are as follows.

Table 1 Comparison of protecting data technique in existing papers

Ref. no.	Author and year	Framework/techniques used	Issues addressed	Security applied on	Benefits
[1]	Vu et al. 2015	SCS, DPaaS	To tackle on access control security problems, less flexibility	Cloud customer	It provided more dynamic or flexibility for protecting data over cloud by using these technique
[2]	Hashi et al. 2016	Intelligent knowledge as a service (IKaaS), live migration	Data contents duplicated from source to destination during migration	Cloud customer as well as cloud provider	It provides suitable data protection and privacy possible while live data migration over the cloud
[3]	Salman et al. 2017	Intrusion detection systems (IDS), machine learning (ML)	To design a secure framework for various security attack types	Cloud customer as well as cloud provider	It provides simple inspection for both security attack detection and categories different attacks in network traffic
[4]	Kapusta et al. 2018	SSMS, AONT-RS	It highlighted the issues that to provide data fragmentation, data protection, information, dispersal, etc.	Cloud customer as well as cloud provider	It provides easy and real way of fragmentation and dispersion of data encryption and data protection
[5]	Colombo et al. 2019	Storage as a service (STaaS)	Mixture of multi-cloud system adding with private cloud and virtual machine data centers	Cloud customer as well as cloud provider	It provides dynamic data storage and control as well data protecting over the cloud
[6]	Cui et al. 2020	Elliptic curve cryptography (ECC)	It faced to ignore vendor lock in and difficult to handle single point failure	Cloud customer as well as cloud provider	It provide robust and extensible authentication scheme

4.1 Description of Different Cloud Framework Used

For providing data security, different authors were used various frameworks in multi-cloud environment. Framework is a systematic process for providing encryption of data at client side by using various encryption algorithms. These various frameworks are as follows:

Secure Cloud Storage (SCS): SCS is a framework for providing data protection using DPaaS to cloud computing users. SCS framework consists of mainly three components: tenant management (TM), cloud platform management (CPM), and data security management (DSM). These frameworks allow user interaction via different functions. These functions operate on three provisioning levels of the framework like tenant account level, VM level, and data levels.

Data Protection as a Service (DPaaS): DPaaS is another framework based on cloud or Web delivered service for protecting data resources. Different companies can utilize this type of service to enhance network security and to build better protection for data in transit and data at rest.

Intelligent Knowledge as a Service (IKaaS): The IKaas framework platform integrated with data store on multi-cloud environment and providing preserving privacy policies simultaneously.

Intrusion Detection Systems (IDS): IDS is a framework in cloud computing system. It monitors a network for malicious activity or policy desecrations. Any malicious activity or violation was typically reported or collected centrally using a security information and event management system.

Secret Sharing Made Short (SSMS): SSMS is a technique for data security and data protection combining symmetric data encryption, information dispersion, and perfect secret sharing techniques.

Storage as a Service (STaaS): STaaS is technique that implements data security solutions that provide assurance for interoperability and enforcement of access restriction across multiple cloud service providers.

Elliptic Curve Cryptography (ECC): ECC is a framework for cloud computing security techniques. It is an approach to public key cryptography based on the algebraic structure of elliptic curves over finite fields. ECC requires smaller keys compared to non-EC cryptography (based on plain Galois fields) to provide equivalent security.

4.2 Issued Addressed

In the above table, there are various authors described various issues for data protection in multi-cloud environments.

Vu et al. [1] addressed issues that less flexibility in existing DEaaS. It did not provide flexibility to users for data protection over the cloud like cloud providers IBM, Amazon, and Google. But it did not provide effective data access control rules or policies to protect their data. Hashi et al. [2] addressed issues that violation of regulations while user data migration from source to destination. It had the least data protection mechanism provided it had violation of regulation of county or organization. Using that regulation violates user data not protected during data movement from source to destination movement. Salman et al. [3] addressed issues that their firewalls and old-style rule-based security techniques were not sufficient to protect user data in multi-cloud environment. It cannot be detected and categorized anomalies in the network traffic. It did not protect against different types of attacks. It decreased performance of cloud services in related of inactivity, load balancing, and scalability. It also faced capital expenditure (CAPEX) and operational expenditures (OPEX) for providing data protection over multi-cloud environments. Kapusta and Memmi [4] addressed issues that data protection and security challenges and opportunities over multi-cloud environments like cloud service deal with a large number of different outsider or external security attacks on day-to-day basis. It required how to slow down external attacks in cloud environments and protect user data. Colombo et al. [5] addressed issues that difficulty in separation of concerns between security and data management like the service is often vender lock in, i.e., the data encryption services is bound to a particular cloud service provider or cloud computing platform. It cannot provide data sharing between users; this is out of scope in DEaaS. Cui et al. [6] addressed issues that escape vendor locked-in and deal with single-point failure problem. It was difficult to provide data sharing between users. The users encounter difficulties in effectively considering one appropriate CSP due to occurrence of several data service providers [7].

In above table, it is shown that security applied by different authors for cloud customer as well as cloud provider.

4.3 Benefits

In above table, it is shown that there were various benefits provided by different authors related to cloud data protection like: It provided dynamicness and custom technique to protect data in cloud computing. It provided suitable data protection and privacy possible while live data migration over the cloud. It provides simple inspection for both security attack detection and categories different attacks in network traffic. It provided easy and real way of fragmentation and dispersion of data encryption and data protection. It provided dynamic data storage and control as well as data protection over the cloud. It provided robust and extensible authentication scheme.

5 Identified Gaps in Literature Survey

Following section contains the various limitation or gaps in earlier frameworks or techniques:

Existing framework or techniques had not implemented mathematical model to provide time complexity, security theorems, and proofs.
Spontaneous classification of data is not done in previous methods.
More dynamic and secure framework should be used in permutation. So, it will provide confidentiality to user data.

6 Mathematical Model

Assumption data size = 1.28 MB
Encryption algorithm AES
Approximate time to encrypt 1 block = 100 ms.

$$T_t = \sum (T_e + T_m) \tag{1}$$

where

T_t Total time for migration of the encrypted data segment.
T_e Time to encrypt particular data segment.
T_m Time to migrate particular encrypted segment.

$$T_e = 100 \text{ ms} \tag{2}$$

1 block of 128 MB requires time is 100 ms

$$T_m = \sum_{P=1}^{n} \frac{Xi}{Yi} \tag{3}$$

where

Xi represent ith data segment.
Yi represent ith channel capacity/bandwidth.

$$T_a = \frac{\sum T_t}{n} \tag{4}$$

where

T_a Average migration.
n No. of blocks.

7 Proposed Methodology or Related Work

Following are the steps to describe Fig. 1 proposed model:

Step1: Client or user login by their user ID and password
Step2: After successful login server send authentication key to client mail ID.
Step3: Client open their mail ID and enter that key to application
Step5: Select n number of records to be processed
Step6: Compressed that semi-structure data if it is large (optional)
Step7: Select n number of records for segmentation
Step8: Select m number of images to be used
Step9: Set counter = 0
Step10: WHILE counter < = n-1

INCREMENT Counter;
WRITE the record in encrypted form
Perform any algorithm here like AES, DES, RSA

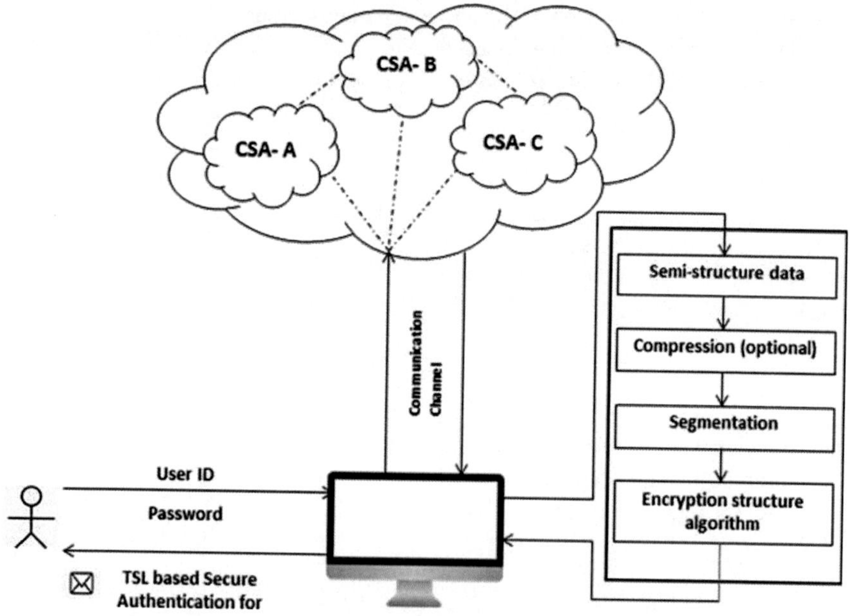

Fig. 1 Secure client system with code hidden features

ENDWHILE
WRITE "The End"

Step11: Stop
Step12: that data will be uploaded over the multi-cloud server like AWS, Google cloud, and Drop box.

In above section, it describes the proposed mathematical model of data protection mechanism at client side. This data protection model is executed by three modules: data protection client, semi-structure data encryption techniques, and data migration over cloud environment. The mechanism can be applied to different types of existing cloud computing system. Figure 1 shows secure client system with code hidden features. This proposed model shows that we proposed data protection model for providing more data protection security than existing system. We proposed such model that includes semi-structure data encryption technique and secure authentication of client or user. The data protection techniques need to protect client data while migrate over the cloud.

Data protection client module use secure client system block structure for protecting client data from different attacks. While protection data as well as client, we proposed to secure client login using TSL based secure authentication for client identification. While semi-structure data encryption technique module, we proposed a separate encryption techniques at client side that includes considering semi-structure data for processing. If size of data will be large, then we proposed compression technique for reducing the size of data for encryption process. This compression technique is optional. After data compression, segmentation will performed on compressed data.

In segmentation, we break the n no. of records in block size before encryption performs. After segmentation, we perform encryption structure algorithm like AES, DES, RSA, etc., we can use any encryption techniques on client data before migration over the cloud. While data migration over the cloud, we process encrypted data migrate over a cloud by using communication channel from client machine to cloud server (like AWS, Google cloud platform, Drop box, etc.). Client can migrate encrypted data over any cloud directly by their choice as well as client encrypted data migrated between cloud providers. Thus malevolent agent programs cannot intrude host or client machine.

8 Conclusion

The cloud service providers and the cloud users must ensure data safety during transition and on the cloud from all the internal threats or malicious, external threats, or attacks and mutual understanding between the customer and provider when it comes to the data protection of multi-cloud environment. The emphasized a mathematical formation along with the component diagram showing the TSL based authentication for the discussed robust system. The proposed mathematical formulation of encrypting the semi-structure data could be implemented in real time. The data protection framework model for data security at client side for live migration of user data over cloud environment could be done using the django or Python scripting connected to appropriate cloud service provider. Furthermore, the existing methodology does not showcase the application on various types of data, the size of the data, and domain of source data.

References

1. Vu, Q.H., Colombo, M., Asal, R.: Secure cloud storage: a framework for data protection as a service in the multi-cloud environment. In: 2015 IEEE Conference on Communications and Network Security (CNS), ISSN (2015). https://doi.org/10.1109/CNS.2015.7346879
2. Hashi, Y., Uchibayashi, T., Hidano, S.: Data protection for cross-border live migration in multi-cloud environment. In: 2016 Fourth International Symposium on Computing and Networking (2016). ISSN 2379-1896
3. Salman, T., Bhamare, D., Erbad, A.: Machine learning for anomaly detection and categorization in multi-cloud environments. In: 2017 IEEE 4th International Conference on Cyber Security and Cloud Computing (2017). ISSN: 978-1-5090-6644-5/17
4. Kapusta, K., Memmi, G.: Enhancing data protection in a distributed storage environment using structure-wise fragmentation and dispersal of encrypted data. In: 2018 17th IEEE International Conference on Trust, Security and Privacy in Computing and Communications/12th IEEE International Conference on Big Data Science and Engineering (2018). ISSN 2324-9013/18
5. Colombo, M., Asal, R., Hieu, Q.: Data protection as a service in the multi-cloud environment. In: 2019 IEEE 12th International Conference on Cloud Computing (CLOUD) (2019). ISSN: 2159-6190/19.
6. Cui, J., Zhang, X., Zhong, H., Zhang, J., Liu, L.: Extensible conditional privacy protection authentication scheme for secure vehicular networks in a multi-cloud environment. IEEE Trans. Inf. Forensics Secur., 1654–1667
7. Vimercati, S.D.C.D., Foresti, S., Jajodia, S., Paraboschi, S., Samarati, P.: Encryption policies for regulating access to outsourced data. ACM Trans. Database Syst. **35**(2), 12:1–12:46 (2010)
8. Yu, S., Wang, C., Ren, K., Lou, W.: Achieving secure, scalable, and fine-grained data access control in cloud computing. In: Proceedings of the IEEE INFOCOM, pp. 1–9 (2010)
9. Huang, Z., Li, Q., Zheng, D., Chen, K., Li, X.: YI Cloud: improving user privacy with secret key recovery in cloud storage. In: Proceedings of the 6th International Symposium on Service Oriented System Engineering (SOSE), December 2011, pp. 268–272
10. Gu, Y., Grossman, R.L.: Sector and sphere: the design and implementation of a high performance data cloud. Philos. Trans. R. Soc. A Math. Phys. Eng. Sci. **1897**, 2429–2445 (2009)
11. Daniel, J., Dimitrakos, T., El-Moussa, F., Ducatel, G., Pawar, P., Sajjad, A.: Seamless enablement of intelligent protection for enterprise cloud applications through service store. In: Proceedings of the 6th IEEE International Conference on Cloud Computing Technology and Science (CloudCom), December 2014, pp. 1021–1026
12. Pawar, P.S., Sajjad, A., Dimitrakos, T., Chadwick, D.W.: Security as-a-service in multi-cloud and federated cloud environments. In: Proceedings of the 9th IFIP International Conference on Trust Management (IFIPTM) (2015)

Prof. Rajkumar Rameshrao Chalse has completed his Master of Engineering in Computer Science with specialization in Wireless Communication and Computing and Bachelor of Engineering in Information Technology. His area of interest lies with cloud computing, data protection and security issues with cloud and multi-cloud environment. He has a good databank of research papers under several headings, which can be fetched from scholar and you can reach him on rajchalse@gmail.com.

Dr. Jay Dave is vastly experienced assistant professor with a demonstrated history of working in the higher education industry since 2009. He is a strong education professional with a Doctorate degree in Computer Engineering. He is having cloud computing as an area of interest. He is a Ph.D. Supervisor with Indus University, Ahmedabad, AWS certified cloud practitioner and AWS Academy accredited educator. Contact email id—jay.dave4u@gmail.com.

Hierarchical Ontology Based on Word Sense Disambiguation of English to Hindi Language

Shweta Vikram

Abstract There are numbers of issues when research comes to question paper translation which should be effectively handled by applying suitable approaches and word sense disambiguation (WSD) algorithms. In order to have a machine translation (MT) system, this could be used for practical purposes. The further study and the analytical work carried in the present research are to develop an efficient MT system. MT would greatly reduce the dependency on human experts in translating questions into different Indian languages for various exams that require bilingual papers. This paper proposes an algorithm based on a hierarchical ontology which uses a hierarchical tree structure. Use of hierarchical structure reduces the time of translation while ambiguity was also reduced. The experiment was done on a real dataset of questions of English language of NCERT and other source.

Keywords Machine translation · World sense disambiguation · Questions · Hierarchical · English and Hindi

1 Introduction

The number of issues arrived during the machine translation process toward translation of English questions in the equivalent Hindi form. It has also been observed throughout this work that a number of researches have been carried in WSD for various languages and have also been successfully applied in MT systems [1–7]. The issue of ambiguity in translation of question paper from English to Hindi is a challenge and needs to be resolved through suitable disambiguous algorithm.

India is a multilingual country with 22 official languages [8], but a majority of people in India are familiar with Hindi language. However, English is still a dominating language as far as the working of the government of India is concerned. Most of the official government documents are still in English, whereas states work and communicate in their regional language as well as in English. This gap

S. Vikram (✉)
Babasaheb Bhimrao Ambedkar Univcrsity, Lucknow, India

H. Sharma et al. (eds.), *Congress on Intelligent Systems*, Advances in Intelligent Systems and Computing 1334, https://doi.org/10.1007/978-981-33-6981-8_24

between central and states requires lots of translations and raises various communication issues. Even most of the examinations conducted in India by various agencies do prefer English as a language for asking questions to the candidates. The states, however, conduct various examinations in their regional languages (including competitive exams). To conduct the examination for various purposes in India, therefore, it requires a number of translations of question paper from English to Hindi and other Indian languages and also between one Indian language to others. The time, energy, and cost involved in translations can be reduced by using a suitable Machine translation (MT) tools.

Translation of questions appearing in various competitive examinations from English to Hindi and other Indian languages are mostly being carried out manually. It includes the timely availability of human specialists in order to correctly translate questions to and from various Indian languages. The translations of question papers using an MT tool may highly help in such circumstances to cut time and energy. Though there are many good Indian languages MT tools available (offline and online) such as Anusaaraka [9, 10], BabelFish [11], Babylon [12], Bing [13], Google [14], they still perform fairly while translating many natural language sentences and the issues such as ambiguity [15], ordering [16], Tense-Aspect and Modality (TAM) [17], gender, synonym aspect [15, 18] often causes translation to become vague. Among these, ambiguity during translation is the most critical aspect. There have been many studies and successful implementations of WSD algorithms to minimize the issue; however, the high-level accuracy of translation still remains a challenging task in MT research.

Word ambiguity is a challenging task in almost all natural language processing (NLP)-based application, and word sense disambiguation (WSD) is a research area which helps inappropriately handling the ambiguity issue. WSD aims to automatically identify the correct sense of a word in a particular context by applying a suitable technique. This problem persists since a long in NLP which lead to many researchers to make machine translation (MT) projects development meaningful.

A large no. of researches has been carried in the area of MT for various language pairs [19, 20]. A number of MT tools have also been developed in India and across the world. These tools are either open domain or are based on the specific area of applications. Many popular MT tools though translate all types of sentences to the varying degree of accuracy, they have not been specifically designed for translation of questions from one natural language to another. Specially in India, when many exam questions are needed to be translated in various Indian Languages, the general domain MT tools often fail to produce the desired accuracy while translating questions from English to Indian languages.

There has not been much effort or study on knowing the issues in automatic translation of questions specially from English to Hindi, though WSD and other important issues have been widely discussed in the literature on various sentences. Therefore, this research aims to specially target various questions sentences to analyze how they behave with the existing MT tools which are otherwise very popular [21–26].

The work carried in this work clearly reveals that there are a number of issues when it comes to question paper translation which should be effectively handled by applying suitable approaches and WSD algorithm in order to have an MT system which could be used for practical purposes. The further study and the analysis of work carried in the present research may help to develop an efficient machine translation system which would greatly reduce the dependency on human experts in translating questions into different Indian languages for various exams that require bilingual papers.

Some previous experimental analysis [27, 28] clearly shows that none of these are capable of appropriately handling the issue raised in our experiments. It is also been observed in many cases that translators while translating the question in Hindi have changed the overall interpretation of the questions. This is due to the fact that these translators could not have appropriately resolved various type of ambiguity that may have occurred in such questions.

In another observation, it is found that as the size of questions increases then translation accuracy decreases [29]. Almost all translators have performed fairly well with category I type (small size) questions, whereas they perform poorly for category III (large size) questions [29].

In another important analysis with different types of questions [30–34] (Wh-questions, objective, match, fill in the blank, and keyword specific), our previous analysis show that among the performances of different types of questions, the objective-questions perform better than other questions and match questions performed poorly [35].

The work described in this paper leads to a number of distinct area for future investigation. Various studies on a different aspect of ambiguity and development of an MT with automated disambiguation can be done for question paper translation from English to the Hindi language. An efficient MT system based on our analysis of WSD impact can help the users a lot, and they can be getting rid of the question translation refinement process to some extent.

2 Proposed Methodology

Whole work is divided into two modules; first, we have developed the ontology [36, 37] of the work in the hierarchical structure where two-level and three-level structures were developed, while second was a testing module where English questions were passed in the trained ontology which will give the relevant word in the output. Whole work is graphically shown in Fig. 1 while an explanation of each step was done step by step.

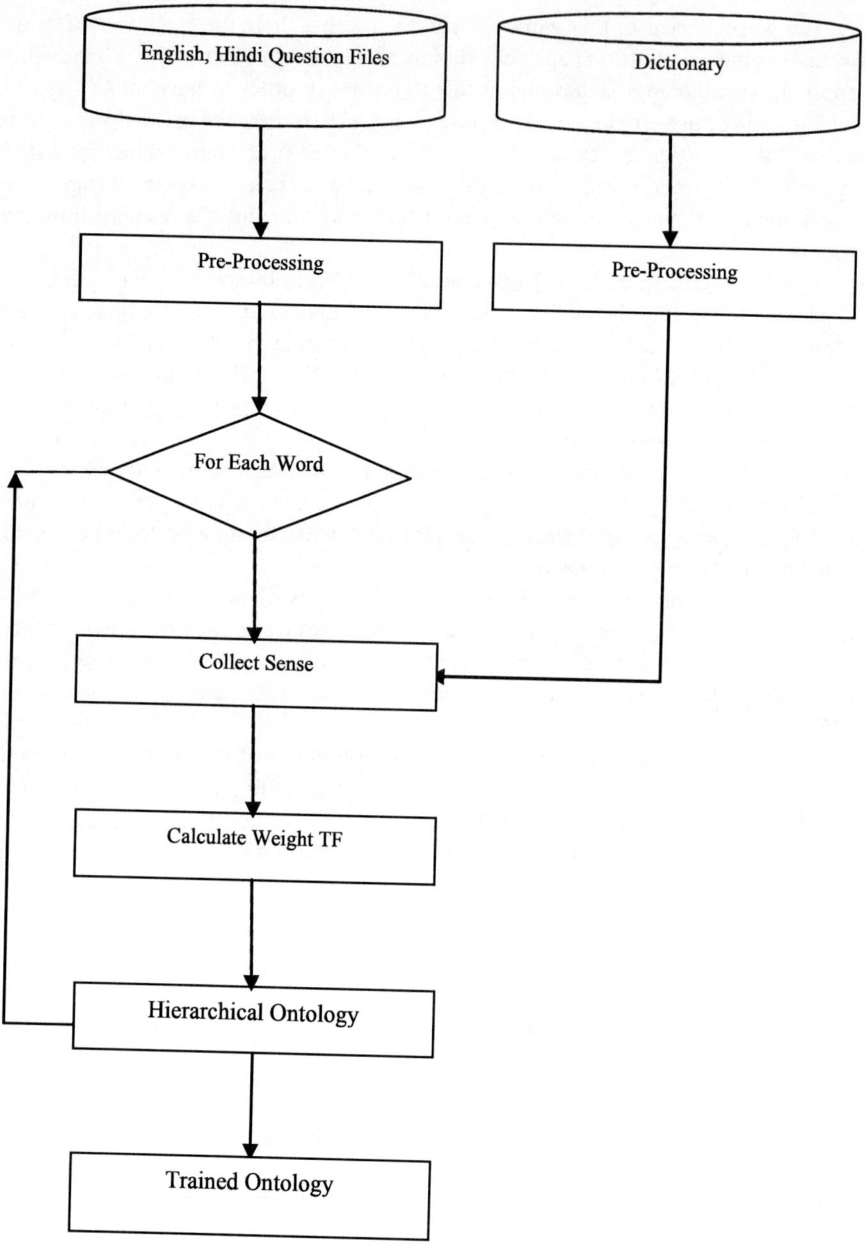

Fig. 1 Proposed work block diagram

3 Preprocessing of English/Hindi Questions File and Dictionary

In this, we discuss methodology for preprocessing of training English data file as well as training Hindi data file, and further we discuss dictionary preprocessing.

3.1 Preprocessing

First, we discuss preprocess for training English data file and training Hindi data file English files are input to the dataset, where each sentence or question is treated separately. Once a file is broken into sentences, each sentence is further segmented into words and special characters. This reduces dictionary searching time for finding the appropriate meaning in the dictionary. This can be understood with an example question file,

Training English (TE): "How does sociology study religion?, What are the strengths and weaknesses of participant observation as a method?".

Now, this is separated into two lines, English Question Vector (EQV).

"How does sociology study religion?".

"Analyze the role of credit for development."

Collect word into English Word Vector (EWV) = ['How', 'does', 'sociology', 'study', 'religion', '?'].

Similar to English files, Hindi file is input to the dataset, where each sentence or question is treated separately. Once the file is broken into sentences, each sentence is further segment into words and special characters. This can be understood with an example question file,

Training Hindi (TH): "समाजशास्त्र धर्म का अध्ययन कैसे करता है?, एक पद्धति के रूप में सहभागी प्रेक्षण की क्या-क्या खूबियाँ और कमियाँ हैं? "

Now this is separated into two lines, Hindi Question Vector (HQV)

"समाजशास्त्र धर्म का अध्ययन कैसे करता है?"

"एक पद्धति के रूप में सहभागी प्रेक्षण की क्या-क्या खूबियाँ और कमियाँ हैं?"

Collect word into (Hindi Word Vector (HWV) = ['समाजशास्त्र', 'धर्म', 'का', 'अध्ययन', 'कैसे', 'करता', 'है', '?'].

3.2 *Dictionary Preprocessing*

In this step, the input dictionary is arranged into a set of an English word and its corresponding Hindi words. Using data matrix (DM), length of the word is to be checked. In order to reduce search time, the words having length more than three in data have been stored in the linear form (in alphabetically). It can be understood as after preprocessing dictionary was transformed into DMqX3 dimension where q is a number of English words in the dictionary and X3 is level of hierarchical ontology.

4 Ontology Development Module

In this section, we discuss three steps that are the collection of senses, the calculation of weight term factor, and at last hierarchical ontology [37, 38].

4.1 *Collection of Senses*

In this step, each word of DM matrix is arranged into hierarchical structure such that all words with the similar prefix of length are bind in a separate tree, as per the size of the dictionary size. Word length will affect search time in the dictionary. In order to reduce the search time, words having length more than three data has been stored in the linear form (in alphabetically).

Now insertion of an element in the tree has to start with a word in the DM, first check the length of the word. If the length of the word is equal or more than m, then this word will be included in the tree of m size prefix. To find out the position of a word in the tree structure Eq. (1) below. Now as per the first m character of the word tree, node position will be selected by using Eq. (1).

$$\text{Pos} \leftarrow \text{Pos} + (C - 96) * 26(m - 1) \tag{1}$$

where

Pos: Position.
C: ASCII value of the English alphabet.
m: Level of the tree.

In the above equation, Pos is node position in the tree as per the input word from DM matrix first m character ASCII numbers.

4.2 Calculation of Weight TF

TF has been evaluated which depends on the input English training file. Each word Ex from the EWV is passed in the tree from the root node, as per Eq. (1), correct position is identified, and after that exact word was compared in the sequential node. Once correct node of Ex was identified, then its corresponding sense in Hindi was collected in S (Sense), now tth HWV of training Hindi file was compared, and this will match with the S. So only one word of HWV is identified in S; hence, TF of that word is increased by 1.

Term frequency is calculated by Eq. (2) [39].

$$\mathrm{TF}_s = \sum_{S=1}^{p} \mathrm{TF}_p + f\left\{ \begin{array}{c} f = 1, S_p \cap \mathrm{HWV}_t \\ f = 0, \sim S_p \cap \mathrm{HWV}_t \end{array} \right\} \tag{2}$$

where

TF: Term frequency,
f is the frequency of Hindi word.
$p = 1, 2, 3 \ldots\ldots n$.
S is Hindi sense of English words,
HWV: Hindi Word Vector.

4.3 Hierarchical Ontology

In this step, term frequency of the words in node was updated by using Eq. (2), for example,

S= { 'अध्ययन', 'परिक्षण करना', 'देखना', ...}, now corresponding, HWV is = ['समाजशास्त्र', 'धर्म', 'का', 'अध्ययन', 'कैसे', 'करता', 'है', '?'], so, 'अध्ययन' word matches in HWV, hence TF of this word gets increased.

In a similar manner other sets of words from the training English, Hindi files are used to increase the weight value of the proposed Hierarchical ontology. This training increases the accuracy of word replacement from the training question files.

5 Proposed Hierarchical Ontology Algorithm

In this step, testing English questions were passed in the system wherein constant time all set of words are identified by the work. Ontology was used which directly gives output inefficient manner, and this efficiency directly depends on two parameters, first, the dictionary is used for training, then second, it is used for training English and Hindi question set.

5.1 Proposed Algorithm

Input: D// Bi-Lingual Dictionary, TE // Training English Question File, TH// Training English Question File.
Output: WO // Weighted Ontology.
EWV ← Pre-Processing(TE) // EWV: English Word Vector.
Loop 1: n // n: number of element in Dictionary.
[E H] ← Fetch-ASCII(D[n]) // Fetch-ASCII: It is a function to Read American Standard Code for Information Interchange Number of E as English and H as Hindi Font.
End Loop.
// Develop Tree Structure for m level.
Loop 1:n
Pos ← 0 // Pos: Position.
Loop 1:m // m: Level of Tree.
C ← E[n, m] // C: Read English ASCII number at m^{th} position.
Pos ← Pos + (C-96)*26(m-1).
End Loop.
WO[Pos, P] ← E[n] // Assign word at Position Pos, P: next linear Position, WO: Weight Ontology.
WO[Pos, P] ← H[n] // Assign word at Position Pos, P: next linear Position.
End Loop.
// Assign Weight to the nodes of the tree as per Training English or Hindi files.
Loop 1: t // t number of training files.
F ← EWV[t] // F: File which we read from tth position of EWV, t: EWV sequence in TE.
Loop 1:x // x: number of words in F file.
S ← Sense(F[x], WO) // Find S: Sense of word at xth position in file F, x: Word position in EWV, S: Sense.
WAssign Weight(S, HWV[t]) // This assigns a weight to the sense present in similar Hindi training file, W: the weight of the Sense, HWV: Hindi Word Vector.
WO ← Update Tree(WO, W, S) // This function assign (W) weight to the (S) sense in WO, WO: Weight Ontology.
End Loop.
End Loop.

6 Experiment and Results

NCERT question sets have been taken as training and testing by category wise (I, II and III) from our previous work [29]. Each category contains English and Hindi question (Table 1).

To evaluate the proposed method, we used bilingual evaluation understudy (BLEU) [40, 41], translation error rate (TER) [42, 43], word error rate (WER) [42, 43], and frequency measure (F-measure) [39] using the following formula

- $$\text{BLEU} = \min\left(1, \frac{\text{Output} - \text{Length}}{\text{Reference} - \text{Length}}\right)\left(\prod_{i=1}^{n} \text{precision}_i\right)^{1/n} \tag{3}$$

  where $\text{Precision} = \frac{\text{Correct}}{\text{Output−Length}}$

- $$\text{F-Measure} = \frac{2(\text{Precision} \times \text{Recall})}{\text{Precision} + \text{Recall}} \tag{4}$$

  where $\text{Precision} = \frac{\text{Correct}}{\text{Output\_ Length}}$,
  $\text{Recall} = \frac{\text{Correct}}{\text{Reference\_ Length}}$

- $$\text{TER (Translation Error Rate)} = \frac{\text{Number of Edits}}{\text{Average Number of Reference Words}} \tag{5}$$

- $$\text{WER (Word Error Rate)} = \frac{S + D + I}{N} \tag{6}$$

  where

  S is the number of substitutions,
  D is the number of deletion,
  I is the number of insertion,
  C is the number of correct words, and
  N is the number of the reference.

Table 1 Training dataset description

English–Hindi	Questions	Words
Category I	24	135
Category II	55	557
Category II	33	554

After evaluation of all the above parameters (BLEU, TER, WER, and F-measure) given Figs. 2, 3, 4, and 5 show all the results. All these below figure shows that the proposed model has increased the BLEU score for 1 g as compared to the translated version used in our previous work [29] by BabelFish MT tool which is found to be best performing MT tool among all translators taken, hereafter referred to as previous work. We used hierarchical ontology with the training module to increase

Fig. 2 BLEU score of previous and proposed model value and proposed

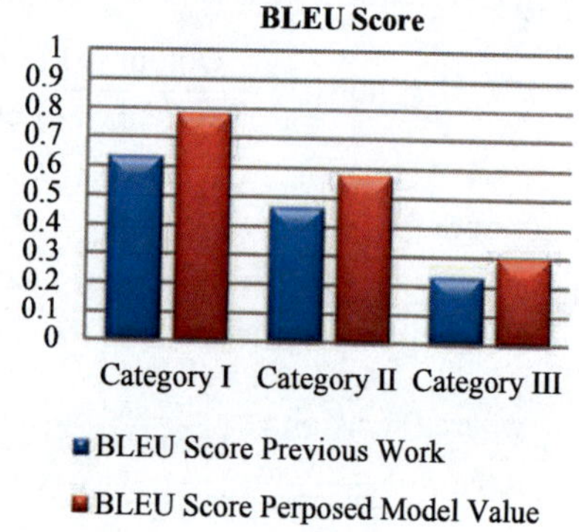

Fig. 3 TER previous model value

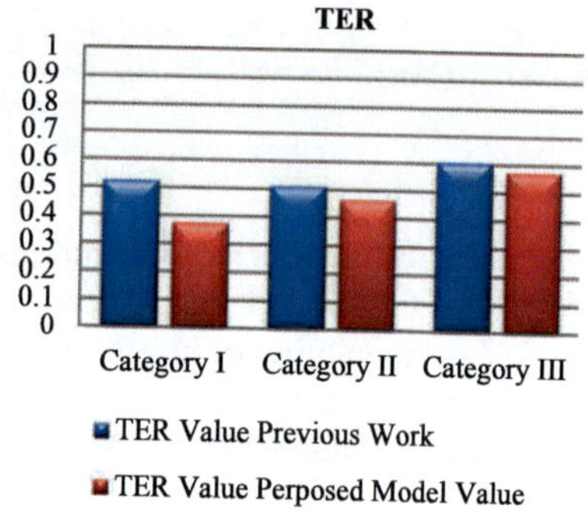

Fig. 4 WER previous and proposed model value

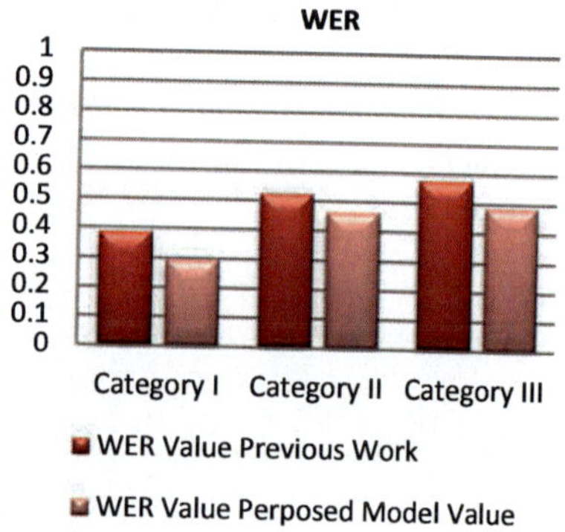

Fig. 5 F-measure previous and proposed model value

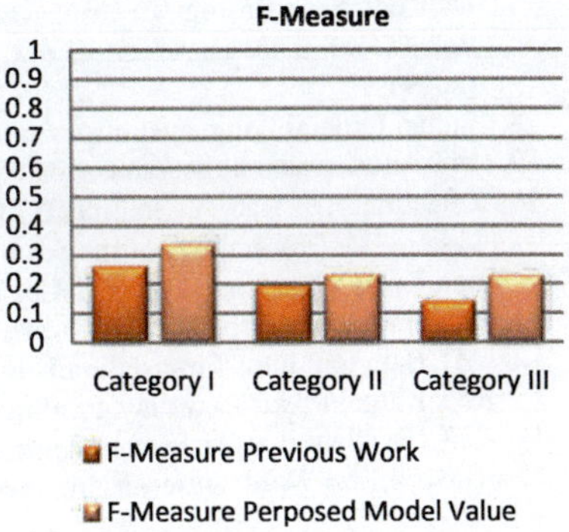

the efficiency of the proposed model. Training of the data increased the weight of those senses which reflect the same as training file and dictionary.

7 Conclusion

A proposed disambiguation technique as that of the disambiguation accuracy of questions can be further improved which may enhance the capability of MT tools to make them more suitable for the automated question translation process. Our approach has established better performance in enhanced WSD technique depending on specific learning sets (Wh-questions). Dictionary was arranged in a hierarchical structure where all set of nodes contain Hindi language sense of English words. Training increases the efficiency of WSD as this involves TF weight for identifying the correct sense as per input English training file. The experiment was done on Wh-questions in English, and results are compared with the reference Hindi Wh-questions by the existing translation.

References

1. Jurafsky, D.: Speech & Language Processing. Pearson Education, India (2000)
2. Mante, R., Kshirsagar, M., Chatur, P.: A review of literature on word sense disambiguation. Int. J. Comput. Sci. Inf. Technol. (IJCSIT), 1475–1477 (2014)
3. Navigli, R., Lapata, M.: Graph connectivity measures for unsupervised word sense disambiguation. In: IJCAI, pp. 1683–1688 (2007)
4. Sinha, M., Kumar, M., Pande, P., Kashyap, L., Bhattacharyya, P.: Hindi word sense disambiguation. In: International Symposium on Machine Translation, Natural Language Processing and Translation Support Systems, Delhi, India (2004)
5. Kunchukuttan, A., Mishra, A., Chatterjee, R., Shah, R., Bhattacharyya, P.: Sata-anuvadak: tackling multiway translation of Indian languages. Pan **841**(54,570), 4–135 (2014)
6. Godase, A., Govilkar, S.: Machine translation development for Indian languages and its approaches. Int. J. Nat. Lang. Comput. (IJNLC), pp. 55–74 (2015)

7. Dwivedi, S.K., Rastogi, P.: Critical analysis of WSD algorithms. In: Proceedings of the International Conference on Advances in Computing, Communication and Control, pp. 62–67. ACM (2009)
8. Indian Official Language: https://rajbhasha.gov.in
9. MT Anusaaraka: https://anusaaraka.iiit.ac.in/drupal/node/2
10. Chaudhury, S., Rao, A., Sharma, D.M.: Anusaaraka: an expert system based machine translation system. In: Proceedings of the 6th International Conference on Natural Language Processing and Knowledge Engineering (NLPKE-2010), pp. 1–6. IEEE (2010)
11. MT BabelFish: https://www.babelfish.com/success
12. MT Babylon: https://translation.babylon-software.com/english/to-french
13. MT Bing: https://www.bing.com/translator/help/#AboutMicrosoftTranslator
14. MT Google: https://translate.google.com
15. Navigli, R.: Word sense disambiguation: a survey. ACM Comput. Surv. (CSUR) **41**(2), 10 (2009)
16. Tromble, R., Eisner, J.: Learning linear ordering problems for better translation. In: Proceedings of the 2009 Conference on Empirical Methods in Natural Language Processing, Association for Computational Linguistics, pp. 1007–1016 (2009)
17. Singh, A.K., Husain, S., Surana, H., Gorla, J., Sharma, D.M., Guggilla, C.: Disambiguating tense, aspect and modality markers for correcting machine translation errors. In: Proceedings of the Conference on Recent Advances in Natural Language Processing (RANLP) (2007)
18. Bhala, R.V., Abirami, S.: Trends in word sense disambiguation. Artif. Intell. Rev., 159–171 (2014)
19. Dave, S., Parikh, J., Bhattacharyya, P.: Interlingua-based English–Hindi machine translation and language divergence. Mach. Transl., 251–304 (2001)
20. Bharti, A., Chaitanya, V., Sangal, R.: Hindi grammar. In: Natural Language Processing—A Panninial Perspective
21. Pechsiri, C., Piriyakul, R.: Developing a why–how question answering system on community web boards with a causality graph including procedural knowledge. Inf. Process. Agric., 36–53 (2016)
22. Zayaraz, G.: Concept relation extraction using Naïve Bayes classifier for ontology-based question answering systems. J. King Saud Univ. Comput. Inf. Sci., 13–24 (2015)
23. Mishra, A., Jain, S.K.: A survey on question answering systems with classification. J. King Saud Univ. Comput. Inf. Sci., 345–361 (2016)
24. Hao, T., Hu, D., Wenyin, L., Zeng, Q.: Semantic patterns for user interactive question answering. Concurrency Comput. Pract. Exp. **20**(7), 783–799
25. Dwivedi, S.K., Singh, V.: Integrated question classification based on rules and pattern matching. In: Proceedings of the 2014 International Conference on Information and Communication Technology for Competitive Strategies, p. 39. ACM (2014)
26. Hao, T., Wenyin, L.: Automatic question translation based on semantic pattern. In: 2008 Fourth International Conference on Semantics, Knowledge and Grid, pp. 372–375. IEEE (2008)
27. Dwivedi, S.K., Vikram, S.: Word sense ambiguity in question sentence translation: a review. In: Information and Communication Technology for Intelligent Systems (ICTIS 2017). Smart Innovation, Systems and Technologies, pp. 64–71. Springer (2017)
28. Vikram, S., Dwivedi, S.K.: Ambiguity in question paper translation. Int. J. Mod. Educ. Comput. Sci. (IJMECS), 13–23 (2018)
29. Vikram, S., Dwivedi, S.K.: Analysis of ambiguity In Wh-question with different machine translation. J. Theoret. Appl. Inf. Technol. (JATIT) (2019)
30. NCERT: https://epathshala.nic.in/e-pathshala-4/flipbook
31. NCERT: https://ncert.nic.in/NCERTS/textbook/textbook.htm
32. Other Type Questions: https://www.jagranjosh.com/articles/uppsc-uppcs-prelims-exam-2017-question-paper (2017)
33. Other Type Questions: https://www.jagranjosh.com/articles/uppsc-uppcs-exam-prelims-question-paper (2017)
34. Other Type Questions: https://iasexamportal.com/upsc-mains/papers

35. Vikram, S., Dwivedi, S.K.: Impact of ambiguity: Wh-questions vs other questions in question paper translation. In: Third International Conference on ICT for Sustainable Development (2018)
36. Gruber, T.R.: A translation approach to portable ontology specifications. Knowl. Acquisition, 199–220 (1993)
37. Ercan, G., Haziyev, F.: Synset expansion on translation graph for automatic wordnet construction. Inf. Process. Manage., 130–150 (2019)
38. Chiang, D.: A hierarchical phrase-based model for statistical machine translation. In: Proceedings of the 43rd Annual Meeting on Association for Computational Linguistics, pp. 263–270 (2005)
39. Wu, H.C., Luk, R.W.P., Wong, K.F., Kwok, K.L.: Interpreting tf-idf term weights as making relevance decisions. ACM Trans. Inf. Syst. (TOIS) **26**(3), 13 (2008)
40. Papineni, K., Roukos, S., Ward, T., Zhu, W.J.: BLEU: a method for automatic evaluation of machine translation. In: Proceedings of the 40th Annual Meeting on Association for Computational Linguistics, pp. 311–318. Association for Computational Linguistics (2002)
41. Ananthakrishnan, R., Bhattacharyya, P., Sasikumar, M., Shah, R.M.: Some issues in automatic evaluation of English-Hindi MT: more blues for BLEU. ICON (2007)
42. Och, F.J.: Minimum error rate training in statistical machine translation. In: Proceedings of the 41st Annual Meeting on Association for Computational Linguistics. Association for Computational Linguistics, pp. 160–167 (2003)
43. Snover, M., Dorr, B., Schwartz, R., Micciulla, L., Makhoul, J.: A study of translation edit rate with targeted human annotation. In: Proceedings of Association for Machine Translation in the Americas (2006)
44. Navigli, R., Velardi, P., Gangemi, A.: Ontology learning and its application to automated terminology translation. IEEE Intell. Syst., 22–31 (2003)

Review Paper: Error Detection and Correction Onboard Nanosatellites

Caleb Hillier and Vipin Balyan

Abstract The work presented in this paper forms part of a literature review conducted during a study on error detection and correction systems. The research formed the foundation of understanding, touching on space radiation, glitches and upsets, geomagnetism, error detection and correction (EDAC) schemes, and implementing EDAC systems. EDAC systems have been around for quite some time, and certain EDAC schemes have been implemented and tested extensively. However, this work is a more focused study on understanding and finding the best-suited EDAC solution for nanosatellites in low earth orbits (LEO).

Keywords Error control · Nanosatellite · FPGA · Radiation · EDAC

1 Introduction

The main focus of this study is preventing and overcoming the effects of radiation in memory onboard nanosatellites. The issues that need to be addressed are the single event upsets (SEUs) and multiple event upsets (MEU) caused by space radiation. This study will take a logical approach to EDAC systems, starting with space radiation (the cause), glitches and upsets (the error), geomagnetism (high probability of occurrence areas) EDAC schemes (solutions), and implementing EDAC systems (implementation).

All research is related and done for nanosatellites orbiting at low earth orbits (LEO). By referencing a number of articles and journals, a board understanding of EDAC systems is established. This review was assembled from a number of papers observations. Using the information and knowledge gained during the completion of review, a thesis topic was identified and motivated. Also included in this paper is the comparison of EDAC systems, based on foundation principals, capabilities, and applications. At the end of this paper, all important aspects relating to EDAC systems will be discussed fully, and the finding will conclude that further work on

C. Hillier · V. Balyan (✉)
Cape Peninsula University of Technology, Cape Town, South Africa

H. Sharma et al. (eds.), *Congress on Intelligent Systems*, Advances in Intelligent Systems and Computing 1334, https://doi.org/10.1007/978-981-33-6981-8_25

Table 1 Effects of charged particle in a space environment [1]

Spacecraft charging	Total ionizing dose	Displacement damage	Single event effects
Surface charging from plasma Deep dielectric from high-energy electrons	Trapped protons and electrons Solar protons	Protons Electrons	Protons: both trapped and solar Heavy ions: both galactic cosmic rays and solar events

EDAC systems in the nanosatellite field is still a needed and interesting field to work in.

2 Methodology

2.1 Space Radiation

Space radiation is a general term given to ionizing radiation found in space. This form of radiation is made up of highly energized particles which are mainly protons and heavy ions. The sources of space radiation are identified by NASA's Space Radiation Analysis Group (SRAG) in the following manner: "There are three naturally occurring sources of space radiation: trapped radiation, galactic cosmic radiation (GCR), and solar particle events (SPEs)" [3]. According to sources on radiation, SPEs have the biggest impact on satellites. The effects of radiation are often minor; however, there are some extreme examples of incidents occurring that have caused space missions to fail [4].

Radiated solar particles are responsible for electronic malfunctions, deterioration of materials, and surface charging and discharging. Due to the severe impact radiation can have on a satellite, radiation is considered to be a fundamental and important factor when it comes to the design and operations of satellites [5]. The effects of charged particles in a space environment are summarized using Table 1. Table 2 shows the three main sources of the radiation in space along with their effects on CMOS devices.

It is important to ensure that all electronic devices and components used to make up any space system have been tested and are able to resist the dose of radiation they might be exposed to. This is essential to ensure data integrity and reliability.

2.2 Glitches and Upsets

A glitch is defined as a sudden, usually temporary malfunction or fault of equipment. Glitches experienced by a satellite can be caused by many factors, such as hardware

Table 2 Effects of radiation on CMOS devices [2]

Space radiation environments and their effects on CMOS devices		
Radiation source	Particle types	Primary effects in devices
Trapped radiation belts	Electrons	Ionization damage
	Protons	Ionization damage; SEE in sensitive devices
Galactic cosmic rays	High-energy charged particles	Single event effects (SEEs)
Solar flares	Electrons	Ionization damage
	Protons	Ionization damage; SEE in sensitive devices
	Lower energy/heavy-charged particles	SEE

problems, corrupted software, or space weather. The exact cause of glitches and upsets is often hard to determine, especially when satellites are exploring unknown areas of space and interplanetary space.

The first warnings for single events upset were first brought to light by Wallmark and Marcus in 1962 [6]. SEUs are errors that occur in electric and digital circuits. According to NASA, these SEU occur: "When charged particles lose energy by ionizing the medium through which they pass, leaving behind a wake of electron-hole pairs" [7]. MEU occurs when two or more bits are upset by a single ion. These upsets are usually soft errors and can be corrected by onboard EDACs or prevented using hardened devices.

There are numerous examples of satellites being affected by glitches and upsets. The Magellan spacecraft on route to Venus suffered both power panel and star tracker upsets after being exposed to a solar flare [8]. Intelsat officials reported on the January 13, 2011 that their Intelsat's Galaxy 15 telecommunications satellite was unable to receive any commands from earth for eight months due to electrostatic discharge that caused a major software error [9].

The source of these types of glitches can normally be determined by analyzing the housekeeping data of the satellite. The referred articles are relevant to this review as solving the problem of glitches and upsets is ultimately the aim of an EDAC system.

2.3 Geomagnetism

Geomagnetism refers to the earth's magnetic field. This field expands into outer space from the earth's core, effecting orbiting spacecraft and also deflects particles from outer space. Figure 1 provides a well laid out illustration of the magnetic fields produced by the earth, as well as the factors that influence it.

Geomagnetism is directly linked to radiation, as its fields cause particles to be trapped, which results in belts like the Van Allen Radiation Belts. The Van Allen

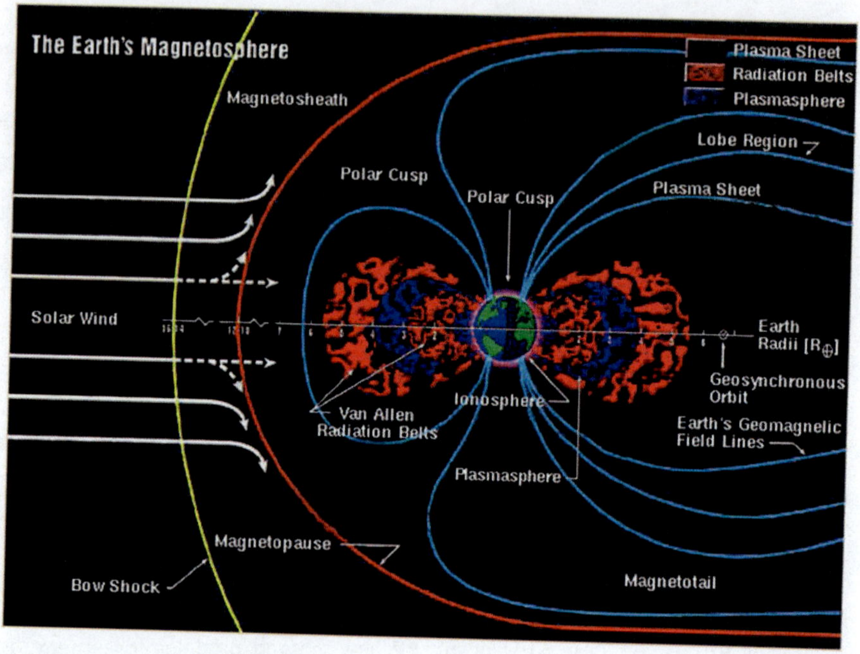

Fig. 1 The earth's magnetosphere [10]

Radiation Belts contain highly energized particles which can have devastating results on unprotected satellites. In 2012, NASA launched two Van Allen Probes spacecrafts with the mission of "study two extreme and dynamic regions of space known as the Van Allen Radiation Belts that surround Earth" [11]. Figure 2 shows both the inner and outer radiation belts.

When discussing radiation with regards to nanosatellites, the South Atlantic Anomaly (SAA) is the main culprit of single error upsets (SEUs) and multiple error upsets (MEUs). The South Atlantic Anomaly (Fig. 3) is the area where the earth's magnetic field is at its weakest. This means it is the area where the Van Allen Radiation belt is the closest to earth.

From an article written by Y. Bentoutou, the effect of radiation on board the Alsat-1 spacecraft is shown clearly in Fig. 4. The orbital location of each upset that occurred from the November 29, 2002 to October 12, 2009 is plotted. It can be noted that the majority (±80%) of SEU fell within the SAA [13].

From the provided information on Geomagnetism, its connection and relevance to nanosatellites design and implementation becomes clear. It is important that methods and schemes are designed that will allow nanosatellites to withstand and operate normally within environments produced by SSA and others.

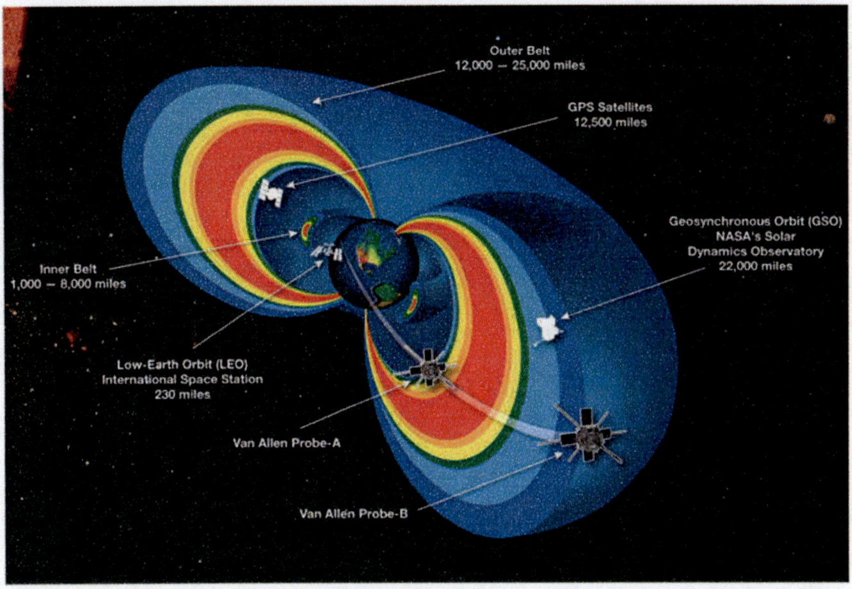

Fig. 2 ERBs including 2 Van Allen Probes satellites [12]

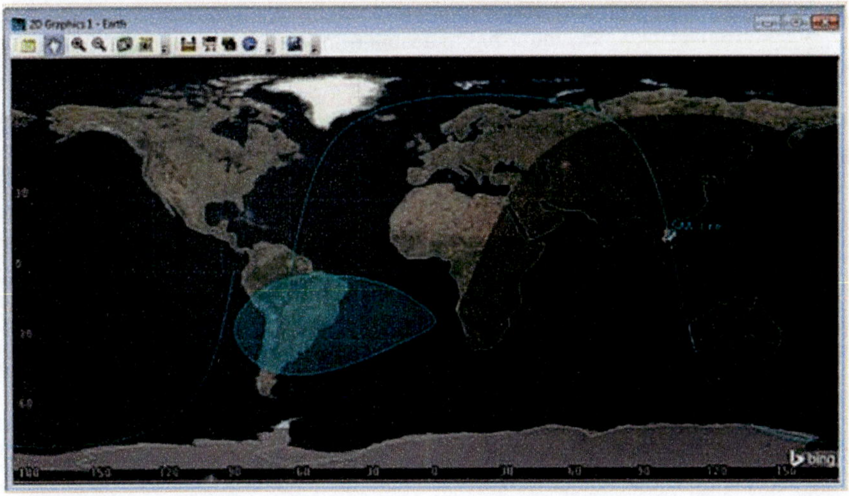

Fig. 3 SAA using STK SEET [14]

2.4 *EDAC Schemes*

EDAC systems are responsible for ensuring reliable data transfer between the satellites onboard computer and its local memory. An EDAC system is a software solution

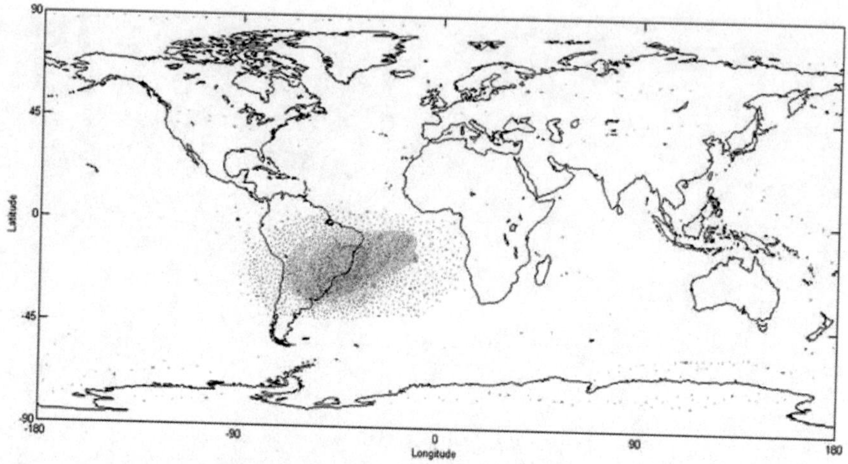

Fig. 4 Orbital position of OBC386 Ramdisk memory upsets [13]

to prevent the effects of radiation on a satellite. There are a number of EDAC schemes that have been developed during the past few years. However, each scheme has its own advantages and disadvantages. These are summarized in Table 3.

Out of all the EDAC schemes mentioned in Table 3, Reed Solomon and Hamming codes are most commonly used in modern day satellites [13].

An additional EDAC technique has to be mentioned as it is normally implemented together with a software EDAC such as Hamming code. This EDAC technique is referred to as triple modular redundancy (TMR). TMR's definition is dependent on the section/hardware for which TMR is implemented. In simple terms, TMR consists of having three identical devices that perform the exact same operation but communicate through a voter. The voter compares all received information to ensure the information matches and that the right result is returned. Figure 5 shows a block diagram of a TMR-based EDAC system which is implemented for RAM [13]. TMR, however, is not the ideal solution as three times the necessary hardware is needed to perform a single operation. This ultimately adds to the satellite's cost, complexity, and computing time.

2.5 Implementing EDAC Systems

Before getting started, it is important to know what EDAC systems are currently being implemented onboard nanosatellites and if additional hardware is needed in order to implement these EDAC schemes.

First, there are mainly two types of memory that need protection against upsets, namely program memory (SRAM) and Ramdisk. SRAM is faster than Ramdisk and is directly linked to the satellite's OBC, as it is typically used as the CPU cache. This

Table 3 Summary of well-known EDAC schemes

EDAC scheme	SED	SEC	DED	DEC	MED	MEC	Description
Parity code [15]	Yes	No	No	No	No	No	This scheme is considered the simplest and most basic error detection scheme. Using a parity bit, the scheme determines whether the string of bits is even or odd. This is done by evaluating the 1 s contained in the string. Creating two variants, even and odd parity bit
2 of 5 code	Yes	No	No	No	No	No	Most popular was the 2-out-of-5 code, which allows decimal digits to be represented using five bits. This code was implemented in barcodes. The *m*-out-of-*n* code makes use of codeword weightings (*m*) and length (*n*) to perform error detection. The weighting value normally represents the sum of the 1 s within a codeword
Berger code	Yes	No	No	No	No	No	This unidirectional error detecting code is only capable of flipping ones into zeroes or only zeroes into ones, such as in asymmetric channels. Used mainly in telecommunications. Berger code counts all the ones or zeroes within the information data (*k* bits long) and then attaches the binary equivalent of the sum to the information forming the codeword (*n* +*k* bits)

(continued)

Table 3 (continued)

EDAC scheme	SED	SEC	DED	DEC	MED	MEC	Description
Hamming code [16]	Yes	Yes	No	No	No	No	This scheme adds additional parity bits (r) to the sent information (k). The codeword can be calculated as $n = 2^r - 1$. This means information data can be calculated by $k = 2^r - r - 1$. Using a parity-check matrix and a calculated syndrome, the scheme can self-detect and self-correct any SEE errors that occur during transmission
Extended Hamming code	Yes	Yes	Yes	No	No	No	The original scheme allows SECSED, but with an addition of one bit, an extended Hamming version allows DECSED
Hadamard code	Yes	Yes	Yes	Yes	No	No	Based on unique mathematical properties namely Hadamard matrixes, this linear code allows both DED and double error correction (DEC). This code was used in 1971 by NASA space probe Mariner 9, to send photos of Mars back to Earth [17]
Repetition code	Yes	Yes	Yes	Yes	No	No	This code is one of the most basic codes as it simply resends a message several times. This result is low performance and transfer rates makes the code less than ideal

(continued)

Table 3 (continued)

EDAC scheme	SED	SEC	DED	DEC	MED	MEC	Description
Four dimensional parity code [15]	Yes	Yes	Yes	Yes	No	No	Also referred to as multidimensional parity-check code (MDPC), this code makes use of multiple parity bits. This means the code basically arranges a message into a grid and then generating parity rows according to horizontal, vertical, and cross diagonally. Ideally used for DDR RAM protection
Golay code [18]	Yes	Yes	Yes	Yes	Yes	Yes	This code is a perfect linear error correcting code and makes use of a look-up table. This codes has the following parameters [8, 12, 24] & [7, 12, 23]
BCH code [19]	Yes	Yes	Yes	Yes	Yes	Yes	This cyclic code is constructed using polynomials over a finite field. This code generally uses a linear-feedback shift register (LFSR) to encode the message block and uses syndromes polynomial to determine the error location during decoding. This results in Bose Chaudhuri Hocquenghem (BCH) codes being complex and difficult to implement while requiring significate processing time

(continued)

Table 3 (continued)

EDAC scheme	SED	SEC	DED	DEC	MED	MEC	Description
Reed Solomon code [20]	Yes	Yes	Yes	Yes	Yes	Yes	This non-binary cyclic code is based on univariate polynomials over finite fields. The error locator polynomial is then found using both the syndrome polynomial and Euclidian algorithm. Errors can then be located or pin-pointed by applying the Chien Search Algorithm. Once the error's location is found, the Forney algorithm is used to correct any errors. This results in Reed Solomon (RS) codes being complex and difficult to implement, while requiring significate processing time

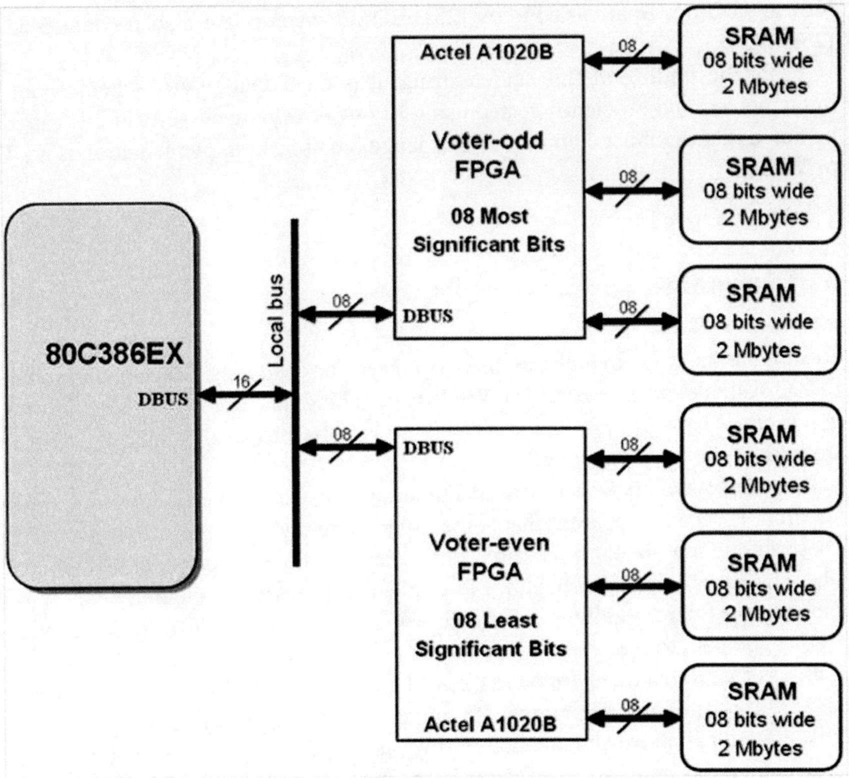

Fig. 5 Block diagram of TMR-based EDAC [13]

implies that the integrity of the information stored within the SRAM is vital to the lifespan and health of the satellite. Ramdisk, on the other hand, refers to the memory that serves as a disk drive, mostly used to store image files and memory-intensive information.

Most commonly used EDAC scheme implemented in nanosatellites for SRAM protection is Hamming code and TMR. These are the most popular schemes because they are relatively easy to implement and have short encoding and decoding delays. RS codes, on the other hand, are more popular for large sets of information, as used in Ramdisk. For this reason, RS codes are known as block codes. RS is popular mainly due to its EDAC capabilities, as it is able to ensure MED-MEC.

From the literature review, it was clear that implementing EDAC systems on FPGA's is quite popular. In a paper by V. Tawar, a four-dimensional parity EDAC scheme was designed, tested, and synthesized on Xilinx FPGA device XC3S500E-4FG320 [21]. A horizontal vertical diagonal (HVD) EDAC was developed for Xilinx [22, 22]. Other examples are orthogonal codes written in Verilog using Altera Quartus-II software [24] and Hamming code implemented on FPGA using Verilog

[16] and Xilinx Spartan-3 FPGA [25], EDAC systems are also implemented in [26, 27].

From the finding of this search string, it is clear that EDAC schemes can be implemented and efficiently tested using FPGA development software. It can also be noted that the most common coding language used in implementation is VHDL or Verilog.

3 Discussion

To summarize, it is obvious that there is a need for designing effective and reliable EDAC systems for nanosatellites. While researching this topic as a whole, information seemed hard to find, however, once broken up into small topics and research areas, the following was found.

The space radiation has caused numerous mission failures. Through further research, it became apparent that some failures are owing to the SEUs and MEUs. These upsets are almost impossible to predict but there are certain areas of space where upsets are more frequent, for example, the Van Allen Radiation belts. Within the context of nano-satellites, which are low earth orbiting (LEO), special attention needs to be paid to the SAA.

It was found that there are a number of EDAC schemes and techniques currently used. Most commonly Hamming, RS codes, and TMR. It was also found that the most effective, non-evasive method of implementing an EDAC system would be to implement the system using an FPGA. Using the above-summarized information as a starting point, this thesis topic will be proposed which will take a more detailed look at the design, development, and implementation of an effective system on chip-based EDAC for nano-satellites.

4 Conclusion

From the literature survey, it is clear that there is a need for research in the area of EDACs. This field is new and constantly evolving as nanosatellites provide a platform from which the boundaries of space and technology are constantly being pushed. As technology advances memory chip, cell architecture is becoming more and more dense, especially with the development of nanotechnology. This creates a growing demand for more advanced and reliable EDAC systems that are capable of protecting all memory aspects of satellites.

Acknowledgements The work is funded by the National Research Fund (NRF), and done in collaboration with the French South African Institute of Technology (FSATI), based at Cape Peninsula University of Technology (CPUT).

References

1. Holbert, K.E.: Space Radiation Environmental Effects (2007)
2. Wall, J., Macdonald, A.: NASA ASIC Guide Title Page. Jet Propulsion Laboratory California Institute of Technology and National Aeronautics and Space Administration (1993)
3. Langford, M.: Space Radiation Analysis Group—NASA, JSC. NASA, JSC (2014)
4. Campbell, K.: Engineering News—Sumbandila provided lessons for next SA satellite. Creamer Media's (2012)
5. Maki, A.: 2-1-5 space radiation Effect on satellites. J. Natl. Inst. Inf. Commun. Technol. **56**(1–4), 49–55 (2009)
6. Wallmark, J.T., Marcus, S.M.: Minimum size and maximum packing density nonredundant semiconductor devices. Proc. IRE **50**(3), 286–298 (1962)
7. NASA/SP: NASA thesaurus volume 1—hierarchical listing with definitions. Natl. Aeronaut. Sp. Adm. **1**, 879 (2012)
8. Odenwald, S.F.: The 23rd Cycle: Learning to Live with a Stormy Star. Columbia University Press, New York (2001)
9. Choi, C.Q.: Software Glitch Blamed for Turning Satellite into Space Zombie. Space News, PARIS (2011)
10. Miller, I.: Geomagnetics 2012—THE SEDONA EFFECT. Sedonanomalies (2012)
11. Zell, H.: Van Allen Probes Mission Overview_NASA. National Aeronautics and Space Administration (2015)
12. Zell, H.: Radiation Belts with Satellites _ NASA. National Aeronautics and Space Administration (2013)
13. Bentoutou, Y.: A real time EDAC system for applications onboard earth observation small satellites. IEEE Trans. Aerosp. Electron. Syst. **48**(1) (2012)
14. System Tool Kit (STK): STK—Ionizing Radiation from the South Atlantic Anomaly (SAA) Using STK SEET (2017)
15. Bilal, Y., Khan, S.A., Khan, Z.A.: A refined four-dimensional parity based EDAC and performance analysis using FPGA. In: International Conference on Open Source Systems and Technologies, 81–86 (2013)
16. Jindal, V.: Design OF hamming code using Verilog HDL. Electron. YOU, no. FEBRUARY, pp. 94–96 (2006)
17. Malek, M.: Coding theory Hadamard codes. Calif. State Univ. East Bay, pp. 1–8.
18. Kanemasu, M.: Golay codes. MIT Undergrad. J. Math., 95–100 (1999)
19. Poolakkaparambil, M., Mathew, J., Jabir, A.M., Pradhan, D.K., Mohanty, S.P.: BCH code based multiple bit error correction in finite field multiplier circuits. In: IEEE—12th International Symposium on Quality Electronic Design, pp. 615–621 (2011)
20. Parvathi, P.: FPGA based design and implementation of Reed-Solomon encoder & decoder for Error Detection and Correction. In: 2015 Conference on Power, Control, Communication and Computational Technologies for Sustainable Growth (PCCCTSG), pp. 261–266 (2015)
21. Tawar, V., Gupta, R.: A 4-dimensional parity based data decoding scheme for EDAC in communication systems. Int. J. Res. Appl. Sci. Eng. Technol. **3**(Iv), 183–191 (2015)
22. Singh, N.P., Singh, S., Sharma, V., Sehmby, A.: RAM error detection & correction using HVD implementation. Eur. Sci. J. **9**(33), 424–435 (2013)
23. Road, H.: FPGA implementation of 4d-parity based data coding technique. IJRET Int. J. Res. Eng. Technol. **4**(3), 593–598 (2015)
24. Reshmi, R., Joseph, S., Praveen, U.K.: EDAC by using orthogonal Codes. Int. J. Adv. Res. Electron. Commun. Eng. **4**(3), 632–635 (2015)
25. Hosamani, R., Karne, A.S.: Design and implementation of hamming code on FPGA using Verilog. Int. J. Eng. Adv. Technol. **4**(2), 180–184 (2014)
26. Hillier, C., Balyan, V.: Error detection and correction on-board nanosatellites using hamming codes. J. Electr. Comput. Eng. **6**, 1–15 (2019). https://doi.org/10.1155/2019/3905094
27. Hillier, C., Balyan, V.: Effect of space radiation on LEO nanosatellites. J. Eng. Appl. Sci. **14**(17), 6843–6857 (2019)

Kerala Floods: Twitter Analysis Using Deep Learning Techniques

Chetana Nair and Bhakti Palkar

Abstract As innovation and Web developed, online networking has gotten significant and unavoidable in our lives. In any event, during emergency or regular cataclysms, individuals utilize social media communication more than any other methods for correspondence. Thus, these online social media platforms contain colossal measure of data identified with such occasions or occurrences. We can utilize this online information to get a great deal of data which can be additionally used to comprehend the occasions through different viewpoints. In this paper, we have extracted Kerala floods related tweets from Twitter and grouped them into different classifications utilizing natural language processing models like bidirectional encoder representations from transformers (BERT), XLNet, Ernie 2.0. It was seen that Ernie 2.0 gave better result as compared with the other natural language processing models utilized for classifying the extracted tweets. These automatic text classification models designed using the above deep learning techniques can be useful in giving data while planning the arrangement and preventive measures for catastrophic events like floods. Investigation of these tweets can likewise give discernment in taking care of such catastrophes in a superior and proficient manner.

Keywords Deep learning · Natural language processing · Text classification · Transformers · BERT · XLNet · Ernie 2.0

1 Introduction

In the present period of Internet and digitization, social media assumes a tremendous role. For each and everything, people resort to social media [1]. The number of individuals using utilizing online networking has expanded exponentially as well. Social networking sites have increased their penetration into all age groups, from cities to rural areas, individuals have social media visibility. As a result, these social networking platforms contain a large amount of data. Researchers use this data for

C. Nair (✉) · B. Palkar
KJ Somaiya College of Engineering, Mumbai, India
e-mail: chetana.nair@somaiya.edu

H. Sharma et al. (eds.), *Congress on Intelligent Systems*, Advances in Intelligent Systems and Computing 1334, https://doi.org/10.1007/978-981-33-6981-8_26

various purposes like sentiment analysis, effectiveness of the schemes implemented, etc.

Twitter is a micro-blogging site, so here users' tweet, i.e., a short sentence, might have a link of an image, video, blog, or article. This textual data can give us lot of insights especially when it is pertaining to a particular event. Twitter generated around 1.2 million tweets during the recent Kerala floods [2]. Here, we are extracting these tweets, i.e., flood-related tweets and performing text classification on them. Text classification can help us to filter out the necessary and useful information related to a crisis like Kerala floods. Kerala flood was a characteristic cataclysm in 2018 which prompted enormous number of deaths and immense financial misfortunes to both the state and the focal government [3]. It was one of the most exceedingly terrible floods in more than 100 years [3]. These floods were followed by landslides which again took many lives [4]. Following summer Kerala confronted dry season like circumstances with temperature taking off up. All these led to decline in economic growth and affected the common man to a great extent. Similar floods occurred in the next year in 2019 with practically same measure of causalities [5]. This is by all accounts a rehashing cataclysmic event which calls for critical answers for forestall it and to deal with or bargain productively during such emergency. Event of flood is because of numerous reasons and consequently the arrangement needs appropriate examination of the circumstance. We have to investigate the past floods, how the fiasco of the executives was performed and how might it be improved, what should be possible to forestall such floods, how to spread mindfulness, what strategies ought to be made, particularly by the focal and state governments like ecological approaches, development of dams, laws against illicit development on waterway banks, and so on. To discover these arrangements, we have to initially investigate the accessible information about the past floods and get bits of knowledge. Subsequently, here the flood-related tweets are gathered and consequently ordered with the goal that we can get some data on finding a superior and proficient answer for taking care of and forestalling normal disasters like floods.

2 Related Work

Text classification is one of the most widely used natural language processing technologies. Common text classification applications include spam identification, news text classification, information retrieval, emotion analysis, and intention judgment. There are many works related to tweet classification, in one of the such works, Maceda et al. [6] have collected earthquake-related tweets, classified them, and manually annotated based on the four labels identified, namely drill/training, earthquake feels, extent of damages, and government measures and rehabilitation [6]. With the standard metrics, the method which obtained the highest rating is the 15 folds validation using SVM. This obtained 82.60% precision metric, recall obtained 82.50%, F-measure resulted to 82.50%, and the correctly classified instances are 82.48% [6]. SVM consistently showed a high evaluation performance as compared to Naïve Bayes and linear logistic regression [6]. Traditional text classifiers based on machine learning methods have defects such as data sparsity, dimension explosion, and poor generalization ability, while classifiers based on deep learning network greatly improve these defects, avoid cumbersome feature extraction process, and have strong learning ability and higher prediction accuracy [7]. Cai et al. [7] have introduced the process of text classification and focuses on the deep learning model used in text classification. Here, recurrent neural network (RNN) was used. The RNN model can deal with this dependence well to achieve significant classification effect, but the RNN model cannot parallelize well, so it is more suitable for short text processing [7]. The training time is too long when processing texts with more than dozens of words [7]. Devlin et al. [8] have introduced a new language representation model called BERT, which stands for bidirectional encoder representations from transformers. It obtains new state-of-the-art results on 11 natural language processing tasks [8]. BERT did give state-of-the-art results in various natural language processing tasks, but it has some drawbacks. Yang et al. [9] have proposed XLNet, a generalized autoregressive pretraining method that enables learning bidirectional contexts by maximizing the expected likelihood over all permutations of the factorization order and overcomes the limitations of BERT [9]. Furthermore, XLNet integrates ideas from Transformer-XL [9], the state-of-the-art autoregressive model, into pretraining. Empirically, XLNet outperforms BERT on 20 tasks, often by a large margin, and achieves state-of-the-art results on 18 tasks including question answering, natural language inference, sentiment analysis, and document ranking [9]. However, BERT and XLNet techniques have not taken into consideration the lexical, syntactic and semantic information while training the language models [10]. Sun et al. [10] have extracted the lexical, syntactic, and semantic information from training corpora and have proposed a continual pretraining framework named Ernie 2.0 which incrementally builds pretraining tasks and then learn pretrained models on these constructed tasks via continual multitask learning [10]. Tripathi et al. [11] presents a novel parallel biogeography optimization-based method to unfold the fake review detection in the big data environment. The experimental analysis is performed on two standard fake review datasets and compared with K-means and four state-of-the-art methods in terms of accuracy. Again Tripathi et al. [12] have introduced a novel MapReduce based K-means biogeography-based optimizer (MR-KBBO) which is proposed to leverage the strength of biogeography-based optimizer with MapReduce model to efficiently cluster the large-scale data [12]. Experimental results demonstrate that the proposed method is efficient in sentiment mining for the large-scale twitter datasets.

From the detailed study of the above-mentioned papers, it has been observed that deep learning models perform better than machine learning models when it comes to natural language processing tasks. Recent developments and research in deep learning techniques have improved the language understanding and thereby increase the output of natural language processing tasks like classification. Transformers form the basis of models with focus on attention to the words in the pretraining process. Models like BERT, XLNet, and Ernie 2.0 also use transformers but have different pretraining tasks. Hence, these models can be used to perform text classification and the appropriate conclusion can be obtained from the results thus obtained.

3 Methodology

Extent of this work is to precisely group the tweets produced during the Kerala floods. These categorized tweets can be utilized to acquire data about the floods with respect to the classified categories. The dataset is obtained from Twitter, for example, the tweets posted by Twitter users during the Kerala Floods. This will be utilized to train the model to accurately classify the tweets into respective categories. The initial step is to precisely gather the flood related tweets from Twitter followed by data preprocessing and labeling. The dataset is then divided into training set and testing set. The training set is then used to fine-tune the pretrained natural language processing models like BERT, XLNet, and Ernie 2.0. The test set is utilized to assess the exhibition of these models. Random tester is likewise used to check if the tweets are accurately classified into suitable classes.

3.1 Data Collection

Here, we have gathered tweets identified with Kerala floods from Twitter utilizing the Twitter scrapper application programming interface (API). The tweets were queried utilizing the famous hashtags utilized during Kerala floods by the Twitter users. Well-known hashtags like #KeralaFloods, #KeralaDonationChallenge, and so on were utilized as question words for looking through the tweets [13]. The tweets so acquired were saved in comma separated format (csv).

3.2 Data Cleaning

The obtained tweets are raw, and to get information from them, it has to be processed. This is done by removing duplicate tweets. All the unnecessary information like the timestamps, user information, etc., are removed. Retweeted tweets are also to be removed. Any special symbols especially at the end of the tweet are removed. Only the column which contains the tweet is extracted from the csv file.

3.3 Data Labeling

The preprocessed tweets are categorized and labeled according to the classification which we want the model to perform. These tweets are manually separated into four classes. The four classes used for classification are:

1. Appreciation/Donation.
2. Help.

3. Destruction/Loss.
4. Political news.

3.4 Data Division and Separation

After the tweets are classified into various categories, they are separated into training and testing set. For separating into training and testing, each category of tweet is divided into four parts, and one part is taken for testing and remaining for training. Once the training and testing set is obtained, we proceed toward data training. Here we have used various natural language processing models like BERT, XLNet, and Ernie 2.0.

BERT. BERT represents bidirectional encoder representations from transformers [8]. As the name proposes, they are gotten from transformers. Transformers comprise of encoders and decoders [8]. Transformers take each word at the same time and create word embeddings; these embeddings are vectors that typify the significance of the words. In the transformer design, the encoder learns the setting of the language. Here, the principle motivation behind the model is to comprehend the language with the goal that it can precisely understand the content. BERT utilizes the heap of encoders to understand the setting of the language. Along these lines, BERT has pretrained layers that get language and we need to fine tune the output layer to the particular task, for example text classification. We have performed supervised training for the output layer using the dataset. Hence, only the output parameters are learnt from scratch, and rest of the parameters are only slightly fined tuned. So, training takes less time.

XLNet. XLNet architecture is similar to BERT, but it has different pretraining procedure [9]. It uses the best of both autoencoding and autoregressive methods. In autoregressive method, the model predicts a word from the words before it. BERT uses autoencoding, i.e., BERT predicts independently and simultaneously. Consider the following sentences as example:

I went to the church to pray.
I went to the gym to exercise.
I went to the library to study.

In the above sentences, if some words are masked like church, pray, gym, exercise, library, and study, then the prediction by BERT will be: I went to the church to exercise or I went to the library to pray. According to BERT, these are correct predictions, but it does not consider the relationship between church and pray, gym and workout, library and study. In order to overcome this flaw, XLNet uses permutation language modeling, two stream self-attention with integrated transformer XL architecture [9].

Ernie 2.0. It is a framework for continuous incremental multitask pretraining [10]. BERT is based on co-occurrence of tokens and sequences while Ernie 2.0 incorporates lexical, syntactic, and semantic information [10]. Like BERT, it also uses transformer encoders, but the difference lies in the pretraining tasks. It has word-aware pretraining where the model understands the meaning of the word using knowledge masking task, capitalization prediction task, and token document relation prediction task [10]. In knowledge masking instead of masking just words like BERT, the named entities, expressions, or phrases are masked together. So, it learns global and local context. Capitalization prediction task makes the model predict if the word was capitalized or not, whereas token document relation prediction task predicts if the given word appears elsewhere in the document, and this helps in understanding the key words and the theme of the document. Ernie 2.0 learns the structure of a sentence using structure aware pretraining task which has sentence reordering task and sentence distance task [10]. It learns the semantics using the semantic aware pretraining tasks like discourse relation task and IR relevance task [10]. Discourse relation task is used to predict the discourse marker, for example:

Kerala received heavy rainfall. (As a result) Kerala suffered from catastrophic floods.

Here, the words ‘as a result’ are the discourse that the model learns from two different sentences (Fig. 1).

4 Implementation and Results

Around 4500 tweets formed the dataset. After this, the dataset cleaned to expel superfluous data like the username, timestamps, and so on. This cleaned dataset was then manually classified into different categories, the four categories being: help (1280 tweets), appreciation or donation (1526 tweets), political (846 tweets), destruction or loss (848 tweets).

The four classes are named utilizing numbers 0, 1, 2, 3, and 4. Here, 0 represents all the tweets in the appreciation or donation classification, 1 for help, 2 for political classification, and 3 for loss or destruction. The quantity of tweets in the previously mentioned classes are not same, for example, the quantity of tweets in the appreciation or donation class is more when contrasted with the tweets in the political classification, henceforth there is an imbalance in the various categories of tweet in the dataset. The distribution of various categories in the dataset can be seen in the bar graph as shown in Fig. 2, where x-axis represents the various categories like 0, 1, 2, 3 and y-axis represent the number of tweets present in the training dataset.

As shown in Fig. 2, there is an imbalance in the distribution of various categories in the dataset. Hence, Mathews correlation coefficient (MCC) was used to evaluate the performance of these models [14] (Table 1).

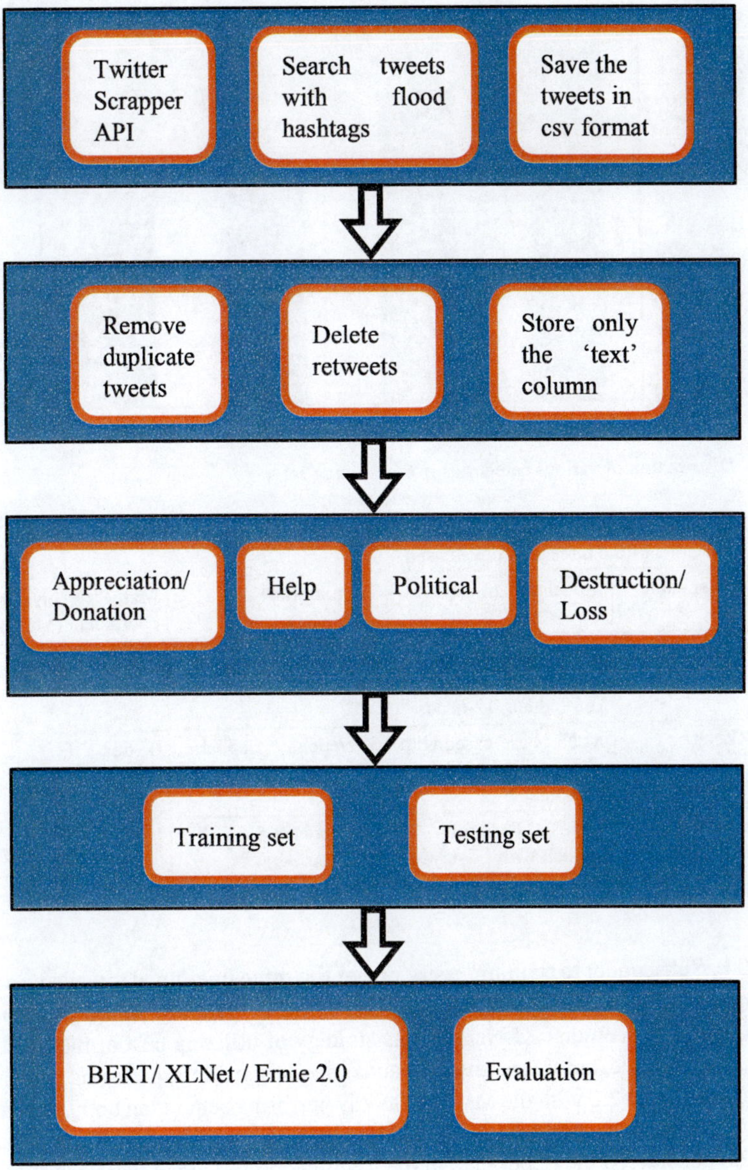

Fig. 1 System design

5 Conclusion

In this paper, we have attempted to perform multiclass text characterization on Twitter information during Kerala floods utilizing various models like BERT, XLNet, and

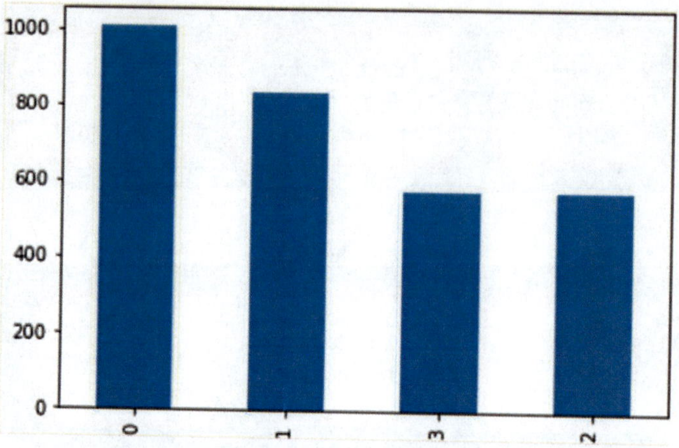

Fig. 2 Distribution of various categories in the training set

Table 1 Results of the three techniques used

	Model name	Model variant used	Training time	Mathew's correlation coefficient (MCC)
1	BERT	BERT-Base, Uncased model with 12-layer, 768-hidden, 12-heads	49 s/epoch * 5 epoch	0.9569
2	XLNet	XLNet-base-cased with 12-layer, 768-hidden, 12-heads	60 s/epoch * 5 epoch	0.9662
3	Ernie 2.0	Ernie 2.0 Base for English with 12 -layer, 768- hidden, 12-heads	44 s/epoch * 5 epoch	0.9769

Ernie 2.0. Subsequent to training, every one of the three fine-tuned models was tried utilizing an arbitrary analyzer, and the models have precisely sorted the random tweets. We have seen that XLNet with its strategy of utilizing best of the two techniques, for example, autoregressive and auto encoding models, performs better than BERT while Ernie 2.0 with the MCC of 0.9769 performs better than both XLNet and BERT. Despite the fact that XLNet considered the constraints of BERT and beat it, Ernie 2.0 considered the richer data in the training corpora along these lines leading to better outcomes. These models have given great outcomes even for a dataset of around 4500 tweets. For future work, the updated or large models of BERT, XLNet, and Ernie 2.0 can be used to improve the results.

References

1. Jahanian, M., Xing, Y., Chen, J., Ramakrishnan, K. K., Seferoglu, H., Yuksel, M.: The evolving nature of disaster management in the internet and social media era. In: 2018 IEEE International Symposium on Local and Metropolitan Area Networks (LANMAN), pp. 79–84, Washington, DC, USA (2018)
2. Kerala floods: Twitterati puts out 2.62 million tweets during deluge. https://www.bgr.in/news/kerala-floods-twitter-2-62-million-tweets-during-august-2018-deluge-686121/. BGR News. Published on 24 Aug 2018. Last accessed 14 Feb 2020
3. Vishwanathan, M.: 2018 Kerala Floods; A journey through death and devastation. https://www.newindianexpress.com/states/kerala/2019/aug/22/2018-kerala-floods-a-journey-through-death-and-devastation-2022362.html. The New Indian Express. Published on 22 Aug 2019. Last accessed 14 Feb 2020
4. India floods: At least 95 killed, hundreds of thousands evacuated. https://www.bbc.com/news/world-asia-india-49306246. BBC News. Published on 10 Aug 2019. Last accessed 14 Feb 2020
5. Kiran, K.P.S.: Kerala floods: Death toll touches 111. https://timesofindia.indiatimes.com/city/thiruvananthapuram/kerala-floods-death-toll-touches-111/articleshow/70701312.cms. The Times of India. Published on 16 Aug 2019. Last accessed 14 Feb 2020
6. Maceda, L., Llovido, J., Satuito, A.: Categorization of earthquake-related tweets using machine learning approaches. In: 2018 International Symposium on Computer, Consumer and Control (IS3C), pp. 229–232, Taiwan (2018)
7. Cai, J., Li, J., Li, W., Wan, J.: Deep learning model used in text classification. In: 2018 15th International Computer Conference on Wavelet Active Media Technology and Information Processing (ICCWAMTIP), pp. 123–126, China (2018)
8. Devlin, J., Chang, M.W., Lee, K., Toutanova, K.: 'BERT: Pre-training of deep bidirectional transformers for language understanding. arXiv:1810.04805, https://arxiv.org/abs/1810.04805 (2018)
9. Yang, Z., Dai, Z., Yang, Y., Carbonell, J., Salakhutdinov, R., Le, Q.: XLNet: Generalized autoregressive pretraining for language understanding. arXiv:1906.08237, https://arxiv.org/abs/1906.08237 (2019)
10. Sun, Y., Wang, S., Li, Y., Feng, S., Tian, H., Wu, H., Wang, H.: ERNIE 2.0: a continual pre-training framework for language understanding. arXiv:1907.12412, https://arxiv.org/abs/1907.12412 (2019).
11. Tripathi, A., Sharma, K., Bala, M.: Fake review detection in big data using parallel BBO. Int. J. Inf. Syst. Manage. Sci. **2**(2) (2019). https://ssrn.com/abstract=3378499
12. Tripathi, A., Sharma, K., Bala, M.: Parallel hybrid BBO search method for Twitter sentiment analysis of large-scale datasets using MapReduce. Int. J. Inf. Secur. Privacy (IJISP), 106–122 (2019). https://doi.org/10.4018/IJISP.201907010107
13. Sidhardhan, S.: #KeralaFloods: Surging on social media. https://timesofindia.indiatimes.com/city/kochi/keralafloods-surging-on-social-media/articleshow/65450526.cms. The Times of India. Published on 20 Aug 2018. Last accessed 14 Feb 2020
14. Luque, A., Carrasco, A., Martín, A., Heras, A.: The impact of class imbalance in classification performance metrics based on the binary confusion matrix. Pattern Recogn. **91**, 216–231 (2019)

An Improved DVFS Circuit & Error Correction Technique

Keshav Raheja, Rohit Goel, and Abhijit Asati

Abstract Dynamic voltage and frequency scaling (DVFS) is useful for low power digital circuit design. The work proposes a novel DVFS module offering any finer clock frequency change to produce an appropriate supply voltage to feed a digital circuit driven by DVFS module. In DVFS with varying supply and clock conditions the chances of setup and hold timing violations in D flip-flop (DFF) circuit may increase. The DVFS module driving a digital circuit utilizing Razor D flip-flop is used to correct errors occurring due to timing violations. The proposed circuit simulation shows that DVFS module driving simple D flip -flop shows error due to timing violations, while the DVFS module driving Razor D flip-flop shows the correct operation. In the digital pipelined circuits any occurrence of timing violations, the Razor DFF uses the error correction mechanism to prevent data loss with a penalty of one additional clock cycle.

Keywords DVFS · Static power · Dynamic power · Transmission gate (TG) · Flip-flop (FF) · Razor DFF

1 Introduction

There are several low power digital circuit design approaches such as pipelining, asynchronous logic circuit design, adiabatic logic circuit design, sub-threshold circuits design, clock gating, power gating, and DVFS. The DVFS reduce dynamic power consumption by reducing the 'power supply voltage' (V_{DD}) and/or 'clock frequency' in digital processor under reduced workload [1, 2]. The dynamic power is proportionally related to frequency of operation of the circuit and also varies proportionally with the square of power supply voltage (V_{DD}) of the circuit. Thus, the combination of supply voltage and frequency scaling has a cubic impact on dynamic power dissipation and suitable for low power Internet of things (IoT) applications (supports low power edge computing). Under lower workload, the power

K. Raheja · R. Goel · A. Asati (✉)
Department of EEE, Birla Institute of Technology and Science, Pilani, Pilani, India
e-mail: abhijit_asati@pilani.bits-pilani.ac.in

H. Sharma et al. (eds.), *Congress on Intelligent Systems*, Advances in Intelligent Systems and Computing 1334, https://doi.org/10.1007/978-981-33-6981-8_27

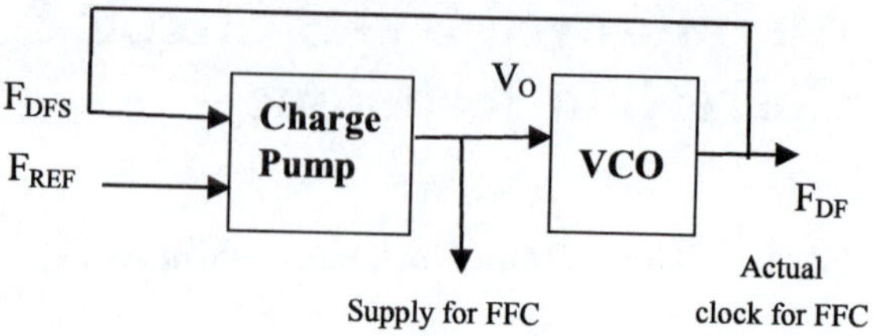

Fig. 1 DVFS circuit

supply and the frequency of digital processor are reduced, while for higher workload power supply and the frequency of digital processor are increased. The workload decomposition method combined with a DVFS technique minimizes the energy consumption [2–7]. In this paper, a novel DVFS circuit is designed using a charge pump circuit and voltage-controlled oscillator (VCO) circuit as shown in Fig. 1. The charge pump output is fed to the VCO (designed using CMOS ring oscillator) as an input. The charge pump produces a constant output voltage under a condition where the frequency of VCO becomes exactly equal to the set value reference frequency (F_{REF}). The output of the charge pump circuit is providing a supply voltage for the digital circuit [for simplicity we choose flip-flop circuit (FFC), so that capacitance can drive this small load] and the output of the VCO becomes the actual clock for the FFC. The FFC uses either transmission gate (TG)-based 'simple D flip-flop' (DFF) or 'Razor DFF' which are generally used in implementation a high speed sequential circuit. The output waveforms of these two circuits are analyzed and compared to check errors under varying data condition. The operation analysis using 'Razor DFF' will also be useful for system on chip with multiple clock domain, in which due to set up and hold time violations meta-stability leaks in to connected logic to output of the flip-flop/register cause huge power dissipation in the digital circuits. Further, the data moving from one clock domain to another clock domain become inaccurate due to timing violations [8, 9]. Conservative DVFS design techniques consider 'legal' voltage and corresponding frequency in a pairs that permits the unfailing operation of the circuit, where supply voltage change applied to VCO adjust its frequency even under fastest safe clock speed to follow all circuit timings (setup, hold, critical path delay etc.) [6]. Thus, the DVS/DVFS circuits designed using razor flip-flop-based error detection are discussed in [10–13] but do not consider circuit operation based on reference frequency setting. Further, DVFS in [5] allows only coarse step frequency change but DVFS architectural details are missing. In this paper, we described the finer reference frequency setting to produce operating frequency using VCO to follow the reference frequency and corresponding correct supply voltage generation for digital circuit (provided the designer ensure that any change in supply voltage ensure the correct circuit timing at changed output frequency) and VCO. During the varying supply and frequency, there is probability of timing violations so DVFS circuit with Razor DFF perform accurate data transfer with a penalty of one clock cycle (in case of error, entire pipeline designed using the razor DFF is re-executed to avoid data loss). Razor DFF is also used to to adjust the supply voltage by observing the error rate [10–13].

Sections 2 and 3 describe the charge pump circuit at transistor level and VCO circuit, respectively, while Sect. 4 describes the DVFS module (combined charge pump and ring oscillator). The TG-based flip-flop (FF) architectures (simple DFF and razor DFF) which are used to implement sequential circuit are described in Sect. 5. The DVFS module driving a TG-based FFC to verify the correctness of operation is explained in Sect. 6, while Sect. 7 concludes the finding in paper.

2 Charge Pump Circuit

The charge pump circuit controls a current in the current arm using the phase detector which is utilizing CKT1 and CKT2 and also uses switches implemented by PMOS as shown in Fig. 2. It uses the two PMOS switches (P_1 and P_2) to govern the output voltage (V_{OUT}) across the large external capacitor ($C_L = 20$ pf). The supply voltage V_{DD} is kept as 1.8 V.

The CKT_1 and CKT_2 of the charge pump circuit with asynchronous 'RESET' input which is fed by a output of NOR gate (i.e., when RESET = 1 then $OUT_1 = OUT_2$ becomes '1' asynchronously), also clock F_{REF} is fed to CKT_1 and clock F_{DFS} is fed to CKT_2. The outputs of CKT_1 and CKT_2 become input for a faster NOR gate.

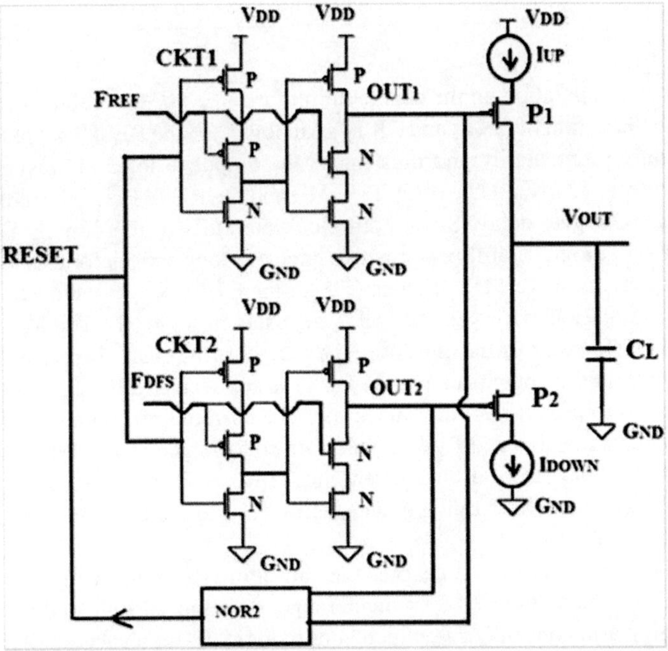

Fig. 2 Charge pump circuit

Table 1 Sequence of operation on the charge pump (For $F_{REF} > F_{DFS}$)

Sequence	Operations in charge pump	OUT_1	OUT_2	V_{OUT}
1	Assume $OUT_1 = OUT_2 = 0$ F_{REF} and F_{FDFS} are both '0'	0	0	I_{up} and I_{down} transistor ON momentarily
	Since $OUT_1 = OUT_2 = 0$, therefore RESET = 1 after NOR gate delay which sets $OUT_1 = OUT_2 = 1$	1	1	Unchanged (P_1 and P_2 are OFF)
2	Since $OUT_1 = OUT_2 = 1$, therefore RESET = 0 after NOR gate delay which keeps $OUT_1 = OUT_2 = 1$ (i.e., OUT_1 and OUT_2 floating, so dynamically preserves previous logic levels)	1	1	Unchanged (P_1 and P_2 are still OFF)
3	@F_{REF} positive edge (arrives first since its faster)	0	1	I_{up}, i.e., P_1 transistor ON (V_{OUT} increases)
4	@F_{FDFS} positive edge (arrives next since its slower)	0	0	I_{up} and I_{down}, i.e., both P_1 and P_2 transistor are ON momentarily (V_{OUT} do not change, but makes RESET = 1 after the NOR gate delay)
5	$OUT_1 = OUT_2 = 0$, therefore RESET = 1 after NOR gate delay	1	1	Unchanged (P_1 and P_2 are OFF)

Sequence of operation on the charge pump for $F_{REF} > F_{DFS}$ is shown in Table 1. When both the output of CKT_1 and CKT_2 go to logic '0', the RESET signal becomes logic '1' only momentarily and thus forces the output to logic '1' (asynchronous RESET). Subsequently, $OUT_1 = OUT_2 =$ '1', and the RESET signal becomes logic '0' after a NOR gate delay. Since both clock are still '0' the outputs OUT_1 and OUT_2 remain at logic '1' till next positive edge of clock arrives for either CKT_1 or CKT_2. After the positive edge of clock (F_{REF}), the OUT_1 sets to logic '0'. After the positive edge of clock (F_{FDFS}), the OUT_2 also sets to logic '0'. The logic at node OUT_1 and OUT_2 controls the current in the current arm (OUT_1 is connected to the gate of PMOS in the upper arm and the OUT_2 is connected to the gate of PMOS in the lower arm) to produce appropriate V_{OUT}. The current arm contains of two fixed value current sources I_{UP} and I_{DOWN}, which are controlled by two PMOS transistors switches to govern current in the current arm ($I_{UP} = 200\ \mu A$, $I_{DOWN} = 200\ \mu A$) and thus charging and discharging of a capacitor (C_L) to produces an output voltage V_{OUT}.

If the $\boldsymbol{F_{REF} > F_{DFS}}$, then it charges the capacitor (since upper or charging arm control OUT_1 goes to logic '0' for a much longer duration as compared to the lower or discharging arm control OUT_2 which goes to logic '0' for a very small duration). The sequences of Table 1 can be observed as shown in Fig. 3, where V_{OUT} increases.

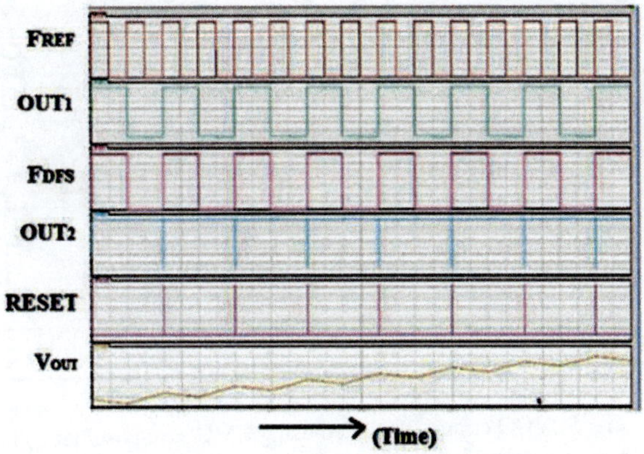

Fig. 3 V_{OUT} increases when $F_{REF} = 50$ MHz and $F_{DFS} = 25$ MHz

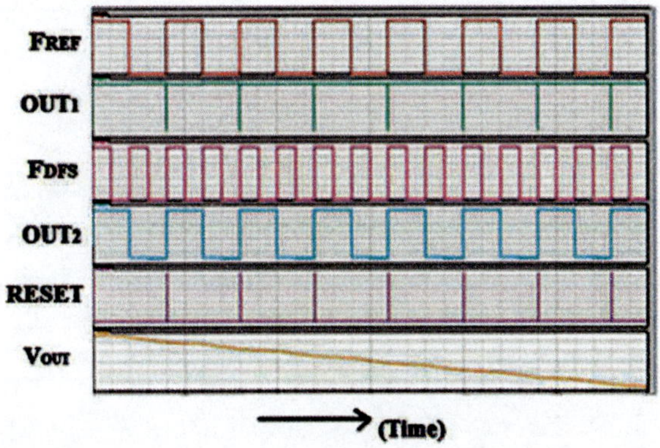

Fig. 4 V_{OUT} decreases at $F_{REF} = 25$ MHz and $F_{DFS} = 50$ MHz

If the $\boldsymbol{F_{REF} < F_{DFS}}$, then it discharges the capacitor (since upper or charging arm control OUT_1 becomes logic '0' for a for a very small duration as compared to the lower or discharging arm control OUT_2 which becomes logic '0' for much longer duration). The V_{OUT} increases as shown in Fig. 4. If the $\boldsymbol{F_{REF} = F_{DFS}}$, then the period of logic '1' of the CKT_1 output and CKT_2 output are same (similarly, period of logic '0' of the CKT_1 output and CKT_2 output are same), thus the capacitor neither charge or discharge as the charging arm and discharging arm both are switched on at the same time having same current value and switch out resistances; hence, maintain the output V_{OUT} at constant value.

3 VCO Circuit

Here, a VCO consisting of ring oscillator having many stages of current starved inverter to introduce a delay is as shown in Fig. 5. The output obtained from the last stage is of VCO fed to the input of the first stage in a ring oscillator. The sustain

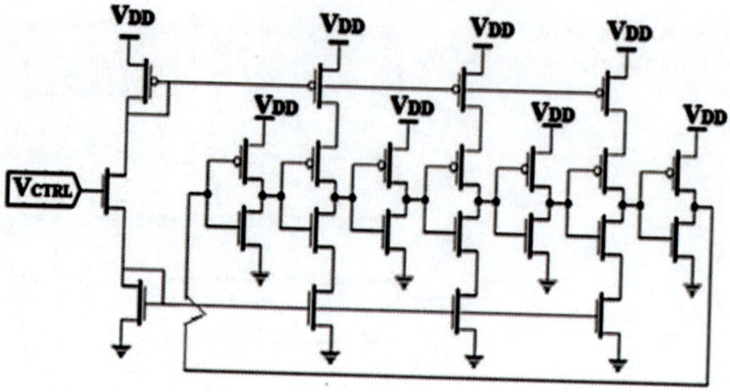

Fig. 5 VCO circuit (designed using CMOS ring oscillator) [14]

Fig. 6 VCO waveform

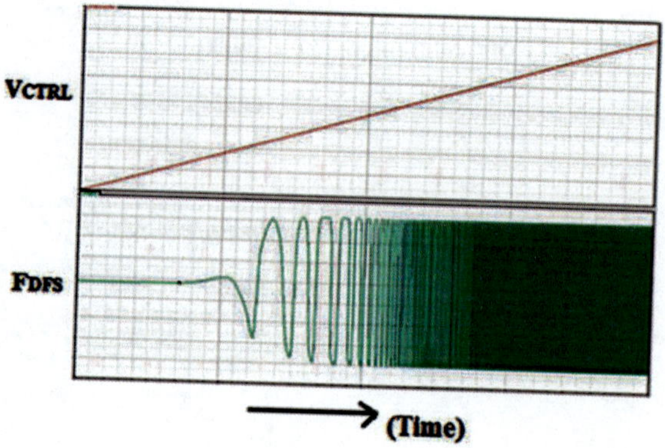

oscillation can be obtained if the ring provides a phase shift of 360° and has a voltage gain of unity at the frequency of oscillation. The supply voltage V_{DD} is kept as 1.8 V.

Figure 6 shows the variation of ring oscillator frequency with the increase in the control voltage (V_{CTRL}) varying from 0 to 1.8 V. After the few oscillations the voltage levels of the oscillator frequency signal varies between 0 and 1.8 V.

4 Combined Charge Pump and VCO: The Complete DVFS Module

The charge pump circuit and VCO are combined by substituting the F_{DFS} fed as the VCO output and output of the charge pump (V_{OUT}) fed as VCO input as shown in Fig. 1. If $F_{REF2} = F_{REF1}$ the V_{OUT} settles to a fixed value. If F_{REF} changed (fed through a 2:1 multiplexer), then the VCO frequency starts to change and keep on changing till the VCO frequency matches with the F_{REF} and V_{OUT} settles to a new voltage level.

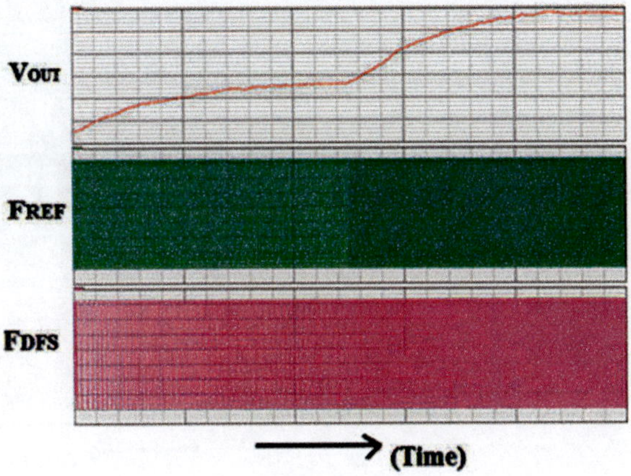

Fig. 7 V_{OUT} and F_{DFS} variation when F_{REF} is switched from (i) F_{REF1} = 8.3 MHz to (ii) F_{REF2} = 16.7 MHz

4.1 $F_{REF2} > F_{REF1}$

When V_{OUT} becomes constant the VCO frequency becomes F_{REF1} if now we change the frequency to F_{REF2} (executed using 2:1 Mux to switch F_{REF} from 8.3 MHz, corresponding to the clock period 120 ns to 16.7 MHz, corresponding to the clock period 60 ns) then the V_{OUT} again increases, which in-turn increases VCO frequency to new value of F_{REF2} as shown in Fig. 7. The V_{OUT} varies from 0.87 to 1.04 V.

4.2 $F_{REF2} < F_{REF1}$

When V_{OUT} becomes constant, the VCO frequency becomes F_{REF1} if now we change the frequency to F_{REF2} (executed using 2:1 Mux to switch F_{REF} 16.7 MHz to 8.3 MHz) and the V_{OUT} again decreases which in-turn increase the VCO frequency to new value of F_{REF2} as shown in Fig. 8. The V_{OUT} varies from 1.04 to 0.87 V.

5 TG FF Circuits

The synchronous digital circuits has fundamental blocks as latches and FFs. A Latch circuits (active low or active high), depending on whether they respond to a '0' or '1' logic level of the clock input. The master–slave architecture is a cascaded combination of active low and active high latches as shown in Fig. 9. The DFF either retain previous output or change its output logic based on the input logic 'D' at the clock edge rising transition [i.e., positive edge triggered (PET)], provided setup and hold times are followed.

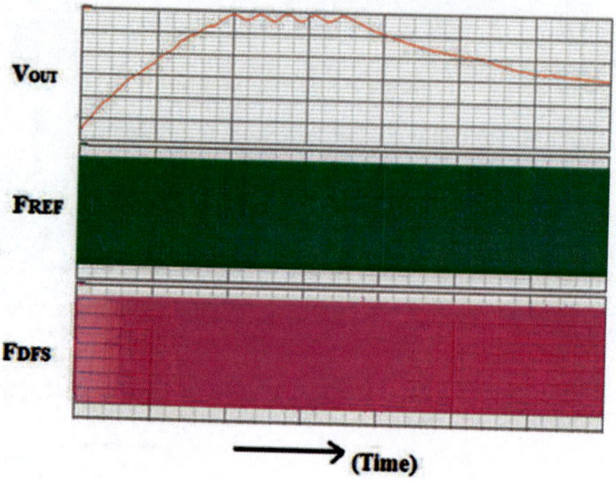

Fig. 8 V_{OUT} and F_{DFS} variation when F_{REF} is switched from (i) $F_{REF1} = 16.7$ MHz to (ii) $F_{REF2} = 8.3$ MHz

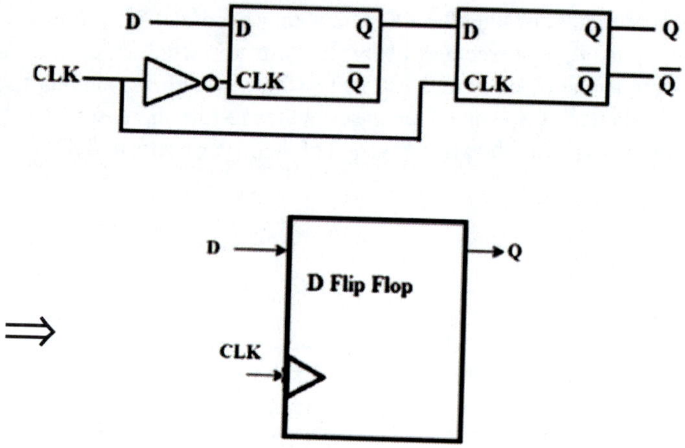

Fig. 9 A Master–slave representation of a PET DFF circuit

The **TG-based** FF architectures (simple DFF and razor DFF) which are generally used to implement sequential circuit are described below.

5.1 *Simple DFF Circuit*

The architecture for the positive edge triggered (PET) simple DFF is shown in Fig. 10.

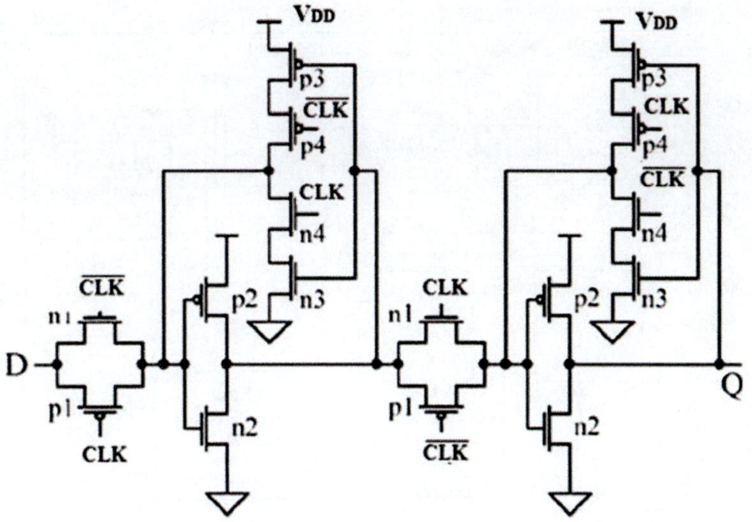

Fig. 10 Simple schematic of a DFF (PET)

5.2 *Razor DFF Circuit [10–13]*

Generally, under varying clocking situations, there is a risk that the FFs operation may fail due to not obeying the setup and hold the timing constraints due to late arrival of data. The Razor DFF utilizing the error correction technique (tolerate setup time and hold time violation leading to meta-stability) to prevent data loss with a one clock cycle penalty. The Razor DFF circuit block diagram is as shown in Fig. 11, which utilizes clock signal 'CLK'.

The main FF is a PET DFF while shadow FF is a negative edge triggered (NET) 'master slave DFF'. The output of both these FF fed to a multiplexer, which is used to select the correct input data value based on the received 'Error_Signal'. Further a combination of XOR and OR gate is used as an error detector (comparing the main FF and shadow FF outputs or meta-stability detection to generate 'Error_Signal'). In case of error or meta-stability, the multiplexer existing at the main FF selects the shadow slave output to generate the correct flip-flop output. In the Razor DFF is required to maintain data for more than half cycle after the positive edge of clock for the correct capture and also allowing error correction.

The waveform for the Razor DFF is explained in Fig. 12. When the timing are met (as indicated for Instr1), then the correct data is latched (i.e., 'Error_signal' = '0'). The cycle 3 shows that the clock edge observe the improper instruction as Instr1 instead of Instr2 due to late arrival of data and hence violating the timing constraints. In this case, shadow latch removes inconsistency, as the shadow latch samples for positive edge of clock; therefore, it samples the correct data. The 'Error_signal' is fed to the multiplexer select line, to select the correct data input. The data missed is

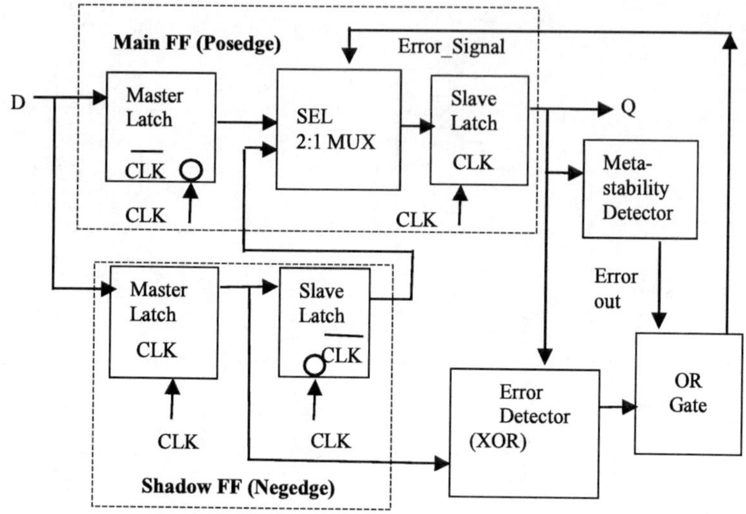

Fig. 11 Razor DFF circuit block diagram

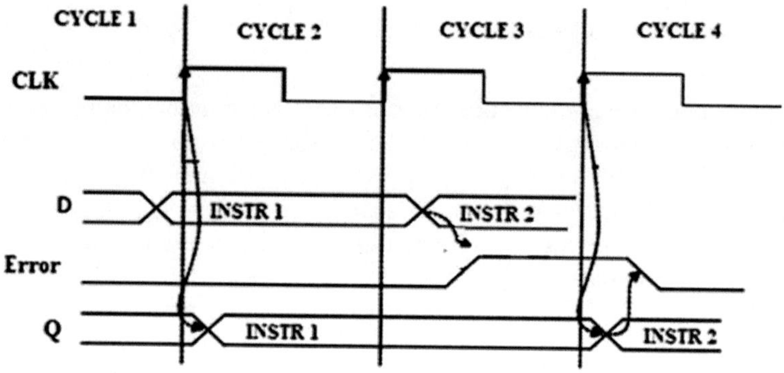

Fig. 12 Waveforms for circuit operation

then sent to the output of flip-flop in the following clock edge which performs the error correction but also adds a one clock cycle penalty.

6 The DVFS Control Circuit Driving a FFC (Simple TG/Razor Type) Load

The DVFS circuit is used to verify the correctness of driving a FFC (i.e., simple TG/Razor type) load. For the simplest FFC-type load, the supply voltage and clock frequency are fed by DVFS circuit as shown in Fig. 13.

Fig. 13 DVFS circuit driving FFC load

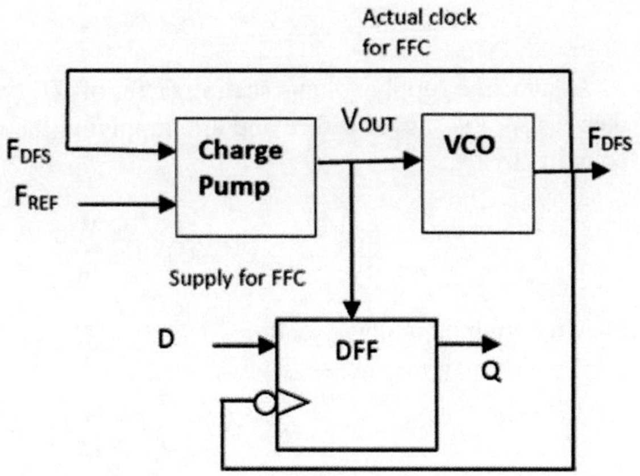

For $L < 0.25\ \mu\text{m}$, short channel approximation is followed. When the critical electrical field is reached (at and after particular drain-source voltage V_{DSAT}), the velocity saturation comes into play (v_{sat} becomes constant) and the current abruptly saturates to I_{Dsat} value.

For a NMOS device, the drain current is as indicated in Eq. 1, similar relationship exist for PMOS device [1].

$$I_{Dsat} = I_D|_{(V_{DS}=V_{DSAT})}$$

$$I_{Dsat} = \mu_n C_{OX}\frac{W}{L}\left[(V_{GS} - V_t)V_{DSAT} - \frac{V_{DSAT}^2}{2}\right]$$

$$I_{Dsat} = \mu_n C_{OX} V_{DSAT}\frac{W}{L}\left[(V_{GS} - V_t) - \frac{V_{DSAT}}{2}\right]$$

$$\frac{\mu_n}{L}V_{DSAT} = v_{sat}$$

$$I_{Dsat} = v_{sat} C_{OX} W\left[V_{GS} - V_t - \frac{V_{DSAT}}{2}\right] \tag{1}$$

where

v_{sat} Saturated velocity of carriers
V_{DSAT} Drain–source voltage at which critical electrical field is reached.

$$I_{Dsat} \propto V_{DSAT}$$

$$V_{DSAT} \propto V_{GS} \text{ or } V_{DD}$$

Therefore, I_{Dsat} variation with supply voltage is given by Eq. (2).

$$I_{\text{Dsat}} \propto V_{\text{DD}} \tag{2}$$

Assume the supply voltage scaling factor of 'U' (where $U > 1$). The propagation delay (τ_{PHL}) is shown before and after applying the voltage scaling is as indicated equation by (3).

$$\tau_{\text{PHL}} \propto C_L \frac{V_{\text{DD}}}{I_{\text{Dsat}}}$$

After applying voltage scaling:

$$\tau'_{\text{PHL}} \propto C_L \frac{V_{\text{DD}}/U}{I_{\text{Dsat}}/U} = \tau_{\text{PHL}} \tag{3}$$

Thus, in short channel device, delays do not vary after voltage scaling due to the velocity saturation. The designer ensures that supply voltage corresponding the highest frequency of operation ensures the correct circuit timing so that timing is also followed at any reduced frequency. Now, assume both the frequency and supply voltage scaling factor of 'U' (where $U > 1$). The dynamic power (P_{dyn}) is shown before and after applying the voltage scaling as indicated equation by (4).

$$P_{\text{dyn}} = f C_L V_{\text{DD}}^2$$

After applying voltage scaling:

$$P'_{\text{dyn}} = \frac{f}{U} C_L \frac{V_{DD}^2}{U^2}$$
$$P'_{\text{dyn}} = \frac{P_{\text{dyn}}}{U^3} \tag{4}$$

Thus, in short channel device, dynamic power reduces inversely proportional with 'U' and shows cubic cubic impact on dynamic power dissipation. Further, the static power (P_{static}) is shown before and after applying the voltage scaling as indicated equation by (5).

$$P_{\text{static}} = V_{\text{DD}} \times I_{\text{Dsat}}$$
$$P'_{\text{static}} = \frac{P_{\text{static}}}{U^2} \tag{5}$$

Thus, scaling voltage will also reduce static power when the circuit elements remain idle for longer period.

The FFC are based on (i) simple DFF (ii) Razor DFF. Figure 14 shows the output waveforms for simple DFF which miss the data and produce erroneous results whenever setup or hold time violation occur due to the late arrival of data, i.e., at clock edge or after the clock edge. When reference frequency is increased, the input data

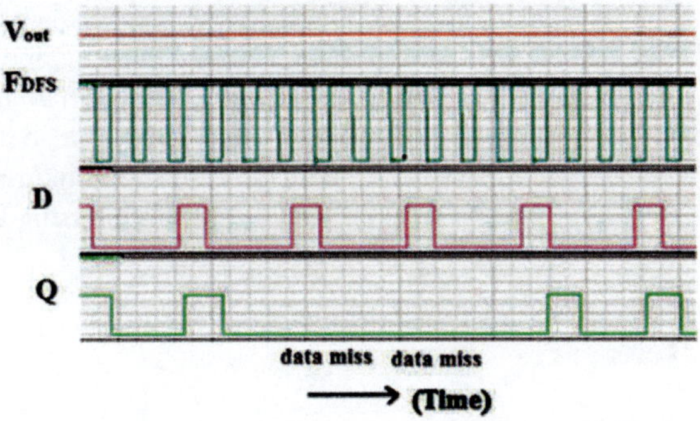

Fig. 14 Output of simple DFF (miss the data)

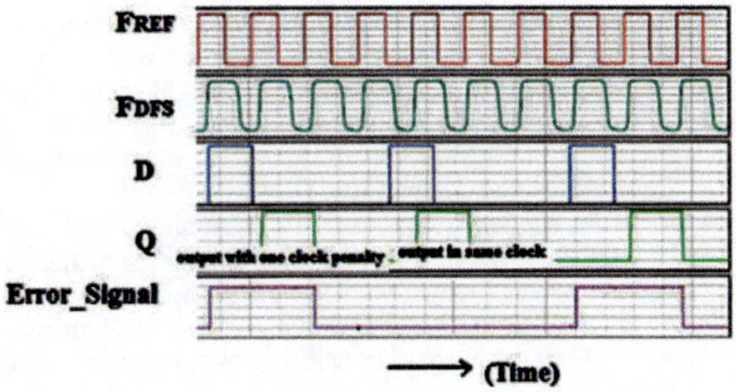

Fig. 15 Output of Razor DFF (does not miss the data)

may arrives late (since the supply voltage to DFF is gradually increasing). Figure 15 shows the output waveforms for Razor DFF which does not miss the data and hence produce correct results with a penalty of one clock cycle. In the Razor DFF since data is maintained more than half cycle after the positive edge of clock since it perform the correct capture and also allow the error correction.

Here, we just verified accurate DVFS operation without considering the power analysis since for detailed power analysis a larger digital circuit as a load will require. The limitation of the implementation is driving strength of FFC supply, but for driving a single DFF, it is sufficient and produce the desired results (since the output voltage (V_{OUT}) of charge pump is feeding a large capacitor ($C_L = 20$ pF) so that a small load of single DFF do not overload it). Thus, V_{OUT} of charge pump acts as 'reference voltage'. We can improve design further by adding a buck/boost converter based power circuit fed with V_{OUT} of charge pump as 'reference voltage' to it and then its controlled dc output voltage feeds to a bigger digital circuit without loading it much.

7 Conclusion

The DFVS circuit helps to reduce the dynamic power under varying workload conditions. The TG-based FF architecture (simple DFF and razor DFF) is fed with DVFS control circuit. The simple DFF circuit misses data hence produces erroneous results while Razor DFF which does not miss the data and hence produce correct results with a penalty of one clock cycle. The DVFS circuit used for Razor DFF-based FFC is best suited for power aware circuit design and also for system on chip with multiple clock domain where every data edge must be captured by destination domain or for designing low power sequential digital circuit using DVFS technique.

References

1. Rabaey, J.M., Chandrakasan, A., Nikolic, B.: Digital Integrated Circuits, 2nd edn. Prentice Hall of India Private Limited (2004)
2. Choi, K., Soma, R., Pedram, M.: Fine-grained dynamic voltage and frequency scaling for precise energy and performance tradeoff based on the ratio of off-chip access to on-chip computation time. IEEE Trans. Comput.-Aided Des. Integr. Circ. Syst. **24**(1) (2005)
3. Choi, K., Dantu, K., Cheng, W., Pedram, M.: Frame-based dynamic voltage and frequency scaling for a MPEG decoder. In: IEEE/ACM Proceedings of International Conference on Computer-Aided Design (2001)
4. Choi, K., Soma, R., Pedram, M.: Dynamic voltage and frequency scaling based on workload decomposition. In: Proceedings of the 2004 International Symposium on Low Power Electronics and Design, pp. 174–179. Newport Beach, California, USA (2004)
5. Beigné, E., Clermidy, F., Miermont, S., Vivet, P.: Dynamic voltage and frequency scaling architecture for units integration within a GALS NoC. In: Proceedings of the Second ACM/IEEE International Symposium on Networks-on-Chip, pp. 129–138 (2008)
6. Lee, Y., Chiu, C., Peng, S., Chen, K., Lin, Y., Lee, C., Huang, C., Tsai, T.: A near-optimum dynamic voltage scaling (DVS) in 65-nm energy-efficient power management with frequency-based control (FBC) for SoC system. IEEE J. Solid State Circuits **47**(11) (2012)
7. Basireddy, K.R., Singh, A.K., Al-Hashimi, B.M., Merrett, G.V.: AdaMD: Adaptive mapping and DVFS for energy-efficient heterogeneous multi-cores. IEEE Trans. Comput. Aided Des. Integr. Circuits Syst. (2019)
8. Zhou, S., Zhang, T., Yang, Y.W.: Cross clock domain signal research based on dynamic motivation model. In: IEEE Fourth International Conference on Dependable Systems and Their Applications (2017)
9. Fan, Y., Xiang, B., Zhang, D., Ayers, J.S., Shen, K.-Y.J., Mezhiba, A.: Digital leakage compensation for a low-power and low jitter 0.5-to-5GHz PLL in 10 nm FinFET CMOS technology. In: IEEE International Solid-State Circuits Conference (2019)
10. Ernst, D., Kim, N., Das, S., Pant, S., Rao, R., Pham, R., Ziesler, C., Blaauw, D., Austin, T., Flautner, K., Mudge, T.: Razor: A low-power pipeline based on circuit-level timing speculation. In: Proceedings of the 36th International Symposium on Microarchitecture (MICRO-36) (2003)
11. Agrawal, R.K., Pandey, N.: Implementation of PFSCL razor flipflop. In: Proceedings of the IEEE 2017 International Conference on Computing Methodologies and Communication (ICCMC) (2017)
12. Kunitakel, Y., Sato, T., Yasuura, H., Hayashida, T.: Possibilities to miss predicting timing errors in canary flip-flops. In: IEEE 54th International Midwest Symposium on Circuits and Systems (MWSCAS) (2011)
13. Rathnala, P., Wilmshurst, T., Kharaz, A.H., Timing error detection and correction for power efficiency: an aggressive scaling approach. IET Circuits, Devices & Systems (2018)
14. Jovanovic, G., Stojcev, M., Stamenkovic, Z.: A CMOS voltage controlled ring oscillator with improved frequency stability. J. Appl. Math. Mech. **2**, 1–9 (2010)

Effective Predictive Maintenance to Overcome System Failures—A Machine Learning Approach

Sai Kumar Chilukuri, Nagendra Panini Challa, J. S. Shyam Mohan, S. Gokulakrishnan, R. Vasanth Kumar Mehta, and A. Purnima Suchita

Abstract As industry is getting advanced day by day incorporating new equipment's on a large scale, there is a need to predict the machine lifetime in order to support the supply chain management. There are many ways of substitutions or upgradations required for any machine over a certain period of time, where its maintenance has become a major challenge. This problem is solved by building an effective predictive maintenance system which provides an intense spotlight for all types of machine industries. The log data is collected from the daily system activity from machines through deployment of various sensors facilitated to monitor the current state of equipment. A huge volume of numerical log data set is analyzed by the system for preparing the time series data for training and analyzing the model. Further steps involve bypassing the anomalies and fetching the clean data. The model is tested focusing on restoration time of any machine. This paper identifies and predicts failures of heavy machines, thus facilitating the predictive maintenance scenario for effective working of the machine at all situations. This work is implemented by LSTM network model for gaining authentic results with numeric data which facilitates major cost savings and offers higher maintenance predictability rate.

S. K. Chilukuri (✉) · J. S. S. Mohan · S. Gokulakrishnan · R. V. K. Mehta · A. P. Suchita
Department of Computer Science and Engineering, SCSVMV, Enathur, Kanchipuram 631561, Tamil Nadu, India

J. S. S. Mohan
e-mail: jsshyammohan@kanchiuniv.ac.in

S. Gokulakrishnan
e-mail: s.gokulakrishnan@kanchiuniv.ac.in

R. V. K. Mehta
e-mail: vasanth.mehta@kanchiuniv.ac.in

N. P. Challa
Department of Information Technology, Shri Vishnu Engineering College for Women, Bhimavaram, Andhra Pradesh, India
e-mail: cnpanini@svecw.edu.in

H. Sharma et al. (eds.), *Congress on Intelligent Systems*, Advances in Intelligent Systems and Computing 1334, https://doi.org/10.1007/978-981-33-6981-8_28

Keywords Predictive maintenance · Machine learning · Data frames · Standardization · Confusion matrix · Forecasting model · Long short-term memory (LSTM) model

1 Introduction

Maintenance is defined as an "entire process of preserving a condition or any situation" by many experts and scientists globally [1]. This is widely applied to different machines related to many domains like automobile, astronomy, household, and many other applications. This is implemented to enable the machine operations and its functions accurately at any point of time. Generally, in industries, workers are responsible for equipment maintenance in a timely manner where the maintenance complexity increased due to the drastic growth in machine utilization over a recent couple of years. The business needs have increased due to growing public demand where the industries focused on product delivery in a reliable manner, which enables the machines to be more specific and accurate. The designers are more concerned about not only solving the failures occurred but also to prevent them. Hence, the computational sector facilitates the machine industry to maintain those complex systems by various strategies.

Currently, many industries are focusing on maintenance activities effectively to optimize their resources by incorporating various computational and data analysis technologies. According to many expert's maintenance is divided into three main categories namely: corrective, preventive, and predictive [2]. The corrective maintenance is defined as solving the problems which specifies or expects the same critical halt of any machine when running a process. This strategy will halt the production, increases the variable costs and hence the repair time cannot be estimated as there might be many failures associated with them in different processes. This type of maintenance is implemented where there is not adverse effect to the production. Thus, preventive maintenance [3] is referred to schedule a planned activity at different time period to reduce the unexpected breakdowns and failures. This is implemented in non-production time by specialists where periodic checks and parts replacement takes place. Some authors presented their strategies of preventive maintenance on high CNC machine centers [4] and gas turbines scheduling system without effecting the production based on reliability and overhaul findings [5]. Predictive maintenance happens with precise formulas based on the data produced by the machine in various time periods. The data produced might be numeric or non-numeric data which can be further analyzed using different measured parameters which ensures cost minimization [6] and any repairs using various deep learning (DL), machine learning (ML), and artificial intelligence (AI) techniques [7].There are several approaches and scenarios which fail at calculating accurate effective prediction at different phases of data acquisition and processing as many systems are unable to handle the large amounts of data generated from a particular machine. Thus, the systems are developed to deal with massive amounts of data produced by the industry 4.0 [8, 9] which are facilitated with a unified framework to acquire, process, and analyze the data to extract the necessary knowledge. This approach facilitates with analyzing the machines lifetime, overcomes the machine failures, and thus predicts the appropriate time [10] for machine maintenance. Some machines adopt the real-time data acquired from different sensors by developing a suitable computational environment which allows automatic data processing (Fig. 1).

The various contributions of this paper are listed as the following. The first part deals with how industry 4.0 [11] is fast advancing and influencing the heavy industry sector. Second part deals with different contributions of various experts and researchers which are analyzed and tabulated accordingly. Third part deals with the work done on data set related to the heavy industrial engines and motors. Thus, the data collected is preprocessed and fed for classification model and the actual machine failure time is calculated facilitating the predictive maintenance based on the forecast model. Fourth part showcases the results obtained from those predictions and tabulated. Fifth part deals with conclusion of this paper abstracting about this work, its benefits, and advantages to the society.

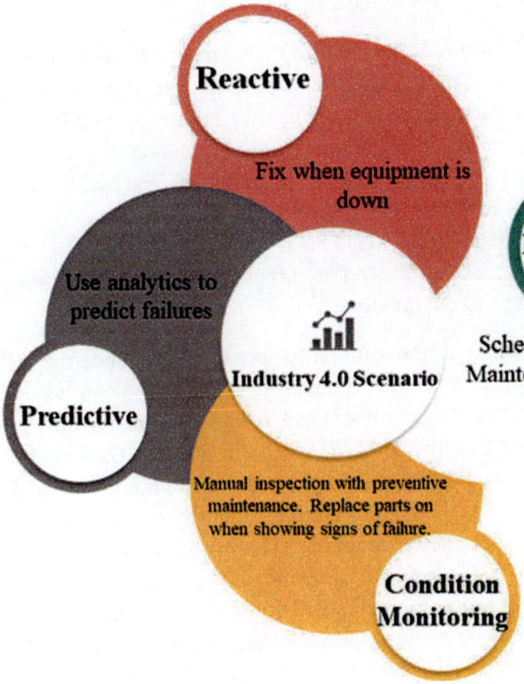

Fig. 1 Industry 4.0 scenario

2 Background and Existing Work

Predictive maintenance (PM) is a form of organized actions which focus on breakdown and failure prevention. This is based on a systematic periodic machine review which is generally applied outside the production time. It is implemented by incorporating different maintenance actions such as periodic checks, parts replacement checks, and many more. Some researchers proposed a computer-aided planning system where maintenance is focused on CNC machines. Scheduling inspection of high speed gas turbines is also implemented using reliability calculations and fault findings [12]. These calculations constitute a part of predictive maintenance which is often an efficient approach for reliability enhancement toward multi-object systems. The main focus here is to predict the equipment failure using predictive models and try to prevent them by performing various maintenance activities. Cost minimization [13] and zero failure rates are being achieved through this maintenance schedule approach. Some researchers have proposed a PM tool based on Web application and R-package which helps engineers and other experts to easily analyze the machine generated data set [14] based on various machine learning algorithms which supports decision support systems [15] for maintenance optimization.

There are different predictions models like

1. Customer Lifetime Value Model (CLVM): This model deals with the customers who are likely to invest more capital into different products and services.
2. Customer Segmentation Model (CSM): This model is based on a group of customers which have similar characteristics and behavior.
3. Predictive Maintenance Model (PMM): This model deals with forecasting the chances of essential machine 9 (equipment) breakdown.
4. Quality Assurance Model (QAM): This deals with identifying and preventing defects by different cost cutting and lay off scenarios when providing different products and services to customers.

Predictive maintenance is a more efficient approach to enhance the reliability of complex multi-object systems. Ideally, maintenance schedule can be optimized to minimize the cost of maintenance and achieve zero failure manufacturing through this approach. Remaining useful life (RUL) prediction of the equipment is the key technology for realizing predictive maintenance. Accurate prediction of RUL plays an important increasingly crucial role in the intelligent health management system for the optimization of maintenance decisions. Therefore, in this paper, we propose a predictive maintenance modeling tool implemented as R-package and Web application (PdM package) that helps engineers and domain experts to easily analyze and utilize multiple multivariate time series sensor data to develop and test predictive maintenance models based on RUL estimation using machine learning algorithms in a rapid for decision support.

Predictive maintenance is a more efficient approach to enhance the reliability of complex multi-object systems. The aim of predictive maintenance is first to predict when equipment failure could occur by using predictive models (based on certain

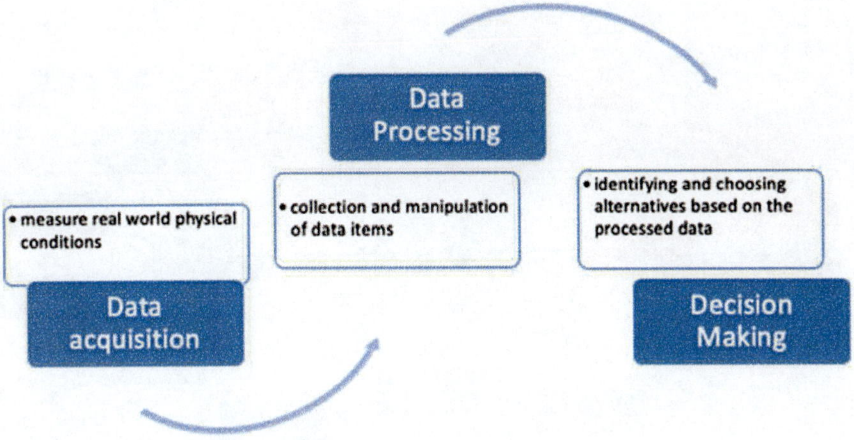

Fig. 2 Steps in development of predictive maintenance

factors), and secondly, to prevent the occurrence of the failure by performing maintenance. Ideally, maintenance schedule can be optimized to minimize the cost of maintenance and achieve zero failure manufacturing through this approach. Remaining useful life (RUL) prediction of the equipment is the key technology for realizing predictive maintenance. Accurate prediction of RUL plays an important increasingly crucial role in the intelligent health management system for the optimization of maintenance decisions. Therefore, in this paper, we propose a predictive maintenance modeling tool implemented as R-package and Web application (PdM package) that helps engineers and domain experts to easily analyze and utilize multiple multivariate time series sensor data to develop and test predictive maintenance models based on RUL estimation using machine learning algorithms in a rapid for decision support. Predictive maintenance plays a major role in cost and time optimization [16] required for any machine. This type of maintenance consists of three different steps, namely data acquisition, data processing [17], and decision making. The first step involves in collecting and storing the data from the machine to perform the predictive maintenance, which is classified in two types namely: event data and condition monitoring data. The former deals with the asset and maintenance estimation related to the machine based on time series data [18]. The later deals with sensor measurement analysis related to its physical assets. The second step deals with data preprocessing where the acquired data consists of missing, inconsistent, and noisy frames in the actual data set. The data preprocessing methods generally consists of different phases like data cleaning, data transformation, and data reduction. The data sets can be containing of wide domain areas such as industries, agriculture, medical, and computer sciences. Data errors [19] are caused due to different factors like sensor faults where eliminating them helps in improving the data quality. This process is referred to as data cleaning which facilitates the maintenance module with accurate data inputs in a timely manner. There are some graphical tools available which are based on human inspection consisting of mean and median values. Many

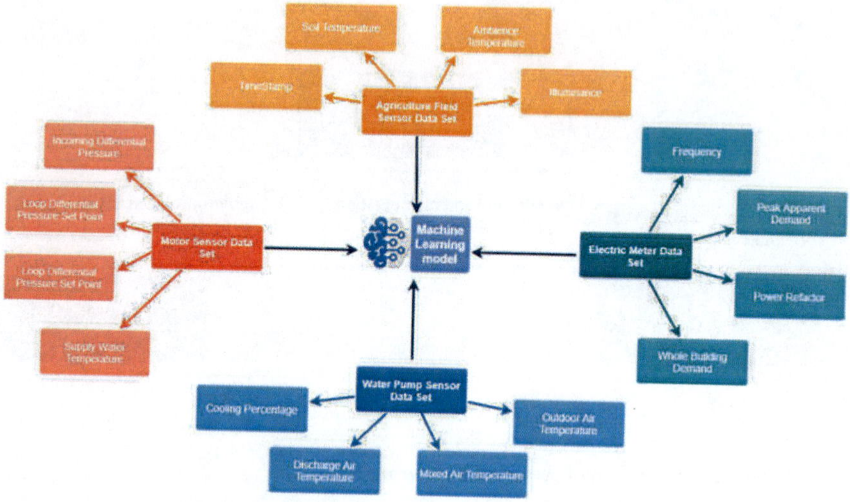

Fig. 3 Different domains using ML model

advanced methods like regression techniques are also used to estimate the missing values. Different methods are proposed for different problems like spectroscopy which is based on values like decomposition, KNN and QRILC data. Various clustering methods [20] are proposed by different authors to eliminate noisy values from data through outlier detection. The implementation of predictive maintenance is to develop an advanced environment which deep computer science and technology for even the small applications and the systems, which are ignored or provided with the less importance. While in real-time perspective, the least important systems like simple automobile engines, computer systems, and other handheld devices do not have any forecasting mechanism [21] and the maintenance of the small systems leads to huge principle amount. Any system should be feasible and have the considerable life-span in order to satisfy the end user. Hence, to ensure the system quality and avoid the scrappy product to the market in order to replace within a short period time, the prediction system can be used. The maintenance of any system or a machine has evolved a lot along with the advancements in every technology. The maintenance approach [22] has changed from reactive method to predictive methods to avoid the unexpected downtime and maintenance cost.

3 Proposed Work

The predictive maintenance carries out various features and advantages. The prediction helps in improving the system and developing an environment to reduce the maintenance cost and time of a particular machine. It helps in improvement of lifetime of the system with a lot more production.

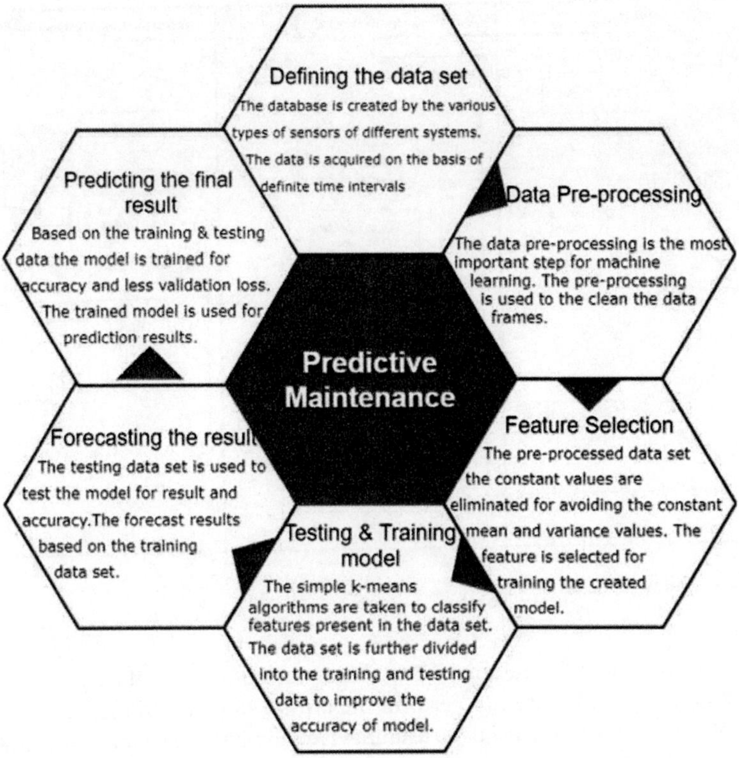

Fig. 4 System architecture

Initially, the data set of any machine is considered and fetched, further it should be changed based on the requirement. The data should be changed into an orderly format before feeding to the model, since the project completely depends on the data which is fed to the model. The data is given at most importance and the variance, anomalies [23] and even the null values are eliminated from the data set. In this project, the model is designed in such a way that it provides the classification analysis which is essential for knowing the most recent failure date of any system.

Unwanted data is removed from the data frame before training the model. The model is organized in a way to yield the time series data. The labeling of the data is performed to declare the target variable for the model to provide the result on the model. The various data frames are brought together and scaling is performed between the range of (0–1) by various scaling algorithms like standard scaling method, min–max scaling method, or the robust scaling method. The intrinsic feature of data alignment is introduced to create the similar data frame configuration.

The time series data is passed down to the created classification model. Based on correlation value (*r* value), the empty cells in the csv data frames are populated with the most accurate values. For schooling the model into the forecast model, massive volume of the statistics is used. Model's accuracy is improved by the amount of

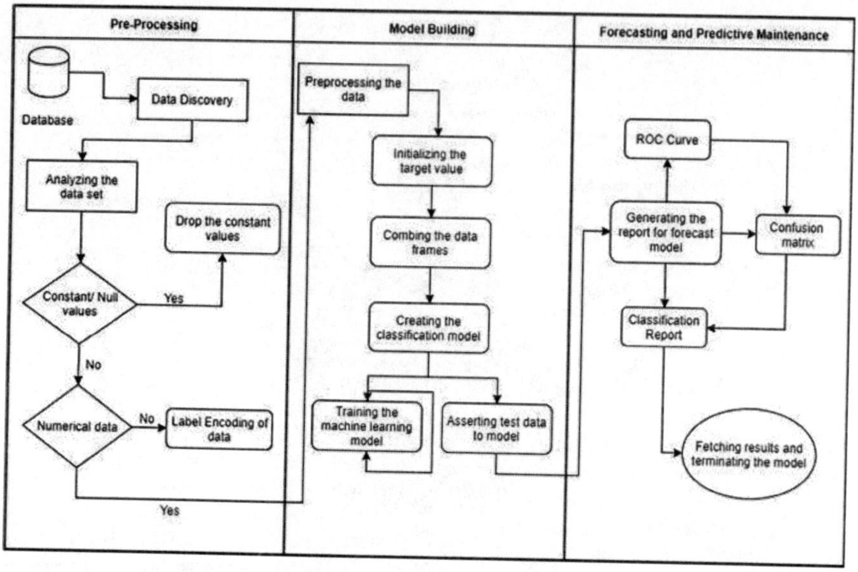

Fig. 5 Workflow of the proposed work

data fed to the model for training. The splitting techniques like the classical shuffle splitting techniques are used to perform the cross validation of the data frames to differentiate the testing data and the training data. Finally, the single data frame with all the appropriate index is used generating conclusion of the model.

4 Database and Data set Configuration

The data sets are acquired from the sensors of the systems which provide the details of performance of the system on the fixed time intervals. The data frames contain the various parameters based on the system configuration. The parameters vary based on the model. Initially, the data set is gone through the label encoding procedure where the data such as text data into the numerical values which can be easily understood by the predictive models. The label encoding depends on the number of parameters present in the particular data set.

4.1 Algorithmic Approach

The algorithm is the finite sequence of well-defined instructions which are used to solve the problem with in the finite number of steps. The algorithm is essential part of project which provides a detailed roadmap for the final requirement of the task. The

```
TimeStamp,IncomingDifferentialPressure_forecast,LoopDifferentialPressureSetPoint_forecast,LoopD
2018-08-01 00:00:10,17.130486107262815,16.0116546385772,17.644669364532216,27.25228837685925,1.
2018-08-01 00:01:10,17.131241626619,16.0113522545869,17.646356063017958,27.214578985124785,1.00
2018-08-01 00:02:10,17.13199655608252,16.01105018871425,17.64805968471132,27.17647257533018,1.0
2018-08-01 00:03:10,17.132750899134667,16.01074847083336,17.649780175301004,27.137971921087352,
2018-08-01 00:04:10,17.13350465939739,16.0104471301507,17.65151747752537,27.09907985753029,1.00
2018-08-01 00:05:10,17.134257840655422,16.010146195576837,17.653271531216443,27.05979927995175,
2018-08-01 00:06:10,17.1350104468171,16.009845695646394,17.655042273250498,27.020133143864307,1
2018-08-01 00:07:10,17.135762481874863,16.009545658522782,17.656829637502156,26.980084465075997
2018-08-01 00:08:10,17.1365139499 2767,16.009246111965965,17.658633554894084,26.939656318039212,
2018-08-01 00:09:10,17.137264855124965,16.008947083343 05,17.66045395331697,26.898851836607808,1
2018-08-01 00:10:10,17.138015201659922,16.00864859961962,17.662290757654887,26.85767421264892,1
2018-08-01 00:11:10,17.13876499375877,16.0083506873409,17.66414388976799,26.81612669575816,1.00
2018-08-01 00:12:10,17.139514235641112,16.008053372637285,17.666013268449298,26.774212593294425
```

Fig. 6 Dataset overview

deep learning algorithms are used to create a classification model for the supervised learning approach. The algorithms play the role for performing various operations like complex matrix multiplication, adding all the types of data frames, and adding the activation functions for the model.

Predictive analytics adopters have easy access to a wide range of statistical [24], data-mining, and machine-learning algorithms designed for use in predictive analysis models. Algorithms are generally designed to solve a specific business problem or series of problems, enhance an existing algorithm or supply some type of unique capability. Clustering algorithms, for example, are well suited for customer segmentation, community detection, and other social-related tasks [25]. To improve customer retention, or to develop a recommendation system, classification algorithms are typically used. A regression algorithm is typically selected to create a credit scoring system or to predict the outcome of many time-driven events.

The algorithms implemented in the project can be classified into two categories for training and testing the model. Generalized algorithm implemented on the data set is given below:

Step: 1—Start the Python program editor.
Step: 2—Describe the time series data frames.
Step: 3—If number_of_dataframes > 1, then create a consolidated data set.
Step: 4—Using the Minmax scaler scale the data set for classification.
Step: 5—Create the target variable using the unsupervised learning algorithmic approach.
Step: 6—Initialize the x_data as the initial data and y_data as the target data for binary classification.
Step: 7—Split the data set into training and test data by stratified shufflesplit.
Step: 8—Initialize the new tensor (third dimension) for the data set.
Step: 9—Create a LSTM model with number of dense neurons, activation method.
Step: 10—Save the model to the directory which contains the least validation loss.
Step: 11—Consider the batch size for the model to update the weights of back propagation.

This algorithm introduced in training the data frame combines both the forward propagation and back propagation. The forward propagation is used to assume the upcoming results and plot the curve based on the data set. Similarly, the backward propagation is an approach where the graph is plotted for the given data frame and the calculation of errors is edited. It is essential to use the backward propagation for the testing of the deep learning model [26] such that the errors in predicting can be verified.

4.2 Model Development

Model development is divided into various steps based on the priority and precedence of each step in creating the model. In this project, the model for forecasting is developed in some important steps which includes analysis of the data, preprocessing, creating the model, training the model, classifying the data into train data as well as the test data and forecasting the recent failure [27] of the system based on the trained model. Every timestamp also has an internal hidden state (ht) and internal memory. The model development is essential in order to cover all the steps which are important in sensing the anomaly in the system, the data frames which helps in classification of the data and providing the index for the data frames.

The model description is used to represent information about the model such as type of model, the number of parameters, and total trainable and non-trainable parameters of the machine learning model.

Forecasting Model Algorithm

Step: 1. Fetch the data set from the directory.
Step: 2. Drop the features which has no effect on the model (constant values).
Step: 3. Using fbprophet and the new data set plot the time series data report.

Fig. 7 Forecasting (LSTM) learning model

$$f_t = \sigma(W^f x_t + U^f s_{t-1})$$

$$o = \sigma(W^o x_t + U^o s_{t-1})$$

$$g_t = tanh(W^g x_t + U^g s_{t-1})$$

$$c_t = f_t \odot c_{t-1} + h_t \odot g_t$$

$$s_t = tanh(c_t) \odot o_t$$

$$h_t = f_t$$

```
Model: "sequential_1"
_________________________________________________________________
Layer (type)                 Output Shape              Param #
=================================================================
lstm_1 (LSTM)                (None, 11, 64)            16896
_________________________________________________________________
lstm_2 (LSTM)                (None, 11, 64)            33024
_________________________________________________________________
lstm_3 (LSTM)                (None, 11, 64)            33024
_________________________________________________________________
lstm_4 (LSTM)                (None, 64)                33024
_________________________________________________________________
dense_1 (Dense)              (None, 64)                4160
_________________________________________________________________
dropout_1 (Dropout)          (None, 64)                0
_________________________________________________________________
dense_2 (Dense)              (None, 1)                 65
=================================================================
Total params: 120,193
Trainable params: 120,193
Non-trainable params: 0
_________________________________________________________________
```

Fig. 8 System model description

The modular implementation provides a particular roadmap for developing an accurate for predictive maintenance.

Prediction algorithm

Step: 1. Load the model accurate model which has less validation loss.
Step: 2. Reading the forecasted data as input to predict the output.
Step: 3. Preprocess the data set for checking the input data requirements.
Step: 4. Run the model with prediction data and fetch the timestamp.

The model development includes three phases which can be termed as:

1. Training the model.
2. Translating the model from trained network.
3. Acquiring the results from the model.

The training method approached in the system is known as offline training approach.

5 Results and Discussion

The outcome for this project is categorized into two divisions, namely:

1. ROC and time series graphs
2. Confusion matrix.

5.1 *ROC and Time Series Graphs*

The receiver operating characteristic is a graphical plot which displays the ability of binary classifier system according to the variable threshold. The ROC curve is created by plotting the true positive rate against the false positive rate at various thresholds. The time series curves are used to plot the timestamp and any other feature present in the data set. The ROC curves are used to display the training and validation losses for a machine learning model [28].

The training and validation are derived from the given data set where the number of epochs used to train the LSTM model. The number of epochs is given by the range starting from 1 to the maximum length of the accuracy. The loss is the measure of bad prediction. A loss is the number indicating how effective the model's predictions are.

epochs = range(1, len(acc) + 1)

The training and validation accuracy are used to define the precision of the model. The model's precision is based on the measure of highest accuracy of the model.

Fig. 9 Training and validation data loss

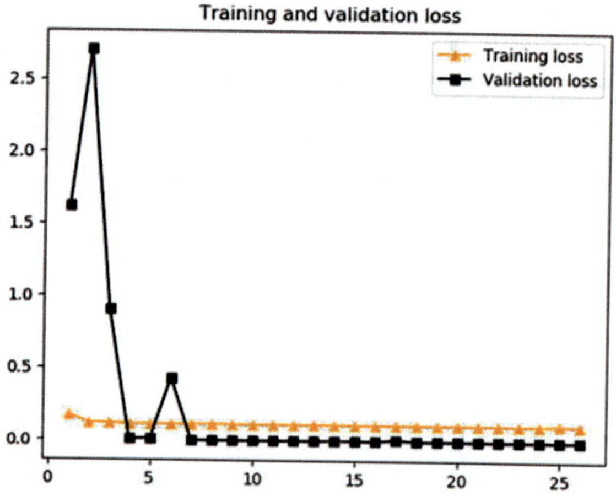

X- axis: Training Data loss (Epoch values)
Y- axis: Validation Data loss (Epoch values)

Fig. 10 Training and validation accuracy

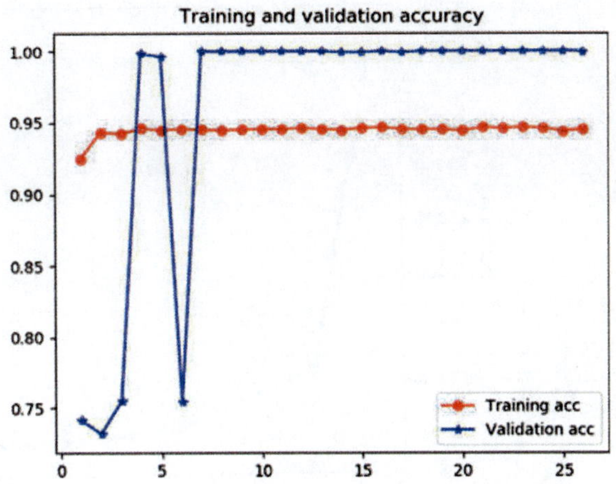

X-axis: Training accuracy (Epoch values)
Y-axis: Validation Accuracy (Epoch values)
Epoch for training model = 40

epochs = range(1, len(acc) + 1)

Similarly, the accuracy of the machine learning model is known as the measurement used to determine which model is best at identifying relationships between variable's and data sets based on inputs and training data.

The graph is used to represent the relation between the number of clusters and the total number of distortions. The clusters are known as the grouping of data points, or features while data points in different groups should have dissimilar properties. The distortions are the theoretical metrics which define finite metric space. This is the time series graphical visualization [29] of the data set which is used to display the forecasted values of the particular selected feature in the data set.

5.2 *Confusion Matrix*

The confusion matrix is the table mostly used to define the behavior of the classification model or classifier on the particular data frame whose true values are known. The confusion matrix provides the visualization [30] of the performance of any model. It is mostly used on the supervised learning algorithms. It is the matrix between the predicted class versus the actual class.

The confusion matrix is given with the x-axis and y-axis for plotting the heat maps based on the matrix data. The precision rate is used to define the model's accuracy and precision. Generally, the binary bits contain two options either true or false, but the basic terminologies of the confusion matrix are given by true positives (TP), true negatives (TN), false positives (FN), and false negatives (FN).

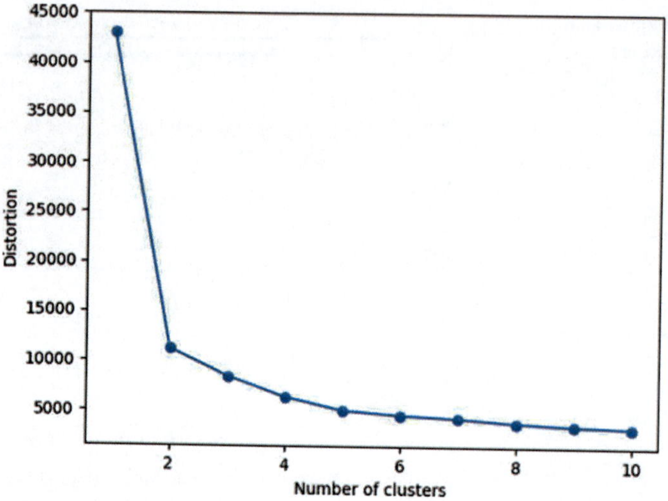

Fig. 11 Number of classes (K-means algorithm)

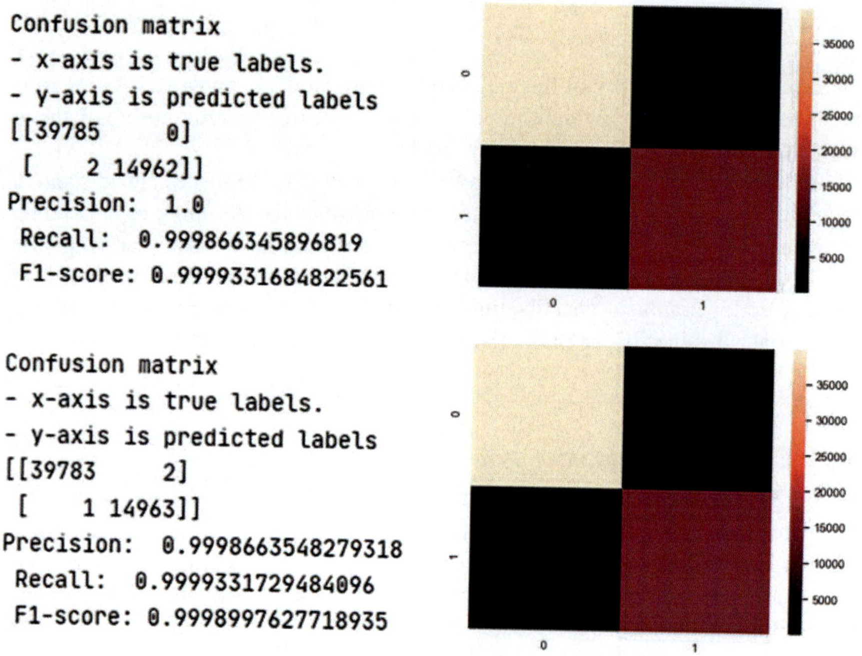

Fig. 12 Confusion matrix for differential pressure feature

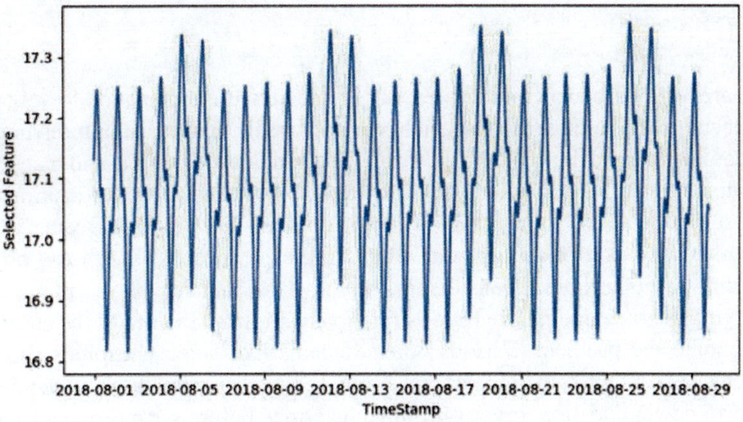

X-Axis: Timestamp
Y-axis: selected feature (differential pressure)

Fig. 13 Final output of PM for differential pressure feature

The classification reports are generated when the test data set containing the true value is provided to the trained classification model.

Table 1 consists data of 8 rows × 12 columns. Hence, the result obtained for this system intimated that they might fail on 2018-08-02, 23:24:10.

Therefore, necessary preventive steps need to be incorporated. Later, the classification report is visualized by the tensor boards. This result facilitates many organizations or industries with major cost savings and offers higher maintenance predictability rate.

Table 1 Classification rate

Incoming differential pressure (feature)	Classification
count	1908.000000
Mean	17.185183
std	0.060343
Min	17.071541
25%	17.138270
50%	17.196218
75%	17.238555
max	17.272476

6 Conclusion

There are various benefits of implementing predictive maintenance for obtaining more accurate and reliable results which are facilitated through various tools/interface providers such as Acxiom, IBM, Microsoft, SAP, Tableau, Teradata, and many more. These tools help the users to use and run predictive models for sales maximization and cost optimization. Some business centric domains like Airlines frequently need to analyze the ticket prices without affecting the past trends. When the physical machining part of airlines is considered, predictive maintenance plays a major role in identifying the machine failure time estimation and further facilitates the authorities to overcome the problem. Various other domains like hotels, hospitals use these maintenance strategies to forecast the room occupancy rate and its revenue. PM is used to detect and stop various criminal activities before serious damages occur ranging from online credit card frauds to cyberattacks in multiple domains.

This paper provides an effective predictive system by analyzing the numeric data generated from heavy machines like Boeing aircraft machine used for identifying the earlier failure of the system based on the trained data frames. The forecast model mentioned in the system uses the numerical data frames for creating the time series data. Based on the input data frame and the target data set, the model provides the results for the target value. This paper is the initial step for the predictive maintenance for the small, medium, or large systems and miniature data frames. The scope of the model can be developed so that the model can accept the alpha-numerical data frames for the model and it can be narrowed to a particular point for the prediction.

References

1. Pal, A., Prakash, P.K.S.: Practical Time Series Analysis. Time Series Data Processing, Visualization and Modelling. Packt Publishing (2017)
2. Marsland, S.: Machine Learning an Algorithmic Perspective, 2nd edn. CRC Press (2015)
3. Lee, J., et al.: Predictive maintenance of machine tool systems using artificial intelligence techniques applied to machine condition data. Procedia CIRP **80**, 506–511 (2019)
4. Yongyi et al.: A survey of predictive maintenance: systems, purposes and approaches. IEEE Commun. Surv. Tutor. (2019)
5. Liao, W., et al.: Data-driven machinery prognostics approach using in a predictive maintenance model. J. Comput. **8**(1), 225–231 (2013)
6. He, Y., et al.: Cost-oriented predictive maintenance based on mission reliability state for cyber manufacturing systems. Adv. Mech. Eng. **10**(1), 1–15 (2018)
7. Cimino, C., Negri, E., Fumagalli, L.: Review of digital twin applications in manufacturing. Comput. Ind. **113**, 103130 (2019)
8. Bousdekis, A., Lepenioti, K., Apostolou, D., Mentzas, G.: Decision making in predictive maintenance: literature review and research agenda for Industry 4.0. IFAC-Papers On Line **52**(13), 607–612 (2019)
9. Ruiz-Sarmiento, J.-R., Monroy, J., Moreno, F.-A., Galindo, C., Bonelo, J.-M., Gonzalez-Jimenez, J.: A predictive model for the maintenance of industrial machinery in the context of industry 4.0. Eng. Appl. Artif. Intell. **87**, 103289 (2020)

10. Pedregal, D.J., et al.: State space models for condition monitoring: a case study. Reliab. Eng. Syst. Safety **91**(2), 171–180 (2006)
11. Thoben, K.-D., et al.: "Industrie 4.0" and smart manufacturing—a review of research issues and application examples. Int. J. Autom. Technol. **11**, 4–19 (2017)
12. Wuest, T., et al.: Machine learning in manufacturing: advantages, challenges, and applications. Prod. Manuf. Res. **4**(1), 23–45 (2016)
13. Moyne, J., et al.: Big data analytics for smart manufacturing: case studies in semiconductor manufacturing. MDPI J. Processes **5**(3), 1–20 (2017)
14. Kaiser, K.A., et al.: Predictive maintenance management using sensor-based degradation models. IEEE Trans. Syst. Man, Cybern. Part A Syst. Humans **39**(4), 840–849 (2009)
15. Kang, H.S., et al.: Smart manufacturing: past research, present findings, and future directions. Int. J. Precise Eng. Manuf. Green Technol. **3**, 111–128 (2016)
16. Abellan, et al.: A review of machining monitoring systems based on artificial intelligence process models. Int. J. Adv. Manuf. Technol. **47**, 237–257 (2010)
17. Janssens, O., et al.: Convolutional neural network based fault detection for rotating machinery. J. Sound Vib. **377**, 331–345 (2016)
18. das Chagas, M., et al.: Failure and reliability prediction by support vector machines regression of time series data. Reliab. Eng. Syst. Safety **96**(11), 1527–1534 (2011)
19. Li, Z., et al.: Failure event prediction using the cox proportional hazard model driven by frequent failure signatures. IEEE Trans. **39**(3), 303–315 (2007)
20. Yang, Z.M., Djurdjanovic, D., Ni, J.: Maintenance scheduling in manufacturing systems based on predicted machine degradation. J. Intell. Manuf. **19**(1), 87–98 (2008)
21. Zhou, Z.J., Hu, C.H., Xu, D.L., Chen, M.Y., Zhou, D.H.: A model for real-time failure prognosis based on hidden Markov model and belief rule base. Eur. J. Oper. Res. **207**(1), 269–283 (2010)
22. Poosapati, V., Katneni, V., Manda, V.K., Ramesh, T.: Enabling cognitive predictive maintenance using machine learning: approaches and design methodologies. In: Proceedings of ICSCSP 2018, p. 388 (2018)
23. Gonzalez, J., Yu, W.: Non-linear system modelling using LSTM neural networks. IFAC-Papers online **51**(13), 485–489 (2018)
24. Mourtzis, D., Vlachou, E., Milas, N., Xanthopoulos, N.: A cloud-based approach for maintenance of machine tools and equipment based on shop-floor monitoring. Procedia CIRP **41**, 655–660 (2016)
25. Bishop, C.M.: Pattern Recognition and Machine Learning for Advanced Information Science and Statistics. Springer (2006)
26. Mobley, R.K.: An Introduction to Predictive Maintenance. Elsevier (2002)
27. Efthymiou, K., Papakostas, N., Mourtzis, D., Chryssolouris, G.: On a predictive maintenance platform for production systems. Procedia CIRP **3**, 221–226 (2012)
28. Carvalho, T.P., Soares, F.A., Vita, R., et al.: A systematic literature review of machine learning methods applied to predictive maintenance. Comput. Ind. Eng. **137**, 106024 (2019)
29. Okoh, C., Roy, R., Mehnen, J.: Predictive maintenance modelling for through-life engineering services. Procedia CIRP **59**, 196–201 (2017)
30. Noman, M.A., Abouel Nasr, E.S., Al-Shayea, A., Kaid, H.: Overview of predictive condition based maintenance research using bibliometric indicators. J. King Saud Univ.—Eng. Sci. **31**(4), 355–367 (2019)

IN-LDA: An Extended Topic Model for Efficient Aspect Mining

Nikhlesh Pathik and Pragya Shukla

Abstract In last decade, LDA is extensively used for unsupervised topic modeling, and various extension of LDA has also been proposed. This paper presents a semi-supervised extension IN-LDA, which uses very few influential words related to the domain for providing supervision in the topic generation process. IN-LDA also improves the performance of the LDA generation process in two ways. First, it deals with multi-aspect terms by passing N-grams vectors, and second simulated annealing-based algorithm is used for tuning hyperparameters of LDA for more coherent output. The experiment is conducted on two popular datasets, movie reviews and 20Newsgroup. IN-LDA is showing improved results when compared with others on coherence value. It also shows a better interpretation of output due to influential words.

Keywords Semi-supervised LDA · Topic modeling · Aspects mining · Hyperparameters tuning · Simulated annealing · Coherence

1 Introduction

Topic modeling first came into focus in 2003 when Beli et al. presented the LDA model for unsupervised clustering from the text in the form of various topics [1]. A large volume of text data can be analyzed with the help of topic models such as LDA. It clusters the documents into various topics. LDA applied to unlabeled data, and it finds clusters of these text documents and grouped them into topics. Different other approaches are applied for aspect extraction. Gou et al. proposed a supervised topic modeling by using the TF-IDF topic frequency method, where the weight of the topic distinguishes the topics. Symmetric and asymmetric Dirichlet prior both have taken as parameters [2]. Shukla et al. proposed various extensions of LDA for efficient aspect extraction from text data [3].

Frequency-based methods fail as different words may be used to represent the same aspect. Rule-based approaches are much dependent on domains and manually

N. Pathik (✉) · P. Shukla
Institute of Engineering and Technology, DAVV, Indore, India

H. Sharma et al. (eds.), *Congress on Intelligent Systems*, Advances in Intelligent Systems and Computing 1334, https://doi.org/10.1007/978-981-33-6981-8_29

designed rules. In this paper, an extended semi-supervised LDA model is presented. The proposed model focused on the issues raised above and tried to make an efficient extension of LDA. The following are the main contributions of the proposed work:-

1. LDA hyperparameter tuning for better coherence value.
2. N-gram vectors for multi-word aspect extraction.
3. Influential words for guiding LDA output and experimentally verify the effectiveness of IN-LDA.

The rest of the paper is organized as follows: Related work is described in Sect. 2. Section 3 discussed the proposed methodology in detail; the experimental setup of the proposed work is explained in detail in Sect. 4. The result and evaluation are discussed in detail in Sect. 5. Section 6 concludes the paper with some future directions.

2 Related Work

Alqaryouti et al. presented a model which aims to help the government to know customers need. They proposed a hybrid approach based on lexicon and semantic rules for aspect extraction and sentiment classification [4]. A supervised extended LDA model was proposed by Zahedi et al. for aspect and sentiment analysis to analyze unlabeled online available opinion datasets [5]. Aziz et al. have used an unsupervised Senti-Wordnet approach for opinion mining. They create a classification model using supervised SVM for defining an opinion [6]. Hughes et al. proposed a training framework for supervised LDA and level prediction. The objective of training generative models is to coherently integrate supervisory signals though a small fraction of training data that are labeled. The model gives high-dimensional logistic regression on text analysis [7].

Fu et al. proposed a topic identification criterion using second-order statistics of the words. This criterion identifies the underlying topics from repulsively violated anchor-word assumptions. They proposed an algorithm for optimization and a primal-dual algorithm [8]. Jabr et al. developed a methodology to represent reviews for each product to measure the reviewer's satisfaction with the extracted aspects. They used machine learning and text mining technique to obtain the product aspects that are important to reviewers from the Amazon dataset [9]. Jin et al. explored an approach to utilize review information for recommendation systems. They named this model as LSTM-Topic matrix factorization (LTMF). LSTM and topic modeling are integrated into the model for review understanding. They proved that LTMF is better for topic clustering than the traditional topic model-based methods [10].

Bagheri proposed a joint generative model sentiment-aspect model (SAM) for aspect mining and sentiment analysis. SAM can detect aspect and their sentiment jointly from the large collection of reviews [11]. Jiang et al. presented a dataset in which each sentence contains multiple aspects with multiple sentiments. They proposed CapsNet and CapsNet-BERT models to learn the complicated relationship between aspects and contexts [12].

Smatana et al. proposed neural networks-based autoencoder known topic AE for topic modeling with input texts. Topic modeling is performed to uncover hidden semantic structures from the input collection of texts. They solved the problem of all three types of topic modeling tasks, e.g., basic topic model, the evolution of topics in time, the hierarchical structure of sub-topics [13]. Bhat et al. proposed two models using deep neural networks. They proposed two variants 2NN Deep-LDA and 3NN Deep-LDA. They have used the Reuters-21578 dataset for experimental evaluation using the support vector machine (SVM) classifier. They explore the modeling of the statistical process of LDA using deep neural network [14].

Luo proposed a text sentiment analysis method based on LDA and CNN to improve the performance of sentiment analysis of public opinion available on the internet. LDA is used for latent semantic representation and CNN as a classifier [15]. Gallagher et al. introduced a correlation explanation approach for topic modeling. This framework separates the most informative topic through anchor words [16]. Jo et al. presented a model with neural network architectures. The model combines two layers of long short-term memory (LSTM) and a latent topic model. The latent topic model trained a classifier for mortality prediction and learned latent topics. They have been designed topic modeling layers for topic interpretability as a single-layer network with constraints inspired by LDA. The LSTM layer captures long-range dependencies in sequential data, trained for mortality prediction [17].

Garcia-Pablosa et al. proposed an unsupervised topic modeling-based model W2VLDA, which performs multilingual and multi-domain aspect-based sentiment analysis [18]. Hai et al. proposed a joint supervised probabilistic model for aspect extraction and sentiment analysis. This extended LDA model represented a review document as opinion pairs and developed an efficient collapsed Gibbs sampling-based inference method for parameter estimation [19]. Lim et al. proposed a model for opinion mining and sentiment analysis based on LDA for twitter opinion data. It influences mentions, emoticons hashtags, and strong sentiment words of tweets [20]. Ramesh et al. explored the use of seeded topic models as Seeded-LDA. It is an extension of topic models that uses a lexical seed set to bias the topics according to relevant domain knowledge. They extract seed words for the COURSE topic from each course's syllabus and capture the sentiment polarity using opinion finder. They explored the massive open online courses (MOOC), a discussion forum platform for student discussions [21]. The majority of research done for improving LDA performance can be summarized on the work done in the following directions:

1. Tuning LDA hyperparameters as a lot of fine-tuning are required [22, 23].
2. Multi-word aspect extraction, because so many aspects, is multi-word.
3. Interpretation of output required human intervention as there is no control over LDA output.

3 Proposed Method

A semi-supervised efficient extension IN-LDA is presented for multi-aspect extraction from text reviews. IN-LDA improves the performance of the LDA topic generation process in two ways. Firstly, it tunes LDA hyperparameter alpha and beta for optimizing LDA performance using simulated annealing-based optimization algorithm SA-LDA. Secondly, for controlling the LDA output, it uses some influential words. N-grams vectors are taken as input to LDA to deal with the multi-word aspect. The detailed description is as follows:

1. The Dirichlet priors α, β significantly affects the LDA performance. SA-LDA algorithm is used for tuning of hyperparameter. SA-LDA works on the principle of simulated annealing and gives the best configuration for the hyperparameter, which increases the performance. Figure 1 represents the flow of the SA-LDA algorithm
2. Aspects are not always unigrams, so for dealing with multi-words aspects, an N-gram approach is used. Bigrams and trigrams are used for the identification of multi-word aspects. The co-occurrence of various words helps in dealing with multi-word and incorrect aspects. It will reduce aspect space, which makes it efficient.
3. We consider only top words (say 100 words) up to three grams. By applying the POS ruleset, we can also find out the main aspects with associated sentiments.
4. There is no control over output in LDA, which makes output interpretation littlie difficult. For these few influential words, say 10–12 words per aspect category is supplied to LDA for some supervision in the topic generation. Aspects are mostly domain-dependent, and sometimes, different words are used for representing the same aspect. Influential words will group these words to identify that aspect. It provides control over topic generation so that co-occurred and relevant words will represent the same topic.
5. The performance of the proposed LDA is compared with LDA and its popular extensions. So, two domains, news and movie, are considered. Top generated words represented the performance improvement based on the coherence value.

Figure 2 represents the flow of the proposed algorithm. Step-by-step description is mentioned below.

3.1 Proposed Algorithms

1. Input: Text review dataset.
2. Preprocessing: Following preprocessing is done on input datasets:

   (a) Tokenization:

      i. Document split into sentences
      ii. Sentences split into words

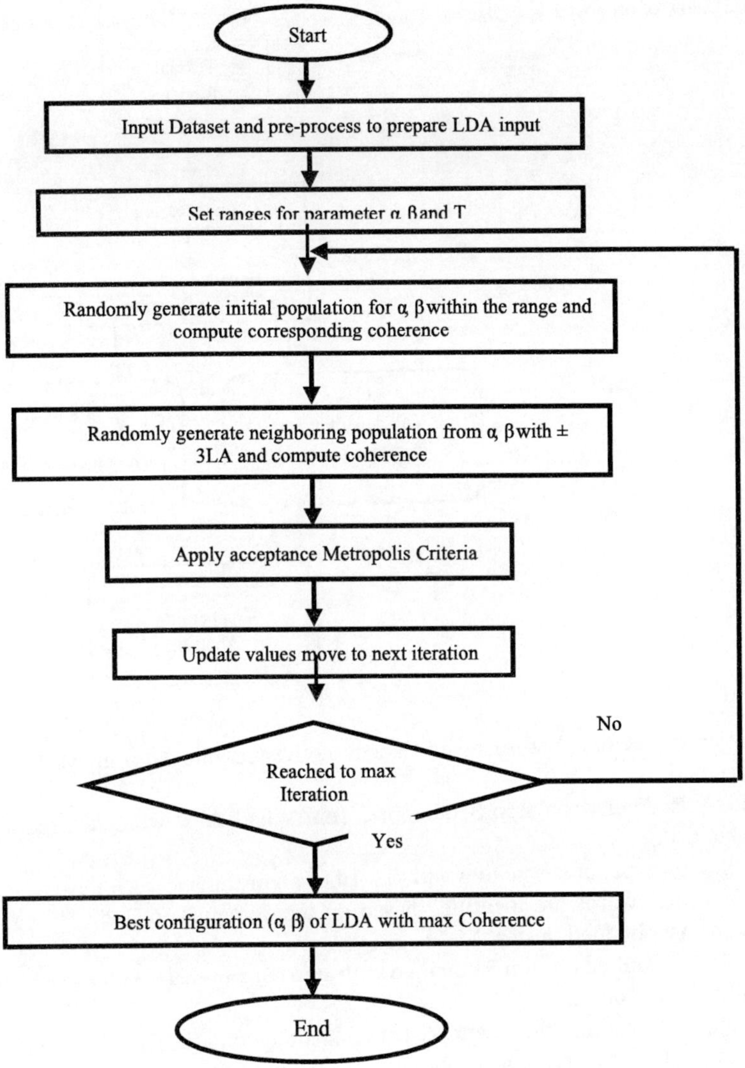

Fig. 1 Flow chart for the SA-LDA algorithm

iii. Words converted into lowercase, and punctuations were removed.

(b) Stop-word elimination: Words like pronouns, preposition conjunction articles, etc. are eliminated. Specific rules are also applied for removing non-useful words. For example, words that have less than three characters are also excluded.

(c) Lemmatization: Similar meaning words are replaced with one word.

(d) Stemming: Words are replaced with their root form.

Fig. 2 Flow of proposed work

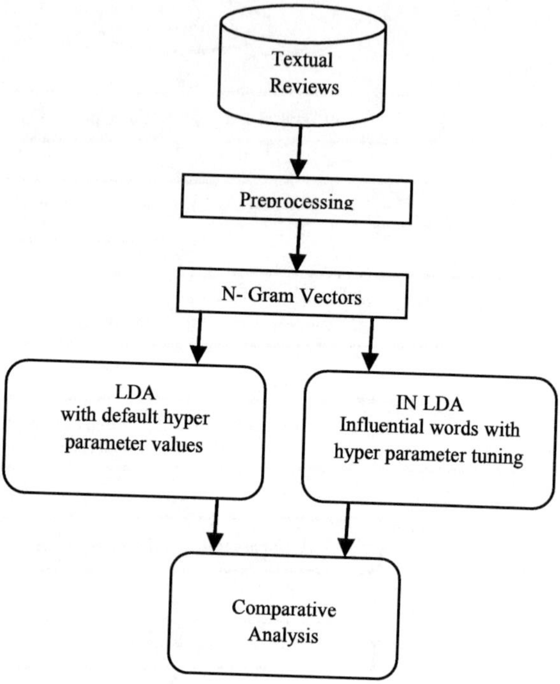

3. N-gram vectors: Preprocessed data is converted into N-gram vectors word vectors.
4. LDA is applied on step 3 outcome. Topics and their word distribution are obtained.
5. LDA hyperparameter tuning with SA-LDA algorithms.
6. Influential words are identified based on the domain knowledge that the top word list obtained in step 4.
7. IN-LDA applied on step 3 outcome with step (5) parameter values and step (6) influential words.
8. Output: Topics and their word distribution along with the Coherence value.
9. Repeat step1 to 8 with the different datasets for performance evaluation.

4 Experiment Setup

We have considered two popular datasets: a movie review from IMDB and a news feed from 20Newsgroup. The movie review dataset contains 2000 reviews with 1000 positive and 1000 negative. It contains 71,532 sentences with 1,583,688 words. 20 Newsgroup data is a collection of around 20 k news posts on 20 different topics. We supply N-gram input vectors along with influential words for guiding the LDA output.

Table 1 Influential words for the movie review and 20 Newgrouop

Domain	Category	Influential words
Movie	Thriller	Murder, psycho, humor, killer, doubt, crime, crime, dual,
	Comedy	Comedy, laugh, joke, funny, entertain, comic, soft, family, entertain, comic,
	Horror	Horror, scary, devil, death, victim, dark, alien, blood, monster, snake, death,
	Love	Romance, love, life, family, wife, husband, father, mother, son daughter, marriage, Romeo, Juliet
	Music	Music, song, singer, soft, soulful, light, cool, fast, old
	Action	Matrix, Jacki, Chan, war, action, Arnold, Steven
	cartoon	Cartoon, Disney, Mickey, Jerri, king, school, young, effect
News	Sports	Game, team, play, player, hockey, season, score, leagu
	Religion	Hristian, Jesus, exist, Israel, human, moral, ISRA, Bibl
	Violence	Kill, bike, live, leave, weapon, happen, gun, crime, hand
	Graphics	Drive, sale, driver, wire, card, graphic, price, apple, software, monitor
	Technology	File, window, program, server, avail, applic, version, user, entri
	Politics	Armenian, public, govern, Turkish, Columbia, nation, president, group
	Space	space, NASA, drive, scsi, orbit, launch, data, control, earth, moon

Influential words provided supervision to LDA. Table 1 represents the influential words for the movie domain related to various topics.

Around 5–10 words per topic are taken for both movies and news domains. Similarly, for other datasets, influential words will be taken. LDA hyperparameters (α, β) tuning is done using simulated annealing (SA-LDA)-based algorithm. SA-LDA gives the best LDA configuration in terms of hyperparameters values, which provide max coherence value.

Gensim and NLTK LDA implementation is taken as reference and further customized as per the proposed method. The experiment is done on I5 4200 M CPU @ 2.5 GHz processor with 8 GB RAM and Windows 8.1 OS. Anaconda environment is used for evaluation. A simple LDA model is run for finding the topic and word distribution for both unigram and bigram. The output is shown in Table 2.

5 Result and Discussion

IN-LDA produces a more coherent output due to influential words and hyperparameter tuning. Influential words guide LDA to produce more coherent and user interpretable output. For the final LDA output, we have taken ten topics per dataset and fifteen words per topic. Table 3 shows IN-LDA top words for the movie review

Table 2 LDA top words list with unigram for the movie review and 20 Newsgroup

Topic	Top words for movie review	Top words for 20 Newsgroup
Topic 0	film, movie, like, character, time, good, scene, go, play, look, come, story, know, year, thing	line, subject, organ, drive, post, universe, card, write, nntp, host, work, problem, need, article
Topic 1	action, chan, film, role, good, perform, jacki, jack, plot, gibson, team, jone, steven, stone, jam	line, subject, organ, write, nasa, space, article, encrypt, chip, clipper, post, like, know, host, acess
Topic 2	school, life, anim, love, comedy, young, high, play, family, girl, father, elizabeth, beauty, disney, voic	Game, team, line, subject, organ, play, hockey, player, univers, write, post, year, article, host, think
Topic 3	music, story, king, battle, life, ryan, world, prince, spielberg, american, visual, voice, john, song, great	window, line, subject, organ, file, write, post, program, host, nntp, problem, univers, thank, know, graphics
Topic 4	drug, girl, spice, arnold, jerri, park, stiller, young, gorilla, tarantino, tarzan, garofalo, grant, disney, schwarzenegg	Write, israel, article, subject, isra, line, Organ, armenian, people, jew, arab, post, say, kill, think
Topic 5	comedy, funny, laugh, joke, humor, play, comic, hilari, eddi, brook, star, love, gag, amus, american	Chrishtian, write, subject, line, people, organ, think, believe, know, article, jesus, say, univers, like, come
Topic 6	life, love, wife, husband, story, white, crime, town, tell, daughter, death, relationship, henri, perform, Michael	Line, subject, organ, write, article, like, post, bike, host, nntp, look, think, universe, know, good
Topic 7	love, perform, wed, carrey, william, truman, family, life, sandler, wife, play, comedy, marry, role, julia	Write, people, line, organ, subject, article, think, right, post, state, like, govern, nntp, know, host
Topic 8	film, effect, human, star, special, action, planet, world, earth, fight, video, game, science, origin, fiction	Organ, subject, line, article, write, post, know, pitt, bank, food, gordon, science, universe, like, think
Topic 9	alien, horror, movie, vampire, ship, know, crew, killer, kill, origin, scream, summer, scari, sequel, blood	Line, organ, subject, write, year, article, game, post, team, think, nntp, baseball, host, player, universe

and 20Newsgroup dataset. Here, we have only considered single word aspects. For multi-word aspects, we have converted out unigram input into bigram.

Table 3 shows more relevant words on the topics. Influential words supervised the topic generation process and produced a better output. Similarly, for generating multi-word aspects, we have taken N-gram(bigram) input vectors along with the N-gram(bigram) influential words.

Table 4 represents the multi-word (bigrams) aspects generated by IN-LDA when bigram vectors are passed as input. We have only shown top bigram words generated by IN-LDA in the movie and news domains. The output is more interpretable as compare to traditional LDA.

Table 3 IN-LDA unigram output top words for movie and 20 Newsgroup datasets

Topic	Top words for movie review	Top words for 20 Newsgroup
Topic 0	life, love, character, story, live, family, perform, wife, turn, beauty, play, father, relationship, work, begin	line, subject, organ, drive, post, universe, card, write, host, nntp, work, problem, know, article
Topic 1	action, plot, chan, fight, jacki, sequenc, hero, jone, chase, gibson, kill, star, scene, stunt, partner	write, line, subject, organ, article, encrypt, chip, post, clipper, space, like, know, host, nasa, access,
Topic 2	anim, king, voic, disney, story, young, children, family, warrior, prince, mulan, gorilla, kid, snake, adventure	game, line, team, organ, subject, write, year, player, article, play, universe, post, think, host, hockey
Topic 3	girl, music, wed, comedy, school, romantic, julia, love, high, play, singer, song, football, team, band	window, line, subject, organ, file, write, post, host, program, nntp, know, thank, universe, graphic, display
Topic 4	drug, reev, live, play, funny, zero, mike, baldwin, matrix, game, comedy, park, keanu, party, tarantino	write, subject, article, people, organ, line, israel, state, isra, post, think, armenian,say, kill, american
Topic 5	comedy, funny, laugh, joke, humor, play, comic, murphi, get, eddy, carrey, think, moment, brook, jack	write, subject, christian, line, people, organ, think, know, believ, articl, say, post, jesus, like, universe, come
Topic 6	crime, harry, stone, director, murder, francy, cage, investig, mystery, killer, detect, shoot, life, neighbor, blood	line, write, subject, organ, article, like, post, think, host, nntp, know, bike, good, universe
Topic 7	film, perform, character, cast, truman, good, carry, perfect, ryan, direct, john, excel, role, joan, steven	Write, people, line, organ, subject, article, think, right, post, state, like, govern, nntp, know, host
Topic 8	film, movie, like, time, character, good, scene, go, look, know, thing, plot, play, come, think	organ, line, subject, article, write, post, universe, host, know, nntp, reply, pitt, bank, food, think,
Topic 9	alien, horror, vampire, human, ship, planet, effect, crew, earth, origin, special, space, kill, scream, killer	game, line, team, organ, subject, write, year, player, article, universe, post, think, host, team, nntp

Table 4 IN-LDA top multi-word (bigram) generated for movie and 20Newsgroup datasets

Domain	Top multi-words
Movie	high_school, romantic_comedy, true_love main_character, best_friend, funny_movie, real_life, love_story, film_star, action_scene, action_sequence hong_kong, special_effect, action_film, million_dollar, motion picture, star_war, true_stori, support_cast, jam_bond, spice_girl, science_fiction, jurass_park, lake_placid, virtual_realiti, aspect_ratio, jacki_chan, urban_legend, video_game
News	comp_window, date_line, comp_hardwar, video_card, gate_comp, window_misc misc_comp, death_day,,imag_process misc_path, crime_street public_road, window_programm comp_graphic, newsgroup_comp kill_women, window_app, graphic_subject, modem_connect, signal_output

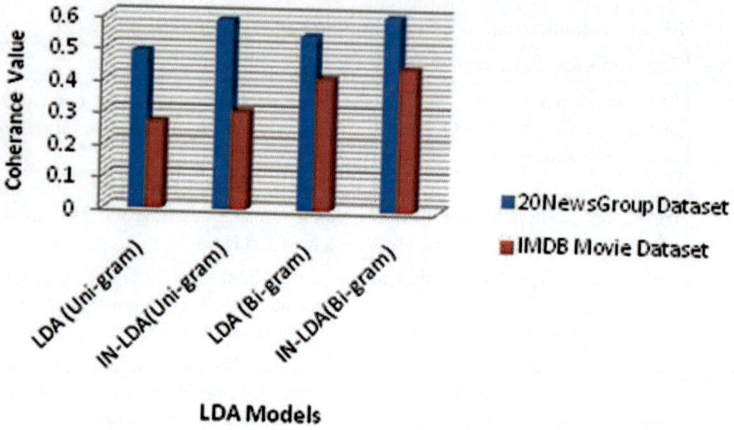

Fig. 3 Coherence plot for IN-LDA with two different datasets

Both LDA outputs are compared based on coherence value, and IN-LDA produced better coherence. With the 20Newsgroup dataset and traditional LDA, value of coherence is 0.488 for unigram, whereas with IN-LDA with the same configuration, coherence is 0.586. When we considered bigram, the value of coherence is 0.54 and 0.60, respectively. Similarly, with movie review, dataset for the unigram value of coherence is 0.27 and 0.304, respectively, for bigram values is 0.407 and 0.44. We can see that influential words guiding LDA for better coherent output. Figure 3 represents the coherence plot for IN-LDA with two above considered datasets.

6 Conclusion

In this paper, semi-supervised IN-LDA is presented for the extraction of multi-word aspects from text reviews. IN-LDA improves the performance of LDA by tuning LDA hyperparameters and providing supervision in terms of influential words. IN- LDA can extract multi-word aspects with better coherence as compare to traditional LDA, which makes output interpretation more clear. The experiment is conducted on two different datasets, which demonstrate the superiority of IN-LDA over LDA. In the future, we will try to combine LDA with neural networks, or a deep learning-based model can be proposed for further performance improvement.

References

1. Blei, D., Carin, L., Dunson, D.: Probabilistic topic models. IEEE Signal Process. Mag. **27**(6), 55–65 (2010)
2. Gou, Z., Huo, Z., Liu, Y., Yang, Y.: A method for constructing supervised topic model based on term frequency-inverse topic frequency. Symmetry **11**(12), 1486 (2019)
3. Pathik, N., Shukla, P.: An ample analysis on extended LDA models for aspect based review analysis. Int. J. Comput. Sci. Appl. **14**(2) (2017)
4. Alqaryouti, O., Siyam, N., Monem, A.A., Shaalan, K.: Aspect-based sentiment analysis using smart government review data. Appl. Comput. Inf. (2019)
5. Zahedi, E., Saraee, M.: SSAM: toward Supervised Sentiment and Aspect Modeling on different levels of labeling. Soft. Comput. **22**(23), 7989–8000 (2018)
6. Aziz, M.N., Firmanto, A., Fajrin, AM., Ginardi, R.H.: Sentiment analysis and topic modelling for identification of government service satisfaction. In: 5th International Conference on Information Technology, Computer, and Electrical Engineering 2018, ICITACEE, pp. 125–130. IEEE (2018)
7. Hughes, M.C., Hope, G., Weiner, L., McCoy, Jr., T.H., Perlis, R.H., Sudderth, E.B., Doshi-Velez, F.: Semi-supervised prediction-constrained topic models. In: AISTATS 2018, pp. 1067–1076 (2018)
8. Fu, X., Huang, K., Sidiropoulos, N.D., Shi, Q., Hong, M.: Anchor-free, correlated topic modeling. IEEE Trans. Pattern Anal. Mach. Intell. **41**(5), 1056–1071 (2018)
9. Jabr, W., Cheng, Y., Zhao, K., Srivastava, S.: What are they saying? A methodology for extracting in-formation from online reviews (2018)
10. Jin, M., Luo, X., Zhu, H., Zhuo, H.H.: Combining deep learning and topic modeling for review understanding in the context-aware recommendation. In: Proceedings of the 2018 Conference of the North American Chapter of the Association for Computational Linguistics: Human Language Technologies 2018, vol. 1, pp. 1605–1614 (2018)
11. Bagheri, A.: Integrating word status for joint detection of sentiment and aspect in reviews. J. Inf. Sci. **45**(6), 736–755 (2019)
12. Jiang, Q., Chen, L., Xu, R., Ao, X., Yang, M.: A challenge dataset and effective models for aspect-based sentiment analysis. In: Proceedings of the 2019 Conference on Empirical Methods in Natural Language Processing and the 9th International Joint Conference on Natural Language Processing 2019, EMNLP-IJCNLP, pp. 6281–6286 (2019)
13. Smatana, M., Butka, P.: TopicAE: a topic modeling autoencoder. Acta Polytech. Hung. **16**(4) (2019)
14. Bhat, M.R., Kundroo, M.A., Tarray, T.A., Agarwal, B.: Deep LDA: a new way to topic model. J. Inf. Optim. Sci., 1–2 (2019)
15. Luo, L.X.: Network text sentiment analysis method combining LDA text representation and GRU-CNN. Pers. Ubiquit. Comput. **23**(3–4), 405–412 (2019)
16. Gallagher, R.J., Reing, K., Kale, D., Ver Steeg, G.: Anchored correlation explanation: topic modeling with minimal domain knowledge. Trans. Assoc. Comput. Ling., 529–42 (2017)
17. Jo, Y., Lee, L., Palaskar, S.: Combining LSTM and latent topic modeling for mortality prediction. arXiv preprint arXiv:1709.02842 (2017)
18. García-Pablos, A., Cuadros, M., Rigau, G.: W2VLDA: almost unsupervised system for aspect-based sentiment analysis. Expert Syst. Appl. **91**, 127–137 (2018)
19. Hai, Z., Cong, G., Chang, K., Cheng, P., Miao, C.: Analyzing sentiments in one go: a supervised joint topic modeling approach. IEEE Trans. Knowl. Data Eng. **29**(6), 1172–1185 (2017)
20. Lim, K.W., Buntine, W.: Twitter opinion topic model: extracting product opinions from tweets by lever-aging hashtags and sentiment lexicon. In: Proceedings of the 23rd ACM International Conference on Information and Knowledge Management 2014, pp. 1319–1328. ACM (2014)
21. Ramesh, A., Goldwasser, D., Huang, B., Daumé, III H., Getoor, L.: Understanding MOOC discussion forums using seeded LDA. In: Proceedings of the 9th Workshop on the Innovative use of NLP for Building Educational Applications, pp. 28–33 (2014)

22. Yarnguy, T., Kanarkard, W.: Tuning Latent Dirichlet Allocation parameters using ant colony optimization. J. Telecommun. Electron. Comput. Eng. (JTEC) **10**(1–9), 21–24 (2018)
23. George, C.P., Doss, H.: Principled selection of hyperparameters in the Latent Dirichlet Allocation model. J. Mach. Learn. Res. **18**(1), 5937–5974 (2017)

Imbalance Rectification Using Venn Diagram-Based Ensemble of Undersampling Methods for Disease Datasets

Soham Das, Soumya Deep Roy, Swaraj Sen, and Ram Sarkar

Abstract Class imbalance is a major problem when dealing with real-world datasets, especially disease datasets. The majority class often consists of non-patient or negative samples while the minority class consists of patient, i.e., positive samples. This imbalance weakens the learning ability of a classifier as supervised learning is governed by classification accuracy, and the classifier disregards the minority class and identifies almost all the samples as belonging to the majority class. Researchers often use undersampling, oversampling, or a combination of both techniques to address this skewness in datasets. In this paper, we use a novel method where we form a Venn diagram-based ensemble of different undersampling algorithms instead of relying on a standalone undersampling algorithm, thereby leading to a more robust model. The proposed method has been evaluated on three class imbalanced disease datasets, namely Indian Liver Patient Dataset, Pima Indian Diabetes Dataset, and Cervical Cancer (Risk Factors) Dataset. Experimental outcomes show that the ensemble approach outperforms some state-of-the-art methods considered here for comparison. The source code along with the relevant datasets are available in the GitHub repository https://tinyurl.com/ycvde8pp.

Keywords Class imbalance · Undersampling · Diabetes · Liver · Cancer · Disease dataset

1 Introduction

Health care is one of the most crucial sectors of our society. Precise and accurate prediction of diseases by doctors based on symptoms and medical conditions can go a long way in saving human lives. Consequently, of late, health care as well as medicine has seen a massive adoption of Data Analytics and Artificial Intelligence (AI) [7, 12].

S. Das (✉) · S. D. Roy
Department of Metallurgical and Material Engineering, Jadavpur University, Kolkata, India

S. Sen · R. Sarkar
Department of Computer Science and Engineering, Jadavpur University, Kolkata, India

H. Sharma et al. (eds.), *Congress on Intelligent Systems*, Advances in Intelligent Systems and Computing 1334, https://doi.org/10.1007/978-981-33-6981-8_30

Owing to our changing lifestyle and habits, various diseases have become commonplace and prevalent. One of the most widespread diseases globally is diabetes: a medical condition arising out of insufficient production of insulin or the inability of the body to use it properly [18]. Liver diseases have also been on the rise primarily due to excessive consumption of alcohol, inhale of harmful gases, intake of contaminated food, pickles, and drugs [16]. Another disease which has plagued women especially in developing countries is cervical cancer [2]. In this work, an attempt has been made to improve the prediction of diabetes, chronic liver diseases, and cervical cancer based on medical indicators.

One of the impediments to successful classification of the said disease data is the severe class imbalance. Mostly, the positive class has very few samples. For instance, the Cervical Cancer Dataset has 803 negative cases and a meager 55 positive cases, the Liver Dataset has 416 non-liver patient records and 167 liver patient records, while the Diabetes Dataset, the least imbalanced of the three, has 500 negative and 268 positive cases. Machine learning algorithms are generally designed to deal with balanced datasets. Hence, to balance the skewed datasets, researchers use oversampling methods which duplicate or create synthetic examples in the minority class, or undersampling methods which delete or merge examples in the majority class. Often a combination of the two is also used.

In this paper, undersampling is preferred to oversampling because synthetic creation of samples may often lead to faulty data points which can be detrimental especially in sensitive domains like disease prediction. Our conjecture is that classification can be improved by using an ensemble of undersampling techniques. The datasets are first undersampled using a multitude of undersampling algorithms out of which the $n - best$ algorithms are chosen in terms of maximum imbalance correction, yielding n $subsets$ of the original dataset. The final set is created by retaining the samples which belong to $atleast$ k of the n subsets. A Venn diagram-based ensemble is chosen as it helps to retain those samples which would have been otherwise excluded by a standalone undersampling method.

The entire paper is organized as follows: Sect. 1 introduces the concept of imbalance correction on disease datasets. Section 2 details some past methodologies used on the datasets considered here. Sections 3 details our proposed machine learning model. In Sect. 4, the results are analyzed, and in Sect. 5, the work is concluded, and some future scope of work in this field is also provided.

2 Literature Survey

Numerous machine learning-based methods have been proposed in the past for accurate prediction of disease from medical data. Some of these research attempts made on the three datasets, namely Pima Indian Diabetes Dataset, Indian Liver Patient Dataset, and Cervical Cancer (risk factor) Dataset, have been briefed.

Deepa et al. [6] proposed a framework for the detection of Type 2 diabetes using Ridge-Adaline stochastic gradient descent (RASGD) classifier. RASGD which adopted an unconstrained optimization model to mitigate the cost of the classifier attained an accuracy of 92%. Karatsiolis et al. [14] suggested region-based support vector machine (SVM) algorithm on cross-validated Pima Indian Diabetes Dataset. The proposed algorithm obtained an accuracy of 82.2%. In [17], Naz et al. presented a method for diabetes prophecy by using several classifiers such as artificial neural network (ANN), naive Bayes (NB), decision tree (DT), and deep learning (DL). DL outperformed the others in all the performance parameters and provided the best results for diabetes prediction with an accuracy of 98.07%.

Adil et al. [1] built a supervised machine learning model using logistic regression (LR) on Indian Liver Dataset. The overall accuracy in this study was about 74%. Gogi et al. [11] utilized SVM, LR, and DT classifiers. They suggested that LR was a better classifier for categorization among all the other classifiers with an accuracy of 95.8% . Auxilia [3] used DT, random forest (RF), SVM, ANN, and NB, and attained 81%, 77%, 77%, 71%, and 37% accuracy scores, respectively.

Gupta et al. [13] assessed the Cervical Cancer Dataset and implemented the Boruta analysis algorithm which was principally built around the RF classifier. In [20], Wu et al. reviewed some cervical cancer risk factors, and three SVM-based approaches (standard SVM, SVM-RFE, and SVM-PCA) were applied to the Cervical Cancer Dataset. Karim et al. [15] applied four classification algorithms: DT, multilayer perceptron (MLP), SVM, and k-nearest neighbor (kNN). SVM with sequential minimal optimization (SMO) followed by bagging yielded the best result.

Most of the aforementioned methods did not take into account the imbalance present in the datasets and used accuracy as a metric to evaluate their model. Our approach involves using an ensemble of various undersampling algorithms like (i) NearMiss, (ii) condensed nearest neighbor (CNN) rule, (iii) cluster centroid, (iv) edited nearest neighbors, and (v) neighborhood cleaning rule. Ensemble approach has been used earlier in different fields due to its usefulness [4, 5, 8–10].

3 Proposed Methodology

In the present work, a Venn diagram-based ensemble of different undersampling algorithms has been formed in order to rectify the class imbalance of the disease datasets. The pipeline of the said model is shown in Fig. 1.

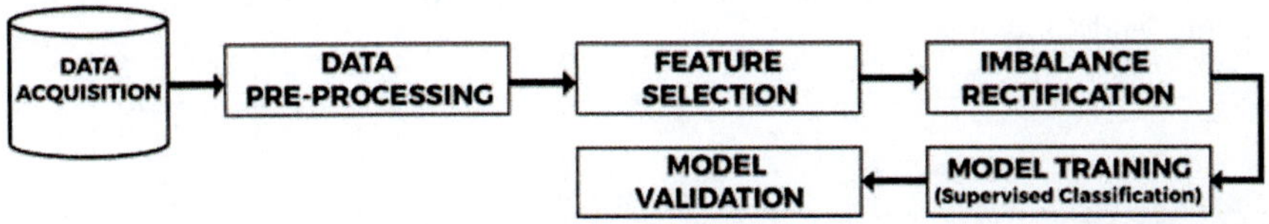

Fig. 1 Pipeline of proposed model

3.1 Dataset Description

Three class imbalanced disease datasets have considered in the present work. For the classification of diabetes, we have used the Pima Indians Diabetes dataset from the National Institute of Diabetes and Digestive and Kidney Diseases. The dataset consists of eight medical predictor variables like *number of pregnancies, BMI, insulin level,* etc., and one target variable, *Outcome*. The liver disease classification is performed on the Indian Liver Patient dataset obtained from the UCI Machine Learning Repository. Ten medical indicators like *total bilirubin, protein, albumin-globulin ratio*, etc. are used as feature variables corresponding to target variable, *Dataset*. We have used the Cervical Cancer (Risk Factor) dataset from UCI Machine Learning Repository, which has 32 feature variables like *age, number of pregnancies, first sexual intercourse*, etc., and target variable which classifies samples into patient and non-patient.

3.2 Data Preprocessing

The three datasets are checked for missing values. The Diabetes Dataset had no missing values, while the Liver and the Cervical Cancer Datasets had multiple missing values. These missing entries are filled with the median values if the variable is numeric and modal values in case of categorical variables.

3.3 Feature Selection and Scaling

Feature selection and scaling are often used to reduce the computational cost of modeling and to improve the performance of the model. Recursive feature elimination with cross-validation (RFECV) is a frequently used feature selection method. RFECV builds a model with an estimator on the entire set of features and computes an importance score for each feature. We have used classification performance of RF as an estimator. After applying RFECV-RF, the optimum number of features for Pima Indian Diabetes, Indian Liver, and Cervical Cancer (Risk Factors) Datasets are found to be 8, 8, and 19, respectively. Using the respective optimal features of the three datasets, we create new datasets and use them for imbalance rectification, model training, and validation.

3.4 *Imbalance Rectification*

All the datasets that we have chosen have varying degrees of class imbalance. To determine the 'degree' of imbalance, we use imbalance ratio (I.R) as the metric. Mathematically,

$$I.R = \frac{\text{\# of samples in Majority Class}}{\text{\# of samples in Minority Class}} \tag{1}$$

Imbalanced datasets are a huge impediment to predictive modeling as most of the machine learning algorithms used for classification are designed for equal distribution of samples between the classes. This yields models having poor predictive performance, especially for the minority class. In many areas, the minority class is more significant. Hence, the problem is more sensitive to classification errors for the minority class than the majority class. To deal with class imbalance, researchers typically choose undersampling methods (NM, CNN, etc.), oversampling methods (SMOTE, etc.) or often a combination of both. This section describes the undersampling techniques traditionally used by researchers and introduces the concept of Venn diagram-based ensemble of undersampling methods used in our model training.

3.4.1 Undersampling Algorithms

Undersampling refers to the technique of balancing the skewed sample distribution between classes by retaining all the minority class samples and removing samples from the majority class based on some algorithm. Let ζ denote the universal set containing all the samples in the dataset, with m samples of the minority class and n samples of the majority class, where $m << n$. Mathematically,

$$\zeta = \{\alpha_1, \alpha_2, ..., \alpha_m, \beta_1, \beta_2, ..., \beta_n\} \text{ and } n(\zeta) = m + n \tag{2}$$

After applying undersampling algorithm on this universal set, we obtain a subset ϕ containing m samples of the minority class and k samples of the majority class where $m \approx k$, i.e. ,

$$\phi = \{\alpha_1, \alpha_2, ...\alpha_m, \beta_1, \beta_2, ...\beta_k\} \text{ and } n(\phi) = m + k \tag{3}$$

where α and β are samples belonging to minority and majority classes, respectively. It is important to note that the k samples are selected from n samples of the majority class based on different algorithms, some of which are as follows:

NearMiss

NearMiss is a collection of undersampling algorithms that retain samples of majority class based on the distance from the samples of minority class. NearMiss-1 chooses majority class samples for which the average distance of the k-nearest minority class samples is the smallest. NearMiss-2 retains majority class samples for which the average distance to the farthest minority class sample is the smallest. NearMiss-3 is a two-step algorithm: first, the m-nearest neighbors for each minority class sample is retained; then, those majority class samples are selected for which the average distance from the k-nearest neighbors is the largest. NearMiss-3 is desirable, as it keeps the majority class examples on the decision boundary. Hence, for the purpose of the paper, we have used NearMiss-3 and future references of NM in this paper will imply the said version.

Condensed Nearest Neighbor (CNN)

In CNN, we start with two blank datasets A and B. The first sample is placed in dataset A with the remaining samples in dataset B. Then, using dataset A as the training set, one instance from dataset B is scanned. If a sample in B is misclassified, it is transferred from B to A. This process is repeated until no point is transferred from B to A.

Edited Nearest Neighbor(ENN)

ENN removes any sample whose class label is different from the class of at least two of its three nearest neighbors. The intuition behind this algorithm is to delete the majority class samples which are near or around the borderline of different classes based on the concept of nearest neighbor.

Neighborhood Cleaning Rule (NCR)

NCR treats majority and minority samples separately while sampling the datasets. NCR uses ENN to remove majority samples. It finds the three nearest neighbors for each sample in the training set. If the sample belongs to the majority class and the three nearest neighbors classify it in the opposite class, then the sample is removed. If the chosen sample belongs to the minority class but is misclassified by its three nearest neighbors, then instead of the minority class instance, the nearest neighbors belonging to the majority class are deleted.

Cluster Centroid-Based Majority Undersampling Technique (CCMUT)

In CCMUT, samples are removed based on the Euclidean distance from cluster centroids. After finding the cluster centroid of the majority class, the sample belonging to the cluster (majority class), which is farthest from the cluster centroid in feature space, is removed while the sample belonging to the majority class, that is nearest to the cluster centroid in feature space, is retained.

3.4.2 Ensemble of Undersampling Algorithms

We apply all the undersampling algorithms discussed above. Each algorithm, in essence, produces a new subset ϕ. Let there be p such subsets corresponding to p undersampling algorithms. Mathematically, we have,

$$\phi_1, \phi_2, ...\phi_p \text{ where } \phi_i \subset \zeta \;\; \forall i \in \{1, 2, .., p\} \tag{4}$$

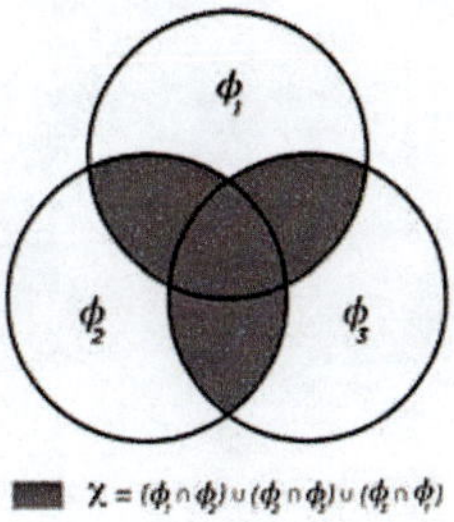

Fig. 2 Venn diagram of the subsets. The shaded area represents χ which is the union of intersections of sets ϕ_1 , ϕ_2, and ϕ_3

Out of the p subsets, we select q subsets which have the least I.R. In this work, $q = 3$, i.e., the three undersampling algorithms corresponding to which the I.R is least are chosen. Using these q subsets, a new set χ is created by retaining those samples which belong to at least r among the q subsets. r is generally chosen to be a number greater than half of q, and in this work, we have used $r = 2$. Using the aforementioned values of q and r, we have,

$$\chi = (\phi_1 \cap \phi_2) \cup (\phi_2 \cap \phi_3) \cup (\phi_3 \cap \phi_1) \tag{5}$$

The same equation can be represented as a Venn diagram as shown in Fig. 2. The process of imbalance rectification can be summarized as under:

1. From the original dataset,
   (a) Create a new dataset by fitting the original dataset on NM
   (b) Calculate the I.R of the new dataset
   (c) Repeat steps *a* and *b* for CNN, ENN, NCR, and CCMUT
2. Select the three datasets corresponding to which the I.R values are the least. Let the selected datasets be ϕ_1, ϕ_2, and ϕ_3.
3. Create a new dataset χ by retaining only those samples which belong to at least two of the selected datasets, as indicated in Eq. 5.

For each of the three disease datasets, we create a subset χ from the original dataset ζ to train our machine learning model. This helps to address the class imbalance as shown in Fig. 3 and simultaneously improves the classification of patients into positive and negative classes.

3.5 Model Training

The machine learning algorithms used for model selection are LR, SVM, with radial basis (RBF) kernel, kNN, DT, RF, XGBoost (XGB), and AdaBoost (AB). The datasets are first trained with the aforementioned algorithms without imbalance rectification, and their performance metrics are noted. The same process is then followed for datasets obtained after imbalance rectification. These results are given in Tables 1, 2 and 3.

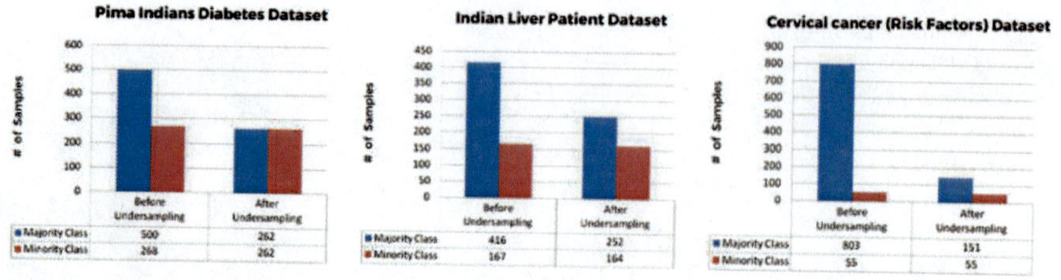

Fig. 3 Change in imbalance ratio after applying the ensemble of undersampling methods on the three disease datasets considered here

Table 1 Performance of different classifiers on Indian Liver Patient Dataset

	Before applying the proposed method						After applying the proposed method					
	LR	SVM	RF	kNN	XGB	AB	LR	SVM	RF	kNN	XGB	AB
ACC	0.71	0.72	0.71	0.65	0.66	0.68	0.85	0.88	**0.99**	0.80	0.90	0.83
P	0.74	0.72	0.75	0.73	0.75	0.74	0.91	0.92	0.99	0.87	0.94	0.94
R	0.91	0.97	0.86	0.82	0.78	0.84	0.87	0.91	0.99	0.74	0.91	0.81
F1	0.81	0.82	0.80	0.77	0.76	0.78	0.89	0.91	0.99	0.80	0.92	0.87
AUC	0.57	0.56	0.58	0.53	0.56	0.57	0.80	0.87	0.94	0.76	0.89	0.84

The boldened cells represent the highest performance for that particular dataset

Table 2 Performance of different classifiers on Pima Indian Diabetes Dataset

	Before applying the proposed method						After applying the proposed method					
	LR	SVM	RF	kNN	XGB	AB	LR	SVM	RF	kNN	XGB	AB
ACC	0.72	0.79	0.79	0.78	0.69	0.69	0.89	0.96	**0.99**	0.85	0.91	0.90
P	0.61	0.73	0.77	0.75	0.56	0.58	0.85	0.94	0.97	0.83	0.85	0.83
R	0.50	0.58	0.62	0.63	0.57	0.53	0.91	0.95	0.99	0.88	0.93	0.91
F1	0.54	0.64	0.69	0.68	0.56	0.55	0.87	0.94	0.98	0.86	0.89	0.87
AUC	0.67	0.74	0.76	0.78	0.66	0.67	0.89	0.91	0.95	0.86	0.90	0.88

The boldened cells represent the highest performance for that particular dataset

Table 3 Performance of different classifiers on Cervical Cancer (Risk Factors) Dataset

	Before applying the proposed method						After applying the proposed method					
	LR	SVM	RF	kNN	XGB	AB	LR	SVM	RF	kNN	XGB	AB
ACC	0.94	0.95	0.96	0.95	0.91	0.92	0.96	0.97	**0.98**	0.90	0.95	0.94
P	0.64	0.66	0.69	0.75	0.63	0.68	0.64	0.72	0.87	0.61	0.68	0.62
R	0.57	0.64	0.71	0.31	0.52	0.56	0.57	0.69	0.87	0.34	0.68	0.78
F1	0.60	0.64	0.69	0.44	0.57	0.61	0.60	0.70	0.87	0.44	0.68	0.69
AUC	0.77	0.80	0.82	0.65	0.74	0.76	0.77	0.79	0.90	0.66	0.82	0.87

The boldened cells represent the highest performance for that particular dataset

4 Result and Analysis

This section describes the outcomes of the proposed ensemble of undersampling methods when applied on the three class imbalanced disease datasets mentioned earlier. Here, we have discussed performance metrics used to evaluate the proposed model followed by a summary of the performance on the three datasets.

4.1 Performance Metrics

For any classification problem, four popular figures of merit measures are used, namely True Positive (TP), False Positive (FP), True Negative (TN), and False Negative (FN).

To evaluate our model, we have used some of the standard performance metrics like precision (P), recall (R), accuracy (ACC), F1 score (F1), and AUC score (AUC) which built on the aforementioned figures of merit. These are defined as under:

$$ACC = \frac{TP + TN}{TP + TN + FP + FN} \tag{6}$$

$$P = \frac{TP}{TP + FP} \tag{7}$$

$$R = \frac{TP}{TP + FN} \tag{8}$$

$$F1 = \frac{2 \cdot P \cdot R}{P + R} \tag{9}$$

The area under curve of receiver-operating characteristic (AUC) curve is a performance metric for classification problems at various thresholds. It is indicative of how much the model is capable of distinguishing between the positive and negative classes. Higher the AUC, better the model is at classifying samples correctly. Hence, higher the AUC, better the model is at distinguishing between patients and non-patients.

For imbalanced datasets, accuracy is often a misleading metric as it is dominated by the majority class and does not take into account how well it classifies the minority class. Hence, we use other measures like P, R, F1 Score, and AUC Score.

4.2 Analysis

As already described, we have used the three datasets, namely Indian Liver Patient Dataset, Pima Indian Diabetes Dataset, and the Cervical Cancer (Risk Factors) Dataset. We trained the datasets on various machine learning algorithms before applying the proposed method and after applying our proposed method. Tables 1, 2, and 3 summarize these results. It is to be noted that regardless of the machine learning model, there is a significant rise in all the performance metrics, especially F1 score and AUC score. This validates our initial hypothesis that model training on a dataset obtained after applying an ensemble of undersampling techniques classifies better than that applied on the original dataset.

RF on Indian Liver Dataset produces an F1 score of 0.99 and AUC score of 0.94. This machine learning model produces an accuracy of 99% for fivefold cross-Vvalidation(CV). RF emerges as the best classifier for Pima Indian Diabetes Dataset with F1 score of 0.98 and AUC score of 0.95. This model has an accuracy of 99% for sevenfold CV. For Cervical Cancer (Risk Factors) Dataset as well, RF is the best machine learning model with F1 score of 0.87 and AUC score of 0.90, attaining an accuracy of 98% for sevenfold CV.

The proposed methodology involving an ensemble of undersampling methods improve the current state-of-the art classification on the same datasets. Table 4 shows a comparative analysis of the performance of the state-of-the-art disease classification available in the literature.

Table 4 Brief comparative study of different methods. Accuracies achieved by the proposed method are highlighted

Work reference	Dataset	Classifier	Classification Accuracy
Adil et al. [1]	Indian Liver Patient Dataset	LR	0.74
Gogi et al. [11]		SVM	0.95
Auxilia [3]		DT	0.81
Present method		**RF**	**0.99**
Deepa et al. [6]	Pima Indians Diabetes Dataset	RASGD	0.92
Naz et al. [17]		DL	0.98
Karatsiolis et al. [14]		SVM	0.82
Present method		**RF**	**0.99**
Wu et al. [20]	Cervical Cancer (Risk Factors) Dataset	SVM	0.94
Karim et al. [15]		SVM	0.98
Rayavarapu et al. [19]		DNN	0.97
Present method		**RF**	**0.98**

The boldened cells represent the highest performance for that particular dataset

5 Conclusion

Class imbalance problem can be commonly found in many real-word datasets such as disease datasets. As a result, classifiers ignore the minority class (often representing the positive patients' information) during the learning process. To this end, this paper describes how using a Venn diagram-based ensemble of different undersampling algorithms improves the F1 score and accuracy of the class imbalanced datasets. We have evaluated our method on three datasets namely the Pima Indians Dataset, Indian Liver Dataset, and Cervical Cancer (Risk Factors) Dataset, which have varying degrees of imbalance. Though our method outperforms some state-of-the-art methods, still there is some room for improvement. For example, we plan to use some recently proposed undersampling algorithms to form the ensemble and use a more intelligent approach in forming the ensemble. Besides, in future, scalability and robustness of our proposed model can be assessed by applying it on higher-dimensional dataset such as microarray dataset.

References

1. Adil, S.H., Ebrahim, M., Raza, K., Azhar Ali, S.S., Hashmani, M.A.: Liver patient classification using logistic regression. In: 2018 4th International Conference on Computer and Information Sciences (ICCOINS). IEEE, August 2018
2. Aswathy, S., Reshma, J., Avani, D.: Epidemiology of cervical cancer with special focus& nbsp ; on india. Int. J. Women's Health, page 405, April 2015
3. Alice Auxilia, L.: Accuracy prediction using machine learning techniques for Indian patient liver disease. In: 2018 2nd International Conference on Trends in Electronics and Informatics (ICOEI). IEEE, May 2018
4. Begum, S., Chakraborty, D., Sarkar, R.: Cancer classification from gene expression based microarray data using SVM ensemble. In: 2015 International Conference on Condition Assessment Techniques in Electrical Systems (CATCON). IEEE, December 2015
5. Chakraborty, A., Rajonya De, Chatterjee, A., Schwenker, F., Sarkar, R.: Filter method ensemble with neural networks. In: Lecture Notes in Computer Science, pp. 755–765. Springer International Publishing (2019)
6. Deepa, N., Prabadevi, B., Maddikunta, P.K., Gadekallu, T.R., Baker, T., Ajmal Khan, M., Tariq, U.: An AI-based intelligent system for healthcare analysis using ridge-adaline stochastic gradient descent classifier. J. Supercomputing, May 2020
7. Esteva, A., Kuprel, B., Novoa, R.A., Ko, J., Swetter, S.M., Blau, H.M., Thrun, S.: Dermatologist-level classification of skin cancer with deep neural networks. Nature **542**(7639), 115–118 (2017). January
8. Ghosh, K.K., Ghosh, S., Sen, S., Sarkar, R., Maulik, U.: A two-stage approach towards protein secondary structure classification. Med. Biol. Eng. Comput., May 2020
9. Ghosh, M., Adhikary, S., Ghosh, K.K., Sardar, A., Begum, S., Sarkar, R.: Genetic algorithm based cancerous gene identification from microarray data using ensemble of filter methods. Med. Biol. Eng. Comput. **57**(1), 159–176, August 2018
10. Ghosh, S., Ghosh, K.K., Chakraborty, S., Bhowmik, S., Sarkar, R.: A filter ensemble feature selection method for handwritten numeral recognition. In: International Conference on Emerging Technologies for Sustainable Development (ICETSD '19), vol. 5-6, pp. 394–398 (2019)

11. Gogi, V.J., Vijayalakshmi, M.N.: Prognosis of liver disease: Using machine learning algorithms. In: 2018 International Conference on Recent Innovations in Electrical, Electronics & Communication Engineering (ICRIEECE). IEEE, July 2018
12. Gulshan, V., Peng, L., Coram, M., Stumpe, M.C., Derek, W., Narayanaswamy, A., Venugopalan, S., Widner, K., Madams, T., Cuadros, J., Kim, Ramasamy, Raman, R., Nelson, P.C., Mega, J.L., Webster, D.R.: Development and validation of a deep learning algorithm for detection of diabetic retinopathy in retinal fundus photographs. JAMA **316**(22), 2402, December 2016
13. Gupta,P., Jindal, I., Goyal, A.: Early detection and prevention of cervical cancer. In: 2019 IEEE 5th International Conference for Convergence in Technology (I2CT). IEEE, March 2019
14. Karatsiolis, S., Schizas, C.N.: Region based support vector machine algorithm for medical diagnosis on pima Indian diabetes dataset. In: 2012 IEEE 12th International Conference on Bioinformatics & Bioengineering (BIBE). IEEE, November 2012
15. Karim, E., Neehal, N.: An empirical study of cervical cancer diagnosis using ensemble methods. In: 2019 1st International Conference on Advances in Science, Engineering and Robotics Technology (ICASERT). IEEE, May 2019
16. Moon, A.M., Singal, A.G., Tapper, E.B.: Contemporary epidemiology of chronic liver disease and cirrhosis. Clin. Gastroenterol. Hepatol., August 2019
17. Naz, H., Ahuja, S.: Deep learning approach for diabetes prediction using PIMA Indian dataset. J. Diabetes Metabolic Disorders **19**(1), 391–403 (2020). April
18. Ogurtsova, K., da Rocha Fernandes, J.D., Huang,Y., Linnenkamp, U., Guariguata, L., Cho, N.H., Cavan, D., Shaw, J.E., Makaroff, L.E.: IDF diabetes atlas: global estimates for the prevalence of diabetes for 2015 and 2040. Diabetes Res. Clinical Pract. **128**, 40–50, June 2017
19. Rayavarapu, K., Krishna, K.K.V.: Prediction of cervical cancer using voting and DNN classifiers. In: 2018 International Conference on Current Trends towards Converging Technologies (ICCTCT). IEEE, March 2018
20. Wu, W., Hao, Z.: Data-driven diagnosis of cervical cancer with support vector machine-based approaches. IEEE Access **5**, 25189–25195 (2017)

Personalized Route Finding System Using Genetic Algorithm

P. Karthikeyan and P. Priyadharshini

Abstract The significant improvement of population and multiple types of vehicles causes traffic congestion and environment problem. The economic prosperity and continuing urbanization have led to quick raise in travel demand in urban areas. Rapid urbanization and motorization in cities cause congestion and an increase in vehicular emission and also increase the accident. Travelers face the practical problem of identifying the most efficient route on their daily journey. Fuel consumption, emission, long travel time, and accident can be direct and indirect consequences of vehicle traffic congestion and rough driving pattern. Personalized driving system needs some specific real-time data like vehicle volume, weather condition of the area, road condition, carbon emission factors, and safety measures. Hence in this work, genetic algorithm (GA) is proposed to find the optimal route for the drivers based on their requirement. In this work, optimal path will be identified from several paths. Route maps of Madurai City, Delhi City, Northern Utah City are considered. The GA result is compared to Bellman–Ford algorithm. The GA finds the best route for 94.6% of accurately, and Bellman–Ford algorithm finds the shortest path for 92.4% of accurately. The results have proven that the proposed GA works better for the personalized route finding system.

Keywords Genetic algorithm · Emission factor · Distance factor · Personalized route finding system · Bellman–Ford algorithm

1 Introduction

The significant increase in number of vehicles and growth of population in urban areas causes many problems. In global scenario, cites in India facing transportation and environment problem regard to traffic congestion, emission of carbon, over speed accidents, and more fuel consumptions in traffic signals. Street fatalities among defenseless street clients, in particular, motorcyclists, cyclists and walkers, and urban

P. Karthikeyan · P. Priyadharshini (✉)
Department of Information Technology, Thiagarajar College of Engineering, Madurai, India

H. Sharma et al. (eds.), *Congress on Intelligent Systems*, Advances in Intelligent Systems and Computing 1334, https://doi.org/10.1007/978-981-33-6981-8_31

street well being are a reason for worry in the area. Street traffic passing among powerless street clients represents 55% of complete street traffic fatalities. Urban transportation network is consisting of four categories such as demand network, organization network, facility network, and route network. In addition, the human interaction in socioeconomic behavior and urban architecture is an intertwined network model. Road traffic network consists of microscopic and macroscopic models. Urbanization also provides more opportunities such as job and education. In current situation, an efficient traffic flow and route optimization are critical challenges in urban cities, because more dense population is living in urban cities. National and worldwide research organizations and their systems must be dispatched to make databases, benchmarks, a lot of center criteria furthermore, focus, just as to screen and report about advancement to national foundation of urban data sharing that ought to be built up in each nation. Many of the research work and project are going on urban planning and smart cities concept. Shortest path distance and minimum travel time are important factors in transportation objective function in urban areas. Shortest route identification alone will not give a feasible solution. Hence, an optimal route finding system is proposed to provide a suitable route for travelers. There are many factors needed to find the optimal route through proposed system. The factors are traffic volume, emission factor, driver condition, driver action, weather condition, road condition, accident data, and safety requirements. It depends upon the driver's options; the optimal route will be suggested based on different types of factors. Intelligent transportation system has one of the research themes, i.e., optimized paths in dynamic environment. The various types of electronic components such as sensors, camera are used to collect different types of traffic parameters. Based on the parameters analysis, processes are carried out through the intelligent computational algorithms. The intelligent route planning system is one of the part of global monitoring systems along with main objective is to monitor, detect, identify the objects in feasible routes.

The related works discussed in Sect. 2, algorithm is discussed in detail in Sect. 3. Section 4 elaborates the methodology of the implementation, and Sect. 5 compares the GA and Bellman–Ford algorithm. Results are discussed in Sects. 6 and 7 which conclude the work.

2 Related Works

it describes the system investigation in geospatial information system (GIS) that gives solid choice help to users in looking through ideal route, finding the closest workplace, and deciding the administration area. Looking through ideal way is a significant propelled check work in GIS. Right now discovering modules, heuristic algorithm calculations have been utilized to complete its hunt methodology. Because of the absence of worldwide examining in the possible arrangement space, these calculations have inspiring probability of being fixed into nearby optima [1]. The intelligent transportation system (ITS) gives drivers route direction and guide data, and it very well may be utilized to help the smart driving support by detecting innovations or by the driving data input from the clients. Intelligent transportation system is the utilization of detecting, investigation, control, and correspondences advancements to ground transportation so as to improve security, portability, and efficiency. Intelligent transportation framework incorporates a wide scope of uses that procedure and offer data to ease blockage, improve traffic the executives, limit ecological effect, and increment the advantages of transportation to business clients and people in general all in all [2]. Route guidance systems (RGSs) are progressively being executed in a wide scope of vehicle route applications. RGSs give pre-trip data about which path to pick, the journey time and whether to hold the excursion; moreover, in transit data can be given about mishaps, clog, and unfavorable climate conditions. The ideal objective point can be gotten to by elective ways utilizing the most limited time, the most secure way, the least fuel consumption, and different variables [3].

3 Algorithm

Genetic Algorithms: It is most generally utilized in improvement issues wherein we need to increase or limit a given target work a reason under a given arrangement of limitations. The way to deal with take care of optimization issues has been featured all through the instructional exercise. Genetic algorithm is bio-inspired operators such as mutation, crossover, and selection [4].

Bellman–Ford algorithm: In this algorithm, it utilizes unwinding to discover single source shortest path on coordinated graph. Furthermore, it likewise contains negative edges. The algorithm will likewise identify if there are any negative weight cycles. In the case of discussing separations on a guide, there is no any negative separation. The essential structure of Bellman–Ford algorithm is like Dijkstra's algorithm. It loosens up all the edges, what is more, does this |V|—1 time, where |V| is the quantity of vertices in the graph. The expense of a way is the entirety of edge loads in the way. Presently, srv is source, and this is single source shortest path issue so find separation structure single source to various goal. This algorithm return esteem that negative cycle is present or not and furthermore returns most brief way. This algorithm finds shortest path in base up way [5, 6].

3.1 Genetic Algorithm Techniques

A calculation is a step-by-step technique for tackling an issue. It is ordinarily utilized for information handling, figuring, and other related personal choice and numerical activities. A calculation is additionally used to control information in different manners, for example, embeddings another information thing, scanning for a specific thing or arranging a thing. Toward the starting an initialization population must be characterized. This should be possible utilizing an arbitrary instatement or can use a few sort of seeding which permits the calculation to work in a detection space where measures are more probable [7, 8]. From this point until a legitimate arrangement is found or the maximal degree of permitted generations is come to, the following steps are performed (Fig. 1).

Population Initialization: Population *P* can also be defined as a set of chromosomes. The initial population *P*(0), which is the first generation, is usually created randomly. They are two part of chromosome to be performed in initialization.

Selection: First we select an extent of the current population to breed a new population. The choice is done on a wellness based methodology where fitter populations are more liable to breed then others.

Reproduction: During the increase stage, the critical edge is made utilizing the two essential techniques, hybrid, and transformation. For each new child, a pair of parent is chosen from which the new child acquires its properties. In the hybrid procedure, genotype is taken from the two parents and consolidated to make another new child.

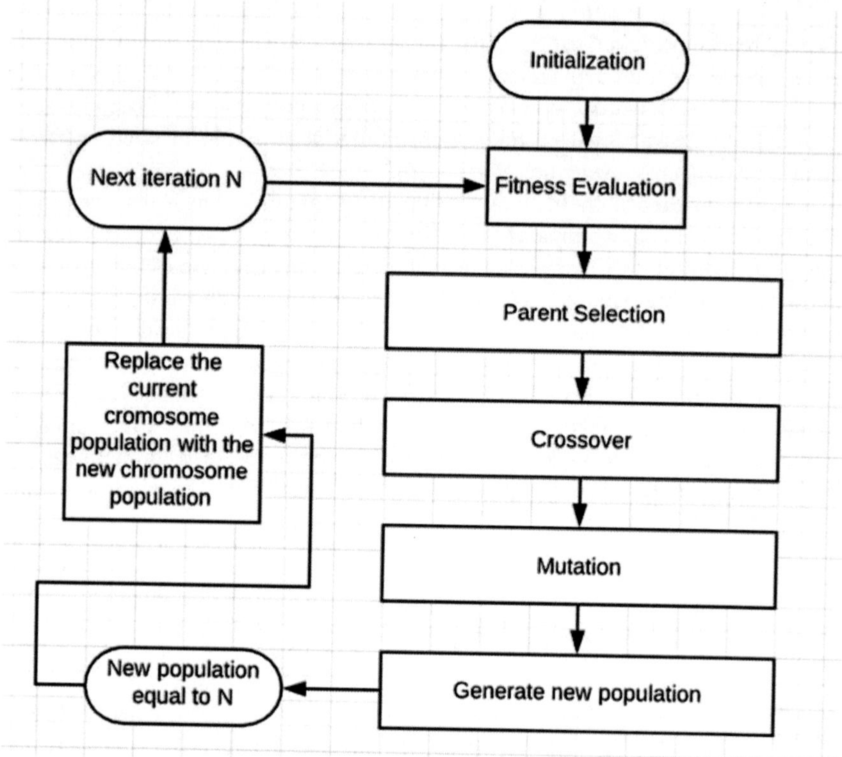

Fig. 1 Flow diagram for genetic algorithm

Chromosome Representation: Chromosome representation the generation of a population in the GA can be viewed as a requested rundown of fake chromosomes where each chromosome speaks to a highway a truck is going to take. Each chromosome contains *K* whole numbers, where *K* is the quantity of qualities a chromosome holds. A quality itself is and whole number too what's more, speaks to the quantity of client.

Crossover: In genetic calculations, hybrid is a genetic operation utilized to change the programming of a chromosome or chromosomes starting with one generation then onto the next. It is a similarity to generation and natural hybrid, whereupon genetic calculations are based. Both actualized hybrids do not do shared trade of genetic material between two guardians. They take data from one individual and addition it in the other to make another new child. The likelihood which hybrid technique must to be utilized can be designed.

Mutation: In genetic calculations, transformation is a genetic operation used to keep up hereditary assorted variety from one age of a populace of chromosomes to the following. It is undifferentiated from natural change. The likelihood which changes will occur, and if transformation happens at all can be designed [9].

The sequence part, two nodes are selected randomly and they are interchanged with each other. There is no change in the topological part of the chromosome.

Fitness Function: The target work is characterized as the wellness capacity of an answer, which is settled by a branch and bound calculation in LINGO tool. In the event that the best wellness work esteem isn't improved over a specific number of successive generations, at that point the calculation is stopped and the best chromosome over all generation is the answer for our freeway service patrol (FSP) issue [10].

4 Methodology

The methodology of this project is in the form of flow diagram, consisting of all the task to be performed in sequential order, starting from the problem identification, literature survey, data collection, and analysis of data from the study area up to the stage were the results are obtained based on that suggestive measures are given (Fig. 2).

4.1 Data Collection and Factor Identification

Dataset collected from around Madurai City selects a stretch that has number of intersections with heavy traffic. We are choosing heavy traffic conjunction and main area for Madurai City that areas like Thiruparankundram and Mattuthavani.

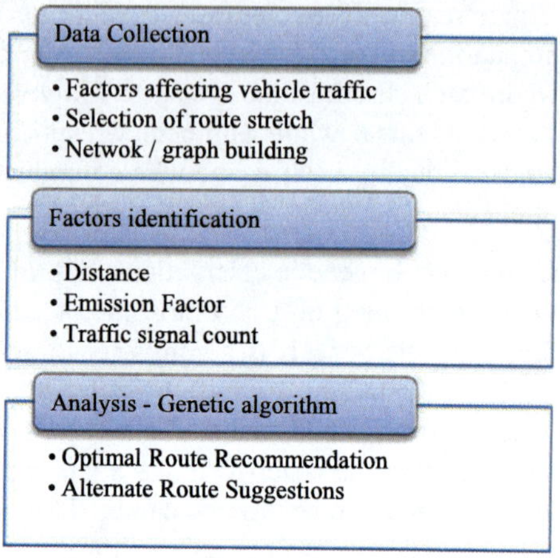

Fig. 2 Model flow diagram

Thiruparankundram is a source node, and Mattuthavani is a destination; they are 22 intermediate nodes (Fig. 3).

We are choosing another two use cases, heavy traffic conjunction, and main area for Delhi City and northern Utah those areas like Ballbhgarth, raj nager, west jarden, and ogden. We have taken the place of ballbhgarth in Delhi City as source node and raj nager as destination. We have put twenty intermediate nodes in between. We have taken the place of west jarden in Northern Utah City as source node and ogden as destination. We have put sixteen intermediate nodes in between (Figs. 4 and 5).

We have selected to prioritize three important factors such as distance, emission factor, and count of traffic signal. We have used emission equation to calculate the emission factor values [7]. Need for some data to calculate the emission values

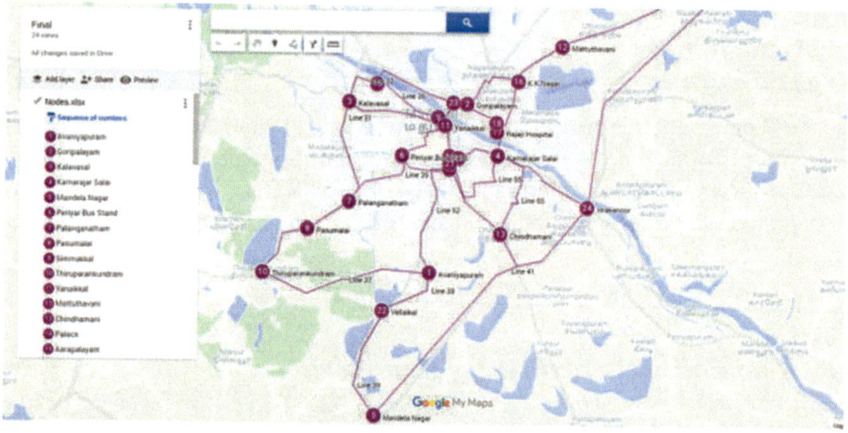

Fig. 3 Selecting of route stretch (Madurai–Thiruparankundram to Mattuthavani)

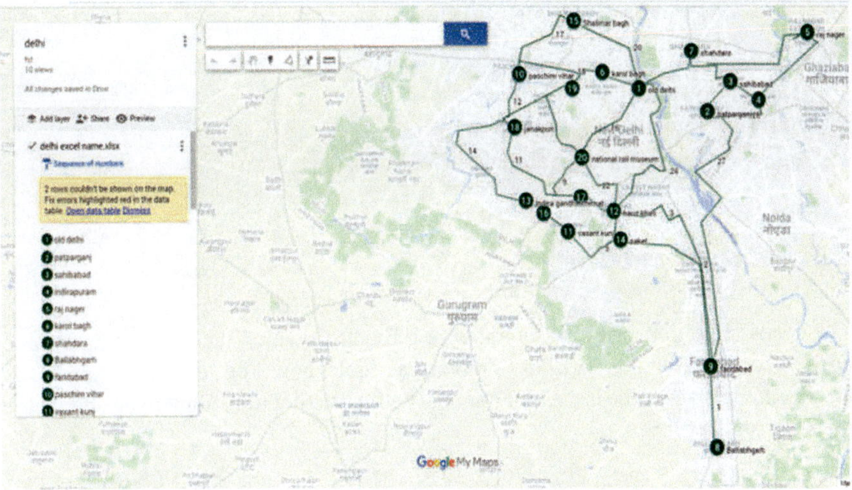

Fig. 4 Selecting of route stretch (use case) Delhi–Ballabhgarh to raj nager

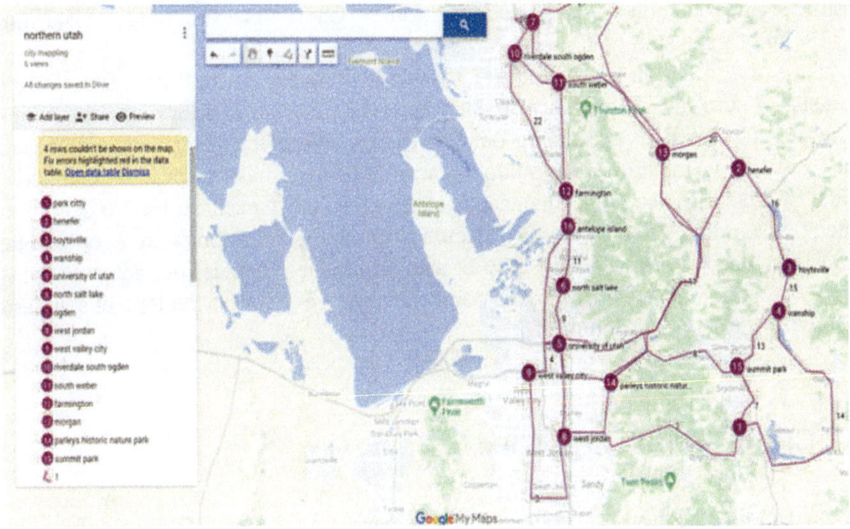

Fig. 5 Selecting of route stretch (use case) Northern Utah–West Jordan to Ogden

like emission of pollutant standard value, number of vehicles, distance traveled in kilometer's, and emission factor for the pollutant of vehicle.

Emission Equation:

$$E = (V \times D) \times (e)$$

where

- E Emission of pollutant
- V Number of vehicles
- D Distance traveled in kilometer's
- e emission factor for the pollutant of vehicle [11].

4.2 Fitness Function

The fitness function just characterized is a function which takes an up-and-comer answer for the issue as info and delivers as yield how "fit" our how "great" the arrangement is as for the issue in thought. The fitness function ought to be adequately quick to register.

$$R = 0.6 * f1 + 0.3 * f2 + 0.1 * f3$$

where EF is emission factor, td is distance between one node to another node, T is traffic signal total count for one node to another node. $f1$ is a sum of td, $f2$ is a sum of EF and $f3$ is a sum of T.

Objective function is devoted to sum of $f1$, $f2$ and $f3$. The fitness function $f1$ consist of sum of total distance in each nodes, $f2$ consist of sum of all the nodes emission factor values and also $f3$ denoted to sum of total traffic signal count in each nodes. We have mentioned the value of each factor in terms of its primary character. First of the entire very short distance route is everyone expects so we are denote to the 60% standard value for distance factor. Second, the emission factor people to be considered so we denote to the 30% of standard value for emission factor. Peoples will eventually take the traffic signal count value so we denote to the 10% of standard value for traffic signal count.

5 Comparison of Algorithms

An in deciding the most limited way data in a course can utilize different algorithm or strategies to decision making. Numerous strategies can be utilized to tackle the most brief way issue in the graph, yet there are numerous contrasts in every strategy or algorithm despite the fact that the equivalent expected objective is to locate the most brief/least direction that is ideal and proficient. Despite the fact that the principle reason for existing is to discover the most limited way, there is additionally another objective that is to discover the briefest way ideal with a quick hunt time. It is very hard to get both. The briefest way of the ideal course takes the most extreme time likewise by the field that numerous ways that can be utilized to look through the briefest way on the graph with the points of interest and detriments of each. The different shortest

Table 1 Comparison to various algorithm performances

Comparison of various algorithm performance strategies	Genetic algorithm	Bellman–Ford algorithm
Accuracy (%)	94.6	92.4
Time complexity	O(g(nm + nm + n))	O(|V||E|)
Negative cycle	No	Yes

path algorithms like genetic algorithm and Bellman–Ford algorithm are used to find the shortest path of a network where the value of the shortest path may vary according to the route. And used it is compared the accuracy, time complexity which is shown in Table 1.

6 Experimental Results

Randomly generate first initial population. The generated trees fitness values are evaluated. If the chromosome is invalid, that chromosome is rejected only random chromosome to be processed. The population size is 10. The best routes identified are illustrated in Figs. 6, 7, 8, and 9 of the selected three different regions (Tables 2 and 3).

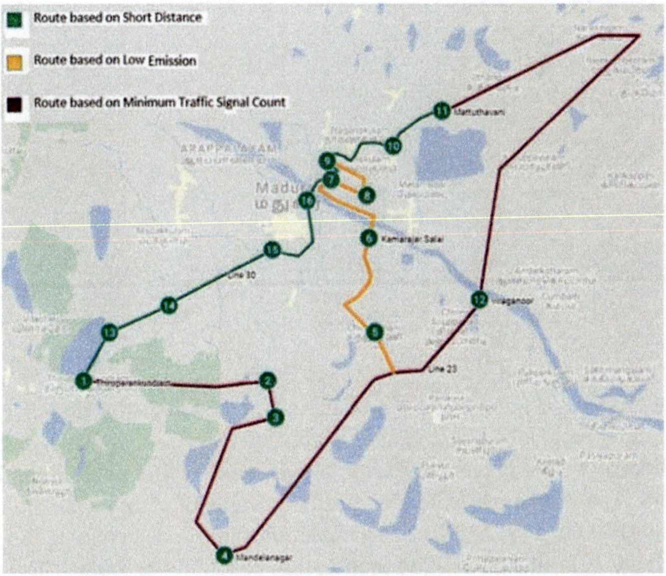

Fig. 6 Madurai City Best Route Mapping using Bellman–Ford algorithm

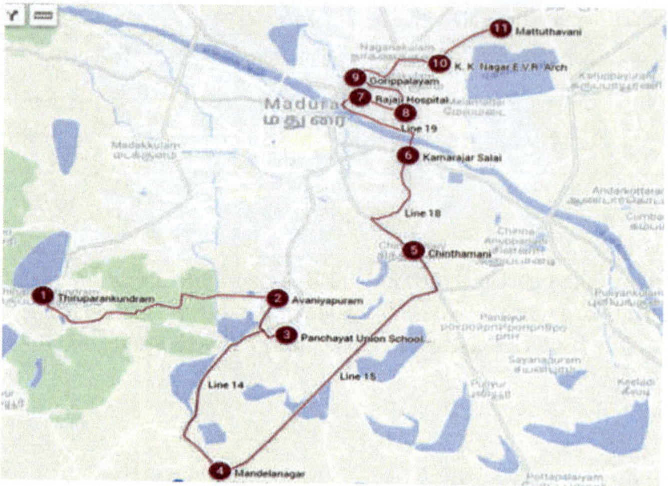

Fig. 7 Madurai City Best Route Mapping using genetic algorithm

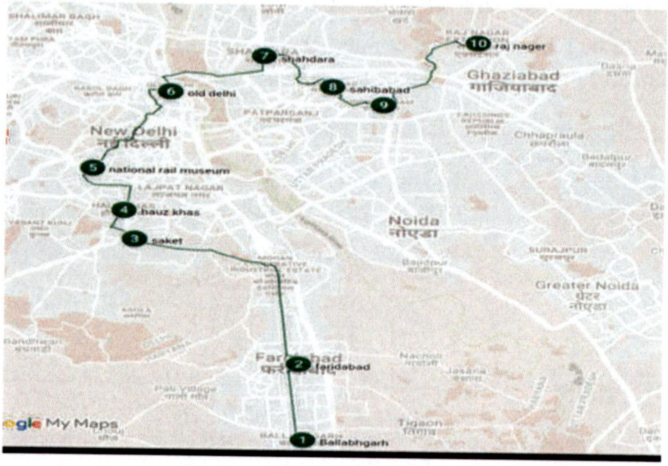

Fig. 8 Delhi City Best Route Mapping using genetic algorithm

Bellman–Ford algorithm single object-based shortest path algorithm so multiple of factors cannot be invoked. But genetic algorithm is taken to find multiple object-based method so we are choosing genetic algorithm. The Bellman–Ford algorithm gives three different types of factors and different type of path. These are the three types of path ID like 5, 7, and 8. But genetic algorithm creates objective function with three factors. In that objective function find best optimal route. The path ID is 7. Figure 6 shows best route identified by Bellman–Ford algorithm.

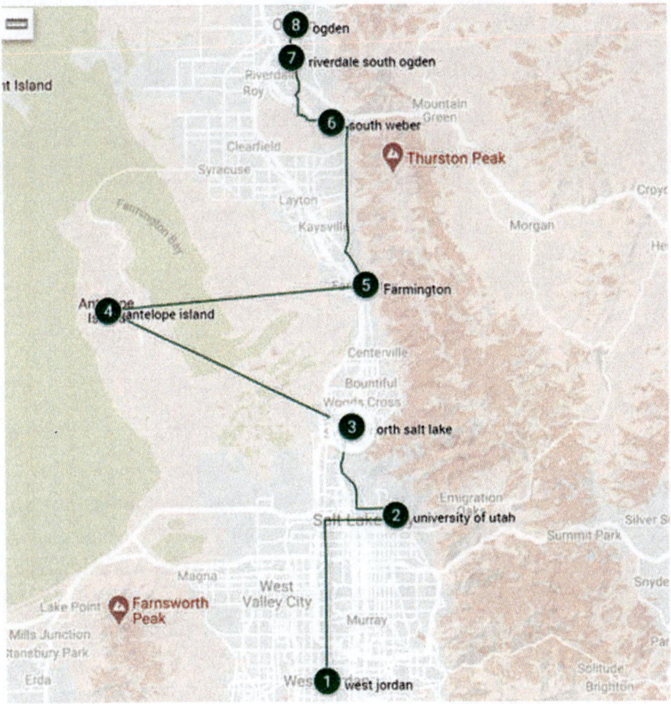

Fig. 9 Northern Utah City Best Route Mapping using genetic algorithm

Table 2 Tabulate from final result for Bellman–Ford algorithm

	Madurai
Population size	10
Source node	10
Destination node	12
Intermediate node	22
Distance value	12.63
Emission value	15.984
Traffic signal count	0
Best ID for distance factor	5
Best ID for emission factor	7
Best ID for traffic signal count factor	8

7 Conclusion

The proposed process of personalized route finding system using genetic algorithm in and around the areas of Madurai City is successfully tested with real-time

Table 3 Tabulate from final result for genetic algorithm

	Madurai	Delhi	Northern Utah
Population size	10	10	10
Source node	10	8	8
Destination node	12	5	7
Intermediate node	22	19	14
Distance value (GA)	56.95100	94.6532	124.0387
Emission value (GA)	11.3340	4.2821	44.7743
Traffic signal count(GA)	8	–	–
Best ID (GA)	7	4	2
Minimum fitness value (GA)	68.2950	98.9353	168.8139

factors. Real-time distance, emission factor, and traffic signal count values to be collected in various methods. Genetic algorithm analyzes to be find the optimal path for frequent route. The results are tabulated in the previous section. Through this proposed process, easily and effectively perform the personalized route finding in and around Madurai City. And also genetic algorithm result is compared to Bellman–Ford algorithm. The genetic algorithm finds best route for 94.6% of accurately, and Bellman–Ford algorithm finds the shortest path for 92.4% of accurately. In future, this method in-turn provides opportunities for smart city implementation, intelligent optimal route finding system, and personalized route finding system application development.

References

1. Zhang, S., Zhang, Y.: A hybrid genetic and ant colony algorithm for finding the shortest path in dynamic traffic networks. Autom. Control Comput. Sci. (3), 1384–1390 (2018)
2. Hu, W., Wang, H., Qiu, Z., Nie, C., Yan, L.: A quantum particle swarm optimization driven urban traffic light scheduling model. Neural Comput. Appl. **4**, 467–477 (2018)
3. Horng, M.-F., Dao, T.-K., Shieh, C.-S., Nguyen, T.-T.: A multi-objective optimal vehicle fuel consumption based on whale optimization algorithm. In: Multimedia Signal Processing, vol. 16, pp. 432–442. Springer, Berlin (2017)
4. Efthymiou, D., Chrysostomou, K., Morfoulaki, M., Aifantopoulou, G.: Electric vehicles charging infrastructure location: a genetic algorithm approach. Int. J. Trans. Sci. Technol. (26), 155–160 (2017)
5. Goldberg, A.V., Radzik, T.: A Heuristic Improvement of the Bellman Ford Algorithm, Technical Report. Stanford University, CA (1993)
6. Lia, Y., Mac, D., Zhua, M., Zengd, Z., Wang, Y.: Identification of significant factors in fatal-injury highway crashes using genetic algorithm and neural network. Accid. Anal. Prevension (34), 43–60 (2017)
7. Ramachandra, T.V.: Emissions from India's transport sector: statewise synthesis. In Atmos. Environ. **43**(34), 5510–5517 (2009)

8. Guo, J., Kong, Y., Li, Z., Huang, W., Cao, J., Wei, Y.: A model and genetic algorithm for area-wide intersection signal optimization under user equilibrium traffic. Math. Comput. Simul. (3), 807–830 (2017)
9. Jordán, J., Palanca, J., del Val, E., Julian, V., Botti, V.: Using genetic algorithms to optimize the location of electric vehicle charging stations. Adv. Intell. Syst. Comput. 3203, 218–224 (2018)
10. Pal, B.B.: Interval-valued goal programming method to solve patrol manpower planning problem for road traffic management using genetic algorithm. In: Multi Objective Optimization, vol. 119, pp. 262–268. Springer, Berlin (2018)
11. Liu, C., Susilo, Y.O., Karlström, A.: Estimating changes in transport CO_2 emissions due to changes in weather and climate in Sweden. Int. J. Transp. Sci. Technol. (5), 136–151 (2016)

Efficient Approach for Encryption of Lossless Compressed Grayscale Images

Neetu Gupta and Ritu Vijay

Abstract To transmit images over a communication medium at lower bandwidth and robustness against differential attacks, compression and encryption processes are used. In this paper, Huffman lossless compression technique is used to compress the grayscale images. In this, different prefix codes of variable lengths are assigned to each pixel. The length of the prefix codes is inversely proportional to frequency of occurrence of characters. Compressed image is encrypted by RSA asymmetrical encryption algorithm. In RSA encryption algorithm, encryption key is kept public and different from decryption key which is kept secret. The efficiency of proposed algorithms is expressed by analyzing the compression parameters as well as the correlation coefficient analysis and entropy analysis. The results are verified on five grayscale test images of 512 × 512 sizes.

Keywords Lossless compression · Asymmetrical encryption · Security · Entropy analysis · Correlation coefficient analysis · Huffman · RSA

1 Introduction

To transmit the information effectively over a communication channel with lower bandwidth, compression techniques are employed during transmission [1, 2]. The compression techniques may be lossy or lossless [3]. During compression process, size of information is reduced by removing the redundant data from original information source. Lossy compression techniques have large compression ratio, but it may lose the important part of information also, whereas lossless compression techniques maintain the integrity and certainty of the information with sufficient amount of compression [4, 5]. Security of the information from various unethical attacks

N. Gupta (✉)
Computer Science and Engineering Department, Banasthali Vidyapith, Banasthali, Rajasthan, India

R. Vijay
Electronics Engineering Department, Banasthali Vidyapith, Banasthali, Rajasthan, India

H. Sharma et al. (eds.), *Congress on Intelligent Systems*, Advances in Intelligent Systems and Computing 1334, https://doi.org/10.1007/978-981-33-6981-8_32

during transmission is also an important aspect. Encryption techniques are incorporated with the information during transmission to provide confidentiality to transmitting data [6]. Encryption techniques can be symmetrical and asymmetrical. Generally, encryption is performed by key generation process. In symmetrical encryption process, the sender key and recipient keys are same which can cause hacking of information during transmission [7]. The security features can be improved by using asymmetrical encryption algorithm in which sender key is public, but recipient key is private [8].

The compression and encryption process both are important for transmission of information at lower bandwidth with security. Compression process may be employed before or after the encryption process. If the encryption is performed on uncompressed image, then it increases the execution time and misuse of encryption process [9]. The possibility of decoding of encrypted image by hackers can be reduced by using compression algorithm before applying encryption algorithms [9, 10].

In this paper, Huffman lossless compression algorithm is applied before applying the RSA asymmetrical encryption technique. The efficiency of reconstruction of image is evaluated by analyzing the compression efficiency parameters like compression ratio (CR), peak signal-to-noise ratio (PSNR), and mean square error (MSE) and comparing them with state-of-the-art methods. The encryption efficiency of proposed algorithm is also analyzed based on entropy and correlation coefficient analysis. The experimental analysis is performed on five standard grayscale test images in.bmp format having size of 512×512.

2 Related Work

In 2017, Yassein et al. [11] have presented a study on different encryption techniques under symmetrical and asymmetrical categories. AES, DES, 3DES, and blowfish symmetrical encryption algorithms are compared with RSA asymmetrical encryption algorithm. Authors have also analyzed the symmetrical and asymmetrical encryption techniques based on different encryption efficiency evaluation parameters and illustrate that RSA encryption technique provides better security against different attacks. In 2016, Galla et al. [12] have also implemented the RSA algorithm for encryption of image data. Authors have demonstrated the dependency of RSA algorithm on factorization of numbers. In 2013, Jamgekar et al. [8] have also implemented and demonstrated the elementary cryptanalysis and shown implementation of RSA algorithm through a public and private key generation. In 2002, Jonsson et al. [13] have presented a TLS handshake protocol-based RSA encryption algorithm and have compared with the partial RSA algorithm. Authors shown that TLS based algorithm reduces the security features which are based on pseudo-random functions. In 2014, Katz et al. [14] have shown the modern cryptanalysis approach which reduces the shortcomings of perfect secrecy by modernizing the computational process. Authors have used private key encryption process and authentication through message signal.

In 2016, Han et al. [15] have addressed the limitations of neural networks by illustrating the deep compression method. This deep compression method consists of pruning, quantization, and Huffman coding. Authors show that the deep compression method reduces the storage requirement without sacrificing the accuracy of reconstructed image. In 2014, Satir et al. [16] have proposed a lossless Huffman compression approach to enhance the compression capacity and security of textual data. In this, stegokeys are used to enhance the security which hides the selected important text. Email has been used as broadcast medium between transmitter and receiver. In 2012, Patel et al. [17] have implemented bzip2-like lossless image compression algorithm with the aim of parallel implementation of MTF, BWT, and Huffman coding. Performance analysis is also carried out to illustrate the strengths and weaknesses of the proposed algorithms. In 2010, Sharma [18] has also analyzed the Huffman compression algorithm with LZW and run-length encoding. Author illustrates the compression ratio, and compression efficiency of Huffman coding is better than other compression schemes.

From literature review, it can be viewed that Huffman coding is most widely used lossless compression technique, and it motivates for the further research to improve the compression ratio and efficiency of reconstructed images.

The aim of this research paper is to implement and analyze the effect of asymmetrical encryption technique over lossless compressed grayscale images. RSA asymmetrical technique is used for encryption purpose while grayscale images are compressed using Huffman lossless technique. The objectives of this paper are to perform the parametric evaluation of compression efficiency in terms of compression ratio, PSNR, and MSE as well as encryption efficiency in terms of correlation coefficient and entropy analysis.

3 Basic Knowledge and Proposed Algorithms

Huffman coding is in category of lossless compression methodology, and it is based on the frequency of occurrence of pixels in image data. In Huffman encoding process, the pixels are arranged in decreasing order of their frequency of occurrence. The frequencies of two pixels having least frequency are combined, and again all the pixels are rearranged in decreasing order of frequency of occurrence during execution of proposed algorithm.

This process will be continuous till only two frequencies of pixels remaining. This process of reducing the number of pixels in Huffman table is called reduction process. Last two frequencies of pixels are encoded by assigning 0 to higher frequency and 1 to lower frequency pixels. Now go back by assigning 0 and 1 to the second digit for the two probabilities that were combined in the previous reduction step till first column of Huffman tree is reached.

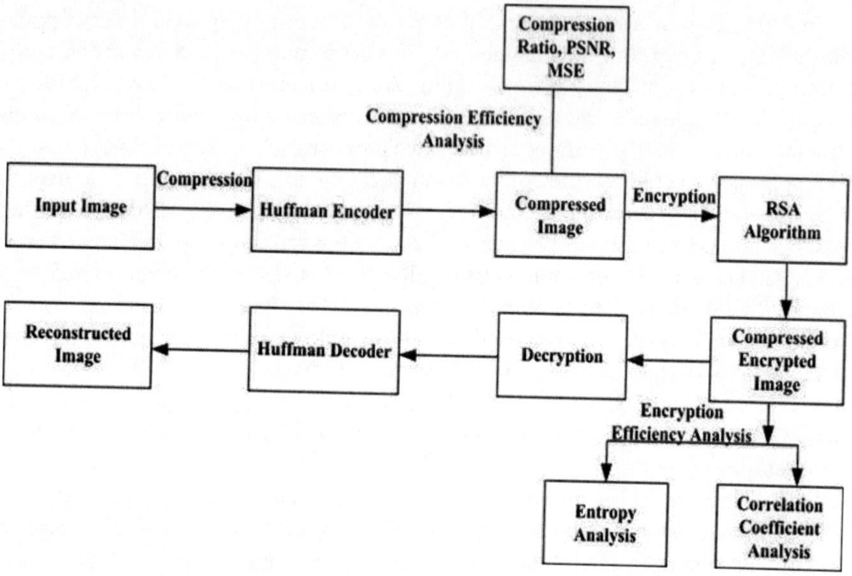

Fig. 1 Block diagram representation of proposed system model

Algorithm 1

 # Each pixel value is represented by 8 bits contains values from 0 to 255 and each value is called as symbol.

1. *Read the corresponding input original image by*
   I = imread(file)
2. *If the original image is colour image then convert it in grayscale image using*
   I = rbg2gray(file)
3. *Calculate the size of image by using*
   [m,n] = size(I), which gives Total count = m x n.
4. *Compute the cumulative probability by using the program*
   Count (cnt) = sum(k(:))
   Pro(cnt) = Count (cnt)/Total count
   Cumpro(cnt) = Sigma + Pro(cnt)

5. Call Huffman code dictionary using

dict = huffmandict(symbols, pro)

6. Array of symbols is converted in vector using

newvec(vec_size) = I(m,n)

7. Huffman encoding is performed using command

hcode = huffmanenco(newvec, dict)

8. Huffman decoding is performed by

dhsig1 = huffmandeco(hcode,dict)

9. Calculate compression ratio, PSNR and MSE.

RSA is an asymmetrical encryption technique in which two separate keys are generated. One key is assigned to sender side, and this key will be public key. Another key is assigned to the receiver, and this key is a private key. Security of this algorithm lies on the selection of two large prime numbers. Public and private keys are generated based on these prime numbers.

RSA encryption algorithm consists of mainly three parts: key generation, encryption, and decryption.

Algorithm 2

 # key generation includes the generation of encryption key and decryption key.

1. Find two distinct but equal bit length integer prime numbers g and h by primer testing process.

2. Calculate key length n in bits by $n = g \times h$

3. Select encryption key e such that it should not be a factor of $(g-1)\times(h-1)$ *and* $1<e<(g-1)\times(h-1)$. *Now (n,e)will form a public key.*

4. Select decryption key d, which is a private key, such that

$(d \times e) \bmod (g-1)\times(h-1) = 1$ *and* $1<d<(g-1)\times(h-1)$.

 # Encryption

5. Decompose the image I into a sequence of blocks I1, I2, . . . , Iq where each Ii satisfies $0 \leq Ii < n$. *Then encrypt these blocks as* $E = I^e \bmod (n)$.

 # Decryption

6. Decrypt the encrypted image E by using $D = E^d \bmod (n)$.

7. Perform the entropy and correlation coefficient analysis to illustrate the efficiency of encryption algorithm.

---

4 Parametric Evaluation of Compression and Encryption Efficiency

4.1 Compression Efficiency Parameters

To analyze the efficiency of Huffman compression technique, compression ratio, peak signal-to-noise ratio (PSNR), and mean square error (MSE) are evaluated. For effective reconstruction of image, PSNR should attain a large value, i.e., above 30 dB, and MSE should be minimum.

$$\text{Compression Ratio} = \frac{\text{Uncompressed image size}}{\text{Compressed image size}} \tag{1}$$

If the dimensions of image data are M and N, then PSNR and MSE can be evaluated as

$$\text{PSNR} = 10\log_{10}\left[\frac{\{M \times N\}^2\}}{\text{MSE}}\right] \tag{2}$$

$$\text{MSE} = \frac{1}{M \times N}\sum_{X=1}^{M}\sum_{Y=1}^{N}[P(x, y) - q(x, y)] \tag{3}$$

where $p(x, y)$ and $q(x, y)$ denote the original and compressed pixel, respectively.

4.2 Encryption Efficiency Parameters

The efficiency of proposed RSA algorithm is illustrated based on the following analysis.

4.2.1 Entropy Analysis

Entropy defines the average information contained in a long sequence of pixels in an image data.

The mean value or entropy of $I(X_i)$ with m different symbols is given by

$$H(X) = \sum_{i=1}^{m} P(X_i) I(X_i) \tag{4}$$

$$H(X) = \sum_{i=1}^{m} P(X_i) \log_2 P(X_i) \tag{5}$$

The entropy $H(X)$ satisfies the following relationship

$$0 \leq H(X) \leq \log 2m \tag{6}$$

4.2.2 Correlation Coefficient Analysis

The correlation coefficient between two horizontal adjacent pixels, two vertically adjacent pixels, and two diagonally adjacent pixels represents the correlation in original image and encrypted image. The correlation coefficient is represented as:

$$C = \frac{\sum_{i=1}^{N} (x_i - \overline{x})(y_i - \overline{y})}{\sqrt{\sum_{i=1}^{N} (x_i - \overline{x})^2} \times \sqrt{\sum_{i=1}^{N} (y_i - \overline{y})^2}} \tag{7}$$

where $\overline{x}$ and $\overline{y}$ are the average or mean values of x and y, respectively, which can be expressed as

$$\overline{x} = \frac{1}{N} \sum_{i=1}^{N} x_i \text{ and } \overline{y} = \frac{1}{N} \sum_{i=1}^{N} y_i \tag{8}$$

The adjacent pixels of original image in horizontal, vertical, and diagonal directions must be strongly correlated and therefore must have high value of correlation coefficient. On the other hand, encrypted image should have loose correlation among pixels and low value of correlation coefficient.

5 System Environments

In this experimental analysis, MATLAB 2018 is used for simulation of results. Window 10 used as operating system. Standard grayscale test images of 512 × 512 sizes with.bmp format from SIPI data base are used to authenticate the results.

6 Results and Discussions

Results for compression parameters using proposed Huffman compression technique are given in Table 1 and also compared with state-of-the-art methods. Compression ratio, PSNR, and MSE are used to represent the compression efficiency. Figure 3 is representing graphical representation of compression efficiency parameters. In Fig. 2, column 2 and 3 are showing original image and corresponding correlation coefficients distribution, respectively. Compressed-encrypted (CE) image and its corresponding correlation coefficients distribution are shown in column 4 and 5, respectively, and column 6 is showing reconstructed image after performing decryption and decompression.

Table 2 is representing the correlation coefficient values of adjacent pixels of original image and encrypted image in horizontal, vertical, and diagonal directions. The values of entropy are also given in Table 2.

7 Conclusion and Future Scope

In this paper, Huffman lossless compression scheme followed by RSA asymmetrical techniques are used for compression and encryption of image data, respectively. From the analysis of results, it is concluded that PSNR of Huffman compression technique is in the range of 30–40 dB which is above the threshold value of PSNR for effective compression while it produces moderate amount of compression ratio because of its lossless nature. MSE lies in the range of 6.85–47.64 which is lower than the other state of art methods. The correlation coefficients in horizontal, vertical, and diagonal directions of encrypted images are having low values, while the entropy is in the range of 3.72–3.93 which shows the robustness of RSA encryption technique against different attacks. The proposed system produces effective compression and encryption of grayscale images. The proposed research can be further implemented and analyzed on grayscale images and color images of different formats and size.

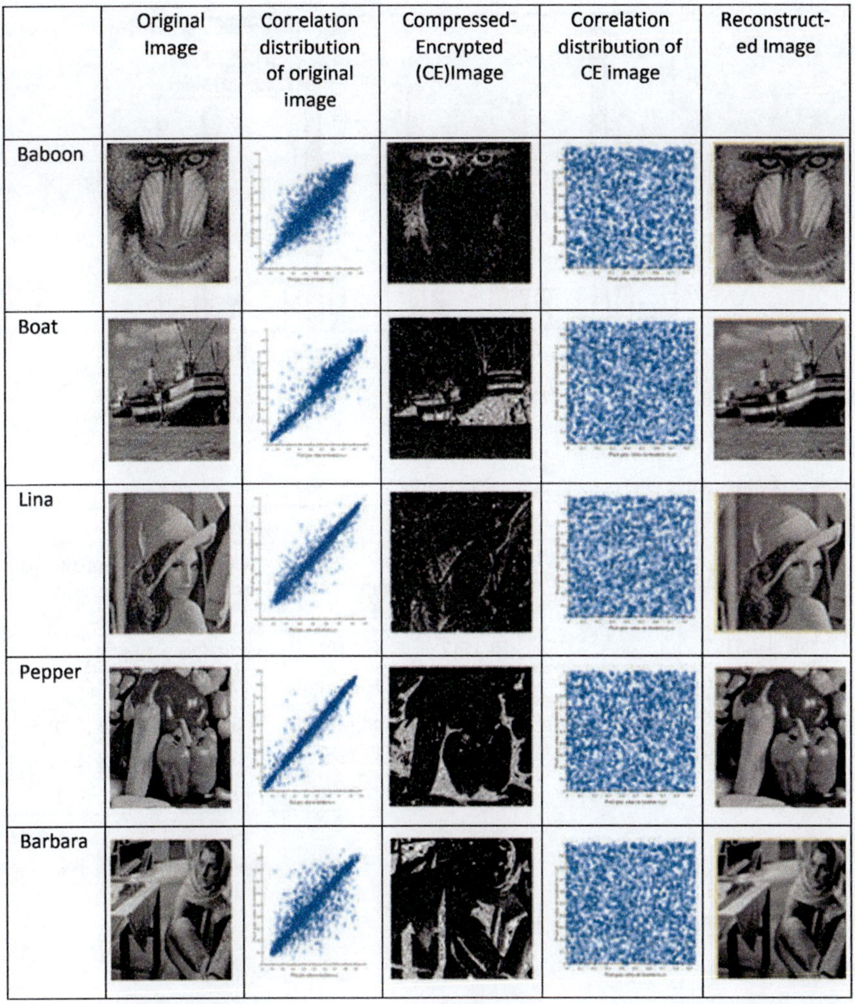

Fig. 2 Original images, compressed-encrypted images, correlation coefficient of original and CE images, reconstructed images

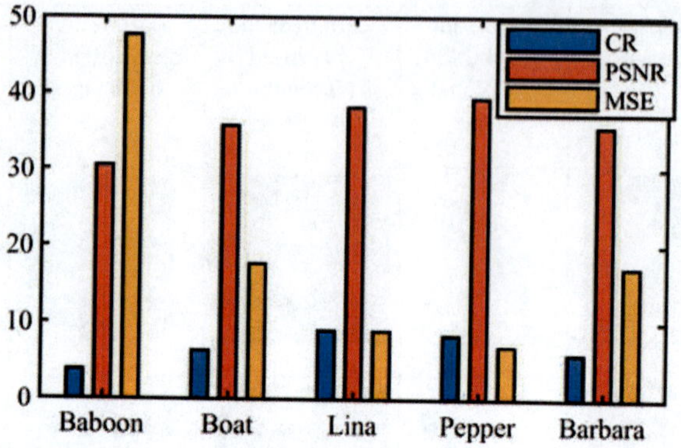

Fig. 3 Graphical representation of CR, PSNR, and MSE

Table 1 Comparative analysis of compression ratio, PSNR, and MSE

Image	CR		PSNR in dB		MSE	
	Our scheme	State-of-the-art comparative method	Our scheme	State-of-the-art comparative method	Our scheme	State-of-the-art comparative method
Baboon	3.8197	X	30.4547	22.19 [19]	47.6420	134.78 [19]
Boat	6.2796	X	35.6674	30.76 [20]	17.6336	X
Lina	8.9828	6.92 [21]	38.1049	30.68 [22]	8.9112	179.68 [23]
Pepper	8.3794	4.0 [24]	39.2817	32.15 [24]	6.8527	39.598 [24]
Barbara	5.8678	X	35.4829	22.92 [19]	17.1231	332.10 [19]

Table 2 Comparative analysis of correlation coefficient and entropy

Image	Vertical		Diagonal		Horizontal		Entropy	
	Our scheme	State-of-the-art method	Our scheme	State-of-the-art method	Our scheme	State-of-the-art method	Our scheme	State-of-the-art method
Baboon	0.0712	0.0093 [25]	0.0687	−0.0251 [25]	0.0886	−0.0225 [25]	3.8652	7.9987 [1]
Boat	0.2294	−0.0064 [1]	0.1920	0.00007 [1]	0.2176	0.010 [1]	3.8863	7.9989 [1]
Lina	0.2110	0.0240 [25]	0.1005	−0.0411 [25]	0.1352	−0.0094 [25]	3.7257	7.9989 [1]
Pepper	0.2521	0.0093 [25]	0.2144	−0.0251 [25]	0.2470	−0.0225 [25]	3.9000	7.9988 [1]
Barbara	0.2894	0.0079 [1]	0.2426	0.0200 [1]	0.2666	0.0122 [1]	3.9308	7.9901 [1]

References

1. Tong, X.J., Chen, P., Zhang, M.: A joint image lossless compression and encryption method based on chaotic map. Multimed. Tools Appl. **76**(12), 13995–14020 (2017)
2. Singh, R.K., Kumar, B., Shaw, D.K., Khan, D.A.: Level by level image compression-encryption algorithm based on quantum chaos map. J. King Saud Univ. Comput. Inf. Sci. (2018)
3. Daubechies, I., Barlaud, M., Mathieu, P.: Image Coding Using Wavelet Transform. IEEE Trans. Image Process. 1(2), 205–220 (1992)
4. Maniccama, S.S., Bourbakis, N.G.: Lossless image compression and encryption using SCAN. Pattern Recognit. **34**(6), 1229–1245 (2001)
5. Zhang, X.: Lossy compression and iterative reconstruction for encrypted image. Inf. Forensics Secur. 53–58 (2011)
6. Al-Khasawneh, M.A., Shamsuddin, S.M., Hasan, S., Bakar, A.A.: An improved chaotic image encryption algorithm. In: 2018 International Conference on Smart Computing and Electronic Enterprise, ICSCEE 2018 (2018)
7. Carpentieri, B.: Efficient Compression and Encryption for Digital Data Transmission, vol. 2018 (2018)
8. Jamgekar, R.S., Joshi, G.S.: File encryption and decryption using secure RSA. Int. J. Emerg. Sci. Eng. (2013)
9. Setyaningsih, E., Wardoyo, R.: Review of image compression and encryption techniques. Int. J. Adv. Comput. Sci. Appl. **8**(2) (2017)
10. Sharma, M., Gandhi, S.: Compression and Encryption : An Integrated Approach. Int. J. Eng. Res. Technol. **1**(5), 1–7 (2012)
11. Yassein, M.B., Aljawarneh, S., Qawasmeh, E., Mardini, W., Khamayseh, Y.: Comprehensive study of symmetric key and asymmetric key encryption algorithms. In: Proceedings of 2017 International Conference on Engineering and Technology, ICET 2017 (2018)
12. Galla, L.K., Koganti, V.S., Nuthalapati, N.: Implementation of RSA. In: 2016 International Conference on Control Instrumentation Communication and Computational Technologies, ICCICCT 2016 (2017)
13. Jonsson, J., Kaliski, B.S.: On the security of RSA encryption in TLS. Lect. Notes Comput. Sci. (including Subser. Lect. Notes Artif. Intell. Lect. Notes Bioinformatics) (2002)
14. Katz, J., Lindell, Y.: Introduction to Modern Cryptography (2014)
15. Han, S., Mao, H., Dally, W.J.: Deep compression: compressing deep neural networks with pruning, trained quantization and Huffman coding. In: 4th International Conference on Learning Representations, ICLR 2016—Conference Track Proceedings (2016)
16. Satir, E., Isik, H.: A huffman compression based text steganography method. Multimed. Tools Appl. (2014)
17. Patel, R.A., Zhang, Y., Mak, J.,Davidson, A., Owens, J.D.: Parallel lossless data compression on the GPU. In: 2012 Innovative Parallel Computing, InPar 2012 (2012)
18. Sharma, M.: Compression using huffman coding. IJCSNS Int. J. Comput. Sci. Netw. Secur. (2010)
19. Raja, S.P., Suruliandi, A.: Performance evaluation on EZW & WDR image compression techniques. In: 2010 IEEE International Conference on Communication Control and Computing Technologies, ICCCCT 2010, pp. 661–664 (2010)
20. Agarwal, C., Mishra, A., Sharma, A.: A novel gray-scale image watermarking using hybrid Fuzzy-BP N architecture. Egypt. Informatics J. **16**(1), 83–102 (2015)
21. Praisline Jasmi, R., Perumal, B., Pallikonda Rajasekaran, M.: Comparison of image compression techniques using huffman coding, DWT and fractal algorithm. In: 2015 International Conference on Computer Communication and Informatics, ICCCI 2015, pp. 1–5 (2015)
22. Zhou, N., Pan, S., Cheng, S., Zhou, Z.: Image compression-encryption scheme based on hyperchaotic system and 2D compressive sensing. Opt. Laser Technol. **82**, 121–133 (2016)
23. Dang, P.P., Chau, P.M.: Image encryption for secure Internet multimedia applications. IEEE Trans. Consum. Electron. (2000)

24. Hu, F., Pu, C., Gao, H., Tang, M., Li, L.: Image compression and encryption scheme based on deep learning. Nauk. Visnyk Natsionalnoho Hirnychoho Universytetu **6**, 142–148 (2016)
25. Zhang, Y.: The unified image encryption algorithm based on chaos and cubic S-Box. Inf. Sci. (Ny) **450**, 361–377 (2018)

Application of Social Big Data in Crime Data Mining

Nahid Jabeen and Parul Agarwal

Abstract The increased fame of the Internet and technology has motivated criminals a lot to commit their evil actions with the help of its various facilities such as social media, e-commerce and so on. Nowadays, almost everything exists on the Internet either to store or to access personal as well as professional data. This helpful nature of the Internet and technology has proved a boon for criminals to find their vulnerable targets. It also provides many options for investigators and police officials to track and catch the criminals. The regular use of the Internet has led to a huge amount of big data that made investigators and police officials hard to analyze and predict crime and criminals. But a combination of crime analysis and computer science or information technology can make it easier and user-friendly for investigators and police officials to fetch useful data from the social big data. By applying data mining approaches, we can help the investigative department to predict and prevent crime based on past crime data. The main focus of this paper is to perform an in-depth analysis of some cyber-crimes that frequently occurred in the Union Territories of India per year from 2014 to 2018. Two open-source data mining tools, namely RapidMiner and Weka, are used with the application of the k-means clustering technique. The results of the two tools are compared, and it is found that though Weka outperforms RapidMiner in execution time, it lacks in accuracy and visualization. But, in general, both help in finding an application for investigation and the police department in predicting patterns of crime occurrences.

Keywords Big data analytics · Criminal · Crime data · Union Territories · Years from 2014 to 2018 · RapidMiner · Weka · K-means clustering · Investigation · Theft · Cyber-bullying

N. Jabeen · P. Agarwal (✉)
Jamia Hamdard, New Delhi 62, India
e-mail: pagarwal@jamiahamdard.ac.in

H. Sharma et al. (eds.), *Congress on Intelligent Systems*, Advances in Intelligent Systems and Computing 1334, https://doi.org/10.1007/978-981-33-6981-8_33

1 Introduction

In the information era, the digitalization of information makes it easily available to everyone. Many people want to exploit this information for their unethical motives. So, every information needs higher security and privacy on the ground of managing sensitive data regarding countries' safety. Security is the prior motto by all political and executive bodies regarding the welfare of every nation. If the Internet and technology are proven as a boon, then cyber-crimes have been proven as a curse for humanity. Crime data helps to predict crime and criminals based on previous criminal records. As no prediction is cent percent accurate, but it can only help to reduce the crime rate to some extent. Any crime contains some data behind it such as:

- Crime spot
- Crime sufferers
- Crime calendar day and date
- Crime suspects
- Crime causes
- Crime moment in time
- Crime tools
- Crime witnesses.

1.1 Social Big Data

In recent years, the term "social media" has become so popular and vague. Social media are affecting the cultural and social life of different people. Social media is a networking platform that enables users to participate and share multimedia content, for example, text, audio, video, images, etc. Nowadays, it is a complete term that is used to describe a group of Internet-based applications. Globalization and technological advancement have led to a quick and complex evolution of crime, but at the same time, it also provided some powerful tools to fight against crime. For example, data mining can synchronize the big data of social media networking for regular updating of criminal activities on it. Social big data is the obtained huge amount of data that is user-generated content on social media Web sites and mobile apps to extract a definite pattern. According to Ishikawa [1], "Analyzing both physical real-world data (heterogeneous data with implicit semantics such as scientific data, event data, and transportation data) and social data (social media data with explicit semantics) by relating them to each other, is called Social Big Data science or Social Big Data for short." The multimedia contents are cloud-based big data content and can be viewed in the form of volume, variety, velocity, veracity, volatility. As social media networking has become a part of economical, political and personal life, there is so much data inside it that can be used for solving crime issues. Social media is the target area for criminals to perform their evil actions which makes social networks very vulnerable. According to Soomro and Hussain [2], many crimes can be committed to the availability of social network data. Some of them are as follows:

1. Cyber-bullying or Cyber-stalking: Cyber-bullying or cyber-stalking is a crime in which harassment is done using digital devices such as computers and mobile phones. This crime is most commonly reported by teenagers due to the expansion and advancement of digital technology. A cyber-bullied person faces anxiety, depression and low self-esteem problems. These are some of the serious problems in our society that result in teen suicide. Under Sect. 354D of IPC, bullying of women or children on electronic devices is a serious crime.
2. Social Engineering and Phishing: Social engineering is a crime in which criminals manipulate victims psychologically to get personal information. In many situations, people get emails or messages from friends for financial help. These kinds of messages may be from criminals who stole your friend's email data. Phishing is a kind of social engineering. Phishing is used in committing disaster fraud. During 2015, Nepal Earthquake trouble made fraudsters send multiple phishing emails in which they request for a donation for the earthquake victims. According to Trend Micro, Facebook is a minefield of scams [2].
3. Identity Theft: Identity theft is about creating fake profiles and posting offensive content, including abusive photographs and videos or to cheat someone via someone else's profile. It is a major offense. Many criminals share their messages among different criminal organizations via emails or any other media. Under Sect. 66-D of IT Act, 2000, it considers the creation of fake profiles to be a serious crime that is liable to punish offenses according to his/her offense.
4. Malware: Malware, any software which is intentionally designed to harm computers, server/client and network. Several kinds of malware exist, such as computer viruses, worms, Trojan horses, ransomware, spyware and adware. Social media is a great platform to spread this malware. Developers hide them in attachments, messages, links, i.e., normal to open by users. Once users respond to them, they spread and infect their system without any information.
5. Burglary through Social Networking: Burglary is also known as house breaking which is illegal entry into a building or any location for the intent of the crime. Usually, it is a kind of theft, but most jurisdictions consider the ambit of burglary. Burglary via social networking became a serious problem for police officials. This is due to the habit of the victim to update their routine on social media, which is a very good indication for burglars that their way of committing the crime is clear. For example, a social media user is having dinner or going somewhere for vacations. Criminals keep an eye on that kind of user to find an easy target where they can get a big-time for burglary.
6. Cyber-casing: Cyber-case deals with a major trend called geo-tagging. Social media played a vital role in promoting this trend. When a person goes out for dinner or shopping, the normal trend is to click pictures, tag them with geographical locations, and shared instantly via social media. This prominent feature of social media used real-world locations using data available in online resources.

This information became the prime element in the cyber-casing process executed by criminals [2].

7. Credit Card Fraud: Credit card fraud or debit card fraud, where an unauthorized person uses someone else's information for withdrawing cash offline or online. This offense can be done either by stealing a card or by getting enough information [2].

1.2 Big Data Analytics

There is a huge number of stored crime and criminal-related data, i.e., online and offline that made investigators hard to find and analyze relevant information. Big data analytics helps evaluate large amounts of data much faster than humans can. So from a crime-solving point of view, data analytics could help police staff to catch criminals who are trying to evade arrest. The law enforcement officials have collected the data, opening new opportunities for digging into information much more efficiently with a significantly reduced human resource demand and previous methods required. In this information age, we need weapons of different kinds to battle crime data, and algorithms and analytics (here k-means clustering algorithm and big data analytics) are the newest addition to the law enforcement to maintain law and order. Big data analytics is a process of analyzing and accessing big data, i.e., a huge amount of data. With the help of big data analytics tools, the collected information can help to prevent crime. Today in many countries, police departments adopted data analytics to track criminal behavior patterns, crime anniversaries and more as they depend on data to help them detect useful patterns that could keep people safer. Big data analytics make tasks easier for police staff to report crime and add it into the system and for information to be provided in real time to frontline staff to help them focus more effectively on prevention of crime. Police staff along with big data can solve and prevent many crimes by the use of an accurate and timely data of crime. Big data analytics can help police departments to do their job with even better results.

1.3 Crime Data Mining

Crime is a social problem that is disturbing our society, both mentally and economically. Crime needs to explore, detect, prevent and solve for better living. Data mining provides an intelligent field to identify pattern trends in crime with the help of crime analysis, machine learning and predictive modeling. With the regular increase in volume and complexity of crime and criminal-related data, data mining provides police officials with a powerful and efficient way to solve the crime at some level. Crime data mining is a way of fetching relevant information from a huge amount of crime and criminal-related data which are mostly in an unstructured format. Crime data mining provides only those information that is based on past crime incidents to investigators and law enforcement officials, which can help them to predict the crime patterns for future use. Data mining is a user-friendly tool for crime analysis and prediction as it does not need previous experience and technical knowledge before committing actions. With the combination of big data analytics and data mining, investigation and police departments can solve crime-related issues using their previously occurred crime-related information. It can predict more efficient and up-to-date future crime occurrences. The user-friendly nature of data mining has been proved as a gift for untrained and non-technical police staff so that they can use it more easily and efficiently for detecting and catching culprits.

2 Background

As crime has become a global issue for all sectors of people, law enforcement officials must find some effective techniques that can help them to catch the culprit. Some many scholars and authors gave their precious time to study different approaches to data mining so that crimes can be solved faster. In this section, we will study different ideas of different scholars based on their notable research papers in the field of crime analysis. In [3], a model using the Weka tool and Google Maps marker has been proposed to analyze various crime datasets in India from 2001 to 2012. In this research paper, the authors had done crime analysis among four selected cases with the help of clustering as well as the classification algorithm. They also used two more algorithms that can help them in performing classification such as random forest and neural networks. They found the proposed model to be 99.93% accurate which is completely verified. The main aim of the authors in this research paper [4] is to produce a model that can help the Libyan Security Committee to identify, detect and solve crime efficiently and earlier than the crime to happen. In this model, two data mining techniques (i.e., association rule mining and cluster algorithm) were applied and compared their applicability to find the best one between them. Various forms of crime are analyzed to identify the diverse nature of criminals utilizing this proposed model. The authors [5] described various types of techniques in their research paper that are used in crime and criminal data mining, for example, K-nearest neighbor (KNN), support vector machine (SVM), classification, clustering, Naïve Bayes and neural network. They also stated some challenges that are faced by researchers and law enforcement officials during the identification and prediction of crime and criminals. The authors used CRISP-DM methodology (stands for cross-industry standard process for data mining) for analyzing huge amounts of heterogeneous criminal data in the criminal investigation.

In [6], the authors applied decision tree and Apriori algorithms for identification and classification of data, respectively. They classified various crimes at both the local level and the national level. In this research paper, information mining and inspection of crime are explained with experiments performed so that a beginner can understand the accessibility of information mining methods. The authors in

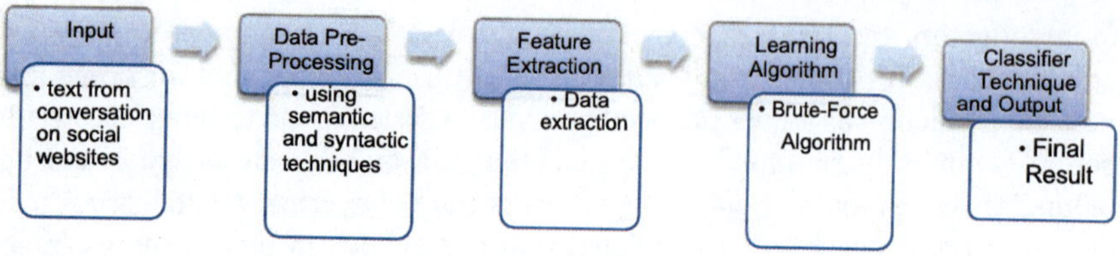

Fig. 1 Systematic architecture of cyber-bullying detection

[7] described the analysis of crime datasets in their research paper using a use-case diagram, activity diagram and various levels of data modeling. They divided the system analysis into functional and non-functional requirements. In this analysis, three algorithms are used such as the autoregressive integrated moving average (ARIMA) model, Bayesian algorithm and artificial neural network (ANN) model of Weka tool for crime analysis. They used k-means for visualization of the output of their project. Sangani et al. [8] emphasize on the importance of data mining techniques, i.e., k-means clustering that can help in analyzing the crime-prone areas. The paper primarily focused on creating a database for criminal's profiles having their complete and detailed criminal records. These intelligent steps can help police officials to detect and prevent any type of crime at any specific and sensitive locations. The main aim of the paper [9] is to present an overview of cyber-bullying on social media like Twitter. The authors proposed a system architecture of detecting cyber-bullying cases using techniques (as shown in Fig. 1).

Bharati and Sarvanaguru [10] clarified how machine learning approaches using Python can help find relationships and patterns among crime datasets. This prediction of crime according to the location can help control crimes. In this paper, the authors also found that the KNN classifier has the highest predictive accuracy and SVM has the least accuracy.

Bharathi and Shilpa [11] discussed the importance of clustering algorithms in the crime domain. In this paper, the authors stated that the narration of crime records having free text is difficult to fetch similar data for the clustering process. The authors also dictated various clustering algorithms to search the best similarity measures that could help the crime branch in better crime prediction. Agarwal et al. [12] primarily focused on crime analysis of crime datasets, i.e., homicide offenses recorded by England and Wales police department (1990 to 2011–2012). In their research paper, crime analysis is done with the help of k-means algorithm using an open-source data mining suite called RapidMiner tool. The authors of the paper [13] primarily focused on some data mining techniques applied during crime prediction. This paper also discussed various research papers based on several classification and clustering methods used in crime analysis. These methods are:

- Text, content and NLP-based method
- Deep learning method
- Crime pattern and evidence-based method
- Spatial and geo-location-based methods

- Prisoner-based methods
- Communication-based methods.

Some research works are mentioned in Table 1.

3 Experiment and Results

There is a need for an open-source data mining tool with the help of which datasets can be implemented and analyzed easily and efficiently. Here, an in-depth analysis is performed on the crime dataset taken from a report file, i.e., in portable document format (PDF) file issued by the National Crime Record Bureau on behalf of the Government of India. The procedure is done with the application of the k-means clustering algorithm in the RapidMiner tool (as shown in Fig. 2).

The following procedures are mentioned below:

1. First of all, extract the required crime dataset.
2. Apply refinement on the dataset as per need.
3. Open RapidMiner tool, read excel file of extracted crime dataset, and execute the operation.
4. Perform the k-means clustering operator and execute the process.
5. Again back to the design page and apply data to the similarity.
6. Apply the performance operator to the data.
7. At last, perform visualization on plot view and get cluster and analysis performed.

3.1 Dataset Used

The crime dataset is taken from a report file, i.e., in PDF format [20] issued by the National Crime Record Bureau (Ministry of Home Affairs) on behalf of the Government of India. The crime rate has been calculated using the actual population census of 2011. The recorded offenses are from the year 2014 to 2018. A sample of the selected crime with their respective dataset is shown in Fig. 3.

3.2 Tool Used

Several open-source tools for data mining are available, e.g., R, Tanagra, Weka, KNIME, Orange and RapidMiner. In this experiment, RapidMiner [21] and Weka both are used in this paper for demonstrating the results of crime-related data.

Table 1 A summary of some research papers along with findings and conclusions of the paper

S. No.	References	Paper title	Findings of the paper	Conclusions
1	[2]	Social Media-Related cybercrimes and techniques for their prevention	The authors discussed various kinds of crime that are strongly connected to social media. They recommended prevention and technique to keep users as far as possible from criminals	The vast amount of big data from social media can be used by law enforcers to fetch criminals and also to prevent crime
2	[14]	Detection of Cyber bullying on social media using data mining techniques	The author applied data mining techniques on Twitter and found that on Twitter cyber-bullying is very high during November–December, 2016	Cyber-bullying via social media can be detected by the application of some useful data mining techniques
3	[15]	Crime data mining for Indian Police Information System	The author presented a model that is intended to extract useful information from the maintained dataset of the National Crime Record Bureau (NCRB). They detect crime using data mining techniques. They used a design model approved by GoI to collect the data, and called and created a database Crime Criminal Information System (CCIS) and an application called Common Integrated Police Application (CIPA) for understanding the work flow of work at primary level under the police system	The proposed tool with the help of a crime analysis tool made it effective for police to respond, detect and prevent crime. The hotspot of crime analysis will give proper geographical information for better results, and this may prove an efficient tool for changing current crime scenarios

(continued)

Table 1 (continued)

S. No.	References	Paper title	Findings of the paper	Conclusions
4	[16]	Crime Analysis and prediction using Data mining	The paper proposed a systematic way to find crime hotspots using several data mining techniques. The authors applied MongoDB for collecting unstructured data, Naive Bayes algorithm for classification, Apriori algorithm for pattern identification, decision tree for predicting crime-prone areas, and visualization is done using a heat map	This research paper concluded that the area that has low security concern is the area of committing crimes like robbery and murder, and the area having any very important person (VIP) appearances causes crime like arson and vandalism. These crimes can be reduced at some extent by night patrolling and fixing burglar alarms
5	[17]	Using Big data Analytics for developing crime predictive model	Crime mapping can be performed in 3 phases—distribution of data geographically that can be implemented using the R tool. On the basis of mapped data, kernel density estimation (KDE) technique estimates clusters for second phase, i.e., cluster analysis. Hadoop platform is required to perform parallel processing on clusters as comparatively faster than traditional data analysis techniques. After the GAMMA test over this phase, the analyzed data is processed by artificial neural network (ANN) using a regression tree for crime pattern prediction	The proposed model helps the investigative authorities to analyze as well as predict crimes. The combination of big data analytics and clustering techniques helps to reduce investigation time and also uncover hidden data

(continued)

Table 1 (continued)

S. No.	References	Paper title	Findings of the paper	Conclusions
6	[18]	Big data analytics: a link between knowledge management capabilities and superior cyber protection	The paper provides evidence that big data analytics enhance cyber-protection with the help of knowledge management. Big data analytics enhances the three processes of knowledge management, i.e., acquisition, conversion and application. It creates a link between knowledge management and business. The research is done using two-dimensional views of cyber-capabilities. They are: • to improve cyber threat detection • to reduce the business impacts of cyber-attacks Steps of research method: • explanatory variables • implication for practices • intermediate outcomes • demonstration of examples	Big data analysis is able to organize and store more data than ever before in this competitive age of business. It made organizations enable knowledge management to better and quickly respond to the changes in global threat
7	[19]	Cyber risk prediction through social media big data analytics and statistical machine learning	In this paper, an algorithm model is designed which is helpful in predicting cyber-risks. Statistical machine learning is used to create algorithms that learn from data and compile a stochastic model to generate predictions and decisions. R software was used to collect the Twitter data, perform data analysis and execute algorithm models. MySQL database is used for vulnerable data storage	The application of the algorithmic model concluded that Apache has the highest cyber-risk. The prediction using algorithms such as Naive Bayes, KNN, support vector machine, decision tree and ANN concluded that the ANN is the most accurate. Cyber-risk monitoring leads to false-negative errors in identification and management when a threat has already occurred. The model used by managers to formulate effective strategies for cyber-risk reduction

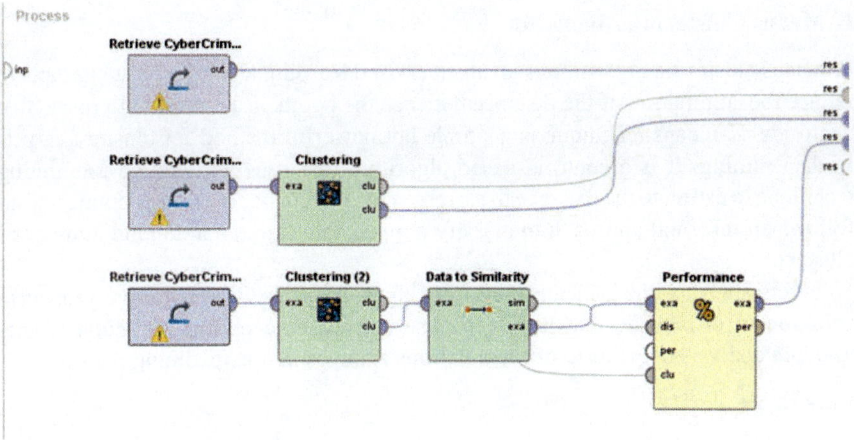

Fig. 2 Procedure for performing k-means clustering algorithm

	A	B	C	D	E	F	G
1	UnionTerr	2014	2015	2016	2017	2018	
2	Andaman	13	6	3	3	7	
3	Chandigar	55	77	26	32	30	
4	Dadar&N	3	0	1	1	0	
5	Daman&D	1	1	0	0	0	
6	Delhi	226	177	98	162	189	
7	Lakshdwe	1	0	0	0	4	
8	Puducher	1	0	2	5	14	
9							
10							
11							
12							

Fig. 3 A snapshot of retrieved dataset

Algorithm and Operators Used

The primary goal of the project is to build a model that can predict cyber-crime based on social networking-related crime records that is more likely to surface some characteristics such as the location of the crime, period of crime and so on. We will also take the help of statistical, graphical and cluster analysis to determine which attributes contribute most.

K-Means Clustering Algorithm

K-means clustering algorithm is an unsupervised learning algorithm that attempts to reduce the summation of the distance between the points in a cluster with respective centroids. K-means technique is a simple but powerful method for cluster analysis in data mining. It is a centroid-based algorithm that uses an iterative fine-tuning technique to estimate the distance to assign points in a cluster. The k-means algorithm follows an informal approach to classify a given data through a definite number of clusters.

In this analysis, the k-means clustering algorithm is used to perform a year-wise variation of crime rate that has been observed via data mining techniques. This in-depth analysis helps us to observe various changes in the upcoming period.

Data to Similarity

The data to the similarity operator calculates the relationship among data of the dataset. It helps to prohibit the same comparison repeatedly. For example, if A is compared to B then B never again compared to A because it results in the same. If there are n items in a dataset, then this operator does not return n^2 similarity comparisons. Instead, it returns $(n)(n - 1)/2$ similarity comparisons. The measure to use for calculating the similarity can be specified through the parameters. Four types of measures are provided: mixed measures, nominal measures, numerical measures and Bregman divergence. We will use mixed measure parameters as measure-type parameters. The available option is the "mixed Euclidean distance" whose range is selected.

3.3 *Result of Graphical Analysis*

In this section, we would look at the results of the graphical analysis conducted which also helped to decide what attribute to include for predictions. Some of the graphs show an attention grabbing pattern in crime, which might not be visible otherwise. As mentioned, both RapidMiner and Weka has been used to analyze the results. Both these tools are used to determine the cluster centroid that can help to locate the distance among clusters.

Figure 4 shows the values of centroid clusters as a table using RapidMiner tool, whereas Fig. 5 shows the same retrieved values of centroid clusters as a table using Weka tool. This similarity among values fetched from the two different data mining tools proves the accuracy of the result. The above-mentioned tools divided the instances into two clusters, i.e., cluster0 and cluster1, where both have same number of instances. For example, cluster0 contains 6 instances and cluster1 contains 1 instance using RapidMiner as well as Weka tool.

Both the figures (Figs. 6 and 7) show the clear visualization of two formed clusters, i.e., cluster0 and cluster1, using RapidMiner tool and Weka tool, respectively. It contains the crime occurrence in seven union territories of India from 2014 to 2018.

Attribute	cluster_0	cluster_1
UnionTerritories	3.833	5
2014.0	12.333	226
2015.0	14	177
2016.0	5.333	98
2017.0	6.833	162
2018.0	9.167	189

Fig. 4 A snapshot of cluster centroid as a table using RapidMiner tool

```
Final cluster centroids:
                                Cluster#
Attribute      Full Data               0            1
                   (7.0)           (6.0)        (1.0)
=====================================================
2014             42.8571         12.3333          226
2015             37.2857              14          177
2016             18.5714          5.3333           98
2017                  29          6.8333          162
2018             34.8571          9.1667          189
```

Fig. 5 A snapshot of cluster centroid as a table using Weka tool

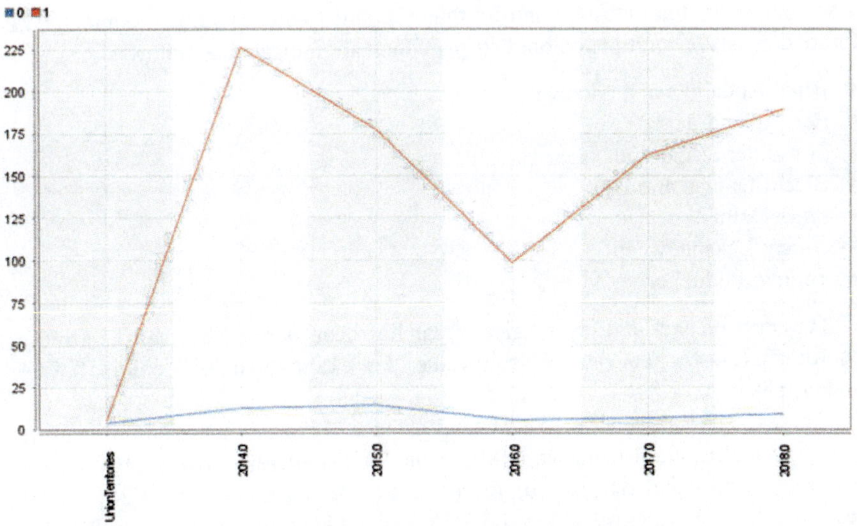

Fig. 6 A snapshot of clusters as a plot view using the RapidMiner tool

Fig. 7 A snapshot of visualization as a plot view using the Weka tool

Figure 6 shows a plot view of centroid using RapidMiner tool that contains two clusters, i.e., cluster0 and cluster1. Cluster0 is indicated by blue line, and cluster1 is indicated by red line where *X*-axis indicates the cluster centroid value and *Y*-axis indicates the selected years for dataset 2014–2018, whereas Fig. 7 shows a visualization of final cluster centroid that contains two clusters, i.e., 0 and 1. This figure consists of multiple colored strips, which are indicated as follows:

- Blue: Andaman and Nicobar
- Red: Chandigarh
- Sky Blue: Dadar and Nagar Haveli
- Black: Daman and Diu
- Pink: Delhi
- Green: Lakshadweep
- Yellow: Puducherry.

The strip of each and every given year has been defined by their centroid as coordinate (cluster name and centroid value). For example, in 2018, pair is (0, 94.5) and (1, 189).

Figures 8 and 9 show collectively the occurrence of cyber-crime in union territories of India with line graph using RapidMiner and Weka tool, respectively. In Weka tool, the graph is sub-labeled as (a), (b), (c), (d) and (e) for each of the years 2014–2018, respectively. The places mentioned on the *X*-axis are in order: Andaman and Nicobar, Chandigarh, Dadar and Nagar Haveli, Daman and Diu, Delhi, Lakshadweep and Puducherry.

Both the figures show the trend of cyber-crime over the years in various union territories of India. On an average, the results shown by both the tools are almost similar. By viewing the plots, it can be observed that the cyber-crime rate in Delhi is on the rise as compared to the UTs: Daman and Diu and Dadar and Nagar Haveli

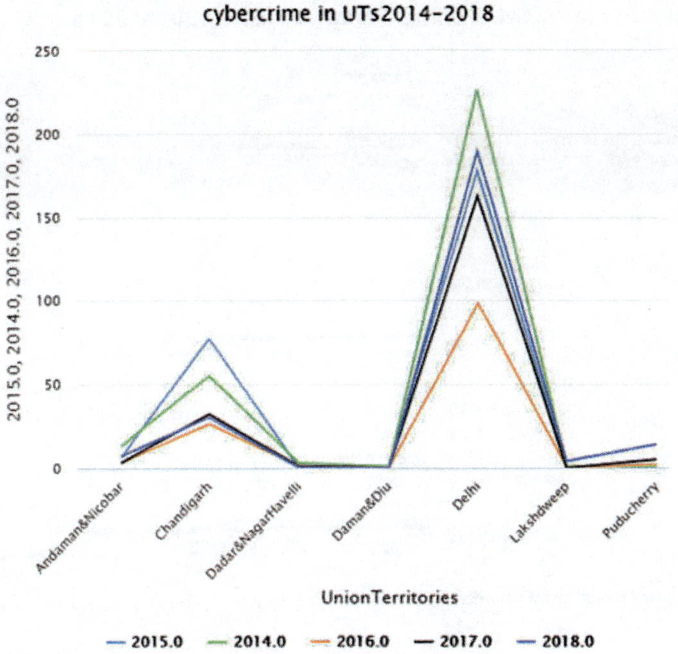

Fig. 8 Rate of cyber-crime per UTs per year using RapidMiner tool

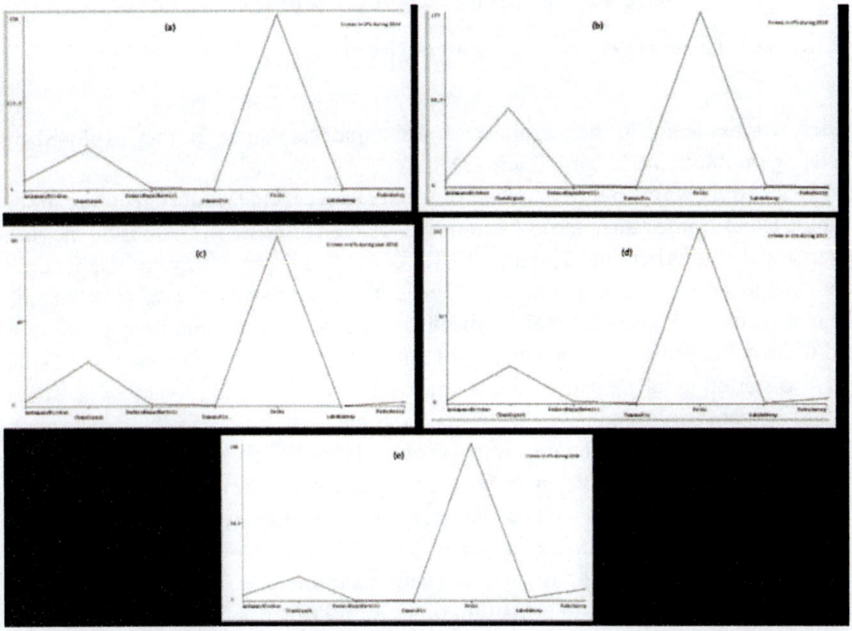

Fig. 9 Cyber-crimes per UTs per year 2014–2018 using Weka tool

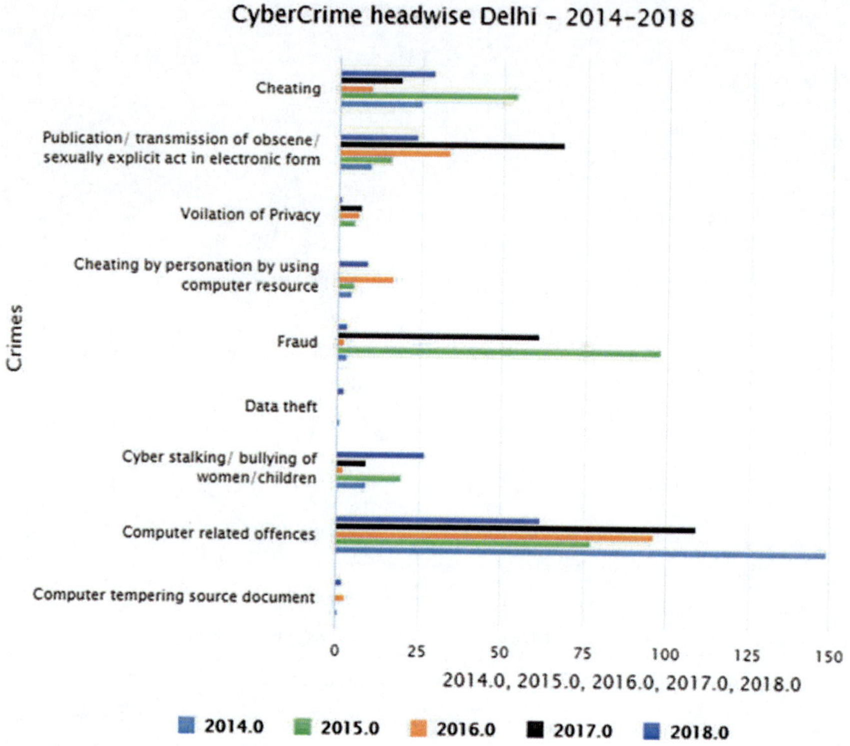

Fig. 10 Rate of crime in Delhi per crime head-wise per year

which witness a fall in their crime rate. This analysis can be further explored by focusing on the crime in particular of Delhi.

Figure 10 shows the trends of cybercrime over the years in Delhi. Here, you can see that the cyber-crime rate in terms of computer-related offenses in Delhi is on the rise, whereas data thefts have increased in 2018 as compared to previous years. Depending upon the previous criminal records of Delhi, computer-related offenses may become the main crime tools for criminals in the upcoming years to commit illegal activities and destroy the privacy and sensitivity of data. The data theft cyber-crime in Delhi has also started to take a path to the destruction of security and privacy of sensitive data. If it is not controlled now, then it may become hard to stop. Delhi may become the center of criminals in the upcoming years to commit illegal activities. Thus, we can see that data mining tools not only help us in analyzing the crime, but also enable us to visualize the results which can be used for combating crime in the future.

Table 2 compares the data mining tools: RapidMiner and Weka, on three parameters. They are accuracy [22], visualization and execution time. Though Weka outperforms the RapidMiner tool in execution time, RapidMiner is far better in two most important features the accuracy and visualization. The accuracy values clearly show the validation performance of RapidMiner over Weka in accurate clustering of

Table 2 Comparison of RapidMiner and the Weka tool

Features	RapidMiner	Weka
Accuracy	0.89	0.81
Visualization	Much better features	Lacks in good visualization of data
Execution time (in sec)	1 s	0.02 s

data into two clusters. Also, as can be seen from the figures above, RapidMiner provides effective visualization of data as compared to Weka as feature histograms are available in RapidMiner.

In this section, we compared our results using RapidMiner and Weka tool. Though both the tools enable us to analyze and provide a visualization of data, yet more effective analytical model needs to be built. The next section proposes an analytical model which shall help in predicting the crime.

4 Proposed Analytical Model

Data mining in crime analysis helped the police department and investigation department to solve many criminal cases. There exists some limitation in solving crime cases due to its less connectivity of data. As we know, the Internet via social media networking has stored a huge amount of data which resulted in big data. The analysis of big data can give much useful information related to criminals and their behavior. The terrorists try to contact innocent people through social media to divert and control their minds for their destructive purpose. With the use of big data analytics and crime data mining techniques such as clustering, classification and aggregation, investigators can stop these social media dedicated crimes. After applying the k-means clustering algorithm with neural networking in crime detection, the police department can make its searching, solving and prediction process easier and effective.

Big data fetched from social media networking such as Facebook and Twitter can become a fine source of evidence. The big data analytics applied to those social big data give a wide overview and statistics regarding crime-related chats, shared media, location, period and thoughts. It can predict much about the activities that criminals and terrorists are planning to do. The application of learning models such as neural networks can make predictions and detection automatically based on both previous and current data. The combination of big data analytics and neural networks can analyze the character or nature of any person. If this proposed system gets connected with the reporting system of social media so if anyone reports a crime or offense, then it can track the issue and also verify it with evidence. On this evidence, the police or investigation department can take action that will be fast and easy to use. For example, if someone is posting an aggressive or violent content on a social media network, then it selects those profiles and will track their behavior and nature. If they are doing so with any dangerous motive, then it will mark them as a threat. It can also assist in tracking and controlling some specific movements of ex-criminals on social media networking so that they will not become a threat to anyone. It can also track how well they are adapting to the society through their social media movements. With the help of our proposed system, a person can also verify a stranger on social media, whether it is safe or not to connect.

5 Conclusion and Future Work

The appropriate collection and intelligent use of data are very significant for the resolution of cyber-crime-related consequences such as cyber-bullying, computer resource tampering and fraud. This paper compares the crime data with respect to most commonly used data mining tools: RapidMiner and Weka. The results clearly show that RapidMiner outperforms the Weka tool in terms of accuracy and visualization though execution time using Weka is less than RapidMiner tool. Also, RapidMiner actually provides significantly more data preprocessing functions and operators than Weka. The integrated use of advanced data mining technology in coordination with crime-dedicated social media opens a new way of exploring crime-related data through data clustering, classification and aggregation. As a part of the future work, the model formed in this project can be trained with big data analytics and neural networks to determine the upcoming incidents based on cloud-based big data. It would be interesting to see the proposed crime detecting model working globally with the help of learning algorithms. This may help law enforcement officials and investigators to detect, solve and predict crimes more easily and effectively.

References

1. Ishikawa, H.: Social Big Data Mining, 1st edn. Taylor and Francis group, CRC Press, Boca Raton (2015)
2. Soomro, T., Hussain, M.: Social media-related cybercrimes and techniques for their prevention. Appl. Comput. Syst. **24**(1), 9–17 (2019)
3. Deepika, K.K., Vinod, S.: Crime analysis in India using data mining techniques. Int. J. Eng. Technol. **7**(2.6), 253–258 (2018). LNCS Homepage, https://www.springer.com/lncs. Last accessed 21 Nov 2016
4. Zubi, Z.S., Mahmmud, A.A.: Crime data analysis using data mining techniques to improve crimes prevention. Int. J. Comput. **8**, 39–45 (2014)
5. Tyagi, D., Sharma, D.S.: An approach to crime data analysis: a systematic review. Int. J. Eng. Technol. Manage. Res. **5**(2), 67–74, Communication, Integrated Networks & Signal Processing (2018)
6. Bhargava, R., Singh, P., Sangwa, R.S.: Analysis of crime data using data mining algorithm. Int. J. Eng. Sci. Res. Technol. **7**(2), 675–681 (2018)
7. Pande, V., Samant, V., Nair, S.: Crime detection using data mining. Int. J. Eng. Sci. Res. Technol. **5**(1), 891–896 (2016)

8. Sangani, A., Sampat, C., Pinjarkar, V.: Crime prediction and analysis. In: 2nd International Conference on Advances in Science & Technology 2019, Mumbai (2019)
9. Ravi, B.N., Kamthe, C.R., Shravan, A.M., Mahesh, K.A.: Cyber bullying detection and prevention for social media using data mining. Int. J. Sci. Eng. Manage. **3**(4), 281–283 (2018)
10. Bharati, A., Sarvanaguru, K.: Crime prediction and analysis using machine learning. Int. Res. J. Eng. Technol. **5**(9), 1037–1042 (2018)
11. Bharathi, A., Shilpa, R.: A survey on crime data analysis of data mining using clustering techniques. Int. J. Adv. Res. Comput. Sci. Manage. Stud. **2**(8), 09–13 (2014)
12. Agarwal, J., Nagpal, R., Sehgal, R.: Crime analysis using K-means clustering. Int. J. Comput. Appl. **83**(4), 1–4 (2013)
13. Sreedevi, M., Reddy, A., Reddy, C.: Review on crime analysis and prediction using data mining techniques. Int. J. Innov. Res. Sci. Eng. Technol. **7**(4), 3360–3369 (2018)
14. Riadi, I.: Hariani: Detection of Cyberbullying On Social Media Using Data Mining Techniques. International Journal of Computer Science and Information Security **15**(3), 244–250 (2017)
15. Gupta M, Chandra B, Gupta MP. Crime data mining for Indian police information system (2006).
16. Sathyadevan, S., M.S, Devan., S. Gangadharan, S.: Crime Analysis and prediction using data mining. ResearchGate (2014).
17. Chauhan,T., Aluvalu,R.: Using Big Data Analytics for developing Crime Predictive Model. In: First International Conference on Research & Entrepreneurship, vol. 1, pp. 1–6.Gujarat, India (2016).
18. Obitade, P.O.: Big data analytics: a link between knowledge management capabilities and superior cyber protection. Springer J. Big Data **6**(71), 1–28 (2019)
19. Subroto, A., Apriyana, A.: Cyber risk prediction through social media bigdata analytics and statistical machine learning. Springer J. Big Data **6**(50), 1–19 (2019)
20. National Crime Records Bureau homepage, https://ncrb.gov.in. Last accessed 24 Feb 2020
21. RapidMiner Homepage, https://en.wikipedia.org/wiki/RapidMiner. Last accessed 24 Feb 2020
22. Huang, Z.: Extensions to the k-Means algorithm for clustering large data sets with categorical values. Data Min. Knowl. Disc. **2**, 283–304 (1998)

IoT Based RGB LED Information Display System

T. Kavya Sree, V. Swetha, M. Sugadev, and T. Ravi

Abstract The Internet of Things (IoT) is an incompatible technology used in various applications. IoT is an interconnected network. Using IoT, we can transfer data without human-to-human interaction. In the present era, in every field, we are using IoT. Normally for the dot matrix display, we are using a single-color LED light or RGB LEDs. It consumes more power, occupies more space, and is heavyweight and not so flexible to program from user perspective. Using RGB LEDs, we can produce different colors by adjusting the intensity of light using the pulse width modulation (PWM) signal. In this paper, the design, implementation, and cost optimization of an IoT controlled RGB display system is presented. In this work, we are using the W60-75 RGB driver board to program the LEDs for the display system using IoT connectivity using Bluetooth and Wi-Fi protocols. This driver board comprises of built-in modules of Wi-Fi modem, U-disk port, and supports GUI based program tool. So instead of a complex programming procedure, it can also be operated by a simpler mobile app called LED-Art for Android operating system. This display system can be used to exhibit any information in transportation, colleges, and in many other places.

Keywords RGB LED · P10 LED module · W60-75 driver board · IoT · LED-Art · HD-2016 software

1 Introduction

Internet of Things (IoT) has gained momentum in all aspects of engineering with the adaptation of Industry 4.0 [1–3]. It is the underlying technology for providing connectivity to all physical things in a cyber-physical environment. Several standards have been defined for IoT devices to meet the challenges in data rate, radio-frequency spectrum, and security [4–6]. Nowadays, LED display systems are very popular to display any information. The LED display systems are mainly preferred because of its better resolution, wider viewing angle, portability, and relatively lower power

T. K. Sree · V. Swetha · M. Sugadev (✉) · T. Ravi
ECE Department, Sathyabama Institute of Science and Technology, Chennai, India

H. Sharma et al. (eds.), *Congress on Intelligent Systems*, Advances in Intelligent Systems and Computing 1334, https://doi.org/10.1007/978-981-33-6981-8_34

consumption. These displays are mostly used in public places like shopping malls, schools, colleges, public transport, railway stations, theatres, etc. The drawback of RGB display system is that it is complex to program and difficulty to alter the messages being displayed by the end user or consumers. It is required for the display controller unit to be connected to a local host computer for debugging or updating the programs [7–10].

Existing systems requires physical access to the communication port of the display unit to alter the display message. The proposed work aims to employ IoT technology in standard RGB display systems to make programming and customizing display content more easier by wireless protocols such as IEEE 802.11 and Bluetooth [6, 11–13].

To make the LED display more portable, the W60-75 driver board is used as the controller circuit. Instead of writing complex programs and the use of host computer for sending a message to the display system, information can be sent through smartphones. Since the W60-75 microcontroller board is integrated with the Wi-Fi module, Bluetooth and U-disk port interface to facilitate easy connectivity with P10 LED modules. In addition, it supports GUI based program tool, which reduces the complexity of writing programs. Due to the availability of GUI based program tool for this particular driver board, it is convenient to create the display information. W60-75 based RGB display makes the system more flexible to display flash news or important notices faster than a wired programmable system.

2 Related Work

Pooja Pawar et al. (2019) have proposed a work on IoT based digital display system using Arduino ATMegaa328. In this paper, they had proposed the work on designing the LED display systems for displaying day-to-day information based on the GSM technology. GSM mobile is used to send the message. A text message can be typed and sent through the GSM-based mobile phone using SMS service. A GSM module connected to the Arduino board is used as the receiver to get the message from the sender and forwards to the Arduino controller unit to display information in the display system [14].

Kunal Verma et al. (2018) have presented an article on the Internet of Things (IoT)-based LED using ESP32. In this, ESP32 is used a microcontroller. Arduino IDE software is used to program the microcontroller. The IP address of the ESP32 is connected to the mobile device. Using ESP32 Wi-Fi in mobile device, we can control the voltage and the brightness of an LED [15].

Amirgaliyev Yedilkhan et al. (2013) have presented a paper on the design of an information display based on several LED matrices and a single microcontroller. 8×8 dot matrix display is constructed using 64 LEDs. LED can be blinked by applying the low signal to the corresponding column terminal and high signal to the corresponding column terminal. This device can be able to display and scroll the text that is provided in the program [16].

Bhawna Saini et al. (2014) have presented a paper on Smart LED display boards. This paper has proposed a work on the smart LED display board based on the GSM module using the AT89S51 microcontroller. The message can be sent through the GSM-based mobile phone. It receives the SMS, validates the sending mobile identification number (MIN), and then displays the information in the display system. Moreover, by using a GSM-based system, these display systems are more flexible to display day-to-day information in schools and colleges at regular interval of time [17].

Sayidul Morsalin et al. presented a work on wireless notice board with improved security features using NodeMCU development board. The display system employs password-based authentication for access to programming or reconfiguration of the unit [18].

A survey on related work reveals that most of the mentioned display systems use Arduino, AT89C51, Raspberry Pi microcontroller boards, PIC16F877A, ESP-8266EX with GSM or Wi-Fi wireless technology or Bluetooth for communication between user and the microcontroller platforms and a high-level programming language like embedded C or Arduino IDE [18, 19].

A common user may not feel convenient with these high-level programming languages and face difficulties for reprogramming of the system. Therefore, a graphical method or GUI based programming is utilized in the proposed work.

3 Proposed Work

The proposed IoT based display system tries to minimize the overheads in program debugging by using a graphical programming tool. In this work, W60-75, an ARM-processor-based development board is chosen as the hardware platform of the display system. P10 RGB LED module is used to build the display panel, which displays the text or images in multicolor. A text message can be typed in LED-Art Android application with different colors, font style, and font size. And, we can also choose the animation effects to display and then send the message to display system from smart phones. Through Wi-Fi, message is delivered to the controller and gets displayed in the LED panel. This system is less cost, robust, and easy to maintain in day-to-day life. Figure 1 shows the block-level representation of the proposed display system.

3.1 Hardware Configuration

In the proposed prototype, W60-75 development board exclusively made for LED display systems is selected because it consists of 32-bit ARM cortex-M3 microcontroller with low programming. It has inbuilt Wi-Fi; the hardware configuration built with ESP-8266 is shown in the block diagram as shown in Fig. 1. The board can be programmed with either LED-Art mobile app or HD2016 programming language

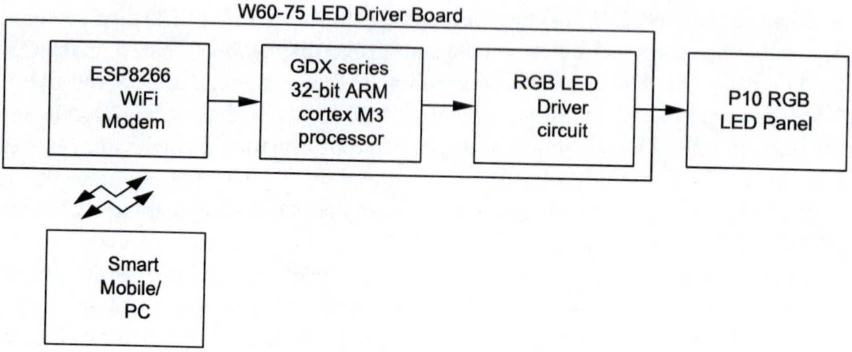

Fig. 1 Block-level representation of proposed RGB display system

based tool. The RGB LED module includes four P10 (128 × 64) display system (each RGB LED 16 × 32). The power supply is connected to all the four RGB LED module and LED driver board which is rated at 5 V, 2A.

3.2 *Power Supply*

In the proposed system, off-the shelf switched mode power supply module N200V5, 5 V power supply is used which is also called G-energy N200V5-A. It is mostly used for LED display system, and it converts AC to DC with the input voltage which is 220–240 V to output voltage which is 5 V. This power supply can be protected during over load, under voltage, and short circuit. It also features built-in air cooling, to bear more heat. Figure 2 shows the power supply module N200V5.

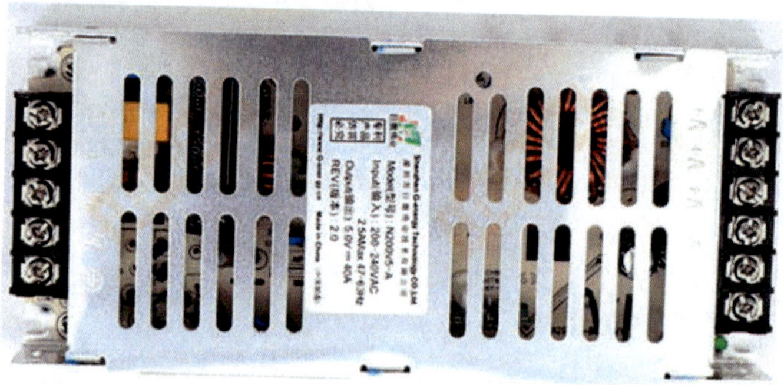

Fig. 2 N200V5-A power supply

3.3 Microcontroller Development Board

W60-75 is used as controller for the display system, and it is the most widely used professional hardware platform for medium resolution LED display systems. W60-75 driver board consists of 32-bit ARM cortex-M3 processor, ESP8266EX Wi-Fi modem, antenna switches, and sensors for temperature and humidity. It also has an inbuilt real-time clock and supports GUI based programming tool. It is operated from a 4.2 to 5.5 V and has 8 M Bytes of memory capacity. It supports different types of display module such as single, dual and tri colors. It consists of two types of communication interface namely U-disk (USB) and Wi-Fi. It has 2HUB75E bus which connects the display to the board.

W60-75 can also be programmed by using embedded C- compiler called HD2016. So for Wi-Fi based programming, LED-Art smart mobile can be utilized, and for USB based programming, HD2016 may be used. Figure 3 shows the W 60-75 development board and its interfaces.

ARM Cortex-M3: The Cortex-M3 processor is aGD32F10x series 32-bit processor that features low interrupt latency and low cost debugger. It is based on the ARMv7 architecture and supports a powerful and scalable instruction set including general data processing I/O control tasks and advanced data processing bit field manipulation. It includes three AHB buses known as I-code, D-code, and system buses and has up to 4 GB memory space. It operates from a 2.6 to 3.6 V power supply.

ESP8266EX: It contains different functional units or modules and communication protocol architectures in a single circuit board such as Wi-Fi module, Bluetooth module, power amplifiers (Radio-frequency), low noise amplifier (LAN), TR switch (Duplexer), WPA (Wi-Fi protected access), serial peripheral interface (SPI), I2C, UART, 10-bit ADC, filters, and power management module. It achieves low power consumption with a combination of proprietary technologies, and we are using Wi-Fi

Fig. 3 W60-75 LED driver board

Fig. 4 ESP8266EX

module in ESP8266EX which delivers highly integrated Wi-Fi SoC, compact design and reliable performance in the Internet of Things industry [4]. Figure 4 shows pin layout of ESP8266EX Wi-Fi modem chip with a SMT package.

3.4 RGB Led

It is a multicolor LED, consisting of three LEDs that can emit red, green, or blue color in a single LED device package. To produce different colors, we need to adjust the intensity of current applied to respective color LED, and resultant color of the output light depends upon the linear combination of intensity of each internal LEDs [12]. So the output of a RGB light is a combination of three primary colors red, green, and blue. By varying intensity of light of each LED using PWM, we can produce 16 million hues in a single RGB LED [13]. Since LEDs are very close to each other, our eyes can see the result of the combination of colors rather than the three colors individually. An RGB LED will have four leads, red, common anode or common cathode, green and blue, respectively [20, 21].

3.5 RGB LED Module

The display panel is built with multiple P10 RGB LED modules with 16×32 pixel size or resolution. These types of LED panels are suitable for indoor or outdoor display systems [2]. In this work, we used four P10 RGB LED modules which are connected in such a way that two LED panels are in each row. The first LED module's output pins (16) are connected to second module input pins (16), such that in the second row is also connected in same way. P10 module has Molex-style header for connecting with the DC power supply of 5 V. Figure 5 shows front and rear panel views of the P10 RGB LED panel.

Fig. 5 **a** Front side of the P10 module, **b** Rear side of the P0 module

3.6 *Software Configuration*

The W60-75 LED driver board is controlled by software called HD2016, and a mobile application called LED-Art. HD2016, shown in Fig. 6, is a program for setting up and managing LED display control cards. The control systems are quite simple, support plenty of operations, have strong function libraries, and are easy to learn. Using this application, we can precisely control the contents of each display pixel-by-pixel. Text and animations can be programmed without any coding, simply drag and drop text-bar and customize the effect, background color, speed of animation, etc. It can be communicated through a serial port or Ethernet port or U-disk.

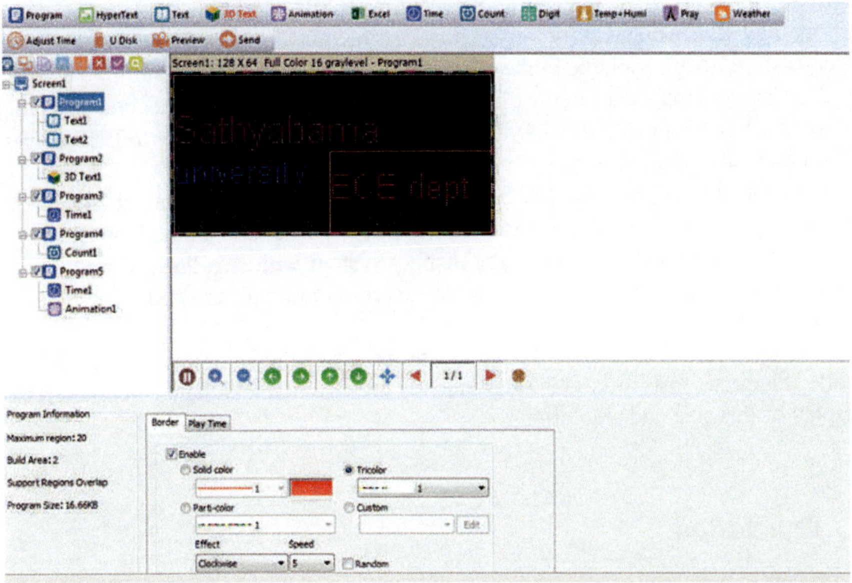

Fig. 6 HD 2016 software screenshot

LED-Art is a software application based on Android operating systems and can be easily installed in smart mobile phones or tablets or any Android OS supporting portable devices. The display system can easily set for single-color or full-color LED display by sending commands from the mobile device through Wi-Fi connectivity. It contains basic setting option like time, brightness, switch, and edit program. In a single screen, we can display different items like text, clock, image, animations, etc., it has a play time, and the program can be displayed in a specified date, time, and day. For each item, there will be an editor section, and the given data can the edited with different effects such as font style, font color, alignment, font size, border effects, etc. Through the mobile phone, the brightness level of the display panel can also be controlled, and we can even switch-off or switch-on the display screen through this mobile app. It is also possible to feed in additional parameters such as data embedded in the image, priority mode, clock frequency, and scan parameters. The screen can be updated as many times as required through the app. Figure 7 shows the screen shot of the LED-Art app installed on a mobile device.

4 Results and Discussion

In this work, an Internet of Things based RGB display system is implemented and successfully tested. Built-in features of W60-75 RGB ARM processor development board make things easier to develop the display system. Programming is done through USB mode in HD2016 as well and LED-Art mobile app with ease and without any complexity. Connectivity between the nodes may also be implemented with Bluetooth protocol available in the controller board.

The above-simplified circuit diagram in Fig. 8 shows the connection between W60-75 LED driver to P10 LED display module and power supply to LED driver board and P10 LED module.

Figure 9 shows completed IoT based RGB display unit with color text characters in the display area with different font style without border and animation effects.

Figure 10 shows the output in the display system with two lines of multicolor border effects which rotates in clockwise direction and with the text characters in the display area.

Table 1 shows the list of components and prices of the components used in this work. It approx. costs around 5000/- which is pretty much less than the commercial display of this size and resolution.

5 Conclusion

In this paper, we proposed a low-cost technique in the design and implementation of a RGB display system with GUI based programming feature. In the proposed work, we used the P10 module and the W60-75 RGB driver which has integrated

Fig. 7 LED-Art application

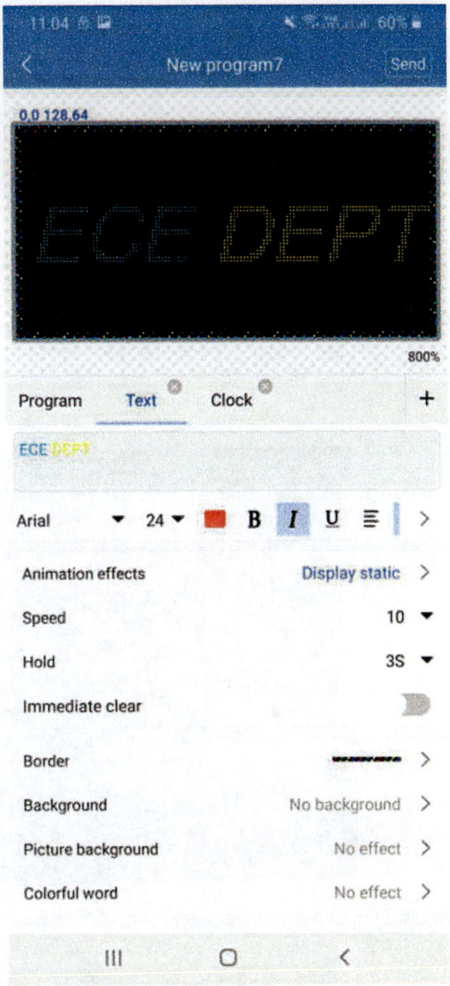

Wi-Fi and Bluetooth modules, and it also has a GUI program tool. So, it reduces the complexity of huge connections and programs, and it also reduces the cost. By establishing Wi-Fi connectivity or Bluetooth connectivity between the W60-75 and smart phone, type the text message or select image which needs to be displayed and select send option in the app. The display unit can be controlled from the mobile phone from a maximum distance of 5 m.

Fig. 8 Hardware connection

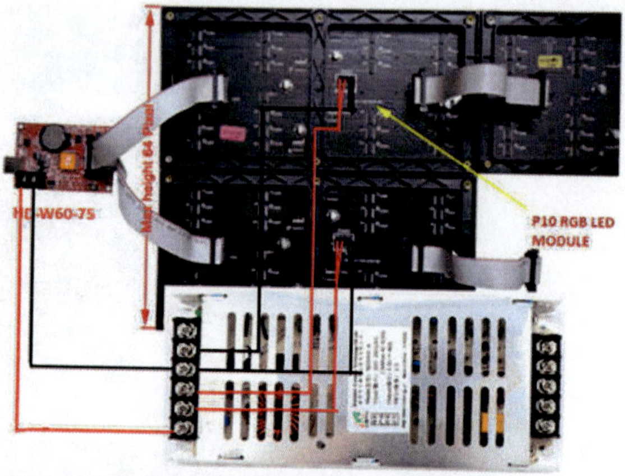

Fig. 9 Output in the display system 1

Fig. 10 Output in the display system 2

Table 1 List of components and prices

Component name	Price
P10 RGB LED module	3020
Module cabinet	500
W60-75 driver board	810
N200V5-A power supply	450
Ribbon cable and wires	200

References

1. Reddy, G.T., Kaluri, R., Reddy, P.K., Lakshmanna, K., Koppu, S., Rajput, D.S,.: A novel approach for home surveillance system using IoT adaptive security. In: International Conference on Sustainable Computing in Science, Technology and Management (SUSCOM) (2019)
2. Maddikunta, P.K.R., Gadekallu, T.R., Kaluri, R., Srivastava, G., Parizi, R.M., Khan, M.S.: Green communication in IoT networks using a hybrid optimization algorithm. Comput. Commun. (2020). https://doi.org/10.1016/j.comcom.2020.05.020
3. Zhou, S.,Wang, J., Wang, F.: A design of outdoor display monitoring system based on NB-IoT. In: International Conference on Information Systems and Computer Aided Education (ICISCAE) (2019). https://doi.org/10.1109/ICISCAE48440.2019.221597
4. Kumar, M.E., Reddy, G.T., Sudheer, K., Reddy, M.P.K., Kaluri, R., Rajput, D.S., Lakshmanna, K.: Vehicle theft identification and intimation using gsm & iot. In IOP Conference Series: Materials Science and Engineering, vol. 263, p. 42062 (2017)
5. Wang, Y.P., Lin, X., Adhikary, A.: A primer on 3GPP narrowband internet of things. IEEE Commun. Mag. **55**(3), 117–123 (2017)
6. Cao, X., Li, Y.: Data collection and network architecture analysis in internet of vehicles based on NB-IoT. In: 2018 International Conference on Intelligent Transportation Big Data & Smart City (ICITBS), pp. 157–160 (2018)
7. Pramanik, A., Rishikesh, Nagar, V., Dwivedi, S., Choudhury, B.: GSM based smart home and digital notice board. In: International Conference on Computational Techniques in Information and Communication Technology (ICCTICT), pp. 340–345 (2016)
8. Rangani, D.G., Tahilramani, N.V.: Smart notice board. In: International Conference on Applied and Theoretical Computing and Communication Technology, pp. 209–214 (2017)
9. Chaurasi, D.M., Sahu, G., VijayaBaskar, V.: A smart informant system for institutional transport system. In: IEEE 2nd International Conference on Innovations in Information Embedded and Communication Systems (ICIIECS), pp. 304–310 (2015)
10. Simha, K., Shreya, Kumar, C., Parinitha, C., Tantry, S.: Electronic notice board with multiple output display. In: International Conference on Signal Processing, Communication, Power and Embedded System (SCOPES), pp. 1558–1561 (2016)
11. Tsai, M.F., Ye, C.K., Sinn, A.: Smart service relying on internet of things technology in parking systems. J. Supercomput. 1–24 (2016)
12. Khera, N., Shukla, D., Avasthi, S.: Development of simple and low-cost android based wireless notice board. In: International Conference on Reliability, Infocom, Technologies and Optimization (ICRITO), pp. 630–633 (2016)
13. Ketkar, P.U., Tayade, K.P., Kulkarni, A.P., Tugnayat, R.M.: GSM mobile phone based LED scrolling message display system. Int. J. Sci. Eng. Technol. **2**(3), 149–155 (2013)
14. Pawar, P., Langade, S., Bandgar, M.: A paper on IoT based digital notice board using arduino ATMega 328. Int. Res. J. Eng. Technol. (IRJET) **6**(3), 7509–7513 (2019)
15. Verma, K.: Internet of Things (IoT) LED Lamp. Electronics for You (2018)
16. Yedilkhan, A., Rassim, S., Andrey, B., Ari, N.: Design of an international display based on several LED matrices and a single microcontroller. In: International Conference on Application of Information and Communication Technologies, pp. 2045–2048 (2013)

17. Saini, B., Devi, R., Dhankhar, S., Haque, M.Z., Kaur, J.: Smart LED display board. Int. J. Electron. Electr. Eng. **7**(10), 1057–1067 (2014)
18. Morsalin, S., Rahman, A., Siddiqe, M.A.B., Saha, P., Halim, M.R.: Password protected multiuser wireless electonic noticing system by GSM with robust algorithm. In: Proceedings of International Conference on Electrical Information and Communication Technology (EICT 2015), pp. 249–253 (2015)
19. Teckchandnri, Y., Perumal, G.S., Mujumdar, R., Lokanathan, S.: Large screen wireless notice display board. In: IEEE International Conference on Computational Intelligence and Computing Research, pp. 709–713 (2015)
20. https://randomnerdtutorials.com/electronics-basics-how-do-rgb-leds-work/
21. https://www.circuitbread.com/tutorials/how-rgb-leds-work-and-how-to-control-color

Determination of Breakdown Voltage for Transformer Oil Testing Using ANN

V. Srividhya, Jada Satish Babu, K. Sujatha, J. Veerendrakumar, M. Aruna, Shaik Shafiya, SaiKrishna, and M. Anand

Abstract Power transformer is an important apparatus in power system. Insulation used in transformer is of solid insulation system and liquid dielectrics. It is important to investigate the cause of the insulation degradation with respect to aging affects. The standalone parameters of transformer oil are analyzed for various breakdown voltages. The absence of a well-defined scheme to correlate the various parameters of transformer oil reinforces the need of this transformer oil testing studies. In this work, a change in the breakdown voltage of insulating oil was observed, and its interrelationship was determined using feed forward neural network (FFNN) trained with quasi-Newton and conjugate gradient learning methods. An expert system is created using FFNN implemented using MATLAB software to accurately determine the sensitivity and also the fault in system due to aging of transformer oil.

Keywords Transformer oil testing · Breakdown voltage · Quasi-Newton algorithm · Conjugate gradient algorithm

V. Srividhya · M. Aruna
Department of EEE, Meenakshi College of Engineering, Chennai, India

J. S. Babu
Department of EEE, REVA University, Bangalore, India

K. Sujatha (✉) · J. Veerendrakumar · M. Anand
Department of EEE, Dr. MGR Educational and Research Institute, Chennai, India

S. Shafiya
Department of ECE, Nalla Narasimha Reddy Education Society, Majarguda, India

SaiKrishna
Department of ECE, Sree Rama Institute of Technology and Science, Khammam, India

H. Sharma et al. (eds.), *Congress on Intelligent Systems*, Advances in Intelligent Systems and Computing 1334, https://doi.org/10.1007/978-981-33-6981-8_35

1 Introduction

Dielectric strength of the insulating oil in power transformers depend on parameters like pressure, temperature, humidity, configuration of the electrode, and characteristics of applied voltage [1–3]. A suitable dielectric material for power transformers can be identified if the breakdown strength analysis is effectively carried out as it would be helpful for high-voltage-related applications. The test kit for breakdown voltage (BDV) comprises of two electrodes with a gap length of 6 mm fixed on the horizontal axis kept in a concealed glass chamber as shown in Fig. 1. Sphere–sphere type of electrodes is used in transformer oil testing kit whose maximum gap length is 9 mm and diameter of the sphere electrodes is 10 mm [4–6].

2 Existing Methods

The existing transformer oil testing apparatus in laboratory is provided with normal oil test container with a translucent envelop and power driven by a step-up transformer whose maximum breakdown voltage value is 60 kV. The test is conducted as per the IS6792 standards, and readings are given in Tables 1, 2, and 3 corresponding to three

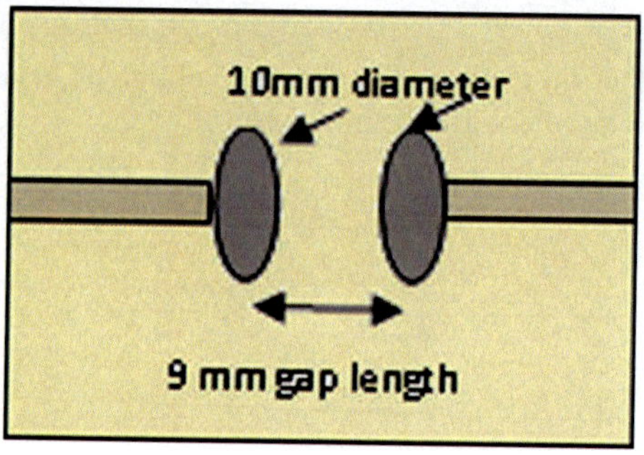

Fig. 1 Electrode configuration in transformer oil testing apparatus

Table 1 BDV analysis for pure heavy oil

Gap length (mm)	Normalized value	Breakdown voltage (KV)	Normalized value
1	0.1	60	0.6
2	0.2	50	0.5
3	0.3	40	0.4
4	0.4	30	0.3
5	0.5	20	0.2
6	0.6	10	0.1

Table 2 BDV analysis for moderately contaminated heavy oil

Gap length (mm)	Normalized value	Breakdown voltage (KV)	Normalized value
2	0.2	60	0.6
3	0.3	50	0.5
4	0.4	40	0.4
5	0.5	30	0.3
6	0.6	20	0.2
7	0.7	10	0.1

Table 3 BDV analysis for highly contaminated heavy oil

Gap length (mm)	Normalized value	Breakdown voltage (KV)	Normalized value
4	0.4	60	0.6
5	0.5	50	0.5
6	0.6	40	0.4
7	0.7	30	0.3
8	0.8	20	0.2
9	0.9	10	0.1

different categories of heavy oil used in the laboratory scale setup [7]. Heavy oil is taken as a standard sample for which the different BDVs are recorded for variation in gap length. Heavy oil used indicates that the oil sample taken for test has fulfilled the test standards under normal working conditions. Initially, a pure quality of heavy oil free from contamination is used for BDV analysis. Thereafter, heavy oil which is contaminated to a medium level referred to as moderate degradation level and the one which is highly contaminated referred to as high degradation level is used for BDV analysis which helps to determine the dielectric strength of the oil. The heavy oil which is over burnt is said to be highly contaminated with pollutants causing internal winding fault.

3 Materials and Methods

Back propagation algorithm includes a group of learning rules to effectively train artificial neural networks (ANNs) following a gradient descent and quasi-Newton approach. The main characteristic of back propagation algorithm is that it is iterative and recursive method for modifying the weights and to improve the performance of the network till the network reaches convergence [8, 9]. Back propagation algorithm uses the first-order derivative of the activation function. Automatic differentiation is a technique that can automatically and analytically provide the derivatives to the training algorithm. In the context of learning, back propagation is commonly used by the gradient descent optimization algorithm to adjust the weight of neurons by calculating the gradient of the loss function. Newton's method often converges faster than conjugate gradient methods. Unfortunately, it is complex and expensive to compute the Hessian matrix for feed forward neural networks. There is a class of algorithms that is based on Newton's method, but which does not require calculation of second derivatives. This is called as quasi-Newton method. The weights update is based on approximate Hessian matrix after every iteration is completed. This update is computed as a function of the gradient.

4 Results and Discussion

The dielectric strength of the transformer oil needs to be estimated to have a proper insulation. This estimation can be done using ANN trained with quasi-Newton and gradient decent algorithm. Figure 2 shows the ANN based training using quasi-Newton algorithm. Tests were carried out on nine samples to investigate various critical characteristics. It is inferred that the quality of the insulating oil used degrades due to contamination of moisture and carbonization. Hence, it is inferred that on repeated usage of the same heavy oil, its viscosity increases, and the BDV value decreases. The decrease in BDV reduces the dielectric strength of the oil. As a result, the oil loses its insulation capacity leading to overheating of transformers. Regression analysis implies that BDV is inversely proportional to the gap length due to repeated use of insulating oil. Thus, the expert system developed for analyzing insulation efficiency helps the operator to have an easy solution about the dielectric strength of the transformer (Figs. 3, 4, 5, 6, 7, 8, and 9).

5 Conclusion

The testing on the transformer oil was carried out, and the results are interpreted using ANN. The ANN trained with quasi-Newton and gradient decent algorithm was found to have effectual monitoring of the power transformers and can give path to find the life time of insulation. This strategy has been designed to integrate trials for detecting faults and unnatural outages in power transformers. It can be accomplished that prognostic monitoring approach will facilitate an expansion in the life time of power transformers.

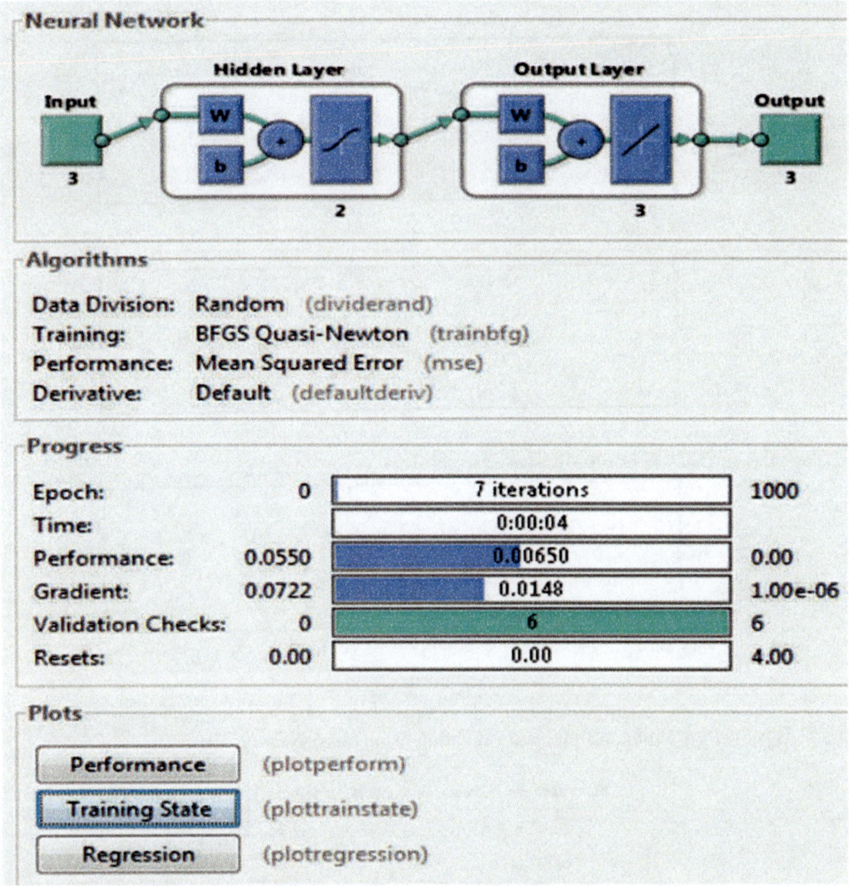

Fig. 2 ANN browser in MATLAB for quasi-Newton algorithm

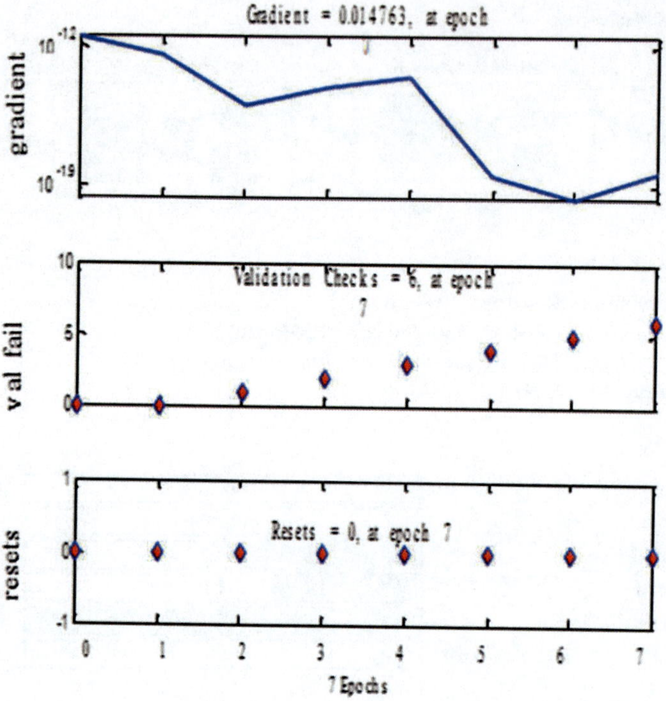

Fig. 3 Results for training, testing, and validation using quasi-Newton algorithm

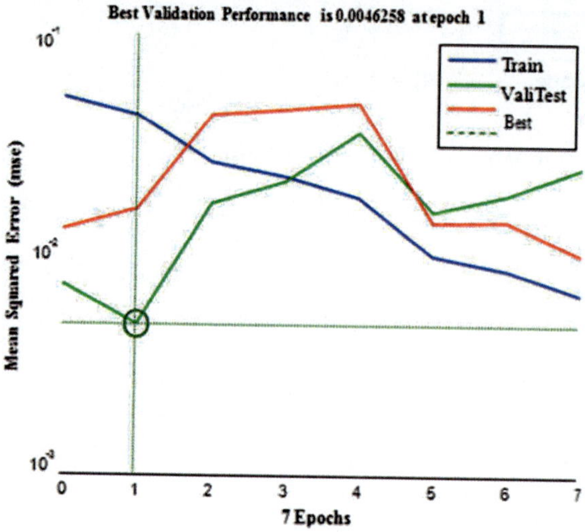

Fig. 4 Performance measure for training, testing, and validation using quasi-Newton Algorithm

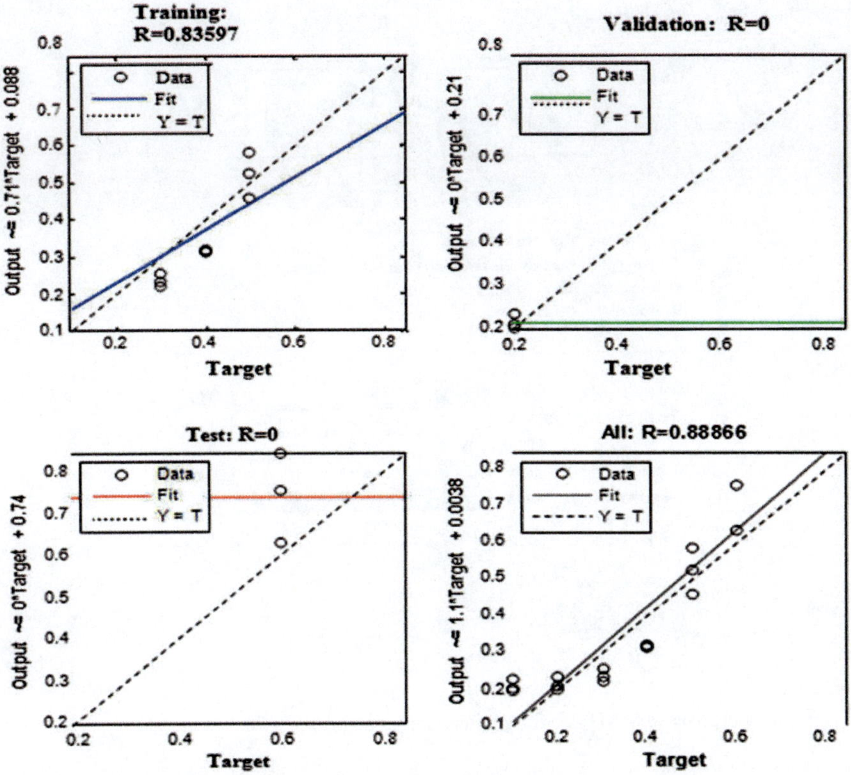

Fig. 5 Variation in actual and target output during training, testing, and validation using quasi-Newton Algorithm

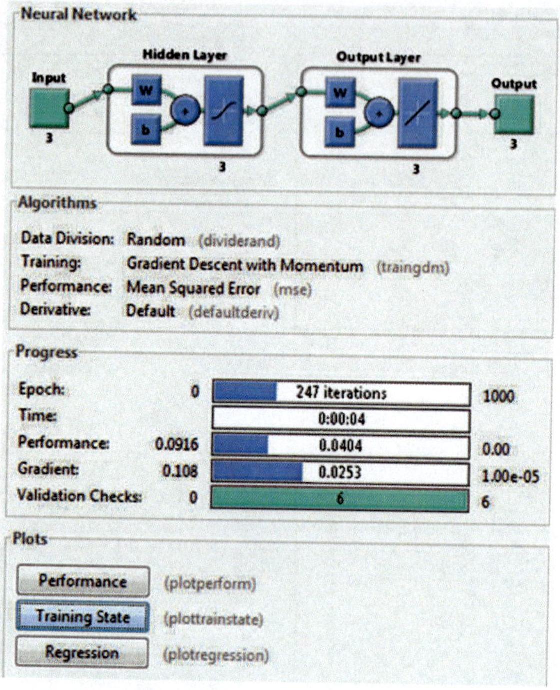

Fig. 6 ANN browser in MATLAB for gradient decent algorithm

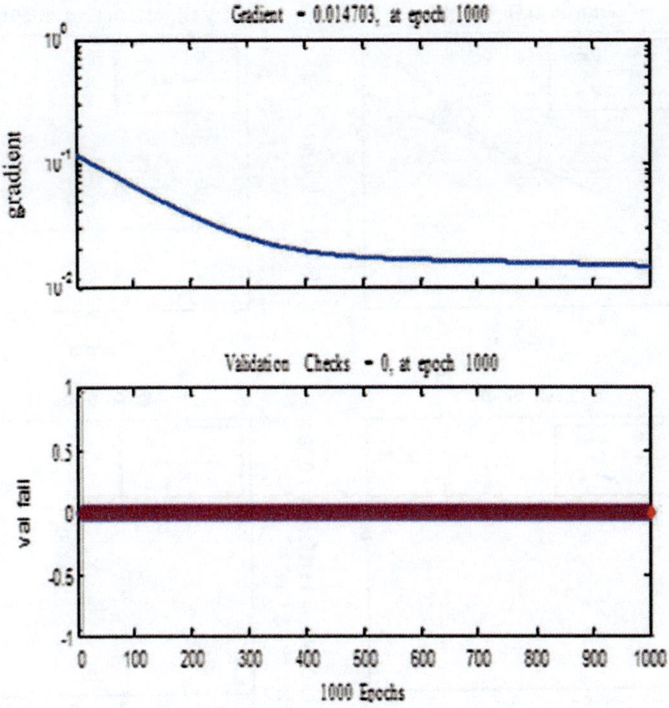

Fig. 7 Results for training and validation using gradient decent algorithm

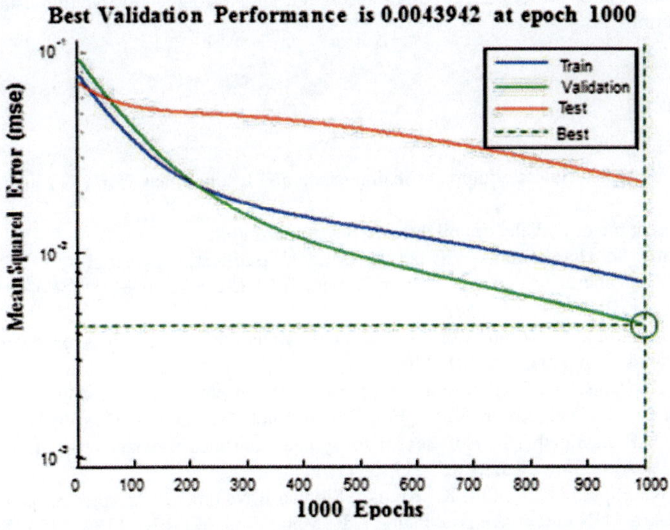

Fig. 8 Performance measure for training, testing, and validation using gradient decent algorithm

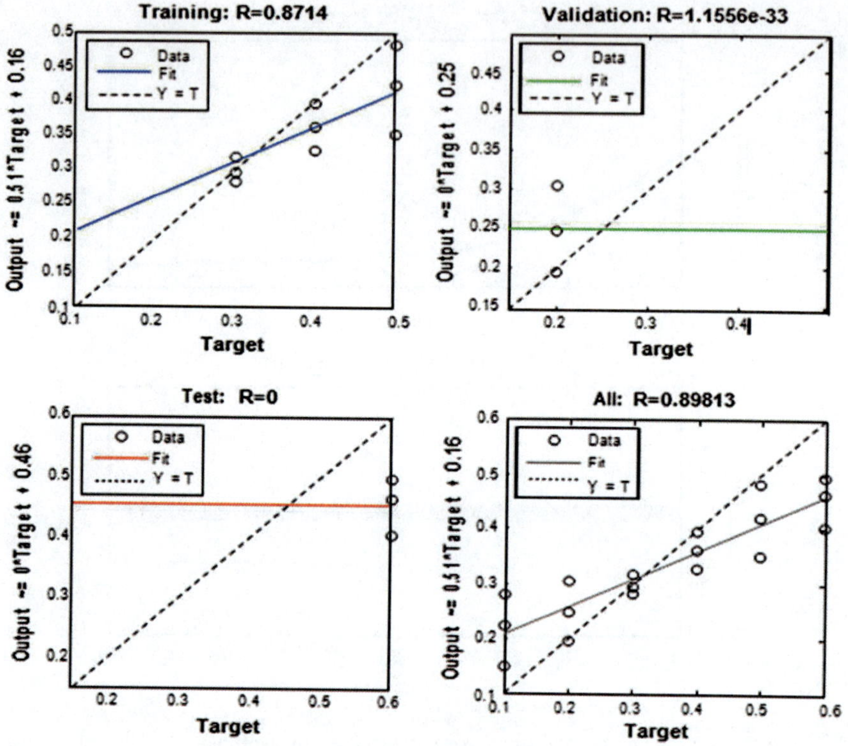

Fig. 9 Variation in actual and target output during training, testing, and validation using gradient decent algorithm

References

1. Gill, P.: Electrical power equipment maintenance and testing. New York, NY [u.a.], Dekker (1998)
2. Loss tangent meter in dielectric oil, http://www.amperis.com/
3. Transformer Oil Detoriation—Why is transformer oil purification essential
4. Rouse, T.O.: Mineral insulating oil in transformers. IEEE Electr. Insul. Mag. **14**(3), 6–16 (1998). https://doi.org/10.1109/57.675572
5. Sujatha, K., Pappa, N.: Combustion monitoring of a water tube boiler using a discriminant radial basis network. ISA Trans. **50**, 101–110 (2011)
6. Sujatha, K., Pappa, N.: Combustion quality estimation in power station boilers using median threshold clustering algorithms. Int. J. Eng. Sci. Technol. **2**(7), 2623–2631 (2010)
7. Bhavani, N.P.G., Sujatha, K.: soft sensor for temperature measurement in gas turbine power plant. Int. J. Appl. Eng. Res. 21305–213169 (2014)
8. Sujatha, K., Pappa, N., Senthil, K., Kumar., Siddharth Nambi, U.: Monitoring power station boilers using ANN and image processing. Adv. Mater. Res. 631–632, 1154–1159 (Trans Tech Publications, Switzerland) (2013)
9. Sujatha, K., Pappa, N., Senthil, K., Kumar., Siddharth Nambi, U., Raja Dinakaran, C R.: Intelligent parallel networks for combustion quality monitoring in power station boilers. Adv. Materi. Res. **699**, 893 -899 (Trans Tech Publications, Switzerland) (2013)

Code Buddy: A Machine Learning-Based Automatic Source Code Quality Reviewing System

Nidhi Patel, Aneri Mehta, Priteshkumar Prajapati, and Jigar Biskitwala

Abstract The growing complexity of software and associated code makes it difficult for software developers to produce high-quality code in a timely fashion. However, this process of assessing code quality can be automated with the help of software code metrics, which is a quantitative measure of code properties. The software metrics consist of several attributes, which describe the source code and this includes lines of code, program length, the effort required, the difficulty involved, cyclomatic complexity, volume, vocabulary, intelligence count, and so on. With the help of these features, code can be classified as a well-written code or badly-written code. This study focuses on evaluating the performance of main classification algorithms: Naïve Bayes, K-nearest neighbors (KNN), logistic regression, stochastic gradient descent (SGD) classifier, support vector machine (SVM), and decision tree (D-Tree) with thirteen of NASA metrics data program (MDP) dataset. The research work also focuses on understanding the math and working of each of the classifiers and the quality of each dataset. The comparison measure for the evaluating classifiers includes confusion matrix and other derived measures, namely F-measure, recall, precision, accuracy, and Matthews correlation coefficient (MCC). The best model is chosen along with the appropriate dataset. In order to allow the developers to use the trained model, we created Code Buddy a SharePoint web-portal; which allows the developers either assess the code quality by sending the review request to any of the colleagues or assess the code automatically using a trained model, which will predict whether the code is well written or badly written. Moreover, if the developer is not satisfied with the results, he/she can send a review request to any fellow colleague who can review the code and provide the review comment on the same.

N. Patel (✉) · A. Mehta · P. Prajapati
Chandubhai S. Patel Institute of Technology (CSPIT), Charotar University of Science and Technology (CHARUSAT), Changa, India
e-mail: 16it085@charusat.edu.in

A. Mehta
e-mail: 16it053@charusat.edu.in

J. Biskitwala
DEMS, Capgemini, Gandhinagar, India

H. Sharma et al. (eds.), *Congress on Intelligent Systems*, Advances in Intelligent Systems and Computing 1334, https://doi.org/10.1007/978-981-33-6981-8_36

Keywords Code review · Cyclomatic complexity · Decision tree · K-nearest neighbors · Logistic regression · Machine learning · Metrics data program (MDP) · Naïve Bayes · Software development life cycles (SDLC) · Stochastic gradient descent (SGD) · Support vector machine

1 Introduction

The procedure of software flaw prediction and detection plays a crucial role in sustaining software quality. Evaluating the code quality in the early stages of software development life cycles (SDLC) reduces the cost involved in the development and maintenance of the software. In addition, this process enables a firm to proliferate the culture of writing clean code amidst its employees. Hence, the software flaw prediction is stirred by the heavy cost involved due to software failures and establishing the culture of developing an error-free code. In order to achieve this, researchers have been using software metrics, which is a quantitative measure of software characteristics that aid in estimating the progress and quality of source code. The software metrics can be used to quantify various software properties. The software metrics consist of several parameters that describe the source code and this includes lines of code, program length, the effort required, the difficulty involved, cyclomatic complexity, volume, vocabulary, intelligence count, and so on. These properties highly associate with the code quality and, hence, helps in defect proneness. The focus of defect detection or prediction is to use code metrics in combination with historic fault data to gain insight into past code metrics. This is achieved using various classification or machine learning algorithms, which can be trained and validated on a historical dataset. From here on, predictions can be made on the quality of future codes using their code metrics, which can be calculated relatively easily and quickly.

The study involves a performance comparison of four classifiers: decision tree, logistic regression, support vector machine, and Naïve Bayes. These classifiers are selected because they have a distinct predictive approach and are widely applied by the machine learning community and researchers. All the classification techniques are applied to the various datasets in order to increase the diversity of code and defects. After evaluating various properties like total instance, class imbalance, erroneous data, and relative performance particular dataset is chosen to work upon. The aim of the software developers is to write clean and high-quality code in a timely fashion. The conventional method of improving code quality such as peer code review is a good example of this. Several features on code reviews discovered by software engineering, such as link graph analysis [1], decision making [2], and sentimental analysis [3].

2 Literature Review

The process of code review has always been in place amongs the software developers in order to check whether the code follows the quality assurance (QA) guidelines, finding defects, search for an alternate or better solution, and increase cohesiveness amongs the developers so on and so forth. The conventional methods include formal inspections, regular change-based code review, over-the-shoulder, email pass around, and walkthroughs. In order to automate the process of code review, static code metrics are taken into consideration that helps in the classification. The software code metrics are quantitative measures of source code characteristics that aid in estimating the progress and quality of source code.

The software metrics used to quantify various software properties. One such publicly available dataset for classification purposes is the NASA MDP dataset, which contains a thirteen-module/function/method level dataset aimed for software metrics research. Each of the thirteen datasets corresponds to a NASA software system containing several instances of software code metrics and fault data for each comprising module. The code metrics include line of code (LOC)-count measure, Halstead measure [4] and McCabe measure [5]. There several influential studies that are carried out in this domain, however, it was Menzies [6] and Lessmann [7] work which was a first big step toward the usage of software code metrics for classification purpose. Both of these are well-cited work of modern defect prediction literature involving several experiments for defects predictions, along with various learning models and usage of publicly available datasets like NASA MDP repository or the PROMISE dataset.

The Menzies' study involved three classifiers while Lessmann's undertook a largescale classifier benchmarking consisting of twenty-two classifier and ten of NASA MDP datasets. Another notable study in this area was done by David Bowes and Jean Petric [8] whereby they have evaluated different classifiers in order to find out whether different defects are found by them or not. These studies along with the publicly available NASA MDP have made substantial contributions and acted as catalysts for the usage of code metrics in order to perform classification of code as good code or bad code. Lastly, Sonali Agarwal and Divya Tomar [9] have done an extensive study on the classification along with feature selection in order to produce optimum results.

Code review is not only constrained to IT-based industries. Modern code review (MCR) is applied at a financial technology company in Stockholm. The company even suggested implementation of high-level conceptual code review (HCCR) where changes in code can be sorted automatically into different commits with every change [10]. There are several approaches of dealing with the review comments such as a set of changed files for the developer of the project team, search for the best set of candidate reviews use the non-dominated sorting genetic algorithm (NSGA-II) [11, 12] used software repositories of different commercialized projects and extracted 15223 review comments to perform their code review process. Integrating MCR with the software development process is beneficial in obtaining software with fewer amount of bugs, improvised code, and easier to maintain software [13].

Several previous research studies have taken into consideration that performed over the years for code reviews [14, 15]. The datasets taken into consideration consist of various attributes relating to the code metrics and have a variable impact on the performance of the classification algorithm. Zhang [16] suggests that a line of code can be an effective attribute in early detection of the defect while others debate that LOC is a poor attribute and has less of an impact on performance than other attributes. As per Malhotra [17], object-oriented metrics contribute much more toward the defect prediction. According to Ostrand [18], the inclusion of developer information in the metrics does not help much in the improvement of predictive performance while Bird [19] reports better performance when the same information used with a socio-technical network of the variable.

In order to increase the diversity and make the results of the study more generic, thirteen of NASA datasets are taken into consideration. Wahono [20] suggests since these datasets are publicly available and easy to access these NASA datasets are most popular for code classification, and the use of PROMISE repository has surged. Datasets have a substantial impact on the performance of the classification algorithms. Some dataset is difficult than others.

For example, the PC2 NASA dataset is one of them since it is highly imbalanced and produces poor predictive results as reported by Menzies [6]. Table 1 represents the statistics of all the thirteen NASA datasets. Arisholm and Briand's [21] studies demonstrate that how different datasets have a considerable impact on the performance of the classifier.

In addition to this, not all the features of the dataset needed to consider for classification purposes. Correlated or repeated attribute barely contribute toward the training of the model. Hence, the feature selection helps in improving the performance of the

Table 1 NSA MDP datastet statistics [22]

Dataset	Language	Total KLOC	No. of Modules	
			Pre training	Post training
CM1	C	20	505	505
KC1	C++	43	2109	2096
KC3	Java	18	458	458
KC4	Perl	25	125	125
MC1	C and C++	63	9466	9277
MC2	C	6	161	161
MW1	C	8	403	403
PC1	C	40	1107	1107
PC2	C	26	5589	5460
PC3	C	40	1563	1563
PC4	C	36	1458	1399
PC5	C++	164	17,186	17,001
JM1	C	315	10,878	7722

model [5, 23]. Another factor that needed to consider for the defect prediction is a balanced number of instances. This is important because many of the NASA MDP datasets are highly imbalanced which results in producing high accuracy during the training phase, however, fails to perform during the unforeseen testing. An example of this is the PC2 dataset, which consists of only 23 data points belonging to defective class out of 5589 data points. An imbalanced dataset affects different classifiers in different manners. For example, the C4.5 decision tree performs poorly with imbalanced data, while fuzzy-based classifiers performed extremely well. The process of data balancing like SMOTE has shown positive signs when used with random forest while there is no substantial improvements with Naïve Bayes [24]. Model performance comparison is a complex task and needs to consider various measures. Often times confusion matrix is used for measurement of predictive performance. In this study, measurement is mainly based on the confusion matrix which reports how a model has done classification compared to the actual classification. Using confusion matrix F-measure, precision, recall, and accuracy is derived which also serves as measures for the study.

F-measure was selected since it is very commonly used by other research studies, hence allowing to compare the result of this studies with other. Added to this, F-measure gives the harmonic mean of both precision and recall. MCC was selected because it is easy to comprehend as range as -1 to $+1$. Additionally, MCC includes all the components of confusion matrix.

3 Proposed System

The 'Code Buddy' provides a platform to rectify common problems during the process of peer code reviewing that a developer or in general, anyone who writes a program comes across in their daily life. This common problem is mostly related to documentation, style, and structure. This kind of issue takes the maximum amount of time, and because of them, they tend to focus less on important aspects such as defects, architectural issues or testing of code.

To address the challenge mentioned above, we propose a machine learning-based code-reviewing system that makes the process of automatic code reviewing more precise, automatic, and easy. The system uses static code analysis and ML algorithm to detect the quality of the code. We have used twelve rules of static code analyzing whose values are then fed into a classifier that predicts the quality of the code to be well written or badly written based on training data provided by peer reviews, public datasets or data collected from some other sources.

The general architecture mainly divided into two phases training and testing (see Fig. 1). The training phase includes training of a pre-processed training dataset. On the other hand, the testing phase uses the classification algorithm and concepts of static code analysis to automatically identify both well-written code and badly-written code. This functionality reduces the work of reviewers up to some extent.

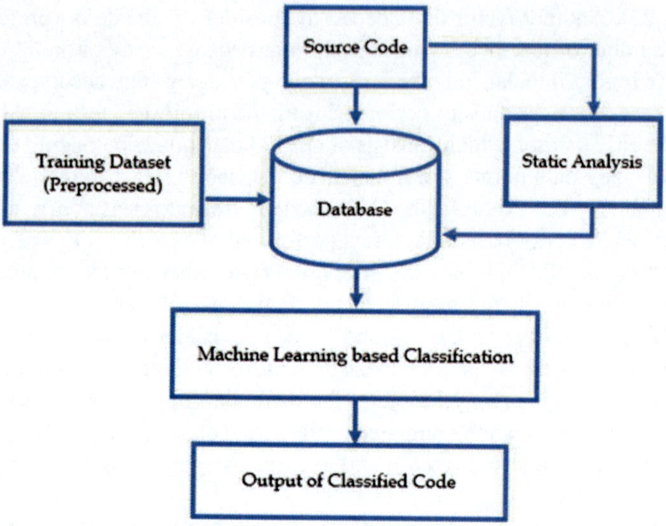

Fig. 1 General architecture of proposed system

Proposed system follows understated steps: data collection, data pre-processing, training, and testing.

3.1 Data Collection

The first step of any machine learning-based project is to collect correct, and a large amount of data that would help the model to learn further providing us a precise result. For Code Buddy, we have used already reviewed code, review pairs, and then pre-processed them accordingly. There are around 11,000 entries in our dataset.

3.2 Data Pre-processing

Pre-processing of the dataset includes transforming code into Halstead's software metrics values. We have used a static code analyzer to get these software metrics value. We have taken into consideration 12 rules of static code analyzing for this purpose mainly cyclomatic complexity, vocabulary, volume, length, difficulty, effort, bugs, time, unique operators, total operators, unique operands, and total operands. Detailed explanation for the above rules given below:

A computer program is an execution of an algorithm considered a group of tokens, which classified as either operators or operands.

Here for metrics purpose, we will consider following as pperators and operands:

Operators: Inbuilt keywords, all types of brackets (( ), , [ ]), +, −, *, /, and all other types of operators (Relational operators, logical operators, etc.)
Operands: Variables we define, constants (Numbers)
To understand every metric, firstly, we need to have some idea about the notations and their meanings, which is as follows:
n1 = Distinct Operators
n2 = Distinct Operands
N1 = Total Operators
N2 = Total Operands

Length: The total number (N) of operators and operand occurrence.

$$N = N1 + N2 \tag{1}$$

Vocabulary: The total number (n) of unique operator and unique operand occurrences.

$$n = n1 + n2 \tag{2}$$

Volume: Proportional to program size; represents the size, in bits, of space necessary for storing the program. This parameter is dependent on specific algorithm implementation. The unit of measurement of volume is the common unit for size 'bits'. It is denoted by V.

$$V = N \times \log_2(n) \tag{3}$$

Difficulty: This parameter (D) shows how difficult to handle the program is. As the volume of the implementation of a program increases, the program level decreases and the difficulty increases.

$$D = (n1 \div 2) \times (N2 \div n2) \tag{4}$$

Effort: Measures the amount of mental activity needed to translate the existing algorithm into implementation in the specified program language. It is denoted by E.

$$E = D \times V \tag{5}$$

Time: Shows time (in minutes) needed to translate the existing algorithm into implementation in the specified program language. It is denoted by T.

$$T = E \div (f \times S) \tag{6}$$

where E is effort, S=18, f=60

Bugs: The number of delivered bugs (B) correlates with the overall complexity of the software. It is an estimate for the number of errors in the implementation. It is denoted by B.

$$B = V \div 3000 \tag{7}$$

Unique Operators: Operators with single occurrence throughout the program.
Total Operators: Total count of operator occurrence in the program.
Unique Operands: Operands with single occurrence throughout the program.
Total Operands: Total count of operands occurrence in the program.

After transforming the code, we have also done some usual pre-processing like dealing with missing values, inconsistent values, etc.

3.3 *Training*

After pre-processing comes the training of the model. For our system, we have tried using five classifiers mainly KNN, SVM, random forest, Naïve Bayes (Gaussian), and stochastic gradient descent (SGD) classifier. Out of all five classifiers, SVM has provided the most accurate results as per our dataset. The comparison measure for the evaluating classifiers includes confusion matrix and other derived measures, namely F-measure, recall, precision, accuracy, and MCC. One can use any classification algorithm suitable for their dataset.

SVM is a supervised classification/regression algorithm in machine learning. The support vector machine is a representation of the training data as points in space separated into categories by a clear gap that is as wide as possible. New examples then mapped into that same space and predicted to belong to a category based on which side of the gap they fall. SVM is a suitable classifier when the number of features and the number of training data is very large. In addition, SVM is best when the sparsity of data is huge. As our data has twelve features and has a large amount of entries, SVM is most suitable for our dataset among other classifiers.

3.4 *Testing*

When a new code arrives, it first goes into the pre-processing stage and the code converted into Halsted's software metrics values. After being converted, the values fed into the model that in return classifies the code into good/needs improvement.

4 Conclusion

Poor quality of code leads to inaccurate or completely wrong results, which further degrades the overall quality of software. Assessing the code quality in the early stages of software development reduces the cost involved in the development and maintenance of the software. In addition, this process enables a firm to propagate the culture of writing clean code amongs its employees. One way of assessing the code quality is through software code metrics, which is a quantitative measure of software characteristics that aid in estimating the progress and quality of source code. This study conducted over thirteen NASA MDP datasets using various classification algorithms, namely Naïve Bayes, logistic regression, SVM, and D-tree. The research work involved understanding the entire mathematics and working behind each of the classification algorithms and later on comparing the statistics with different datasets. It turned that out of all thirteen datasets, JM1 turned out to be a balanced dataset with enough instance to work upon. In addition, in the case of the classification algorithm, it was SVM turned out to be the best performer, which later trained on JM1 and integrated into Code Buddy.

References

1. Agarwal, S., Tomar, D.: A feature selection based model for software defect prediction. Assessment **65** (2014)
2. Shi, S.T., Li, M., Lo, D., Thung, F., Huo, X.: Automatic code review by learning the revision of source code. In: Proceedings of the AAAI Conference on Artificial Intelligence vol. 33, pp. 4910–4917 (2019)
3. Arisholm, E., Briand, L.C., Johannessen, E.B.: A systematic and comprehensive investigation of methods to build and evaluate fault prediction models. J. Syst. Software **83**(1), 2–17 (2010)
4. McCabe, T.J.: A complexity measure. IEEE Trans. Softw. Eng. **4**, 308–320 (1976)
5. Bowes, D., Hall, T., Petrić, J.: Software defect prediction: do different classifiers find the same defects? Softw Qual. J. **26**(2), 525–552 (2018)
6. Menzies, T., Greenwald, J., Frank, A.: Data mining static code attributes to learn defect predictors. IEEE Trans. Softw. Eng. **33**(1), 2–13 (2006)
7. Lessmann, S., Baesens, B., Mues, C., Pietsch, S.: Benchmarking classification models for software defect prediction: a proposed framework and novel findings. IEEE Trans. Softw. Eng. **34**(4), 485–496 (2008)
8. Bowes, D., Hall, T., Petrić, J.: Software defect prediction: do different classifiers find the same defects? Softw Qual. J. **26**(2), 525–552 (2018)
9. Agarwal, S., Tomar, D.: A feature selection based model for software defect prediction. Assessment **65**, (2014)
10. Hirao, T., McIntosh, S., Ihara, A., Matsumoto, K.: The review linkage graph for code review analytics: a recovery approach and empirical study. In: Proceedings of the 2019 27th ACM Joint Meeting on European Software Engineering Conference and Symposium on the Foundations of Software Engineering. pp. 578–589 (2019)
11. Hossain, S.S., Arafat, Y., Hossain, M.E., Arman, M.S., Islam, A.: Measuring the effectiveness of software code review comments. In: International Conference on Advances in Computing and Data Sciences, pp. 247–257. Springer (2020)
12. Lessmann, S., Baesens, B., Mues, C., Pietsch, S.: Benchmarking classification models for software defect prediction: a proposed framework and novel findings. IEEE Trans. Softw. Eng. **34**(4), 485–496 (2008)
13. Malhotra, R.: A systematic review of machine learning techniques for software fault prediction. Appl. Soft Comput. **27**, 504–518 (2015)
14. McCabe, T.J.: A complexity measure. IEEE Trans. Softw. Eng. **4**, 308–320 (1976)
15. Menzies, T., Greenwald, J., Frank, A.: Data mining static code attributes to learn defect predictors. IEEE Trans. Softw. Eng. **33**(1), 2–13 (2006)

16. Olausson, A., Louca, S.: Reviewing code review: defining and developing high-level conceptual code review at a financial technology company (2020)
17. Malhotra, R.: A systematic review of machine learning techniques for software fault prediction. Appl. Soft Comput. **27**, 504–518 (2015)
18. Rodriguez, D., Herraiz, I., Harrison, R., Dolado, J., Riquelme, J.C.: Preliminary comparison of techniques for dealing with imbalance in software defect prediction. In: Proceedings of the 18th International Conference on Evaluation and Assessment in Software Engineering, pp. 1–10 (2014)
19. Saini, N.: Using machine intelligence to prioritise code review requests (2020)
20. Wahono, R.S.: A systematic literature review of software defect prediction. J. Softw. Eng. **1**(1), 1–16 (2015)
21. Arisholm, E., Briand, L.C., Johannessen, E.B.: A systematic and comprehensive investigation of methods to build and evaluate fault prediction models. J. Syst. Software **83**(1), 2–17 (2010)
22. Siow, J.K., Gao, C., Fan, L., Chen, S., Liu, Y.: Core: Automating review recommendation for code changes. In: 2020 IEEE 27th International Conference on Software Analysis, Evolution and Reengineering (SANER), pp. 284–295. IEEE (2020)
23. Wahono, R.S.: A systematic literature review of software defect prediction. J. Softw. Eng. **1**(1), 1–16 (2015)
24. Zhang, H.: An investigation of the relationships between lines of code and defects. In: 2009 IEEE International Conference on Software Maintenance, pp. 274–283. IEEE (2009)

A Software Reusability Paradigm for Assessing Software-as-a-Service for Cloud Computing

Deepika and **Om Prakash Sangwan**

Abstract Cloud computing is a mode of computing in which there is on demand availability of IT or computer system resources over the network as an online service. Virtualization is the core driving technology of cloud computing. With the increase in utilization of infrastructure, it speeds up IT operations and reduces cost. SaaS is generally operated and it gives numerous advantages to assist end users. To accomplish these advantages, it is important to calculate the reusability of SaaS and there is requirement to develop a reusability paradigm for evaluating SaaS for cloud computing. In this paper, we suggested an inclusive paradigm for assessing reusability of SaaS. We first describe key characteristics of SaaS. And then, we defined reusability attributes from the key characteristics, and derived metrics for the reusability attributes. Correlation measure is used for validating the reusability paradigm of SaaS. This proposed paradigm can be used for evaluation and quality management of SaaS by service providers.

Keywords Software-as-a-service (SaaS) · Reusability · Reusability attributes · Metrics

1 Introduction

Cloud computing comes forward as an efficient reuse criterion, where hardware or software resources are provided in the form of cloud services through the Internet [1]. Software-as-a-service (SaaS) is one category of the cloud service areas, which provides on demand access to software services or application software. It provides business services such as enterprise resource planning (ERP), customer relationship management, sales, and management of documents, social networking and mail services. Cloud application SaaS is advantageous in the manner that it accelerates the process of software adoption and upgradation by reducing the cost required for development of software [2]. SaaS provides complete product; we do not have to think about underlying infrastructure or maintenance. SaaS provide various benefits by

Deepika (✉) · O. P. Sangwan
Guru Jambheshwar University of Science and Technology, Hisar, Haryana, India

H. Sharma et al. (eds.), *Congress on Intelligent Systems*, Advances in Intelligent Systems and Computing 1334, https://doi.org/10.1007/978-981-33-6981-8_37

maintaining commonality, ease of access via Internet, and supporting multi-tenant's access. Services are developed and set up among various applications with common and reusable features. Reusability is the key characteristic of SaaS. It is an important criterion for evaluation of services quality. High return on investment (ROI) is provided by a service if it is highly reusable. Reusability is the degree with which we can use the previously acquired concepts in new situation [3]. A well-defined reusability model is required to be produced for evaluation of reusability of SaaS. The desirable properties of a SaaS are high reusability, availability, and scalability.

Therefore, there is need for developing a reusability model for evaluating SaaS cloud services. In this paper, we have proposed an inclusive reusability model for estimating SaaS. Key characteristics of SaaS are defined first and then we derived reusability attributes from these key characteristics. Metrics for these reusability attributes are defined. Validation is conducted by assessment in term of correlation measure. By using the proposed reusability model, evaluation and quality management of SaaS can be done by service providers.

The rest of this paper is prearranged as follows. Part II represents related work and Part III defines the key characteristics of SaaS. In part IV and V, attributes related to reusability and metrices are explained, respectively. Assessment is conducted in part VI. And part VII represents conclusion and future work.

2 Related Works

In this section, we explored and analyzed works related to SaaS services management. However, little work is done related to SaaS quality management. But on the basis of past work, we defined various metrics related to SaaS features which will help in SaaS services reusability evaluation.

Cho and Kim [4] done the evaluation of reusability of services by proposing a quality model for service-oriented architectures (SOA). They divided services into two types on the basis of granularity. Atomic and composite services. As atomic service implements service components only, and composition of services is implemented with composite services. Key features of service areas in SOA are explained and defined quality attributes of reusability on the basis of key characteristics. Applicability of the model is shown by case study. They concluded that by using suggested model, reusability of SOA can be measured efficiently and effectively.

A comprehensive quality model has been proposed by Lee et al. [5] for the evaluation of quality of SaaS in cloud computing. First, important characteristics of SaaS were defined and then attributes of quality related to these key features are derived. Metrics related to these key characteristics were defined and validated by conducting assessment on the basis of IEEE 1061. This quality model predicts ROI of SaaS and will also act as a SaaS quality indicator. The validation criteria used are correlation, consistency, and discriminative power. La and Kim [6] proposed methods with the help of which high quality SaaS can be developed. They defined intrinsic characteristics of SaaS and develop SaaS with desired properties for defining SaaS cloud services development process. They also highlighted the importance of modelling commonality and variability to increase reusability. Designing of meta model is also presented to develop high quality SaaS. La et al. [7] defined metamodels for SaaS and mash-up cloud services and also discussed related technical issues. They discussed four problems, i.e., scalability, availability, commonality, and adaptability and also provided solutions for these problems. Proposed solutions for the technical problems will make cloud services more reliable and reusable.

A framework is introduced for simplified and easier development of SaaS-EasySaaS by Tsai et al. [8]. This framework provides two different alternatives to tenants. In first alternative, a consumer-centric platform is developed for SaaS. In which the tenants publish their requirements of application with related specifications of the required service and on the basis of that SaaS providers search actively for the tenants' needs. In second alternative, tenants can customize and search services that are reusable according to their requirements. Testing support is also provided in this proposed EasySaaS framework. Oh et al. [9] proposed evaluation model for measuring reusability which taken into account quality attributes and metrics for cloud services. Characteristics of cloud services are explained and then quality attributes and metrics for service reusability are defined. Results show that proposed evaluation model can improve reusability by assessing the quality of cloud services. Singh and Singh [10] proposed cloud computing reusability (CRC) model. They used component-based development for cloud computing and proposed a new dimension for interacting with reusability. Reverse engineering is used as basis for the proposed model.

Tsai et al. [11] provided a summary of SaaS structural design and main technical matters. They four main SaaS structural designs, i.e., database oriented, middleware oriented, PaaS based, and service oriented. Technical issues such as customization, multitenancy and scalability related to above-mentioned architectures have been discussed and compared. A quality evaluation model is proposed by Dhanamma and Sunita [12] that uses constraint based clustering (CBC), a data mining technique which focuses on service user's requirement and quality attributes. On the basis of business and provider's perspective, Samir and Darwish [13] explored SaaS reusability quality attributes and metrics. They also explained the methodology of extracting useful attributes related to SaaS reusability and provided metrics related to these attributes. A detailed survey of cloud computing services is presented by Hassan et al. [14] which includes latest trends, challenges associated, and solutions related to security of cloud computing services. They evaluated the security threats and risks associated and also presented a brief overview of regulatory bodies for security of cloud. Mohamed and Yousif [15] proposed a model that reduces gap between service oriented architectures and SaaS by getting benefit from SOA reusability and SaaS scalability. There are scalability issues in SOA and for tackling these issues, this model is developed. Healthcare system is used as an evaluation criterion for the proposed model.

However, little work is done for reusability and SaaS as we explained above, but no work is done for developing a model for SaaS reusability evaluation. So, in our work, we developed a model on the basis of SaaS features, reusability attributes related to those characteristics, and defined metrics for the attributes as explained in next sections.

3 Features of SaaS

To describe a reusability paradigm for SaaS, it is essential to recognize key characteristics of SaaS. On the basis of detailed evaluation on our current SaaS references [5, 6, 13], key features identified are as displayed in Fig. 1.

Supporting Commonality: By giving common functionality and features, SaaS provides various benefits to consumers or service users. These commonality of services gives easier reusability, which in turn provides high return on investment.

Accessibility via*Internet*: All cloud services including SaaS are easily accessible as these are available online and can be assessed through internet.

Providing Complete Functionality: SaaS provides complete functionality of a software through internet services.

Supporting Multi-tenants': SaaS installed on service provider's side make services available to public. SaaS has to be designed to support multi-tenants'. There is no requirement of prior notice by consumers to access the services.

Thin Client Model: SaaS services are provided to consumers on pay-as-you-go basis. In thin client model most of the work is done by server. Consumers uses browser to access the services that runs on SaaS provider's side.

Fig. 1 Features of SaaS [6]

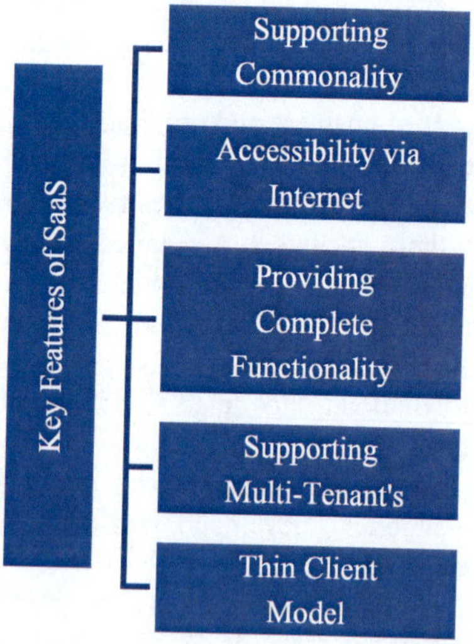

4 Reusability Attributes

On the basis of various studies, the attributes of software reusability which covers specific features of SaaS is identified. Derived from the key features of SaaS, understandability, coupling-ability, flexibility, variability, commonality, efficiency, customizability, availability, and accessibility are anew specified.

Figure 2 represents the mapping between SaaS features and reusability attributes. Figure 3 represents sub-attributes of customizability. Definition of each attribute and their significance with respect to SaaS is as following.

Commonality: This attribute states common features and attributes across SaaS application. Services of SaaS which are having high commonality or sharing of common characteristics would yield high return on investment (ROI). Increased commonality directly relates to increased ROI and increased reusability.

Accessibility: This attribute defines the ease of use of services of SaaS and provides a way for users to interact with available services. Accessibility depends on the complexity and availability of services. Increased accessibility gives easier service usability via internet which in turn increases reusability and yield high return on investment (ROI) for SaaS.

Availability: This attribute gives SaaS service uptime. It measures the percentage of time for which the service is available. Increased availability maximizes the usage

Fig. 2 Mapping from SaaS features to reusability attributes

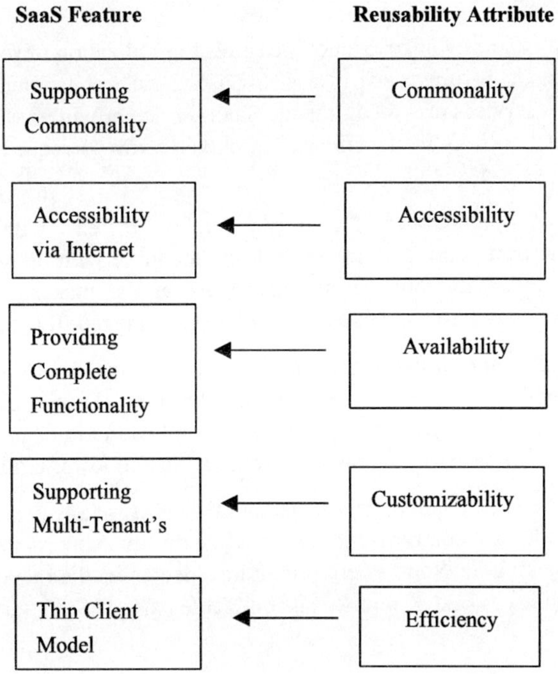

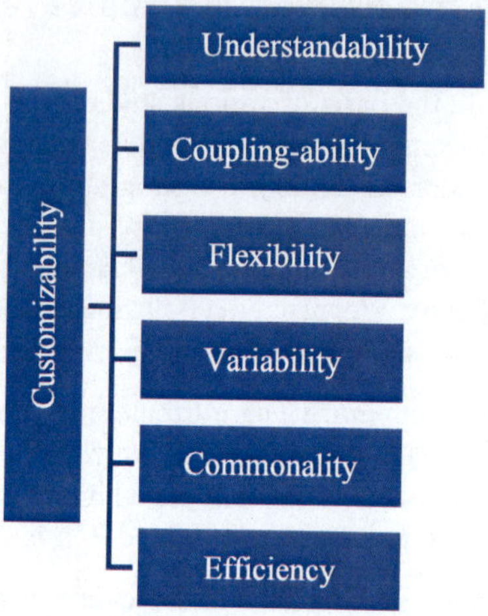

Fig. 3 Customizability reusability sub-attributes [13]

potential of SaaS services by providing complete functionality. For the successful execution of SaaS, there should be high availability of services.

Customizability: This attribute defines the ability of a service to get modified to perform required task. It describes something that can be changed according to required usage. Providing services that better serves user's need require high customizability. SaaS should have high customizability in order to support multi-tenants.

Efficiency: This attribute measures the utilization of resources by SaaS. Thin client model is supported by SaaS. In SaaS, service consumers or users cannot have control of services provided. It is the SaaS service providers which deploy and run services. For a SaaS to be efficient, it should provide required functionality under given conditions.

Understandability: This attribute defines the ease with which a software or service becomes easier to understand and integrate with other services or system. Understandability enhances software reuse. It also makes a system or software easier to integrate or change so that it can be customized. It increases inheritance.

Coupling-ability: This attribute specifies the degree of interdependence between modules of software. It specifies that how changes in one module affect other. It should be low, so the ability of an application to change other application is lower which result in increased customization and lower maintenance.

Flexibility: This attribute specifies the ability of a system to adapt or change according to future requirements. Increased flexibility supports multi-tenants feature of SaaS, making SaaS more reusable. It also decreases the rate of coupling giving increased ability to add or remove customizable parts of SaaS services.

Variability: This metric defines the ability of a system or service to get customized or changed according to a particular framework. It specifies the point or location where change can be introduced in a system. Increased variability directly increases customizability and affects reusability of SaaS.

On the basis of above-defined reusability attributes, reusability metrics related to each attribute are defined in next section for the evaluation of SaaS reusability.

5 Reusability Metrics

In this part, we describe metrics related to every attribute and also explain formula, interpretations, and limit of every single metric. Figure 4 exemplifies the relationship between reusability attributes and their metrics.

Commonality: The commonality is assessed by coverage of similarity (CoS) metric. Coverage of similarity can be measured through dividing the number of applications sharing common features by total number of applications.

$$\text{Cos} = \frac{\text{Number of application sharing common features}}{\text{Total Number of applications in the domain}} \quad (1)$$

If every application in the domain shares common features, then the value of CoS will be 1, given the limit of Cos is 0…1 which indicates more commonality of the applications in SaaS giving better reusability.

Accessibility: The accessibility is measured by ease of usage (EoU) metric. This metric measures a ratio of number of accessible services to the total number of services.

$$\text{EoU} = \frac{\text{Number of accessible services}}{\text{Total number of services in the domain}} \quad (2)$$

The limit is 0…1. Better value of EoU indicates better accessibility of SaaS.

Availability: The availability is assessed by openness of service (OoS). This metric measures availability through dividing the time for which a service is available to the total time for which the service is expected to be available. Availability is generally calculated in percentage form.

$$\text{OoS} = \frac{\text{Time for which service is available}}{\text{Total time for which the service is expected to be available}} \quad (3)$$

Higher value of metric indicates higher availability of SaaS applications given the limit is 0…1.

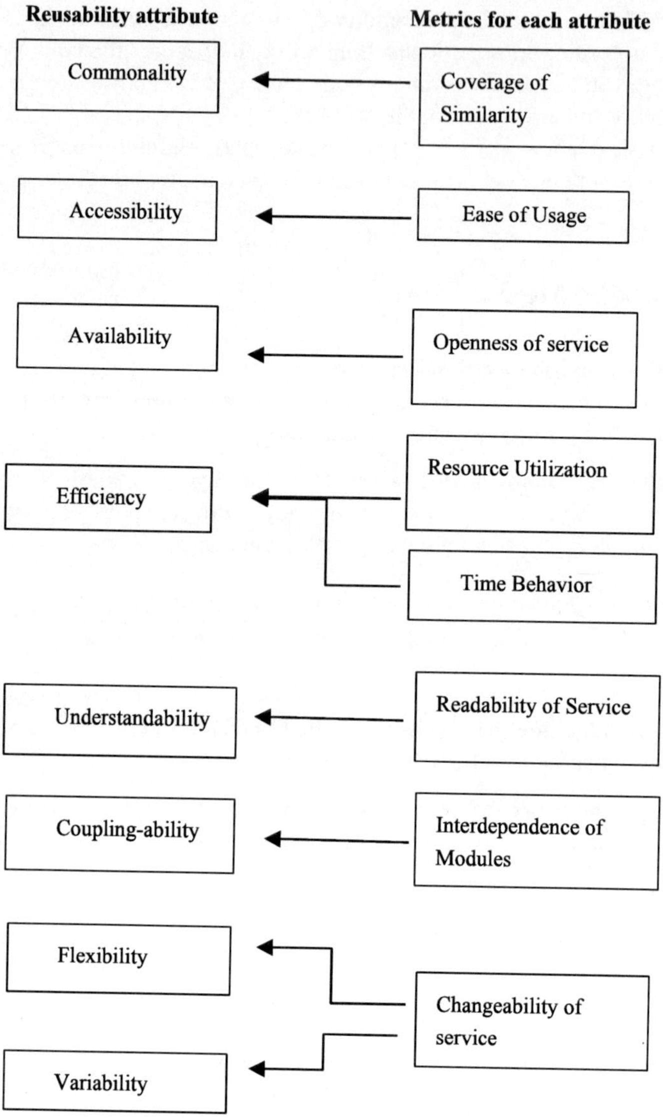

Fig. 4 Metrics for reusability attributes

Efficiency: Efficiency of SaaS can be measured by calculated two metrics i.e., Resource Utilization (RU) and Time Behaviour (TB) [5].

$$\mathrm{RU} = \frac{\text{Amount of resource allocated}}{\text{Actual amount of resources}} \tag{4}$$

Higher value indicates high resource utilization of SaaS given limit value between 0...1.

$$\text{TB} = \frac{\text{Time for service execution}}{\text{Total time for service establishment}} \quad (5)$$

Higher value indicates higher time efficiency of SaaS given limit is 0...1.

Finally, efficiency is calculated as some of both metrics:

$$\text{Efficiency} = W_{\text{RU}} \cdot \text{RU} + W_{\text{TB}} \cdot \text{TB} \ [5] \quad (6)$$

where W_{RU} and W_{TB} represent the weight of each metric, which sum 1. The weight value is allocated according to significance of each metric.

Higher value of efficiency indicates higher effectiveness of SaaS which contributes in increasing reusability also. Limit is 0...1.

Understandability: Understandability can be measured by readability of service (RoS). This metric measures understandability through dividing number of areas with acceptable understandability by total number of areas.

$$\text{RoS} = \frac{\text{Number of areas with acceptable understandability}}{\text{Total number of areas}} \quad (7)$$

The limit is 0...1. More value of RoS implies higher understandability of SaaS.

Coupling-ability: It can be measured by interdependence of modules (IoM) metrics. This metric measures coupling-ability through dividing number of modules with low coupling by total number of modules.

$$\text{IoM} = \frac{\text{Number of modules with low coupling}}{\text{Total number of modules}} \quad (8)$$

The limit is 0...1. More value of IoM implies low coupling of modules which leads to higher reusability of SaaS.

Flexibility and Variability: Both attributes can be collectively measured by changeability of service (CoS) metric. This metric measures flexibility and variability through ratio of number of changeable or variation points supported to total number of variation or changeable points in the domain.

$$\text{CoS} = \frac{\text{Number of chnageable or variation points supported}}{\text{Total number of changeable points in the domain}} \quad (9)$$

If each application in the field supports changeability, then the value of CoS will be 1. It would be advantageous if the value of CoS will be high. The limit of metric is 0...1.

6 Assessment

In this portion, we evaluate feasibility of 8 metrics suggested in this article by using correlation criteria. The metrics suggested in this work are obtained from reusability attributes. Between reusability attributes and metrics, there is strong linear association. Reusability of SaaS is measured by CoS, EoU, OoS, efficiency, RoS, IoM, and CoS.

According to formula

$$\text{Reusability} = \text{CoS} + \text{EoU} + \text{OoS} + \text{Efficiency} + \text{RoS} + \text{IoM} + \text{CoS} \tag{10}$$

Hence, each value will affect calculated result of reusability. Different weights ranging from 0…1 can be assigned by evaluator to each metric of reusability. The sum of metrics value will determine the level of SaaS reusability. Higher value will indicate higher reusability of SaaS.

7 Conclusion and Future Scope

SaaS provides various benefits to service consumers by supporting commonality, easier accessibility with the help of Internet, availability, and efficiency. It supports thin client model and have multi-tenants' access. Therefore, for successful management of SaaS, the evaluation of reusability of SaaS becomes necessary.

In this article, we presented a reusability paradigm for evaluation of SaaS service area in cloud computing. First, we extricated characteristics of SaaS, related to that characteristic's reusability attributes are derived. Then, we defined metrics related to those reusability attributes. To demonstrate the authenticity of model, correlation criteria is used for evaluating reusability of SaaS services.

All these metrics defined will cover essential features of SaaS and will help service providers in selecting reusable services of SaaS. It may also help service providers for predicting their return on investment and these results can be further used for SaaS quality evaluation.

As there is very little work done in the field of SaaS reusability evaluation, so in the future, we will further extend our work to develop a model using machine learning techniques for SaaS reusability evaluation.

References

1. https://aws.amazon.com/what-is-cloud-computing/.
2. Abuhussein, A., Alsubaei, F., Shiva, S.: Toward an effective requirement engineering approach for cloud applications. In: Ramachandran, M., Mahmood, Z. (eds.) Software Engineering in the Era of Cloud Computing. Computer Communications and Networks. Springer, Cham (2020)

3. Deepika, Sangwan, O.P.: Neuro-fuzzy based approach to software reusability estimation. In: International Conference on Sustainable Computing Techniques in Engineering Science and Management (IJCTA), pp. 3811–3891 (2016)
4. Choi, S.W., Kim, S.D.: A quality model for evaluating reusability of services in SOA. In: 2008 10th IEEE Conference on E-Commerce Technology and the Fifth IEEE Conference on Enterprise Computing, E-Commerce and E-Services (2008)
5. Lee J.Y., Lee J.W., Cheun D.W., Kim S.D.: A quality model for evaluating Software-as-a-Service in cloud computing. In: Seventh ACIS International Conference on Software Engineering Research, Management and Applications, pp. 261–266 (2009)
6. La, H.J., Kim, S.D.: A systematic process for developing high quality SaaS cloud services. Cloud Comput. 278–289 (2009).
7. La, H.J., Choi, S.W., Kim, S.D.: Technical challenges and solution space for developing SaaS and mash-up cloud services. In: 2009 IEEE International Conference on e-Business Engineering (2009)
8. Tsai, W.-T., Huang, Y., Shao, Q.: EasySaaS: a SaaS development framework. In: 2011 IEEE International Conference on Service-Oriented Computing and Applications (SOCA) (2011)
9. Oh, S.H., La, H.J., Kim, S.D.: A reusability evaluation suite for cloud services. In: 2011 IEEE 8th International Conference on e-Business Engineering (2011)
10. Singh, S., Singh, R.: Reusability framework for cloud computing. Int. J. Comput. Eng. Res. 169–177 (2012)
11. Tsai, W., Bai, X., Huang, Y.: Software-as-a-service (SaaS): perspectives and challenges. Sci. China Inf. Sci. **57**(5), 1–15 (2014)
12. Dhanamma, J., Sunita, M.: CBC approach for evaluating potential SaaS on the cloud, vesit.edu, vol. 2, pp 43–49 (2014)
13. Samir, A., Darwish, N.R.: Reusability quality attributes and metrics of SaaS from prespective of business and provider. Int. J. Comput. Sci. Inf. **14** (2016)
14. Hassan, W., Chou, T., Li, X., Appiah-Kubi, P., Tamir, O.: Latest trends challenges and solutions in security in the era of cloud computing and software defined networks. Int. J. Inform. Commun. Technol. 162–183 (2019)
15. Mohamed, A.S., Yousif, A.: Bridging the gap between service oriented architecture and cloud software as a service. In: 2019 International Conference on Computer, Control, Electrical, and Electronics Engineering (ICCCEEE), Khartoum, Sudan, pp. 1–6 (2019)

Personal Assistant for Career Coaching

Arbaz Khan, Vinit Masrani, Anoop Ojha, and Safa Hamdare

Abstract There has been an increasing hype and usage of chat bots nowadays. If chat bots are used as career assistants who can guide an individual to the right track, then this will motivate people to learn and acquire new skills. Currently, there is no standard and proper resource which can prepare an individual pertaining to industry requirements. In this paper, we propose a model to develop a standard platform for assisting people in career coaching. The users of this system can get their career-related queries solved and can also obtain information needed for making them industry ready. The proposed system is developed in order to develop a tool for communicating with job seekers to help them find jobs according to their skills. This system will also recommend courses that will enhance their skills in their field of interest for professionals. The proposed system is developed in order to develop a tool for communicating with job seekers to help them find jobs according to their skills. This system will also recommend courses that will enhance their skills in their field of interest for professionals.

Keywords Artificial intelligence · Career · Chat bot · Natural language processing

1 Introduction

When production and supply diverge, industry and academia need to work together, much as they do in any other sector, as they form a "market—supply—value chain." Educating job prospects and choices often involves identifying human desires, educating them about the value of setting targets, and eventually presenting them with a demanding atmosphere. In many professional courses, there is always need for full filling the gap between educational and professional requirements. Due to this large gap, many professionals are unemployed which lead to affecting the country's growth. It is just because of lack of proper guidance and resources.

A. Khan (✉) · V. Masrani · A. Ojha · S. Hamdare
Computer Engineering Department, St.Francis Institute of Technology, Borivali, Mumbai, India

S. Hamdare
e-mail: safahamdare@sfit.ac.in

H. Sharma et al. (eds.), *Congress on Intelligent Systems*, Advances in Intelligent Systems and Computing 1334, https://doi.org/10.1007/978-981-33-6981-8_38

The term Career Guidance in Education means giving students the important information, qualities, and abilities which are needed to take proper educational decision and finally to lead the career path to a domain. More specifically the term Career Guidance means that a person who has more ace or efficient guides or instructs on the general topics. In this modern world of communication, chat bots have become essential. Chat bots are nowadays used in many applications as virtual assistants and also have been seen all over the messenger platforms.

Career Guidance chat bot is a tool which can interact with job seekers in an interactive manner. Users can ask questions pertaining to their working domain, such as which area to pick, the current trends in order to clarify their work needs. Chat bots can even understand the meaning and intent behind the user asked questions in those cases also where they the users have failed to compose their queries properly and the chat bots reframe the questions on its own. It may also help users to find jobs according to their skills, their desires, and their work experience based on their needs. However, such systems could be used further to recommend courses to users to upgrade their skill or pursue their dreams. Also chat bots can be developed in different languages to attract wide range of users from different regions.

The primary goal behind making this system is that we make an expert in educational domain which assist them in preparing for the interview, give customized suggestion regarding career pathways, and suggest the resources and answer to the FAQs of students. This system will take into account the user interested domain and properly suggest the jobs in that domain only. This system is unique from the rest as it does not require the students to browse anywhere and then also, they will get answers to their questions instantly which will make the life of the students easy. This system will assist the students in evolving in them entrepreneurship skills and employability skills. Also, in order with assisting them with planning along with choosing the right career, the system is of utmost importance. Thus, we are preparing an AI-powered chat bot which will assist the user to guide in their career path and enhance their skills for industry.

2 Literature Survey

The main or primary function of the chat bots are to have customized talks with the users of the system, get in touch and interact with them in order to connect with them and answer their queries. The chat bot technologies improve people-service communications, improving company experience. At the same time, they give new incentives to businesses to boost their customer experience and organizational performance by reducing the traditional customer care expense. So, apart from using chat bot just for conversation, they can be used in different fields like for marketing, virtual assistants, etc. One such trending area of interest where chat bots can do wonder is education.

Chat bots are nowadays used as virtual teaching assistant, used for answering inquiries of the students concerning the course module, schedule of the courses, activities, and time limits. One more use of chat bots is to make ready and mentor the user to advise them in their paths ahead in the career, answer the FAQs and to acquire the new skills. Also, the system will be providing the timely notification to the user of their status on the course and thus keep the student glued to study [1]. In order to develop an educational chat bot, the bot has to achieve a high level of accuracy in identifying the intent of the user. However if intent is not identified, then the bot cannot understand the user correctly. The chat bots have the most significant feature of being able to determine the intent of the user. The chat bot derives the necessary details from the questions asked and would try to response based on the intent identified. This makes them more interactive and easy to use by the user.

Rahman et al. [2] discuss the various types of chat bot which are developed and the complexity involved in it. The paper mentions that there are various challenges in developing a chat bot w.r.t natural language processing (NLP) and machine learning (ML). Also, the chat bots are low cost-effective solution by providing a better encounter with the system of any business by increasing the effectiveness of reaching to higher masses. Also, with the increasing complexity of chat bot, the challenges to develop the chat bot increase while a basic chat bot can be easily made.

Bocklisch et al. [3] a proposal has carried out to know about the open-source RASA framework of developing chat bots. This paper discus about in depth technology of RASA and provides insights of how RASA could be implemented w.r.t Actions, Intent, etc. The paper proposes a framework which is divides into two parts: RASA NLU and RASA CORE.

Adam et al. [4] discuss about the importance of chat bots in the user interaction field and have experimented about the same and concluded that both the verbal anthropomorphic design indication and foot-in-the-door technique has hindered the effective user experience of the user with chat bot. The paper explains an experiment, where 308 people were researched out of which only 153 participants were included in the survey. Through the experiment, author discussed about the effectiveness of the chat bot backing it up with a theory and going on with the detailed step-wise study. He also concluded that though chat bots have been a very important tool for interaction with various system, than also, many users fail to meet up the satisfactory experience while interacting with the chat bot.

Miner et al. [5] author of this paper has taken the light from the recent pandemic of COVID-19 and talked about how important is a chat bot during a pandemic. The authors start by mentioning about the CDC new launched chat bot for handling COVID-19 called Clara and during these difficult time, people tend to hide the critical information from human and tend to share with chat bot more frankly and this can lead to effective way to help people during a pandemic. But along with great position comes responsibility so here, the main responsibility is providing proper information only is the main challenge along with handling the psychological illness during the pandemic and proper guiding of the user to the precautionary steps. If this can be handled, then we can reach out to large masses to help in curbing the pandemic faster.

Smutny et al. [6] this paper gives a survey about the presence of educational chat bots on Facebook messenger platform. They found that there are 89 unique chat bot related to this field out of which only 47 chat bots meet the standard. The authors have discussed about the history of chat bot, language in which user prefer to chat and also how its use in educational field can help to reach out to masses through Facebook platform. Through this paper, the authors have clearly mentioned that even though chat bots can be used in the educational domain, it still had many scope of improvement.

Author Kasilingam [7] has mentioned about the use of chat bot in mobile ecommerce domain. He has presented the study of the newly emerging field of the use of the cha bot in the mobile e-commerce and this he has backed by starting with the fact that almost all the users spend more than 90% of their time of using mobile using emails and messaging services. And also with the chat bot nowadays being deployed in various platforms like Facebook Messenger, What's app, Skype, etc., the reach and the use of chat bot will increase. The paper also gives a detailed study and found that there is more than 72% chance that the users will adapt chat bot technology in any application for enhancing user experience.

3 Challenges Identified

Some of the challenges identified while building a chat-bot are as follows:

(1) Making chat-bot user retainable: The greatest limitation of any conversation bot is to receive and maintain the attention of the user until completion.
(2) Understanding user sentiments and emotions in case of voice bot: Speech bots can go really wrong; if the feelings of the recipient are not detected and can be deceptive. It's not easy for the bot to be educated in true human sounds, in order to build the right tone and pitch, to differentiate consumer problem from the voice, split them into substantial expectation and to react with the right voice to respond to the right feelings. It takes a lot.
(3) Building language chat bot for different languages: Because in many cases, talents get hide just because of the language barrier. To make user experience more interactive.
(4) Challenges in adopting multilingual approach: Nonstandard language, attitude and context language specialization.

4 Methodology

This proposes an educational agent system which can be used to get recommended resources, personalized professional suggestions, and help to study in different interviews. To create a well-developed virtual assistant, we should have a provision of having it to handle different input in by the user in many varied languages like English, Hindi, Gujrati, and Marathi thus making our system multilingual conversation agents. Different methods of machine learning are used to create and train the system model that can perform the following steps [8].

Step 1: The conversation of the user and a virtual assistant is started by welcoming the user by system, and goes on ahead with welcoming the user with different recommendation questions through voice or text which is based on the mode of operation selected by the user (Text or Audio) in respective language (Hindi, Marathi and Guajarati).

Step 2: The user conversation is being analyzed to detect the input language and converted to English again with the Google translate API.

Step 3: All the tokens in the sentences are corrected using a custom spell checking module. Custom Spell-Checking module is fine-tuned on data from Web sites like Quora and Reddit.

Step 4: Data is embedded into a dense vector form using word2vec vectorizations.

Step 5: The system will use cosine similarity to identify the domain of the question based on the context of the question asked by the user.

Step 6: Question is further classified into intent using context of question in order to get the clarity of intent.

Step 7: Using the NER technique, the system can extract the various entities from once the Intent Classification step is completed.

Step 8: As per the defined stories relevant action gets triggered using LSTM model and corresponding answer will be fetched and delivered using NodeJS and cloud interfaces.

Step 9: These whole stories which are nothing but pairs of intent and action mapping are saved and later referred for model revaluation to build it stronger and thus increasing the accuracy of the system (Fig. 1).

5 Implementation

This system comprises of two modules: Solving the FAQs of the user and recommending the right technology stack to a user based on his skill sets.

5.1 Implementation of Module 1

Module 1 is responsible for handling frequently asked question by users. In order to provide answer for frequently asked questions by students, we are using Rasa NLU, used to classify the intent of asked questions. Here, Rasa Core is used for providing answers from predefined lists. For example, to process a query of students as "What is the best source for getting reliable data?" the implementation of answering the FAQ's is done using following steps:

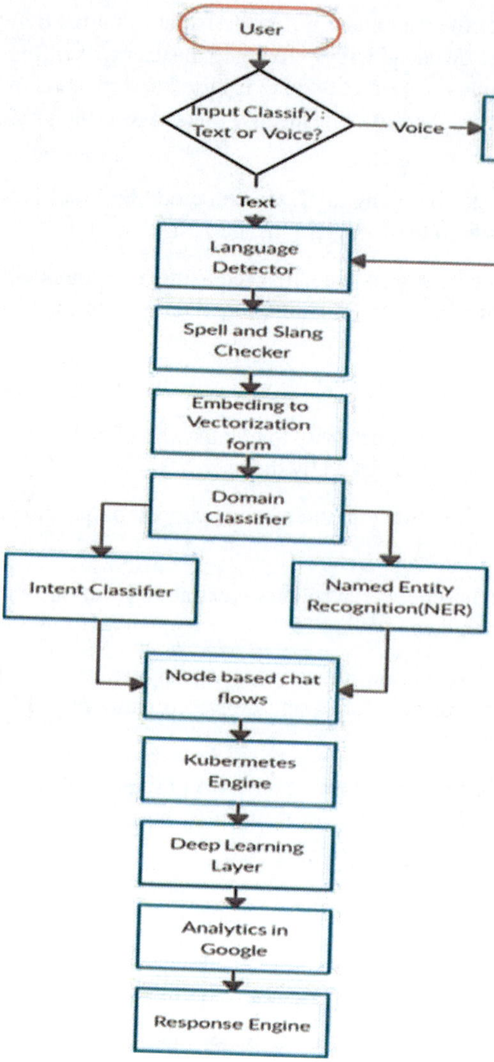

Fig. 1 System flow diagram of the system

1. **Input Conversion and Language Translator**: This step checks in which form the user input is provided in voice or text, and if that is in voice, then the system will convert the voice to text using the Google API's before processing it further. So, here, the question will remain the same as the original query was in English and in text, "What is the best source for getting reliable data?"
2. **Custom Spell Checker**: Now, using the custom checker for spelling check of that language which is trained on the data available in NLU and scraped data from Quora, reddit for different domains like Data Science, Full Stack, Machine

Learning, etc. This is implemented using Peter Norvig Method Since, there is no spelling mistakes in the query this module will also not be used. "What is the best source for getting reliable data?"

3. **Intent Classification and Entity Recognition**: Intent classification is done using the Word2vec library to load pre-trained language models, which are used as a word embedding for any word in the user document. The vector representations of words by Word embedding means that each word is converted to a thick numeric vector. The capture semantic and syntactic elements of words through Word embedding. Similar words should therefore be depicted with similar vectors.
4. **Intent**: This shows us what the user needs to do. The intent classification model is as shown in Fig. 2.
5. **Entities**: This helps in getting details of each task done by the user. The Entity classification module is shown in Fig. 3.
6. **Confidence Score**: This is a distance metric that indicates how the NLU can classify the result into the intents list.
7. **Node Based Chat Flows**: NodeJS is used as a backend framework and ReactJS is used as a front-end framework for the management of dialogue flow.

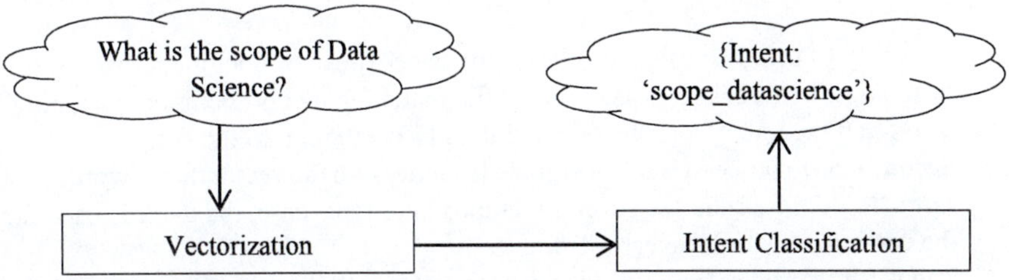

Fig. 2 Intent classification module

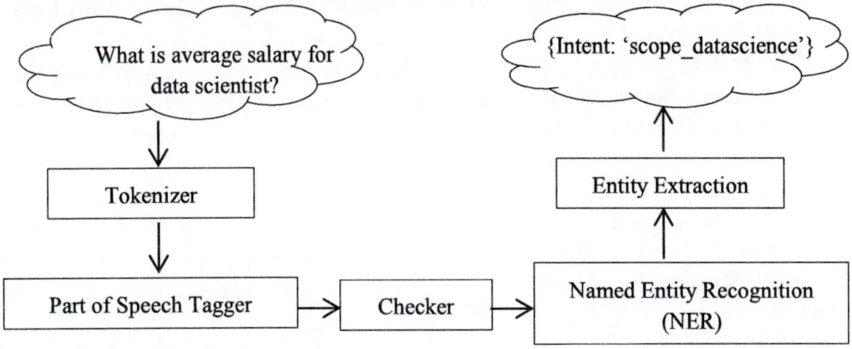

Fig. 3 Entity Identification module

8. **Google Kubernetes Engine**: Google Kubernetes Engine (GKE) is a Docker container and container cluster management and orchestration system that runs in public cloud Google services. Project Kubernetes Engine is based on Kubernetes, the container management system of Google's open source. It facilitates device delivery [9].

5.2 Implementation of Module2

Module 2 is responsible for recommending the right technology stack to a user based on his skill sets.

1. **Cosine Similarity**

The cosine similarity is measure which compute the angular cosine between the two vectors on which are calculating it. This metric does not signify the magnitude but it tells about the orientation; it can be regarded as a comparison of documents in a normalized space, because the magnitude of each count of documents, but the angle between the papers, is not taken into account [10]. It is processed using the following steps:

a. First step to use the cosine similarity for the purpose of document or sentence matching (in this case it is used for profile matching that is document matching) is to fetch the letter from the document and save them into the list.
b. Second step to proceed is to convert the list letter into the vector that is word2vect form. To do this, there are several methods but in this case, the tf-idf is used for the conversion word to vector form.
c. The tf-idf is used to get a vector for each letter, with zero and non-zero values. Zero values represent words that are common to all letters, and thus do not have any predictive power. Non-zero values are words that are not present in all letters, but maybe only a few. The same reasoning can be applied for any topic. After converting this into vector form.
d. After getting the vector, cosine similarity is calculated using (1) with the testing or incoming document. Based on that, profiles are segregated and the profile document (in this case resume) resume for career guidance recommendation [11].

$$\textbf{Similarity} = \cos(\theta) = \frac{A \cdot B}{||A||||B||} = \frac{\sum_{i=1}^{n} A_i B_i}{\sqrt{\sum_{i=1}^{n} A_i^2}\sqrt{\sum_{i=1}^{n} B_i^2}} \tag{1}$$

This system recommends the right learning track for users using this cosine similarity based on the given skill set of the user. Let us take the given recommended field as a data scientist. Then, based on the type of the level: Beginner, intermediate, and expert, it will give the user information in that field. A suggestion to the user falling in these levels is as below:

1. So, for the beginner data scientist user level is the one who has heard about the domain and has a little bit of skill set for that domain so the system gives as suggestion the prerequisite courses to the user like probability, statistics, etc.
2. For the users who have intermediate level of skills sets in data scientist, the system gives the list of suggested courses which are the advanced level proper course for the field like reinforcement learning, using AI for image processing, etc.
3. For the expert level data scientist type of user, system gives the list of suggested jobs and internships, and projects he/she should do.

6 Result Analysis

To better search the correct answer and extract the most useful information, we have evaluated the performance of the chat bot to see the correctness or its accuracy with which it is able to answer the question asked by the user. This testing is done based on the feedback provided by the users. And these users include various professionals who tested the chat bot and along with that many normal day users of our system, the college going students. The chat bot was tested by almost 6700 times (Fig. 4).

We found out that the most answers fall under the Good Answer (46%) meaning the answers provided by the chat bot was not the actual expected answer but which in a way answers the user. The second maximum category is the Related (22%) which means the chat bot could not find the answer for the question asked by the user so it showed a list of similar question which might be related to the question asked by him/her. Third category is Exact Answer (21%) indicating that the user has got the exact answer to the question asked by him/her. The category having the least percentage is Unrelated (11%) which includes the questions of user is not found by chat bot so it replies gives default reply that sorry answer not available.

7 Conclusion

There has been an increasing hype and usage of chat bots nowadays. If chat bots are used as career assistants who can guide an individual to the right track, then this will motivate people to learn and acquire new skills. This paper addresses one such multilingual career guidance chat bot with an objective of bridging gap between industry

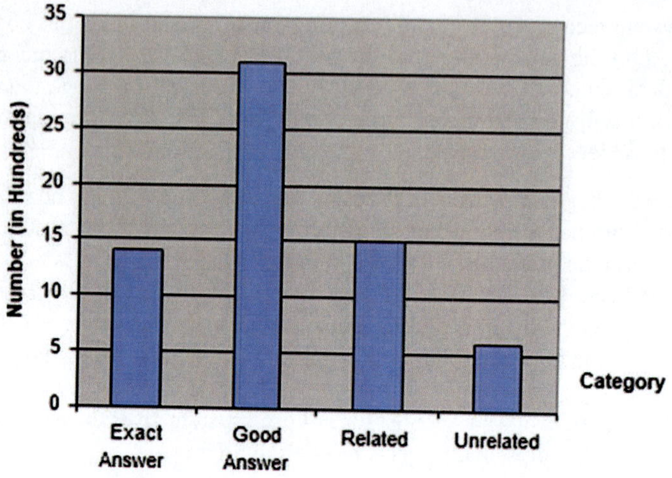

Fig. 4 Graph of performance of chat bot

requirements and students skill set in order to solve the problem of un-employability. The main aim of developing such system is to make chat bot understand the intent of the user even though if questions are not syntactically correct. This system focuses on responding to career related user queries and also recommends some courses required for profile building. In future this chat bot could be used as self-learning agent, by evaluating its performance through user feedback and improves itself.

Appendix

See Table 1.

Table 1 Abbreviation

Word	Abbreviations
ECG	Educational and Career Guidance
NLP	Natural Language Processing
AI	Artificial Intelligence
FAQ	Frequently Asked Questions
NLU	Natural Language Understanding
API	Application Programming Interface
LSTM	Long Short Term Memory
GKE	Google Kubernetes Engine

References

1. What is a Chatbot and How to Use It for Your Business, Anadea, https://medium.com/swlh/what-is-a-chatbot-and-how-to-use-it-for-your-business-976ec2e0a99f [Online]
2. Rahman, A.M., Al Mamun, A., Islam, A.: Programming challenges of chabot: current and future prospective. IEEE (2017)
3. Bocklisch, T., Faulkner, J., Pawlowski, N., Nichol, A.: Rasa: open source language understanding and dialogue management. In: NIPS 2017 Conversational AI Workshop, Long Beach, CA, USA (2017)
4. Adam, M., Wessel, M., Benlian, A.: AI-based chatbots in customer service and their effects on user compliance. In: Electronic Markets, pp. 1–19 (2020)
5. Miner, A.S., Laranjo, L., Kocaballi, A.B.: Chatbots in the fight against the COVID-19 pandemic. NPJ Digit. Med. 3(1), 1–4 (2020)
6. Smutny, P., Schreiberova, P.: Chatbots for learning: a review of educational chatbots for the facebook messenger. Comput. Educ.103862 (2020)
7. Kasilingam, D.L.: Understanding the attitude and intention to use smartphone chatbots for shopping. Technol. Soc. 101280 (2020)
8. Rasa Docs, Official Rasa Webpage, https://rasa.com/docs/ [Online]
9. Basic Chatbots Using RASA, Lishi Patel, https://medium.com/acmvit/basic-chatbots-using-rasa-d15efd4ec1f7 [Online]
10. Top Research papers in Conversational AI for Chatbots and Intelligent Agents, Mariya Yao, https://www.topbots.com/most-important-conversational-ai-research/ [Online]
11. Chatbot Report 2019: Global Trends and Analysis, BRAIN [BRN.AI] White Label Chatbots, https://chatbotsmagazine.com/chatbot-report-2019-global-trends-and-analy-sis-a487afec05b [Online]

Detection of Cardiac Stenosis Using Radial Basis Function Network

G. Indira Priyadarshini, K. Sujatha, B. Deepa Lakshmi, and C. Kamatchi

Abstract Worldwide, the cardiac stenosis is the number one cause of death among human community. The countries which are below the poverty line and having mediocre income experience nearly 80% of deaths due to cardiovascular disease. If the present status continues, approximately, about 20 million people will die due to coronary artery block by 2030. Effective decision making is the only solution the physicians can offer to detect the occurrence of heart disease so that the patients can be relaxed and can undergo treatment at appropriate time. The heart attack mainly occurs due to decreased blood and oxygen supply to the heart due to deposition of cholesterol on the sidewalls of the arteries causing heart disease. Conversely, there is a shortage in effective scrutiny tools to identify the concealed relationship in the data pattern. This proposed technique intends to provide a study of existing techniques to form the knowledge base which will guide the cardiologists to take an effective decision. The objective of this scheme is to diagnose the presence of various levels of block in coronary artery using the features extracted from the computed tomography (CT) angiogram. Seven features with radial basis function network were used to measure the percentage of blocks in the coronary artery. This investigation proved that the data mining approaches like RBFN were used to measure the blocks in the coronary artery.

Keywords Cardiac disease · Coronary artery stenosis · Computed tomography angiogram · Radial basis function

G. I. Priyadarshini · K. Sujatha (✉) · B. D. Lakshmi
ECE Department, Dr. MGR Educational and Research Institute, Chennai, India

EEE Department, Dr. MGR Educational and Research Institute, Chennai, India

ECE Department, Ramco Institute of Technology, Rajapalayam, India

C. Kamatchi
Department of Biotechnology, The Oxford College of Science, Bangalore, India

H. Sharma et al. (eds.), *Congress on Intelligent Systems*, Advances in Intelligent Systems and Computing 1334, https://doi.org/10.1007/978-981-33-6981-8_39

1 Introduction

The heart in the human body is only one fist zone. It is the strongest muscle in the human body. The heart starts to beat in the uterus long before birth, usually by 21–28 days after conception. The average heart beats about 100,000 times daily or about two and a half billion times over a 70 year lifetime. With every heartbeat, the heart pumps blood around the body. It beats approximately 70 times a minute, although this rate can double during exercise or at times of extreme emotion [1, 2].

From the left chamber, blood is pumped out of the heart to all other parts in the human body. The arteries carry oxygenated blood to all parts of the body. The arteries finally branch as capillaries supplying oxygenated blood to all tissues, like skin and other major organs present in the human body. The deoxygenated after supplying oxygen and nutrients enters the heart for purification through the right chamber via the veins. When the oxygenated blood is circulated through the liver, the waste products are eliminated. Hence, if there is any breakdown in this network, the entire system collapses and may prove to be fatal to humans [3, 4].

When there is a hindrance in the blood vessels, there is a reduction in blood supply to the heart causing coronary heart disease. The major factors contributing to coronary heart disease are high blood pressure, high saturated blood cholesterol, tobacco and alcohol use, unhealthy oily diet, physical inactivity, increase in blood sugar level, and any other genetic disorders. The other factors include malnutrition due to poverty, lack of awareness, mental depression, and swelling of blood vessels [5, 6].

Rheumatic heart disease damages the cardiac muscles and valves due to rheumatic fever, caused by streptococcal bacteria. Genetic factors contribute to congenital heart disease like hole in the heart, malfunctioning of heart valves, and chambers. All these defects arise due to alcohol consumption, usage of medicines like warfarin during the pregnancy period, infections like rubella, malnutrition, and consanguinity. Stroke occurs due to interrupted blood supply to the brain which may happen due to blockage (ischaemic stroke) or bursting of a blood vessel (haemorrhagic stroke). Deaths from cardiovascular diseases (CVD) and advancing age prove to be challenge in the field of medicine [7].

2 Literature Survey

Block in coronary artery is a common and frequently occurring heart disease causing heart attack. Presently, angiogram is used to diagnose block in coronary artery blockage and other heart diseases. Md. Ashraful Alam et al. (2019) has used computerized fully automated scheme for detection of block in coronary artery using image processing techniques so as avoid the manual intervention by the physician. Effective image processing and intelligent algorithms facilitate a speedy and trustworthy detection of the reduced area of the wall of coronary arteries due to the condensation by blocking agents. Either a 64 or 128slice CT image of the heart is used as input from which the region of interest is acquired. Two phases are involved, which include pre-processing and decision. The pre-processing phase involves feature extraction which is used by AI algorithms to make a decision regarding the coronary heart disease. Hence, early detection of coronary artery blockage is achieved by segmentation, quantification, and identification of degree of blockage causing heart attack [8].

Angiography technique used for the detection of coronary artery blockage leads to a variety of complications like artery dissection, arrhythmia, and also death. Nayana Mohan et al. (2016) has proposed an image processing method for identifying the stenosis in regions of the artery. Digital Imaging and Communications in Medicine (DICOM) images of the heart is used for the detection of stenosis. The region of interest (coronary artery) is segmented using vessel enhancement diffusion filter and also usage of morphological operations like dilation and erosion. The artery is segmented along the medial line using fast marching-based method. The diameter of the artery is measured along the medial line location in order to diagnose the blockage. The sudden decrease in the artery's diameter indicates stenosis. This method is evaluated for a set of images with an accuracy of 86.67% for diagnosing the stenosis [9].

3 Methodology

The CT images of the heart for 132 patients were reviewed for identification of block in the heart. Firstly, 120 CT scan images were analysed to identify and measure block in the arteries present in the human heart and 12 more CT scan images for validation. Randomly, the samples are selected with various percentages of blocks in the coronary artery. This meant that nearly, 40 patients had 10–50% of coronary artery disease, secondly, 40 patients had 50–70% of coronary artery block, and third set is with 40 patients having greater than 70% of coronary artery block. The CT images shown in Fig. 1 correspond to various levels of coronary artery block obtained from the open source database (https://openheart.bmj.com/).

4 Results and Discussion

The detection of blockage in coronary artery at various levels is substantiated with quantitative results for detection supported by training and validation of the data collected which is provided in Table 1. At the detection stage by the RBFN, a sensitivity of 97% and a positive predictive value (PPV) of 96% are achieved for block >70%, a sensitivity of 96% and a positive predictive value (PPV) of 94% are achieved for a block in the range 50–70%, a sensitivity of 92% and a positive predictive value (PPV) of 93% are achieved for a block in the range 10–50%. These results are

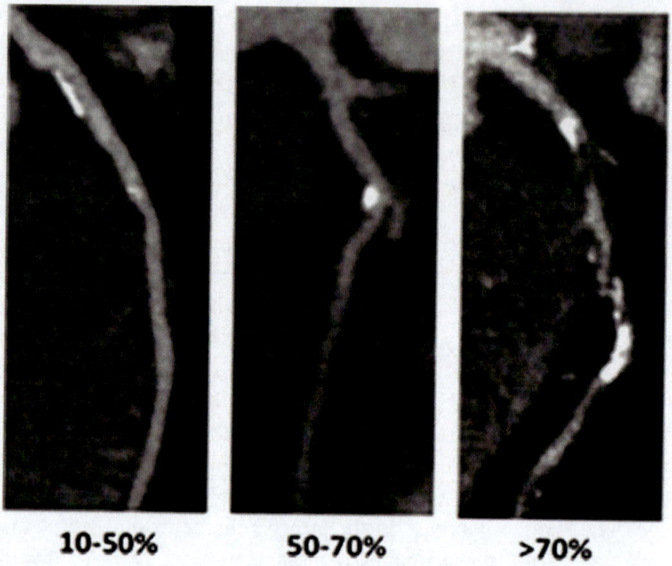

Fig. 1 CT images with various levels of block in coronary artery

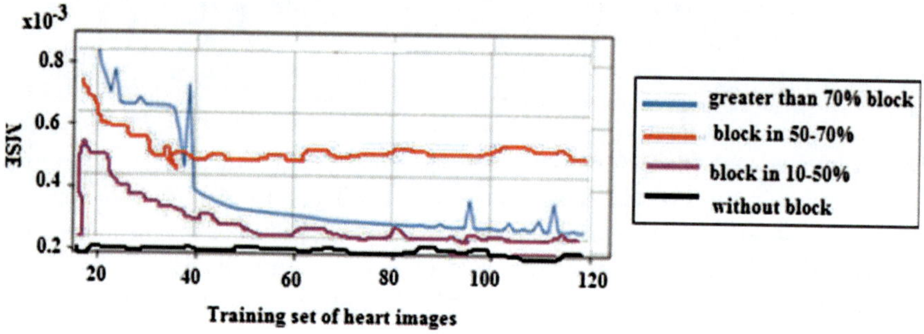

Fig. 2 Training of RBFN

Table 1 Quantitative results for coronary heart disease detection

S. No	Name of the algorithm	Number of images with 10–50%		Number of images with 50–70%		Number of images with >70%	
		% sensitivity	% PPV	% sensitivity	% PPV	% sensitivity	% PPV
1	BPA	18	24	16	22	12	23
2	RBFN	92	93	96	94	97	96

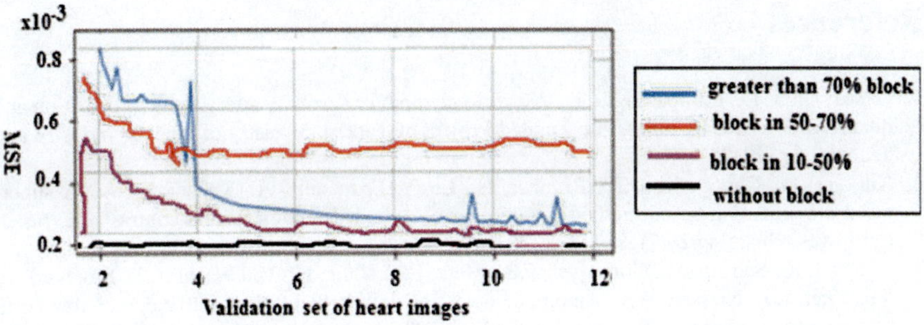

Fig. 3 Validation of RBFN

compared to BPA, while a sensitivity of 18% and a PPV of 24% for block >70%, a sensitivity of 16% and a positive predictive value (PPV) of 22% are achieved for a block in the range 50–70%, a sensitivity of 12% and a positive predictive value (PPV) of 23% are achieved for a block in the range 10–50% are obtained as compared to RBFN. However, the blockage in coronary artery is quantified with an averaged absolute difference (AAD) of only 2.7% for RBFN as compared to BPA. Thus, although the algorithm performed relatively well for detection of 50% constriction in coronary artery, it remains to be improved for better categorization of the block identification with mild, moderate, severe, and occlusion categories. Finally, the average execution time for RBFN on a CT image with 3.4 GHz processor is just about 9.9 s.

The algorithm presented here is capable of a new automated system for detection of coronary artery stenosis and quantification of the blocks. An acceptable PPV was obtained with slightly less sensitivity on quantitative evaluation. The user interaction and detection time can be improved by making small changes with respect to the parameters in the algorithm so that the identification accuracy is maximized with reduction in AAD. Figures 2 and 3 show the results for training and validation by RBFN for various conditions of block in coronary artery.

5 Conclusion

In this research paper, a RBFN-based classification method is presented to detect the coronary artery stenosis from CT A1ngiography scan images. It is inferred that a RBFN classifier is more robust as compared with BPA, so as to reduce the false positive maintaining a high accuracy in the identification rate. At the first stage, the method discriminates the arterial block followed by the density of block present within the regions of the heart. Thus, a classification-based approach has the potential to ease the difficulty of the physicians by reducing the rate of manual interaction which is eventually required to, eliminate them completely.

References

1. Oksuz, Unay, D., Kadipasaoglu, K.: A hybrid method for coronary artery stenosis detection and quantification in CTA mages. In: Funded Project of Turkish Ministry of Science Industry and Technology (2012)
2. Kurkure, U., Chittajallu, D.R., Brunner, G., Le, Y.H., Ioannis, A., Kakadiaris, A.: A supervised classification-based method for coronary calcium detection in non-contrast CT. Int. J. Cardiovasc. Imaging **26**(7), 817–828 (2010)
3. Shahzad, R., Schaap, M., Goncalves, F.B., Metz, C.T., Tang, H., Van Walsum, T., Moelker, A., Van Vliet, L.J., Niessen, W.J.: Automatic detection of calcified lesions in the descending aorta using contrast enhanced CT scans. In: 2012 9th IEEE International Symposium on Biomedical Imaging (ISBI), pp. 250–253 (2012)
4. Dirksen, M.S., Lamb, H.J., Doornbos, J., Bax, J.J., Wouter Jukema, J., De Roos, A.: Coronary magnetic resonance angiography: technical developments and clinical applications: CORONARY ANGIOGRAPHY. J. Cardiovasc. Magn. Reson. **5**(2), 365–386 (2003)
5. Goyal, P., Goyal, K., Gupta, V.: Calcification detection in coronary arteries using image processing. Lancet **3**(2277), 599–603 (2013)
6. Isgum, I., Prokop, M., Niemeijer, M., Viergever, M.A., Van Ginneken, B.: Automatic coronary calcium scoring in low-dose chest computed tomography. Med. Imaging IEEE Trans. **31**(12), 2322–2334 (2012)
7. Westin, C.-F., Wigström, L., Loock, T., Sjöqvist, L., Kikinis, R., Knutsson, H.: Three-dimensional adaptive filtering in magnetic resonance angiography. J. Magn. Reson. Imaging **14**(1), 63–71 (2001)
8. Ashraful Alam, M., Shakir, M.B., Pavel, M.I.: Early detection of coronary artery blockage using image processing: segmentation, quantification, identification of degree of blockage and risk factors of heart attack. In: Proceedings, Micro- and Nanotechnology Sensors, Systems, and Applications XI, vol. 10982 (2019)
9. Mohan, N., Vishnukumar, S.: Detection and localization of coronary artery stenotic segments using image processing. In: International Conference on Emerging Technological Trends, IEEE Xplore (2016)

Telugu Scene Text Detection Using Dense Textbox

Srinivasa Rao Nandam, Atul Negi, and D. Koteswara Rao

Abstract This paper presents an end-to-end trainable Scene Text Detector and a Script Classification Network to filter Telugu text. The Scene Text Detector named Dense Textbox efficiently makes accurate predictions using Densenet as a base architecture. It introduces K-Means clustering to set aspect ratios of default boxes and predicts at vertical offsets [−0.125, −0.25, 0.125, 0.25] similar to Textboxes and at a resolution of 512×512 to produce more text candidate proposals which are filtered using non-maximum suppression. SSD Focal loss is used instead of traditional SSD loss to accurately differentiate text regions from the background. Dense Textbox outperforms other similar frameworks on the IndicSceneText dataset using no preprocessing and standard ICDAR2013 dataset using a single pass at a single resolution. The framework has no problem in detecting text containing various scripts like the images present in IndicSceneText dataset. Traditional Scene Text Detection frameworks fail to classify the type of script Ex: Telugu, English present in an Image. We introduce a Script Classification network that is used to filter images that contain Telugu script from a non-Telugu script based on CRNN network, which uses both convolutional and Recurrent networks to classify the type of script present. Classifying and filtering text regions helps to send the text regions to appropriate Scene Text Recognition modules. The network identifies the type of script with 92.4% accuracy.

Keywords Computer vision · Deep learning · Scene text detection

S. R. Nandam · A. Negi (✉) · D. Koteswara Rao
University of Hyderabad, Hyderabad Central University Rd, CUC, Gachibowli, Hyderabad, Telangana 500046, India
e-mail: atulcs@uohyd.ernet.in

D. Koteswara Rao
e-mail: dkoteswararao_cse@mgit.ac.in

H. Sharma et al. (eds.), *Congress on Intelligent Systems*, Advances in Intelligent Systems and Computing 1334, https://doi.org/10.1007/978-981-33-6981-8_40

1 Introduction

Text plays a major role in everyday life, like communication and expressing ideas. Scene Text refers to the text that occurs in natural images as opposed to text that occurs in a document. Scene Text contains high-level semantic information that can help us understand the world around us. Some of the areas where a text extraction can be useful are license plate recognition, reading labels, industrial automation, etc. Traditional OCR is much simpler when compared to Scene text Detection because of varying lightening conditions, contrast, the orientation of the text.

Traditional Text Detection methods tend to be complex because of multiple levels of processing involved. These architectures [1–6] are hard to train because of different modules present, the number of parameters to train, the hand built features which have to be selected based on prior knowledge. The architecture for Scene Text Detection proposed is end-to-end trainable which makes it easy to train and deploy. The architecture proposed predicts word level bounding boxes similar to Object Detection frameworks [7–12]. These Object Detection frameworks form the basis of multiple Scene Text Detection frameworks based on deep learning [13–18].

Dense Textbox predicts text bounding boxes and is based on Textboxes framework [13], and it can predict text accurately and with less time compared to other Scene Text Detection methods. It replaces VGG-16 [19] with Densenet [20] as its base architecture to extract better feature representations. Utilization of multiple vertical offsets [−0.125, −0.25, 0.125, 0.25] to make more predictions at a feature cell instead of just a couple of offsets [−0.25,0.25] and a higher resolution of input 512×512 helps make more accurate and dense predictions [13]. Traditional SSD loss cannot account for the imbalance present between foreground text instances and background instances. The number of text instances is low when compared to background instances. To tackle this problem, we introduce SSD Focal loss proposed in Focal Loss for Dense Object Detection [21] to account for this imbalance. The default bounding box aspect ratios are set using K-Means clustering, unlike Textboxes, which uses predefined aspect ratios similar to YOLOV3 [12]. This helps us to use the prior information of the data to make better predictions. The text predictions happen at multiple levels of feature map hierarchy, and these predictions are offsets from default bounding boxes. There are multiple default boxes with different aspect ratios. When offsets from these default boxes are predicted at different levels of feature hierarchy, they represent different scales. These predictions of offsets for each default box at every feature map level happen at each cell in the feature map.

The predictions contain the coordinate offsets for horizontal bounding boxes combined with the conditional probability of the region represented by coordinates containing text given an input feature map. The Dense Textbox is trained from ground up without using transfer learning of weights learned on Imagenet like other similar networks [13, 14, 16]. To account for the sparsity of text objects along the vertical axis, we predict the text at different vertical offsets similar to Textboxes [13]. We use vertical offsets [−0.125, −0.25, 0.125, 0.25] and input size of 512×512 [13]. The Dense Textbox achieves better accuracy by utilizing predictions at a single scale and in a single forward pass compared to similar frameworks. The presence of a variety of scripts has no effect on detection performance, as shown in Fig. 5.

To identify the type of script which traditional Scene Text Detection frameworks fail to perform, we introduce a Script Classification module. Script Classification network uses a CRNN [22] based architecture. A CRNN is composed of a convolution network followed by a recurrent network to detect the type of script present. Script Classification helps to filter outputs based on a script and send them to appropriate Scene Text Recognition modules, which simplifies those modules. Convolution networks by themselves cannot find the inter-dependencies present in different characters of a script which recurrent networks can identify easily. It is difficult to train Convolutional networks so as to identify the type of script, and they require large datasets to produce better results. In contrast a CRNN based approach produces accurate results in classifying the type of Script present in images.

2 Related Work

Scene text detection frameworks based on traditional methods focused on hand-built features to detect and recognize the text. They use Connected Components [1, 2] or Sliding window [4–6] based approaches. Sliding window approaches use windows of varying sizes, and these windows are moved over image to generate text candidate regions. These regions are used to filter non-text regions using handcrafted rules or machine learning techniques based on predefined features. Some methods [5, 6] extract features based on different properties of text to recognize text in natural images. Label encoding which encodes a string to image information is used by other methods [23].

Deep learning greatly increases the accuracy of scene text detection frameworks. Deep learning has its benefits in learning a feature representation automatically. Various frameworks based on Object detection frameworks like SSD [10], YOLO [11], Faster RCNN [8], Mask RCNN [9] are customised to detect text present in an image. Some methods try to simplify the architecture [13, 14, 17] and some try to focus on utilizing different segments present in an image to detect text [24]. Yao et al. [25, 26] are some of the procedures that apply multi-step procedures to localize text which have complex post-processing steps. Jiang et al. [13, 14, 16, 24] are some of the simpler methods, these methods use simple post-processing and can be trained end-to-end. The above methods have made some changes to bounding box prediction and region proposal networks. Some of the methods are based on either SSD [10] or Faster RCNN [8]. SSD is only a single stage architecture, and Faster RCNN has two stages.

3 Proposed Methodolgy

Our motive is to detect horizontal text present in a given image and to filter regions containing only Telugu text. We modify SSD [10] to detect horizontal text and also utilize the procedure of predicting at multiple vertical offsets similar to Textboxes [13]. Densenet [20] is used to extract feature maps instead of VGG-16 used in Textboxes [13]. To penalize easily recognizable backgrounds and to accurately identify foreground text, SSD Focal Loss is also introduced [21]. To make dense predictions, we use predictions at multiple vertical offsets and higher resolution inputs. We utilize K-means clustering to set default bounding box aspect ratios [12], unlike Textboxes, which utilizes predefined aspect ratios. Figure 1 shows an overview of the architecture.

3.1 Dense Textbox Architecture

This architecture is based on Textboxes [13] and we use Densenet [20] as the base network. Textbox layers are attached at multiple levels of feature map hierarchy to predict the bounding box offsets. Traditional Convolution networks which have L layers have L connections. Each layer is connected to the previous one and the one after that. DenseNet has $\frac{L(L+1)}{2}$ connections. Each feature map is connected to all the previous layers. This reduces vanishing gradient problem, makes for better feature propagation, and uses fewer parameters to make better predictions. Traditional Convolutional networks take input from a layer and pass it to the next one. If the number of layers increases, this may lead to a vanishing gradient problem. To tackle this, Renets [27], Highway Networks [28] use identity connections to make feature propagation possible to lower layers of the architecture. In contrast, Densenets concatenate all previous layers to the current ones, which produce multiple channels that contain previous and current layers. This makes it so that the no of parameters to be learned becomes less compared to similar architectures like Resnet that has to add multiple filters at each layer to increase channel count.

Each layer has gradients directly passed to them by loss functions, and inputs are also directly connected. This leads to better training because of direct supervision.

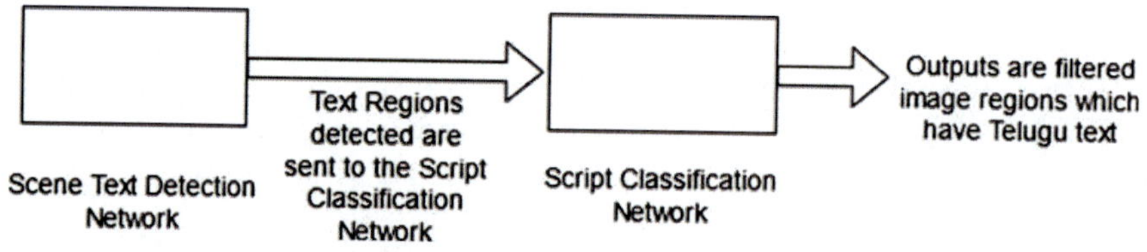

Fig. 1 Overview of Architecture for Telugu Scene Text Detection. Scene Text Detection Network detects text regions which contains both Telugu, English Text. The regions predicts are filtered to give only Telugu text regions by Script Classification network

If x is the input at layer L then H_l is applied, which is non-linear transformation. H_l is a composite function in which Batch Normalization is applied. This is followed by ReLU and a 3×3 convolution. The input to the function is the concatenation of previous layers. All the layers do not have the same size in a convolutional network. Whenever we want to concatenate layers of different sizes in Densenet, a pooling layer is used to change the size of the layer to match the lower layer. At layer L the number of filters is given by $k_0 + k \times (l - 1)$ where k_0 is the input layer channel size, and k is the growth rate. The number of channels before applying a 3×3 convolutional filter is large. To reduce them, we use a bottleneck layer. Each bottleneck layer outputs 4k channels by applying Batch Normalization, followed by 1×1 convolution.

3.2 Textbox Layers to Predict Text

Textbox layers [13] are generally 1×5 convolution filters that are applied to feature maps at multiple levels of a base network to detect text bounding boxes offsets and corresponding confidence scores. At each feature map of size $(w_{\text{map}}, h_{\text{map}})$ we predict bounding box offsets from a default box. The default box generally has aspect ratio set using K-Means clustering and generally, multiple bounding boxes are used. At position (x_i, y_i) we have the default box that is represented by (x_i, y_i, w_0, h_0) where (x_i, y_i) is the center of the box and (w_0, h_0) are the width, height which depend on the aspect ratio of the default box. At the position (x_i, y_i) we predict bounding box offsets and a confidence score as $(\Delta x, \Delta y, \Delta w, \Delta h, c)$. The formula to calculate the bounding boxes given offsets and default boxes are given by Eqs. (1), (2), (3) and (4).

$$x = x_i + w_0 \Delta x, \tag{1}$$

$$y = y_0 + h_0 \Delta y, \tag{2}$$

$$w = w_0 \exp(\Delta w), \tag{3}$$

$$h = h_0 \exp(\Delta h) \tag{4}$$

We use six default boxes with aspect ratios calculated by K-Means clustering[12]. The text is sparse along the vertical direction. To remedy this, we try to predict two offsets for each default box with vertical offsets [−0.125, −0.25, 0.125, 0.25]. We use these vertical offsets and the input size of the image is set to 512×512, unlike Textboxes [13], which uses vertical offsets [−0.25, 0.25] because of smaller input size 300×300. There are a total of seven levels of feature maps used to predict these offsets, and they represent predictions at different scales because of variation of the size of feature maps at different levels of a feature pyramid (Figs. 2 and 3).

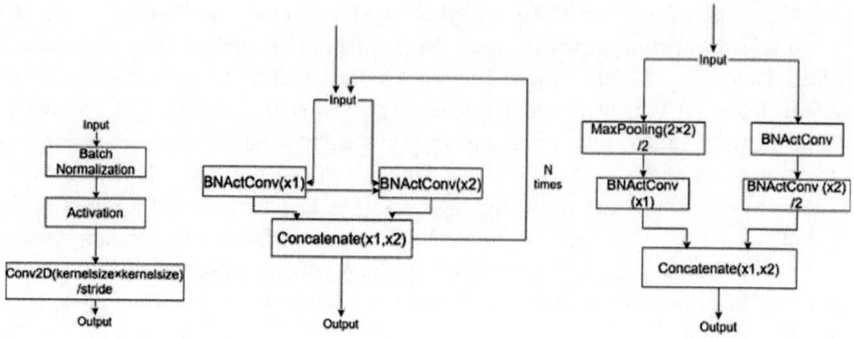

Fig. 2 BNActConv Layer applies batch normalization, activation and a convolution of size 'kernelsize' with stride 'stride', Dense layer contains multiple levels of BNActConv layers which are concatenated to form a dense block, Downsampling layer applies a maxpooling operation and a BNActConv to input parallely and concatenate them which helps to downsample the input

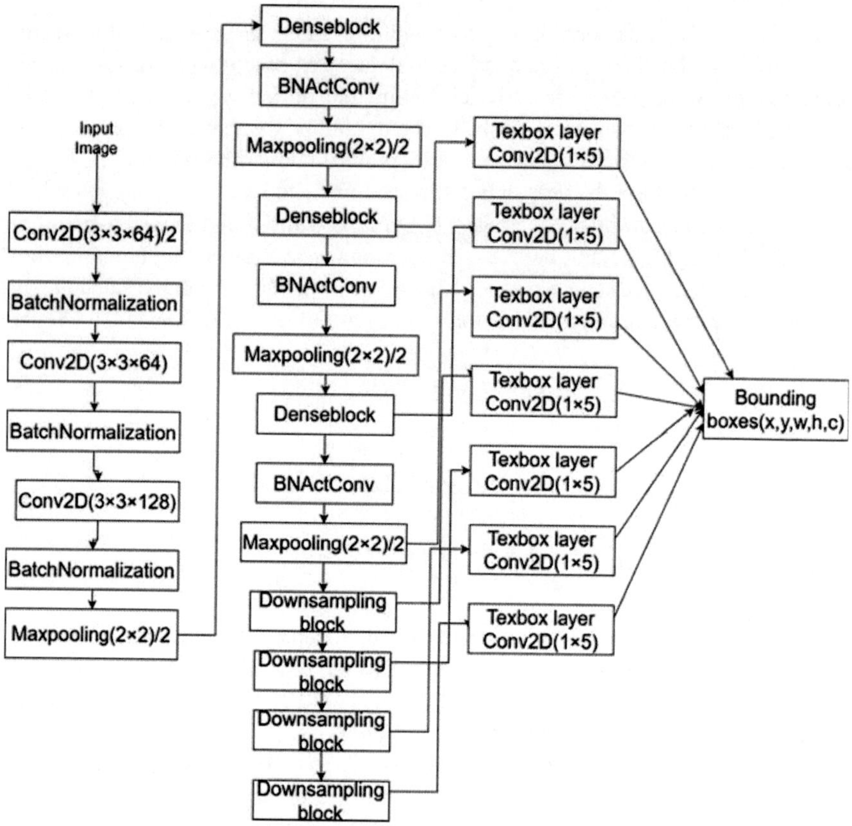

Fig. 3 Dense textbox architecture for detecting text in natural images

3.3 Loss Funtions

SSD Focal Loss function [21] has a composition of two functions one for confidence score and other for bounding box regression. The confidence score calculated by cross-entropy loss function gives easily classifiable background instances more weight during training. To reduce this, we use the focal loss presented in RetinaNet [21]. We multiply cross-entropy loss function with $(1 - p_t)^{\gamma}$. Equation 6 shows balanced cross-entropy loss 'CE' with balance factor α_t and Eq. (7) shows the modified focus cross-entropy function 'FL' with the term $(1 - p_t)^{\gamma}$ added where p_t is the input probability. Let 'x' is a matrix that represents if the default box and ground truth have a match, that is, if they are intersecting or not above a certain threshold, g be the ground truth location, c represents confidence score, l be the predicted location. Then the loss function is given by Eq. (5)

$$L(x, c, l, g) = \frac{1}{N}(L_{\text{conf}}(x, c) + \alpha L_{\text{loc}}(x, l, g)) \tag{5}$$

$$\text{CE}(p_t) = -\alpha_t \log(p_t) \tag{6}$$

$$\text{FL}(p_t) = -(1 - p_t)^{\gamma} \log(p_t) \tag{7}$$

Here L_{conf} is a focal cross-entropy loss function, and L_{loc} is a smooth L1 loss function that is used regress bounding box offsets. N is the no of bounding box predictions, and α is a hyperparameter.

3.4 Script Classification Network

Script Classification network detects if the type of script is Telugu or not. Convolutional layers are used to extract features from an input image which are passed onto recurrent layers to classify input images. The architecture used is similar to the architecture used in CRNN-CTC [22]. We replace CTC loss function with Softmax loss, and the recurrent layer output at last step is used for Script prediction instead of predicting at each step like CRNN-CTC framework mentioned in the paper [22]. The architecture consists of five convolutional layers, followed by two Bidirectional LSTM layers. The last step prediction of the LSTM layer is passed to a 2-class softmax layer.

4 Experiments

4.1 Results of Scene Text Detection Framework

The Scene Text Detection framework is implemented using Keras framework. We show the results on ICDAR 2013 [29] and InidicSceneText2017 [30] dataset. SynthText [31] is used to first train the data. It consists of 800 k synthesized text images that contain text blended with natural images. ICDAR 2013 consists of real-world images, and the text present in these images are mostly in a horizontal orientation. This dataset consists of 229 training images and 223 test images. It is a standard dataset used to compare Scene Text Detection Frameworks. These images contain various complexities, including lightening, contrast. IndicSceneText2017 dataset consists of around 1000 images containing Malayalam, Telugu, Devanagari, English text instances. 80% is used for training and the rest for testing. It also contains mostly horizontal oriented text instances, and these are also of varying complexity. It is used to train network to show detection performance when images contain various scripts.

The input size of the images is 512×512, which is passed to the input layer. SSD Focal Loss [21] is used as a loss function. During training for ICDAR, IndicSceneText aspect ratios are calculated using KMeans clustering [12]. KMeans clustering is applied to the dataset using the height and width of the image normalized to the input image size, which is 512×512. This data is then used to calculate the aspect ratio of each image and then cluster these into clusters of size 6. Weight decay is set to 5×10^{-4}. The learning rate is set to 5×10^{-4}, and Momentum is set to 0.9. The training is done on Synthtext for 60 k iterations. For IndicSceneText, ICDAR 2013 datasets, the weights trained on Synthtext are used as the basis, and they are fine tuned to those particular datasets. Instead of training on single datasets for training, we use a mix of IndicSceneText, ICDAR 2013, ICDAR 2011 datasets to increase precision and recall.

As shown in Table 1, we get 83% F-measure on ICDAR 2013 dataset and IndicSceneText datasets without using multiple passes at different resolutions like Textboxes. Dense Textbox results are shown for a single pass of an image at a single resolution of 512×512.

Table 1 Comparision of different frameworks on ICDAR, IndicScenetext2017 dataset

Architecture	ICDAR 2013			IndicSceneText2017		
	Precison	Recall	F-measure	Precison	Recall	F-measure
Textboxes	0.74	0.86	0.80	0.70	0.81	0.75
Faster RCNN	0.56	0.60	0.59	0.53	0.59	0.55
Proposed method	0.836	0.826	0.830	0.838	0.841	0.839

Fig. 4 Outputs of Text Detection Network on different IndicSceneText test set images with mostly horizontal text instances. Inputs are the same images without the highlighted boundaries and masks. The predicted box is represented by a rectangle with a green boundary and Groundtruth is highlighted with a green mask. The network detects almost all text instances without any issue even when a variety of scripts are present in an image and has some difficulties with oriented and soft text instances

ICDAR 2013 dataset, IndicSceneText dataset, is used for comparison of Dense Textbox and Textboxes, as shown in Fig. 5. This dataset is a standard dataset used by all the architectures to measure text detections. The results show that the Dense Textbox identifies the text regions with greater accuracy. The architecture produces good results, as seen in Fig. 4, it has detected all the text instances without any issue. The instances present are mostly horizontal. Texts of various script types are also detected, as shown in Figs. 4 and 5. As shown in Fig. 4, some of the images shown are difficult for the network to detect because of various complexities. The network finds it challenging to detect text with orientation. The orientation of text makes it difficult to localize text. The network also finds it difficult to detect soft text instances, as shown in Fig. 4.

4.2 Results of Script Classification Network

Script Classification network uses CRNN [22] attached to a 2-class softmax layer. The data used is from ICDAR 2013, IndicSceneText2017 cropped text instances. As previously stated IndicSceneText2017 and ICDAR 2013 dataset contain text instances in natural images.

Learning rate is set to 0.001 and Adam is used as an optimizer. The input given to the image is of size 160×32. We use a cross-entropy function as a loss function to train a network. Some of the results are shown in Fig. 6. The network can distinguish the text script present in the images. The accuracy obtained is 92.4%. The overall flow of inputs through the network is shown in Fig. 6.

Fig. 5 **a**, **d** are input images of standard ICDAR 2013 test data used to compare horizontal text detection, **g**, **j** are IndicSceneText dataset images that contain Telugu, Malayalam, Devanagari **b**, **e**, **h**, **k** are output of Proposed Dense Textbox, **c**, **f**, **i**, **l** are outputs of Textboxes architecture. The predictions of Dense Textbox shown by the green border rectangle show that they accurately predict the regions failed to be detected by Textboxes. A green mask is used to highlight ground truth boxes

Fig. 6 Overview of working of architecture. Scene Text detection produces detections as shown in the second image with green borders showing detections. The detections are sent to the classification network which classifies them as containing either Hindi or Telugu text

5 Conclusion

We present a precise and effective Scene Text Detection framework and a Script Classification network to filter text regions. The framework proposed detects text accurately, even when the image has multiple scripts like in IndicSceneText Dataset. Extracting features using Densenet [20] as base architecture helps produces better feature representations. By predicting at various vertical offsets [13] at higher resolutions and using the dense predictions produced, we can accurately predict the text bounding boxes. SSD Focal Loss [21] helps with false positives of background instances, which produces more precise bounding boxes. K-Means clustering used for setting aspect ratios also further helps predictions [12]. The addition of Script Classification Network can simplify the process of Scene Text Recognition, making it possible to apply different Text Recognition modules to different scripts. The results show accurate detection and classification of text present in an image. The Scene Text Detection framework can be extended to oriented text detection and also produce segmentation masks.

References

1. Epshtein, B., Ofek, E., Wexler, Y.: Detecting text in natural scenes with stroke width transform. In: 2010 IEEE Computer Society Conference on Computer Vision and Pattern Recognition, pp. 2963–2970 (2010)
2. Huang, W., Lin, Z., Yang, J., Wang, J.: Text localization in natural images using stroke feature transform and text covariance descriptors. In: Proceedings of the 2013 IEEE International Conference on Computer Vision, ICCV'13 (USA), pp. 1241–1248, IEEE Computer Society (2013)
3. Jain, A.K., Yu, B.: Automatic text location in images and video frames. In: Proceedings. Fourteenth International Conference on Pattern Recognition (Cat. No.98EX170), vol. 2, pp. 1497–1499 (1998)
4. Coates, A., Carpenter, B., Case, C., Satheesh, S., Suresh, B., Wang, T., Wu, D.J., Ng, A.Y.: Text detection and character recognition in scene images with unsupervised feature learning. In: 2011 International Conference on Document Analysis and Recognition, pp. 440–445 (2011)
5. Shi, C., Wang, C., Xiao, B., Zhang, Y., Gao, S., Zhang, Z.: Scene text recognition using part-based tree-structured character detection. In: 2013 IEEE Conference on Computer Vision and Pattern Recognition, pp. 2961–2968 (2013)

6. Yao, C., Bai, X., Shi, B., Liu, W.: Strokelets: a learned multi-scale representation for scene text recognition. In: 2014 IEEE Conference on Computer Vision and Pattern Recognition, pp. 4042–4049 (2014)
7. Girshick, R., Donahue, J., Darrell, T., Malik, J.: Rich feature hierarchies for accurate object detection and semantic segmentation. In: 2014 IEEE Conference on Computer Vision and Pattern Recognition, pp. 580–587 (2014)
8. Ren, S., He, K., Girshick, R., Sun, J.: Faster r-cnn: towards real-time object detection with region proposal networks. In: Proceedings of the 28th International Conference on Neural Information Processing Systems—Volume 1. NIPS'15, (Cambridge, MA, USA), p. 91–99, MIT Press (2015)
9. He, K., Gkioxari, G., Dollár, P., Girshick, R.: Mask r-cnn. In: 2017 IEEE International Conference on Computer Vision (ICCV), pp. 2980–2988 (2017)
10. Liu, W., Anguelov, D., Erhan, D., Szegedy, C., Reed, S.E., Fu, C.-Y., Berg, A.C.: Ssd: single shot multibox detector. In: ECCV (2016)
11. Redmon, J., Divvala, S., Girshick, R., Farhadi, A.: You only look once: unified, real-time object detection. In: 2016 IEEE Conference on Computer Vision and Pattern Recognition (CVPR), pp. 779–788 (2016)
12. Redmon, J., Farhadi, A.: Yolov3: an incremental improvement (2018) CoRR abs/1804.02767
13. Liao, M., Shi, B., Bai, X., Wang, X., Liu, W.: Textboxes: a fast text detector with a single deep neural network. In: The Association for the Advancement of Artificial Intelligence (AAAI) (2017) abs/1611.06779
14. Liao, M., Shi, B., Bai, X.: Textboxes++: a single-shot oriented scene text detector. IEEE Trans Image Proc **27**(8), 3676–3690 (2018)
15. Tian, Z., Huang, W., Tong, H., He, P., Qiao, Y.: Detecting text in natural image with connectionist text proposal network. ECCV **9912**, 56–72 (2016)
16. Jiang, Y., Zhu, X., Wang, X., Yang, S., Li, W., Wang, H., Fu, P., Luo, Z.: R2CNN: rotational region CNN for orientation robust scene text detection (2017) CoRR abs/1706.09579
17. Zhou, X., Yao, C., Wen, H., Wang, Y., Zhou, S., He, W., Liang, J.: East: an efficient and accurate scene text detector. In: 2017 IEEE Conference on Computer Vision and Pattern Recognition (CVPR), pp. 2642–2651 (2017)
18. Lyu, P., Liao, M., Yao, C., Wu, W., Bai, X.: Mask textspotter: an end-to-end trainable neural network for spotting text with arbitrary shapes (2018) ArXiv, abs/1908.0820
19. Simonyan, K., Zisserman, A.: Very deep convolutional networks for large-scale image recognition (2015) CoRR abs/1409.1556
20. Huang, G., Liu, Z., Weinberger, K.Q.: Densely connected convolutional networks. In: 2017 IEEE Conference on Computer Vision and Pattern Recognition (CVPR), pp. 2261–2269 (2016)
21. Lin, T.-Y., Goyal, P., Girshick, R., He, K., Dollár, P.: Focal loss for dense object detection (2017)
22. Shi, B., Bai, X., Yao, C.: An end-to-end trainable neural network for image-based sequence recognition and its application to scene text recognition. IEEE Trans Pattern Anal Mach Intell **39**, 2298–2304 (2017)
23. Rodriguez-Serrano, J., Gordo, A.: Label embedding: a frugal baseline for text recognition. Int J Comp Vis **113**, 193–207 (2015)
24. Shi, B., Bai, X., Belongie, S.: Detecting oriented text in natural images by linking segments (2017)
25. Yao, C., Bai, X., Sang, N., Zhou, X., Zhou, S., Cao, Z.: Scene text detection via holistic, multi-channel prediction (2016)
26. Zhang, Z., Zhang, C., Shen, W., Yao, C., Liu, W., Bai, X.: Multi-oriented text detection with fully convolutional networks (2016) CoRR abs/1604.04018
27. He, K., Zhang, X., Ren, S., Sun, J.: Deep residual learning for image recognition. In: 2016 IEEE Conference on Computer Vision and Pattern Recognition (CVPR) (2016) pp. 770–778
28. Srivastava, R.K., Greff, K., Schmidhuber, J.: Highway networks (2015) ArXiv abs/1505.00387
29. Karatzas, D., Shafait, F., Uchida, S., Iwamura, M., Bigorda, L.G.I., Mestre, S.R., Mas, J., Mota, D.F., Almazàn, J.A., de las Heras, L.P.: Icdar 2013 robust reading competition. In: Proceedings

of the 2013 12th International Conference on Document Analysis and Recognition, ICDAR'13, (USA), pp. 1484–1493, IEEE Computer Society (2013)
30. Mathew, M., Jain, M., Jawahar, C.V.: Lbenchmarking scene text recognition in devanagari, telugu and malayalam. In: ICDAR-MOCR Workshop (2017)
31. Gupta, A., Vedaldi, A., Zisserman, A.: Synthetic data for text localisation in natural images. In: 2016 IEEE Conference on Computer Vision and Pattern Recognition (CVPR), pp. 2315–2324 (2016)

Elastic Optical Networks Survivability Based on Spectrum Utilization and ILP Model with Increasing Traffic

Suraj Kumar Naik and Santos Kumar Baliarsingh

Abstract Communication always demands growth in technologies to fulfill the requirement of bandwidth in a network. Requirements like multimedia, cloud services, high use of Internets which are exponentially increasing with respect to time. Due to this high requirement of bandwidth, it is always a challenge to maintain the survivability of the existing optical networks for a defined period. The existing network can handle the traffic within its certain limits on spectrum efficiency and data rate. These two parameters are most important to predict the survivability of the network for a long period to handle the traffic. Recent technologies are to provide better quality of service with flexible spectrum utilization and adaptive data rate in a network. The network with flexi grid in spectrum and the adaptive networking properties called elastic optical network (EONs) can improve spectrum efficiency. The paper focuses on parameters EONs by considering the spectrum utilizations and the presence of guardbands using integer linear program (ILP). The ILP returns the utilized slots with and without guardband at increased request demand in the networks. Finally, the methodology to calculate the spectrum utilization of the elastic networks discuss the survivability of the elastic optical network with guardband. The value of spectrum utilization factor with respect to the scaled traffic demand allows predicting the survivability of the existing networks.

Keywords Elastic optical network · DWDM · Flexi grid · ILP · Spectrum utilization

S. K. Naik (✉)
Siksha 'O' Anusandhan University, Bhubaneswar, Odisha, India
e-mail: surajnaik@soa.ac.in
URL: https://www.soa.ac.in

S. K. Baliarsingh
IIIT Bhubaneswar, Bhubaneswar, Odisha, India
e-mail: c115011@iiit-bh.ac.in

H. Sharma et al. (eds.), *Congress on Intelligent Systems*, Advances in Intelligent Systems and Computing 1334, https://doi.org/10.1007/978-981-33-6981-8_41

1 Introductions

Wavelength division multiplexing (WDM) is the traditional communication network for fixed grid spectrum utilization. Increased traffic requests require more bandwidth and the fixed grid technologies of the existing network can handle it with limitation. The elastic optical network changed the scenario of the spectrum handling capacity by introducing technologies like bandwidth variable transponder (BVT), wavelength selective switch WSS, and superchannels. EONs efficiently utilize the spectrum of the existing network, which overcome the drawback of a WDM network. In other word, EONs enhance the existing network spectrum handling capability to fulfill the requirement demand and increase the longevity of the networks [1]. In WDM optical networks due to the rigidness of the spectrum, it is not possible to utilize the available bandwidth and there is wastage of the spectrum. EONs support flexibility in center frequency and adaptive spectrum utilization, which prevent the wastage of ITU-T fixed grid and support the survivability beyond the 100 Gb/s era [2]. The prior concern of routing assignment in an EONs is to reduce the spectrum utilization of the existing networks to fulfill the traffic request demands [3]. The dense wavelength division multiplexing (DWDM) is used upon WDM, which enhances the spectrum capacity of the optical networks. Guardband supports uninterrupted transmission of the signal by providing spectrum space between the two channels. It avoids the interference between the data signal of the adjacent path during routing assignment. The orthogonal frequency division multiplexing enhances the spectral efficiency by orthogonally modulating the adjacent subcarrier frequency spectrum. The transmission separated by guardband and its effect in spectrum utilization factors is considered with the traffic. The spectrum utilization without guardband and the effect of the guardband in an EONs are considered together with respect to the increasing traffic and increasing guardband. On the hardware side to fulfill the traffic demand, an EONs uses wavelength (bandwidth) variable transponder, wavelength selective switch, and advanced technologies like add-drop multiplexer.

This paper focuses on trade-offs between the spectrum utilization factors and the effects of guardbands to provide a better quality of service with increased traffic demand. In the further section, DWDM and EONs network architecture model are discussed. Also, the spectrum allocation in an optical network is explained. The ILP models are elaborated in the Sect. 3. The part-4 explains the result with discussion. Finally, the last section concludes the paper. DWDM network model and elastic optical networks.

Wavelength division multiplexing (WDM) is a technique to allocate required frequencies to an optical signal. The allocated frequencies are in the range of light with a certain frequency band. WDM transmits each signal in such a way that, there is no interference with each other and each channel has its different independent bandwidth. Dense wavelength division multiplexing (DWDM) uses closely spaced frequencies, which results in handling high signal as compared to the WDM technique. The WDM/DWDM technique uses fixed grid architecture. These are considered as circuit-based switched fixed grid optical networks. In this traditional network,

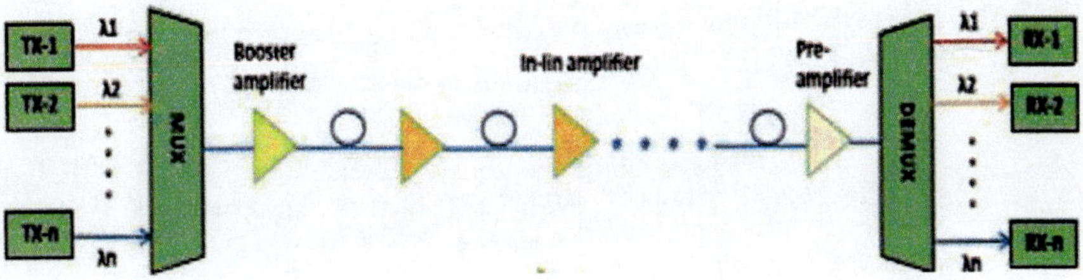

Fig. 1 DWDM system

which is nothing but a light path is transmitted through an optical channel by an optical link of the path. Basically in this, a circuit is executed in between the two optical nodes.

DWDM technology is used in a fixed grid network. Figure 1 shows a DWDM system. Here, the transmitter is an optical transmitter that generates optical signals of multiple wavelengths [4]. The different wavelength signals are then pass through a WDM multiplexer to generate a DWDM signal (Fig. 1). At this stage, the signal is in an optical domain which is then passed through the optical fiber and the optical amplifiers. The optical amplifier at different stages regenerates the signal and compensates for the fiber loss. At the receiver, the signal is passed through a demultiplexer and generates the original data signal. It provides better efficiency and increases network performance.

Nowadays, the optical network needs to be more efficient to fulfill the required demand. WDM technology support low bandwidth, which reduces the network efficiency due to fixed granularity. [4] The network efficiency will depend on how efficiently the bandwidth is used. In the fixed grid, a connection is established by using a full dedicated path between the end to end-users. Also due to the wavelength continuity constraints, the optical light path will occupy the single wavelength along with the entire fiber link.

Due to this, the fixed grid results an inefficient utilization of spectrum and reduces the spectrum utilization of the network. The flexible-grid concept overcomes this limitation and utilizes the full wavelength of the spectrum. This technology, according to the demand request efficiently utilizes the full wavelength of the network. This supports the transmission beyond 100 Gb/s and beyond [2, 5].

The requirement of efficient utilization of the spectrum can be possible by using flexi grid technology and the generated network is called elastic optical network. It fulfills the demand request and efficiently utilize the spectrum of the network. In other words, according to the requirement, it utilizes the spectrum of the network (Fig. 2).

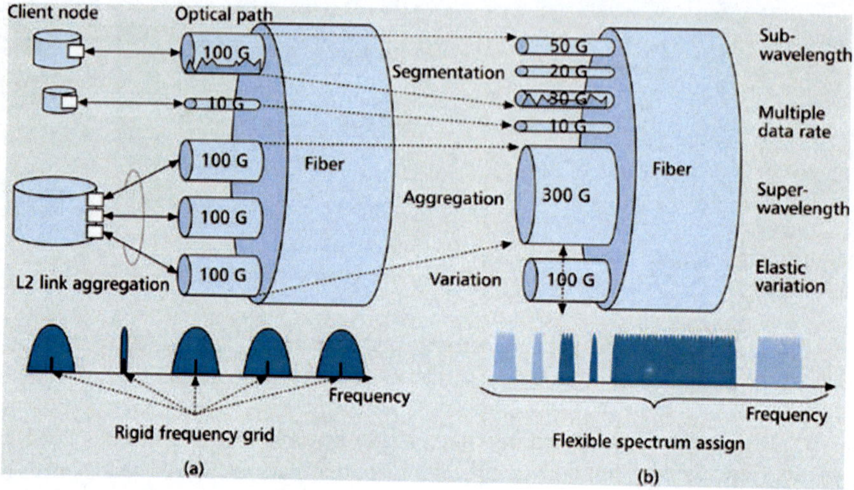

Fig.2 **a** WDM and **b** EONs technologies [3]

1.1 Spectrum Allocation

The concept of EON is to allocate the properly required wavelength (Optical bandwidth) to an end-to-end optical path. This generates the flexible-grid technologies and is different from the fixed grid WDM technology; flexible grid allow the division of optical spectrum flexibly and to generate a path with variable bitrates. According to the request demand, the EONs efficiently utilize the optical path [6, 7].

The present ITU-T frequency grid specified in G.694.1 [8] equals to 193.1 THz and it handles different spacing of 12.5, 25, 50, and 100 GHz Fig. 3.

The use of 25, 12.5 GHz (Low-frequency span) results in a reduction in transmission of signal due to unfit of request on the 12.5 GHz/25 GHz spectrum width. For example, for a traffic request of 100 Gb/s, it is not possible to aggregate the request with a 12.5 GHz spectrum. Similarly, for a higher frequency span, there is always a chance of loss due to the unused spectrum.

Figure 3 shows the spectrum allocation for a WDM network, EON with a single-slice approach, and EON with a double-sided half slice approach [3]. The first one is the single-sliced approach and the frequency segment is between 193.1 + (fn ± I?2)fcs Where fcs = Channel spacing and fn = frequency number. Here, the middle frequency of the segment deviates from the wider channel spacing frequency grid by i/2fcs. The second one is the double-sided half sliced [9] interpreted as two spectral slices at both sides of the frequency *f*. Here, the slice width of each frequency is ½ fcs. The frequency segments with various width can be represented in the same way as the single sliced on the grid approach. The half slice represents the frequency segment designated by wider channel spacing frequency grid better than the segment compatible with the ITU-T G.694.1 due to the half-width of the slot [9].

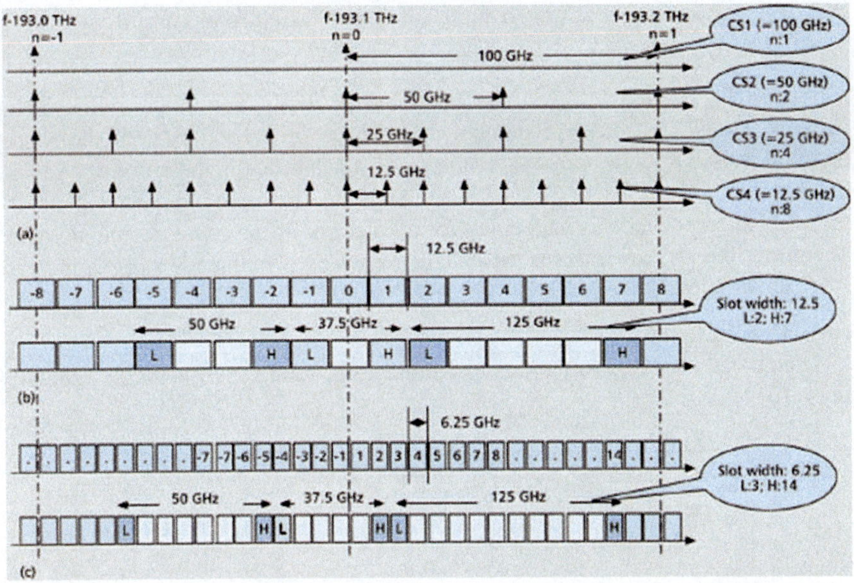

Fig. 3 Spectrum allocation: **a** WDM networks, **b** EON (single slice), **c** EON (double-sided half slice) [9]

2 ILP Model for RSA and Problem Formulation

2.1 ILP Model Without Guardband

In a communication network, such as mesh network, it is always a challenge to route the entire request without changing the traffic with maintaining all the frequency slots constraints. In this model, we first elaborate on the objective function, which is the basic model used to minimize the network frequency slots [10]. The routing and spectrum allocation (RSA) has constraints like single-path routing, spectrum continuity, and contiguity constraints are explained. The ILP takes care of the availability of the first frequency slot, frequency slot continuity constraint, and the requested demand in the link of the route.

The ILP parameters are: *S*—Source node, *d*—Destination node, *f*, *k*—Frequency slot, *r*-Route in the network, traf[*i*][*j*] = Traffic matrix for the network, and *E*[*i*][*j*] = Adjacency matrix for the network.

$\delta^{sd}_{ij,r}$	It is a binary that is set in case link is *ij* is used in route r to meet demand (*s*, *d*), and zero otherwise.
Dist^{sd}_{r}	Number of hops in route *r* for *sd*.
N	Indicate the value of frequency slots in link (*i*, *j*).
$X_{ij,k}$	Binary; 1 if frequency slot '*k*' is used in link (*i*, *j*), else '0'.

C_{rf}^{sd} Binary output; 1 if the frequency slot'*f*' is the highest slot reserved to between request (*s*, *d*) on route '*r*', else '0'.

In traf[*i*][*j*],'*s*' and '*d*' are the source and the destination of the traffic of the network. Dist_r^{sd} indicates the number of the hops in the route '*r*' of the source destination *s-d*. C_{rf}^{sd} Denotes '*f*' is the first slot that has highest priority for the request *s-d* on the route '*r*'. As we are minimizing the use of frequency slots, the process indicates that it is happening for all the link of the route, for the network. Therefore, the relation ensures the calculation of used frequency slots to increase spectrum efficiency in each link.

Objective Function:

The objective of this formula is to minimize the total use of network frequency slots [10].

$$\text{Min} : \sum_{w} \sum_{r} \sum_{f} \text{traf}[s][d] \times \text{Dist}_r^{sd} \times C_{rf}^{sd} \tag{1}$$

Constraints:

$$\sum_{r} \sum_{f} C_{rf}^{sd} = 1 \tag{2}$$

$$C_{rf}^{sd} \leq X_{ij,k}^{r}; \begin{cases} \forall s, d/s \neq d \& r \in R \\ f = 1, \ldots(N - \text{traf}[s][d]) \\ \forall i, j/E[i][j] = 1 \\ k = f, \ldots.(f + \text{traf}[s][d]) \end{cases} \tag{3}$$

$$\sum_{sd} \sum_{r} \sum_{f} \delta_{ij,r}^{sd} \times C_{rf}^{sd} - X_{ij,k} \leq 0; \begin{cases} \forall k/k \geq f \& \\ k \leq f + \text{traf}[s][d] \end{cases} \tag{4}$$

$$\sum_{k} X_{ij,k} \leq N; \forall (i, j) \in E \tag{5}$$

The constraints (2) indicate that the first frequency slot is available for each demand. Constraints (3–5) ensure for frequency slot continuity and maintain the minimum frequency slot allowed.

2.2 *ILP Model for EONs with Guardband*

Objective of this formula is to minimize the total use of network frequency slots with gaurdband [11]

Given:

Tr[s][d]	Requested traffic for the network in the form of matrix
Adja[i][j]	Adjacency matrix of the network model.
$D_{ij,r}^{sd}$	It is binary '1′ if the ij link is used in the route 'r' of demand sd and zero otherwise.
Dist_r^{sd}	Number of hops in route r to meet demand sd.
N	Available link frequency slot.
U	Frequency slot for guardband.
$X_{ij,k}$	'1', if the slot 'k' is used in link (I, j) else '0'.
C_{rf}^{sd}	'1' if 'f' is the 1st frequency slot in route 'r' of the request sd and zero otherwise.

Objective Function:

$$\text{Min}: \sum_{sd}\sum_{r}\sum_{f}[T[s][d]+U]\times \text{Dist}_r^{sd}\times C_{rf}^{sd} \tag{6}$$

Constraints:

$$\sum_{r}\sum_{f} C_{rf}^{sd}=1 \tag{7}$$

$$C_{rf}^{sd}\le X_{ij,k}^{r};\begin{cases}\forall s,d/s\ne d\&r\in R\\ f=1,\ldots(N-(Tr[s][d]+U)+1)\\ \forall i,j/\text{Adj}[i][j]=1\\ k=f,\ldots.(f+Tr[s][d]+U)\end{cases} \tag{8}$$

$$\sum_{sd}\sum_{r}\sum_{f} D_{ij,r}^{sd}\times C_{rf}^{sd}\le X_{ij,k};\begin{cases}\forall k/k\ge f\&\\ k\le f+(Tr[s][d]+U)\end{cases} \tag{9}$$

$$\sum_{k} X_{ij,k}\le N;\forall(i,j)\in \text{Adaj}[i][j] \tag{10}$$

Here, the goal of the objective formulation is to find the total frequency slot used by the network by considering the guardband.

Constraint formula (7) indicates that each connection has only one first frequency slot.

Constraint formula (8) assures for each request (s, d), the first frequency slot is within (N-(Tr[s][d] + U) + 1). In other words, the first frequency slot of the request (s, d) cannot be greater than the maximum frequency slot minus the ((Tr[s][d] + U) + 1) quantity of slot requested by request sd.

Constraint formula (9) ensures the frequency slot continuity constraints.

Constraint formula (10) indicates that total used in the link (I, j) is within the maximum available frequency slot.

2.3 Spectrum Utilization Model

Using the simple algorithm with the above ILP, it can find out the frequency slot allocation for the node pairs of the network. This can help in calculating the spectrum utilization, which is used to predict the traffic and survivability of the present network. The overall spectrum utilization for a network is expressed as

$$\eta_{su} = F/F_{max}$$

With F = Used frequency slot calculated using ILP and F_{max} indicates the complete number of FS across the network on all the fiber links. F_{max} is found by multiplying the total number of links in the network and the number of available frequency slots in each fiber. Here,

$$F_{max} = MN_E$$

where

M Number of available frequency slots in each fiber link.
N_E Total number of fiber link.

The value of F indicates the minimized value for the objective function. It is nothing but the total number of the frequency slots used in the network fora given traffic. It is always convenient to consider the fiber link as bidirectional and the value of 'M' is calculated by multiplying two with the number of available frequency slots in each fiber link. The traffic requirement is increasing drastically due to the increased use of multimedia and the Internet. The survivability of the existing network depends on the value of 'F' due to the growing demand for traffic to allow the existence of a network for a certain limit. The ILP allows us to measure the used frequency slot for a certain limit and it is decided by the available request of network traffic. At a certain increased value of the 'F', the ILP will stop working indicating that the existing network will handle the traffic for a certain limit. It is a challenge to find at what point of multiplied traffic, the ILP will stop working and its need to upgrade the network (Fig. 4).

2.4 Minimum Allowable FS Estimation Model

The traffic is generated between the source and destination in the network according to the traffic demand. For each generated request, the model search for a free-frequency slot (FS) in an available route. If the route is not free, the request is blocked and if the frequency slot is not free, the request is blocked. On the other hand, if the frequency slot for a route is free, then the estimation of the used frequency slot 'F' is calculated using the ILP model. The ILP model estimates for a particular value of the requested

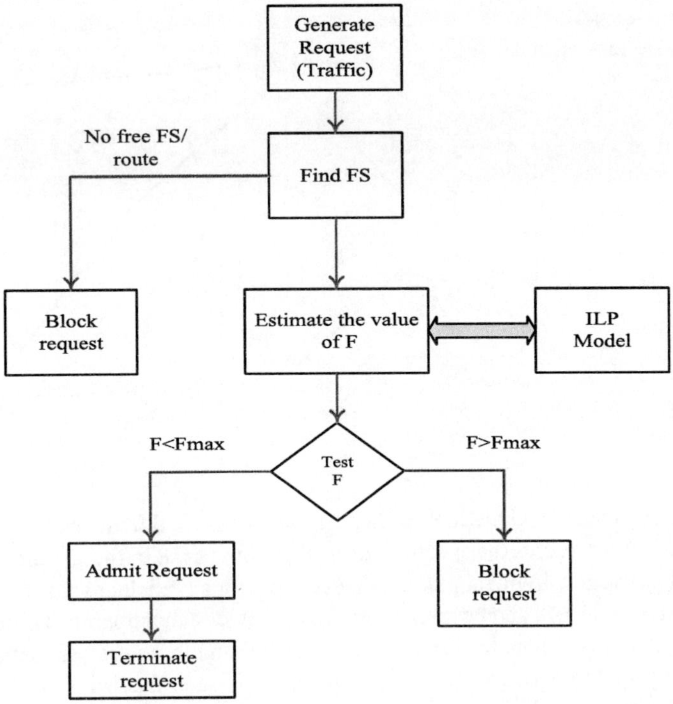

Fig. 4 FS estimation workflow

traffic and gives the value of '*F*'. It also takes care of the number of used fiber link with the request progress in the fiber link. Once the value of the used frequency slots in a network is estimated, a decision is required to admit the request or block the request. If the estimated value of the frequency slot is exceeding the Fmax, it blocks the request else, the request is accepted. The process is repeated for different traffic demand and the spectrum utilization of the network is then calculated using the above formula. The scale-up traffic demands required up-gradation in the network model. The spectrum utilization with respect to the scaling of the traffic gives a clear estimation to predict the admit traffic and the estimation of spectrum utilization can predict the required frequency slot and upgrade the elastic optical network.

3 Results and Discussion

In this work, we have used a planar network model of six nodes as in Fig. 5. The number in the link indicates the link distance in kilometers. A guardband is very essential to avoid interference between signals of the adjacent channel. The model is used to find the engaged frequency slots to calculate the spectrum utilization factor.

Fig. 5 6-node EONs network model (link distance is in 1000 s of km)

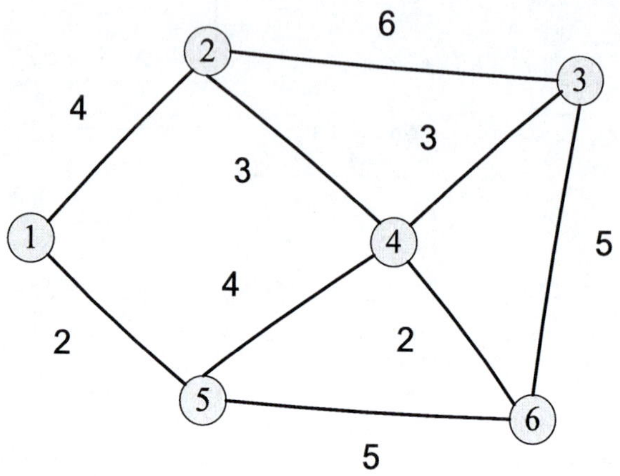

Using a traffic matrix and the ILP model, the number of used frequency slot is calculated. Figure 6 represents the spectrum utilization versus the traffic scaling factor for the value of available frequency slots. It is seen that for a particular value of available slots, the ILP will work and beyond a certain limit, it stop functioning. This indicates that the network will handle certain demand traffic and it is possible to predict the survivability of the network. Using guardband, it is seen that with the increasing value of the guardband, the spectrum utilization factor will decrease and the spectrum utilization reaches the saturation at a certain value of traffic demands [Fig. 7]. The two parameters, efficient spectrum utilization and guardband are essential for predicting the survivability of the existing networks.

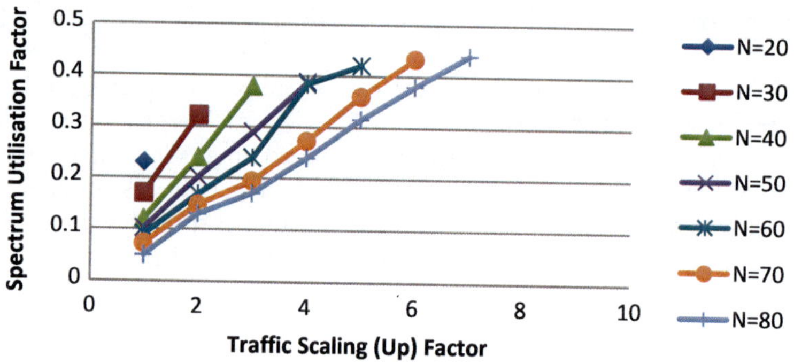

Fig. 6 Spectrum utilization factor versus traffic scaling factor

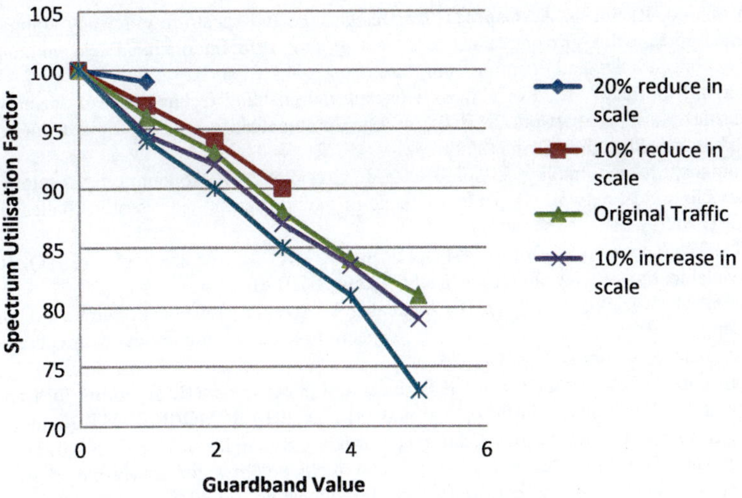

Fig. 7 Spectrum utilization versus guardband for different traffic

4 Conclusion

The paper gives a brief study of the survivability elastic optical networks by considering spectrum utilization with increasing traffic demand and the guardband. The spectrum utilization factor gives the capacity of the optical network to handle the traffic. The ILP model optimizes the network to find the minimized frequency slots. The result shows that the capacity of handling the traffic by the network depends on how efficiently the spectrum is utilized. The survivability of the network depends on the traffic handling capacity to provide a better quality of service. It is seen that the spectrum utilization factor saturates at the value of the traffic scaling factor and decreases with the increase of the guardband. So there is a tradeoff between the spectrum utilization factor and the guardian for the survivability of the elastic optical network.

References

1. Gerstel, O., Jinno, M., Lord, A., Yoo, S.B.: Elastic optical networking: a new dawn for the optical layer? IEEE Commun. Mag. **50**(2), s12–s20 (2012)
2. Jinno, M.: Elastic optical networking: roles and benefits in beyond 100-Gb/s era. J. Lightwave Technol. **35**(5), 1116–1124 (2016)
3. Jinno, M., Takara, H., Kozicki, B., Tsukishima, Y., Sone, Y., Matsuoka, S.: Spectrum-efficient and scalable elastic optical path network: architecture, benefits, and enabling technologies. IEEE Commun. Mag. **47**(11), 66–73 (2009)
4. Gerstel, O.: On the future of wavelength routing networks. IEEE Network **96**(11), 14–20 (1996)

5. Monteiro, P., Sousa, A., Lopes, C.B.: Design cost and spectrum efficiency comparison of fixed-grid and flex-grid optical networks with grooming. In: International Telecommunications Network Strategy and Planning Symposium, pp. 1–4 (2014)
6. Jinno, M., Takara, H., Kozicki, B.: Concept and enabling technologies of spectrum-sliced elastic optical path network (SLICE). In: Asia Communications and Photonics Conference and Exhibition, Shanghai, China (2009)
7. Tomkos, I., Azodolmolky, S., Sole-Pareta, J., Careglio, D., Palkopoulou, E.: A tutorial on the flexible optical networking paradigm: state of the art, trends, and research challenges. Proc. IEEE **102**(9) 1317–1337 (2014)
8. ITU-T.: Spectral grids for WDM applications: DWDM frequency grid (2012). [Online]. Available: https://www.itu.int/rec/T-REC-G.694.1-201202-I/en
9. Jinno, M., Kozicki, B., Takara, H., Watanabe, A., Sone, Y., Tanaka, T., Hirano, A.: Distance-adaptive spectrum resource allocation in spectrum-sliced elastic optical path network. IEEE Commun. Mag. **48**(8), 138–145 (2010)
10. Capucho, J.H.L., Resendo, L.C.: ILP model and effective genetic algorithm for routing and spectrum allocation in elastic optical networks. In:2013 SBMO/IEEE MTT-S International Microwave & Optoelectronics Conference (IMOC), Rio de Janeiro, pp. 1–5 (2013)
11. Adhikari, D., Datta, R., Datta, D.: Design methodologies for survivable elastic optical networks with guardband-constrained spectral allocation. In: 2018 3rd International Conference on Microwave and Photonics (ICMAP), Dhanbad, pp. 1–6 (2018). https://doi.org/10.1109/ICMAP.2018.8354464

Open MV-Micro Python Based DIY Low Cost Service Robot in Quarantine Facility of COVID-19 Patients

S. Yogesh, B. Prasanna, S. Parthasarathi, and M. A. Ganesh

Abstract COVID 19 is a pandemic which is affecting the entire world. Numerous efforts are being taken by the governments to maintain social distancing. But in quarantine facilities it is difficult to maintain these norms. Nowadays, service robots are being used in quarantine facilities to avoid spread of the virus. The robots used are generally expensive and manufacturer oriented. In this paper we propose a Micro Python based DIY low cost service robot with an algorithm to operate in quarantine facility of COVID-19 patients. By using OpenMV M7 Cam as their vision and controller element, we implement vision based navigation. OpenMV IDE is used with a PID control for smooth steering of mobile robot. Since it is a Do It Yourself robot, it can be built easily and can be used in rural quarantine facilities. For implementation, we assign a unique colour tag to each patient bed in the ward. When the required colour is selected by the operator the robot detects and tracks that colour tag on the bed. This facilitates effortless setup of service robots to assist health care workers in treating the affected patients. The paper also compares the proposed robot with the existing robots.

Keywords COVID19 · OpenMV · Micro Python · Quarantine · DIY service robot

S. Yogesh · B. Prasanna · S. Parthasarathi · M. A. Ganesh (✉)
Thiagarajar College of Engineering, Madurai, India
e-mail: ganeshma2015@tce.edu

S. Yogesh
e-mail: yogeshsrinivasanv@gmail.com

B. Prasanna
e-mail: prasannab@student.tce.edu

S. Parthasarathi
e-mail: parthasarathi_s@tce.edu

H. Sharma et al. (eds.), *Congress on Intelligent Systems*, Advances in Intelligent Systems and Computing 1334, https://doi.org/10.1007/978-981-33-6981-8_42

1 Introduction

1.1 Service Mobile Robots

The autonomous navigation of service mobile robots are based on GPS, LIDAR and Vision. GPS offers absolute location, but the bandwidth and precision are limited. It is difficult to detect small or flat objects on the ground. Additional sensors therefore have to be used to gather ground surface information [1]. Mobile robots based on vision have the potential to manoeuvre autonomously in their environment, suggesting that they can navigate an unknown environment in real time [2]. Most of the vision based robots use object tracking methods for navigation. These mobile robots use colour identification, line follower, HSI colour space based detection, shape detection and colour segmentation image processing techniques to detect the target object. By using various sensors like webcam, CMOS camera, IR sensor, LDR, ultrasonic sensor and proximity sensor [3–5]. For tracking motion they use complex fuzzy logic control, FPGA control using VHDL and neural network control algorithms are implemented using Matlab, Arduino, Microcontroller and Verilog programming software [6, 7]. Recently Raspberry pi and pi cam based object tracking mobile robots are developed and are capable of tracking objects using color in virtual grid using openCV [8]. Agribot works on voice command given by the user through Bluetooth medium will sow, furrow, watering and cut the harvest in the farm [9].

1.2 Service Mobile Robots in Hospitals

In healthcare lot of research are done by using latest technologies, such as healthcare analysis by An AI based intelligent system [10] and Early Detection of Diabetic Retinopathy Using Deep Learning Model [11]. Hospital service robots can be used for transporting medicines and food [12]. Drip-Stand Patient Follower Robot uses off-the-shelf RGB and pinhole cameras along with template matching and colour histogram comparison technique for visual tracking of markers placed on patients clothing [13]. Currently COVID 19 outbreak quickly spread all around the world. Affected people are isolated in quarantine facilities. In the quarantine facility the virus spread to the health care staff must be reduced. As a result, service robots are implemented for delivering materials and medicines to patients, cleaning, garbage disposal and sanitizing the area.

1.3 Service Mobile Robots for COVID 19

Countries like China, USA are using service mobile robots for providing the safe quarantine facilities to the patients and people working in those isolation wards.

UVD Robots disinfect patient rooms and treatment rooms in hospitals by using powerful short wavelength ultraviolet-C lights [14]. Keenon's disinfection robot uses Lidar, machine vision, and multi-sensor fusion to autonomously navigate around the quarantine area [15]. Tug robots are used to help nurses and doctors by autonomously delivering food and drugs to patients [16, 17]. NELBOT is designed to carry food, medicines, medical reports and even newspapers for patients [18]. CO-BOT can hold a tray that can carry a necessary material and deliver it to the patients by use of remote control [19].

The service robots in the literature are expensive and are not open source. In quarantine facilities of rural areas these robots are not affordable. Thus there is an immediate need to develop low cost DIY open source service robots for effective implementation of social distancing norms. In most of the existing service robots uses camera and computer to process separately and it take more time for processing.

In this paper we propose a Micro Python based DIY low cost service robot with an algorithm to operate in quarantine facility of COVID-19 patients. By using OpenMV M7 Cam as their vision and controller element, we implement vision based navigation to deliver necessary materials to the patients by maintaining social distancing. OpenMV IDE is used with a PID control for smooth steering of mobile robot. It is programmed using Micro Python programming language. This can process the image at the rate of 60 FPS and hence increasing the processing speed of the mobile robot. As we implement a small in built frame wok by using OpenMV M7 cam as an on- board controller and camera. Since it is a Do It Yourself robot, it can be built easily and can be used in rural quarantine facilities. For implementation, we assign a unique colour tag to each patient bed in the ward. When the required colour is selected by the operator the robot detects and tracks that colour tag on the bed. Thus, a robot model which is open source, do it yourself (DIY) is experimented. This model can be easily built and deployed in rural quarantine facilities. A colour tag is placed for identification of each quarantine bed and OpenMV camera is programmed on Micropython to control the robot.

2 Experimental Setup

2.1 Hardware Configuration

2.1.1 OpenMv

The OpenMV Cam is a small, low power, micro controller board which allows you to easily implement applications using machine vision in the real-world. It can be programmed with high level Python scripts instead of C/C++. This makes it easier to work with the complex outputs and high level data structures of machine vision algorithms. The OV7725 image sensor in the OpenMV M7 cam can capture 640 × 480 8-bit grayscale images or 16-bit RGB565 images at 75 FPS when the resolution is 320 × 240 and 150 FPS when resolution is 160 × 120. In this paper we configure the image sensor to capture 160 × 120 RGB565 images at 120 FPS by selecting the QQVGA mode. It can detect up to 16 colours in a frame and each colour can have any number of distinct blobs (detected coloured object). The microcontroller can control the I/O pins in real time using Python scripts. Thus OpenMV M7 Cam is choosen as a vision and controller element.

2.1.2 Two Wheeled Platform

The Two Wheeled Platform consists of two wheels mounted on a chassis (20 cm * 10 cm) in a common axis and castors wheel at the front stabilize the platform without compromising movements. The wheels are driven by a 5 V geared DC motor. This is the base of the mobile robot upon which the camera, controller, motor driver and container to carry the materials are to be mounted.

2.1.3 Motor Drive

L293D Motor driver is used to drive the wheels in differential drive configuration. It is a popular 16-Pin motor driver IC. The IC works on the principle of Half H-Bridge to run the motors. It can control a set of DC motors simultaneously in any directions as instructed by the controller.

2.1.4 Ultrasonic Sensor

In this paper, HC-SR04 sensor is used to measure the distance of obstacles from the robot. It takes Trigger Signal of 10 micros TTL impulse as an input and gives out Echo Signal as Output TTL PWM Signal. It has a sensing range of 2cm to 450cm.

2.1.5 Power Source

The power bank of 10000mAh capacity is used as a power supply to the robot. The power bank has two outputs as (5 V and 1 A) and (5 V and 2.1 A) which can provide power supply to the controller and the motor.The prototype of the proposed COVID19-BOT is shown in Fig. 1.

2.2 *Assembly Process*

By using these steps the proposed mobile robot prototype is assembled and circuit connections are implemented.

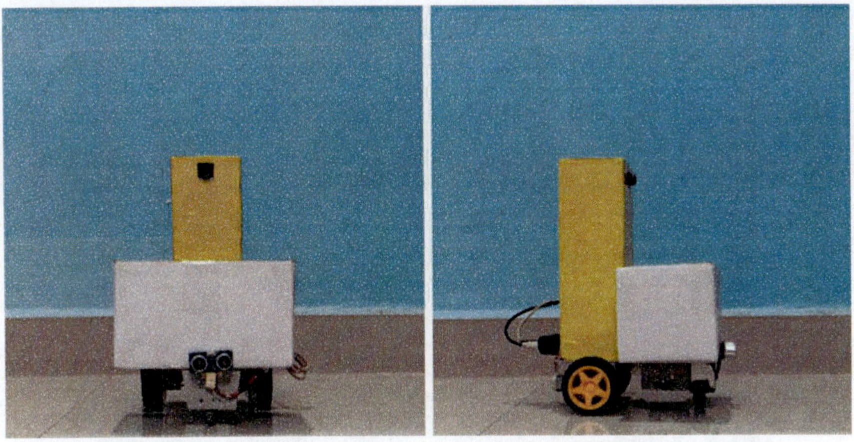

Fig. 1 COVID19- service robot prototype

- The two 5 V geared motor is fixed to the back side of our chassis.
- The castors wheel is fixed to stabilize the chassis without compromising movements.
- The cubical box of height 23 cm, width 8 cm and length 8 cm approximately is made using a foam sheet.
- The cubical box has 3 compartments each of height 7.5 cm and placed at the back side of the chassis.
- In top compartment the OpenMV M7 smart camera is placed facing the front side of the Mobile robot.
- The motor driver is placed in the middle compartment so that connection can be easily established.
- The ultrasonic sensor is placed at the front side of the mobile robot so that it can detect the obstacles and stops the robot.
- The power bank is placed at the last compartment so that it can supply power to the camera board and motor driver.

The maximum payload of the materials which the robot can carry is approximately 500 g.

2.3 Implementation Environment

Figure 2 shows a local quarantine ward for COVID 19 patients with three beds following social distancing. At each bed there is a colour tag for identification. The robot will detect and track the selected colour tag and reach the correspond- ing bed. Thus, it can deliver the materials for the corona patients such as water bottles, food packages, medicines and other necessary items and return to the operator.

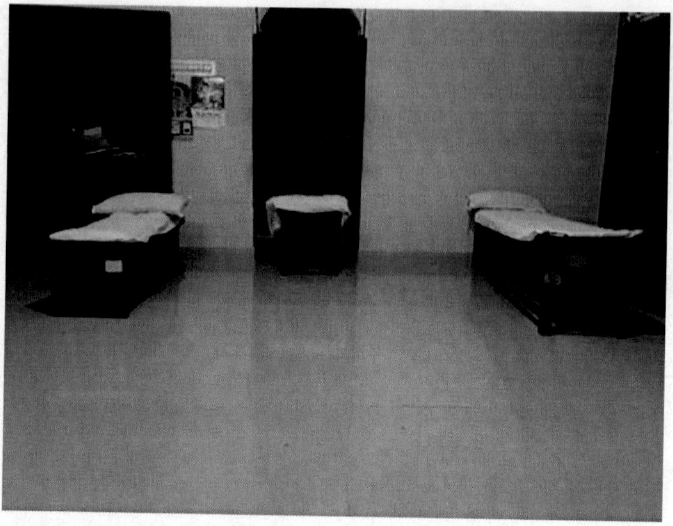

Fig. 2 Environment setup

3 Software Configuration

The implementation of the process includes communication between Micro python based OpenMV IDE, L293D motor driver and HC-SR04 sensor.

3.1 Micro Python Based OpenMV IDE

Micro Python is a lean and efficient implementation of the Python 3 programming language that includes a subset of the Python standard library and is optimised to run on micro-controllers and in constrained environments. It is compact enough to fit and run within just 256 k of code space and 16 k of RAM. Hence, it is used to program the OpenMV M7 cam board. It has a Machine Vision Support, image processing library implements a fairly comprehensive set of image processing algorithms and supports multiple image formats. By using this module we can easily detect the required colour by skipping the major image processing steps to identify the required threshold values in the image.

3.2 Selecting Threshold Values of Colour Tag

The OpenMV M7 cam can detect various colours if threshold values are provided as input. The threshold value of the colour to be detected can be obtained from

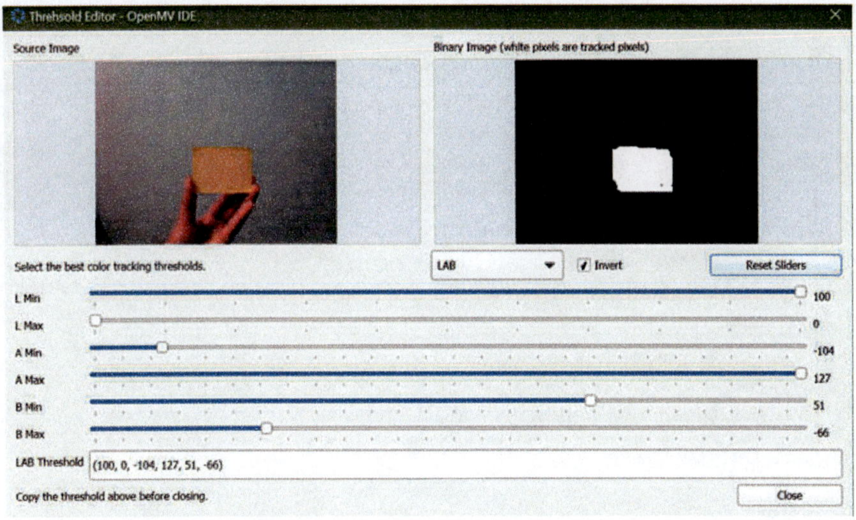

Fig. 3 Threshold editor (machine vision tool)

the OpenMV IDE by using the Machine vision tool. Select ⇒ Tools ⇒ Machine Vision ⇒ Threshold Editor ⇒ Frame Buffer to capture the image of the colour tag. The threshold editor displays it in the source image section and black and white image in the binary image section. Adjust the slides to change the L Min, L Max, A Min, A Max, B Min, B Max values till the colour to be detected appears white and others appear black in the binary image. The threshold editor is shown in Fig. 3. The threshold value of the detected yellow object is (100, 0, −104, 127, 51, −66).

The proposed algorithm used for bed detection is briefed in algorithm 1

Algorithm 1: Bed Detection Algorithm

Result: Color Blob Detection
initialize OpenMV Camera;
while *True* **do**
  Start Capturing Frames;
  if *Check For colour Blobs* **then**
    if $Threshold == obtained$ **then**
      Blob detected;
    else
      Blob not Detected;
    end
  else
    Rotate Bot in Clockwise direction;
  end
end

The proposed algorithm used for robot navigation is briefed in algorithm 2

Algorithm 2: Colour Tag tracking Algorithm

Result: Colour Tag tracking and Bot Movement
Colour tag Detected;
while *No Obstacle* **do**
  Calculate *h_error, x_error*;
  if $h\_error < 0$ **then**
    if $x\_error < 0$ **then**
      Move Left;
    else
      Move Right;
    end
    if $x\_error == 0$ **then**
      Move Forward;
    end
  else
    Idle;
  end
end

Here the x_error and h_error are given to the PID control operation and the suitable PID constant values (Kp, Ki, Kd) are identified by using trial and error method. The obtained values are Kp = 0.5, Ki = 1, Kd = 0 for x_error and Kp = 0.05, Ki = 0.1, Kd = 0 for h_error By using the PID control for these error values we get xoutput and houtput values that can be feed to the motor driver to vary the pulse width percentage input of the motors. Correspondingly it varies the speed of left and right and generates a smooth and precise movement of the robot towards the blob.

4 Results and Discussion

4.1 Bed Detection

Once the mobile robot is turned ON, the live video is streamed by the smart camera which is converted to image frames at the rate of 120 fps. By performing the colour thresholding process, binary images for each frame in the live video is generated. When the blob is detected the controller will assign a white rectangle around the blob by using the contour detection process and a cross symbol (+) at the centre of the rectangle. In this project, green, yellow and pink colour tags are used to identify the position of bed. The threshold value of green colour is (34, 46, −70, −25, −67, 127), yellow colour is (100, 0, −104, 127, 51, −66) and the pink colour is (41, 58, 33, 127, −4128, 127). These values are obtained by the Threshold editor in special

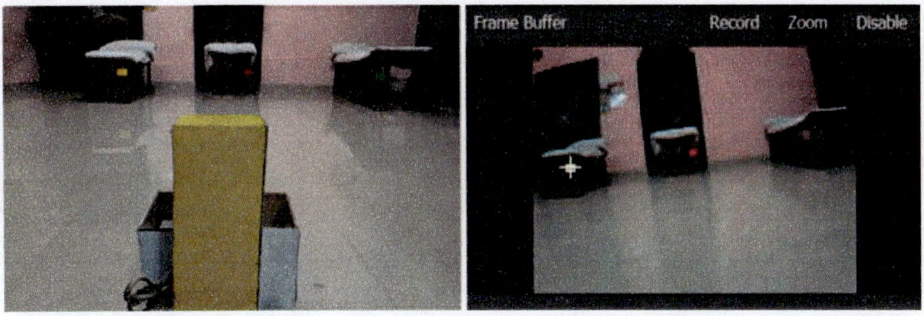

Fig. 4 Bed detection

machine vision tool of OpenMV IDE. However the robot would be able to detect a particular colour tag which is selected. The detecting of the yellow tagged bed is shown in Fig. 4.

4.2 Colour Tag Tracking

The proposed mobile robot starts to track the selected tag once it is in the field of view. x_error and h_error values which are calculated is converted to left and right wheel pulse width percentage. This varies the wheels speed with PID controller. This results in precise tracking movement of the robot.

If the position of the tag is left to the centre of the camera screen (i.e.) x_error value is negative. Thus, the robot turns left to minimize the error. Similarly, if the tag is at right to the centre of the camera screen (i.e.) x_error value is positive. Thus, the robot turns right to minimize the error. If the tag is at the centre of the camera screen (i.e.) the x_error value is zero. This drives the robot in a forward direction. The tag shown in Fig. 5 has a pixel area of greater than 2000 pixels (i.e.) the x_error and the h_error values are zero. As a result, the controller stops the mobile robot at approximately 20 cm from the colour tag. During an implementation trial, the operator wants to provide medicines to the patient in the yellow tagged bed. The bot reaches near the bed as shown in Fig. 5.

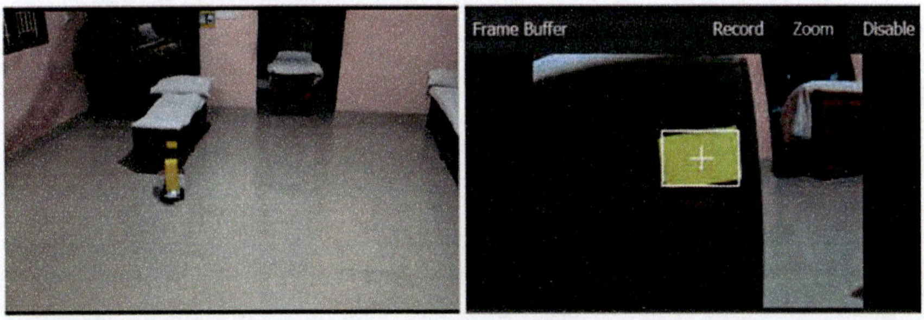

Fig. 5 Colour tag tracking

4.3 Obstacle Detection

When the mobile robot is used in quarantine facilities it can be hit by the dynamic obstacles. These obstacles can be detected by the ultrasonic sensor aboard the robot. The ultrasonic sensor detects any obstacles within 10 cm and immediately stops the robot. After detection of clear vicinity, it will continue the tracking movement without any deviation from the path.

4.4 Performance Analysis

In this project, PID control is implemented for precise tracking motion of the robot by using x_error and y_error values. The following graphs represent the error reduction and movement input of the wheels. The robot detects and tracks the yellow colour tag on the bed and reaches close to it. The distance between the initial position of the robot and the bed is approx. 3 m. It takes nearly 28560 ms (approx. 30 s) to reach that bed.

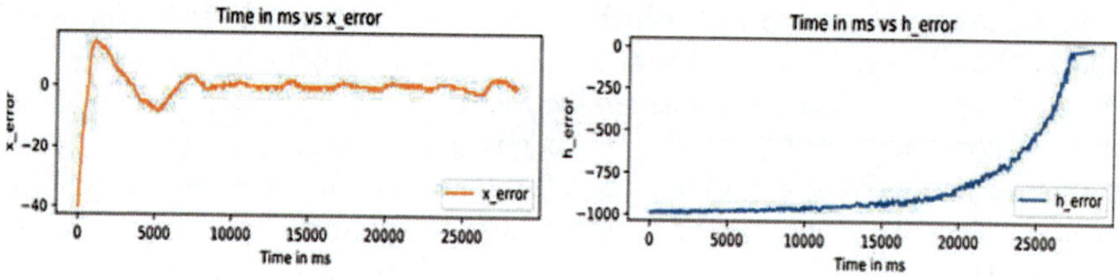

Fig. 6 Real time error correction

The graph as shown in Fig. 6 (left) is plotted between the Time and x_error value. x_error value is plotted for every 30 ms. This shows the damping line that indicates error reduction trend using PID control.

From this graph we infer that the x_error value is initially −40 indicating that the tag is at left to the camera screen. The robot turns left to minimize the error value. At 1470 milli second the error value reaches 14 indicating that the tag is at the right to the camera screen. The robot turns right to minimize the error. This left and right deviation of the tag on the screen continues until the error value reaches zero (i.e.) tag at the center of the screen. Due to the implementation of PID controller we obtained this damping line. The graph as shown in Fig. 6 (right) is plotted between the Time and h_error value. h_error value is plotted for every 30 ms. This shows the curve that indicates error reduction trend using PID control.

From this graph we infer that the h_error value is initially −1000 indicating that the pixel area of the tag is less than 2000. The robot moves towards the tag based

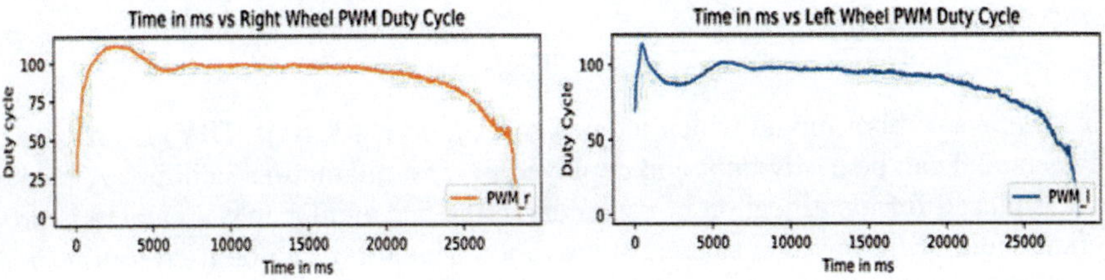

Fig. 7 PWM generation by PID algorithm

on the x_error value. At 27060 milli second the h_error value nearly reaches zero. Then the robot stops when the error value reaches zero.

Based on these x_error and h_error values the corresponding pulse width percentage for the left and right wheel is generated by the PID controller. This drives both wheels based on the generated pulses. Graph is plotted between Time and left right wheel pulse width percentage as shown in Fig. 7. From these graphs we infer that at 1000 ms the speed of the leftwheel reduces and the speed of right wheel increases. The robot turns left as the tag is on the left of the screen. At 2000 ms the speed of the right wheel increases and speed of left wheel reduces. The robot turns right as the tag is in right of the screen. Thus the left and right wheel speed keeps changing based on the position of the tag. When the errors are zero the robot stops.

4.5 Scalability and Comparison with Robots Used in Quarantine Facility

Our proposed service robot is designed to carry a maximum payload of 500 g. The total cost of the developed robot is Rs. 7500/-. To increase payload capacity of the robot to 2 kg, 12 V geared DC Motor of 2 Kgcm rated torque should be fixed at metal chassis of dimension 30 cm × 15 cm × 7cm powered by a 12 V rechargeable battery pack. The total estimated cost is Rs. 9000/-.

From the literature [16–19] it is observed that the various service mobile robots implemented in various quarantine and hospitals with its price as Rs. 114,390/- at lease per month, Rs. 1,800,000/-, Rs. 250,000/-, Rs. 250,000/- respectively. These robots are unaffordable and difficult to be implemented in rural quarantine facilities. The proposed service robot can be adapted to various sizes by simply changing its chassis, motor and power supply capacity based on the application and load to carry. This proposed prototype is suitable for small quarantine facilities. Colour tracking is difficult to scale up as it is complex to assign unique colour to all patients and maintain constant lighting. Hence to implement this in large quarantine facilities April tags or QR code tags [20] can be used instead of colour tags. OpenMV M7 camera have inbuilt open source libraries for April tags and QR code tags detection. So it is technically feasible and scalable to implement April tags and QR code tags for the identification of patient's bed.

4.6 Conclusion

In this paper, robot model which is open source, do it yourself (DIY) is proposed. This model can be easily built and deployed in rural quarantine facilities. A colour tag is placed for identification of each quarantine bed and OpenMV camera is programmed on Micropython to control the robot. This robot can be used to transport medicines, food and other necessary items to the COVID 19 patients thus implementing social distancing norms. Our robot is compared with the existing service robots in COVID 19 hospitals and it is observed that the proposed robot is open source and less expensive.

Acknowledgments We would like to sincerely thank Thiagarajar College of Engineering which provided the environment and facility to complete this project in Centre for Robotics, Department of Mechatronics, Thiagarajar College of Engineering, Madurai.

References

1. Murtra, C., Trulls, E., Sandoval, O., Perez-Ibarz, J.: Autonomous Navigation for Urban Service Mobile Robots. Int. Conf. Intell. Rob. Syst. (2010). https://doi.org/10.1109/IROS.2010.5649151
2. Chatterjee: Vision Based Autonomous Robot Navigation Studies in Computational Intelligence. Springer-Verlag Berlin Heidelberg (2013). https://www.springer.com/gp/book/9783642339646
3. Das, P.K., Mandhata, S.C., Patro, C.N., Patro, S.N.: Vision based object tracking by mobile robot. Int. J. Comput. Appl. **45**(8), 0975–8887 (MAY-2012). http://citeseerx.ist.psu.edu/viewdoc/download?doi=10.1.1.734.7780&rep=rep1&type=pdf
4. Luo, B., Luo, K., Ding, Y., Liu, J.: Research on mobile robot path tracking based on color vision. Chinese Automation Congress (CAC), IEEE (2015). https://doi.org/10.1109/CAC.2015.7382527
5. Bold, S., Jeong, M.A., Jeong, S.M., Ro Lee, S.: FPGA based real time embedded colour tracking mobile robot. In: International Conference on IT Convergence and Security (ICITCS), IEEE (2013). https://doi.org/10.1109/ICITCS.2013.6717835
6. Kim, S., Lee, S., Kim, S.: Object tracking of mobile robot using moving colour and shape information for the aged walking. In: FGCN '08: Second International Conference on Future Generation Communication and Networking vol. 02, pp. 293–297 (2008). https://doi.org/10.1109/FGCN.2008.180
7. Soans, R.V., Hegde, A., Kumar, A., Ranjith, C.S.: Object tracking robot using adaptive colour thresholding. in: 2nd International conference on Communication and Electronic System (OCT-2017). https://doi.org/10.1109/cesys.2017.8321192
8. Gurram, R., SweathaSuresh, B., Sneha, B.R., Sushmitha, R.: Object tracking robot on raspberry Pi using opencv. Int. J. Eng. Trends Technol. (IJETT) **35**(4) (2016). https://ijettjournal.org/archive/ijett-v67i10p209
9. Poluru, R.K., Reddy, M.P.K., Kaluri, R., Lakshmanna, K., Reddy, G.T.: Agribot: In Architecture and Security Issues in Fog Computing Applications. IGI Global, pp. 151–157 (2020)
10. Deepa, N., Prabadevi, B., Maddikunta, P.K., Gadekallu, T.R., Baker, T., Khan, M.A., Tariq, U.: An AI based intelligent system for healthcare analysis using ridge adaline stochastic gradient descent classifier. J. Supercomput. (2020). https://doi.org/10.1007/s11227-020-03347-2

11. Gadekallu, T.R., Khare, N., Bhattacharya, S., Singh, S., Maddikunta, P.K. R., Srivastava, G.: Early detection of diabetic retinopathy using Deep learning model", electronics **9** (2020). https://www.mdpi.com/2079-9292/9/2/274
12. Mettler, T., Sprenger, M., Winter, R.: Service robots in hospitals: new perspectives on niche evolution and technology affordances. Eur. J. Inf. Syst. (2017). https://doi.org/10.1057/s41303-017-0046-1
13. Wu, Z., Payandeh, S.: Research article: toward design of a drip-stand patient follower robot. Hindawi, J. Rob. Article ID 9080642 (Feb 2020). https://doi.org/10.1155/2020/9080642
14. Ackerman, E.: Autonomous Robots Are Helping Kill Coronavirus in Hospitals. IEEE Spectrum, (11 Mar 2020). https://spectrum.ieee.org/automaton/robotics/medical-robots/autonomous-robots-are-helping-kill-coronavirus-in-hospitals
15. The Robot Report (Exploring the business and application of robots) Keenon rolls out disinfection robot to hospitals in China and beyond"(May 1,2020). https://www.therobotreport.com/keenon-rolls-out-disinfection-robot-china-covid-19/
16. WIRED: Tug, the busy little robot nurse, will see you now (11.10.2017). https://www.wired.com/story/tug-the-busy-little-robot-nurse-will-see-you-now
17. MIR a better way Mobile robot transports sterile goods in hospital (June 2018)
18. The Hindu: COVID-19 Robot to assist doctors at Nellore government hospital (Apr 2020). https://www.thehindu.com/news/national/andhra-pradesh/covid-19-robot-to-assist-doctors-at-nellore-government-hospital/article31457740.ece
19. The Hindu: ITI Cuttack develops low-cost robots to combat COVID 19 (Apr 2020). https://www.thehindu.com/news/national/other-states/iti-cuttack-develops-low-cost-robots-to-combat-covid-19/article31463832.ece
20. Wikipedia: The Free Encyclopedia, QR Code, Geneva: ISO/IEC (2000). https://en.wikipedia.org/wiki/QR_code

Educator's Perspective Towards the Implementation of Technology-Enabled Education in Schools

Gopal Datt and Naveen Tewari

Abstract This study gives an insightful view of applications of technology-enabled education in the schools of the Uttarakhand region. As we know the Uttarakhand is a hill state where Internet connectivity is a big issue in hilly areas of the region. The government of India has launched its National Mission on Education through Information and Communication Technology in the year 2009 to produce the chance for all the learners and educators within the country to pool their collective knowledge. The policymakers are keen to use the technology in education toward strengthen the mission on technology-enabled education, to form the standard education accessible to any or all meriting learners. We have conducted an extensive survey to identify awareness, importance, and hurdles to applying the applications of technology-enabled education in the region. We have received 148 total responses about 29 survey questions, where some of them are of type qualitative and others are of quantitative by nature. The opinion of the respondents analyzed and the end results are presented. The study further states to various aspects and phases of implementation and hurdles of technology-enabled education (TEE) in schools.

Keywords Technology-enabled education (TEE) · Information and communication technology (ICT) · Online learning initiatives · NMEICT · MOOC

1 Introduction

Technology-enabled learning is the way to connect with people and able to create interaction among them regardless of time, place, and person. The advantages of ICT to enrich today's teaching–learning process with multimedia such as audio, video, pictures, animation and graphics, and Web resources [1]. In this era, the importance of technology-enabled education has been recognized across the world by the governments, learners, and educators. The Ministry of Human Resource and Development

G. Datt (✉)
School of Vocational Studies, Uttarakhand Open University, Haldwani, India

N. Tewari
School of Computing, Graphic Era Hill University, Bhimtal, India

H. Sharma et al. (eds.), *Congress on Intelligent Systems*, Advances in Intelligent Systems and Computing 1334, https://doi.org/10.1007/978-981-33-6981-8_43

(MHRD, India), Government of India (GoI) has launched its National Mission on Education through Information and Communication Technology (NMEICT) [2].

This mission has initiated with several projects related to accessing multimedia content, hands-on-training (virtual training), digital library, etc. Popular NMEICT initiated projects to promote technology-enabled education in the country are—Virtual Laboratories (www.vlab.co.in) [3], Talk to Teacher (AVIEW—Amrita Virtual Interactive e-Learning World) [4], Spoken Tutorial (www.spoken-tutorial.org) [5], Consortium for Educational Communication (www.cec.nic.in) [6], E-Yantra (www.e-yantra.org) [7], Digital Library Inflibnet (www.inflibnet.ac.in) [8], Quantum and Nano Computing (www.dei.ac.in/dei/quantumNano) [9], Oscar++ (Open Source Courseware Animations Repository), Free Open Source Software for Education project (www.Fossee.in) [10], E-Kalpa (www.dsource.in) [11], Pedagogy Project (www.ide.iitkgp.ernet.in/Pedagogy1/pedagogy_main.jsp) [12], E-PG Pathshala (epgp.inflibnet.ac.in) [13], and etc., Vidwan-Expert Database and National Researcher Network (vidwan.inflibnet.ac.in) [14], Indian Research Information Network System (irins.org/irins) [15], Virtual Laboratories project is a paradigm shift that has been set up for remote-experimentation where too many Web-enabled experiments are designed for remote-operation and viewing. Virtual Laboratories do not require any additional infrastructural setup for conducting experiments at user premises; as its name implies virtual, it facilitates simulation-based experiments and that can be accessed remotely via the Internet from anywhere.

A-VIEW is an indigenously e-multimedia and e-learning platform that is almost as good as a real classroom experience. Spoken Tutorial project is an educational content portal where anyone can learn various Free and Open Source Software by his own, and all the courses are of self-paced and multi-lingual, so that anybody can learn from anywhere with a smart device with an Internet connection [16]. Consortium for Educational Communication (CEC) is an Inter-University Centers which is set up by the University Grants Commission (UGC), India. The goal of CEC is to address the needs of Higher Education through the use of powerful medium such as television along with the appropriate use of emerging Information Communication Technology (ICT). The e-Yantra is an educational initiative to spread embedded systems and robotics among the learners. The objective of this initiative is to provide hands-on learning to higher secondary/engineering students those who have limited access to laboratories and mentors. The MHRD, Govt. of India has also formed e-ShodhSindhu merging three consortia initiatives, namely (i) UGC-INFONET Digital Library Consortium, (ii) National Library, and (iii) Information Services Infrastructure for Scholarly Content (NLIST) and Indian National Digital Library in Engineering Sciences and Technology (INDEST)-All India Council for Technical Education (AICTE) Consortium for archiving online e-resources. The project Quantum & Nano computing provides an environment for scientists and mathematicians to explore fundamentals of quantum systems. The project Indian Sign Language education and recognition platform (ISLERS) is aimed to develop an interactive and feature rich audio-visual mode system through automatic Indian Sign Language education for hearing impaired learners. Open Source Courseware

Animations Repository (OSCAR) is a repository of learning objects, such as animations and other teaching learning resources. The project, Free Open Source Software for Education (FOSSEE) is aimed to promote the use of Open Source Software in education to improve the accessibility and quality of education. The E-Kalpa project is basically initiated for creating a digital resource database on design. This project also focuses on digital content for design, knowledge accumulation, and dissemination. The Pedagogy project is designed with the objective to develop learner-centric curricula. The aim behind to design this type of curricula is outcome-based learning. The E-PG Pathshala project is subdivided into e-Adhyayan, e-Pathya, and UGC-MOOCs [17]. The e-Adhyayan and e-Pathya are a repository of digital contents for graduate and post graduate courses whereas UGC-MOOCs are the Massive Opne Online Courses for the aspirant learners. The project Vidwan is a database of scientists, researchers, and faculty members those are involved in teaching and research in India [18]. The project, Indian Research Information Network System (IRINS) is a Web-based Research Information Management system. This project involves the development of organizations, research, and academic activities. *[Sources—Various Web Portals of Govt. of India, National Mission on Education through Information and Communication Technology].*

Technology-enabled learning (TEE) tools in teaching and learning are not bounded. The term TEE, has a variety of similar other names as—Technology Enhanced Education, Technology Supported Learning, E-Learning, Online Learning, Learning Through Internet, Blended Learning, Technology-Mediated Education and etc. The Open Educational Resources (OERs) are the innovations in teaching–learning which are especially relevant for achieving equitable access to quality education. OERs are learning materials and open content to freely accessible to anyone [19]. As per UNESCO, Open Educational Resources are any type of educational materials that are in the public domain or introduced with an open license. The nature of these open materials means that anyone can legally and freely copy, use, adapt, and re-share them. OERs range from textbooks to curricula, syllabi, lecture notes, assignments, tests, projects, audio, video, and animation. Anyone can modify and enhance the OERs because their creators have permitted to do the same. OERs include materials like presentation slides, podcasts, syllabi, images, lesson plans, lecture videos, maps, worksheets, and even entire textbooks. OERs allows educators to work together to ensure consistency and quality of matter [20].

The government of India (GoI) also initiated the OERs and designed a National Repository of Open Educational Resources (NROER) [21]. It is a collaborative platform, which brings together everyone interested in school and teacher education. The term Open Educational Resources (OERs) was created at the United Nations Educational, Scientific, and Cultural Organization (UNESCO) in the year 2002. The OER campaign was initiated to make freely adaptable educational content. This campaign gradually has geared up and incorporated by the global educational community into their teaching–learning practices. The impact of adopting UNESCO's commitment to the free exchange of ideas and knowledge, toward the OER; the number of educational institutions offers free educational content in form of audio, video, presentations, e-books, MOOCS and etc.

UNESCO organized world OER congress in the year 2012 with the objectives that the products of public-funded work in support of education should launch under open licenses. This was the first world OER congress which brought together the leading researchers, expert practitioners, policymakers, and ministers of education and human resource and development. The second OER Congress was organized during September 2017, Slovenia, with the following recommendations [22]. As- Building the capacity of users to find, re-use, create and share OER; Language & Cultural issues; Ensuring inclusive and equitable access to quality OER; developing sustainability models; and developing supportive policy environments.

The Massive Open Online Courses (MOOCs) are gradually establishing global recognition as equivalent to a regular university degree because of learners' interests. The MOOCs is conceptualized as an online course with an unlimited number of enrolments and opens to all through the Internet globally. Some popular MOOC platforms across the world are—EdX, Coursera, Khan Academy, Udemy, Canvas, Future Learn, Udacity, Open Education Europa, The Open University, SWAYAM, NPTEL, etc.

In light of Open Educational Resources initiatives by UNESCO and world community; the Govt. of India has also initiated several projects for technology-enabled education. One of the popular project is National Digital Library of India (NDL, India). The NDL India is a pilot project supported by MHRD and developed by the Indian Institute of Technology (IIT) Kharagpur, India. NDL India holds thousands of e-learning contents in different formats and languages as—audio, video, animation, text, etc. NDL featured with a rich searching facility and supports several languages for all academic levels including researchers and life-long learners, all disciplines, all popular forms of access devices, and also open for differently abled learners.

National Council for Educational Research and Training (NCERT) initiated several projects to broaden the use of technology-enabled education over traditional education, such initiatives are—Curricula for ICT in education, Access textbooks online, E-pathshala, Tamanna, National initiative for school heads' and teachers' holistic advancement (NISHTHA), and National Repository of Open Educational Resources (NROER) [21].

2 Related Work

2.1 Existing Research

Aktaruzzaman et al. conducted a study on trends and issues to integrate information and communication technology (ICT) in Teaching Learning for the Future World of Education. In his study he found that children retain 20% of what they hear, 40% of what they see and hear and 75% of what they see and do. So, one of the key reasons to introduce the latest educational technology tools in teaching–learning is to impart the global needs of the education system. He also suggested using several ICT tools and technologies, such as—Internet, video, audio, graphics, text, images, info-graphics, etc.; to offer students a better experience of learning. He further stated that we have created a new global economy that is powered by technology, fuelled by information, and driven by knowledge [23]. The emergence of this new global economy will have serious implications in the future education system, as the access to information continues to grow exponentially, schools cannot remain mere venues for the transmission of a prescribed set of information from teacher to student over a fixed period, rather than that they should promote the acquisition of knowledge and skills that makes learning continuous.

Cinganotto et al. conducted a study on "TEAL as an Innovative Teaching Model (Insights from Educational Avant-Garde movement in Italy)". They found the Technology-Enabled Active Learning (TEAL) is the most powerful and most challenging which insights rethinking of learning pathways in Schools in Italy. Author(s) also stressed to implement TEAL in Schools through which learning becomes more attractive, engaging, and also reduces school drop rate [24]. McCoy et al. studied "The Digital Learning Movement: How Should Irish Schools Respond?" In their study, they discussed Information and Communication Technology (ICT) in education-related policy in Ireland as well as they also stressed on enabling the use of ICT in Schools, especially for teaching–learning purposes. They also suggested to harnessing the potential of technology-enabled education to support learning and teaching approaches and prepare students for twenty-first century skills [25].

Chen studied "Exploring the Digital Divide: The Use of Digital Technologies in Ontario Public Schools"; in his study, he discussed the patterns of ICT integration in Ontario's publicly funded schools and their impacts on student academic achievement and empowerment. He also highlighted the role and importance of parent's awareness and involvement in ICT for student's achievement [26].

Sultan et al. conducted a study on "A Constructivist Approach for Digital Learning: Malaysian Schools Case Study"; in their study, they discussed influences of Constructivist Learning Environment in Malaysian schools. The author(s) emphasized the integration of multimedia in learning and also identified various barriers to the teacher's face [27]. Rajesh and Nair have conducted a study on "Indian School Teachers' Perspective on Globalization of Education—A Case Study of Atomic Energy Education Society School Teachers"; in their study, they reveal on directions of the forces affecting globalization of the education. They emphasized the challenges towards globalization of education from the school teachers' perspective, such challenges are—adaptation to the latest techniques of teaching–learning, e-learning, online education, etc. [28]. Halverson and Graham studied "Learner engagement in blended learning environments: A conceptual framework"; in their study, they explored the academic achievement and satisfaction regarding learner engagement. The author(s) focused on learner engagement tactics and identifies most relevant constructs to learning in a blended learning environment, and also suggested a conceptual framework for learner engagement in technology-mediated learning contexts [29].

The US Department of Education discussed the national technology competency standards for students of secondary schools, therefore throughout the country; the educators/teachers/policymakers believe technology literacy is a critical factor to equip students with the skills and knowledge necessary to succeed in the twenty-first century. The National Council for Accreditation of Teacher Education (NCATE) of the USA is seriously working on implementing technology competencies for teachers. The International Society for Technology in Education (ISTE) also recommends some set of standards for educators which can help the students to become empowered learners. These standards are—promote collaborative learning; rethink traditional approaches of teaching–learning, to become a leader, etc. The International Society for Technology in Education (ISTE) Standards for Educators is the guidelines for strengthening the role of technology in teaching–learning. Such standards include—Learner, leader, citizen, collaborator, designer, facilitator, and analyst. The ISTE standards ensure that we should set professional learning goals to explore and apply pedagogical approaches, to pursue professional interests by creating and actively participating in local and global learning networks and to stay with research that supports improved student learning outcomes.

The teachers/educators should seek out opportunities for leadership and also shape, advance and accelerate a shared vision towards equitable access to educational technology tools. To participate and contribute in a digital education world, the teachers should act a model for their pupil and colleagues to explore and evaluate new digital learning resources.

3 Methodology

We have collected primary data set from teachers across the state Uttarakhand, India. The Uttarakhand is a hill state, where Internet connectivity is a big issue in itself, the state is subdivided into 13 districts. We have designed a survey form and spread among the stakeholders via WhatsApp/e-mail/ other social mediums.

We have received total of 148 responses from the respondents. The first part of this questionnaire was designed to gather the information of the respondents and the next part of this questionnaire is about to their use of ICT in teaching related activities. The data was collected and analyzed to achieve the intended objectives.

4 Results and Discussion

4.1 Quantitative Results

The respondents are teachers who are employed in either government or private schools. Figure 1 shows that the 60 respondents are working in government schools (either Hindi or English medium) rest 88 respondents are working in private schools.

We have categorized the schools as per the physical location where the school is situated. Figure 2 shows the school locations, as—a village or rural area where fewer than 3 thousand people are living, a small town where more than 3 thousand to 15 thousand people are living, a town where more than 15 thousand to 1 lakh people are living, a city where more than 1 lakh to 10 lakh people are living and a large city where over 10 lakh people are living.

As per recorded responses, Fig. 3 about class/section teaching; 13 respondents (teachers) are teaching pre-primary section which includes playgroup up to class 2; 30 respondents are teaching primary section which includes class 3 up to class 5; 40

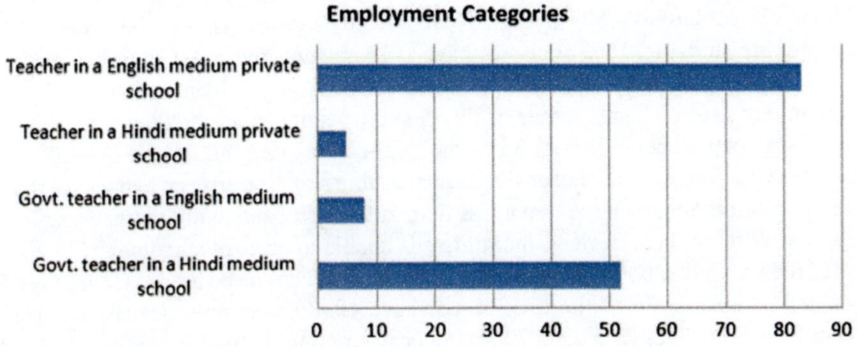

Fig. 1 Respondent's employment categories

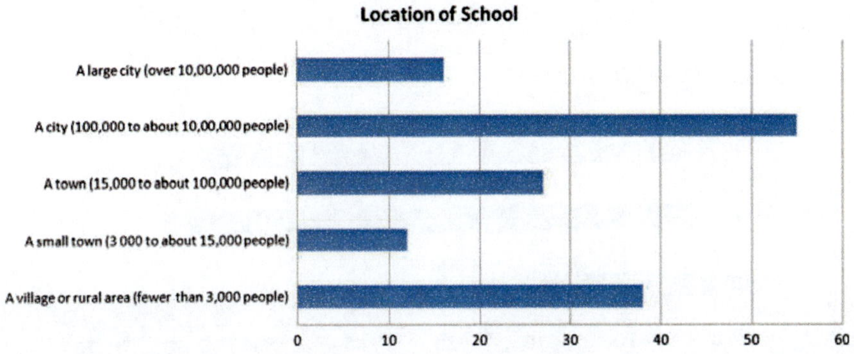

Fig. 2 Location of schools

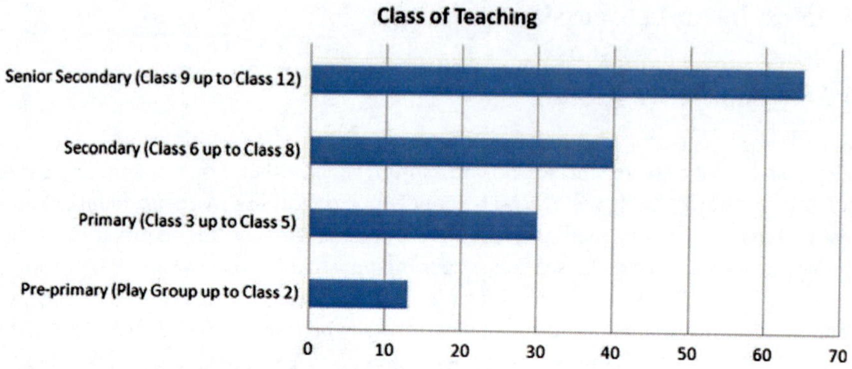

Fig. 3 Class of teaching

respondents are teaching secondary standard which includes play class 6 to class 8; and rest 65 respondents are teaching senior secondary standards which include class 9 to class 12.

61% of classes have 21–40 students enrolled in their classes. Figure 4 shows that 57% of classes have 41–60 students. There are only 11% of classes where above 60 students are studying. 19% of classes have 1–20 students present in their classes.

An important aspect of technology-enabled education is to identify internet facilities among users. The responses of Fig. 5 show details about the internet-enabled devices among students. About 35 of the total classes have almost all (81–100%) students with Internet-enabled devices. An alarming situation arises when we see that around 52 of the total classes have less than 20% of students with internet-enabled devices. Without these devices, the student is unable to access technology (Fig. 6).

12 Schools have above 150 of electronic devices, which can be used in technology-enabled education. 36 of the total school have 40–80 electronic devices. 21 The school has 80–150 devices, and 79 schools have around 0–40 electronic devices in

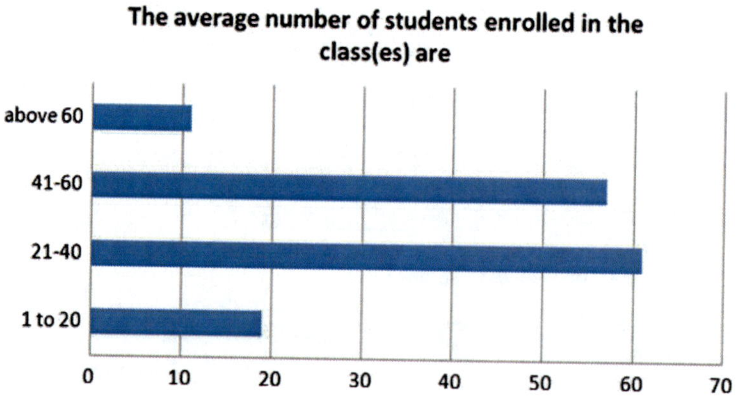

Fig. 4 Number of students enrolled in the classes

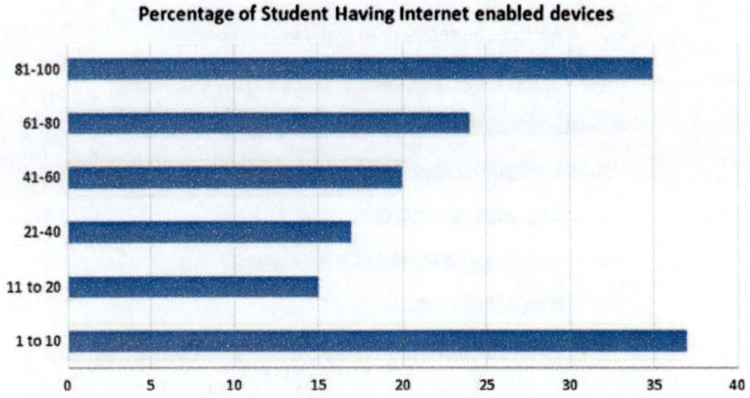

Fig. 5 Student having internet enabled devices

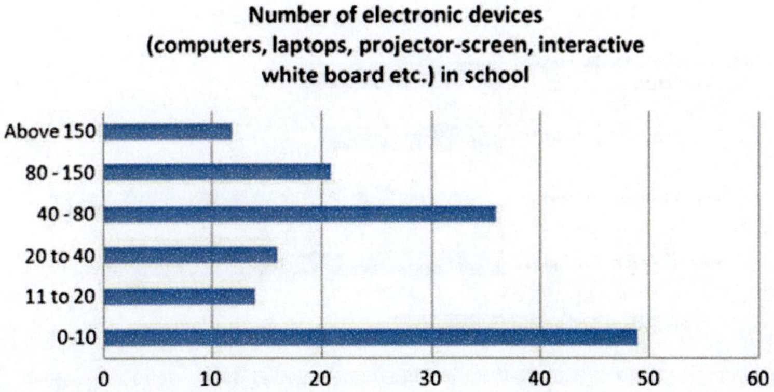

Fig. 6 Total number of electronic devices in school

their school. This shows that school is quite ready for delivering technology-enabled education to its students.

25 Schools have 80–100 Devices which are Internet enabled. Responses in Fig. 7 show that 35 schools have 40–80 Internet-enabled devices. There are 26 schools, which have no Internet connection available. Around 43 schools have 1–20 devices which are Internet enabled.

A very important aspect of delivering content through the Internet is its speed. Figure 8 shows that 50 schools have connectivity with 6.1 mpbs-17 mbps speed. 12 schools have poor connectivity, whereas 22 schools have no connectivity at all. 20 schools have greater than 20 Mbps speed.

When asked that if the teacher thinks that a computer facility should be made available to students in the school, 106 teachers give a positive response in favour (Fig. 9).

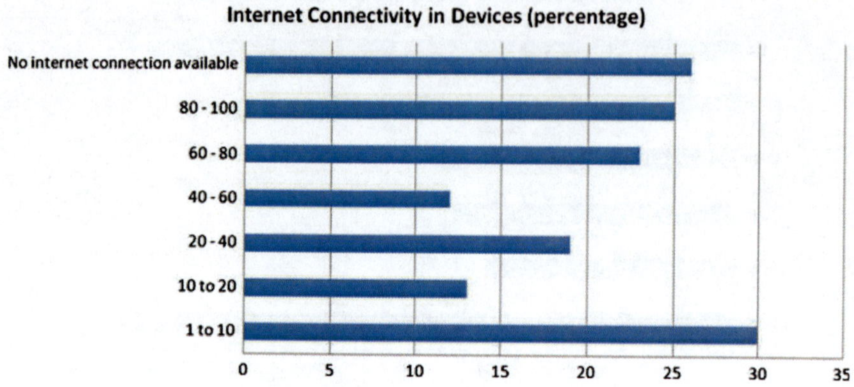

Fig. 7 Devices with internet connectivity

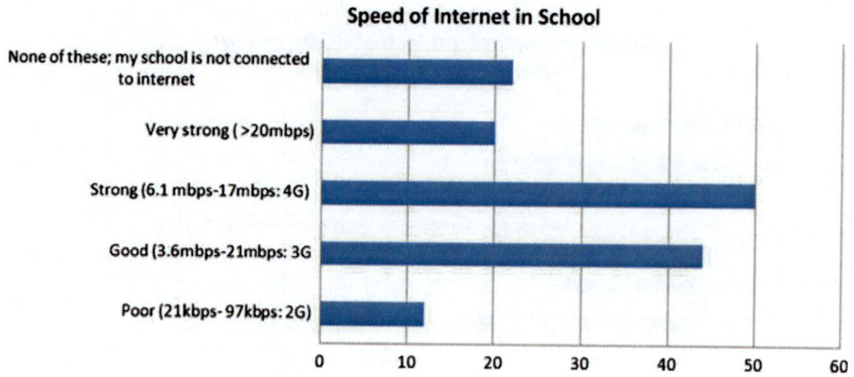

Fig. 8 Internet speed in schools

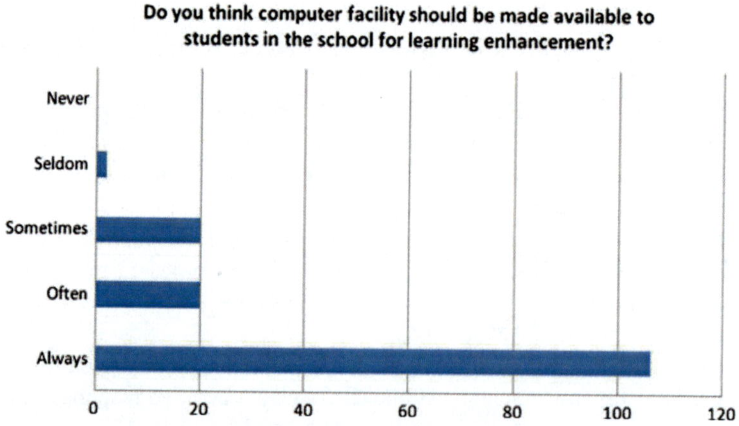

Fig. 9 Teacher responses about availability of computer facility

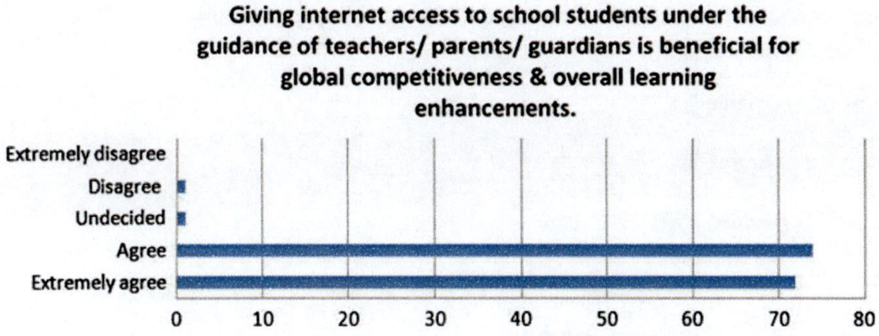

Fig. 10 Teacher responses about giving internet access to school students

146 teachers in Fig. 10 agree in giving Internet access to school students under the guidance of elders. No one disagrees in this respect.

There are around 106 teachers attended ICT-related workshops/ hands-on training in the last 10 years (Fig. 11). 42 responses never attend such training at all.

Almost all teachers agree that after attending ICT workshops, their teaching–learning method is enhanced (Fig. 12).

142 teachers agree that training on ICT tools should be organized from time to time. This will enrich academic performance. Only 1 disagrees with this (Fig. 13).

43 teachers say that they do not have adequate facilities or they are not comfortable with operating computers/ Internet, but they are interested in technology-enabled teaching. 88 responses in Fig. 14 show they are always interested in this type of ICT-related tools.

Over 137 teachers know computers well (Fig. 15), Only nine responses think that they are poorly computer literate. Computer literacy is not an issue as seen by the responses.

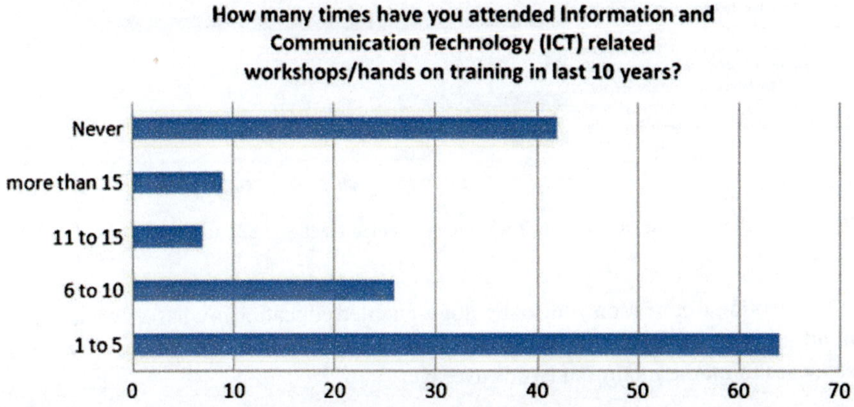

Fig. 11 Teacher responses about ICT workshops/ training in last 10 years

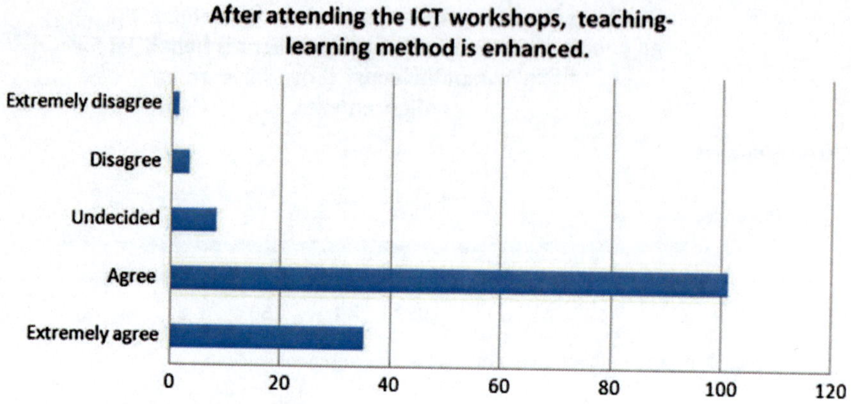

Fig. 12 Enhancement of teaching-learning methods after attending workshops

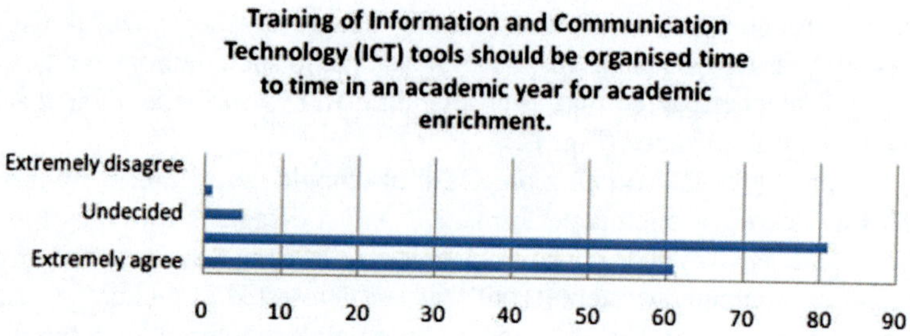

Fig. 13 Teacher responses about ICT training

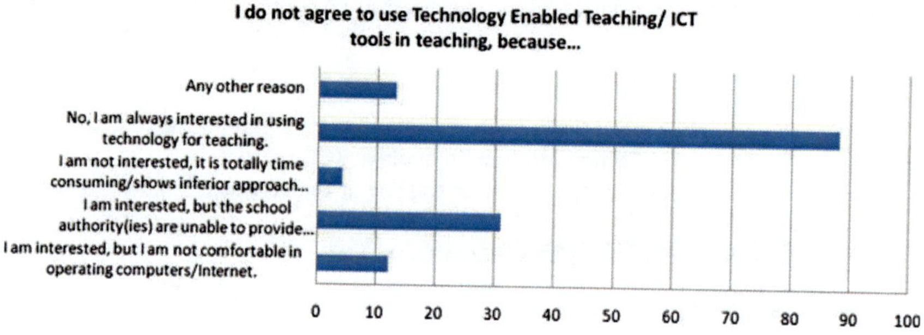

Fig. 14 Teacher's reason about using Technology enabled teaching/ ICT tools

Conventional education with technology-enabled education is what we want today for all students. 80 responses in Fig. 16 say it is excellent to blend the above two. 68 responses think the mixing of two is average.

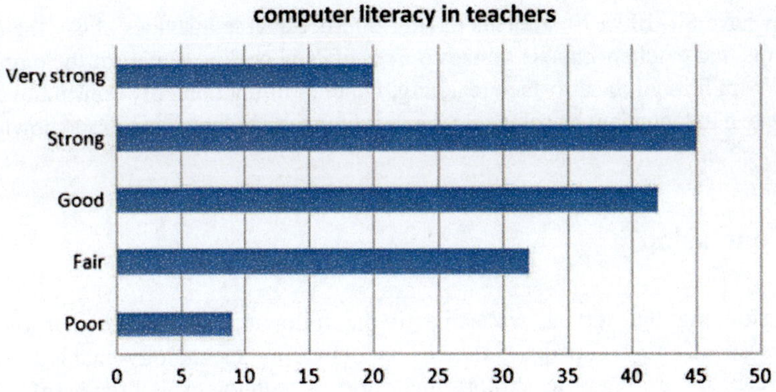

Fig. 15 Computer Literacy in teachers

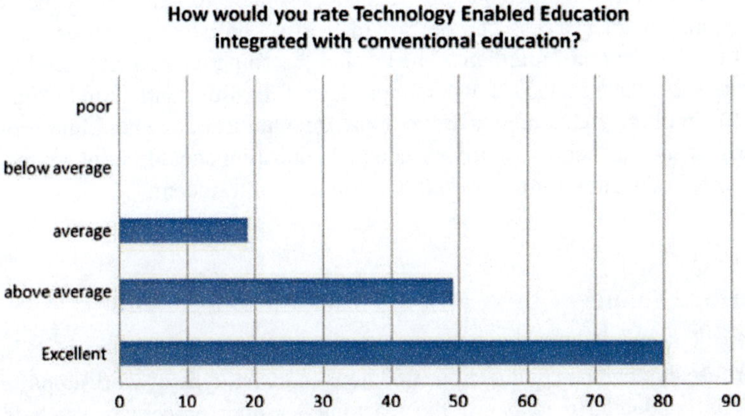

Fig. 16 Teacher's rating about TEE education

5 Evaluation

From the survey, we can see the complete scenario of technology-enabled education from the teacher's perspective. Teachers are ready and quite enthusiastic to use technology in education. They know that technology will make them confident by making them globally competent. And, technology will also make the teaching process quite easy resulting in an increase of understanding by the students. Almost all the teachers 94% are computer literate, which shows that teachers are already aware of technological advancement. They only need is to understand various modes/tools of technology-enabled education. Furthermore, 60% of total respondents already interested and using technology for teaching. 85% of teachers use one or the other means/tools in their teaching purposes. Many students do not have Internet-enabled devices; this can cause a problem in providing e-content to them. Only 35 of the total classes have 80–100% of students having Internet-enabled devices. Also, the result confirms that teachers choose to use courses offered on-line as part of the emphasis on their traditional, face-to-face teaching, which will undoubtedly contribute to the increase in independent learning and the deepening of students' acquired knowledge.

6 Conclusion

After studying the survey, we came to the following conclusion that Internet and its speed are the major concerns in delivering technology-enabled education. Teachers are ready to explore the world of technology if proper infrastructure is provided to them. 81% of schools do not have proper infrastructure for technology-enabled teaching or they do not provide proper training facilities to their teachers. Computer facilities should be made available to students in the school for learning. Internet access can be provided to school students under the guidance of teachers/parents/guardians. The teaching learning method enhanced in 98% of teachers after attending ICT workshops. Therefore, Information and Communication Technology (ICT)-related workshops/hands-on-training should be provided regularly to the teachers. Technology-enabled education should be integrated with conventional education for the overall development of a student.

7 Future Scope

Further research can be conducted to find how and what type of workshop/training should be conducted to enhance the teaching–learning process in teachers and students. We can also check through the study that how the enhancement can be made in the technological infrastructure of a school.

References

1. Department of School Education and Literacy, Ministry of Human Resource Development, Government of India. National Policy on Information and Communication Technology (ICT) in School Education (2012)
2. National Mission on Education Through Information & Communication Technology, Government of India (Mission Document)
3. Virtual Labs: Available at www.vlab.co.in. Last Accessed 15 Dec 2019
4. Talk to Teacher: Available at www.aview.in. Last Accessed 15 Dec 2019
5. Spoken Tutorial: Available at www.spoken-tutorial.org. Last Accessed 15 Dec 2019
6. Consortium for Educational Communication: Available at www.cec.nic.in. Last Accessed 18 Dec 2019
7. E-Yantra: Available at www.e-yantra.org. Last Accessed 18 Dec 2019
8. Digital Library Inflibnet: Available at www.inflibnet.ac.in. Last Accessed 18 Dec 2019

9. Quantum & Nano Computing: Available at www.dei.ac.in/dei/quantumNano. Last Accessed 19 Dec 2019
10. Available at www.Fossee.in, (Free Open Source Software for Education project). Last Accessed 21 Dec 2019
11. Available at www.dsource.in, (E-Kalpa). Last Accessed 21 Dec 2019
12. Pedagogy Project: Available at www.ide.iitkgp.ernet.in/Pedagogy1/pedagogy_main.jsp. Last Accessed 21 Dec 2019
13. E-PG Pathshala: Available at epgp.inflibnet.ac.in. Last Accessed 21 Dec 2019
14. Vidwan-Expert Database and National Researcher Network: Available at vidwan.inflibnet.ac.in. Last Accessed 21 Dec 2019
15. Indian Research Information Network System: Available at irins.org/irins. Last Accessed 25 Dec 2019
16. Rogers, D.L.: A paradigm shift: technology integration for higher education in the new millennium. Educ. Technol. Rev. (2000)
17. Rajasingham, L.: Will mobile learning bring a paradigm shift in higher education? Educ. Res. Int. (Hindawi Publishing Corporation) (2011)
18. Das, A.: Emergence of open educational resources (OER) in India and its impact on lifelong learning. Emerald Group Publishing Limited, pp. 10–15 (2011) https://doi.org/10.1108/07419051111163848
19. Carnoy, M. (2004). ICT in Education: Possibilities and Challenges. https://www.uoc.edu/inaugural04/dt/eng/carnoy1004.pdf
20. White, G. (2008). ICT Trends in Education. Australian Council for Educational Research
21. National Repository of Open Educational Resources: Available at nroer.gov.in/welcome. Last Accessed 25 Dec 2019
22. Yusuf, M.O.: Information and communication technology and education: analysing the Nigerian national policy for information technology. Int. Electron. J. **6**(3), 316–321 (2005)
23. Aktaruzzaman, Md., Shamim, Md., Huq, R., Kum, C.C.: Trends and Issues to integrate ICT in teaching-learning for the future world of education. Int. J. Eng. Technol. **11**(3), 114 (2011)
24. Cinganotto, L., Panzavolta, S., Garista, P., Guasti, L., Dourmashkin, P.: TEAL as an innovative teaching model (insights from educational avant-garde movement in Italy). J. e-Learn. Knowl. Soc. **12**(2), 115–126. ISSN: 1826–6223, eISSN: 1971–8829 (2016)
25. McCoy, S., Marcus-Quinn, A., Hourigan, T.: The digital learning movement: how should Irish schools respond? Econ. Soc. Rev. **50**(4), 767–783 (2019)
26. Chen, B.: Exploring the digital divide: the use of digital technologies in Ontario public schools. Can. J. Learn. Technol. (CJLT) **41**(3), 1–23 (2015)
27. Sultan, W.H., Woods, P.C., Koo, A.-C. (2011). A constructivist approach for digital learning: Malaysian schools case study. J. Educ. Technol. Soc. **14**(4), 149–163. https://doi.org/10.2307/jeductechsoci.14.4.149
28. Rajesh, M., Nair, S.P.: Indian school teachers' perspective on the globalization of education. Turk. Online J. Distance Educ.-TOJDE. **10**(4), 12–26. ISSN 1302–6488 (2009)
29. Halverson, L.R., Graham, C.R.: Learner engagement in blended learning environments: a conceptual framework. Online Learn. J. **23**(2), 145–178 (2019). https://doi.org/10.24059/olj.v23i2.1481

An Automatic Digital Modulation Classifier Using Higher-Order Statistics for Software-Defined Radios

Nikhil Marriwala and Manisha Ghunsar

Abstract Automatic modulation recognition (AMR) plays a vital role in various applications such as software-defined radios (SDRs), cognitive radio (CR) receivers, and surveillance systems. This paper is based on the recognition and classification of digital modulation schemes. The designed system is capable of recognizing nine digital modulation schemes, namely 2FSK, 4FSK, 8FSK, 2PSK, 4PSK, 8PSK, 16QAM, 64QAM, and 256 QAM in additive white Gaussian noise (AWGN) environment. The system includes two steps: feature extraction and classification. In feature extraction step has used seven higher-order cumulants, namely C_{40}, C_{42}, C_{44} C_{51}, C_{53}, C_{62}, and C_{80} as features for statistical analysis of signals. In the classification, the step has used the principal component analysis technique on the feature set for compression of data and for classification of different modulation schemes. The simulation results show that the presented system has 100% classification accuracy at 16 dB SNR.

Keywords Software-defined radio (SDR) · Automatic modulation classification (AMC) · Higher-order cumulants · Higher-order moments · Signal-to-noise ratio

1 Introduction

Early radios were hardware-defined radios (HDRs), correlates to the classical view of radio, consisting of only hardware components that cannot be modified. On the other hand, software-defined radios (SDRs) are known to be more flexible, interoperable, reconfigurable, capable, and efficient radios. The HDRs lack all these qualities and this is the reason why they were overtaken by SDRs. The Wireless Innovations Forum with the Institute of Electrical and Electronics Engineers (IEEE) P1900.1 group has worked to found a definition of SDR that provides stability and a clear outline of the technology and its associated advantages [1].

N. Marriwala (✉) · M. Ghunsar
Electronics and Communication Engineering Department, University Institute of Engineering and Technology, Kurukshetra University, Kurukshetra, India

H. Sharma et al. (eds.), *Congress on Intelligent Systems*, Advances in Intelligent Systems and Computing 1334, https://doi.org/10.1007/978-981-33-6981-8_44

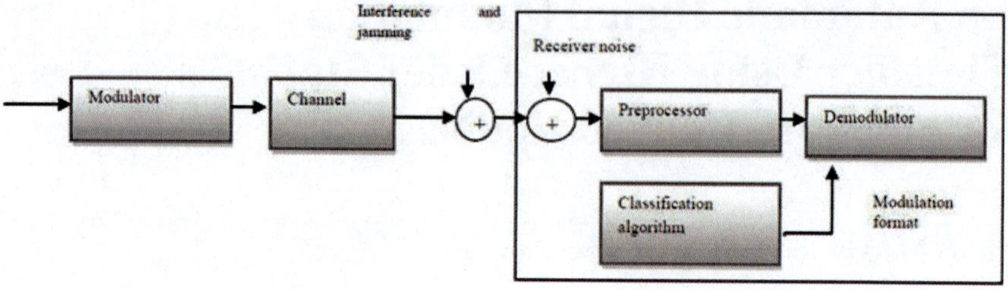

Fig. 1 Block diagram of AMC [12]

Definition: SDR is a radio in which some or all of the physical layer functions are software defined. Significant advantages of SDRs like interoperability, efficient reuse of resources, quick technology up-gradation, flexibility, reconfigurability, reduced logistics cost, multiple band, and multimode operations and many more, have made it so powerful, popular, and widely acceptable. It plays a vital role in today's modern wireless communication [2, 3]. To make these technologies possible, SDR is becoming increasingly commonplace. Modulation recognition is very important in communication intelligence (COMINT) applications for perfect monitoring of the intercepted signal. Digital modulation schemes have been used widely in many fields [4]. The technique to classify the modulation format of the noisy signals received from the fading channels is termed as automatic modulation classification (AMC). Modulation recognition is an intermediate step between signal detection and data demodulation and plays a very important role in civilian and military applications [5]. Figure 1 shows a block diagram of an AMC. AMC algorithms can be divided into two categories: Likelihood-based method (LB) [6–8] and feature-based (FB) method [9–11].

The classification is performed from a decision test based on maximum likelihood estimation (MLE) [8, 13]. FB approach is based on the extraction of features of received signal like a signal envelope, DFT, statistical features, instantaneous features, etc. It is seen from the literature survey done that the radio functions can be applied with multiple software modules for various standards. Under these circumstances, the identification of transmitted signal modulation type has to turn out to be vital to demodulate for further processing. Various researchers have investigated and proposed many modulation classification modules using signal envelope [9, 14], statistical moments [12, 15, 16], wavelet transform, and DFT techniques [11].

From the techniques already mentioned, it is observed that their ability to recognize the modulation schemes is not very efficient under low SNR conditions [16, 17]. Therefore, it is required to develop an automatic classification module or system that is proficient in classifying most of the digital modulation scheme using well-defined signal features even at low SNR values [18]. The main goal of this work is developing such a system which can facilitate certain possibilities like establishing automatic digital modulation identification in SDRs even at low SNR. In this paper, an attempt has been made to develop a system that can classify nine (2FSK, 4FSK, 8FSK, 2PSK, 4PSK, 8PSK, 16QAM, 64QAM, and 256 QAM) modulations out of a variety of digital modulation schemes using the FB approach. Thus, the work is divided into two main modules: a feature extraction module and a classification module. The feature extraction module uses higher-order moments (HOMs) and higher-order cumulants (HOCs) as features for differentiating among digital modulation schemes. The classification module includes the application of principal component analysis (PCA) for classifying digital modulation schemes for SDR receivers. The rest of the paper is organized as follows: Section 2 presents a discussion about feature extraction, Sect. 3 presents classifier implementation, Sect. 4 carries the simulation results and performance evaluation of the presented work, Sect. 5 presents the performance comparison and finally Sect. 6 concludes the paper and discusses the future scope.

2 Feature Extraction Module

There mainly exist four types of digital modulation schemes: ASK PSK, FSK, and QAM [19–21]. Every modulation scheme has different characteristics. Hence, the problem is to find the suitable features to recognize or classify different modulation schemes. This paper considers nine digital modulation schemes out of a variety of schemes. The nine digital modulation schemes considered are 2FSK, 4FSK, 8FSK, 2PSK, 4PSK, 8PSK, 16QAM, 64QAM, and 256 QAM. Based on extensive study and research, HOMs and HOCs are presented as effective features. These features can be used to describe the distinctiveness of the probability density function. The deciding factor for choosing HOMs and HOCs as features for classification is that they show robust behaviour to a variety of transformations which is a useful property while characterizing the signals in the systems. Single effect on the received signal during transformation is a change in its mean but the variance, HOMs, and HOCs remain unaffected. When signal undergoes multipath distortion or other corruptions, the rotation of constellation occurs, though some other parameters remain invariant to rotation such as covariance matrix and Eigenvalues. The following subsections briefly describe these features.

3 Higher-Order Statistics (HOMs and HOCs)

A. *Moments*

Statistical moments are the expected value of a random variable with the order of the moment as the powers of it. The mean of the variable is denoted by the first-order moment and the measurement of power is denoted by the second-order moment. Equation 1 describes the moment of a random variable [15].

$$E_{x,a,b} = E\left[x^a(\bar{x})^b\right] \tag{1}$$

where,

x the variable
$\bar{x}$ its complex conjugate.
$a + b$ the order of this moment.

To normalize the moments measure of signal power is taken as the order of the moment in this paper.

B. ***Cumulants***

Equations (2) and (3) give the characteristic functions of the cumulants of a random variable x [16].

$$\Phi(s) = \int_{\infty}^{-\infty} f(x)e^{sx}\,dx \tag{2}$$

$$\text{And } \Psi(s) = \ln \Phi(s) \tag{3}$$

Equation (4) gives the nth-order cumulant of x which is the nth derivative of the second characteristic function and is evaluated at $s = 0$:

$$\lambda_n = \frac{d^n\ \Psi(0)}{ds^n} \tag{4}$$

For consistency, the cumulants have been represented by $C_{x,a,b}$. The cumulants used in the simulations have been expressed as functions of equal and lower ordered moments, to make the calculation simple [15, 22].

4 Classifier Implementation

A. ***Feature Selection***

The FSK signal spectrum is enough to distinguish it from other digital modulation schemes such as the PSK and QAM. It is also easier to differentiate among M-ary FSK schemes because of its signal spectrum. A simple way to automate the analysis of the spectrum is, first, the FFT of the signal is computed and then the peak values in obtained transform are compared to each other to find whether the signal is FSK and if yes, then what type [23]. For PSK and QAM signal, HOMs and HOCs are chosen as features. Through experimentation, it is seen that the value of HOMs and HOCs remains constant for FSK modulated signals which makes hard to distinguish it, that is why spectrum-based classification is chosen for FSK signals.

B. *Signal Generation and Corruption*

In this paper, the signals have been simulated at baseband. For PSK and QAM signals, this suggests that they have one sample per symbol as the output. The spectrum of the FSK signal is centred at 0 Hz as FSK signal is generated using different frequencies with several samples per symbol thus down converting the digital signal to baseband signal from its carrier frequency. For PSK and QAM signal which are into consideration, it signifies that the symbol rates have been obtained. Standard AWGN noise is added in all the digital signals in the channel. Figure 2 shows the generated signals for the training phase of the classifier.

C. *Classifier Module*

In this paper, the unsupervised machine learning algorithm has been used. The classifier is trained using seven higher-order cumulants, namely C_{40}, C_{42}, C_{44}, C_{51}, C_{53}, C_{62}, C_{80} as features for the classification of PSK and QAM signals. In the presented scheme, applied PCA to the data which seeks the best representation of the data in the least-squares sense [24].

The classifier decomposes the data covariance matrix into its Eigenvectors which are chosen from the projection matrix. The classifier very intelligently constructs the feature space by random selection of the most significant Eigenvectors as basis vectors [6]. Then, it calculates the centroid value for each modulation scheme. To determine the centroid of each modulation scheme we project the training data or we can say centred matrix into space and is given by Eq. (5).

$$T = W^{\mathrm{T}}\bar{x} = \left|W^{\mathrm{T}}\bar{x}_1\right|W^{\mathrm{T}}\bar{x}_2|\ldots|W^{\mathrm{T}}\bar{x}_p| \tag{5}$$

where W^{T} is the transpose of a matrix made of the most significant eigenvectors and $\bar{x}$ centred training profile matrices of training data.

The centroid of any modulation scheme ‘l’ will be given by Eq. (6)

$$g_l = \frac{1}{n}\sum_{i=1}^{n} W^{\mathrm{T}}\bar{x}_i \tag{6}$$

The projection of the test profile ‘t’ is computed into the feature space by the proposed classifier. Further the proposed classifier chooses the centroid with the smallest Euclidean distance to the point. The point having minimum distance with the compared centroid value of the modulation scheme is chosen as that modulation type.

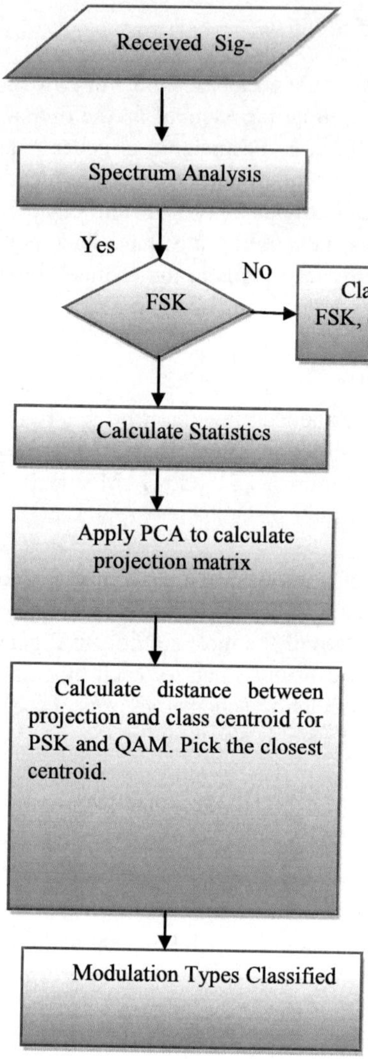

Fig. 2 Classification flowchart for test signals

5 Simulation Results and Discussion

This section discusses the results achieved from the proposed classifier after the simulation. The digital signals (messages) are randomly generated for every test to make sure results are independent of the transmitted signal. Estimation (theoretically) of moments and cumulants value for considered PSK and QAM modulation types is discussed in Sect. 2. First 50 training signals of 20,000 symbols are generated for each modulation type and estimates for the seven selected cumulants are computed for each one. Generated signals are corrupted using AWGN channel. In each trial of the testing phase, 1000 signals are generated and a modulation scheme is chosen at random for each one. To ensure that the proposed classifier produces statistically significant results the proposed classifier is simulated for a minimum of 100 trials at SNR (20 dB). The analysis of each received signal is done to observe that it matches one of the FSK signal types. The matching is done by computing FFT on a minimum of 4096 samples of the signal. The ratios are calculated for the signal peaks from 2nd to 3rd peak, 4th to 5th peak, and 8th to 9th peak. These calculated rations are then further compared with the threshold values estimated in a separate simulation. The computation is on the basis that if the ration of the 2nd to 3rd peak of the signal is greater than the threshold value as specified the signal is classified as BFSK signal by the classifier. With the similar comparisons, the classifier is able to determine whether the signal is 4-FSK or 8-FSK. If the signal peak is below the threshold value, the signal is not FSK signal according to the proposed classifier output. The classifier calculates the projects the cumulants into the new feature space on the basis of the projection matrix formed by the proposed training algorithm for the system. The proposed classifier selects the modulation type of the signal by comparing it with the class centroids. The closest one of the class is then shown as the modulation type of the signal. To analyse the performance of the classifier for different types of signal, a confusion matrix is used. Table 1 shows the ideal confusion matrix obtained from simulation using the presented scheme as all entries are zero except the diagonals. During the simulation, we generated 1000 unknown modulated test signals and identified the modulation scheme of each generated unknown modulated signal. We found the presented scheme correctly identified and classified all of the 1000 unknown test signals and put them into their respective modulation class which is shown in the form of the confusion matrix. Also, the confusion matrix is shown in graphical form in Fig. 3 at 16 dB SNR.

Sum of nine columns (peak value of each column) shows total is equal to 1000, which signifies each testing signal generated has been correctly recognized and classified by its modulation type, giving 100% accuracy. The proposed classifier is trained at 20 dB and produces better accuracy while handling test signals at 20 dB than 5 dB. And on testing, simulation result shows that our classifier gives 100% accuracy even at 16 dB SNR. Table 3 shows the presented classification algorithms accuracy at different SNR values during the testing stage (Fig. 4 and Table 2).

6 Performance Comparison

Table 2 represents the comparison of the proposed work with the important previous work done in this field. As it is seen from paper [14] recognition accuracy is very less and several schemes it includes are also less. In paper [18], it is seen that the authors used statistical measures for classification but our work achieves the same result at low SNR value (16 dB) while they attained the same at 20 dB. In paper [16] the scheme used gives a result at a very high value compared to our work. Based on

Table 1 Confusion matrix

	BFSK	4-FSK	8-FSK	BPSK	QPSK	8-PSK	16-QAM	64-QAM	256-QAM	Total
BFSK	127	0	0	0	0	0	0	0	0	127
4-FSK	0	110	0	0	0	0	0	0	0	110
8-FSK	0	0	100	0	0	0	0	0	0	100
BPSK	0	0	0	111	0	0	0	0	0	111
QPSK	0	0	0	0	107	0	0	0	0	107
8-PSK	0	0	0	0	0	106	0	0	0	106
16-QAM	0	0	0	0	0	0	113	0	0	113
64-QAM	0	0	0	0	0	0	0	114	0	114
256-QAM	0	0	0	0	0	0	0	0	112	112
Total										1000

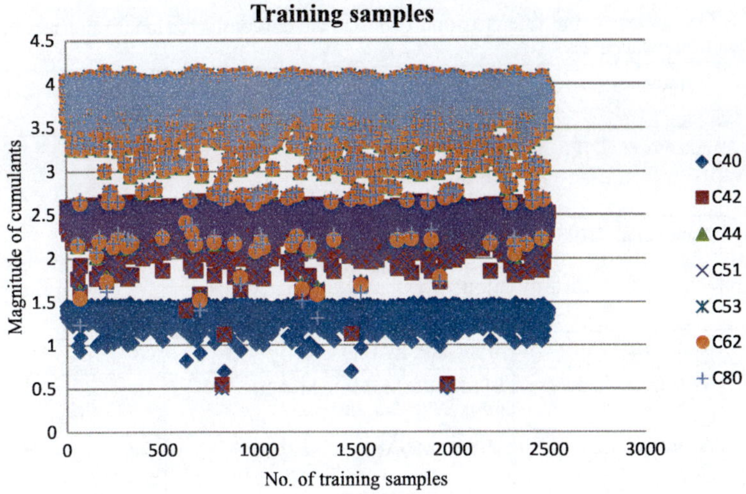

Fig. 3 Training samples

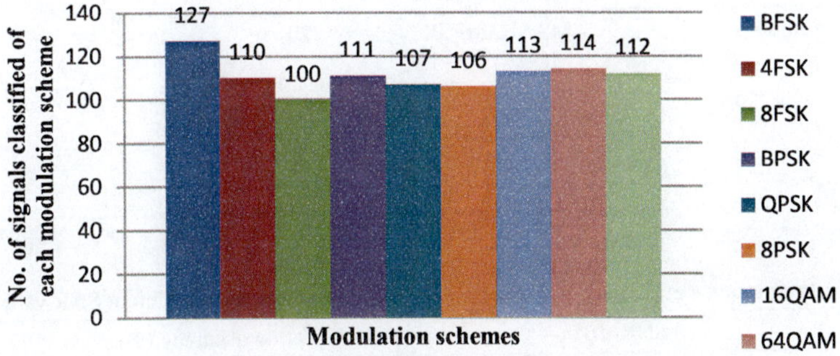

Fig. 4 Bar graph of the generated confusion matrix

comparison, we can say that our scheme works well in low SNR scenarios and is less complex. Table 3 highlights the performance evaluation of the presented classifier algorithm at different SNR values.

7 Conclusion

In the literature survey, we study various methods for AMC and based on that, the criteria for selecting higher-order statistics for the classification of nine modulation schemes (2FSK, 4FSK, 8FSK, 2PSK, 4PSK, 8PSK, 16QAM, 64QAM, and 256 QAM) has been discussed. This work presented the implementation of PCA and

Table 2 Comparison of the different works in the case of the considered modulation required SNR and classification accuracy

S. No	References	Considered modulation schemes	SNR (dB)	Recognition accuracy (%)
1	Azzouz et al. [9]	2ASK, 4ASK, 2PSK4PSK, 2FSK, 4FSK	10	90
2	Nandi et al. [10]	AM, DSB, VSB, LSB,USB, FM, 2PSK, 4PSK, 2ASK 4ASK, 2FSK, 4FSK	15	96
3	Mobasseri [25]	4PSK, 8PSK, 16QAM	5	90
4	Swami and Saddler [26]	BPSK, 4PAM, 16QAM 8PSK	10	96
5	Abdelmutalab et al. [15]	MPSK, MQAM	20	100
6	Jithin jagnnath et al. [27]	BPSK, QPSK, 8PSK 16QAM, 32QAM,CPFSK, GFSK, GMSK	20–25	>99
7	Jajoo [28]	MASK, MPSK, MQAM	20	100
8	Our work	2FSK, 4FSK, 8FSK, 2PSK, 4PSK, 8PSK, 16QAM 64QAM, 256QAM	16	100

Table 3 Performance evaluation of the presented classifier algorithm at different SNR values

S.No	SNR(dB)	Classification accuracy (%)
1	16	100
2	15	99.8
3	14	98.8
4	13	95.8

k-means classifier on the feature set for classification. The presented work works well in AWGN channel conditions. The classification accuracy achieved by our proposed algorithm is 100 per cent at 20–16 dB SNR range. We have checked accuracy for different SNR values also. The results show that the higher-order statistics are highly efficient and effective in classifying the digital modulation schemes and the presented system works well even at low SNR values and is less complex. In the future, the presented work can be further investigated for studying its behaviour under different fading channels conditions. Another challenge is to identify a smaller set of moments/cumulants for modulation classification and that shows robust behaviour under low SNR conditions.

References

1. Pucker, L.: SDR architecture. SDR Forum, pp. 1–6 (2006)
2. Tribble, A.C.: The software defined radio: fact and fiction. In: 2008 IEEE Radio Wireless Symposium, pp. 5–8 (2008)
3. Mitola, J.M.: Software radios survey, critical evaluation and future directions. IEEE Aerosp. Electron. Syst. Mag. **8**(4), 25–36 (1993)
4. Godbole, B.B.: Performance improvement by changing modulation methods for software defined radios. **1**(6), 72–79 (2010)
5. Barbero, L.G., Thompson, J.S.: Performance of the complex sphere decoder in spatially correlated MIMO channels. Eng. Technol. **1**(1), 122–130 (2007)
6. Dubey, H.C., Tiwari, A.K.: Blind modulation classification based on MLP and PNN. In: 2012 Students Conference on Engineering and Systems, IEEE, pp. 1–6 (2012)
7. Ibrahim, S.A., Alias, M.Y., Ahmad, N.N.: Performance of adaptive modulation scheme for adaptive minimum symbol error rate beamforming receiver. Wirel. Pers. Commun. **71**(2), 873–886 (2013)
8. Xu, J.L., Su, W., Zhou, M.: Software-defined radio equipped with rapid modulation recognition. IEEE Trans. Veh. Technol. **59**(4), 1659–1667 (2010)
9. Azzouz, E.E., Nandi, A.K.: New techniques for the baud duration estimation. in 8th European Signal Processing Conference (EUSIPCO 1996), Trieste, Italy, pp. 1–4 (1996)
10. Nandi, A.K., Azzouz, E.E.: Algorithms for automatic modulation recognition of communication signals. IEEE Trans. Commun. **46**(4), 431–436 (1998)
11. Hu, Y.Q., Juan, L., Tan, X.H.: Digital modulation recognition based on instantaneous information. J. China Univ. Posts Telecommun. **17**(3), 52–59, 90 (2010)
12. Green, R.J.J., Joshi, H., Higgins, M.D.D., Leeson, M.S.S.: Recent developments in indoor optical wireless [optical wireless communications]. Eng. Technol. **2**(1), 3–10 (2008)
13. Jagannath, J. et al.: Design and evaluation of hierarchical hybrid automatic modulation classifier using software defined radios. In: 2017 IEEE 7th Annual Computing and Communication Workshop Conference, CCWC 2017 (2017)
14. Jurian, M., Lita, I., Visan, D.A., Oprea, S., Cioc, I.B.: LabVIEW application for systems with automatic modulation recognition. In: 2008 31st Int. Spring Semin. Electron. Technol. Reliab. Life-time Predict. ISSE 2008, pp. 639–643 (2008)
15. Abdelmutalab, A., Assaleh, K., El-Tarhuni, M.: Automatic modulation classification based on high order cumulants and hierarchical polynomial classifiers. Phys. Commun. **21**, 10–18 (2016)
16. Ghodeswar, S., Poonacha, P.G.: Modulation recognition techniques for improving communication efficiency in SDR networks (2013)
17. Reddy, K.P.K., Yeleswarapu, Y.: Performance evaluation of cumulant feature based automatic modulation classifier on USRP testbed, pp. 393–394 (2017)
18. Li, S.: Automatic modulation classification of MPSK signals using high order cumulants, pp. 6–9 (2006)
19. Marriwala, N., Sahu, O.P., Vohra, A.: 8-QAM software defined radio based approach for channel encoding and decoding using forward error correction. Wirel. Pers. Commun. **72**(4), 2957–2969 (2013)
20. Marriwala, N., Sahu, O.P., Vohra, A.: Design of a hybrid reconfigurable software defined radio transceiver based on frequency shift keying using multiple encoding schemes. Egypt. Inform. J. **17**(1), 89–98 (2016)
21. Marriwala, N., Sahu, O.P., Vohra, A.: Novel design of a low cost flexible transceiver based on multistate digitally modulated signals using Wi-Fi protocol for software defined radio. Wirel. Pers. Commun. **87**(4), 1265–1284 (2016)
22. Davidson, K.L., Goldschneider, J.R., Cazzanti, L., Pitton, J.W.: Featijnc-based modulation classification using circular, pp. 765–771 (2004)
23. Mustafa, H., Doroslova, M.: Digital modulation recognition using support vector machine classifier, pp. 2238–2242 (2004)

24. Duda, R.O., Hart, P.E., Stork, D.G.: Pattern Classification, p. 680. Wiley, New York (2011)
25. Mobasseri, B.G.: Digital modulation classification using constellation shape. **80**, 251–277 (2000)
26. Swami, A., Member, S., Sadler, B.M.: Hierarchical digital modulation classification using cumulants. IEEE Trans. Commun. **48**(3), 416–429 (2000)
27. Jagannath, J., Connor, D.O., Polosky, N., Sheaffer, B., Foulke, S.: Design and evaluation of hierarchical hybrid automatic modulation classifier using software defined radios (2017)
28. Jajoo, G., Kumar, Y., Kumar, S., Adhikari, B., Kumar, A.: Blind signal modulation recognition through clustering analysis of constellation signature. Expert Syst. Appl. **90**, 13–22 (2017)

An Improved Machine Learning Model for IoT-Based Crop Management System

Harish Sharma, Ajay Saini, Ankit Kumar, and Manish Bhardwaj

Abstract Smart farm is a management system that influences farmers in the highest demand for agriculture, which predicts the future of the field instead of the fruit of the field by learning the appropriate machine in your area, which is what the farmers want to produce, the year they want to know. This paper is aimed at the development of an IOT based smart agriculture system which can help and provide all information to farmers from their fields on mobile and Web server; this system is helpful for farmers to increase quality product and preserve the corps. The purpose of the article is to analyze the environmental parameters such as agricultural production zones, annual production rates, and crop location to produce, which influences crop yields and establishes relationships among these parameters. Typically, the farmers are still using traditional methods for monitoring their crop and agriculture field. To meet the end of the growing world needs, the farmers have to use new techniques of monitoring and tracking their crops which in turn help them by improvement in yield, reduction in farming cost, reduction in destruction to the environment, and increase in the quality of the produce. Existing systems and methods of monitoring crops use wireless sensor network to collect data from the different sensors deployed at various nodes and send it through the wireless protocol. In this study, regression analysis (R.A.) is used to stress environmental factors and crop yields. In this search, the random forest classification method for the deduction and precision model used is 87.04%; therefore, improve the accuracy of the prediction of the hybrid model,

H. Sharma
Department of Computer Science & Engineering, Engineering College Kota, Kota, India
e-mail: hsharma@rtu.ac.in

A. Saini (✉)
Department of Computer Science & Engineering, Arya Institute of Engineering and Technology, Jaipur, India

A. Kumar
Department of Computer Science & Engineering, Swami Keshvanand Institute of Technology, Management & Gramothan, Jaipur, India

M. Bhardwaj
Department of Computer Science & Engineering, Poornima Institute of Engineering & Technology, Jaipur, India

H. Sharma et al. (eds.), *Congress on Intelligent Systems*, Advances in Intelligent Systems and Computing 1334, https://doi.org/10.1007/978-981-33-6981-8_45

which is a combination of linear regression and random matching, and 93.72% of the efficiency model.

Keywords IoT · Machine learning · Model · Linear regression · Random forest model · Accuracy

1 Introduction

For a country, one of the most important development circles around their ability to produce food aspects. For decades, agriculture has been associated with the production of essential food crops. The urbanization rate is currently the focus of our much higher civilization. By doing this, we reduce ignorance, our capacity for agriculture, especially in terms of land and fertility. We will focus on making the most of what we have as the amount of land will not be increasing in this era of urbanization and globalization, Because of this, we need the device new ways to cultivate arable land and extract the absolute maximum from the limited land resources [1]. In this age of technology and science, if applied correctly, the agricultural sector can also be greatly affected. It is true that the farmer is the best to decide cultures and cultures. However, machine learning techniques can be applied in this area for greater accuracy and stability of the selection. In this research, we tried to find techniques that lead us to choose suitable crops depending on the specific condition, specific district, station, and other environmental aspects. In this research, we tried to find techniques that lead us to choose suitable crops depending on the specific condition, specific district, station, and other environmental aspects. Machine learning techniques can be applied in this area for greater accuracy and stability of the selection. However, the U.S. 80% of farmers use certain types of SFT in Europe which does not exceed 24% [2]. From the perspective of farmers, smart agriculture must provide better prices for farmers and more efficient operation and management operations. This includes system support disease (DSS) for the complete management of complete instructions for customizing return tickets farms, while saving resources, which is enabled by the latest generation of GPS, amplitude image GNSS tires unmanned aircraft and super structural images provided by the Sentinel satellites. Create a local variable map that can be measured (such as crop yield, soil conditions/tile, organic matter content, moisture content, and nitrogen content). Automation and Robotics agriculture: application of robotics process, automatic control and artificial intelligence techniques at all levels of agricultural production [3]. Smart not only agricultural applications of large agriculture and conventional operating targets, but may also lead to the new lever to increase other general trends or held increasingly, that family farms (small space or complex, the specific culture and/or livestock, high-quality savings and a variety of species organic farms, and European consumers, according to social consciousness and respected market and transparent agricultural growth. Smart agriculture seeks to make the best production on farms using the most modern methods of sustainable resources, increase production, and provide the best quality products and to return the maximum. A wide range of used areas including IoT sensors, laptops, GPS services, drones, robots and drones provide real-time data for systems that support the production line and decision support. This allows less waste and maximum efficiency.

- **Information Management System**: A system designed to collect, process, store, and distribute the forms necessary to run the operations and functions of livestock.
- **Precision Agriculture**: To reduce the impact of input and environmental impact, and regional management of the temporal variability to improve income.

1.1 Problem Definition

An intelligent system to utilize Internet of Things (IoT) to monitor a plurality of crops in an agriculture field and further enables a farmer to manage the crops, and the agriculture field in real-time is provided substantially, as shown in and/or described in connection with at least one of the figures, as set forth more completely in the claims. The intelligent system comprises a plurality of sensor units, a wireless unit, a server unit, a database, a computing device, a software application, a water motor and water management unit, a buzzer unit, a CCTV camera, an unmanned-aircraft vehicle system (UAVs), and a pesticides control unit and crop quality unit. The plurality of sensor units is configured with the crops to periodically detect a plurality of parameters pertaining to the crops and the agriculture field. The software application provides the detected data in an analog form and enables the farmer to check the quality of their corps [4]. In an aspect, the water motor and water management unit automatically supplies water to the soil on receiving a wireless command from the computing device of the farmer. The ultrasonic sensor monitors and controls the water level in water for pouring the water onto the soil when required.

1.2 Motivation

In smart agriculture system, it shows the real-time data by sensing the agriculture field data using Wi-Fi as allows user to control the system and monitor their agriculture from anywhere of the world as the application is connected to the cloud where the data from the sensor is stored and allow the application to connect to the hardware. They use various sensors like DHT (temperature and humidity) sensor for sensing the local atmosphere data, ultrasonic sensor for sensing the distance or obstacle in the field as it provides security to the agriculture field as well as for water level indicator in the tank, soil moisture and humidity sensor for sensing the current soil condition, water motor for supplying the water to the soil automatically by sensing through the sensor or whenever the user wants, fencing of the field for security to protect the field from animals or theft as it will send a notification to the farmers mobile and buzzer will start buzzing, CCTV camera and drone technology for continuously monitoring of farm through the application, pesticides control system which provides the perfect solution from insects for their valuable crops. The proper ratio of pesticides is given to the farmer, so that they can act quickly and accurately, and the application is used to control and monitor data sensed from the sensor and send it to the server than to use application which is connected to hardware through Internet system, the Android application is used to provide data storage. It also shows data in analog form.

2 Literature Review

Global demand for cereals increases every year. The only effective solution to meet food needs is to improve the production of food crops. In order to increase agricultural production, it is necessary to increase the productivity of field crops and crops, one of at least two. Due to lack of land for farming and/or lack of resources to harvest available land, the growth of this region is very difficult because of the increase in crop and productivity is determined by various factors such as genetic potential crop varieties, soil, climate, and management variables that vary widely over time and place. Early prediction of crop yields is important for various policy decisions and planning adoption. Many countries use conventional strategies. Using these methods is subjective, expensive, and time-consuming. Empirical models for predicting crop yields were developed from the data and also suffered from several problems. Remote sensing data is used for satellite monitoring as well as for crop monitoring and yield forecasting.

Boydell and McBratney [5] studied using field data and satellite combined for identification and estimation of the cultivated area in County Durham, North East England. Research has shown the ability of Landsat TM data (Thematic Mapper) to discriminate between agricultural crops in the study area. The results showed that satellite data could be used for identification of the largest agricultural crops geographic areas with small field sizes and environmental and physical characteristics. The study produced a land cover type map thirteen classifications with over 80% accuracy. The classification accuracy was assessed quantitatively using the information of land use obtained from the known sample units visited during the field study.

Bramley and Ouzman [6] evaluated several narrowband clues EO-1 Hyperion images to discriminate sugarcane areas affected by the disease of leaf rust "against the healthy growing area". Forty spectral vegetation index (PSI), focusing on the strips pigment bonded sheets, the internal structure of the sheet, and the water content of the sheet is generated from an acquired image along with Mackay, Queensland, Australia. The results showed that the Hyperion images could be used to detect rust disease leaves in the sugar cane crops, while some evidence generated using visible bands in the near-infrared (VNIR) offered only reparability.

Donohue et al. [7] water needs of crops are estimated for the Pontian plain, central Italy, with R.S. based soil classification system and the application of a simple water balance in a GIS environment.

3 Proposed Work

3.1 IOT Hardware Used in Smart Agriculture Framework

Hall Effect Water Flow Sensor

The Hall effect water flow sensor is a sensor that detects the current flow of water passing through it [8]. It tells the outgoing rate of water or the speed per unit of time at which the water is flowing (Fig. 1).

The water flow sensor has three wires, i.e., red for Vcc, black for GND, and yellow for the output pulse. Its interfacing to the Arduino is quite easy [9]. The sensor's pulse frequency is 4.5Q, where Q is the flow rate in liters/minute. Since the sensor works on Hall effect, therefore, the flow rate is directly proportional to the pulse rate. In other words, we can say that the calibration factor of the sensor is 4.5.

FC28 Soil Moisture Sensor

The soil moisture sensor is also an analog sensor, it can work in digital mode also, but in the digital mode, it can only tell the presence or absence of moisture instead of quantifying the moisture content (Table 1).

Relay: A relay is an electromechanical device that uses movable electromagnetic contacts to switch positions between open and closed. A relay is an electrically operated switch, and it is generally employed at a place where a low power circuit is

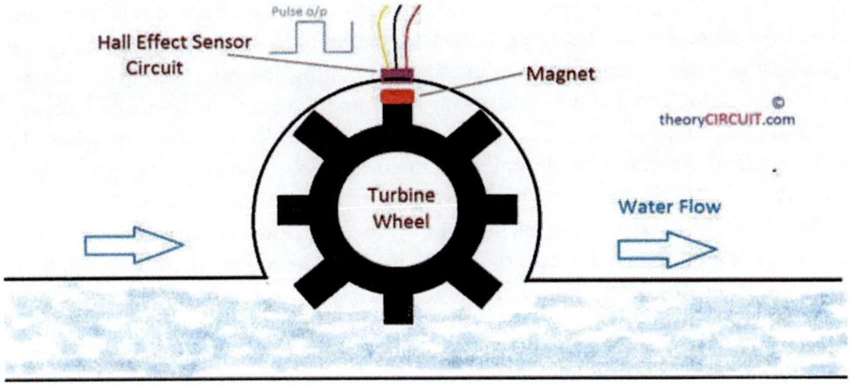

Fig. 1 Working of water flow sensor

Table 1 FC28 soil moisture sensor

	Moisture (%)	Voltage	Corresponding digital value
Minimum	0	550 mV	1023
Maximum	100	10 mV	0

used to control another high-power circuit such as heaters, lamps, motors, or other AC circuits. A relay is an output device (actuator) which switches the AC circuits, that is, why a relay is used to build a home automation system [10].

ULN 2003 Motor Driver I.C.: It is one of the most popular and commonly used motor driver IC. This I.C. comes in use when we need to drive high current loads using digital logic circuits like operator maps, timers, gates, Arduino, PIC, ARM, etc. For example, an motor that requires 9 V and 300 mA to run cannot be powered by an Arduino I/O; hence, we use this I.C. to source enough current and voltage for the load [11, 12].

3.2 *Software/Platform*

Arduino IDE is used for all the programming of microcontrollers as it is free to use, powerful, robust, and easily integrated with many types of microcontrollers.

3.3 *Machine Learning Used in Agriculture*

Linear Regression

Methods involving the regression are essential for data analysis models that attempt to describe the association between a response variable and a number of predictor variables. Like all regression analyses, linear regression is a predictive analysis. Linear regression is used to describe the data and explaining the relationship between a binary dependent variable and one or more nominal, ordinal, or levels of independent variables interval ratio. This algorithm works with binary data, or when the event occurs, represented by "1" or the event does not occur, represented by 0. Then, given a certain "*X*" feature, which asks whether an event occurs "I do not". So "and" maybe "0" or "1" to be or. if the event occur value of and logic is 1 otherwise it should be 0. If the event does not occur, the "and" is given the value "0". For example, if "and" if it indicates a particular culture of a variety of cultures and "and" is "1" if the culture develops [13, 14].

Random Forest

We have successfully achieved high accuracy in the application in our data set using the random forest classifier [15, 16]. Random forest is a supervised learning algorithm. As the name implies, this algorithm generates a forest and uses precision techniques at random. Based on the "forest", a set of decision trees, most of whom, are trained in a "pocketed" manner. The general idea of the bagging method is that the combination of learning styles enhances the overall outcome [17]. To put it simply: Random forests grow multiple decision trees and come together to get more accurate and stable predictions [2]. One of the great benefits of random forest is the classification problem, and two regressions can be used. We have randomly established forest classification because the classification is sometimes seen as the basis of machine learning, and it was a common fact that this aspect of the algorithm, based on the type of research we have made. We are required to use on top. Figure 3.3 shows representative random forests.

3.4 *Proposed Model*

We proposed the hybrid model is a combination of linear regression and random forest model because we have seen that most of the classification technique is not suitable to predict agricultural production the ratio of the Data is not defined that is to say the agricultural production of the year will depend on the data predicting the harvest of next year's total external dependent we want to produce huge amount crops for that we require to know that quality of soil and level of soil parameter for mass production of crops. Thus, only a classification technique cannot provide good accuracy; we have designed the hybrid model. The work proposed a new model to integrate the strengths of both model linear regression and random forest. All the data used in the search for open access to Government of India records are obtained. It was 1998–2015 [18] for the agricultural production season. Great set of preliminary data, for this survey, a limited number of important factors that have the greatest impact on crop production were selected. The diagram below shows the parameters selected for this study [19].

Precipitation (mm): Total rainfall during the Kharif season (June to November) for each year of each district is calculated from the monthly average rainfall of a district in this year.

Minimum, average agricultural production, maximum: Agricultural production must be affected by changes throughout the year. For the minimum results of production, the average and maximum crop yields per year in each district were considered for this survey. The average production of the season this year is calculated from the average monthly temperature for the minimum, medium, and maximum temperatures for each district [20].

Reference Evaporation (mm): Reference evaporation was calculated based on the monthly average for the entire season for each district.

- (H) The paddy area was considered for this study in all the selected districts of each state of Rajasthan

3.5 Proposed Methodology

Following the inclusion of all the set data from this study in the CSV file, the following steps were followed to prepare the data for processing.

Step 1: Acquisition of each parameter (rainfall, minimum, average, maximum yield, and reference evaporation) Average monthly records for each district in the Government of India 2002–2005.

Step 2: Calculate total rainfall, minimum average yield, average and maximum reference evaporation, and the average for each district in each season in the state of Rajasthan.

Step 3: Then, the raw data set is collected on the same sheet with the following columns in the CSV file: Mo no., district name, year, rainfall, minimum, average, maximum reference evaporation, region, production, and yield.

Step 4: District or production data of the year were not available for some of the given parameters, so the data is not used for the actual study. The record was added to each record

Step 5: To prepare the data set for the application of the selected technology, the columns should not be omitted. They are sr. no., the district name, and year.

Step 6: The information below is classified according to the area. No less than 100 ha were considered in this investigation and so excluded records

Step 7: Then, all data folders are sorted on a yield basis to categorize lower, middle, and upper.

Step 8: The yield was calculated on the basis of the zone, and therefore, these two columns of production are omitted.

Step 9: Then, this data set will be saved in a CSV format to apply more machine learning techniques.

3.6 *Hybrid Model*

Algorithm 1: Proposed Hybrid Algorithm

Input

Input: Training sample a: number of input instance to be used to generate classification
UT: Total number of classification Feature

O.T.: Classification Output from each Feature

Output

Output: V: Hybrid Classification set

Procedure Begin

Training
Given a set S of data on each iteration i, a set of training is sampled If replacing S (i.e., bootstrapping)
The classifier is taught by If each
Classification: Given a sample X invisible
Each classifier returns Ci class prediction
the bagging classifier H and count the votes assigns the class X more votes
Regression: It can be applied prediction of continuous values, taking the mean value for each prediction.
random forest (R.F.) are a combination of predictors of trees
Each tree depends on the values of a random vector of the sample independently
The generalization error depends on the strength of individual trees and the correlation between
Using a random selection of performance characteristics favourable to vote, and are more robust noise WRT
Execution Model
Given a training set S
For i = 1 to k do:
If the construction subset by sampling with replacement S
If Ti Learning Tree
On each node:
Choose a better division of random subset of features F
Each tree grows without expanding the size
They are making predictions by the majority vote of all the trees k.

Table 2 Prediction matrix

	Observation	
To discover	True	False
True	TP	F.P
False	Tennessee	F.N

4 Result and Analysis

4.1 Evaluation of Performances

Each example is classified into two classes in the binary classification model. Both classes are false and true. The number of correct predictions is negative in a case. This can be presented as a confusion matrix that is given in Table 2 as the Contingency.

Ranking for an event observed against a model's expected classification in a confusion matrix. Along the long diagonal of the table, this ranking table is the correct classification as true positive and true negative. Model errors are intended for other fields. Only real true positive and negative fields will be filled in by a perfect model, and other fields are set to zero. From the confusion matrix, several performance measures can be modeled. The most common metric is defined as the overall success rate of the accuracy [21] classification and calculated as (Fig. 2).

$$\text{Accuracy} = (\text{TP} + \text{TN})/(\text{TP} + \text{FN} + \text{FP} + \text{TN})$$

Table 3 shows the annual production of all cultures of the year-wise state of Rajasthan

We separated the production of crops in the district wise rate to the production of crop into district wise (Fig. 3).

The above chart displays the observation points between production area and total production for various crops like wheat, bajra, maize, rice, and barley from the previous year's Statistical dataset [22]. Different colors are used to show the

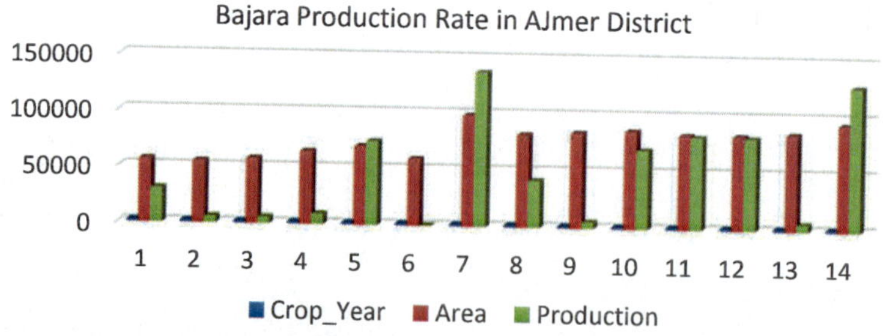

Fig. 2 Bajra production in Ajmer district

Table 3 Production versus year surface

S. No	Crop year	Zoned	Production
0	1997	45,500.0	156,800.0
a	1998	48,730.0	176,455.0
two	1999	52,097.0	139,835.0
3	2000	31,501.0	115,721.0
4	2001	44,550.0	176,784.0

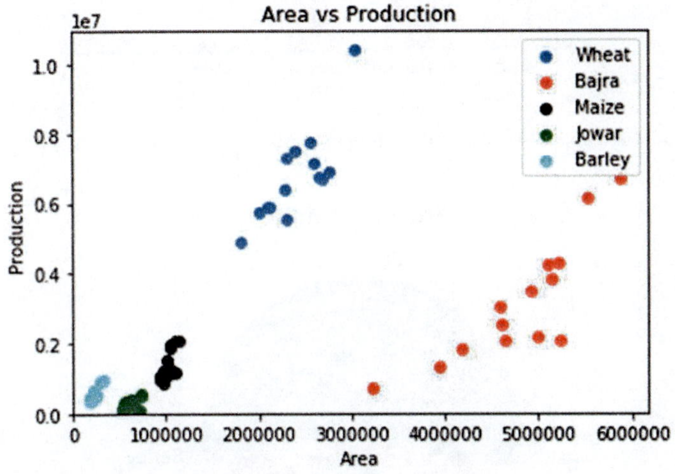

Fig. 3 Area versus production rate of crop

observation points for different crops. In the statistical analysis of existing data, we can see that wheat and bajra need a larger production area as compared to other crops (Fig. 4).

Figure 5 displays the relation between crop and their production for various crops like wheat, bajra, maize, barley, etc., from the previous year's dataset. The statistical analysis of existing data in the above chart shows that the production of wheat, bajra, and maize is highest than other crops that is why we can say that wheat, bajra, and maize crops have the more suitable environment in Rajasthan.

Description: above pie chart simply visualizes that wheat, bajra, and maize crops have the larger production area as well as larger crop production than other crops; thus, the analysis of existing dataset shows that these crops have the most suitable environment in Rajasthan. This production may vary at the various places in Rajasthan, which we can analyze in this research work and can predict the crop yield in the specific district on the basis of previous years data and the farmer can be informed about the same predicted crop production in his/her district so that the total production of various crops can be kept balanced in order to fulfill the need of people.

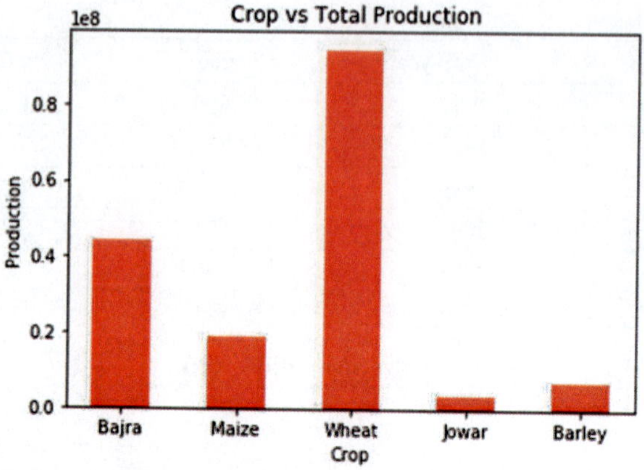

Fig. 4 Crop versus total production rate of proposed work

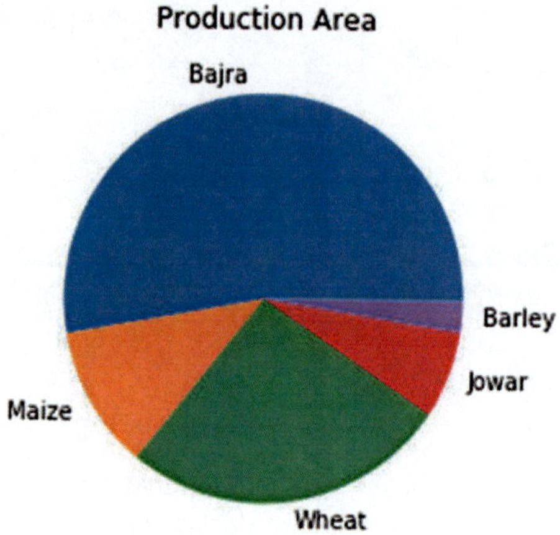

Fig. 5 Wheat, bajra, and maize crops have the larger production area

5 Conclusion

An ambitious crop management program for farmers and government would be our solution to this issue. We forecast in this method the average (upper) concomitance by machine learning of many crops using open-source data on crop production and consumption from previous years. Although IoT is still the way we use technology in the field of agriculture in India, we should be optimistic. When farmers undergo adequate technological instruction with a knowledgeable smartphone in their pockets, they can carry out much of their farming duties without ever getting there. In theory, it allows farmers to stay connected from everywhere. It also helps to reduce human effort by increasing productivity, while at the same time boosting farmers' economies. Therefore, the agriculture industry will have a better future for the next decade with fully developed apps and the Internet of Things and make India great over the next few days. With this exacting forecast, we are assigning each farmer to a cultivation and quantity.

References

1. Balaghi, R., Tychon, B., Eerens, H., Jlibene, M.: Empirical regression models using NDVI, rainfall and temperature data for the early prediction of wheat grain yields in Morocco. Int. J. Appl. Earth Obs. Geoinf. **10**, 438–452 (2008)
2. Bishop, T., Horta, A., Karunaratne, S.: Validation of digital soil maps at different spatial supports. Geoderma **241–242**, 238–249 (2015)
3. Bishop, T., Lark, R.: A landscape-scale experiment on the changes in available potassium over a winter wheat cropping season. Geoderma **141**, 384–396 (2007)
4. Bishop, T.F.A., McBratney, A.B., Laslett, G.M.: Modelling soil attribute depth functions with equal-area quadratic smoothing splines. Geoderma **91**, 27–45 (1999)
5. Boydell, B., McBratney, A.B.: Identifying potential management zones from cotton yield estimates. Precis. Agric. **3**, 9–23 (2002)
6. Bramley, R.G.V., Ouzman, J.: Farmer attitudes to the use of sensors and automation in fertilizer decision-making: nitrogen fertilization in the Australian grains sector. Preci. Agric. (2018). https://doi.org/10.1007/s11119-018-9589-y
7. Donohue, R.J., Lawes, R.A., Mata, G., Gobbett, D., Ouzman, J.: Towards a national, remote-sensing-based model for predicting field-scale crop yield. Field Crops Res. **227**, 79–90 (2018)
8. Huete, A., Didan, K., Miura, T., Rodriguez, E.P., Gao, X., Ferreira, L.G.: Overview of the radiometric and biophysical performance of the MODIS vegetation indices. Remote Sens. Environ. **83**, 195–213 (2002)
9. Jones, J.W., Hoogenboom, G., Porter, C.H., Boote, K.J., Batchelor, W.D., Hunt, L.A., et al.: DSSAT cropping system model. Eur. J. Agron. **18**, 235–265 (2003)
10. Kantanantha, N., Serban, N., Griffin, P.: Yield and price forecasting for stochastic crop decision planning. Journal of Agricultural, Biological, and Environmental Statistics, 15, 362–380. Eating, B.A., Carberry, P.S., Hammer, G.L., Probert, M.E., Robertson, M.J., Holzworth, D., et al.: An overview of APSIM, a model designed for farming systems simulation. Euro. J. Agron. **18**, 267–288 (2003)
11. Lewis, A., Oliver, S., Lymburner, L., Evans, B., Wyborn, L., Mueller, N., et al.: The Australian geoscience data cube—foundations and lessons learned. Remote Sens. Environ. **202**, 276–292 (2017)
12. Censimento Della Agricoltura. Available online: https://censimentoagricoltura.istat.it. Accessed on 3 May 2018
13. ClassificazioneDelleAttività Agricole. Available online: https://www.codiciateco.it/coltivazioni-agricole-eproduzione-di-prodotti-animali--caccia-e-servizi-connessi/A-01. Accessed on 15 Aug 2018
14. Sunseeker, H., Verdouw, C., Wolfert, S., PrezFreire, L:. Internet of food and farm 2020. In: Digitizing the industry—Internet of Things connecting physical, digital and virtual worlds, vol. 2. River Publishers, Gistrup, Denmark (2016)
15. Wolfert, S., Ge, L., Verdouw, C., Bogaardt, M.-J.: Big data in smart farming a review. Agric. Syst. **153**, 69–80 (2017)
16. Biradarand, H.B., Shabadi, L.: Review on IoT based multidisciplinary models for smart farming. In: Proceedings of the 2nd IEEE International Conference on Recent Trends in Electronics, Information Communication Technology (RTEICT), Bangalore, India, pp. 1923–1926 (2017)

17. Ramya, R., Sandhya, C., Shwetha, R.: Smart farming systems using sensors. In: Proceedings of the 2017 IEEE Technological Innovations in ICT for Agriculture and Rural Development (TIAR), Chennai, India, pp. 218–222 (2017)
18. Yoon, C., Huh, M., Kang, S.G., Park, J., Lee, C.: Implement farm with IoT technology. In: Proceedings of the 20th International Conference on Advanced Communication Technology (ICACT), Chuncheon-si Gangwon-do, Korea, pp. 749–752 (2018)
19. Arkeman, Y., Utomo, H.A., Wibawa, D.S.: Design of web-based information system with green house gas analysis for palm oil biodiesel agroindustry. In: Proceedings of the 3rd International Conference on Adaptive and Intelligent Agroindustry (ICAIA), Bogor, Indonesia, pp. 238–244 (2015)
20. Amanda, E.C.R., Seminar, K.B., Syukur, M., Noguchi, R.: Development of expert system for selecting tomato (Solanum lycopersicum L.) varieties. In: Proceedings of the 3rd International Conference on Adaptive and Intelligent Agroindustry (ICAIA), Bogor, Indonesia, pp. 278–283 (2015)
21. Nurulhaq, N.Z., Sitanggang, I.S.: Sequential pattern mining on hotspot data in Riau province using the prefixspan algorithm. In: Proceedings of the 3rd International Conference on Adaptive and Intelligent Agroindustry (ICAIA), Bogor, Indonesia, pp. 257–260 (2015)
22. Murphy, F.E., Popovici, E., Whelan, P., Magno, M.: Development of an heterogeneous wireless sensor network for instrumentation and analysis of beehives. In: Proceedings of the 2015 IEEE International Instrumentation and Measurement Technology Conference (I2MTC), Pisa, Italy, pp. 346–351 (2015)

A Study on Intelligent Control in PV System Recycling Industry

A. S. L. K. Gopalamma and R. Srinu Naik

Abstract Solar is one of the accessible green energy resources, but to make it really green, we should focus on the end life management of it. It is a peremptory aspect for the industries and government to work together for a green energy waste management plan. Increased volume of PV panels might create environmental issues, and there exists unprecedented offers to chase new economic opportunities. The proper way of recycling and reuse of material can unlock significant challenges behind. Three Rs play a major role in any production to end the circular economic cycle and they are *Reduce, Reuse, and Recycle*. The priority is to reduce waste, reuse it with minor modifications, and lastly, for recycling. For the recycling process, proper infrastructure, machinery, and its control are needed. Waste treatment plants for automobile industries, composite materials like steel, glass, and metal, plastic manufacturing/recycling units are having similar machinery, control logic, and electrical installations. In this work, we proposed energy flow in each and every unit and its control in the PV recycling Industry. In industrial scenario, any industry associated with equipment like conveyors, hopper, crusher systems to transport, and to handle the materials processed. We have taken each unit of individual system as the combination of a drive, programmable logic controller. The conveyor logic thus executed with proper interlocks and permissives in codesys simulation software and the same logic executed in siemens S71200 PLC and G120 drive.

Keywords Codesys software · G120 drive · Intelligent control · Programmable logic controller (PLC) · Recycling · S71200 PLC

1 Introduction

In general, the materials we are using now will become scrap tomorrow; it may destroy the environment or else limit the living area. Sometimes, we should compromise with the cost for quality products, time to execute, and environmental responsibility. When we interlink the advancement in technologies-to-waste management,

A. S. L. K. Gopalamma (✉) · R. Srinu Naik
Andhra University College of Engineering (A), Visakhapatnam, A.P, India

H. Sharma et al. (eds.), *Congress on Intelligent Systems*, Advances in Intelligent Systems and Computing 1334, https://doi.org/10.1007/978-981-33-6981-8_46

it gives us fruitful results. "Do it Right Now" is not only just a slogan, but it implicates the essentiality towards the prevention and prediction of the upcoming waste Tsunami that now we are inventing. One should aware of the end of life of the products we are creating, and society also should take responsibility towards it before inviting.

Solar is one of the accessible green energy sources, but to make it green, we have to look after the end life management of the material used in the process. It is a peremptory aspect for the industries and government to work together for a green energy waste management plan. India targeted to have 100 GW of solar energy by the year 2022. Bridge to India estimated the waste of solar components volume to grow 1.8 million tonnes (nearly) by 2050, and this volume is almost equal to the total e-waste volume being annually generated waste in India. Solar modules manufacturing process involves dangerous materials, which include lead, polymers, and cadmium compounds. Any work with the proper way of doing will give us good results, but an inappropriate way of doing it results in negative impacts. In case the disposal process of solar modules not done in a proper way, it led to severe environmental issues and health impacts. As of now, India does not have appropriate policy guidelines, and the infrastructure facilities and technologies involved in recycling. As per the E-Waste Management and Handling Rules, 2011, there is no provision on the solar PV waste. Even the conventional recycling facilities for glass (laminated) and e-waste are unavailable. The process of recycling the solar module is not attractive commercially. Compared to the materials we are getting from the process, the cost is less to the capital cost involved [1]. He explains the recent research contributions towards PV recycling in Europe, and he justifies the importance of pilot program to regulate solar E-waste, which increasing rapidly. He introduces low recycling costs and high recycling rates in the field of PV recycling [2]. He well demonstrated the process involved in PV panel recycling and the need for quality assurance checking at the panel reuse centre and the necessity of optimizing the energy flow and weight flow as per the recycling process involved. He demonstrated the similarity within the existing recycling units and the chemical, and features of photovoltaic generation stations [3, 4].

Waste treatment plants for automobile industries, composite materials like steel, glass, and metal, plastic manufacturing/recycling units are having similar machinery and control logic, and the electrical installations. But the point is to save the energy at each level in the PV recycling unit and fulfilling the objectives involved, such as low cost for recycling and a high recycle rate with maximum efficiency [5–7].

The recycling techniques are still improving in each stage, and some of the recyclers have started to develop new recycling technologies on time. If the tailoring process ready for this product will make the process with large quantities simpler. The recycling percentage also increases shortly [8, 9]. Even though the present PV panel lifetime is nearly 20 years, inexorable waste created when panels come to an end.

The research contribution to PV recycling is little, and so many authors propose a way to recycle. In this paper, we focused on energy control, saving monitoring, and control at each stage. The proposed schematic can enrich the new buds in this field.

2 PV Panel Recycling System

2.1 PV System

PV system consists mainly of solar panels (Thin film/Si-based) and energy storage systems like batteries. As we are facing global warming, to reduce it, we are adopting solar energy. It is the right choice this time. The worth mentioning matter is how long it is adaptable? The number of solar panels using now after 20–25 years what happens? When we are thinking about these questions, suddenly, our mind says why we take prevention and prediction of the future. We noted to adopt a waste management plan from the manufacturing level. We can reduce the number of panels by improving its efficiency. We can design the panel in such a way to reuse after the end of life, finally design Implementations of Recycling Procedures and Infrastructural Facilities and Machinery and Management.

2.2 PV System Boundaries

Figure 1 explains the boundaries of the PV system from the manufacturing level to end of life. The recycling process and its approach depend on the market forces and

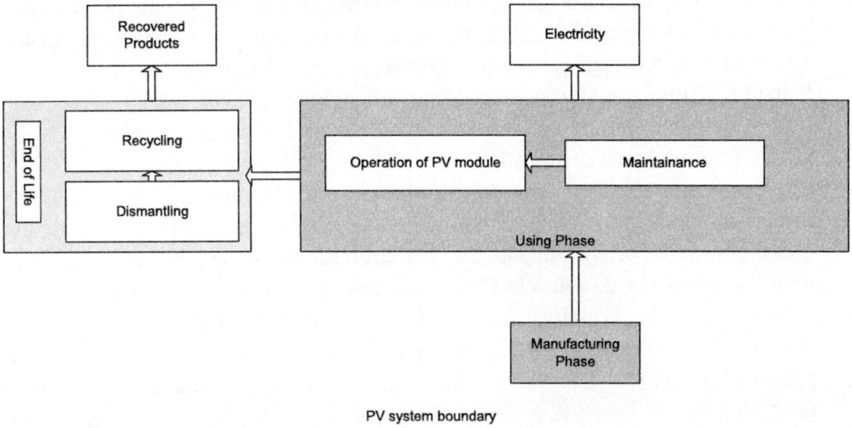

Fig. 1 PV system boundaries

involvement of so many industries and government. The process involves material-specific energy management systems, which include material handling, transport, and recovery facilities. If we wish to learn the process of recycling, first of all, we should acquire knowledge and experience from similar products recycling. Then, we can formulate the new material-specific recycling infrastructure, machinery, and supervisory control. But the issue involved here is the economic feasibility.

2.3 *Reduce, Reuse, and Recycle*

Three Rs play a significant role in any production to end the circular economic cycle. They are Reduce, Reuse, and Recycle. In the case of solar panels, reduce the number of hazardous materials and reuse the materials when their life ends and recycle using the best method to extract more materials and, most crucial impact on the environment. The priority is to reduce and next for reuse and then last for recycling. For the recycling process, proper infrastructure is needed [4, 10, 11].

1. PV panel cleaning: Ample cleaning mixtures, high-pressure sprays which are capable of restore 98% of its reflectance to the surface, and those are safe, low cost and biodegradable, and the dirt cleaned with proper care and without damaging panel coating. If we adopt any algorithm for cleaning these panels, especially for large PV arrays, make the energy-saving task simple. An accurate monitoring system also gives good results for us to know when to clean based on energy in/outflow and its control.
2. Proper orientation: one needs to ensure the appropriate orientation for better efficiency of the PV system. Optimization of tilt/Azimuth angles based on the energy output will provide us with adequate information regarding proper orientation.
3. Panel failures: Panels cracking and decolourization of PV panels reduces the panel efficiency incredibly. We should check these failures at manufacturing levels, and with proper care, we can overcome this challenge.
4. Hotspots: High temperatures sometimes cause hotspots. One such hot spot can lead to damage to multiple cells in the PV panel. The shading must neglect to get rid of it, and proper amended electrical contacts are a must.
5. Environmental effects: Snow, wind, soiling are the leading causes of damage/low production. In fact, due to that, we are getting reduced efficiency. In dusty areas, we should orient the panel with a large tilt angle, which avoids frequent maintenance. One should add the cleaning of the panel also in a warranty for a large PV array. Usually, wind patterns are nonlinear and can damage the corners of the panel, which are the most sensitive portion of the array. With proper experience and prediction of wind simulations, one can overcome this problem [12].

2.4 PV Panel Recycling Process

There are two types of solar panels that require different approaches to recycling in industrial processes [1, 4].

1. Silicon-based
2. Thin film-based

The worth mentioning point is that though some of the research contributions towards recycling reach 96% efficiency by adopting new technologies, the targets are to raise the adoption of PV recycling soon, it needs proper infrastructure and transportation facilities, optimum values of energy inflow and outflow. Not only is it eco-friendly but also witnessing new economic opportunities. Constant continuous solar energy adoption results in tremendous opportunities in the PV recycling sector, and thus we make solar energy truly green with proper recycling policy adoption.

2.5 Materials Extracted from Recycling Process

Recycling should enhance the life of material and its profitability and reduces the raw inputs required for manufacturing. Prioritize the reusing material with minimum modifications. As per the author's suggestion by the combination of both physical and chemical technologies, it is possible to get zero waste. Step by clever step separation of materials using the optical nano-type process [13–15].

3 Power Flow and Its Control in Recycling Plants

To achieve the eco-friendly energy-saving PV recycling system, we should focus on the details of the selection of energy-saving equipment. As the resemblance between other manufacturing/ material handling units to PV recycling provokes us to study the process of the existed material handling plants and proposed the energy control and protection at each level. Usually, material handling units have similar machinery and controls to recycling units. In industries, most commonly used loads are either fan/pump loads, variable, and constant load, as shown in Fig. 2.

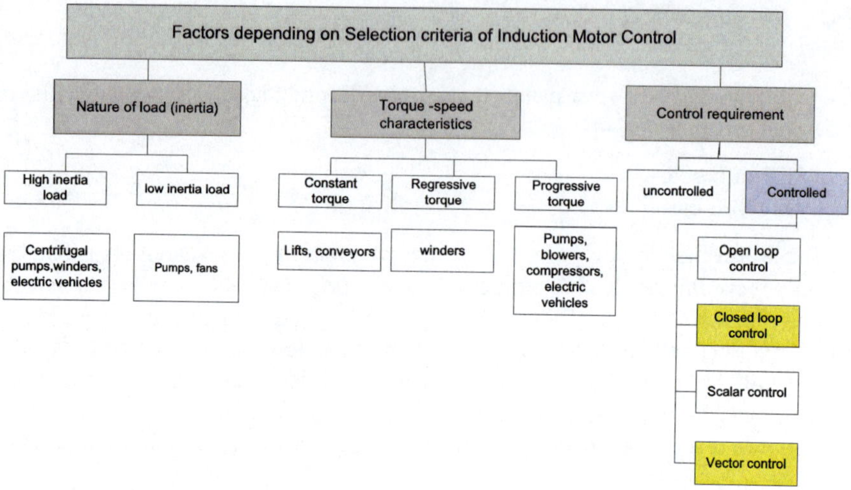

Fig. 2 Classification of loads in industries

3.1 Types of Loads

3.2 Selection of Drives and Its Specifications

1. **Step 1**. Collecting the data regarding the following aspects
   - Voltage (*v*) and frequency (*f*) in volts and Hz, respectively
   - Starting, breaking torques in N-m
   - Load torque and its characteristics for different loads
   - Speed range of VFD (rev/min)
   - Requirement of acceleration
   - Load and motor moment of Inertia
2. **Step 2**. Selection of the number of poles and type of motor
   - In modern VVVF drives, frequency range available is almost 400 Hz, but practically, a few systems are using above 100 Hz
   - For constant torque, loads where maximum speed should be in the range of 50–100 Hz. But in the case of pump/Fan drives, torque is proportional to the square of speed. So, the maximum speed is to be chosen so that it occurs at 50 Hz only.
3. **Step 3**. Selection of power rating of the motor based on the load capability curves of the drive as shown in Fig. 3
   - An essential requirement of the motor to be selected for the desired application is, it does not become stalled/overloaded under any circumstances over the entire speed range.

- It can withstand the thermal runaway, modern IEC motors having these abilities, and having internal thermal reserve to serve this purpose.
- Below base speed-constant torque region
- Above base speed-constant power region

4. **Step 4**. Selection dependency on Inertia of load
   - During acceleration, dynamic torque is added, which defined as the effect of change in kinetic state from one speed to another.
   - Moment of Inertia and time for acceleration decides the maximum torque of the motor, size of motor, and its cost.
   - The torque due to dynamic acceleration expressed as

$$T_A = J * 2\pi \frac{dn}{dt} \ Nm \tag{1}$$

$$T_A = J_T * \frac{2\pi}{60} \frac{d(n_2 - n_1)}{dt} \ Nm \tag{2}$$

$$T_{\max} \geq T_L + T_A \tag{3}$$

$$t = J_T * \frac{2\pi}{60 * T_A} \frac{d(n_2 - n_1)}{dt} \tag{4}$$

$$T_{st} = 1.5 * \frac{I_{\text{converter}}}{I_{\text{motor}}} * T_{\text{Max}} \tag{5}$$

where dn represents the change in speed. During acceleration, dt represents change in acceleration time; J is the moment of inertia; T_{st} represents starting torque, and it represents acceleration time.

5. **Step 5**. The selection of suitable converter and control
   - We should take care about the values of rated currents and converter current. It must follow this as follows $I_{\text{converter}} > I_{\text{motor}}$ and the type of duty involved

4 Results and Discussions

Though the process involved in PV recycling is different, the machinery and control logic is similar to material handling units. Every unit, whether it is conveyor, shredder, or any other unit equipped with a motor, drive unit, controller. Here, the basic conveyor system and hopper system logic control are performed. Figure 4 gives a brief idea about the process control in industries. Crushers, conveyors are constant load drives. Every unit has its local mode of control and remote mode of control.

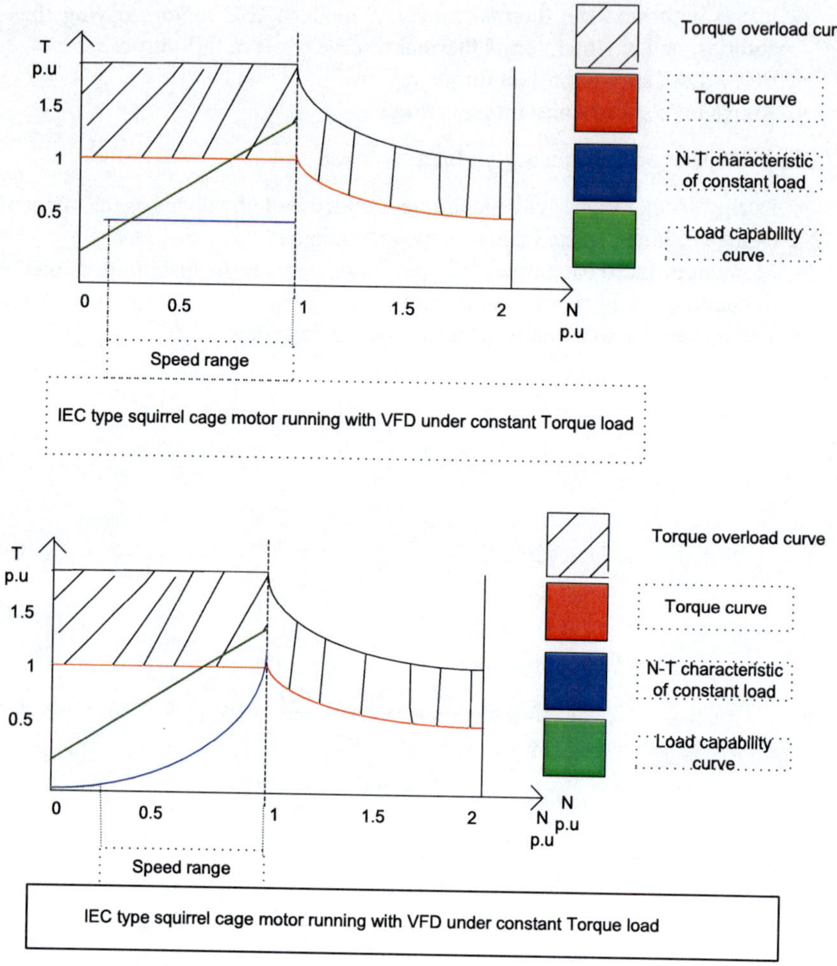

Fig. 3 IEC-type induction motor characteristics for constant/variable torque load

There should be proper communication between individual unit controller and the main controller through interlocks and permissions (Fig. 5).

4.1 *Logic Screens in Codesys Software*

Figure 6 represents the variables used to explain the energy control in the individual unit and overall system. Assumptions made in this study are (1) Each unit has one programmable logic controller (PLC) and variable frequency drive (VFD) (Figs. 7 and 8).

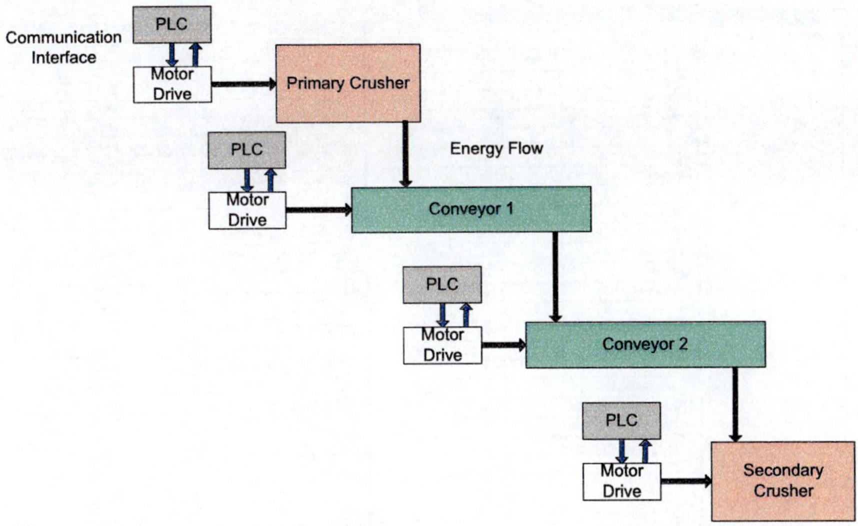

Fig. 4 Schematic diagram of industrial process system

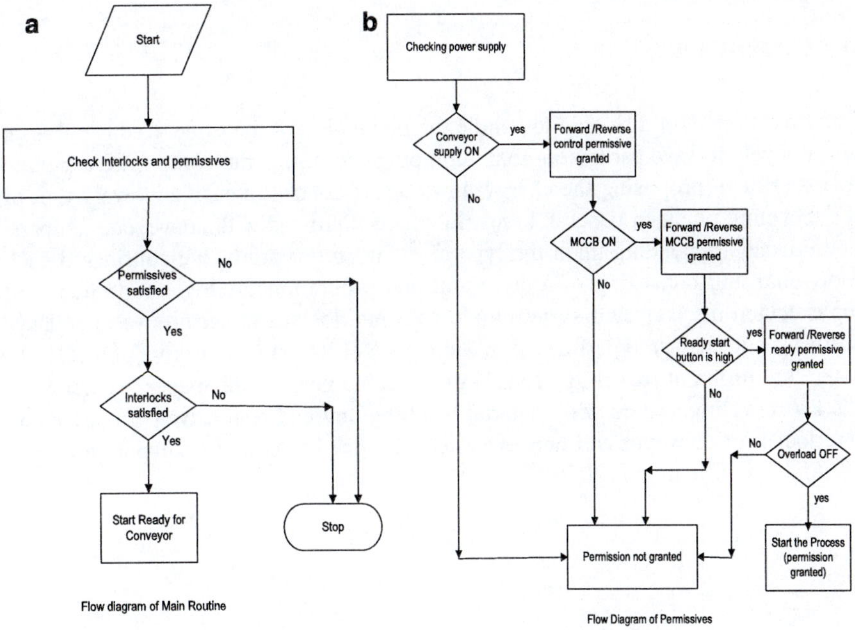

Fig. 5 **a** Main routine flow diagram. **b** Permissions/interlocks flow diagram

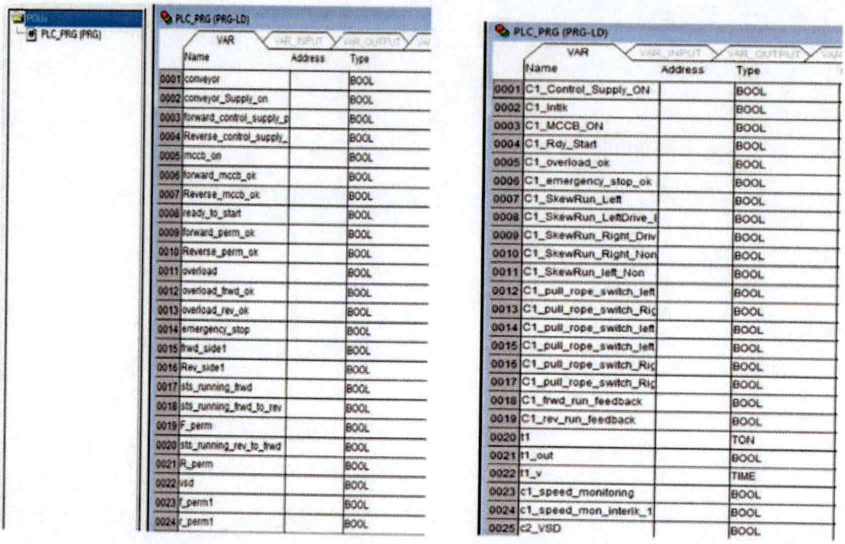

Fig. 6 Variables and their states assigned during process control

5 Conclusions

The invention should be able to handle the prevention of the same. Reduce, Reuse, and Recycle to save earth from toxic dumping. From this motivation, we executed an experiment proposing the energy flow and its control in each and every unit of the recycling plant. In general, industrial plants equipped with conveyors, hoppers, and crushers depending upon the type of plant. In this study, we considered each individual unit consisting of a drive unit and programmable logic controller, and we executed the conveyor system logic with interlocks and permissive in codesys software and the same performed in Siemens S71200 PLC and drive unit (G120). There are different recycling procedures available now; in this paper, we analysed the PV recycling system with material handling unit and respective permissive and interlocks for conveyor and hopper system designed in codesys software and then verified the same in s71200 PLC.

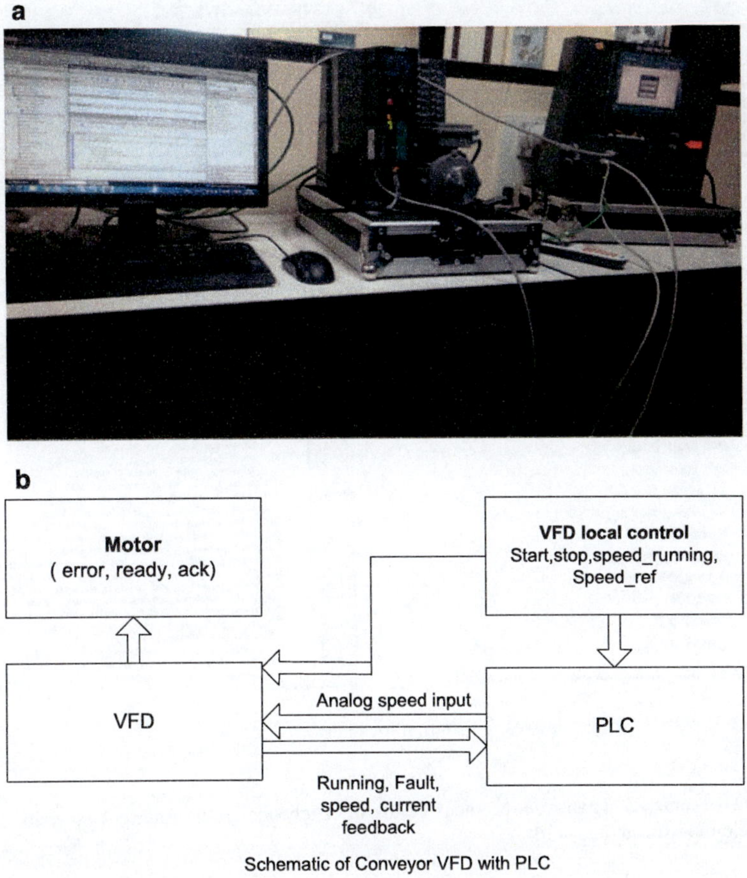

Fig. 7 **a** Experiment setup of individual unit (S71200 PLC, G120 Drive). **b** Schematic of conveyor VFD with PLC

a

```
PROGRAM PLC_PRG
VAR
    conveyor: BOOL;
    conveyor_Supply_on: BOOL;
    forward_control_supply_permission_ok: BOOL;
    Reverse_control_supply_permission_ok: BOOL;
    mccb_on: BOOL;
    forward_mccb_ok: BOOL;
    Reverse_mccb_ok: BOOL;
    ready_to_start: BOOL;
    forward_perm_ok: BOOL;
    Reverse_perm_ok: BOOL;
    overload: BOOL;
    overload_frwd_ok: BOOL;
    overload_rev_ok: BOOL;
    emergency_stop: BOOL;
    frwd_side1: BOOL;
    Rev_side1: BOOL;
    sts_running_frwd: BOOL;
    sts_running_frwd_to_rev: BOOL;
    F_perm: BOOL;
    sts_running_rev_to_frwd: BOOL;
    R_perm: BOOL;
    vsd: BOOL;
    f_perm1: BOOL;
    r_perm1: BOOL;
END_VAR
```

b

```
PROGRAM PLC_PRG
VAR
    C1_Control_Supply_ON: BOOL;
    C1_Intlk: BOOL;
    C1_MCCB_ON: BOOL;
    C1_Rdy_Start: BOOL;
    C1_overload_ok: BOOL;
    C1_emergency_stop_ok: BOOL;
    C1_SkewRun_Left: BOOL;
    C1_SkewRun_LeftDrive_End: BOOL;
    C1_SkewRun_Right_Drive_End: BOOL;
    C1_SkewRun_Right_Non: BOOL;
    C1_SkewRun_left_Non: BOOL;
    C1_pull_rope_switch_left: BOOL;
    C1_pull_rope_switch_Right: BOOL;
    C1_pull_rope_switch_left1: BOOL;
    C1_pull_rope_switch_left2: BOOL;
    C1_pull_rope_switch_Right1: BOOL;
    C1_pull_rope_switch_Right2: BOOL;
    C1_frwd_run_feedback: BOOL;
    C1_rev_run_feedback: BOOL;
    t1: TON;
    t1_out: BOOL;
    t1_v: TIME;
    c1_speed_monitoring: BOOL;
    c1_speed_mon_interlk_13: BOOL;
    c2_VSD: BOOL;
    Iterlocks_intrlk_15: BOOL;
    c1_mtr_maint: BOOL;
    Iterlocks_intrlk_16: BOOL;
    Iterlocks_local_mode_intrlk_16: BOOL;
    c1_local_sel: BOOL;
    c1_rem_sel: BOOL;
    c1_sts_ovld: BOOL;
    c1_Iterlocks_intrlk_15: BOOL;
    C1_Intlk_Intlk00: BOOL;
    C1_Intlkinp_mccb_Intlk01: BOOL;
    C1_Intlk_RDY_Start_Intlk02: BOOL;
    C1_Intlk_overload_ok_Intlk03: BOOL;
    C1_Intlk_overload_ok_Intlk04: BOOL;
    C1_Intlk_Emergency_Stop_ok_Intlk05: BOOL;
    C1_Intlk_skew_ru_left_de_ok_Intlk06: BOOL;
    C1_Intlk_skew_ru_Right_de_ok_Intlk07: BOOL;
    C1_Intlk_skew_ru_Right_Nde_ok_Intlk08: BOOL;
    C1_Intlk_skew_ru_left_Nde_ok_Intlk09: BOOL;
    C1_Intlk_pullrope_left_switch_Intlk10: BOOL;
    C1_Intlk_pullrope_left_switch_Intlk11: BOOL;
    C1_Intlk_pullrope_Right_switch_Intlk12: BOOL;
    C1_Intlk_pullrope_Right_switch_Intlk13: BOOL;
    c2_VSD_Out_Run: BOOL;
    C1_SkewRun_left_NonDrive_End: BOOL;
    C1_SkewRun_left_Non_Drive_End: BOOL;
    C1_SkewRun_Right_Non_Drive_End: BOOL;
END_VAR
```

Fig. 8 **a** Program for permissives. **b** Program for interlocks

Acknowledgements Thanks for Siemens Centre of excellence group, Andhra University, for the permission to execute this work.

References

1. Libby, C., Shaw, S., Heath, G., Wambach, K.: Photovoltaic recycling processes. In: 2018 IEEE 7th World Conference Photovoltaic Energy Conversion, WCPEC 2018—A Jt. Conference 45th IEEE PVSC, 28th PVSEC 34th EU PVSEC, pp. 2594–2599 (2018). https://doi.org/10.1109/PVSC.2018.8547376
2. Komoto, K., Oyama, S., Sato, T., Uchida, H.: Recycling of PV modules and its environmental impacts. In: 2018 IEEE 7th World Conference Photovoltaic Energy Conversion, WCPEC 2018—A Jt. Conference 45th IEEE PVSC, 28th PVSEC 34th EU PVSEC, pp. 2590–2593 (2018). https://doi.org/10.1109/PVSC.2018.8547691
3. Weckend, S., Wade, A., Heath, G.: End of life management solar PV panels (2016)
4. Gómez, V., Lange, A., Clyncke, J.: Collective implementation of the take-back and recycling obligation for end-of-life PV panels: experience of the European system. Electronics Goes Green 2012+, ECG 2012—Jt. International Conference Exhibition Proceedings, pp. 1–4 (2012)
5. Palitzsch, W., Loser, U.: Integrated PV-recycling-more efficient, more effective. In: 2017 IEEE 44th Photovoltaic Specialist Conference, pp. 2272–2274 (2018). https://doi.org/10.1109/pvsc.2017.8521517

6. Einhaus, R. et al.: Recycling and Reuse potential of NICE PV-Modules. In: 2018 IEEE 7th World Conference Photovoltaic Energy Conversion, WCPEC 2018—A Jt. Conference 45th IEEE PVSC, 28th PVSEC 34th EU PVSEC, no. 1, pp. 561–564 (2018). https://doi.org/10.1109/PVSC.2018.8548307
7. Urashima, N., Izumina, M., Arita, A., Matsumoto, K.: Research & development on recycling technology. In: 3rd World Conference Photovoltaic Energy Conversion, pp. 1985–1987 (2003)
8. Burduhos, B.G., Visa, I., Duta, A., Neagoe, M.: Analysis of the conversion efficiency of five types of photovoltaic modules during high relative humidity time periods. IEEE J. Photovolt. **8**(6), 1716–1724 (2018). https://doi.org/10.1109/JPHOTOV.2018.2861720
9. News, T., Editions, S., Edition, S.: India needs a policy on solar PV waste management (2019)
10. Tasnia, K., Begum, S., Tasnim, Z., Khan, M.Z.R.: End-of-life management of photovoltaic modules in Bangladesh. In: ICECE 2018—10th International Conference on Electrical and Computer Engineering, pp. 445–448 (2019). https://doi.org/10.1109/ICECE.2018.8636782
11. Ito, M., Doi, T.: PV module recycling solution and module defects in the field. In: 2018 IEEE 7th World Conference Photovoltaic Energy Conversion, WCPEC 2018—A Jt. Conference 45th IEEE PVSC, 28th PVSEC 34th EU PVSEC, pp. 709–711 (2018). https://doi.org/10.1109/PVSC.2018.8547313
12. Collins, K., Anctil, A.: 2014 Photovoltaic waste characterization with environmental considerations. In: 2014 IEEE 40th Photovoltaic Specialist Conference (PVSC), pp. 1419–1423 (2014).https://doi.org/10.1109/PVSC.2014.6925183
13. Kadiyan, N., Pandey, K., Prakash, A.: Issues and maintenance of photo-voltaic cell-based system. In: 2018 International Conference on Power Energy, Environment and Intelligent Control (PEEIC), 2018, pp. 77–81 (2019). https://doi.org/10.1109/PEEIC.2018.8665671
14. Sinha, P., Wade, A.: Assessment of leaching tests for evaluating potential environmental impacts of pv module field breakage. IEEE J. Photovolt. **5**(6), 1710–1714 (2015). https://doi.org/10.1109/JPHOTOV.2015.2479459
15. Urashima, N., Izumina, M., Arita, A., Matsumoto, K.: Research & development on recycling technology of photovoltaic power generation systems—a social system for PV recycling. In: Proceedings of 3rd World Conference on photovoltaic Energy Conversion, Osaka, vol. 2, pp. 1985–1987 (2003)

Malware Attacks on Electronic Health Records

ShymalaGowri Selvaganapathy and Sudha Sadasivam

Abstract Technological advancements along with the surge of high-speed Internet has enabled healthcare organizations (HCO) to provide enhanced medical treatment and better patient support. Equivalently, cybercriminals are lured by the sensitive patient data available which urges them to launch attacks to their benefits. Artificial intelligence systems involving machine learning and deep learning techniques are progressively utilized for medical diagnosis and predictions. Attackers try to disrupt the working of the learning techniques to cause misclassification or launch privacy attacks. Attackers daunt the health sector to seek personally identifiable information (PII) causing Health Insurance Portability and Accountability Act (HIPAA) violation. These attacks can have severe effects in a critical clinical environment. This article tries to detail on the possibility of adversarial attacks on medical reports by considering the portable document format (PDF) file format. Electronic health records (EHR) and DICOM reports supporting PDF file format are under constant threat to malware attacks. The results demonstrate the validity of the suggested hypothesis and provides security guidelines for health sector to thwart such attacks in realtime.

Keywords Electronic health records · Cyberattacks · Malware · DICOM reports

1 Introduction

Information and communication technology (ICT) has facilitated enhancements in all facets of computing systems enabling benefits to various sectors like education, health care, commerce, and transportation, to name a few. These progressions have resulted in transformations across patient care. Healthcare providers are connected over the Internet and are able to offer better services to their patients through Internet of Medical Things (IoMT). Patient data monitored by smart wearable devices and other sensors are collected and transmitted for intelligent analysis to provide bet-

S. Selvaganapathy (✉)
Department of Information Technology, PSG College of Technology, Coimbatore, India

S. Sadasivam
Department of Computer Science and Engineering, PSG College of Technology, Coimbatore, India

H. Sharma et al. (eds.), *Congress on Intelligent Systems*, Advances in Intelligent Systems and Computing 1334, https://doi.org/10.1007/978-981-33-6981-8_47

ter inferences and personalized treatments. Hospitals enable maintenance of EHR containing patient health information to facilitate timely access and easy sharing of reports along with patient medical histories.

In India, initiatives by government sectors like the NatHealth [1] are on the rise to promote digital health care. Telemedicine schemes like the Social Endeavor for Health and Telemedicine (SEHAT) enable facilities for rural people in India to consult with doctors online [2]. Online diagnostic services are offered by NH healthcare service provider in partnership with CISCO to rural and urban sectors of India [3].

The adverse blowback is the widespread rise of cybersecurity concerns affecting healthcare sector such as theft of sensitive patient electronic data, manipulation of the EHR contents, denial of service (DoS) to access health records, hindering medical transcript processing and embedding malware to cause havoc.

This has a major impact on the healthcare sector as it deals with patient safety and may lead to loss of reputation and trust among patients. Healthcare organizations [4] have been haunted by malware attacks like WannaCry and NotPetya ransomware attacks in 2017. Phishing attack in 2018 on the Florida's Agency of Healthcare Administration uncovered 30,000 Medicaid enrolled patients data. The hacked network of the Oklahoma State University Center for Health Sciences exposed around 2,79,000 Medicaid billing data details.

Medical images and Digital Imaging and Communications in Medicine (DICOM) format reports are susceptible to malware attacks [5]. Along with storage and transmission of images, the DICOM standard supports other formats for handling reports and signals such as PDF, objects, and HL7 Clinical Document Architecture (CDA) as shown in Fig. 1. Embedded malware in the DICOM supported formats can get executed on the picture archiving and communication system (PACS) server or the diagnostic workstation to cause havoc [5].

PDF file format is notorious for malware attacks due to its ability to insert contents such as executable JavaScript or image objects [6]. Arbitrary code can be effected by prompting bugs like buffer overflows in an application through manipulated PDF files. This article explains the working of PDF malware to cause misclassification on the considered learning models. Adversarial machine learning (AML) [7] attempts to exploit vulnerabilities inherent in machine learning techniques to divert them to cause misclassification through the insertion of adversarial samples. HCO's and medical device manufactures configure anti-malware endpoint detection and response (EDR) software to scan medical images and protected EHR with least priority. This has served as the motivation for manipulating medical images and protected EHR as focal attack vectors for AML. The contribution of this article are as follows:

- Identify the possibility of malware threat vectors in DICOM report format and in EHR.
- Uncover the PDF malware exploitation on medical machine learning techniques used for diagnosis and prediction.
- Assess the misclassification caused by adversarial attacks on the AI medical support system using a case study on PDF malware.
- Suggest suitable guidelines to avoid HIPAA violation caused by malware.

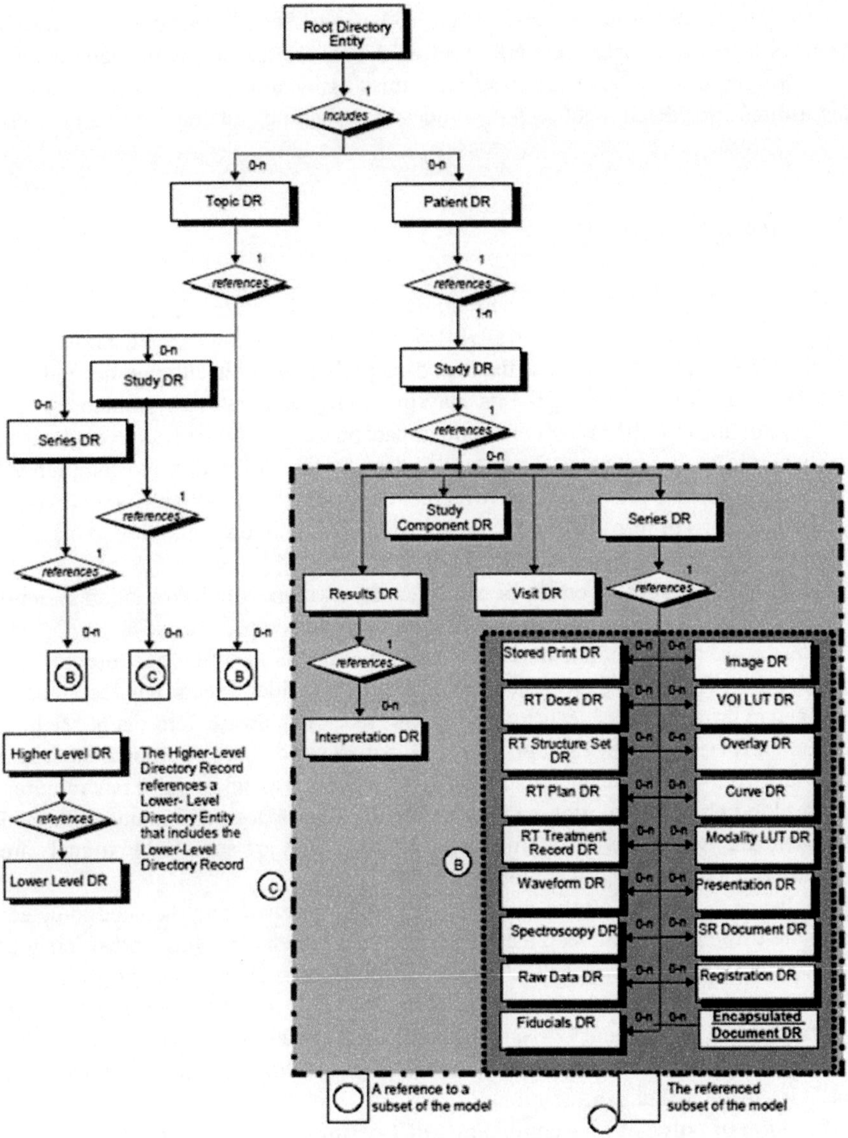

Fig. 1 DICOM report format supporting PDF file format adapted from [8]

The rest of this article is articulated as follows: Sect. 2 elucidates the related work on security attacks for EHR. Section 3 delivers the proposed methodology. Section 4 performs evasion attacks to cause misclassification on the machine learning algorithms considered. Section 5 concludes by providing suitable future directions.

2 Literature Survey

The usage of EHR [9] started with its validation in the Institute of Medicine's (IOM) report entitled as the Computer-Based Patient Record: An Essential Technology for Health Care. This brought about the idea of digitalization of health records and provide better quality service to patients. Hayrinena [10] aggregates a systematic review on the structure of EHR, its contents, and impact on the quality of service provided by medical information systems. The usage of deep learning techniques to comprehend the EHR contents to provide effective medical diagnosis and prediction is surveyed in [11]. The details of information extraction from EHR and representation to machine learning algorithms is elaborated in detail.

Mishra [13] advocates on the need of security awareness in information systems used by healthcare sectors with a HIPAA compliance perspective.

Luna et al. [14] elaborates on the cybercrime threats haunting healthsector like identify theft caused by data breach, cybersquatting, insider threats, and recommend the need to bring in better security provisions to tackle malware. Internet of Medical Things [15] has brought the opportunity for medical devices to interconnect and exchange confidential medical data. Smart devices such as smart pulse rate monitor, smart blood pressure monitor, smart glucometer, smart pacer, and many connected wearable are available with Internet connectivity. These are susceptible to malware attacks which is profoundly explored in [15].

Malware attacks are on the rise in all sectors of software and the need to detect malware using intelligent deep learning techniques which are robust and efficient is expedited in [16] Adversarial attacks tampering 3D medical imagery using generative adversarial nets (GANs) have been exhibited in [17]. Attackers can use learning techniques like conditional GANs to insert or delete manifest of medical conditions in 3D scan reports. This shows the mammoth of possibilities inherent in cyberattack space for medical data perturbations.

Creation of polyglot files combining DICOM image and malware Windows PE file has been elucidated by Ortiz in [18]. The article details on the ability to create specification-compliant DICOM images viewable from any DICOM software to be simultaneously capable of being a Windows malware. Sun [19] propose an adversarial attack leveraging deep learning models to identify susceptible locations in EHR at patient level and cohort level.

Thus, the above literature motivates the need to understand malware attacks on EHR, triggers the need to comprehend their working better and provide better security solutions. In addition, the learning models deployed usually serve as a black box model yielding good performance but poor interpretibility. Reasoning behind the results predicted is a mystery which can be resolved by utilizing the descriptive accuracy which can be computed from the post-hoc analysis as part of the data science life cycle [20]. Explainable artificial intelligence (XAI) model interpretability technique's like local interpretable model-agnostic explanations (LIME), shapley additive explanation (SHAP), sub-modular LIME (SP LIME), integrated gradients, and DeepLift, to name a few can be incorporated to improve the confidence on the model's outcome. Also, the problem of handling sparsity in the feature space extracted from the malware samples can be tackled by usage of dictionary learning techniques like class-specific dictionary learning, Ccentralized class-specific dictionary learning, and maximum correntropy-based dictionary learning framework to name a few. Sparse coding techniques [21] can be used to address the same, and a relative comparison between the performance offered by the deep learning models against sparse coding and dictionary learning methods can be assessed [22].

3 Proposed Methodology

This article coheres into security attacks in healthcare sectors. The validity of possible malware attacks on either EHR using PDF file format or DICOM defined encapsulated documents in PDF format is to be assessed. This hypothesis is backed up by the dataset collection comprising 5000 malware PDF files and 5000 goodware PDF files from the contagio dump [23]. The Adobe PDF file consists of the header which contains PDF file version information, the body which contains a number of objects that defines the file, embedded data operations such as text, images, and scripting code along with the cross-reference table (x-ref) which lists the offset of each object in the file of the reader. PDF files can be analyzed [12] using static or dynamic analysis techniques to exploit information from the features extracted as shown in Fig. 2. Static method analyzes and tests the files by evaluating their code alone, without actually running it. Dynamic analysis explores its actions and behavior during runtime. In addition, hybrid analysis can combine both static and dynamic analysis to extract features [6].

A PDF file is composed of objects marked by one or more tags. These tags represent features for characterizing the file. Indirect object contents are retrieved to assess the nature of the file. For instance, the presence of the keyword /JavaScript can help to infer that the file contains some scripting contents which could be malicious. In this article, static features are extracted using the keyword-based analysis by deploying PDFiD tool and tree-based analysis using the PEEPDF tool. PDFiD extracts a set of 21 features along with their occurrence by performing textual analysis of the dictionaries contained in the file as shown in Fig. 3.

PeePDF tool performs a tree-based analysis by searching for the trailer object/Root, then progressively following the references to find the/Object list recursively to build a tree structure as shown in Fig. 4. Suspicious actions can be identified by performing a tree search. Around 31 features are extracted using this method to build the feature matrix. Figure 5 lists the feature matrix constructed using PeePDF tool with

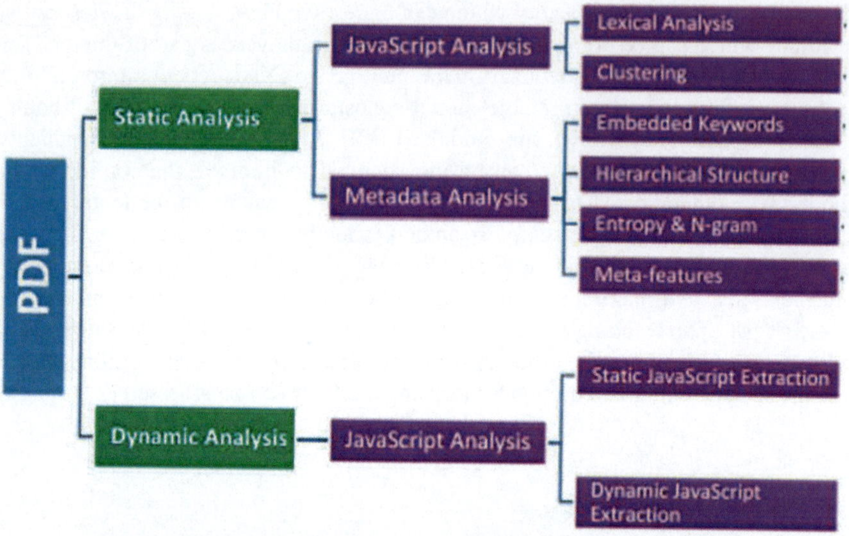

Fig. 2 Taxonomy of PDF feature extraction techniques adapted from [12]

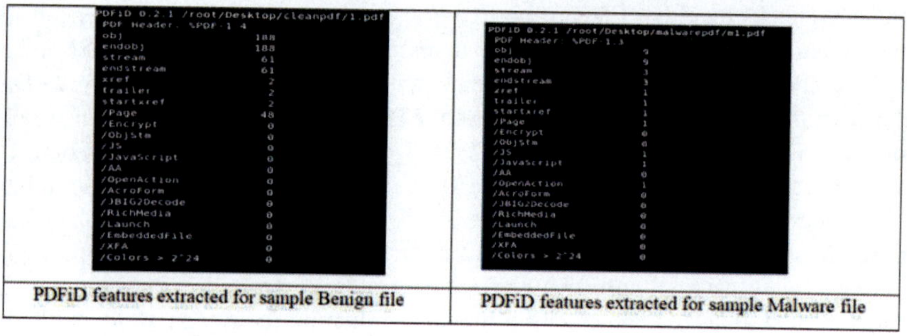

Fig. 3 Keyword-based analysis using PDFiD tool

```
PPDF> tree

/Catalog (1)
        /Pages (3)
                /Page (4)
                        /Pages (3)
                        stream (5)
                        /Font (6)
        /Action /JavaScript (9)
        /Filespec (7)
                stream (8)
        /Outlines (2)

PPDF>
```

Fig. 4 Tree-based analysis using PeePDF tool

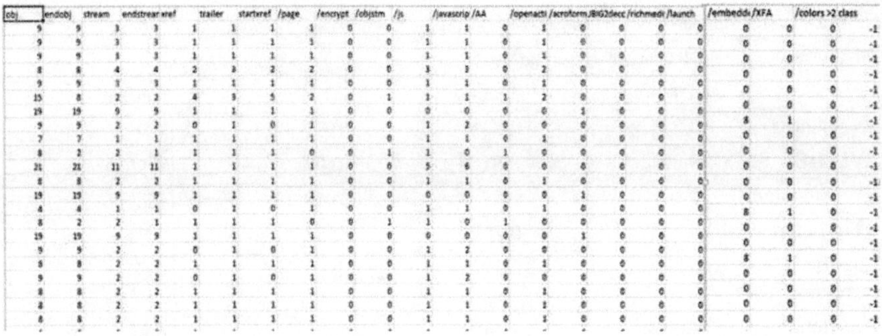

Fig. 5 Feature matrix constructed using PeePDF tool

class labels associated as −1 for malware and +1 for benign files. Standard machine learning models from the scikit learn library [24] is used to build the model. Linear SVM along with other kernel SVM variants like RBF, sigmoid, polynomial followed by logistic regression, decision tree, naïve Bayes, ensemble learners such as random forest for bagging-based and gradient boosting for boosting-based learners are considered. The dataset is split into 80:20 for training and testing purposes.

The confusion matrix is computed from which performance metrics are calculated as follows.

Positive (P): Observation is positive.
Negative (N): Observation is not positive.
True Positive (TP): File is malware, and is predicted to be malware.
False Negative (FN): File is malware, but is predicted benign.
True Negative (TN): File is benign, and is predicted to be benign.
False Positive (FP): File is benign, but is predicted malware.
Accuracy = TP +TN / (TP+TN+FP+FN).

Table 1 infers that the classifiers decision tree, random forest classifier, gradient boosting algorithm, and naïve Bayes have better accuracy than sigmoidal and polynomial SVM for both the analysis methods.

4 Performing Evasion Attacks

The trained model is to be attacked by crafting the adversarial samples using the good word attack [25]. The considered naïve attack tries to cause misclassification of malware samples into benign samples to evade detection. The machine learning (ML) model considered for the attack are SVM, logistic regression, decision tree, naïve Bayes techniques, and ensemble-based approaches like Rrandom forests for the parallel bagging methodology and gradient boosting for the sequential boosting technique. For malware scenario, insertion only modification is feasible as altering

Table 1 Performance of ML classifiers

Classifier	Keyword-based analysis		Tree-based analysis	
	Training	Testing	Training	Testing
SVM linear	0.83	0.82	0.98	0.998
SVM RBF	0.82	0.81	0.97	0.95
SVM sigmoid	0.50	0.51	0.68	0.69
SVM polynomial	0.75	0.74	0.82	0.79
Logistic regression	0.81	0.83	0.99	0.99
Decision tree	**0.98**	**0.95**	**1.0**	**0.99**
Random forest	0.98	0.94	1.00	0.99
Gradient boosting	0.98	0.95	0.99	0.99
Naïve Bayes	0.81	0.83	1.00	0.99

Bold indicates best results

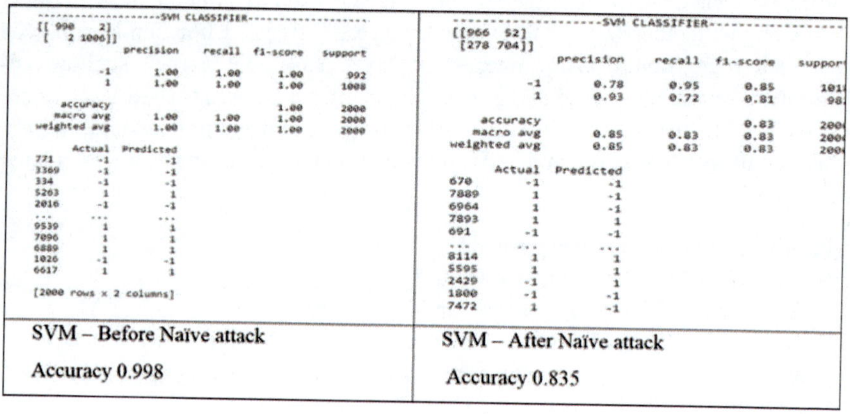

Fig. 6 Illustration of misclassification caused on SVM by naïve attack

feature values or removing them can cause to malware to breakdown. So, in the naïve attack, the malware test samples are taken and modified by randomly inserting features to the test sample until the considered ML model misclassifies the sample as a benign sample. Thus, the amount of objects in the infected PDF files is increased until the model misclassifies them as clean PDF files. This can be used to assess the vulnerability of the learner deployed in the anti-malware EDR software. A white box-based attack is considered in which it is assumed that the attacker knows everything about the data used and learner deployed. Here, the naïve attack shall be considered where the attacker does a brute force attack by iteratively adding one feature at a time to the feature space until misclassification happens. The misclassification caused for linear SVM under the naïve attack for tree-based analysis is shown in Fig. 6. The misclassification caused for all the learners considered is given in Table 2. It can be inferred from the tabulation that among the ML models considered, the model that is

Table 2 Tabulation of evasion attack performance on ML models

Classifier	Keyword-based analysis test accuracy		Tree-based analysis test accuracy	
	Before native attack	After native attack	Before native attack	After native attack
SVM linear	0.82	0.681	0.998	0.835
Logistic regression	0.83	0.71	0.99	0.88
Decision tree	**0.95**	**0.45**	**0.99**	**0.53**
Random forest	0.94	0.62	0.99	0.64
Gradient boosting	0.95	0.87	0.99	0.84
Naïve Bayes	0.83	0.78	0.99	0.86

Bold indicates best results

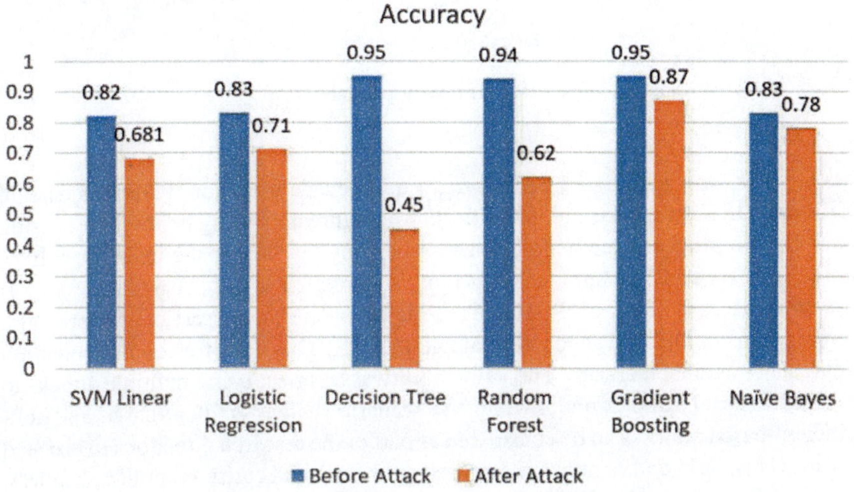

Fig. 7 Misclassification by naïve attack for keyword-based analysis

most vulnerable to the considered evasion attack is decision trees. Figure 7 illustrates the misclassification caused by the naïve attack for keyword-based analysis. The naïve attack behavior for tree-based analysis causing misclassification is shown in Fig. 8. This affirms the fact that learning models can be easily manipulated to cause misclassification by embedding adversarial samples in EHR using PDF file format or usage of DICOM defined encapsulated formats for documents in PDF format.

5 Conclusion and Future Work

Nowadays, HCOs are bound to medical IT and medical ML techniques to provide efficient service to patients. Medical data are prone to security attacks, and the obtained results demonstrate the validity of the above hypothesis. Malware attack-

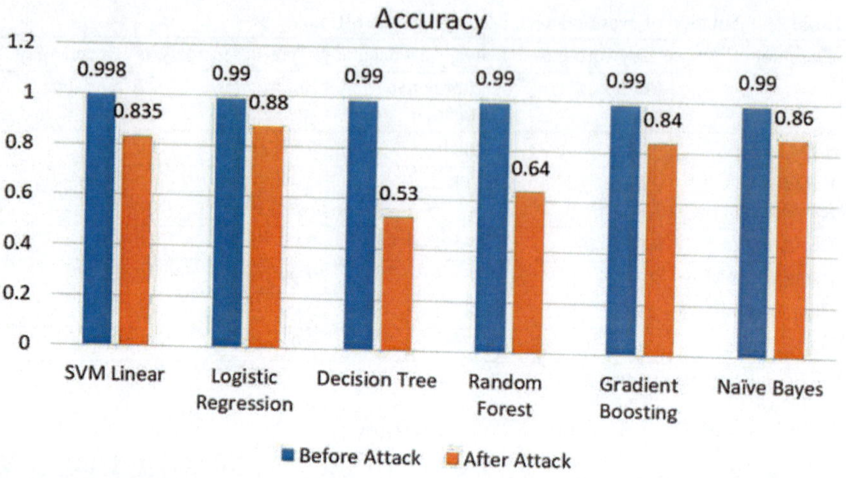

Fig. 8 Misclassification by naïve attack for tree-based analysis

ers lure machine learning-based anti-malware to evade detection. PDF files used in EHR and DICOM reports are under the constant threat of being exposed to evasion attacks. The proposed methodology tries to elaborately assess the inherent vulnerability of usage of machine learning models deployed in the AI medical support system. The case study on PDF malware is considered as it serves as a potential threat vector in DICOM report format and in EHR. This calls for robust protection approaches against malware. The vulnerability assessment using multiple attacks to craft adversarial samples under white box scenario such as gradient-based and non-gradient-based attacks can be considered as part of the research directions to proceed further. Hospitals and healthcare sector should ensure security compliance. NIST, PACS Cybersecurity for the Healthcare Sector [26] provides guidance for DICOM users, device manufactures, and other stakeholders which can serve as an initial measures for improved security. More attention should be given to enhance the security and robustness of machine learning models to become immune to adversarial attacks.

References

1. nathealthindia. https://www.nathealthindia.org/. Last accessed on 5 Aug 2020
2. Social Endeavor. https://en.wikipedia.org/wiki/Social_Endeavor_for_Health_and_Telemedicine. Last accessed on 5 Aug 2020
3. narayana-health-partnersHealthcare Sector. https://cio.economictimes.indiatimes.com/news/enterprise-services-and-applications/narayana-health-partners-with-cisco-to-offer-remote-diagnostics-services/54049553. Last accessed on 5 Aug 2020
4. Banerjee, U., et al.: Achieving information security in healthcare information systems: an essential challenge for the future of nursing informatics. In *SINI: 28th Summer Institute in Nursing Informatics* (2018)

5. Eichelberg, M., Kleber, K., Kämmerer, M.: Cybersecurity Challenges for PACS and Medical Imaging. Academic Radiology (2020)
6. Maiorca, D., Biggio, B.: Digital investigation of PDF files: unveiling traces of embedded malware. IEEE Secur. Privacy **17**(1), 63–71 (2019)
7. Papernot, N., et al.: Towards the science of security and privacy in machine learning (2016). ArXiv preprint arXiv:1611.03814
8. DICOM. ftp://medical.nema.org/medical/dicom/final/sup104_ft.pdf. Last accessed on 5 Aug 2020
9. Dick, R.S., Steen, E.B., Detmer, D.E. (eds.): The Computer-Based Patient Record: An Essential Technology for Health Care. National Academies Press (1997)
10. Häyrinen, K., Saranto, K., Nykänen, P.: Definition, structure, content, use and impacts of electronic health records: a review of the research literature. Int. J. Med. Inf. **77**(5), 291–304 (2008)
11. Shickel, B., et al.: Deep EHR: a survey of recent advances in deep learning techniques for electronic health record (EHR) analysis. IEEE J. Biomed. Health Informat. **22**(5), 1589–1604 (2017)
12. Nissim, N., et al.: Detection of malicious PDF files and directions for enhancements: a state-of-the art survey. Comput. Secur. **48**, 246–266 (2015)
13. Mishra, S., et al.: Security awareness for healthcare information systems: a HIPAA compliance perspective. Issues Inf. Syst. **12**(1), 224–236 (2011)
14. Luna, R., et al.: Cyber threats to health information systems: a systematic review. Technol. Health Care **24**(1), 1–9 (2016)
15. Wazid, M., et al.: IoMT malware detection approaches: analysis and research challenges. IEEE Access **7**, 182459–182476 (2019)
16. Vinayakumar, R., Alazab, M., Soman, K.P., Poornachandran, P., Venkatraman, S.: Robust intelligent malware detection using deep learning. IEEE Access **7**, 46717–46738 (2019). https://doi.org/10.1109/ACCESS.2019.2906934
17. Mirsky, Y., et al. CT-GAN: malicious tampering of 3D medical imagery using deep learning. In: 28th USENIX Security Symposium (USENIX Security 19) (2019)
18. Desjardins, B., et al.: DICOM images have been hacked! Now what? Am. J. Roentgenol. **214**(4), 727–735 (2020)
19. Sun, M., et al.: Identify susceptible locations in medical records via adversarial attacks on deep predictive models. In: Proceedings of the 24th ACM SIGKDD International Conference on Knowledge Discovery and Data Mining (2018)
20. Murdoch, W.J., et al.: Interpretable machine learning: definitions, methods, and applications (2019). arXiv preprint arXiv: 1901.04592
21. Oja, E., Hyvärinen, A., Hoyer, P.: Image feature extraction and denoising by sparse coding. Patt Anal. Appl. **2**(2), 104–110 (1999)
22. Mathews, S.M.: Dictionary and Deep Learning Algorithms with Applications to Remote Health Monitoring Systems. University of Delaware, Diss (2017)
23. contagio dump. http://contagiodump.blogspot.com/2013/03/16800-clean-and-11960-malicious-files.html. Last accessed on 5 Aug 2020
24. scikit https://scikit-learn.org/. Last accessed on 5 Aug 2020
25. Lowd, D., Meek, C.: Good word attacks on statistical spam filters. CEAS (2005)
26. Healthcare Sector. https://www.nccoe.nist.gov/sites/default/files/library/sp1800/hit-pacs-nist-sp1800-24-draft.pdf. Last accessed on 5 Aug 2020

Question Answering System Using Knowledge Graph Generation and Knowledge Base Enrichment with Relation Extraction

K. Sathees Kumar and S. Chitrakala

Abstract Question answering system is one of the predominant application that used to get the answer for the input natural language question. It reduces the search time to find the answer for particular question. In order to answer a natural language question, question understanding plays a vital role. Understanding a question semantically, leads to get the intention of the input question accurately. In order to understand the question, semantic query graph will be constructed to analyze the question phrases in a structured way by extracting entity and relation phrases. If there is no relevant relation phrases in knowledge base, then query graph cannot be constructed. In order to overcome this unavailability of relation phrases, using *n*-gram-based attention model and joint learning, relation phrases will be extracted. Using encoder-decoder model, extracted relation phrases along with that corresponding entity and predicate phrases will be converted as triples with unique IDs. Finally, using top-k sub–graph matching algorithm, the answer with the top score will be selected as the final answer. In this proposed **question answering system with relation enrichment (QARE)** using two datasets such as QALD5 and web questions benchmark, comparative analysis had been carried out with two existing systems. Experimental results show that proposed system achieves better performance than existing systems.

Keywords Question answering · Semantic matching · RDF · Graph database

1 Introduction

Search engines present relevant results based on the various factors such as popularity measures, keyword matching, and frequencies of asking questions on a particular domain. Likewise, in question answering system, answer will be retrieved based on the phrases relatedness present in the input question. Nowadays, data is being stored in various formats. Each format belongs to separate schema. Resource description framework (RDF) performs the data merging when data belong to the different schemas.

K. Sathees Kumar · S. Chitrakala (✉)
Department of Computer Science and Engineering, Anna University, Chennai, Tamil Nadu, India

H. Sharma et al. (eds.), *Congress on Intelligent Systems*, Advances in Intelligent Systems and Computing 1334, https://doi.org/10.1007/978-981-33-6981-8_48

RDF provides the linking structure that provides the relationship between the text phrases. It follows the standard format referred to as a triple. Using this format, data of structured and semi-structured will be mixed. This format forms the directed labeled graph where each node represents the resources and edges represent the named link between those resources. RDF data model is similar to entity relationship model. Its idea is making a common expression of the form subject-predicate-object of web resources. For example, consider the sentence, Who is the president of India? In this sentence who denotes the subject, president of, denotes predicate and India denotes the object. Henceforth, RDF converts data from different formats into this centralized format so that data can be manipulated easily.

SPARQL is the standard query language to access RDF data. Understanding syntax of the SPARQL is complex. In order to answer questions efficiently along with hiding the complexity of SPARQL, in this system RDF questions answering system has been implemented. Question answering involves two stages that are question understanding and query execution. Question understanding determines how phrases in the question related to each other and phrase linking, i.e., mapping similar phrases that matches with the input question. While linking phrases ambiguity may occur. For example, list the movies directed by James Cameron? In this sentence, the person James Cameron may refer to different personalities such as actor, writer, producer, and director. Among these different personalities, correct phrase should be mapped to find the answer. Query execution analyzes selecting appropriate answer for the input question. The advantage of graph data driven approach in the RDF data model helps to evaluate the question by constructing query graph since data representation will be carried out easily. Generally, question answering system can be categorized into two types; open domain question answering system and restricted domain question answering system. Open domain QA system deals with multiple domains whereas closed domain QA system is answering limited number of domain's questions. In this system, questions related to areas such as books, music, films, history, and places are handled. Additionally, the categories of the questions that one system can answer are wh-type questions, factoid questions, etc. In the proposed system, three types of questions are answered that are wh-type questions, true or false questions, and list-type questions. These types can be categorized into two types that are explicit type questions and implicit type questions. Explicit type questions are easy to answer comparatively with the implicit questions. In implicit type questions, relation phrases will be presented in the indirect manner. In this system, implicit type questions have been handled. Generally, question answering system is categorized into two types that are semantic parsing based and information retrieval based. In the semantic parsing-based methods such as [1] finds the semantic representation of the input questions and finds the logical predicates for input question. Proposed system differs by updating knowledge base by adding triple with extraction of relation phrases. Reference [2] addresses graph indexing problem in large networks. But approximation of graph query has not been supported. Reference [3] analyzes the usefulness of semantic analysis and semantic distance metrics. Even though, [3] still needs further improvement in named entity recognition and representing a sentence in logical form. [4] uses semantic similarity to map a natural language question to a SPARQL query that can be improved by using better methods to select the answers. Reference [5] provides solution to the disambiguation while analyzing natural language question, but it does not handle the complex question such as, In what country do people speak Spanish? Reference [6] understands natural language questions by generating binary templates by performing two-level disambiguation, i.e., entity level ambiguity and structure level ambiguity. Reference [7] constructs the model that learns low-dimensional embedding of words using knowledge base using which candidate answers will be selected to natural language questions. Reference [8] represents the approach to retrieve the answer from textual resources by determining the entities and relational phrases by effectively searching structured web data. Subgraph matching is the technique used to find the relevant answers. Reference [9] identifies the predicate arguments and semantic features using probabilistic inference using extracted relations. Reference [10] deals with the how to interface query with RDF linked data which is large and heterogeneous knowledge base. References [11, 12] presents proposal based on semantic information and semantic role for the answer extraction for open domain question answering system.

2 Methodologies

2.1 Detailed System Architecture

In the proposed **QARE**, unavailabilty of relation phrases issue will be resolved efficiently by finding more relationship between entities and predicate phrases that does not exist in knowledge base. In order to handle the issue of unavailability of relation phrases, knowledge base enrichment has been carried out by extracting additional relation phrases from textual resources. For that collection of sentences will be converted to triples in the format of ⟨subject, predicate, object⟩.

In order to understand the question, initially using stanford parser dependency tree will be constructed for the input question. Entity mention dictionary and relation mention dictionary (datasets) will be used to make the comparison with the dependency tree to extract the entity phrases, relation phrases. Semantic query graph will be constructed using the extracted phrases. In this proposed **QARE**, nodes will be introduced first after that edges will be added. Nodes represent entity phrases and edges represent relation mention phrases. Query graph will be constructed based on the candidate list which is generated from the dataset, i.e., extracted entity and relation phrases. Each candidate from the list will be replaced with the nodes and edges of semantic query graph. Entities and wh-words will be replaced with the nodes. Relation phrases will be replaced with the edges. Then, match will be predicted by making the comparison with RDF graph. Initially, query graph will be constructed with the single node. If the match is found, then graph will be extended. A match with the top score will be considered as the answer for the given question. Figure 1 represents the overall system architecture of the proposed **QARE**. In some cases,

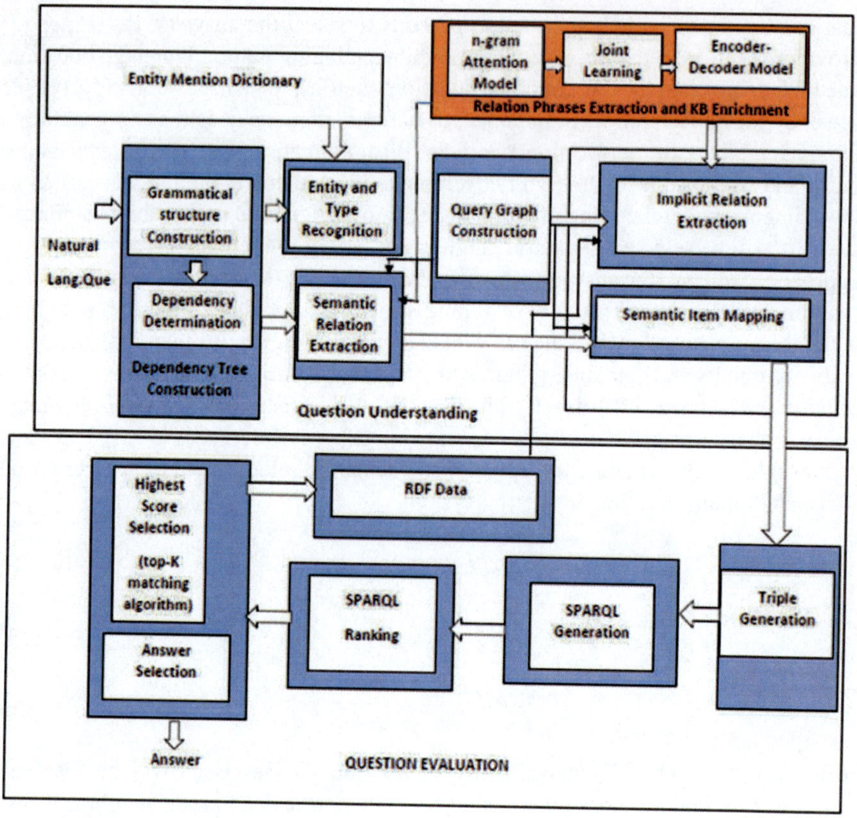

Fig. 1 Overall proposed QARE architecture

while extracting relevant phrases to the question from knowledge base, there may not be similar phrases. In a such situation, query graph cannot be constructed which leads to difficulties in finding the answer. Each sentence in the corpus will be converted to sequence of triples and that will be added to the RDF knowledge base.

2.2 *Extraction of Entity Phrases, Type Phrases, Semantic Relation Phrases*

Initially, natural language question will be given as input. Using the stanford parser [13] grammatical structure of the sentence will be generated. It is used to understand the grammatical relationship between the words in the sentence. Each word will be associated with the part of speech (POS) tags using which grammatical relation will be identified. Each node represents the words in the sentence and edges represent the pos tags. It represents the dependency among the sentence. Based on the dependency relationship between the words in the sentence, dependency tree will be constructed.

Each word from the question will be analyzed in order to remove the unnecessary stop words such as and, many, list, etc. Comparison will be carried out with each POS tag as well in order to ignore such words. Among the remaining words, entity and type words need to be identified. From the remaining words, each word's entity ID and type ID will be retrieved by making comparison with the entity fragment and type fragment (datasets) where collection of entities and type phrases would be stored. For the input question, what is the movie directed by James Cameron?, all the entities related to the name James Cameron will be retrieved and for the word movie relation phrases such as film, direct by, writer, producer, director will be retrieved. While generating triples for the each question entity id, relation ID will be assigned for the each question. Based on those ids, related phrases will be retrieved.

2.3 Knowledge Base Enrichment

Generally, knowledge base consists of the knowledge graph which would have transformed in the triple format with the entities and relation phrases. For example, consider this sentence "Abdul Kalam was the president of India," in which Abdul Kalam and India are the entity phrases. "President of" is the relation phrase which connects the entity phrases. After processing the text data, the sentence will be transformed into triple such as ⟨Abdul Kalam, president of, India⟩. Let us say, "Abdul Kalam" and "India" are recognized and but "president of" might not be recognized by the algorithm since it was not available in the knowledge base. As a result of this, query graph would not have been constructed properly that will lead to the improper answer to the input question. If insufficient relation phrases are in the knowledge base, then query graph cannot be constructed. In order to address this issue, neural encoder–decoder model [14]-based relation extraction has been carried out.

In this proposed approach, WIKI dataset is used to update the existing dataset. WIKI dataset consists of 3,30,005 triples and 158 predicate phrases. In order to process, this much data efficiently joint learning of word and entity embedding has been used to find out the similarity relationship among the enitties and relation phrases. Words that have similar meaning will be represented in the near by positions. Joint learning will be performed using margin-based objective function using the knowledge graph. For each entity and relation phrase, ID will be assigned in order to make the fixed representation of the phases using anchor context model [14]. For example, ⟨Entity name, Entity ID⟩; ⟨Abdul Kalam, Q20203⟩. Word embedding model will be trained in order to generate the vectors.

N-gram-based attention model is used to resolve the issue of wrongly mapping of entities. For example, New York University and New York Times are two different entities. While assigning ID, these two entities have to be assigned with dissimilar ones. Otherwise, it will lead to the improper alignment of triples. In order to avoid this issue, *n*-gram-based attention model will learn all the words of the sentence input. It will assign the weight based on the *n*-gram combinations using which context vector will be assigned. Randomly three phrases will be taken from the output produced by the encoder-decoder model using which triples will be generated. Modified beam search will verify entity IDs and re-rank them if they are not in order. Triple classifier will be used to evaluate the validity of the triple and filter out the valid triples.

Pseudo Code for Proposed EDRE:
Input: *Collection of sentences, paraphrase dictionary*
Output: *Collection of triples*
for each *sentence in corpus* ***do***
perform coreference solution to find implicit relation between entities
for each *predicate$_i$ in paraphrase dictionary* ***do***
remove sentences which doesn't have any relation phrases
for each *sentence in corpus* ***do***
perform word embedding to assign unique IDs to entities and predicates
perform encoding to convert sentence into n-gram tokens and decoding to convert it as sequence of triples
build classifier to determine validity of triples

2.4 Implicit Relation Extraction

Dependency tree words will be considered to find out the modifier and modified words which means, for example "Chinese movie," in these two words, movie is the modifier which modifies the word Chinese. Likewise collection of modifiers list will be created. In order to introduce the first node of graph, target word needs to be identified. Usually, wh-words will be the target word. If the first node is a wh-word, then its corresponding represented word needs to be identified. For example, who is the president of India? In this sentence, the word who represents the president. In order to expand the query graph, DFS technique is used. Each word from the dependency tree will be checked whether it is an entity or type phrase after that only those words will be introduced as nodes in the query graph. Two semantic units from the query graph will be taken. Shortest path phrases of those two nodes will be extracted from the dependency tree. Each shortest path phrase will be compared with paraphrase dictionary, and then all the similar words that matches will be listed out. From that list, each word will be compared with the sub-tree of dependency tree. If any word matches with the sub-tree, then it will be considered as the relation phrase.

Words for which relation could not find will be identified. Among those words, entity and type phrases will be identified using the modifiers list. After finding out the entity, top-k entities will be retrieved that are similar with the passed entities. Predicates that lie between these two entities and type phrases will be listed. After that, corresponding out edge of type, i.e., entity which appears next to this type phrase will be identified and then predicate which is in between these two entities will be considered as the implicit relation. In order to construct the SPARQL, triple needs to be formed. Identified words along with identified relation phrase will be used to form the triple, and then it will be added with the triples list to construct the SPARQL.

2.5 *Answer Selection*

Initially, semantic relations will be passed. Among those constant nodes will be ignored, and only entities and relations will be added to entity and predicate phrases list. All entities and predicate phrases will be combined. For each entity and predicate, corresponding entity and predicate phrases list will be added. In order to form the triple subject, predicate, object needs to be determined. In the semantic relation arg1, i.e., subject is entity phrase, then its corresponding id, word, and score will be retrieved otherwise arg2; i.e., type phrase will be introduced as subject, and entity will be the object. After forming the triples, question focus, i.e., target word needs to be identified in order to construct the SPARQL. Target word would be identified in the query graph construction module which will be considered for the construction of SPARQL. After finding the target word, type of query will be determined whether the count of the target word is needed or the answers alone. Then, triples will be added with the question focus that will be passed to external interface gstore to find the top-k possible matches. Finally, the matches that have the top score will be considered as the final answer for the given question. Score will be calculated for each match with RDF subgraph. Answers that have the top-k score will be selected as the final answer for the input question. According to the above-mentioned equation, score will be calculated as represented in Eq. 1 for each match with RDF subgraph. Answers that have the top-k score will be selected as the final answer for the input question. Answers will be generated for the input question, give me all movies directed by Francis Ford Coppola.

$$\text{Answer Score} = \sum_{vi \epsilon V(Q)} \text{match probability}((v_i, \text{ent}_i)) + \sum_{vi,vj \epsilon E(Q)} \text{match probability}\left(\left(\text{rel}_{v_i v_j}, P_{ij}\right)\right) \quad (1)$$

v_i Vertex in RDF Graph
ent_i Entity phrase in RDF Graph
rel Relation phrase

P_{ij} Predicate phrase.

```
Pseudo Code for Answer Selection:
Input: Edges E(Q) and Vertices V(Q) of of Query Graph Q
Output: Answer for the input question
for each E(Q) and V(Q) do
    get entity, type words and find target word
    construct triple with target, entity and type phrases
for each triples do
    construct the sparql queries
for each E(Q) and V(Q) do
    if match is found in RDF Graph
        add the match with score in answers list
```

3 Experiment

3.1 Experimental Setup

DBPedia A dataset which contains over 1.9 billion triples. Entity fragments and relation paraphrases files will be used for the entity recognition and relation extraction. These files are already in the pre-processed format. Entities are actually mapping of predicate's, entity's and type phrase's name to integers and fragment files based on ID files. Entity fragment of a given entity A contains information of the predicates, entities, and types related to A. Type fragment records the predicate and entity related to a given type. Predicate fragment records the types that a given predicate may accept as object or subject. Using the files that are listed below, proposed system has been implemented and evaluated. **QALD-6** and **Web Questions** A dataset from which 150 questions have taken for testing. **WIKI** A dataset contains 3,30,005 triples and 158 predicate phrases which is used for the knowledge base enrichment.

4 Result Analysis

The QA system was evaluated with respect to precision, recall, and F-measure. Table 1–3 shows the evaluation results of the proposed QARE.

4.1 Precision, Recall, and F-Score

Precision is the ratio of number of questions answered correctly and total number of questions tested.

$$P = \frac{\text{number of questions answered correctly}}{\text{total number of questions}} \tag{2}$$

Using Eq. 2, from the given input questions, precision gives the measure of the ratio of the number of questions correctly answered with that of the total number of questions

Recall is the ratio of number of correct answers for the given question and number of relevant answers in answer key. Using Eq. 3, recall measure is the ratio of number of questions answered correctly with that of number of questions in the answer key.

$$R = \frac{\text{number of questions answered correctly}}{\text{number of questions in the answer key}} \tag{3}$$

F-score is the weighted average of the precision and recall. On the basis of these measures, overall precision and recall values as well as an overall F-measure value were computed as Eq. 4, the average mean of the precision, recall, and F-score values for all questions.

$$F\text{–Score} = \frac{2 * (\text{Precision}(q) * \text{Recall}(q))}{\text{Precision}(q) + \text{Recall}(q)} \tag{4}$$

4.2 Effectiveness Evaluation

Results on QALD-6

Given questions are tested from two bench mark datasets that are QALD-6 and web questions benchmark and checked for their accuracy. The comparison done against answer keys in benchmark with the system and system generated that how many questions answered correctly from the benchmark. The performance details are given in Table 1. As mentioned in Table 1, benchmark questions are divided into three

Table 1 Performance score on types of questions handled

Types of questions	Precision (%)	Recall (%)	F-score (%)
List-type	57	84	67
Wh-type	57	92	70
True or false	63	77	69

types list, wh-type, and true or false questions. Performance of each type questions is clearly given in Table 1. As shown in Figs. 2 and 3, evaluation of proposed **QARE** is carried out with two benchmark datasets that are QALD-6 and web questions. 100 questions from QALD-6 benchmark and 50 questions from web questions benchmark have taken for the evaluation. In [15], more analysis was not done in the relation extraction. Proposed system outperformed [16] which answers questions via template decomposition and could not understand the question through knowledge graph. In [17], semantic items could not be generated as there was difficulty in identifying relations.

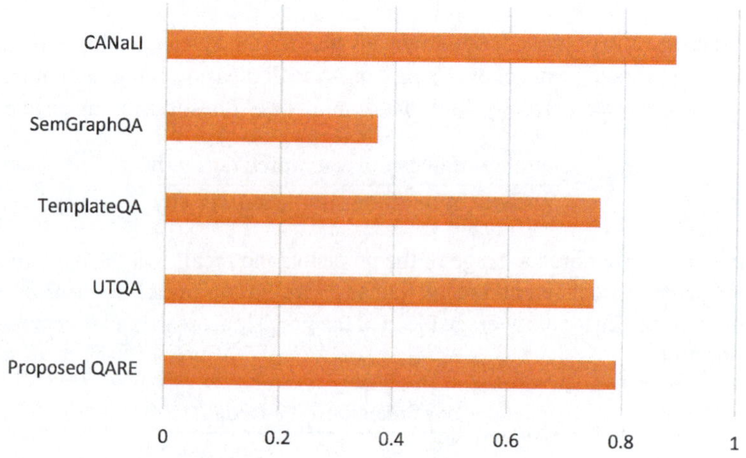

Fig. 2 Comparative analysis on QALD-6

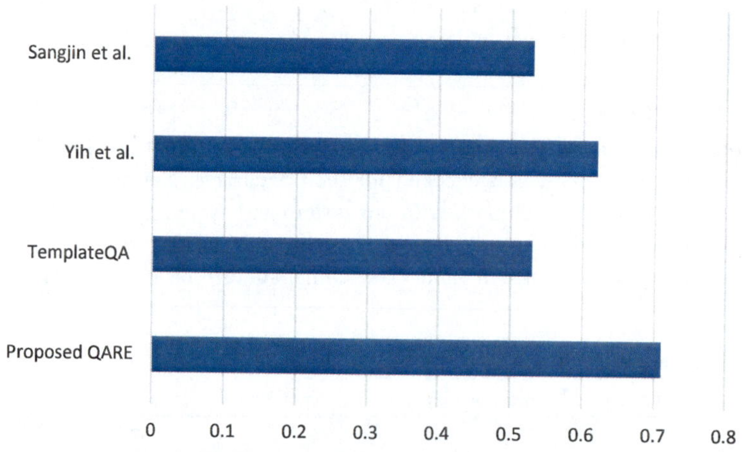

Fig. 3 Comparative analysis on web questions

Table 2 Comparative analysis with existing systems on QALD-6

Methodologies	*F*-score
Proposed QARE	0.71
TemplateQA (2018) [16]	0.53
Yih et al. (2016) [19]	0.62
Sangjin et al. (2018) [20]	0.53

Table 3 Comparative analysis with existing systems on web questions

Methodologies	*F*-score
Proposed QARE	0.79
UTQA (2016) [15]	0.75
TemplateQA (2018) [16]	0.76
SemGraphQA (2015) [17]	0.37
CANaLI (2016) [18]	0.89

Results on Web Questions

Proposed systems has been compared over various existing methods. References [18–20] could not construct the semantic query graph for the complex multiple entity questions. Proposed QARE outperformed these systems with the better F1-score.

Tables 2, 3 and Figs. 2, 3 show the comparison of performance measure of existing and proposed systems for the answer selection. It can be seen that the performance of proposed system which is higher than the existing systems.

5 Conclusion

Proposed **QARE** selects the answer for the given input question properly using knowledge graph. Different from existing solutions, proposed system solves the two key issues that are representing the query intention of the natural language of the question in a structural way. Second one is that knowledge base enrichment by extracting additional relation mention phrases. In order to speed up query evaluation, top-k subgraph matching algorithm is used to find out the match with RDF graph. Proposed **QARE** addresses the issue of unavailability of relation mention phrases. Collectively, this graph data-driven approach is the efficient way to handle the RDF question answering systems. Future work will be handled with complex aggregation and open-minded questions.

References

1. Hu, S., Zou, L., Xu Yu, J., Wang, H., Zhao, D.: Answering natural language questions by subgraph matching over knowledge graphs. IEEE Trans. Knowl. Data Eng. **30**(5), 824–837 (2018). https://doi.org/10.1109/tkde.2017.2766634
2. Zhao P., Han, J.: On graph query optimization in large networks. Proc. VLDB Endow. **3**(1), 340–351 (2010). https://doi.org/10.14778/1920841.1920887
3. Alfonseca, E., De Boni, M., Jara-Valencia, J.-L., Manandhar, S.: A prototype Question Answering system using syntactic and semantic information for answer retrieval. In: Proceedings of the 2001 Conference on question answering, January 2001
4. Park, S., Shim, H., Geunbae Lee, G.: ISOFT at QALD-4: semantic similarity-based question answering system over linked data. In: Proceedings of the International Conference on question answering, pp. 1236–1248 (2014)
5. Hazrina, S., et al.: Review on the advancements of disambiguation in semantic question answering system. Article in Information Processing and Management, Elsevier, pp. 1–18 (2016). https://doi.org/10.1016/j.ipm.2016.06.006
6. Bordes, A., Weston, J., Chopra, S.: Question answering with subgraph embeddings. In: Proceedings of the 2014 Conference on Computation and Language, October 2014
7. Yahya, M., Berberich, K., Elbassuoni, S., Weikum, G.: Robust question answering over the web of linked data. ACM Trans. Inf. Syst. (2013). https://doi.org/10.3115/v1/D14-1067
8. Narayanan, S., Harabagiu, S.: Question answering based on semantic structures. In: Proceedings of the 20th International Conference on Computational Lingistics, pp. 693–701 (2004). https://doi.org/10.3115/1220355.1220455
9. Lopez, V., et.al., Evaluating question answering over linked data. In: Web Semantics: Science, Services and Agents on the World Wide Web, pp. 1–10 (2013). https://doi.org/10.1016/j.websem.2013.05.006
10. Moreda, P., Llorens, H., Saquete, E., Palomar M.: Combining semantic information in question answering systems. Article in Information Processing and Management, Elsevier, pp. 870–885 (2011). https://doi.org/10.1016/j.ipm.2010.03.008
11. Yih, W., He, X., Meek, C.: Semantic parsing for single-relation question answering. In: Proceedings of the 52nd Annual Meeting of the Association for Computational Linguistics, pp. 643–648 (2014). https://doi.org/10.3115/v1/p14-2105
12. Reddy, Siva, Tackstrom, Oscar, Collins, Micheal, Kwaitkowski, Tom, Das, Dipanjan, Steedman, Mark, Lapata, Mirella: Transforming dependency structures to logical forms for semantic parsing. ACL Trans. Assoc. Comput. Linguist. **4**, 127–140 (2016). https://doi.org/10.1162/tacl_a_00088
13. de Marneffe, M.-C., Dozat, Silveria, N., Haverinen, K., Ginter, F., Nivre J., Manning, C.D.: Universal Stanford Dependencies: a cross-lingual typology, Christopher D. Manning, "Stanford Typed Dependencies Manual", In: Proceedings of the Ninth International Conference on Language Resources and Evaluation (2014)
14. Trisedya, B.D., Weikum, G., Qi, J., Zhang, R.: Neural relation extraction for knowledge base enrichment. In Proceedings of the 57th Annual Meeting of the Association for Computational Linguistics, pp. 229–240 (2019). https://doi.org/10.18653/v1/p19-1023
15. Veyseh, A.P.B.: Cross-lingual question answering using common semantic space. In Proceedings of the 2016 Workshop on Graph-Based Methods for Natural Language Processing, NAACL-HLT 2016, pp. 15–19 (2016). https://doi.org/10.18653/v1/w16-1403
16. Zheng, W., Xu Yu, J., Zou, L., Cheng, H.: Question answering over knowledge graphs: question understanding via template decomposition. In: Proceedings of the VLDB Endownment, vol. 11, pp. 1373–1386 (2018). https://doi.org/10.14778/3236187.3236192
17. Beaumont, R., Grau, B., Ligozat, A.-L.: SemGraphQA@QALD-5: LIMSI participation at QALD-5@CLEF, CLEF (2015)
18. Mazzeo, G.M., Zaniolo, C.: Answering controlled natural language questions on RDF knowledge bases. In: 19th International Conference on Extending Database Technology, pp. 608–611, France (2016). https://doi.org/10.5441/002/edbt.2016.60

19. Wen-tau Yih, Ming-Wei Chang, Xiaodong He and Jianfeng Gao, 'Semantic Parsing via Staged Query Graph Generation: Question Answering with Knowledge Base', In Proceedings of the Joint Conference of the 53rd Annual Meeting of the ACL, Vol. 1, pp. 1321–1331 (2015). https://doi.org/10.3115/v1/p15-1128
20. Shin, S., Jin, X., Jung, J., Lee, K.-H.: Predicate constraints based question answering over knowledge Graph, pp. 445–462. Elsevier (2018). https://doi.org/10.1016/j.ipm.2018.12.003

Delivering Newspapers Using Fixed Wing Unmanned Aerial Vehicles

Varun Agarwal and Rajiv Ranjan Tewari

Abstract The advent of unmanned aerial vehicles (UAVs) or drones, as they are popularly known as, has opened a plethora of areas where they can be used and efficiently deployed to perform tasks that are otherwise redundant to humans. Companies across the world are contemplating the applications of UAVs in the delivery of items. We, in this paper, plan to incorporate a class of UAVs called fixed wing UAVs for the delivery of newspapers. Our aim is to determine the exact location where a newspaper parcel needs to be dropped from a UAV so that it reaches its intended destination. Forces of gravity and air drag have been considered in our calculations. Based on the permissible flying ranges, the weight of the newspaper parcel and the drag experienced, our equations calculate the point of parcel drop.

Keywords Fixed wing UAVs · Newspaper delivery · Article delivery using UAVs · Drones · Drone-based newspaper delivery

1 Introduction

In an era of continuous advancement and research in the field of unmanned aerial vehicles (UAVs), backed by their ever evolving uses, we propose the use of UAVs in the delivery of newspapers. So far, there has not been any study in this area. We propose the use of fixed wing UAVS for this delivery process [1]. Fixed wing planes resemble the passenger airplanes in that they require a run way to take off and land. Also, they need a minimum speed during flight (often called the stall speed). They differ from rotary wing UAVs [2], which resemble the helicopters in their operation. Rotary wing UAVs can hover, i.e., come to a standstill in mid air, enabling their use in parcel delivery.

What differentiates newspaper delivery from other items that can be delivered using UAVs is, firstly, the fact that a newspaper is a light, unbreakable object. It can be dropped from a reasonable height with the expectation that the object will not suffer any damage. This particular property of newspaper allows the use of fixed

V. Agarwal (✉) · R. R. Tewari
University of Allahabad, Prayagraj, India

H. Sharma et al. (eds.), *Congress on Intelligent Systems*, Advances in Intelligent Systems and Computing 1334, https://doi.org/10.1007/978-981-33-6981-8_49

wing UAVs. Secondly, newspaper delivery involves delivering the same type of parcel (consisting of the same set of newspapers) to the same set of locations at a fixed time of the day. In case of weekly or fortnightly newspapers, deliveries will be taking place accordingly; however, the schedules and times would be fixed and known in advance. This differs from UAV deliveries of parcels and food which has a dynamic nature.

In any form of literature and practical use, we have not found fixed wing UAVs being used for deliveries. It is always the rotary wing UAVs that are used for delivery purpose, simply because they can hover and can even deliver items right into the recipient's hands. Our approach suggests using fixed wing UAVs to drop newspaper parcels on rooftops of houses or open areas. The possibility of using fixed wing UAV presents the opportunity to harness its advantages, which are as follows [3]:

1. Longer flight time: The average flight duration of a fixed wing UAV is 45 min. This is in contrast to the flight time of a rotor wing UAV that averages 10–15 min. A longer flight time would translate into more number of newspapers delivered.
2. Cover a larger area: A normal flight of a fixed wing UAV covers an area of around 10 km^2. Compared to this, a rotor wing UAV covers a much lesser area. Larger area coverage would imply newspaper deliveries to a greater number of recipients over sparsely populated locales.
3. Better aerodynamic performance: Fixed wing UAVs can fly at wind speeds of up to 45 km/h with minimal turbulence. This translates into greater likelihood of unhindered newspaper delivery activity on all days.
4. Higher flight safety: In case of a technical snag, fixed wing UAVs have a greater tendency to complete their flight in a controlled way, as compared to rotor wing UAVs.

2 Proposed Approach

Steps in the newspaper delivery process:

1. Choose an appropriate UAV—design considerations are area of newspaper delivery, number of newspaper parcels to be delivered, average weight of newspaper parcel and hence the capacity required. The weight of the parcel and the aircraft should be within the maximum take off weight (MTOW) of the UAV [4]. MTOW is the maximum sum total of the aircraft's weight and the weight of the payload that the aircraft can have, in order to fly successfully.
2. Feed the route to be taken by the UAV in delivering the newspapers. The route should be static, with changes taking place in it infrequently.
3. Manual feeding of newspaper parcels in the UAV.
4. The UAV traces the chosen path and reaches near the location where a delivery has to be made. The use of a fixed wing UAV for delivery implies that the UAV cannot reach the actual location where the newspaper parcel has to be delivered. It has to reach a point at a given speed and altitude which is at a certain distance

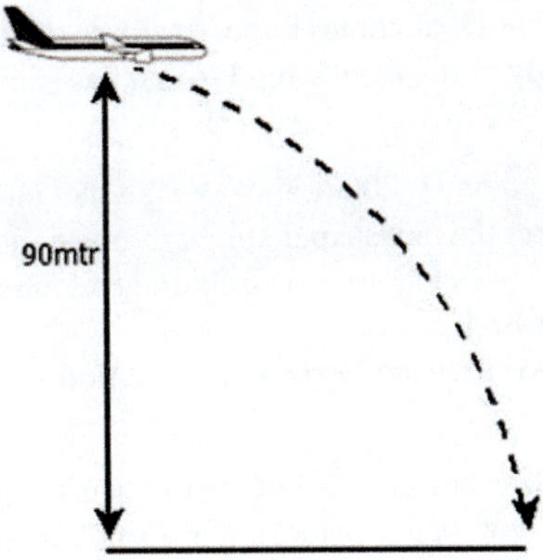

Fig. 1 Showing the projectile path traced by a parcel

before the actual location of delivery and drop the parcel at that point. The parcel, having the velocity of the UAV, would trace a projectile path and deliver the parcel at the scheduled spot. Our problem is focused on finding the point where the UAV needs to drop the parcel in order to reach the desired destination. Figure 1 should make the problem clearer.

5. UAV descends to a height of 90 meters and reduces its speed to 12.9 m/s. Federal aviation administration (FAA) guidelines indicate that UAVs are permitted to fly below a height of 400 ft only [5]. This translates into 121.92 m. So, the UAV cannot fly at a height greater than that. However, we assume 90 m as it a safe height above the ground to avoid any obstacles like high rise buildings, trees, power transmission poles, etc. So, the UAV would be flying at a height between 90 and 121.92 m in normal course and descending to an altitude of 90 m when it needs to drop its parcel. Also, we keep the speed of the UAV as close as possible to the stall speed of the UAV, when the packet has to be dropped. Here, we assume it to be 12.9 m/s [6, 7]. The stall speed of an aerial device is the minimum speed at which it can fly.
6. It drops its parcel at the point based on the below computation.
7. UAV regains the flight altitude and continues to step 4.

Step 6 in detail.

Before stating our actual calculations, we would like to describe our newspaper release technique. We adopt a store release system for dropping the newspaper parcel. It consists of a support member, a newspaper store facility, and an ejection mechanism. The support member is a mechanical device to hold and release the newspaper parcel. The newspaper store contains the payload, the set of newspapers that are to be delivered in a given flight. The ejection mechanism consists of a hydraulic piston that moves the support member vertically. When the UAV reaches the point where a newspaper has to be dropped, the piston should move the support member from a retracted position to the extended position. This action should take place along the vertical axis. The extension of the support member should be such that when the newspaper is released from it, it comes under the force of gravity and falls down and not impacted by the body of the fixed wing UAV. It has been proved in [8, 9] that the above approach has the following advantages:

1. The size of the cargo bay width or the opening that allows the support member to exit the UAV with the newspaper stored to it is greatly reduced.
2. The newspaper-to-UAV clearance, an important parameter to facilitate safe UAV navigation, is increased.
3. A reduction in UAV induced horizontal deviation of the newspaper parcel is reduced.

Our aim is to calculate how much before the actual delivery point should the UAV actually drop the newspaper parcel for it to fall at the delivery point. When the UAV drops the newspaper parcel, it has to consider its initial velocity and the applicable forces on the parcel. These forces govern the path traced by the parcel, which ultimately determines the actual location where the parcel reaches the ground. We assume that the UAV was flying in a path parallel to the earth's surface, thus the initial vertical velocity (v_{if}) of the parcel is zero and the initial horizontal velocity (v_{hf}) is same as that of the UAV at the time of dropping the newspaper parcel.

For our first scenario, let us ignore the affect of air drag. This scenario is a simplified version of our actual problem and is practically possible only in vacuum.

We have assumed:

$$\Delta y = -90\,\text{m} \tag{1}$$

The height from which we are dropping the parcel has been assumed to be 90 m. The parcel will fall from this height to the ground, i.e., displacement will be in the downward direction. Since, Δy is a vector, we have assigned it a negative value. Also, the vertical acceleration is negative.

$$a_y = -9.81\,\text{m/s}^2 \tag{2}$$

$$v_{iy} = 0 \tag{3}$$

Using the above given values, we solve for the vertical motion of the parcel.

$$\Delta y = v_{iy}\Delta t + \left(\frac{1}{2}\right)a_y\Delta t^2 \tag{4}$$

$$-90 = 0.\Delta t + \left(\frac{1}{2}\right)(-9.81) * \Delta t^2$$

$$\Delta t = \sqrt{2 * \frac{(-90)}{(-9.81)}}$$

$$\Delta t = 4.284\,\text{s} \tag{5}$$

This calculation of time, which we obtained considering the vertical motion of the parcel, holds valid for the parcel's horizontal motion also. Applying the calculated value of time in the horizontal motion,

$$v_x = 12.9\,\text{m/s} \tag{6}$$

$$a_x = 0 \tag{7}$$

$$\Delta x = v_{ix}\,\Delta t + \left(\frac{1}{2}\right) a_x\,\Delta t^2 \tag{8}$$

$$\Delta x = 12.9 * 4.284$$

$$\Delta x = 55.26\,\text{m}$$

The above calculation states that when a UAV, traveling at a speed of 12.9 m/s and 90 m above the ground, drops a newspaper parcel, the parcel will trace the path of a projectile and fall at a point on the ground which is 55.26 m ahead of the point of release from the UAV, provided the affect of air drag is ignored. In other words, if the delivery point is known, the UAV should drop the parcel at a distance of 55.26 m before reaching that point so that the newspaper parcel is correctly delivered at its intended destination.

However, any object, when dropped in air, is subjected to the air drag, apart from the other forces, as discussed in the above scenario.

Next, we consider the affect of air drag in our newspaper delivery problem. Figure 2 depicts the vacuum trajectory, which shows the first scenario explained above. Also, the figure shows the actual trajectory, which is the path traced by the newspaper parcel when air drag affects its projectile path. The target marks the point where the parcel will fall in this scenario.

Parameters considered:

$$\text{Density of air } \rho_{\text{air}} = 1.275\,\text{kg/m}^3 \tag{9}$$

This is the density of air at sea level and at zero degrees centigrade temperature [10, 11]. This value decreases marginally with increase in temperature. It has been stated in [12] that the density of air remains more or less constant, reducing by about 3% at every 305 m increase in altitude. Since, our flight altitude is much less than that; we assume the density of air at sea level for our calculations.

$$\text{Drag Coefficient } C_d = 0.04 \tag{10}$$

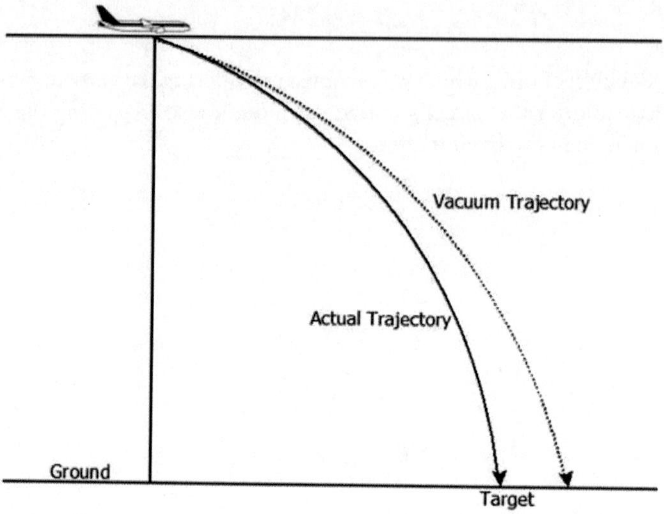

Fig. 2 Effect of air drag

For our experiments, we have decided to roll the newspaper into a cylinder, creating parcels of the same before feeding them into the fixed wing UAV (Fig. 3).

When the parcel is rolled in this form and dropped from the UAV, it would face the least amount of air resistance, with air passing from above and below it, as shown in Fig. 4.

Also, the drag faced by the newspaper parcel would be similar to that experienced by a streamlined body. Hence, we assume the drag coefficient for our parcel to be 0.04 [15, 16], and it is a unitless quantity.

$$\text{Horizontal velocity } v_{ix} = 12.9\,\text{m/s} \tag{11}$$

The minimum velocity at which a UAV can fly, also known as the stall speed.

Fig. 3 A rolled newspaper [13]

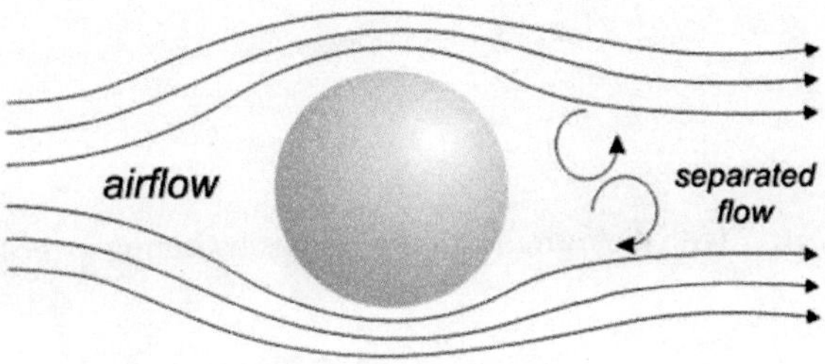

Fig. 4 Air drag around the parcel [14]

$$\text{Area of newspaper parcel (A} = 0.0208\,\text{m}^2 \tag{12}$$

The area, in consideration, is the one exposed to air when the object is in its projectile motion. Hence, instead of using the area of the cylinder formula ($\pi r^2 h$), we use the length * width of the newspaper parcel. Assuming the length to be 26 cms and width 8 cms, the area comes out to be 0.0208 m^2.

$$\text{Mass of newspaper parcel (m)} = 0.12\,\text{kg} \tag{13}$$

While calculating the mass of a newspaper parcel, there were multiple factors [17] we considered, namely:

- Quality of paper used—Most of the newspapers are using 35–50 gsm paper.
- Area of a page—Larger the paper size, more the weight of the newspaper.
- Number of pages in the newspaper—Weight is proportional to the quantity of pages the paper contains.
- Ink quality—Although miniscule, it can still not be ignored.
- Ink quantity—If the paper contains more words, the overall weight would increase proportionately.
- Moisture—Newspaper page material can absorb a great amount of moisture, affecting the total weight.

Considering the various parameters and their roles, we conclude an average weight of 0.120 kgs for a newspaper parcel. This has been calculated assuming a single parcel having two newspapers, each containing an average of 22 pages. The weight of a single page has been assumed to be 0.096 oz or 2.72 gms.

We aim to use the force equation ($F = ma$) while modifying our original calculations to incorporate the effect of drag. Here also, we split our calculations into the horizontal and vertical components.

For the horizontal calculations,

$$F_x = ma_x \tag{14}$$

The only force acting on the falling parcel in the air is the drag force, in the direction opposing the motion of the parcel. Hence,

$$F_x = -F_{\text{drag } x} \tag{15}$$

$$-F_{\text{drag } x} = ma_x \tag{16}$$

Replacing $F_{\text{drag } x}$ with the formula for drag and substituting in the above equation,

$$-\frac{1}{2}\rho_{\text{air}} v_{ix}^2 C_d A = ma_x$$

$$a_x = \frac{-\rho_{\text{air}} v_{ix}^2 C_d A}{2m}$$

$$a_x = -\frac{1.275 * 12.9 * 12.9 * 0.04 * 0.0208}{2 * 0.12}$$

$$a_x = -0.73553\ \text{m/s}^2$$

We substitute this calculated value of horizontal acceleration (or rather, deceleration, since the value is negative) in the equation of motion

$$v_{fx} = v_{ix} + a_x * t \tag{17}$$

where v_{fx} denotes final velocity, v_{ix} denotes initial velocity and t is taken as 0.01 s. The reason of using the above equation is to find out the final velocity of the newspaper parcel at the end of 0.01 s when the initial velocity is known, here 12.90 m/s.

$$v_{fx} = 12.9 - 0.73553 * 0.01$$

$$v_{fx} = 12.8926\ \text{m/s} \tag{18}$$

Thus, at the end of 0.01 s, the effect of air drag has resulted in a reduction in velocity to the above-calculated quantity. We will, later, write a C program where the above formulae will be fed and the depreciating velocities, at intervals of 0.01 s will be calculated. Keeping a value of $t = 0.01$ s will help us in getting values at very short intervals of time, thus giving accurate results at the end of the experiment.

Now, we use another equation of motion to calculate the horizontal displacement of the parcel after an interval of $t = 0.01$ s.

$$x_{\text{f}} - x_{\text{i}} = v_{ix}t + \left(\frac{1}{2}\right)a_x t^2 \tag{19}$$

where x_{f} is the final horizontal position of the parcel, x_{i} is its initial position, here assumed to be 0, v_{ix} denotes initial velocity, t is taken as 0.01 s and a_x

$$x_{\mathrm{f}} = x_{\mathrm{i}} + v_{ix}t + \left(\frac{1}{2}\right)a_x t^2$$

$$x_{\mathrm{f}} = 0 + 12.9 * 0.01 + \left(\frac{1}{2}\right)(-0.73553)(0.01)^2$$

$$x_f = 0.129 + \left(\frac{1}{2}\right)(-0.73553)(0.01)^2$$

$$x_{\mathrm{f}} = 0.12896\,\mathrm{m}$$

The above calculation shows that at the end of time = 0.01 s, for the initial values of acceleration and velocity, the distance covered in horizontal direction is as calculated above. Again, we would be using this formula and calculating the horizontal displacement (x_{f}), in our C program, at time intervals of 0.01 s each.

Now, let us repeat the above set of formulae for the vertical motion of the newspaper parcel.

$$F_y = ma_y \tag{20}$$

Now, in the vertical motion, the newspaper parcel is subjected to two forces, the force of gravity, acting in the downward direction and the drag force acting upwards. Thus, the two forces oppose each other. Hence,

$$F_y = F_{\mathrm{drag}y} - F_g \tag{21}$$

Replacing $F_{\mathrm{drag}y}$ with the formula for drag in the above equation.

$$ma_y = \frac{1}{2}\rho_{\mathrm{air}} v_{\mathrm{iy}}^2 C_d A - mg \tag{22}$$

Since initial vertical velocity v_{iy} is zero,

$$ma_y = -mg$$

$$a_y = -g$$

Repeating the calculations in the equations, we used for vertical motion, we substitute this acceleration value in the equation of motion

$$v_{\mathrm{fy}} = v_{\mathrm{iy}} + a_y * t \tag{23}$$

$$v_{\mathrm{fy}} = v_{\mathrm{iy}} - g * t$$

where v_{fy} denotes final velocity, v_{iy} denotes initial velocity and t is taken as 0.01 s. The reason for using the above equation is to find out the final velocity of the newspaper parcel at the end of 0.01 s when the initial velocity is 0.

$$v_{fy} = 0 - 9.81 * 0.01$$

$$v_{fy} = -0.0981\ \text{m/s}$$

Thus, at the end of 0.01 s, the effect of gravity has resulted in the velocity to have a value stated above with the negative sign (due to the negative value of gravity used in the formula). The effect of drag could not be found as the initial velocity was zero; however, since the final velocity gets a value other than zero, if we calculate for the next 0.01 s time interval, the drag effect will be observable. The resulting table of our program depicts the varying final vertical velocities under the opposing influences of gravity and air drag.

Now, just like we performed in the horizontal motion scenario, we find the vertical displacement of the parcel after time $t = 0.01$ s.

$$y_f - y_i = v_{iy}t + \left(\frac{1}{2}\right)a_y\, t^2 \tag{24}$$

where y_f is the final vertical position of the parcel, y_i is its initial position, here assumed to be 0, v_{iy} denotes initial velocity, t is taken as 0.01 s and a_y is the acceleration in vertical direction.

$$y_f = y_i + v_{iy}t + \left(\frac{1}{2}\right)a_y\, t^2$$

$$y_f = 0 + 0 * 0.01 + \left(\frac{1}{2}\right)(-9.81)(0.01)^2$$

$$y_f = 0.0004905\ \text{m}$$

The above calculation shows that at the end of time = 0.01 s, for the initial values of acceleration and velocity, the distance covered in vertical direction is as calculated above. Again, we would be adding this formula and calculating the vertical displacement (y_f) at time intervals of 0.01 s.

We, thus, reach the end of the first run of our experiment at time $t = 0.01$ s. Using the formulae discussed above, we create a C program to calculate the values at recurring intervals of 0.01 s. We are attaching the flowchart of the program we have written, shown in Fig. 5. Also, Table 1 shows the values calculated from the formulae discussed above.

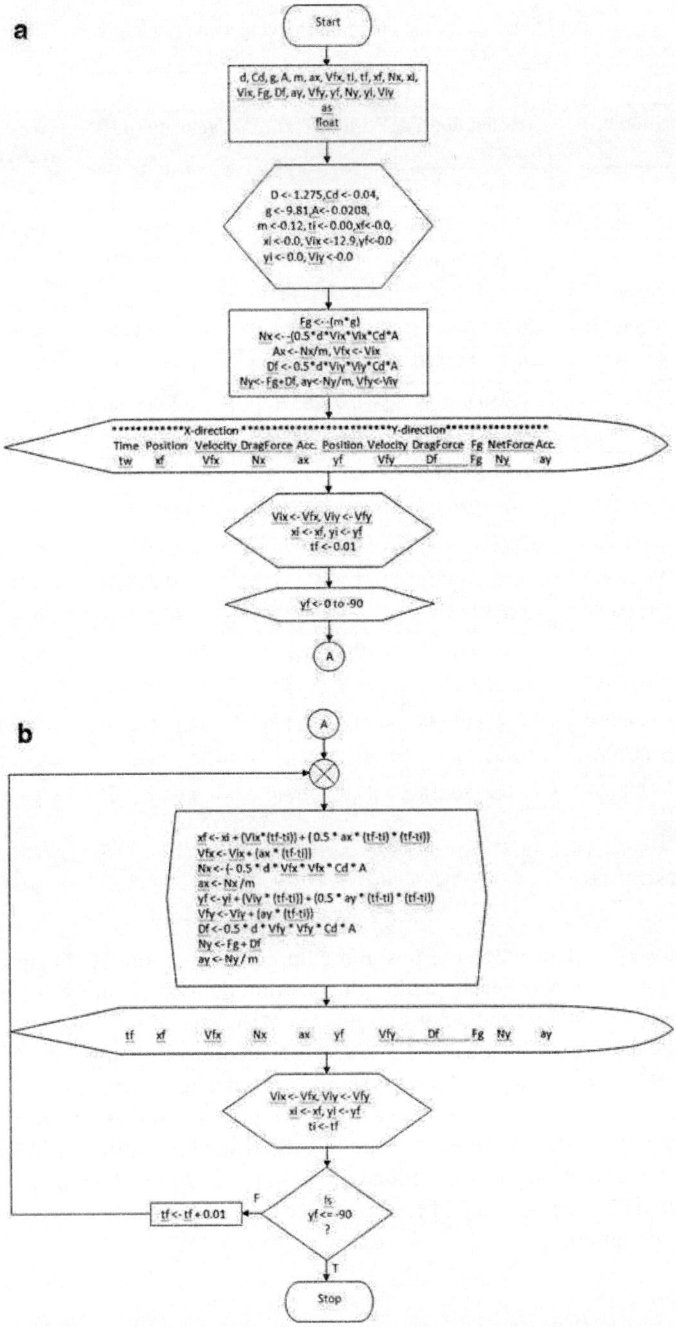

Fig. 5 **a** Flowchart of our C program. **b** Flowchart of our C program

Table 1 A snapshot of the results of the program, showing some random values obtained by our calculations

x-direction			*y*-direction		
Time (s)	Position (m)	Acceleration (m/s^2)	Position (m)	Velocity (m/s)	Acceleration (m/s^2)
0	0	−0.7355322	0	0	−9.81
0.01	0.128963223	−0.734693668	−0.0004905	−0.0981	−9.809957464
0.02	0.257852935	−0.733856569	−0.001961998	−0.196199575	−9.809829855
0.03	0.38666922	−0.7330209	−0.004414485	−0.294297873	−9.809617178
0.04	0.515412161	−0.732186658	−0.007847945	−0.392394045	−9.809319439
0.05	0.644081841	−0.731353839	−0.012262351	−0.490487239	−9.808936646
0.06	0.772678344	−0.730522441	−0.01765767	−0.588576606	−9.808468813
0.07	0.901201754	−0.729692461	−0.02403386	−0.686661294	−9.807915954
0.08	1.029652153	−0.728863894	−0.031390869	−0.784740453	−9.807278086
0.09	1.158029624	−0.728036738	−0.039728637	−0.882813234	−9.806555232
0.1	1.28633425	−0.727210989	−0.049047097	−0.980878787	−9.805747415
0.5	6.359719657	−0.695299553	−1.224127172	−4.887874709	−9.704400369
1	12.54545099	−0.658280396	−4.870638899	−9.672645896	−9.396464452
2	24.43130433	−0.592591197	−19.08279252	−18.5662128	−8.286407181
3	35.72348373	−0.536266934	−41.55594549	−26.13028443	−6.7920604
4	46.47849694	−0.487608338	−70.82472496	−32.14824064	−5.241888558
4.5	51.67007967	−0.465726164	−87.52441603	−34.590447	−4.521474314
4.57	52.38748267	−0.462780618	−89.99676211	−34.90405345	−4.425145172

3 Conclusion

It can be observed that when the position column (*y*-direction) in the spreadsheet shows −90 m (−89.9967 in the table), the position column (*x*-direction) reads the value 52.387 m. This implies that the effect of UAV speed, gravity, and air drag results in the packet reaching a point 52.387 m ahead of its dropping point, i.e., if the delivery destination is known, the UAV should drop the newspaper parcel 52.387 m before it, so that the parcel may trace a projectile path and deliver the parcel at the correct destination. Another observation from the results is that the air drag causes the newspaper parcel to reach the ground 55.26 m–52.387 m = 2.87 m before. (The value 55.26 m had been obtained in our first set of calculations when the affect of air drag had been ignored.)

4 Future Work

We have ignored the affect of cross winds in our experiments. They may play a critical role, considering the parcel weigh is less. We shall, in later work, try to incorporate the affects of cross winds in our future experiments. Another area that has attracted our attention is the fact that the route chosen for the UAV remains fixed for multiple tours, as the deliveries are fixed. This facilitates the use of a route selection algorithm like traveling salesperson problem (TSP). We would like to include in a later research paper, the appropriate route selection algorithm for our delivery mechanism.

References

1. Advantages of Fixed Wing UAV than Rotary Wing, viewed 4 July 2019 https://www.cleverdronemaps.com/advantages-of-fixed-wing-uav-than-rotary-wing/?lang=en/
2. Chao, H., Cao, Y., Chen, Y.: August. Autopilots for small fixed-wing unmanned air vehicles: a survey. In: 2007 International Conference on Mechatronics and Automation, pp. 3144–3149. IEEE (2007)
3. Dalamagkidis, K., Valavanis, K.P., Piegl, L.A.: On unmanned aircraft systems issues, challenges and operational restrictions preventing integration into the National Airspace System. Prog. Aerosp. Sci. **44**(7–8), 503–519 (2008)
4. Free Rolled Newspaper Cliparts, viewed 15 Sept. 2019. http://clipart-library.com/rolled-newspaper-cliparts.html/
5. Grosse, M.: Newspaper vending machine. U.S. Patent 3,917,114 (1975)
6. Hassanalian, M., Abdelkefi, A.: Design, manufacturing, and flight testing of a fixed wing micro air vehicle with Zimmerman planform. Meccanica **52**(6), 1265–1282 (2017)
7. Hauchecorne, A., Chanin, M.L.: Density and temperature profiles obtained by lidar between 35 and 70 km. Geophys. Res. Lett. **7**(8), 565–568 (1980)
8. Holtrop, J.W.: US Secretary of Navy, 1984. Constrained store ejector. U.S. Patent 4,441,674
9. Jaimes, A., Kota, S., Gomez, J.: An approach to surveillance an area using swarm of fixed wing and quad-rotor unmanned aerial vehicles UAV (s). In: 2008 IEEE International Conference on System of Systems Engineering, pp. 1–6. IEEE (2008)
10. Jakubowski Jr., T., Foster, J.K.: McDonnell Douglas Corp, 1999. Constrained store release system. U.S. Patent 5,904,323
11. Kidder, F.E.: The Architects' and Builders' Handbook: Data for Architects, Structural Engineers, Contractors, and Draughtsmen. Wiley (1921)
12. Lasso, I.A., Weidman, P.D.: Stokes drag on hollow cylinders and conglomerates. The Phys. Fluids **29**(12), 3921–3934 (1986)
13. Levine, B.D., Stray-Gundersen, J., Mehta, R.D.: Effect of altitude on football performance. Scand. J. Med. Sci. Sports **18**, 76–84 (2008)
14. Small Unmanned Aircraft Systems Advisory Circular, AC 107-2, viewed 14 July 2019 https://www.faa.gov/uas/resources/policy_library/media/AC_107-2_AFS-1_Signed.pdf/
15. Thurrowgood, S., Moore, R.J., Soccol, D., Knight, M., Srinivasan, M.V.: A biologically inspired, vision-based guidance system for automatic landing of a fixed-wing aircraft. J. Field Robot. **31**(4), 699–727 (2014)
16. Tran-Cong, S., Gay, M., Michaelides, E.E.: Drag coefficients of irregularly shaped particles. Powder Technol. **139**(1), 21–32 (2004)
17. Understanding Rocket Aerodynamics, viewed 22 Sept. 2019 http://learn.sigmarockets.com/lessons/understanding-aerodynamics-with-the-educator-rocket/?print=print/

A Study on Design Optimization of Spur Gear Set

Jawaz Alam, Srusti Priyadarshini, Sumanta Panda, and Padmanav Dash

Abstract In this article, a profile modification approach is adopted for the design of a spur gear set to minimize the contact stress and optimize the weight. An altered tooth sum method is used to reduce the hertizian contact stress along the path of contact. Four case studies are performed for the spur gear set with specific center distance and tooth sum of 90 ($\pm$5). A multi-objective optimization problem is formulated with contact stresses along path of contact and weight of gear set as the design objectives. This nonlinear-constrained optimization problem has been addressed by mean of the particle swarm optimization algorithm. Six design variables related to gear geometry and material property are used in this optimization and all the imposed constraints are satisfied at the optimal solution. Gear and pinion surface temperatures are below flash temperature, indicating protection against scoring wear. Specifically, a lighter gear with less contact stress and adequate scoring resistance is reported in this study. Furthermore, a CAD model is developed by means of optimized design parameters so as to check the geometric interference and practical feasibility of the design.

Keywords Spur gears · Contact ratio · Profile shift · Contact stress · Scoring

1 Introduction

The optimum design of a spur gear with lower contact stress across path of contact, lightweight, and excellent wear resistance is a very cumbersome task. So, it is necessary to improve the design of spur gear by considering different aspect like gear contact stress, vibration level, profile modification, and adequate lubrication. Previously, a lots of research work has been done to improve the design of spur gear. In a recent study, Sachidananda et al. [1] reported one of the important approach to reduce the contact stresses in gear by using addendum modification. In the study, SEM images of the gear are analyzed to study the morphology of gear wear. The optimum weight design (OWD) problem was first introduced by Yokota et al. [2]

J. Alam (✉) · S. Priyadarshini · S. Panda · P. Dash
Department of Mechanical Engineering, VSSUT, Burla, Odisha 768018, India

H. Sharma et al. (eds.), *Congress on Intelligent Systems*, Advances in Intelligent Systems and Computing 1334, https://doi.org/10.1007/978-981-33-6981-8_50

while focusing on a single-stage spur gear and applying an improved genetic algorithm method. The OWD problem of single-stage spur gear was again investigated by Savsani et al. [3] using two advanced optimization algorithms, i.e., particle swarm optimization (PSO) and simulated annealing (SA) to determine an optimum set of design variable for minimum weight of a spur gear set. Furthermore, Panda et al. [4] minimize weight of a spur gear train using an evolutionary based algorithm, i.e., differential evolution (DE). In fact, the authors have performed FEA on to optimized gear set to predict the most critically stressed region. İn a similar study, Rao [5] considered two design cases as reported by [3] and investigated it using the teaching learning-based optimization (TLBO) while comparing against other algorithms. Rao [5] demonstratively reasoned that the parameter less feature of the TLBO algorithm resulted in the improvement of the solution. In a longitudinal study, the OWD problem was investigated using the gray wolf optimizer (GWO), a nature inspired algorithm by Dörterler et al. [6]. Lately, Maputi and Arora [7] uses multi-objective teaching learning based design optimization (MOTLBO) for the same OWD problem with an extension by including power output maximization.

The extensive literature review reveals that no work related to simultaneous optimization of weight and contact stress along the path of contact on multi-objective basis has been reported. So, in this study, an attempt is made to optimize the weight of spur gear set and contact stress along the path of contact. Some constraints related to bending fatigue, surface fatigue, scoring, etc., (as reported in [3]) are imposed on the optimization problem to prevent the failure of the spur gear set against these modes of wear. A CAD model is also developed using the optimized geometry of the spur gear, so as to check the geometric interference.

2 Problem Synthesis

The basic spur gear geometry considered for the weight optimization is illustrated in Fig. 1 as reported in [2].

Where, $D_r = m(aZ_1 - 2.5)$, $l_w = 2.5m$, $D_i = D_r - 2l_w$, $b_w = 3.5m$, $d_0 = d_2 + 25$, $d_p = 0.25(D_i - d_0)$, $D_1 = mZ_1$, $D_2 = amZ_1$, $N_2 = N_1/a$, $Z_2 = Z_1 * D_2/D_1$, $\gamma = \pi D_1 N_1/60,000$, $b_1 = 1000P/\gamma$, $b_3 = 48.68e6P/(N_1\tau)$, $b_3 = 48.68e6P/(N_2\tau)$, $F_s = \pi K_v K_w \sigma bmy$, $F_p = 2K_v K_w D_1 b Z_2/(Z_1 + Z_2)$, $a = 4$, $P = 7.5$, $n = 6$, $\rho = 8$, $\sigma = 294.3$, $y = 0.102$, $b_2 = 0.193$, $\tau = 19.62$, $K_w = 0.8$, $K_v = 0.389$.

PSO is a population-based optimization algorithm initially exhibited by Kennedy and Eberhart [8]. The PSO is based in changing the velocity (i.e., accelerating) of each particle toward its 'p Best' and 'g Best' locations (global version of PSO) at each step. The acceleration is weighted by a random term with separate random numbers being generated for acceleration toward 'p Best' and 'g Best' locations. The velocity and position updates of the particles are accomplished as per Eqs. 1 and 2. The detail steps of PSO algorithm is reported in [8].

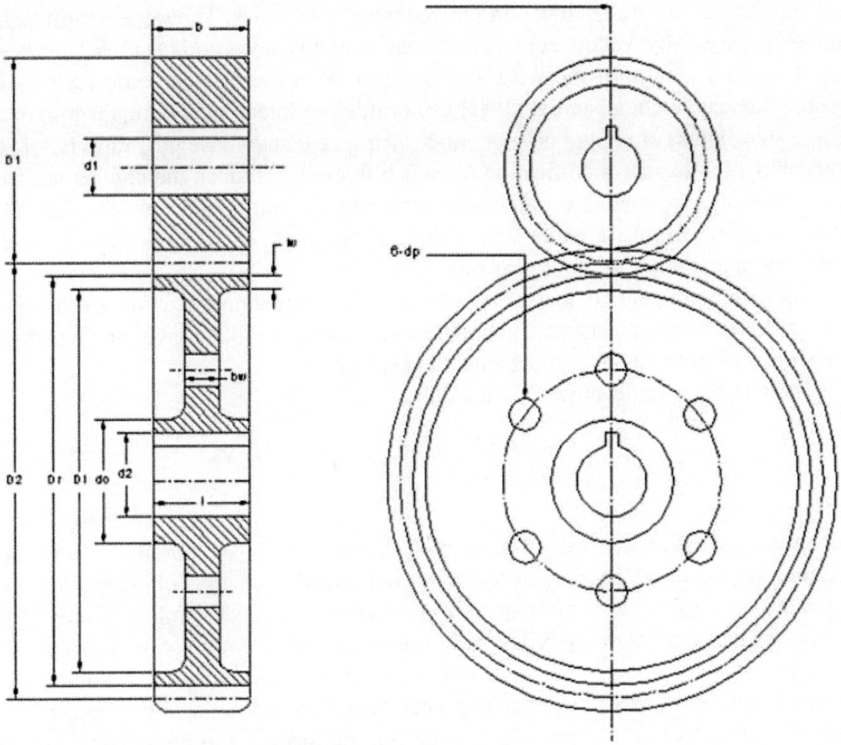

Fig. 1 Basic single-stage spur gear geometry Yokota et al. [2]

$$V_j(i) = V_j(i-1) + c_1 r_1 [P_{\text{best},j} - X_j(i-1)] + c_2 r_2 [G_{\text{best}} - X_j(i-1)];\quad j = 1, 2, \ldots, N \tag{1}$$

$$X_j(i) = X_j(i-1) + V_j(i);\; j = 1, 2, \ldots, N \tag{2}$$

where c_1 and c_2 are the cognitive (individual) and social (group) learning rates, respectively, and r_1 and r_2 are uniformly distributed random numbers in the range 0 and 1. The parameters c_1 and c_2 denote the relative importance of the memory (position) of the particle itself to the memory (position) of the swarm. The values of c_1 and c_2 are usually assumed to be 2 so that $c_1 r_1$ and $c_2 r_2$ ensure that the particles would overfly the target about half the time.

This research on the gear design is mainly focused on dimensional design of the gear drives and to reduce the contact stress along the path of contact. A number of design and calculation procedures were developed by researchers for the gear drives design. In a compact design of the gear box, weight of the gear drive plays a pivotal role. In this work, a single-stage spur gear train weight and contact stress along the path of contact are considered as design objectives on a multi-objective basis with ten non-linear constraints involving mixed integer variables. The same optimization model proposed by Yokota et al. [2], Savsani et al. [3], and Panda et al. [4] has been used for formulation of the weight of the gear set by using geometric features of gear. Whereas Sachidanand et al. [1] use profile modification techniques to reduce the contact stress along the path of contact for a gear set whose gear ratio is 1:1. In this study, the profile modification approach is used to reduce the contact stresses along the path of contact through altering the tooth sum of the spur gear set. The main objective of this research is to minimize the weight of a single-stage spur gear train and the contact stress of gear sets

The basic spur gear geometry considered for the weight optimization is illustrated in Fig. 1. The multi-objective function for optimization problem involving weight of gear and contact stresses is represented in Eq. (3).

The nonlinear multi-objective function

$$\text{Min } F(x) = w_1 \times \frac{W(x)}{W_0(x)} + w_2 \times \frac{\sigma(x)}{\sigma_0(x)} \tag{3}$$

where w_1 and w_2 are the weight constant such that $w_1 + w_2 = 1$ and $W(x) =$ weight of gear, $W_0\,(x) =$ weight of gear when it is optimized as a single objective, $\sigma\,(x) =$ contact stress, $\sigma_0\,(x) =$ contact stress when it is optimized as a single objective.

On the basis of operating requirements, different objective function for spur gear may be proposed. One of the most important design requirement is the minimum contact stress at the path of contact. In normal operating condition, pitting and scoring are the main mode of failure in spur gear. So, the formulation of contact stress as reported in [1].

$$(\sigma_{\max})_X = \text{con}\sqrt{\frac{QX}{(\text{RTR})_X}} \tag{4}$$

where, $\text{con} = \sqrt{\frac{0.35E}{2}}$ and Q is load, X is profile shift, $(\text{RTR})_X$ radius of curvature at different points, E is Young's Modulus of elasticity.

$$(\text{RTR})_X = \frac{T_1 X (T_1 T_2 - T_1 X)}{T_1 T_2} \tag{5}$$

The profile shift "X" can be calculated

$$X = \frac{Z_e(inv\phi_w - inv\phi)}{2\tan\phi} \tag{6}$$

where $\O_w$ is the working pressure angle and Z_e is the altered sum of tooth.

$$\sigma_x = \frac{\sigma_A}{\sigma_{A_0}} + \frac{\sigma_B}{\sigma_{B_0}} + \frac{\sigma_{B2}}{\sigma_{B20}} + \frac{\sigma_C}{\sigma_{C0}} + \frac{\sigma_D}{\sigma_{D0}} + \frac{\sigma_{D2}}{\sigma_{D20}} + \frac{\sigma_E}{\sigma_{E0}} \tag{7}$$

where $\sigma_A, \sigma_B, \sigma_{B2}, \sigma_C, \sigma_D, \sigma_{D2}, \sigma_E$ are the values of contact stress at point A, $B.B_2$, $C.D$, D_2, and E respectively and $\sigma_{A0}, \sigma_{B0}, \sigma_{B20}, \sigma_{C0}, \sigma_{D0}, \sigma_{D20}, \sigma_{E0}$ are the values of contact stress at corresponding points when it is optimized as single objective.

The formulation of weight of the gear as reported in [3] is given by Eq. (8)

$$\text{Weight} = \text{F}(x) = (\pi\rho/4000)\left[bm^2 Z_1^2(1+a^2) - \left(D_i^2 - d_0^2\right)(l - b_w) - nd_p^2 b_w - \left(d_1^2 + d_2^2\right)b\right] \quad (8)$$

The design variables considered in this optimization problem are facewidth (b), pinion shaft diameter (d_1) gear shaft diameter (d_2), hardness (H) pinion profile shift co-efficient (x_1), and gear profile shift co-efficient (x_2). The gear ratio is 4:1 for the spur gear set considered in this study.

Mainly, three types of constraints are used in this optimization problem, i.e., design, geometric and control parameter constraints as reported in [4]. The formulations of these constraints reported in [4] are considered in this optimization problem.

3 Result and Discussion

In this study, the addendum modification is performed with standard tooth sum of 90 as well as with negative and positively altered tooth sum (±5), i.e., 85 and 95 tooth sum. The profile modification approach is adopted to reduce the contact stress at different points, namely A, $B.B_2$, $C.D$, D_2, and E along the path of contact (Can be seen in Fig. 2). The gear ratio is assumed to be 4:1. In this work, four cases are considered, a standard gear with 90 tooth sum, a profile modified gear with 90

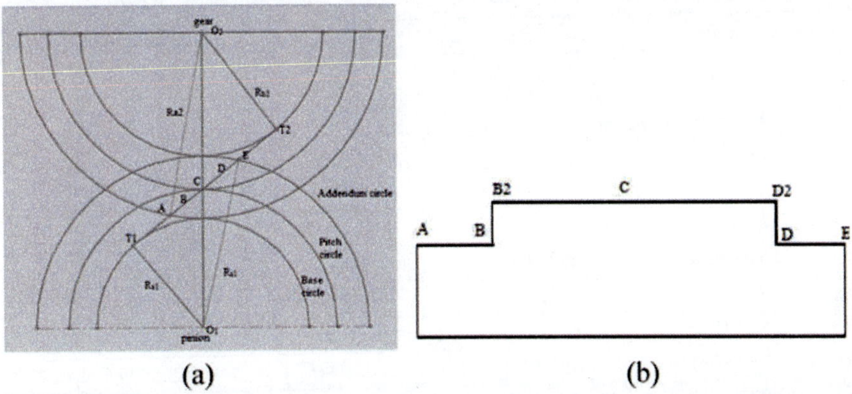

Fig. 2 Path of contact from beginning to end described as A, B, C, D, and E. A, beginning of contact; B, end of two-pair mesh; B_2, beginning of single-pair mesh; C, pitch point; D_2, end of single-pair mesh; D, beginning of two-pair mesh; E, end of contact Sachidanand et al. [1]

tooth sum, negatively altered profile modified gear with 85 tooth sum, and positively altered profile modified gear with 95 tooth sum. The weight and contact stress along path of contact are optimized for all cases, so that a lightweight gear with fairly low contact stress along path of contact can be designed. The optimum geometries along with the material property obtained through the PSO algorithm for standard and other gear set are reported in Table 1.

It is observed from Table 2 that the weight of profile modified gear for tooth sum 90 is 4.976% more than the standard gear set. Similarly, the weight of positively altered tooth sum of 95 it is 27.906% greater than the standard tooth sum. However,

Table 1 Optimum dimensions of spur gear sets

Z_e	85 ($X_1 = 1.33, X_2 = 1.4662$)	90 ($X_1 = 0, X_2 = 0$)	90 ($X_1 = 1.1874, X_2 = -1.1874$)	95 ($X_1 = 0.800, X_2 = -2.9455$)
b	24.0096	21.3241	20.5	20.5
D_1	20	20	20	20
D_2	36.4	36.4	36.4	36.4
m	2	2	2	2
a	90	90	90	90
R_{11}	17	18	18	19
R_{12}	68	72	72	76
R_{a1}	21.62	20	22	21.88
R_{a2}	71.8	74	76	71.4
R_{d1}	17.7	15.5	17.5	18.1
R_{d2}	67.82	69.5	67.5	67.62
R_{b1}	15.4	16.31	16.31	17.21
R_{b2}	61.62	65.25	65.25	68.88
Cont. Ratio	1.0034	1.4809	1.1947	0.9401
Hardness	386.5	357.4920	343.9871	327.2928

Table 2 Comparisons of contact stress in the gear sets

Z_e	85 ($X_1 = 1.33, X_2 = 1.4662$)	90 ($X_1 = 0, X_2 = 0$)	90 ($X_1 = 1.1874, X_2 = -1.1874$)	95 ($X_1 = 0.800, X_2 = -2.9455$)
σ_A	577.334	877.5740	621.6522	660.8847
σ_B	576.9891	682.4049	604.7149	665.1639
σ_{B2}	977.4843	1147.3	948.9034	920.133
σ_C	813.1218	888.4024	906.1738	1044.0
σ_D	750.8531	854.40	821.4437	953.5499
σ_{D2}	877.2945	984.0831	795.2336	757.4041
σ_E	327.0412	345.3264	427.9347	593.5618
Weight	1316.4832	1487.2	1561.00	1902.22

the maximum contact stresses along path of contact at lowest point of single tooth contact are less as compared to the standard tooth sum. Optimum weight of negatively altered tooth sum of 85 is 11.479% less as compared to the standard one. The weight of the optimized gear sets obtained in this study is less as compared to the results reported in [4] for all the cases except altered tooth sum of 95. Also it is observed that the facewidth of the optimized gear set for 85 tooth sum is more as compared to other cases; however, still it satisfies the scoring constraint and hence it is resistance against scoring wear [9]. However, the reported result in Table 2 reveals that the contact stress along path of contact is less as compared to standard gear except at point C and E for the case 90 tooth sum with profile modification and at point C, D and E in the case of 95 tooth sums gear set. The above mention fact is due to $T_1C > T_1B$, $T_1C < T_1E$ and the radius of curvature $(\mathrm{RTR})_x$. This condition reveals that C is a single-pair mesh. Whereas the contact stresses along the path of contact of the 85 tooth sum is reduced at all seven points. Further, it can be predicted that wear near pitch line will be more in the case of 90 profile modified and 95 profile modified gear set. The contact ratio in modified teeth is approaching 1.2, 0.9401 and 1.0034 for tooth sum of 90, 95, and 85, respectively. One of the significant achievements of this study is that all the imposed constraints are satisfied at optimal solution point. It is also found that the hardness of the profile modified gear sets are approaching toward the higher limit which indicates toward wear and pitting resistance.

4 Convergence Study

Case 1:- Standard tooth sum ($Z_e = 90$)

The convergence characteristics of contact stresses along path of contact are depicted in Fig. 3. It can be clearly observed from Fig. 3 that the contact stress at point A (σ_A) shows an initial fluctuation in its value and finally, it converges to a constant value of 877.574 MPa after 39th iteration. Similarly, convergence plot for contact stress at point $B(\sigma_B)$, $B_2(\sigma_{B2})$, $C(\sigma_C)$, $D(\sigma_D)$, $D_2(\sigma_{D2})$ and $E(\sigma_E)$ follows the same pattern and finally converges to stable value 682.404 MPa, 1147.300 MPa, 888.402 MPa, 854.400 MPa, 984.083 MPa, and 345.326 MPa, respectively. Figure 4 shows the convergence characteristics of weight of gear set, and the weight of the gear converge quickly to a value of 1487 gms after 39th iteration. Figure 5 shows the convergence characteristics of some of the important parameter such as facewidth, (b) diameter of pinion (D_1), gear shaft (D_2), hardness of the material (H) required, and contact ratio (C.R). The facewidth after showing initial fluctuation is finally converged to a value of 21.3241 mm, while diameter of pinion shaft and gear shaft converge at 20 mm and 36.4 mm, respectively. The hardness of the material initially shows variation but after 41th iteration, it starts converging and its finally converge to a value of 357.4920 BHN. Contact ratio is constant in this case with a the value of 1.4809. Figure 6 illustrates the convergence of profile shift factor and it shows no variation because no addendum modification is adopted in this case. Similarly, the convergence characteristics of T_1C and T_2C also constant throughout the iteration.

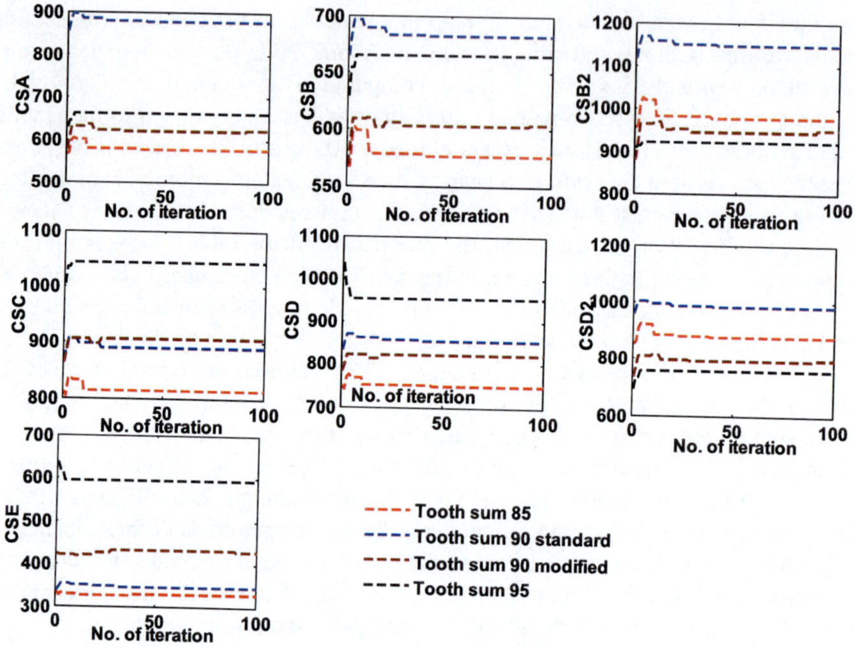

Fig. 3 Convergence of contact stresses along path of contact for spur gear sets

Case 2:- Tooth sum ($Z_e = 90$) with profile shift

The convergence characteristics of contact stress along the path of contact are depicted in Fig. 3. It can be clearly observed from Fig. 3 that the contact stress at point A (σ_A) shows an initial fluctuation in its value and after 89th iteration, finally, it converges to a constant value of 621.652 MPa. Similarly the convergence plot for contact stress at point $B(\sigma_B)$, $B_2(\sigma_{B2})$, $C(\sigma_C)$, $D(\sigma_D)$, $D_2(\sigma_{D2})$ and $E(\sigma_E)$ follows the same pattern and finally converges to stable value 604.714 MPa, 948.903 MPa, 906.173 MPa, 821.443 MPa, 795.233 MPa, and 427.934 MPa, respectively. Figure 4 shows the convergence characteristics of weight of gear set. The convergence characteristic of weight of gear set shows a quick convergence of the weight to a value 1561 gms after 89th iteration. Figure 5 which shows the convergence characteristics of some of the important parameter such as facewidth, (*b*) diameter of pinion (D_1), gear shaft (D_2), hardness of the material (*H*) required, and contact ratio (C.R). The facewidth after showing initial fluctuation is finally converged to a value of 20.5 mm, while diameter of pinion shaft and gear shaft converge at 20 mm and 36.4 mm, respectively. The hardness of the material initially shows variation but after 89th iteration, it starts converging and it is finally converging to a value of 343.9871 BHN. The convergence characteristic of contact ratio indicates an initial fluctuation around a value of 1.2 and ultimately it converges to a value of 1.1947. Figure 6 illustrates the convergence of profile shift factor of pinion (x_1) and gear (x_2) which initially varies and finally converge to the value of 1.1874 and −1.1874, respectively. Whereas

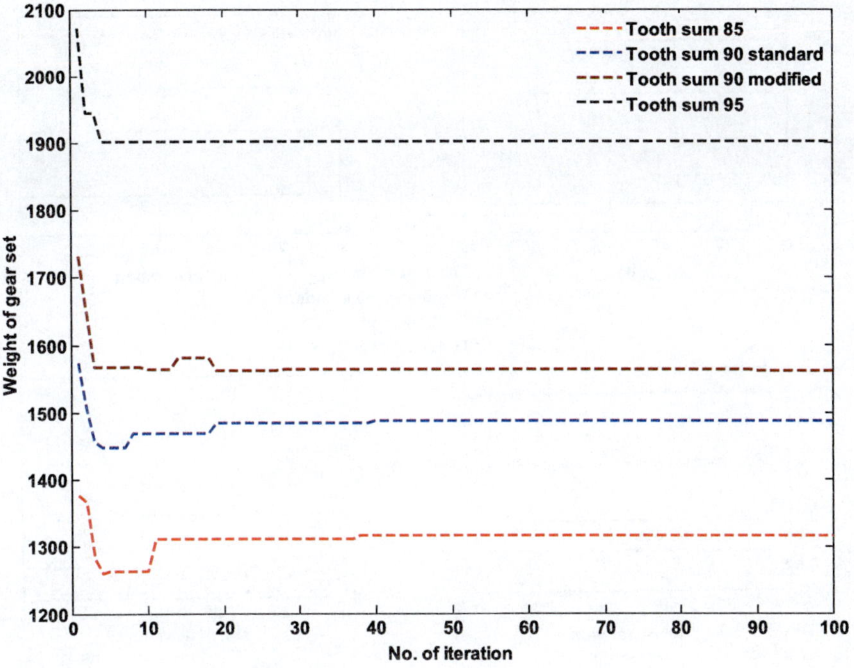

Fig. 4 Convergence of weight for spur gear sets

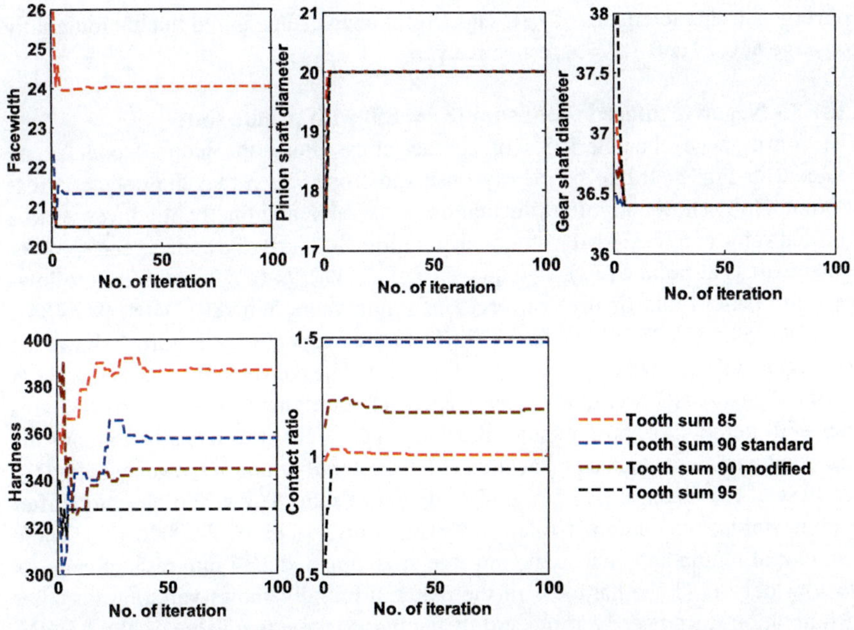

Fig. 5 Convergence of important parameters (b, d_1, d_2, H, C.R) for spur gear sets

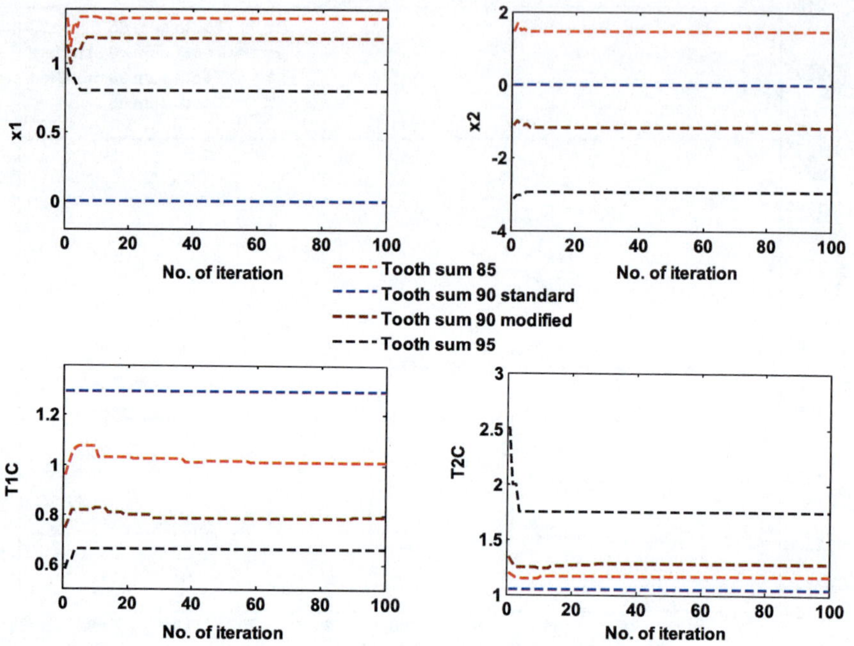

Fig. 6 Convergence of profile shift factor and T_1C, T_2C for spur gear sets

convergence characteristics of T_1C and T_2C showing some initial fluctuation finally converge at 0.79 and 1.2775, respectively.

Case 3:- Negative altered tooth sum ($Z_e = 85$) with profile shift
The convergence characteristics of contact stress along the path of contact are depicted in Fig. 3. It can be clearly observed from Fig. 3 that the contact stress at point $A(\sigma_A)$ shows an initial fluctuation in its value and finally, it converges to a constant value of 577.334 MPa after 95th iteration. Similarly, the convergence plot for contact stress at point $B(\sigma_B)$, $B_2(\sigma_{B2})$, $C(\sigma_C)$, $D(\sigma_D)$, $D_2(\sigma_{D2})$ and $E(\sigma_E)$ follows the same pattern and finally converges to stable value 576.9891 MPa, 977.4843, 813.218, 750.8531, 877.2945, and 327.0412 MPa, respectively. Figure 4 shows the convergence characteristics of weight of gear set. The convergence characteristic of weight of gear set shows a quick convergence of the weight to a value 1316.4832 gms after 95th iteration. Figure 5 shows the convergence characteristics of some of the important parameter such as facewidth, (*b*) diameter of pinion (D_1), gear shaft (D_2), hardness of the material (*H*) required, and contact ratio (C.R). The facewidth after showing initial fluctuation it finally converged to a value of 24.0096 mm, while diameter of pinion and gear shaft converge at 20 mm and 36.4 mm respectively. As we consider Fig. 5 the hardness of the material initially shows variation but after 95th iteration it starts converging and its finally converge to a value of 386.5 BHN. The convergence characteristic of contact ratio indicates an initial fluctuation around

a value of 1 and ultimately it converges to a value of 1.0034. Figure 6 illustrate the convergence of profile shift factor of pinion (x_1) and gear (x_2) which initially varies and finally converge to the value of 1.33 and 1.4662, respectively. Whereas the convergence characteristics of T_1C and T_2C which initially fluctuate and finally converge to the value of 1.0135 and 1.1727 respectively.

Case 4:-Positive altered tooth sum ($Z_e = 95$) with profile shift
The convergence characteristics of contact stress along the path of contact are depicted in Fig. 3. It can be clearly observed from Fig. 3 that the contact stress at point $A(\sigma_A)$ shows an initial fluctuation in its value and after 3rd iteration, finally converges to a constant value of 660.8847 MPa. Similarly, the convergence plot for contact stress at point $B(\sigma_B)$, $B_2(\sigma_{B2})$, $C(\sigma_C)$, $D(\sigma_D)$, $D_2(\sigma_{D2})$ and $E(\sigma_E)$ follows the same pattern and finally converges to stable value of 665.1639 MPa, 920.133 MPa, 1044.0 MPa, 953.5499 MPa, 757.4041 MPa, and 593.5618 MPa, respectively. Figure 4 shows the convergence characteristics of weight of gear set. The convergence characteristic of weight of gear set shows a quick convergence of the weight to a value 1902.22 gms after 3rd iteration. Figure 5 shows the convergence characteristics of some of the important parameter facewidth (b), diameter of pinion (D_1), gear shaft (D_2), hardness of the material (H) required, and contact ratio (C.R). The facewidth after showing initial fluctuation is finally converged to a value of 20.5 mm, while diameter of pinion and gear shaft converge at 20 mm and 36.4 mm, respectively. The hardness of the material initially shows variation but it starts converging and its finally converge to a value of 327.2928 BHN after 4th iteration. The convergence characteristic of contact ratio indicates an initial fluctuation around a value of 1 and ultimately it converges to a value of 0.9401. Figure 6 illustrate the convergence of profile shift factor of pinion (x_1) and gear (x_2) which initially varies and finally converge to the value of 0.8 and −2.9455, respectively. Whereas the convergence characteristics of T_1C and T_2C which initially fluctuate and finally converge to the value of 0.7073 and 1.5774, respectively.

5 CAD Model

The CAD model was developed in solidworks 16 by using the optimized geometry of the spur gear set as in Table 1. The developed CAD model for all the gear set, i.e., 90 tooth sum without and with profile modification as well as positive and negative altered tooth sum, i.e., 85 and 95 tooth sum with profile modification is depicted in Fig. 7. Further, it is observed that in all the geometric modeling there is no interference of geometric dimension. So, it is evident that the optimized dimensions are practically feasible.

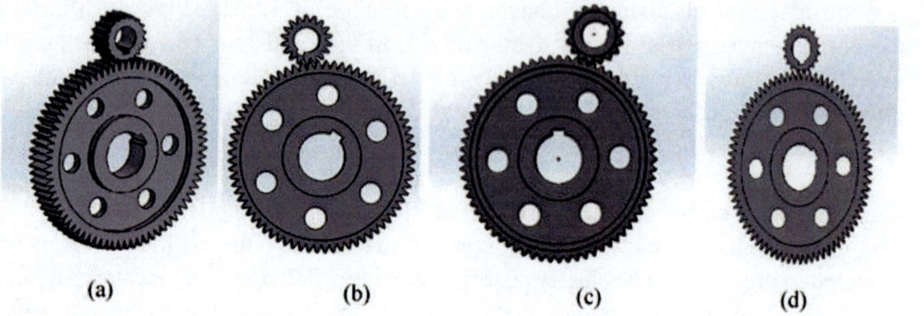

Fig. 7 CAD model of optimum gear sets **a** 90 standard, **b** 90 modified, **c** negatively altered tooth sum 85 **d** positively. Altered tooth sum 95

6 Conclusion

In this study, the weight and contact stress along the path of contact by doing addendum modification with standard tooth sum of 90 as well as with negative and positively altered tooth sum (±5), i.e., 85 and 95 tooth sum is optimized. The estimation of hertizian contact stress at seven different points A, $B.B_2$, $C.D$, D_2, and E along the path of contact is performed for standard and profile modified gearing using PSO algorithm. From the computation and analysis, the following conclusions are drawn.

- The optimized weight obtained by using PSO algorithms shows significant improvement over the results obtained by Panda et al. [4] using a differential evolution (DE) algorithm.
- For the profile modified tooth sum of 90, it is observed that the profile shift factor for pinion and gear is same; however, in case of gear, it is negative. As we consider negative altered tooth sum (−5), i.e., 85 tooth sum more profile shift is given to the gear. While in the case positively altered tooth sum (+5), i.e., 95 tooth, sum positive profile shift is for pinion and negative profile shift is for gear is preferred.
- Minimum contact stress is obtained at end point contact (E), this is due to maximum radius of curvature.
- The hardness of the 90 profile modified and 95 profile modified gear set is less indicating that a low cost material can be used to prevent failure of the gear set against scoring.
- It is observed that through profile modification the contact point can shifted along the path of contact, i.e., optimum addendum modification will result in shift of contact point from single-pair mesh to two pair mesh.
- Negatively altered tooth sum with positive profile shift on pinion has minimum contact stress and weight as compared to standard tooth sum.

Acknowledgements The authors deeply acknowledge the financial support provided by TEQIP-3 and the computational facilities provided in CAD lab. of Dept. of Mechanical Engineering, VSSUT, BURLA.

References

1. Sachidananda, H.K., Raghunandana, K., Gonsalvis, J.: Design of spur gears using profile modification **58**, 736–744 (2015)
2. Yokota, T., Taguchi, T., Mitsuo, G.A.: solution method or optimal weight design problem of the gear using genetic algorithms. J. Mech. Mach. Theory **35**, 523–526 (1998)
3. Savsani, V., Rao, R.V., Vakharia, D.P.: Optimal weight design of a gear train using Particle swarm optimization and simulated annealing algorithms. Mech. Mach. Theory **45**531–541 (2010). https://doi.org/10.1016/j.mechmachtheory.2009.10.010
4. Panda, S., Biswal, B., Jena, S., Mishra, D.: An approach to weight optimization of a spur gear. Proc. Inst. Mech. Eng., Part J: J. Eng. Tribol. **231**(2), 189–202 (2016). https://doi.org/10.1177/1350650116650343
5. Rao, R.V.: Teaching learning based optimization algorithm (2016). https://doi.org/10.1007/978-3-319-22732-0
6. Dörterler, M., Şahin, İ., Gökçe, H.: A grey wolf optimizer approach for optimal weight design problem of the spur gear. Eng. Optim. **51**(6), 1013–1027 (2019). https://doi.org/10.1080/0305215X.2018.1509963
7. Maputi, E.S., Arora, R.: Multi-objective spur gear design using teaching learning-based optimization and decision-making techniques. Cogent Eng. **6**, 1665396 (2019) https://doi.org/10.1080/23311916.2019.1665396
8. Kennedy, J., Eberhart, R.: Particle swarm optimization. In: Proceedings of ICNN'95—International Conference on Neural Networks. https://doi.org/10.1109/icnn.1995.488968
9. Winter, H., Michaelis, K.: Scoring load capacity of gears lubricated with EP-oils. AGMA Paper 219.17,1983

Hybrid Cryptography Algorithm for Secure Data Communication in WSNs: DECRSA

Amine Kardi and Rachid Zagrouba

Abstract In this paper, we propose a dynamic hybrid cryptosystem based on the two famous cryptographic schemes RSA and ECC dedicated to Wireless Sensor Networks (WSNs). Our proposal well adapted to the specific characteristics of the sensors, benefits from the advantages of ECC and improves it by the RSA system to ensure a high security level in WSNs. This hybrid solution overcomes the weakness of existing solutions identified in this work and offers the possibility of adjusting the security level according to the needs of the application. The impact of the Hello flood and the Jamming attacks is highlighted through simulation conducted under the NS3 simulator. Obtained results show that the network lifespan for a WSN based on the famous LEACH routing protocol is on the wane by more than 30% in the presence of Hello flood attacks and decreases by more than 45% when the Jamming attack is launched.

Keywords DECRSA · Hybrid encryption · Security · Wireless Sensor Networks

1 Introduction

A WSN is an ad hoc network consisting of several tiny devices called sensors. These small and lightweight sensors are able to collect and transmit data autonomously from a sensing area to the Base Station (BS) [1]. Each node includes various units such as sensing, processing, storage, wireless, and power units and can integrate some additional modules such as GPS [2].

In addition to civil applications, WSNs are used for military, health, and environmental projects and led the scientific community toward the interconnection between

A. Kardi (✉)
Faculty of Mathematical, Physical and Natural Sciences of Tunis, University of Tunis El Manar, 2092 Tunis, Tunisia
e-mail: Amine.kardi@fst.utm.tn

R. Zagrouba
College of Computer Science and Information Technology, Imam Abdulrahman Bin Faisal University, P.O. Box 1982, Dammam 31441, Saudi Arabia
e-mail: rmzagrouba@iau.edu.sa

H. Sharma et al. (eds.), *Congress on Intelligent Systems*, Advances in Intelligent Systems and Computing 1334, https://doi.org/10.1007/978-981-33-6981-8_51

the Internet and all objects under the name of the Internet of Things (IoT) [3–5]. From here, communications security becomes one of the major constraints of WSNs [6, 7]. Indeed, in addition to their specific characteristics, sensors are often deployed in critical locations such as battlefields or for vital reasons such as controlling the human body. Therefore, they require a very high security level [8, 9].

The rest of this paper contains six sections organized as follows: Sect. 2 summarizes the best known attacks and security mechanisms in WSNs while Sect. 3 presents a performance evaluation of routing attacks and asymmetric algorithms in WSNs. Then, we present our contribution in Sect. 4. Section 5 analyzes the obtained results and presents the main findings while Sect. 6 concludes the paper and specifies the main headlines of our future work.

2 Routing Attacks and Security Mechanisms in WSNs

As with other types of networks, WSNs are vulnerable to a wide variety of attacks. In the following, we describe the best known attacks and present their effect on the routing task.

2.1 *The Best Known Attacks*

The specific characteristics of WSNs make them an attractive target for miscellaneous forms of routing attacks based on several strategies and techniques. Among the most famous attacks threatening the WSNs, we cite:

- **Jamming attack**: This attack targets wireless transmission in a form of noise. It benefits of the weaknesses of the wireless nature of the used communication medium to create interference thus forming a denial of service in the targeted network [10, 11]. The phenomenon proceeds as follows; a malicious node emits unnecessary information in the same frequency of the target network in order to impede regular communications sent and received between legitimate nodes. Several parameters are used to detect the presence of a jammer in WSNs such as Packet Send Ratio (PSR), Packet Delivery Ratio (PDR), Packet Loss Ratio (PLR), Bit Rrror Rate (BER), and Received Signal Strength Indicator (RSSI) [12]. Jamming attack comes in many forms:
- *Constant*: If it aims to target and corrupt data packets in transit.
- *Deceptive*: It consists in drowning the network in a constant data stream sent in the network.
- *Random*: The injected data streams can be dispersed on various separate periods instead of being constant over time.
- *Reactive*: A special jam signal will be always sent following the detection of traffic in the network.

Many defense techniques have been proposed to ward off Jamming attack involving spread-spectrum techniques allowing to use and to switch betwixt many frequency channels and ensuring various authentication mechanisms serving to detect suspicious and replayed packets in addition to the creation of secure medium access control protocols… [13–15].

- **Hello Flood attack** [16]: Hello flood attack is a network layer attack in which the attacker benefits of the signal strength of its attacking device in order to broadcast erroneous information of an optimal non-existent route to all the nodes. Consequently, sensors will update their local tables, and as a result, the network becomes disoriented. In fact, following the update of its table, a targeted node will not be able to communicate using the broadcasted route because the imaginary neighbor is out of range. Successive communication attempts result in a deep waste of energy which led to a malfunction of the network [17].

To counter this type of attacks, nodes should verify the bi-directionality of a path before taking action received over it to ensure the existence of a legitimate neighbor verified through authentication mechanisms.

2.2 *Security Mechanisms in WSNs*

Several types of attacks can threaten the stability, data integrity, and communication security of WSNs ranging from eavesdrops to resource depletion [18].

As a solution, several security methods have been proposed in the literature [19]. These solutions, despite their proven effectiveness, have shown their inadaptability in the context of WSNs because of their energy and memory needs in addition to their algorithmic complexity and the problems of implementation on devices having low processing capabilities [20]. These security methods can be classified in different ways. In the following, we detail the most known security systems applicable in WSNs.

Symmetric Cryptography in WSNs

A symmetric encryption algorithm transforms a plaintext M to an encrypted message C and vice versa using a single secret key K. This invertible encryption function has the advantage of relatively short keys but it presents a problem of key distribution and high rates of data throughput. The most known symmetric algorithms use either a block cipher or a stream cipher. Block cipher algorithms use fixed-size data blocks, e.g., 64-bit or 128-bit contrary to cipher stream algorithms where the data is processed by bit [21]. Among the popular cryptography algorithms for WSNs, we quote:

- **Advanced Encryption Standard (AES)**: It is a block cipher-based encryption algorithm. The data is processed in 128-bit blocks using three common secret key lengths of 128, 192, or 256 bits, hence the three variants AES-128, AES-192, and AES-256 [22]. Encryption and decryption schemes use functions such as the BYTE_SUB function (Byte Substitution) which is a nonlinear function that substitutes each byte by another according to a lookup table. The SHIFT_ROW function that shifts to the left a certain number of steps and the MIX_COL that mixes and transforms each byte by a linear combination of bytes.

- **Rivest Cipher 4 (RC4)**: It is a stream cipher algorithm [23]. It uses the XOR operator (⊕) to combine a series of bytes (or bits) *K*, produced from the key, with the bytes (or bits) of plain text *M* to give the cipher text *C*, using Eq. 1:

$$C = M \oplus K \tag{1}$$

Equation 1: RC4 encryption formula.
This operation is reversible, and the initial message is found through the formula:

$$M = C \oplus K \tag{2}$$

Equation 2: RC4 decryption formula.

Asymmetric Cryptography in WSNs
Asymmetric cryptography, or public key cryptography, integrates two encryption keys, a public key communicated without any restriction to be used for encryption and a private key used for decryption [24]. Among the asymmetric cryptosystems applicable in WSNs, we find:

- **Rivest-Shamir-Adleman (RSA)**: the most used asymmetric cryptographic algorithm today. RSA exploits the mechanism of factorization of large prime numbers. It is composed of three phases, namely key generation, encryption process, and decryption process [25, 26]. Encryption and decryption algorithms are based on one single modular exponentiation. Each node disposes of a public key (*n*, *e*) used for encryption and a private key (*n*, *d*) used for decryption constituted as follows:

  - p, q, two prime numbers
  - $n = p * q$, $\varphi(n) = (p\text{-}1)(q\text{-}1)$
  - e, with gcd $(\varphi(n), e) = 1$; $1 < e < \varphi(n)$
  - $d \equiv e^{-1} \bmod \varphi(n)$.

To encrypt a plaintext *M*, node *A* uses the public key (*n*, *e*) of node *B* to generate the cipher text *C* using Eq. 3:

$$C \equiv M^e \text{mod} n \tag{3}$$

Equation 3: RSA encryption formula.
Upon receipt of the cipher text *C*, node *B* decrypts it using its private key (*n*, *d*) to retrieve the plaintext *M* using Eq. 4:

$$M \equiv C^d \text{mod} n \tag{4}$$

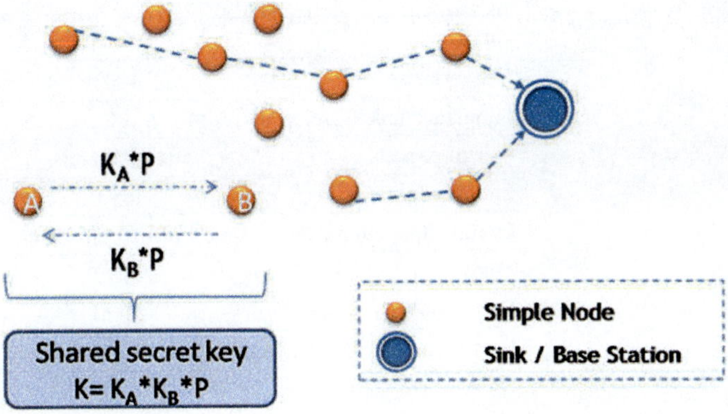

Fig. 1 ECDH under WSNs

Equation 4: RSA decryption formula.

- **Elliptic Curve Diffie-Hellman (ECDH)**: This algorithm adopts the concept of elliptic curves to Diffie–Hellman encryption mechanism [27]. As shown in Fig. 1, the classical ECDH runs as follow: Initially, two communicating nodes agree on a particular elliptic curve E (a, b) and a point P. Public keys are generated by multiplying P with private keys K_A and K_B secretly chosen. After sharing their public keys, they generate a shared secret key by multiplying the public keys received by their private keys. Hence, the secret shared key is $K = K_A * K_B * P$.

Hybrid Cryptography in WSNs
Hybrid cryptography systems combine and benefit from symmetric and asymmetric cryptosystems. In fact, the role of asymmetric encryption will be limited to the exchange of a secret key which will be used as part of symmetric cryptography because asymmetric scheme are rapacious in term of time and resources use [28].

3 Performance Evaluation of Routing Attacks and Asymmetric Algorithms in WSNs

3.1 Analysis of Routing Attacks in WSNs

To highlight the effects of the attacks presented previously in this paper, a sensing area of 100 m * 100 m containing an unlimited power sink placed in the center (50, 50) with 100 randomly deployed nodes following a centralized architecture is simulated under the NS3 simulator using the configuration shown in Table 1. The famous LEACH protocol is used in the simulated network, and the confidence interval is taken to get precise plots.

Table 1 Network parameters

Parameters	Values
Network size	100 m * 100 m
Simulated nodes	100
Size of packets	4000 bits
Initial energy	0.5 J
Energy dissipation (Efs)	10 pJ/bit/m^2
Transmission energy (ETx)	50 nJ
Reception energy (ERx)	50 nJ
Energy for data aggregation	5 nJ/bit/report

To show the effect of the attacks mentioned above, we evaluate the network in term of network lifetime and residual energy.

- **Network lifetime**

The network lifetime, usually defined by the time taken by the network from the moment of the beginning until the death of the last sensor, is one of the main used metrics to evaluate the efficiency of routing protocols in WSNs.

Figure 2 shows the network lifetime of the simulated WSN based on the LEACH protocol in cases of presence and absence of attacks mentioned above.

As shown in Fig. 2, the network lifetime is about 1560 rounds in the absence of attacks. This network lifetime has decreased by more than 30% in the presence of Hello flood attacks and decreased by more than 45% when the Jamming attack is launched. This is explained by the doubled effort exerted by the sensors in the presence of attacks. This leads to an excessive consumption of energy and subsequently the rapid death of the network.

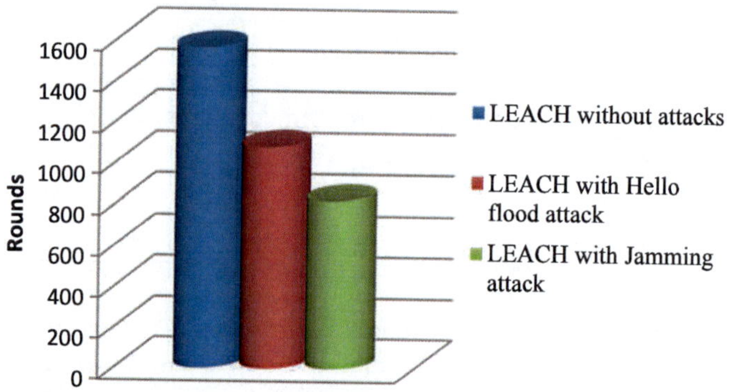

Fig. 2 Network lifetime

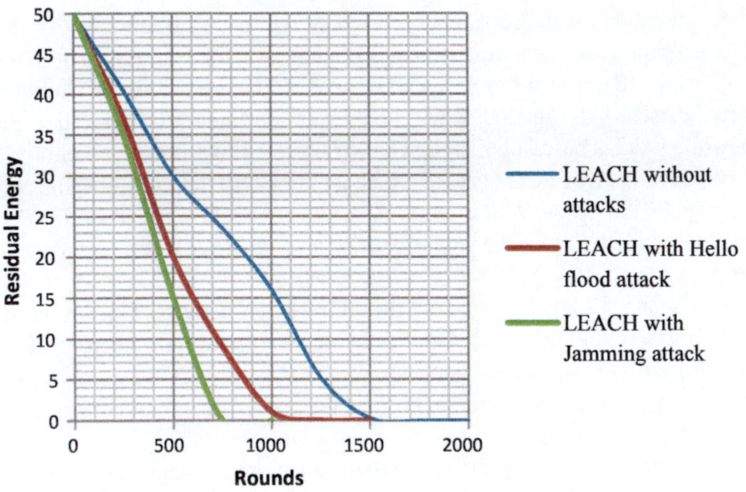

Fig. 3 Residual energy versus time

- **Residual energy**

Figure 3 presents the evolution of the residual energy in the network for the three simulated scenarios.

As for the evolution of the network lifetime, the presence of the attacks led to the rapid exhaustion of the batteries of the sensors. This results in the death of several sensors and subsequently the formation of several isolated zones which largely affects the efficiency of the network.

3.2 *Performance Evaluation of Asymmetric Algorithms in WSNs*

All cryptographic research agree that symmetric encryption algorithms are the fastest and the least energy greedy but they are more vulnerable to various types of attacks and present key distribution problems especially in the case of WSNs where sensors are often disseminated in conflict areas and publicly accessible locations [29]. Using two encryption keys, the asymmetric cryptography offers more security and overcomes key distribution issues but leads to the appearance of time and energy consumption problems [30]. Seen that authenticity and confidentiality are the main cryptographic objectives, the cryptographic algorithm must ensure the complete protection especially in WSNs. Therefore, we opted for security in the first place, while using adaptive methods to WSNs in terms of energy, memory, and processing capabilities, that is why we founded our proposal on asymmetric algorithms.

Due to memory restrictions, encryption key size is still important. As shown in Table 2, asymmetric schemes differ in terms of key size, for example, to offer the same security level, RSA algorithm needs a key size of 512 bits, while ECC only require 112 bits key size [31]. Indeed, RSA is based on the Integer Factorization Problem (IFP), while ECC is based on Elliptic Curve Discrete Logarithm Problem (ECDLP). The main advantage of ECC over RSA is that the ECDLP takes full exponential time contrary to IFP of RSA, which requires sub-exponential time. Therefore, smaller parameters with ECC give the same security level as RSA when larger parameters are used.

Figures 4 and 5, developed based on the research done by Quirino et al. [32], compare, respectively, the processing time for encryption and decryption and the number of used memory pages between RSA-2048 and ECC-224 (El-Gamal [33]) for encrypting and decrypting the same data under the platform sa1core used in ARM simulator via arm-linux-gcc compiler.

Figure 4 shows that ECC is faster and therefore more efficient than RSA in WSNs. In fact, to encrypt and decrypt the same data, RSA needs around 32,592 ms while ECC requires only 356.6 ms.

Figure 5 shows that ECC algorithm occupied only 90 memory pages while RSA requires 178 pages.

Obtained results, as well as several research, show that both solutions are adaptable in the context of WSNs but ECC solution is the best [34, 35]. In contrast, researchers have shown that the ECC has some weaknesses. In fact, several finite fields including

Table 2 Key size comparison for ECC and RSA

ECC key size (bits)	RSA key size (bits)
112	512
160	1024
224	2048
256	3072
512	15,360

Fig. 4 Total time for encryption and decryption (RSA vs. ECC)

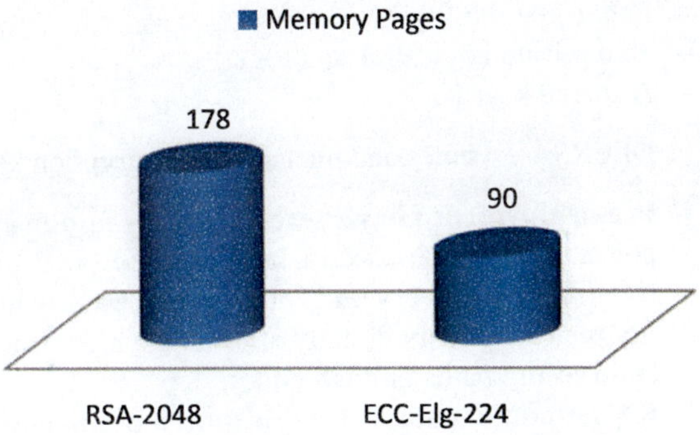

Fig. 5 Used memory pages (RSA vs. ECC)

F_2^{210}, F_2^N with $N \in [160, 600]$ are weak for ECC and vulnerable to different types of attacks such as GHS attack seen that any instance of the ECDLP for any elliptic curve over these fields can be solved in significantly less time than it takes Pollard's rho method to solve the hardest instances [36].

Other research proved that ECC could be vulnerable to the man-in-middle attack when the secret integer chosen by the communicating nodes is equal to one [37]. Hence, it will be easy to unravel the shared secret code through the public key sent without encryption.

Designing a reliable security solution for WSNs become a challenging task due to their special characteristics where increasing the security level automatically affects the sensor resources. Hence, we need to found tradeoffs between these two parameters: security level and resource requirements in WSNs. Our work presented in the following section presents a hybrid method that aims to achieve this compromise.

4 Proposed Algorithm: Dynamic EC-RSA (DECRSA)

Our proposal is based on the concept of flexibility. Indeed, we ensured a maximum level of protection through the combination of RSA and EC cryptosystems (ECDH) while offering the possibility of integrating a symmetric cryptographic mechanism where the system presents energetic constraints or including El-Gamal ECC instead of ECDH for authentication reasons.

The combination of RSA and ECC is based on several research and aims to counter the majority of attacks and weaknesses of ECC, previously detailed, through the encryption of its public keys with RSA which will be used only once during the network initialization phase and therefore does not affect the energy resources. In this way, the maximum energy consumption using DECRSA is approximately the same as with ECC, which is applicable in the context of WSNs. As well, our proposal ensures the update of encryption keys in a secure way.

DECRSA is based on four encryption keys where one is public and three are private:

- Public key: PbK
- Two private keys: PK1 and PK2
- A shared key: K.

DECRSA executes and manages the encryption keys in three phases, namely

- **Pre-deployment phase**: several security information (elliptic curve parameters, points…) are pre-loaded in the sensors before deployment.
- **Neighbors discovery and routing path establishment phase**: The sensors identify their neighbors directly after deployment (random/non-random) in order to build secure paths through nodes.
- **Keys update phase**: The sink or the BS ensures the update (destruction, renewal…) of encryption keys in order to maintain a high level of security.

The detailed execution, shown in Fig. 6, is as follows:

- **Pre-deployment phase**: Before deployment, each node is loaded with:
  - A prime number q.
  - An elliptic curve E (a, b) defined over a finite field Fq. E consists of a set of points (x, y) forming an abelian group such that $E = \{(x, y) \epsilon Fq \times \overline{Fq}\} \text{U}\{\text{O}\}$. (O is the point at infinity).
  - A point P $(x_p, y_p) \in E(Fq)$ of order w.
- **Neighbors discovery and routing path establishment phase**:
  Directly, after deployment, each node:
  - Generates a second prime number r distinct from q (pre-loaded) through a probabilistic algorithm
  - Calculates $N = q * r$
  - Calculates $\varphi\ (N) = \varphi\ (q)\ \varphi\ (r) = (q - 1)\ (r - 1)$
  - Chooses e such as e and $\varphi\ (N)$ are coprime
  - Generates d such as: $d \equiv e^{-1}\ (\text{mod}\ N)$
  - Generates an integer $v < w$
  - Calculates the point $v.P$.

  Thus, the obtained keys are:

- The values (e, N) represent the public key of the node (PbK)
- The values (d, N) represent the private key 1 of the node (PK1)
- The point $v.P$ represents the private key 2 of the node (PK2) (not encrypted under ECC).

To establish a secure link with node B, node A proceeds as follows:

- A sends his PbK_A (e_A, N_A) to B
- B checks if A is a node of the network (checks if N_A is a multiple of q)
- B generates an integer $v_B < w$
- B calculates its $\text{PK2}_B = v_B.P$

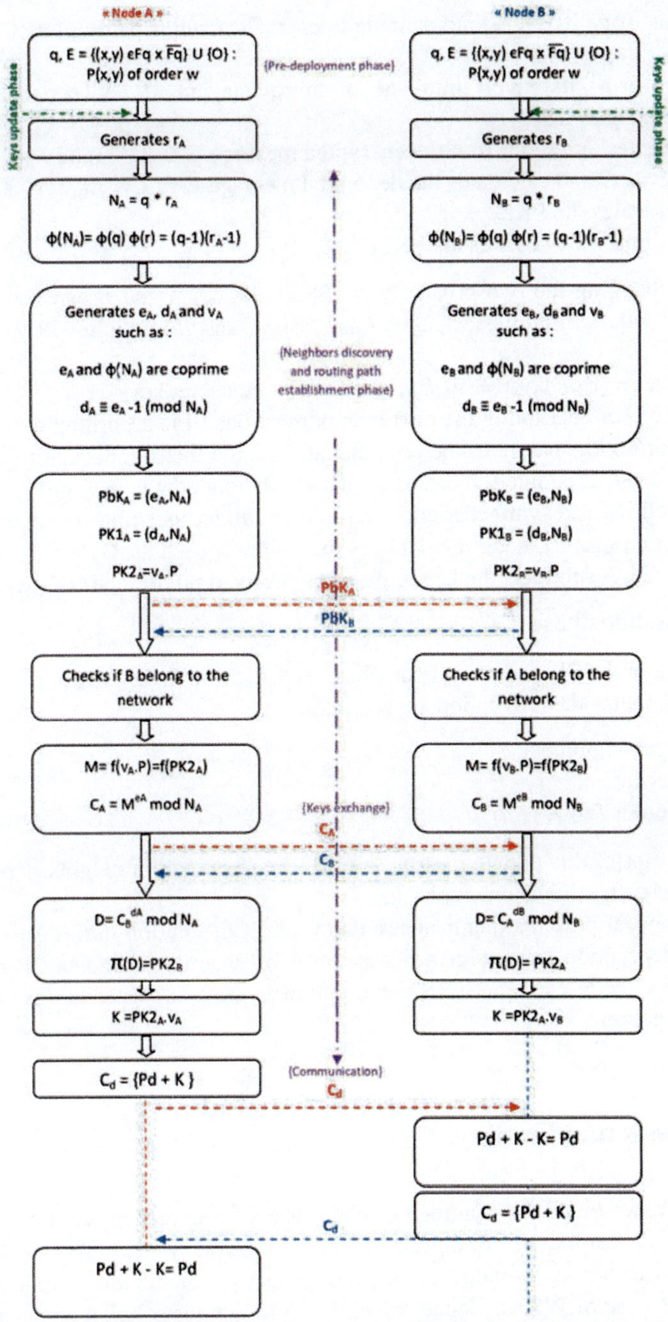

Fig. 6 DECRSA algorithm

- *B* Transforms its $PK2_B$ into a single integer The Cantor pairing function (π: $N^n \rightarrow N$)
- B transforms its PK2B into one or more integers *M* between 0 and N_A-2 (Differently to RSA)
- *B* calculates and sends to *A* the encrypted message $C \equiv M^{eA} \bmod N_A$
- *A* receives *C* and calculates the decrypted message $D \equiv C^{dA} \bmod N_A$
- *A* regenerates the $PK2_B = \pi^{-1}(D)$
- In the same way, A sends his $PK2_A$ to *B*.

By exchanging the two private keys ($PK2_A$, $PK2_B$) *A* and *B* can communicate safely. Indeed, A multiplies $PK2_B = (v_B.P)$ by v_A and *B* multiplies $PK2_A = (v_A.p)$ by v_B.

Hence $K = (v_B.P).v_A = (v_A.P).v_B$ is the shared secret key.

K is used for encrypting the next communications. It is according to the type of the application, the nature of the possible attacks, and the sought security level that the secret key *K* is adopted. It can be used partially (e.g., the nodes agree on a part of *K* to be used for next symmetric encryption) or in full for an elliptic curve encryption (to encrypt a point P_d, node *A* calculates the point C_d such as: $C_d = \{P_d + K\}$ and to decrypt C_d, *B* subtracts the key *K* from the received point: $P_d + K - K = P_d$).

- **Keys update phase**:

The sink or the BS, periodically or when needed, through the diffusion of a special message, updates the encryption keys including:

- A new prime number *q*
- A new elliptical curve *E*
- A new point $P \in E(Fq)$

The receipt of this update message causes the re-launch of neighbors discovery and routing path establishment phase.

Our proposal presents identical results to the ECC solution in terms of encryption and decryption time with a higher level of security as indicated in Fig. 7 because as already explained, RSA encryption is used to secure the exchange of keys information.

5 Analysis and Findings

In this work, we found that symmetric solutions are faster and consume less energy than asymmetric solutions. In contrast, asymmetric cryptosystems offer much higher level of security while symmetric algorithms present key distribution problems especially in the case of WSNs. Hence, we chose to found our solution on asymmetric algorithms in order to ensure a total protection while adopting this solution to the specific characteristics of WSNs so that limited resources will not be affected.

Fig. 7 Total time for encryption and decryption (ECC vs. DECRSA)

Then, we found that ECC is more suitable to WSNs and gives better results than RSA solution but it presents some weakness (weak finite fields, vulnerability to man-in-the-middle attack…). Therefore, the proposed solution is to combine ECC with RSA in order to overcome these problems. RSA will only be used once and, therefore, will not affect the resources of the nodes.

Our proposal is flexible and configurable so that the user has the choice to integrate symmetric cryptographic mechanism if the system presents energetic constraints or including other ECC variants instead of ECDH such as El-Gamal-ECC.

6 Conclusion and Future Work

WSNs offer today many solutions for numerous problems. The specific characteristics of sensors make the security constraints a top concern of the scientific community and make the creation of a perfect security solution for WSNs a challenging task. In this paper, we proposed a hybrid encryption algorithm combining RSA and ECC. Our solution is well adapted to the characteristics of the WSNs and offers a total protection and flexibility facilitating its configuration and use according to the needs of the applications and usage scenarios.

As our future work, we will evaluate the DECRSA algorithm for different application of WSNs using various technical configurations.

References

1. Tehseen, M., Javed, H., Shah, I.H., et al.: A lightweight key negotiation and authentication scheme for large scale WSNs. In: Recent Trends and Advances in Wireless and IoT-enabled Networks, pp. 225–235. Springer, Cham (2019)
2. Al-Turjman, F.: The road towards plant phenotyping via WSNs: an overview. Comput. Electron. Agric. (2018)
3. Basha, A.R., Sivakumar, G.: An energy and delay aware optimal data aggregation using three fold algorithm for WSN. Int. J. Bus. Inf. Syst. **33**(3), 299–319 (2020)
4. Lele, A.: Internet of Things (IoT). In: Disruptive Technologies for the Militaries and Security, pp. 187–195. Springer, Singapore (2019)
5. Kardi, A., Zagrouba, R.: RaCH: a new radial cluster head selection algorithm for wireless sensor networks. Wirel. Personal Commun. pp. 1–14 (2020)
6. Sharma, K., Sharma, S.: Issues in wireless sensor networks with an emphasis on security. In: International Conference on Communications and Cyber Physical Engineering 2018, pp. 419–425. Springer, Singapore (2018)
7. Sujanthi, S., Kalyani, S.N.: SecDL: QoS-aware secure deep learning approach for dynamic cluster-based routing in WSN assisted IoT. Wirel. Personal Commun. pp. 1–35 (2020)
8. Sharma, C., Vaid, R.: Analysis of existing protocols in WSN based on key parameters. In: Proceedings of 2nd International Conference on Communication, Computing and Networking, pp. 165–171. Springer, Singapore (2019)
9. Farhat, A., Guyeux, C., Makhoul, A., et al.: On the coverage effects in wireless sensor networks based prognostic and health management. Int. J. Sens. Netw. **28**(2), 125–138 (2018)
10. Adil, M., Almaiah, M.A., Omar Alsayed, A., et al.: An Anonymous channel categorization scheme of edge nodes to detect jamming attacks in wireless sensor networks. Sensors **20**(8), 2311 (2020)
11. Rose, S.G.H., Jayasree, T.: Detection of jamming attack using timestamp for WSN. Ad Hoc Netw. **91**, 101874 (2019)
12. Meenalochani, M., Sudha, S.: Jammed node detection and routing in a multihop wireless sensor network using hybrid techniques. Wirel. Personal Commun. **104**(2), 663–675 (2019)
13. Shabana, K., Fida, N., Khan, F., et al.: Security issues and attacks in wireless sensor networks. Int. J. Adv. Res. Comput. Sci. Electron. Eng. (IJARCSEE), **5**(7), 81–87 (2016)
14. Zhu, J., Zou, Y., Zheng, B.: Physical-layer security and reliability challenges for industrial wireless sensor networks. IEEE Access (2017)
15. Bhavathankar, P., Sarkar, S., Misra, S.: Optimal decision rule-based ex-ante frequency hopping for jamming avoidance in wireless sensor networks. Comput. Netw. (2017)
16. Pruthi, V., Mittal, K., Sharma, N., et al.: Network layers threats & its countermeasures in WSNs. In: 2019 International Conference on Computing, Communication, and Intelligent Systems (ICCCIS). p. 156–163, IEEE (2019)
17. Srinivas, T.A.S., Manivannan, S.S.: Prevention of hello flood attack in IoT using combination of deep learning with improved rider optimization algorithm. Comput. Commun. (2020)
18. Pawar, M., Agarwal, J.: A literature survey on security issues of WSN and different types of attacks in network. Indian J. Comput. Sci. Eng. **8**, 80–83 (2017)
19. Mugheri, A.A., Siddiqui, M.A., Khoso, M.: Analysis on security methods of wireless sensor network (WSN). Sukkur IBA J. Comput. Math. Sci. **2**(1), 52–60 (2018)
20. Nabavi, S.R., Mousavi, S.M.: A review of distributed dynamic key management schemes in wireless sensor networks. JCP, 13(1), 77–89 (2018)
21. Hamici, Z.: Towards genetic cryptography for biomedical wireless sensor networks gateways. IEEE J. Biomed. Health Inf. 22(6), 1814–1823 (2018)
22. Hung, C.-W. Hsu, W.-T.: Power consumption and calculation requirement analysis of AES for WSN IoT. Sensors 18(6), 1675 (2018)
23. Sasi, S.B., Sivanandam, N., et al.: A survey on cryptography using optimization algorithms in WSNs. Indian J. Sci. Technol. **8**(3), 216 (2015)

24. Senthilkumar, M.K., Senthilkumaran, U.: Review of asymmetric key cryptography in wireless sensor networks. Int. J. Eng. Technol. **8**(2), 859–862 (2016)
25. Leelavathi, G., Shaila, K., Venugopal, K.R.: Design and implementation of montgomery multipliers in RSA cryptography for wireless sensor networks. In: Proceedings of First International Conference on Smart System, Innovations and Computing, pp. 565–574. Springer, Singapore (2018)
26. Fotohi, R., Firoozi Bari, S., Yusefi, M.: Securing wireless sensor networks against denial-of-sleep attacks using RSA cryptography algorithm and interlock protocol. Int. J. Commun. Syst., 33(4), e4234 (2020)
27. Iqbal, U., Shafi, S.: A provable and secure key exchange protocol based on the elliptical curve Diffe–Hellman for WSN. In: Advances in Big Data and Cloud Computing, pp. 363–372. Springer, Singapore (2019)
28. Abdullah, K.M., Houssein, E.H., Zayed, H.H.: New security protocol using hybrid cryptography algorithm for WSN. In: 2018 1st International Conference on Computer Applications & Information Security (ICCAIS), pp. 1–6. IEEE (2018)
29. Singh, P., Chauhan, R.K.: A survey on comparisons of cryptographic algorithms using certain parameters in WSN. Int. J. Electr. Comput. Eng. **7**(4) (2017) (2088-8708)
30. Adnan, S.F.S., Isa, M.A.M., Hashim, H.: Energy analysis of the AA β lightweight asymmetric encryption scheme on an embedded device. In: 2016 IEEE Industrial Electronics and Applications Conference (IEACon), pp. 116–122. IEEE (2016)
31. Khan, A., Shah, S.W., Ali, A., et al.: Secret key encryption model for Wireless Sensor Networks. In: 2017 14th International Bhurban Conference on Applied Sciences and Technology (IBCAST), pp. 809–815. IEEE (2017)
32. Da Silva Quirino, G., Moreno, E.D.: Architectural evaluation of algorithms RSA, ECC and MQQ in arm processors. Int. J. Comput. Netw. Commun. **5**(2), 153 (2013)
33. Kardi, A., Zagrouba, R., Alqahtani, M.: Performance evaluation of RSA and elliptic curve cryptography in wireless sensor networks. In: 2018 21st Saudi Computer Society National Computer Conference (NCC), pp. 1–6. IEEE, (2018)
34. Verma, D., Jain, R., Shrivastava, A.: Performance analysis of cryptographic algorithms RSA and ECC in wireless sensor networks. IUP J. Telecommun. **7**(3), 51 (2015)
35. Sharma, C., et al.: Performance analysis of ECC and RSA for securing CoAP-based remote health monitoring system. In: Ambient Communications and Computer Systems, pp. 615–628. Springer, Singapore (2018)
36. Menezes, A., Teske, E., Weng, A.: Weak fields for ECC. In: Cryptographers' Track at the RSA Conference, pp. 366–386. Springer, Berlin, Heidelberg (2004)
37. Huang, X., Shah, P.G., Sharma, D.: Protecting from attacking the man-in-middle in wireless sensor networks with elliptic curve cryptography key exchange. In: 2010 Fourth International Conference on Network and System Security, pp. 588–593. IEEE (2010)

A Perspective of Security Features in Amazon Web Services

Harsha Surya Abhishek Kota, J. S. Shyam Mohan, and Nagendra Panini Challa

Abstract In the recent times, increase in ICT tools has made significant progress in various sectors. As a result, data storage has become one of the most challenging tasks. Cloud computing has resolved this problem by providing on-demand computing power like virtual servers, storage space, function as a service, and many more. Cloud computing indisputably provides everlasting benefits and is cost-effective model. However, the major drawback in cloud computing is security. In this paper, we discuss about shared responsibility model, global infrastructure and compliance program, and network security features in cloud. These will act as a base for all the services provided by the Amazon Web Services (AWS). We have explored the security features in Amazon EC2, Amazon EBS, Amazon ELB, Amazon VPC, Amazon Route 53, AWS Direct Connect.

Keywords Cloud computing · Amazon Web Services · Cloud security

1 Introduction

With usage of ICT tools, Internet usage and storage of data increased in largescale posed a challenging task for companies for storing and securing data. Cloud computing provided solution for storing and securing data in a single location or at multiple locations. Cloud computing services are given below that provides security and accessing data easily [1].

Software-as-a-Service (SaaS): The vender hosts In SaaS module software and the user based on pay-per-use policy access it. In this type of module, user cannot host their own application in cloud.

H. S. A. Kota (✉) · J. S. Shyam Mohan · N. P. Challa
Department of CSE, SCSVMV, Kanchipuram 631561, India

Department of IT, SVECW, Bhimavaram, Andhra Pradesh 534202, India

J. S. Shyam Mohan
e-mail: jsshyammohan@kanchiuniv.ac.in

H. Sharma et al. (eds.), *Congress on Intelligent Systems*, Advances in Intelligent Systems and Computing 1334, https://doi.org/10.1007/978-981-33-6981-8_52

For example, Google mail, Dropbox, Google Apps, etc.

Infrastructure-as-a-Service (IaaS): In IaaS module networking, storage capacity, virtual computing power, and many more are provided as a service for the end user based on pay-as-you-go. It also provides the scalability option for the user to scale and shrink the resources as needed [1]. For example, Amazon Web Service (AWS), Microsoft Azure, Google Compute Engine.

Platform-as-a-service (PaaS): In PaaS, the cloud venders will provide a platform for the users to deploy their own applications, softwares, and codes in cloud. Users can use Java, .Net, Python for their application Server and MySQL, Oracle for Database Servers. For example, Force.com, OpenShift.

1.1 Deployment Models in Cloud

To deploy and host an application in cloud, first we should know the deployment options present in cloud. Cloud computing has four types of deployment models for different types of cloud users [2]. Various cloud deployment models are:

1. Public cloud
2. Private cloud
3. Hybrid cloud
4. Community cloud.

AWS has a vast network area of around 190 countries and more than 1 million active users, which includes 2000 government agencies, 5000 educational institutions. The main reason for AWS to stand as best cloud provider is because of its global architecture, low cost, agility, instant elasticity, and it is secure. Its infrastructure is built with AWS regions and availability zones. Security, reliability, cost optimization, performance efficiency, and operational excellence are the main key frames of a well-architected framework. The public cloud market has increased rapidly in the past few years. In this cloud competition market, many major corporate companies like Microsoft, Google, Amazon, and many more had launched their cloud platforms. Among these entire cloud providers, the Amazons Cloud platform called Amazon Web Services (AWS) has occupied the majority of public cloud market share of nearly 47% as of survey conducted by Goldman Sachs in 2019. Security services and the features will be the main point of view of the customer when he tries to deploy an application or stores his important information in the cloud. Occupying the majority of the public cloud market share what are the security features and services provided by Amazon Web Services (AWS) to ensure that the data of their customers is highly secured is the main aspect of this paper (Fig. 1; Table 1).

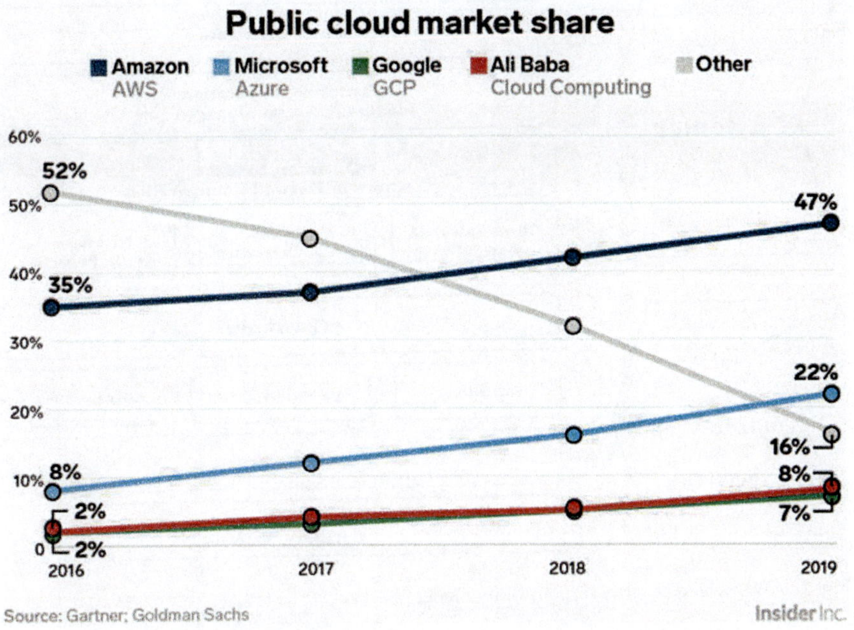

Fig. 1 Market share of public cloudity-Image adopted from Gartner, Goldman Sachs

Table 1 Global security services

Global security services	
Security responsibility model	Amazon Web Services (AWS) has a shared responsibility model in which both AWS and customer has equal level of responsibility for securing the data
AWS compliance program	Amazon Web Services (AWS) had certified with many non-profit organization like ISO, FISMA, and many more in security aspect
Network security	Many network and firewall devices are used in Amazon Web Services (AWS) to enhance their security levels
AWS access	Customer has full control access permissions for AWS account

2 Security Responsibility Model

Amazon Web Services (AWS) security is a shared security responsibility model in which 50% of the security is taken care by AWS and remaining 50% should be taken care by the customers. Though Amazon Web Services (AWS) takes care of providing and maintaining the emphasize cloud infrastructure, customers should perform several security configurations to protect their data present in the cloud. AWS provides security via. Hypervisor also called virtual machine monitor (VMM). Hypervisor is installed on the Hardware layer that helps for paravirtualization of the

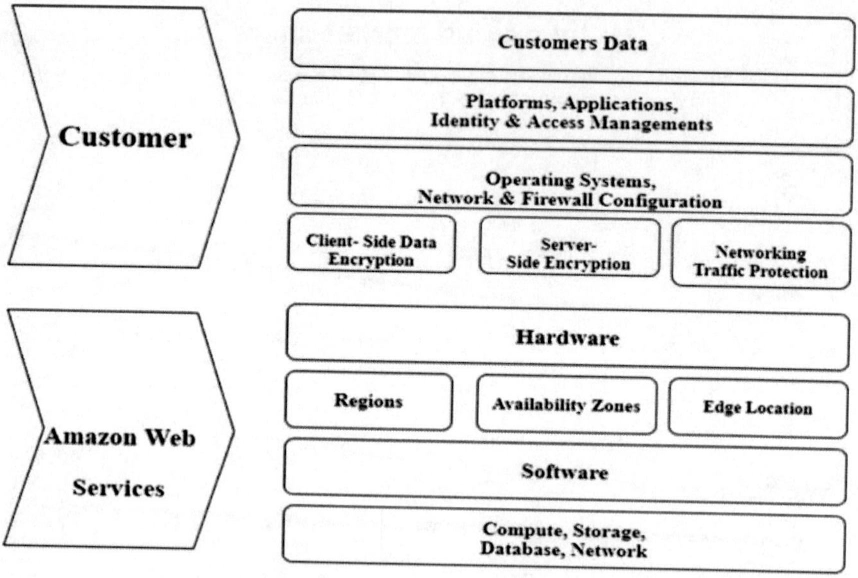

Fig. 2 AWS shared security responsibility model

hardware used for the services in AWS. AWS is also responsible for the security of hardware, software, facilities, and networks that run all products and services [3, 4] (Fig. 2).

AWS services like Amazon EC2 and Amazon S3 are the services in which AWS can only provide the security features but it cannot configure it. In those cases, customer has to configure their own security. In Amazon EC2, customer should configure security options like which OS should we select for the work done in that instance, configuring the firewalls provided by the AWS, and what are the software application should be installed on the instances. These Amazon EC2 and Amazon S3 services are categorized in semi-managed services by AWS. We also had fully managed services like Amazon Relational Database Service (RDS), Amazon Redshift, and Amazon WordDocs. In these types of services, Amazon will take care of launching and maintaining the instances and it will also take care of data backups, security configurations, data replications, and firewall configurations. Features like Identity and Access Management (IAM) user account and credentials, SSL for data transformation, and user logging activity are used by the customer for the security. All these security features should be configures be the customer.

3 AWS Global Infrastructure Security

The AWS worldwide cloud infrastructure incorporates the offices, system, equipment, and operational programming (e.g., have operating system, virtualization programming, and so forth.) that help in provisioning and usage of assets. The AWS infrastructure security is the best practices as overseen and adopted in security guidelines. An AWS client is guaranteed that they are building Web models using the AWS secure framework.

3.1 AWS Compliance Program

AWS has effectively finished numerous reviews, confirmations, and accreditations. AWS appropriates a Service Organization Controls SOC 1 report, disseminated under both the SSAE 16 and the ISAE 3402 master standards, as SOC 2-Security and SOC 3 Report. Furthermore, AWS has accomplished ISO 9001, ISO 27001, ISO 27017, and ISO 27018 affirmations, has been effectively approved as a Level 1 specialist co-op under the Payment Card Industry (PCI) Data Security Standard (DSS), and presently offers HIPAA Business Associate Agreements to secured elements and their business partners subject to HIPAA. In the area of open location confirmations, AWS has completed FedRAMP consistence, has gotten approval from the US General Services Administration to work on the FISMA Moderate level, and is moreover the level for packages with Authorities to Operate (ATOs) beneath the Defense Information Assurance Certification and Accreditation Program (DIACAP). NIST, FIPS 140-2, CJIS, and DoD SRG Levels 2 and 4 are a portion of different affirmations AWS. Some of the Accreditations AWS provides is imparting to its customers with variety of IT security standards and fine practices in industry, includes:

1. SOC 1/SSAE 16/ISAE 3402 (formerly SAS 70)
2. ITAR
3. HITRUST
4. MTCS Level 3
5. PCI DSS Level 1
6. FISMA, DIACAP, and FedRAMP
7. MTCS Level 3
8. ITAR.

AWS is also providing several industry-specific standards to their customers to deploy their solution, including [3]

1. Criminal Justice Information Services (CJIS)
2. Health Insurance Portability and Accountability Act (HIPAA)
3. Cloud Security Alliance (CSA)
4. Motion Picture Association of America (MPAA).

3.2 *Availability*

AWS data centers are working in groups in different worldwide regions. All data centers are on the Web and serving clients. AWS provides the customers to store their data in multiple geographical regions and across multiple availability zones in the same region. This will helps the customer's deployed data or applications to secure even though there are any failure modes like natural disaster and system failure [5].

4 Network Security

Network security in AWS is a well designed in such a way that customers can configure their own level of security and can modify it as per their workloads. AWS had built a high-caliber network infrastructure which helps their customers to build geographical distribute, fault-tolerant Web architecture with cloud resources and all these are deliberately observed and overseen [6]. Networking elements and Firewall devices are used to monitor and control, hels to set the rules, access control lists (ACL), and configure the specific incoming traffic. Amazon Information Security endorses ACLs tendon strategies [7].

4.1 *Secure Access Points*

AWS allows only limited access points for monitoring of incoming and outgoing communications and network traffic. These user access points are referred to as API endpoints, and they allow access to HTTP and HTTPS, which allows customers to setup a secure communication connection to storage and compute instances present in the AWS. AWS GovCloud (US) are FIPS 140-2-predictable are used for assisting clients [8].

4.2 *Transmission Protection*

Clients can connect to an AWS access point via. HTTP or HTTPS using Secure Sockets Layer (SSL), a cryptographic protocol that is designed to shield in opposition to eavesdropping, tampering, and message forgery. Customers requiring an additional layer of security, AWS provides Amazon Virtual Private Cloud (VPC) that helps securing network.

4.3 Network Monitoring and Protection

Amazon Web Services (AWS) uses a wide range of automated monitoring system for the ease of the customers. These monitoring systems are designed in such a way that all surprising or unapproved exercises and conditions at ingress and egress communication points. These automated systems are used to monitor server and network usage, unauthorized intrusion attempts, port scanning activities, and application usage. These tools can set custom execution measurements limits for bizarre movement.

5 AWS Access

To get access into the Amazon Web Services (AWS) console client needs user authorization credentials that depends upon user IDs, passwords, and Kerberos. For programmatic access user need to have SSH public-key authentication through a bastion host.

5.1 Account Review and Audit

For every 90 days, Amazon Web Services (AWS) accounts are reviewed for the re-approval or they are automatically closed. If Amazon's Human resources system terminated an employee's record, the access for that particular person is automatically abolished. In that, particular case Amazon's permission management system will disable the Windows and UNIX accounts of that user from all the systems.

5.2 Background Checks

AWS policies and procedures will help to establish and maintain the minimum standards for logical access to AWS console and host infrastructure. These policies will help to identify the functional responsibilities of security and logical access.

5.3 AWS Credentials

Amazon Web Services (AWS) will provide several types of credentials like Passwords, cryptographic keys, digital signatures for authentication to make sure that only authorized users can access their account. It is also providing two-step verification, which is called as Multi-Factor Authentication (MFA). AWS also provides security deep into its services available on that infrastructure. This helps the customers to protect their sensitive data and applications more extensively on AWS.

6 Compute Services

Amazon Web Services provides wide range of computing services that can incorporate wide range of compute instances that are capable of scaling easily.

6.1 Amazon Elastic Compute Cloud Security

Elastic compute cloud is also called Amazon EC2 that is a scalable computing resource using server instance in AWS's data centers. The main aspect of Amazon EC2 is to make the computing process simpler by empowering customers to acquire and arrange computing power with minimal efforts. Customer can launch instance in Amazon EC2 by adding more hardware and software applications if required [9].

6.1.1 Multiple Levels of Security

Amazon EC2 is built with multiple layers of securities like operating system (OS), virtual instance OS, firewalls, and other API calls. All these security levels are interdepended on each other.

6.1.2 Hypervisor

Hypervisor is software, which runs and creates virtual machines on the hardware layer. This helps the hardware to distribute its resources to the host system OS equally or in a customized manner. Currently, Amazon EC2 is making use of highly customized version of Xen hypervisor and exploiting paravirtualization (in Linux). Para-virtualized guests depend on the hypervisor because it provides support for the operations, which normally required privileged access. These guest OS do not have access to the CPU. Normally, CPU provides four stages of separate privilege methods called as rings. These rings will be varying from 0 to 3. Zero is the most privileged and three the least. Host OS will be running on ring 0, which is the most privileged, guest OS will be running on the ring 1, and all the other applications will be running on the ring 3, which is the least privileged. This virtualization of the physical resources will be the separate of guest and hypervisor, bringing about extra security detachment between the two [10]. Generally, hypervisor helps to safeguard the physical hardware and bios, virtualize the CPU, storage, networking, and helps

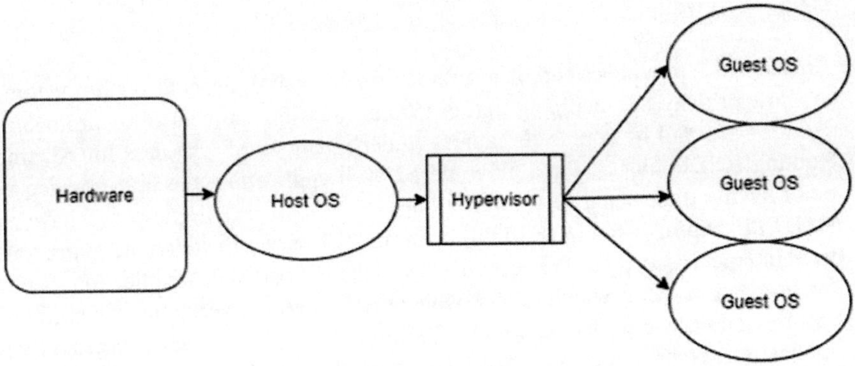

Fig. 3 Hypervisor

to customize the capabilities. With the help of nitro systems, we are able to divide these functions and join them with dedicated hardware and software and it will help us to reduce the cost by providing all the resources of the server to the customer instance (Fig. 3).

6.1.3 Instance Isolation

Many non-identical instances will be running on the same physical hardware machines and these instances are isolated from each other via Xen hypervisor. AWS firewall will be present inside the hypervisor layer for securing the data between the instance's virtual interface and physical network. All the data packets of individual instances should pass through this layer, and in this way, another instances will not have any access to that instance. Hence, all the instances will be treated as there are on separate physical hosts. Using similar mechanisms, physical RAM is also separated.

6.1.4 Guest Operating System

Guest OS is present on the virtual instances that are completely controlled by the customer. Customer will have full access for administration control on accounts, services and applications. To login securely into customers UNIX/LINUX EC2 instances AWS will also support for using Secure Shell (SSH) network protocol. In this, SSH network protocol user is given with a public/private key pair by the AWS, which will help to reduce the unapproved users to access your instance. For windows instances, user can access it by Remote Desktop Protocol (RDP) with a generated certificate for your instance.

6.1.5 Firewall

Amazon EC2 provides a complete solution for firewall. It depends on the customers to allow or deny the incoming traffic. Incoming traffic can control by protocol, by service port, and by source IP address (independent IP of Classless Inter-Domain Routing (CIDR block). For a three-tiered Web application, the best practice will be of having three separate firewall groups. The first firewall group will have port 80(HTTP) and/or port 443 (HTTPS) open for Web server in which incoming traffic from Internet is accepted. The second one will having port 8000 (application specific) for application server which is accessible only for Web server group. The third one will be of database servers in which 3306 port (MySQL) will have access only for application server group. Administrative access for all these groups will be on port 22 (SSH).

6.2 Amazon Elastic Block Store (Amazon EBS) Security

Amazon Web Services (AWS) is providing block-level storage called as Amazon Elastic Block Store (Amazon EBS). In this, customer can create a storage volume from 1 GB to 16 TB and this storage volume can be mounted to Amazon EC2 instance as a device. We can restrict the access of the Amazon EBS volume operations to the AWS root account or the AWS IAM user account. Information stored in EBS volumes will be copied in multiple physical locations within the same availability zone. This process does not need any permission from the customers. For making the data, more secure AWS recommends the customer to take a snapshot of the volume for long haul data durability [11, 12]. The good practice of the securing the data is to encrypt it. AWS provides the option of encryption to EBS volumes and EBS volume snapshots with Advanced Encryption Standard—256. This type of encryption supports the largest bit size of 256. This encryption is done in the server hosted in the EC2 instance. This process is done when the data is transferred between EC2 instance and EBS storage. This encryption feature is present only in EC2s powerful instance types (Figs. 4 and 5).

7 Networking Services

Amazon Web Services (AWS) has a wide range of networking services, which helps the customers to set up a private network connection and to segregate there network with other networks.

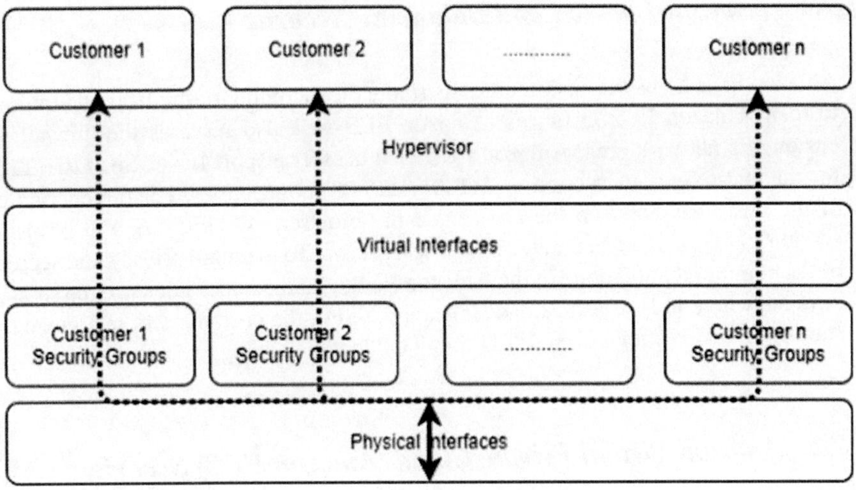

Fig. 4 Amazon EC2 multiple layers of security

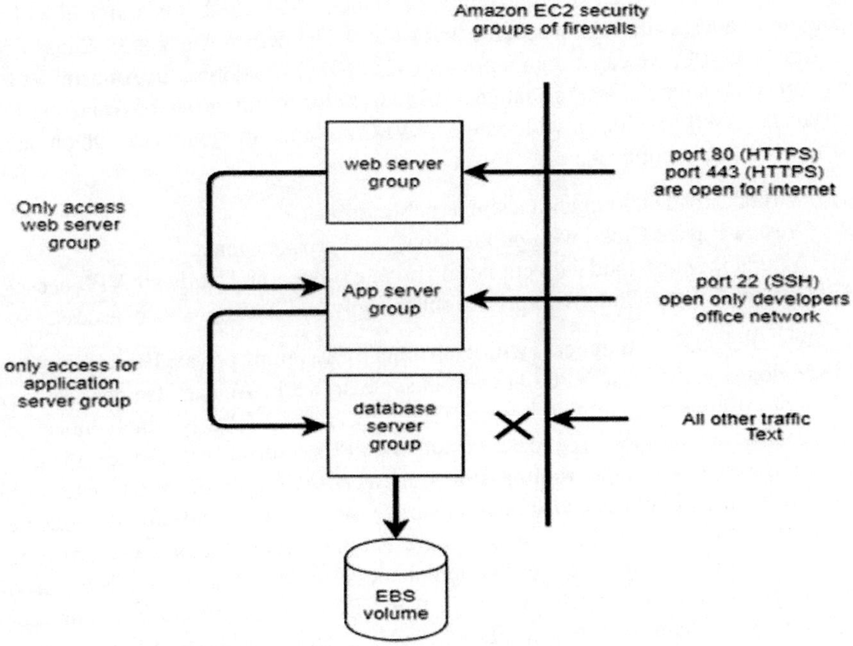

Fig. 5 Amazon EBS security group

7.1 Amazon Elastic Load Balancing Security

Amazon ELB helps the customer to distribute the incoming traffic from the Internet to their Amazon EC2 instances. Amazon ELB will also serve as the firewall for customer's network area. Amazon ELB also uses Transport Layer Security (TLS) for the end-to-end encryption on HTTP (HTTPS) connections. For the encryption of the messages between the server and the browser, HTTPS/TLS will be using session's keys, that are generated by the secret keys. To maintain privacy and secrecy of the communication done in the Amazon ELB, Amazon does not store the session keys anywhere [13]. Elastic Load Balancing allows the customers to recognize the starting IP address regardless HTTPS or TCP load balancing.

7.2 Amazon Virtual Private Cloud (Amazon VPC) Security

Virtual Private Cloud (VPC) helps the customer to create isolate there cloud environment form others. When an Amazon EC2 instance is launched, it is assigned with a random IP address that is not specific, but when we launch an Amazon EC2 instance in Amazon VPC, we can assign a private (RFC 1918) IP address. In Amazon VPC, we can create subnets, assign a range of IP address for similar group of Amazon EC2 instances. AWS provides a wide range of VPC architecture templates, which have different level of public access.

1. Virtual Private Cloud with a single public subnet.
2. Virtual Private Cloud with public subnet and private subnet.
3. Virtual Private Cloud with public and private subnets and hardware VPN access.
4. Virtual Private Cloud with private subnet only and hardware VPN access.

Two VPCs can communicate with each other by assigning private IP address. This can be done only if these VPCs are within same network. We can also connect two VPCs of different account within same region by creating VPC peering connection between those two VPCs. To make customers VPCs more secure, Amazon is also providing security groups, routing tables, network ACLs, and external gateways. Each of them will help the customers to isolate their network and provide a secure connection. In Fig. 6, all the components present in the AWS region are arranged in a topological manner in which each component has its importance. VPCs are used in a single region but subnets present in that VPCs can be present in a single or multipliable availability zone. Similarly, the VPC network shown in Fig. 6 is in a single region but the subnets present inside them are present in two availability zones A and B, respectively. Public subnet present in availability zone B is connected to the Internet through Internet Gateway (IGW). This Internet Gateway (IGW) will help us to ingress and egress the Internet access. It helps to access the outside world to VPC for communication purposes. Internet will not have the access for private subnet present in the availability zone A. Only the Network Address Translation

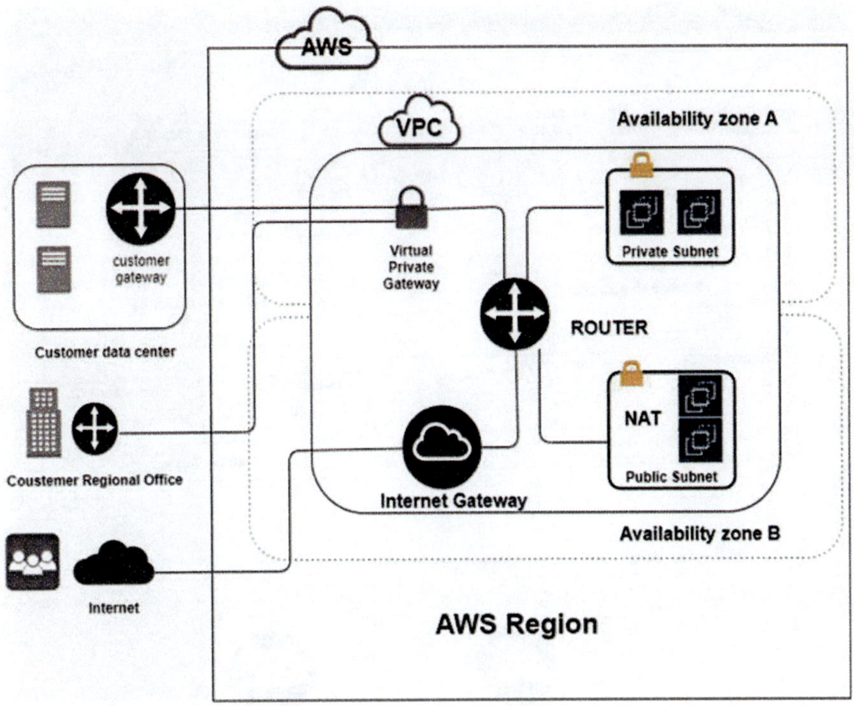

Fig. 6 Amazon VPC network architecture

(NAT) instance present in the public subnet can have access to the private subnet. Client or Customer should get through two gateway from which one will be their gateway and another comes from AWS, which is called as Virtual Private Gateway (VPG) to access AWS VPC.

7.3 Amazon Route 53 Securities

Amazon Route 53 helps systems to communicate with each other by decoding domain name into IP addresses. Route 53 helps the customer to manage the incoming traffic by providing the IP address a specific name. Route 53 manages global traffic of various routing types like, Geo DNS, Latency Based Routing (LBR), and Weighted Round- Robin (WRR) [14]. AWS also provides Domain Name Registrations for the customers. Route 53 is has an option that monitors customer's website regularly.

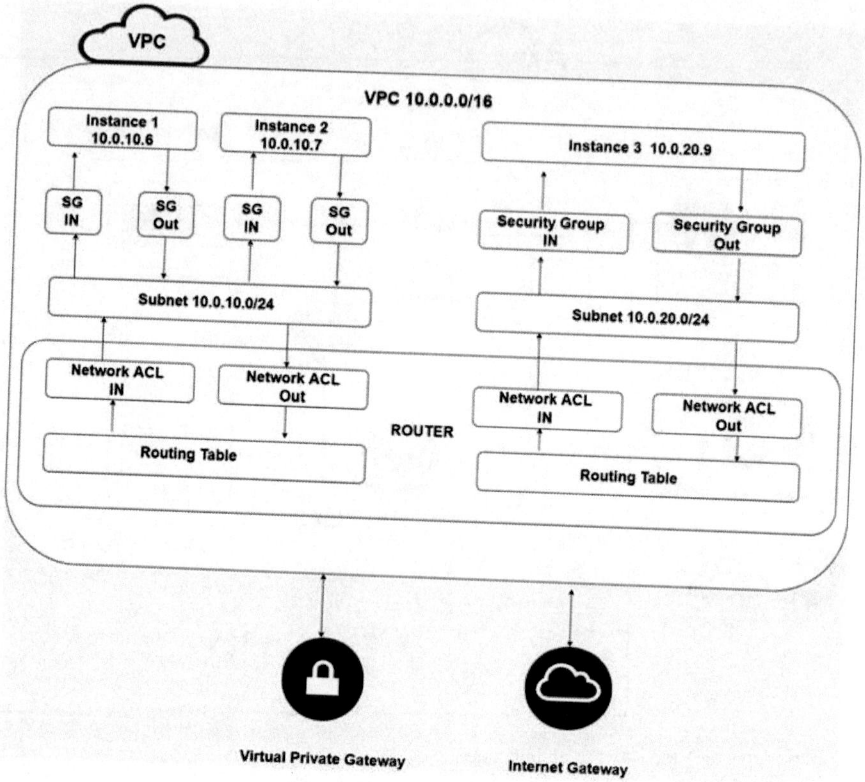

Fig. 7 Flexible network topologies

7.4 AWS Direct Connect Security

Customers can connect to internal network to Aws region using AWS Direct Connect and thereby reduces the risk and improves the latency factors. Amazon Direct Connect use Border Gateway (BGP) protocol and Autonomous System Number (ASN) for security of the data. For creation of virtual interface and message authorization, it uses MD5 cryptographic key [15] (Fig. 7).

8 Conclusion

All these services will help the customer to build a well architecture solution in Amazon Web Services (AWS) cloud. Amazon Web Services (AWS) is providing all the features and options to maintain privacy and protect their sensitive data from external cyber attacks. Shared security responsibility model will say that not only Amazon Web Services (AWS) but also customers should take care of the security of their data or applications. Amazon Web Services (AWS) has completed many audits, attestations, and certification from many International organizations. Every service provided by the Amazon web Services (AWS) has high level of security.

References

1. Srivastava, P., Khan, R.: A review paper on cloud computing. Int. J. Adv. Res. Comput. Sci. Softw. Eng., Res. Gate, 17–20 (2018)
2. Tadapaneni, N.R.: Cloud computing security challenges. Int. J. Innov. Eng. Res. Technol. 1–5 (2020)
3. Amazon Web Services (AWS) Whitepapers: Amazon Web Services Risk and Compliance (2017)
4. Amazon Web Services (AWS) Whitepapers: Overview of Amazon Web Services (2020)
5. Anthony, A.: Mastering AWS Security, Packt Publishing (2017). ISBN: 9781788293723
6. Mosca, P., et al.: Cloud security: services, risks, and a case study on Amazon Cloud Services. Int. J. Commun., Netw. Syst. Sci. 529–535 (2014)
7. Kanikothottu, H.: AWS Security Cookbook. Packt Publishing (2020). ISBN: 9781838826253
8. Amazon Web Services (AWS) Whitepapers: Overview of Security Processes (2020)
9. Kulkarni, G.: Cloud computing-infrastructure as service AMAZON EC2. Int. J. Eng. Res. Appl. 117–125 (2012)
10. Barrowclough, J.P.: Securing cloud hypervisors: a survey of the threats, vulnerabilities, and countermeasures. Secur. Commun. Netw. (2018)
11. Amazon Web Services (AWS): https://aws.amazon.com/eds/
12. Ashalatha, R., Evaluation of auto scaling and load balancing features in cloud. Int. J. Comput. Appl. **117**(6), (2015)
13. Amazon Web Services (AWS) Whitepapers: Amazon Virtual Private Cloud Connectivity Options (2020)
14. Soni, M.: Practical AWS Networking. Packt Publishing (2018)
15. Amazon Web Services (AWS): https://aws.amazon.com/directconnect

Sensorless Speed Control of Induction Motor Using Modern Predictive Control

N. P. G. Bhavani, M. Aruna, K. Sujatha, R. Vani, and N. Priya

Abstract This paper uses model predictive control that is superior technique to handle sensorless speed control of induction motor driven by a solar system. The proposed system, solar photovoltaic (PV) array connected with voltage source inverter (VSI) and predictive control system, is MRAS. A MPPT algorithm based on perturbation and observation is used for harnessing the full power from a PV series. System predictive control strategy is a predictive control approach to standard optimization. For less dependency on model parameters, MPC can achieve sufficient control efficiency and has been widely used in process control systems. MRAS speed estimation method was implemented to achieve sensorless operation to boost the device reliability and cost reduction of hardware. In MATLAB/Simulink environment, the desired configuration is modeled and simulated.

Keywords Speed sensorless control · Model predictive control (MPC) · Photo voltaic (PV) · Perturb and observe algorithm (PO) · Model reference adaptive system (MRAS)

1 Introduction

In the present development, because of the fast exhaustion of non-renewable resources, many countries are promoting renewable energy resources to meet the growing demand for electric power. The age of sunlight-based PV-based vitality

N. P. G. Bhavani (✉) · M. Aruna
Department of Electrical and Electronics Engineering, Meenakshi College of Engineering, Chennai, India

K. Sujatha
Department of Electrical and Electronics Engineering, Centre for Electronics, Automation and Industrial Research (CEAIR), Dr. MGR Educational and Research Institute, Chennai, India

R. Vani
Department of ECE, SRM University, Chennai, India

N. Priya
Department of EEE, Easwari Engineering College, Chennai, India

H. Sharma et al. (eds.), *Congress on Intelligent Systems*, Advances in Intelligent Systems and Computing 1334, https://doi.org/10.1007/978-981-33-6981-8_53

has come up as a significant option for a few purposes. The acceptance engine has supplanted DC engines in light of its mechanical straightforwardness, strength, dependability, minimal effort, better, and less upkeep. A solar PV system is used here to feed induction motor drive using predictive control model. Single power peak is a main feature in photovoltaic, and maximum power attainment is an important feature of PV system. A bother and perception dependent on the maximum power point tracking (MPPT) calculation is utilized to control and watch the greatest force. P&O algorithm improves monitoring time and produces increased energy when the irradiation varies greatly, which is operationally simple and requires less hardware and easy implementation.

To achieve maximum motor output, induction motors (IM) permit one DC–DC converter and one voltage source inverter (VSI). Therefore, VSI performs regulation of the DC link voltage itself. The system, however, needs at least seven switches from the power converter and thus increases the switch losses [1]. This also includes a DC–AC conversion performed in three-phase IM. Therefore, it is important to use single-stage control drive which in turn reduces the quantity of switches and misfortunes. A VSI must keep up the MPP in a solitary stage design, and it likewise controls the voltage of the DC connect. Field-oriented control (FOC) and direct torque control (DTC) are two particular engine control frameworks for their enlistment. These two plans are basic, simple to execute, and have a quick powerful reaction. Yet, they are model based, and furthermore delicate to evolving boundaries. Framework prescient control rises under this kind of interest. Thus, MPC has gotten the third decision of an elite drive framework notwithstanding the FOC and DTC speed control frameworks, creating considerable interest in research.

Model reference adaptive system (MRAS) is one of the speed observator regularly utilized for acceptance engine drives. MRAS speed eyewitness depicts two models, a reference model and a movable model which yields and looks at two related yield signals, for example, motion, back emf, or stator current. A rule which is sensitive to speed moves the output error to null. It is shown that, according to the simulation results, the combined MPC, MRAS framework shows better speed control execution and state estimation execution over a wide speed range and burden torque run.

2 Methodology

The design of sun-powered PV exhibit that took care of sensorless engine acceptance drive with prescient control of the system is shown in Fig. 1. The proposed framework is a PV cluster followed by a VSI-controlled three-stage enlistment engine drive. MRAS figures the engine speed, which is determined by the voltage of the DC connection and the motor currents. For MPPT, a perturb and observe algorithm is used to produce VSI switching pulses, maintaining specification accuracy. Figure 1 shows a straightforward schematic of a three-stage enlistment engine, 415 V, used to drive the framework controlled by a sun-based PV cluster of max. 8.7 kW. A 8700 W PV exhibit is intended to drive a 7.5 kW acceptance engine. The rating of

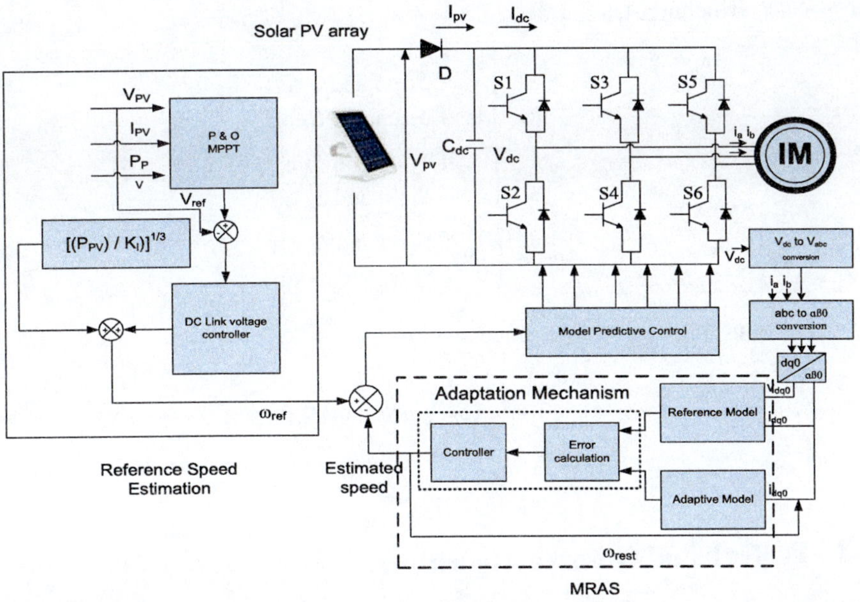

Fig. 1 PV fed induction motor drive using model predictive control

the PV exhibit is picked instead of the rating of the engine, with the goal that the productivity of the engine stays unaffected by misfortunes of the engine and the converter. A PV exhibit is introduced by associating 34 PV modules in open circuit voltage (V_{oc}) arrangement equivalent to 734 V and 25 modules corresponding to the short out current (I_{sc}) identical to 15.5 A.

The voltage of the DC connected to the motor is

$$V_{dc} = \sqrt{2} \times V_L = \sqrt{2} \times 415 = 587\,\text{V}$$

DC link voltage = 600 V.

$$\frac{1}{2} \times C_{dc} \times \left(V_{dc}^2 - V_{dc1}^2\right) = 3aV_pI_t = 3 \times 1.2 \times 239.6 \times 13.5 \times 0.005$$

DC link capacitor is 2.5 nF where V_{dc} is the voltage of the DC connection and V_{dc1} is the minimum permissible voltage of the DC connection during the transient state, t is the time needed to recover the minimum permissible voltage of the DC link, current of the motor phase is I, and voltage of operation is V_p. Hence, the condenser value is selected as 2.5 nF. The inverter comprises of three modules with half extension; at moments of a given time, the upper and lower power switches of every unit join turning on and off. Contingent upon the condition of the switches in the individual half extension, every one of the three yield terminals can either be wired to the positive DC-interface voltage potential or to the negative potential.

Such switching states are

$$S_a = \begin{cases} 1, \text{ if } S_1 \text{ on and } S_2 \text{ off} \\ 0, \text{ if } S_1 \text{ on and } S_2 \text{ off} \end{cases}$$

$$S_b = \begin{cases} 1, \text{ if } S_3 \text{ on and } S_4 \text{ off} \\ 0, \text{ if } S_3 \text{ on and } S_4 \text{ off} \end{cases}$$

$$S_c = \begin{cases} 1, \text{ if } S_5 \text{ on and } S_6 \text{ off} \\ 0, \text{ if } S_5 \text{ on and } S_6 \text{ off} \end{cases}$$

In vector form, this can be expressed as:

$$S = \frac{2}{3}\left(S_a + aS_b + a^2 S_c\right)$$

3 Perturb and Observe Algorithm

The single-stage system includes MPPT to extract maximum power through inverter. Here, P&O algorithms for maximum power point tracking (MPPT) are implemented. VSI switching using model predictive control and speed estimation with MRAS. Consideration is given to the direction of perturbation according to the power transition [2]. If there is a positive power change, then the voltage increases in the direction of the right side, and in the event that it is negative or diminishes, at that point the voltage unsettling influence increments in course of the contrary left, and the aggravation way is resolved if the voltage at present is higher than the voltage at the past one, which would then be able to be estimated as a result of this adjustment in the PWM. At starting, overshoot shows up as indicated by this condition and bit by bit diminishes until it arrives at a steady state, and when the yield power surpasses its full qualities, the control activity stop (Fig. 2).

From the flowchart, it is observed that: On the off chance that force is on the left half of the most extreme force point, at that point, the voltage increments, and in like manner when force is on the correct side of the greatest force point, at that point the voltage diminishes (Fig. 3).

4 Speed Estimator Using MRAS

The MRAS velocity estimation structure essentially consists of a reference model, unique model, and a versatile system as shown in Fig. 4. Independent of rotor speed, measures the state variable, X, against the voltage and current of the terminal. The adjustable model which depends on the speed of the rotor estimates the variable of

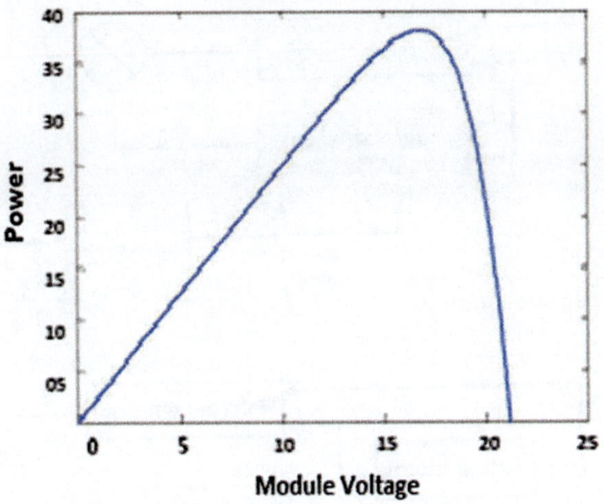

Fig. 2 Power–voltage curve of the solar

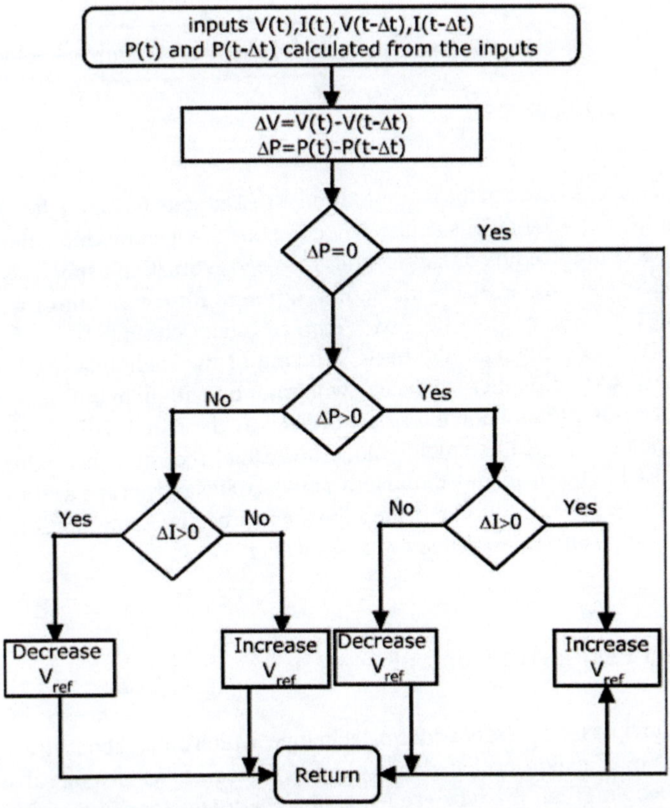

Fig. 3 Flowchart of perturb and observe algorithm

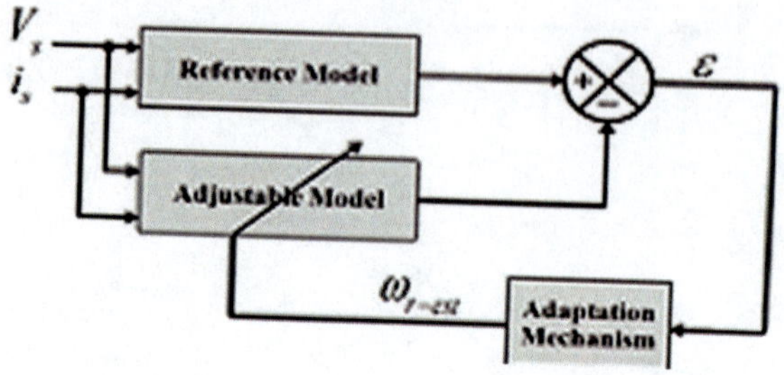

Fig. 4 MRAS estimation structure

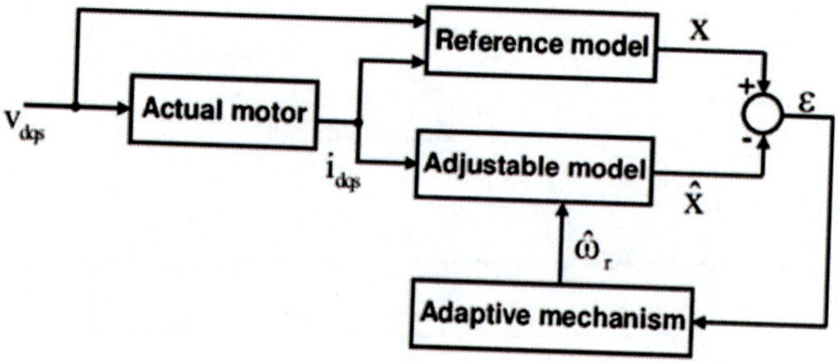

Fig. 5 Rotor flux-based MRAS

state, i. The error s between the measured and assessed state factors is then utilized by tf to drive the flexible model's adjustment component which produces the evaluated speed, tV. It should be noted that methods of speed estimation using MRAS can be divided according to the state variable into different forms. The most widely used are MRAS based on the rotor flux, back emf, and stator current.

As approximate amounts, the basic structure of the traditional style rotor flux dependent MRAS is identical to the one with rotor transition. In this plan, the enlistment engine stator circuit conditions are utilized by the speed-free reference model normally communicated as voltage model to extract the rotor flux value. Usually expressed as the current model, the adaptive model relies on rpm and uses rotor circuit equations of the induction motor. In the adjustment procedure, a proportional-integral (PI) controller is utilized to gauge rotor speed (Fig. 5).

5 Model Predictive Control

IN MPC principles, we present a control technique, which thinks about all exchanging successions over a short exchanging skyline N, alluded to as the control skyline in the MPC people group [3]. The sequence of semi-conductor switch positions S is defined as a switching sequence, from time stage 0 to time step N–1 after some time span N. The MPC plot figures the drive reaction for each exchanging succession in a subsequent stage, for example, the advancement of yield factors over the exchanging skyline N, in view of the nonlinear, discrete-time forecast model. To reproduce a since a long time ago yield skyline, the "promising" yield directions are extrapolated, and the quantity of time-steps are resolved when the primary yield variable hits a hysteresis limit. The expenses related with each exchanging arrangement are determined by partitioning the complete number of exchanging changes in the arrangement by the length of the extrapolated direction. Limiting this punishment permits the ideal grouping of changing to be applied to the inverter and the following ideal switch position [4–6]. The model prescient plan can be worked in two different ways, with $N > 1$ and $N = 1$, frequently varying in the level of opportunity for the exchanging successions, consequently the computational weight and the productivity.

6 Results and Discussion

Simulations of the model predictive control scheme were performed with MATLAB-Simulink to determine its effectiveness [7–10]. The comparative results of MPC and vector control are given in Table 1 (Figs. 6, 7, 8 and 9).

Table 1 Comparison result of MPC and vector control

Control method	Speed (RPM)	Settling time (ms)
MPC	305	2.25
Vector control	305	3.5

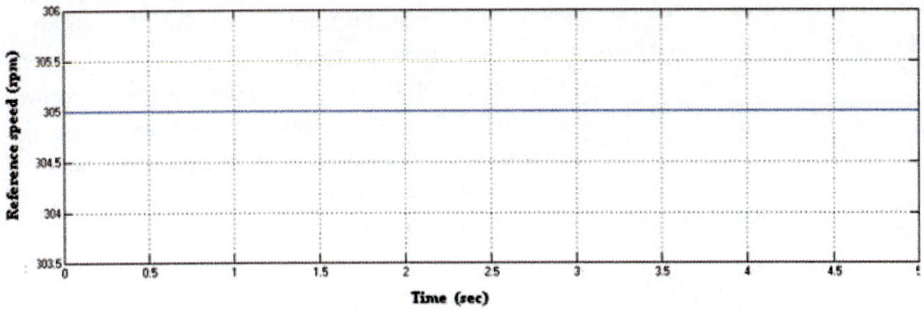

Fig. 6 Rotor reference speed 305 RPM

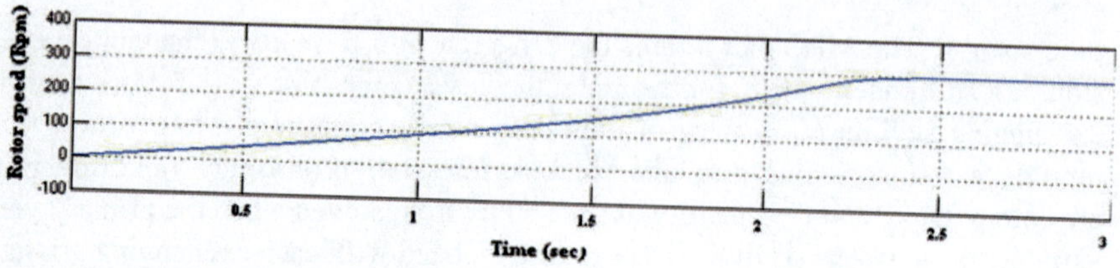

Fig. 7 Rotor speed by using MRAS 305 RPM

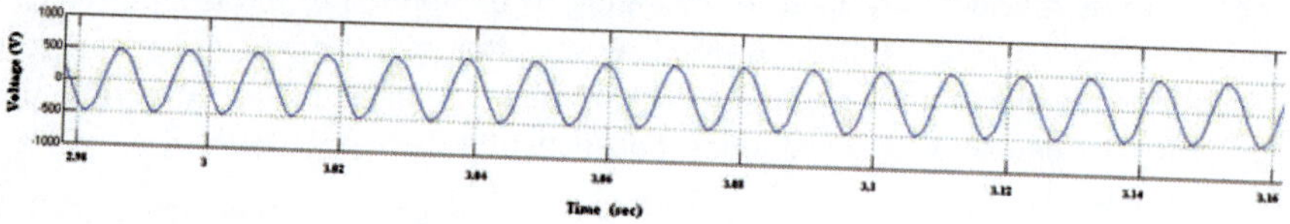

Fig. 8 Stator voltage

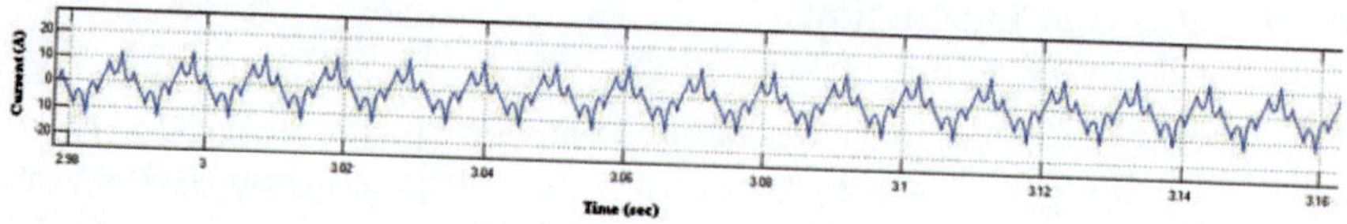

Fig. 9 Stator current

7 Conclusion

Speed control using MPC fed sensorless IM was found to be very satisfactory, and it is suitable for electric vehicles. The engine takes off easily. The reference speed is produced by the DC interface voltage controller that manages the voltage at the DC association alongside the speed estimator. The intensity of the PV exhibit is held at full force point at the hour of progress in irradiance, which was possible through MPPT algorithm based on P&O. Combining the model predictive control with the MRAS speed estimator for high-performance induction motors is feasible. The simulation results obtained without using sensors show better performance. PV system is combined with sensorless induction motor which joins with the MPC current controllers and the MRAS speed estimation.

References

1. Sahoo, S.K., Bhattacharya, T.: Field weakening strategy for a vector-controlled induction motor drive near the six-step mode of operation. IEEE Trans. Power Electron. **31**(4), 3043–3051 (2016)

2. Zhang, S., Xu, Z., Li, Y., Ni, Y.: Optimization of MPPT step size in stand-alone solar pumping systems. In: IEEE Power Eng. Society General Meeting, June 2006
3. Kennel, R., Linder, A.: Predictive control of inverter supplied electrical drives. In: Proceedings of IEEE Power Electronics Specialist Conference (PESC'00), Galway, Ireland, 2000, (CD-ROM).
4. Elgendy, M.A., Atkinson, D.J., Zahawi, B.: Experimental investigation of the incremental conductance maximum power point tracking algorithm at high perturbation rates. IET Renew. Power Gener. **10**(2), 133–139 (2016)
5. Singh, B., Bhuvaneswari, G., Garg, V.: A Novel polygon based 18-pulse AC–DC converter for vector controlled induction motor drives. IEEE Trans. Power Electron. **22**, 488–497 (2007)
6. Kumar, R., Das, S., Syam, P., Chattopadhyay, A.K.: Review on model reference adaptive system for sensorless vector control of induction motor drives. IET Elec. Power Appl. **9**, 496–511 (2015)
7. Holtz, J.: The induction motor—a dynamic system. In: Proceedings of Annual Conference of the IEEE Industrial Electronics Society (IECON'94) vol. 1, pp. P1–P6, Bologna, Italy, (1994)
8. Camacho, E., Bordons, C.: Model Predictive Control. Springer, London, UK (1999)
9. Zhang Y., Yang, H.: Model predictive control for speed sensorless induction motor drive. In: Proceedings of the CSEE, vol. 34, no. 15, pp. 2422–2429 May 25, 2014 (in Chinese).
10. Rodriguez, J., Cortes, P.: Predictive Control of Power Converters and Electrical Drives. Wiley-IEEE Press, Chichester, United Kingdom (2012)

Air Quality Monitoring System Using Machine Learning and IoT

B. R. Varshitha Chandra, Pooja G. Nair, Risha Irshad Khan, and B. S. Mahalakshmi

Abstract Monitoring air pollution is of increasing concern today. People are suffering from health problems as a result of prolonged exposure to polluted environments. This project aims to develop an air quality monitoring system using machine learning with Internet of Things (IoT), an Internet server network of physical nodes. This system consists of three sections: an air pollution detection model developed in python and built using machine learning algorithms, random forest and support vector machine, a low-cost air monitoring device comprising of a hardware unit that detects different pollutants like CO, NO_x, PM2.5, and an IoT Cloud, ThingSpeak, acting as a middleman for the captured data between the hardware component and the algorithm for air pollution classification. The final output displays the predicted air quality index (AQI) and provides a comparison between the two algorithms used, random forest and support vector machine, in terms of accuracy and various other statistical data. The accuracy depicted by random forest exceeds 95% and that of support vector machine is 85%.

Keywords AQI · Pollution · IoT · Random forest · Support vector machine

1 Introduction

Atmospheric pollution is capable of causing heart and lung associated diseases along with deadly diseases such as cancer and premature deaths of babies as reported in the European Union [1]. Some substances that contribute to such a pollution, for example, cadmium, arsenic, polycyclic fragrant hydrocarbons, and nickel are human genome toxic cancer-causing agents. They do not pose a risk if it is below an identifiable threshold.

Air contamination also severely damages soil quality, underground water systems or water bodies also harms biological systems through eutrophication (abundance

B. R. Varshitha Chandra (✉) · P. G. Nair · R. I. Khan · B. S. Mahalakshmi
B. M. S. College of Engineering, 1908, Bull Temple Road, Bangalore, Karnataka 560019, India

B. S. Mahalakshmi
e-mail: bsmahalakshmi.ise@bmsce.ac.in

H. Sharma et al. (eds.), *Congress on Intelligent Systems*, Advances in Intelligent Systems and Computing 1334, https://doi.org/10.1007/978-981-33-6981-8_54

nitrogen contamination) and corrosive downpour [2]. Thus, it affects horticulture and woods, as well as materials and establishments. Air pollution has originated from many sources and means but it is mainly from energy creation, ventures, transport, and agro-business. Air pollution in certain parts has declined in recent times but the long-haul objective is still at risk, namely to accomplish air quality levels that do not have critical negative effects on human well-being and the earth. Air contamination norms are often infracted, especially in urban areas (the "hotspots"). Fine PM2.5, NO_2 and ground-level ozone are the most problematic pollutants today [3].

1.1 Objective

The main goal of this venture is to implement an air pollution monitoring model utilizing the concept of machine learning with Internet of Things which facilitates the monitoring of air pollution by reading the values of gases such as PM2.5, NO, NO_2, NO_x, SO_2, benzene, along with parameters such as toluene, xylene, CO, and ozone from the environment using an embedded system. Air pollution classification model will be developed in python and built using the best-suited machine learning algorithms. These measured data values are collected by the web server for the IoT cloud service and transferred to the air pollution classification model, which processes the data and provides the accurate air quality index to the current instance of the atmosphere. Thus, reducing the health risks for the people living in a given area by providing accurate data thus allowing them to take the required measures to improve the air quality and in turn improve their quality of life.

1.2 Existing System

Reparation of tools and assessment of air contamination observational stations is necessary to ensure data accuracy, quality control, and following guidelines for observing and adjustment. But manual monitoring does have certain limitations. Various measuring techniques and instruments at the same location yield varying results on the environmental concentration. Studies have stated that the median error for PM monitoring ranges between ten to twenty-six percent, mainly due to inaccurate flow measurement and calibration. If the power supply is not regular and shows fluctuation of voltage, then it also often affects the results of observation. In the case of gas contaminants, the inspecting time frame, test weakening, and inversion monitoring are fundamental to guarantee that testing is done appropriately. The type of quality sensors used is also a major in determining the accuracy.

Three vital concepts in the field of city air computation are introduction, estimation, and characteristic examination of the fine air quality. The outcomes of the concepts can be extremely useful in using that information for controlling air contamination and thus produce amazing social and technological impact. Much of the recent work approaches these three problems using various methods separately.

The explanation of the missing variables is associated with the instruments that control air pollutants. To be specific, each region has only one control unit that has to be monitored at particular time intervals, and when the systems re-calibrated, repaired, etc., there will be no outputs for the station.

1.3 Proposed System

In this proposed system depicted in Fig. 1, we collect data from the environment using a hardware component that comprises of an Arduino board, breadboard, NodeMCU, gas sensors, namely MQ135 for sensing all the NO compounds and benzene, and MQ7 and a dust particulate sensor for PM2.5. Then, we store this data consisting of PM2.5, NO, NO_2, NO_x, CO, on a ThingSpeak cloud server. This data is sent to the air prediction model built using random forest and SVM in order to predict the AQI of the current atmosphere. The method of forecasting the concentration of each type of compound or gas will then be invoked, and the outcome will be displayed to the user. The proposed method of selection and analysis of features shows the significance of specific input features to neural network predictions, thus providing the potential to expose any internal component of black-box models, which is not constrained to counteraction and regulation of air pollution, but it may also be widened to a lot of different applications, for instance, such as terrorism and medical diagnosis. This

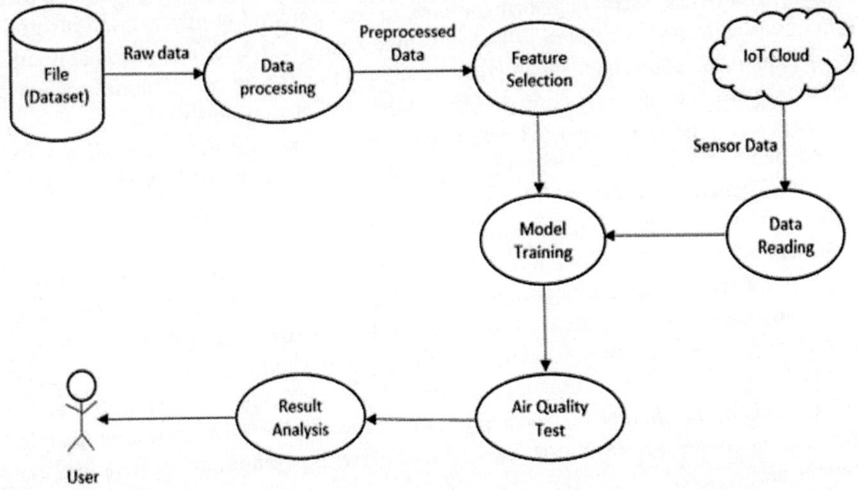

Fig. 1 Flowchart of the proposed system

action cannot be taken based on the predictions of a model without knowing whether it is accurate and trustworthy, else the effects may be disastrous.

2 Related Work

Monia et al. [4] implemented a model that predicts the monitoring and publishing of pollution levels in the air as a web service in various major cities in India. A study was conducted in various predictive analytics, primarily in applications of machine learning such as regression and deep learning.

Xie et al. [5] focused on the deep neural network for reliably calculating the concentration of PM2.5 emissions, based on multiple learning. The multiple learning approach defines meteorological variables. Secondly, the deep belief network (DBN) was developed for learning the input. Lastly, modeling of the deep characteristics on network is done. The study shows that the deep learning in concentration forecasting for PM2.5 is a promising technique based on the multiple learning.

Ayele et al. [6] proposed an IoT-based air pollution prediction monitoring program that can be used in any area. The proposed program would concentrate on tracking air pollution by integrating IoT with an algorithm of machine learning known as the recurrent neural network commonly known as LSTM.

Kök et al. [7] inspired to use methods of deep learning to predict IoT results and produce a novel model of deep learning to analyze IoT smart city data. The results of the model proposed are deemed positive and showed that the model can be used for other problems of smart city prediction.

Soh et al. [8] aim to predict the quality of air until 48 h using a combination of multiple neural networks which includes a convolutionary neural network, an artificial neural network, and a long-term memory to derive spatial-temporal relationships. It involves patterns from several locations, meteorological data from the past hours, and terrain data. Experiments using datasets reveal that the proposed model achieves excellent performance and is outperforming the methods currently used.

Zeng et al. [9] focused on an architecture to visualize the air quality forecasted using big data implementing a platform-based framework based on ETL (Extract-Transform-Load), which consists of storage. The results show that a large visualization of the data system would work fine for air quality analysis.

Khaefi et al. [10] say the fire and events in South Asia have increased in recent years along with land clearing, rapid deforestation, and extreme dry seasons and have transboundary consequences. Because of this, aerosols are released from the fires at high concentrations, worsening air quality and reducing visibility, causing economic, social, environmental, and health problems. Results show that information about the visibility can increase the accuracy of forecasts over a baseline approach. In addition, the model applies to the forecast model, with potential to improve disaster prevention.

Hu et al. [11] presented an implementation, design, and optimization of the network sensing our own air quality, which can provide the monitoring region with fine-grained and a real-time air quality chart. The main aim is to reduce the average common error of the already developed real-time air quality chart. Simulations tested the performance of the suggested approaches and demonstrated a major performance benefit when implementing the above two strategies.

Ao et al. [12] introduced a model, a cross of air quality prediction that uses deep neural networks and K-Means clustering. It is found that the proposed model consisting of K-means and hybrid algorithm used on test data brings a more precise prediction of the air quality.

Madaan et al. [13] forecasted a fine-grained information on the ambient air quality based on two types of data-historical, real-time air quality, and meteorological data reported by the central pollution control board. This is taken from five major areas in Delhi. The proposed LSTM network based on adaptive attention outdoes multiple baselines for models of regression and classification with an accuracy of 15–20%.

Rijal et al. [14] estimated concentrations of particulate matter 2.5 through the usage of exterior door images through a requested set of deep nerve-neural network-based regression that comprises a feedforward neural network. The experimental outcome showed that this proposed model can provide a more correct estimate of particulate matter 2.5 in comparison to all the three individual deep learning frameworks that were used.

Qi et al. [15] suggests a common extensive and efficient method altogether in a single model named deep air learning (DAL), to solve the three problems. The offered model encompasses a deep convolutionary neural network (CNN) to categorize innate figures. The outcomes of the experiment show that our model is reasonable with datasets that use images for PM2.5 concentration calculation.

Ghoneim et al. [16] say clean air is one of the most critical environmental requirements for human well-being. To increase prediction model accuracy, he developed the concept of using models of deep architecture tested on large data of weather and pollution. This model of deep learning is used to learn the ozone level features, and it is trained by using a technique called grid search and shows that the proposed method has best performance for predicting the levels of ozone.

Tao et al. [17] have done a short-term forecasting model based on deep learning proposed for PM2.5 with an aerodynamic diameter of less than or equal to 2.5. Comparing the predictive results with the traditional results, it is proved that the CBGRU model error is lower and the performance of predictions is better.

3 Implementation

The raw data from the file consisting of the dataset undergoes cleaning to remove the data with null values and any special characters. This is called as data pre-processing. The pre-processed data will undergo training in order to give the accurate result. This is then saved for future use. The data flow diagram given below (Fig. 2) represents the processes involved in the software component.

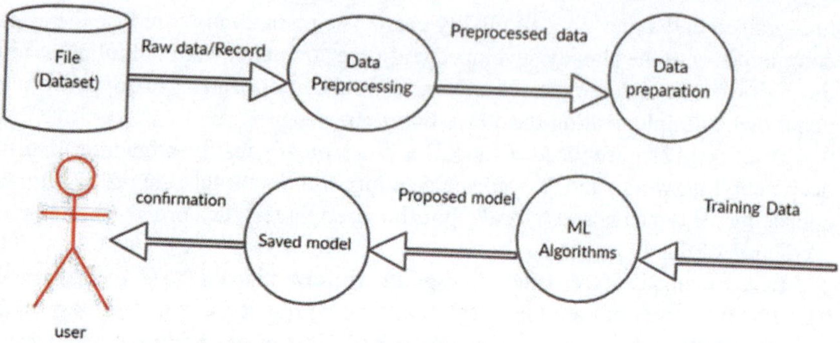

Fig. 2 Data flow diagram of the software component

3.1 System Architecture

The system architecture (Fig. 3) consists of the following: Microcontroller is referred to Arduino Uno microcontroller unit which has sensors for various gases like PM2.5, CO, NO_x, by performing the telemetry operation. Wi-Fi transceiver module transmits

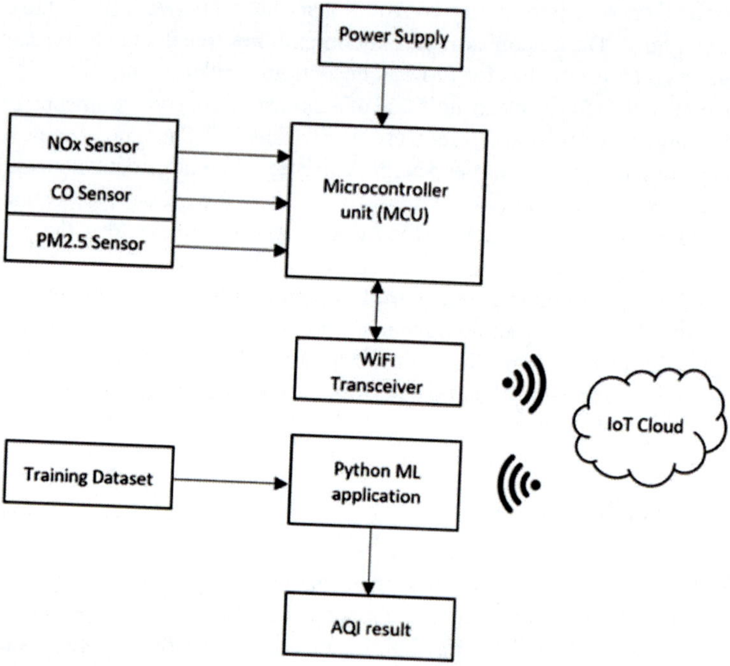

Fig. 3 System architecture

all the parameters that are sensed to the cloud. Captured data from the cloud is collected and sent to the web application. Clients (MCU) and Python application are based on HTTP PROTOCOL. The users of the Python application must feed the training dataset which will be used to build machine learning models. This system uses machine learning algorithms that are fed by the user to build the model. The Python application is built using anaconda framework with Jupyter Notebook IDE.

3.2 Hardware Components

i. Arduino Uno
ii. ESP8266 Wi-Fi module
iii. MQ gas sensors.

- MQ-135 gas sensor
- MQ-7 gas sensor
- PM2.5 GP2Y1010AU0F dust smoke particle sensor.

The components are configured using the Arduino code that runs on the Arduino IDE. It is used to set up the Arduino and configure the Wi-Fi setting for esp8266, and the sensors to sense the data according to the given time interval. This captures the readings from the sensors that are displayed on the serial monitor. The data is then sent to ThingSpeak. The specified channel captures the data and plots graphs accordingly as represented in Fig. 4.

3.3 Training and Testing the Dataset

The dataset contains data on air quality and air quality index (AQI) at various stations at the hourly and daily levels across multiple cities in India. The above-mentioned dataset has been considered for training and testing purposes. The most common type of pollutants, namely PM10, PM2.5 NO, NO_2, NH_3, NO_x, CO, O_3, SO_2, benzene, xylene, and toluene, are considered. "Label" column indicates the AQI index given on a scale from 0 to 5 (0 being the best).

The dataset is pre-processed and cleaned eliminating all the null values after which the length of the data reduces from 26,219 to 20,290. The original dataset is now divided into two parts: the training module and testing module. X_train and X_test consist of training data, and Y_train and Y_test contain the output values. Test size has been set to 0.20, indicating 20% of the dataset is being utilized for testing, while the remaining 80% is utilized for training the machine learning models.

We calculate the confusion matrix in order to show the accuracy. A confusion matrix is defined as a table that is used to characterize a classification result on a set

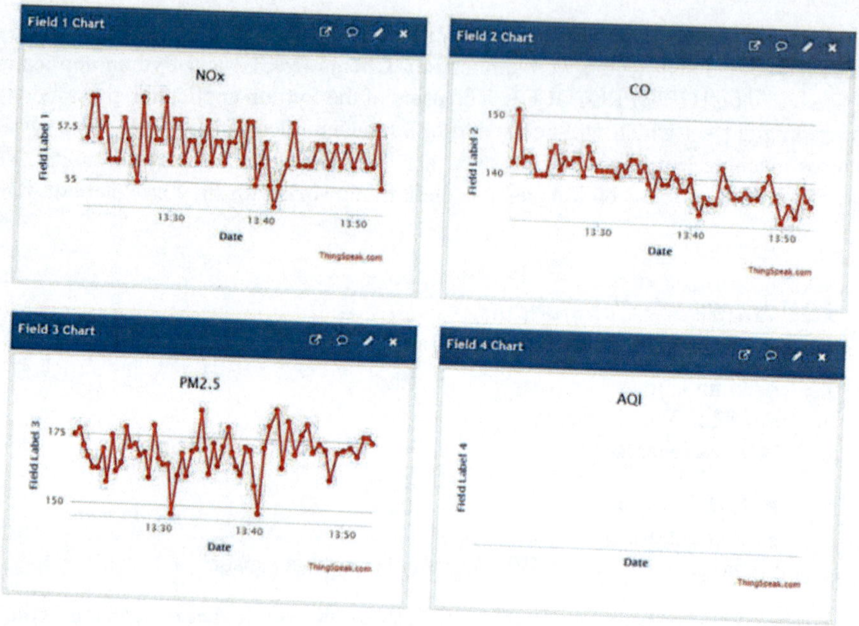

Fig. 4 ThingSpeak capturing the various levels of the respective gases

of test data for which the true values and false values are compared. It enables the output of an algorithm to be visualized.

The AQI categories have been categorized into the following ranges and associated with indices starting from 0 to 5, 0 relating to "good air quality" and 5 relating to "severe/unhealthy air quality." A direct comparison is plotted (Fig. 5) in the form of a bar graph to depict the accuracy of the two algorithms indicating that RF has an accuracy of ~99.9%, and SVM has an accuracy of ~85%.

Testing of the proposed system has been done taking into account several test cases as depicted in the tables below. Two scenarios of test cases have been considered. One being of the air pollution predicting model (Table 1) and the other of the low-cost hardware unit (Table 2).

3.4 Results

AQI is classified into the following categories, as shown in Fig. 6, based on the level of the various gasses present in the atmosphere:

- Good (0–50): Minimal impact.
- Satisfactory (51–100): Minor breathing discomfort to sensitive people.
- Moderate (101–200): Breathing discomfort to people with lung diseases, heart diseases, young children, and adults.

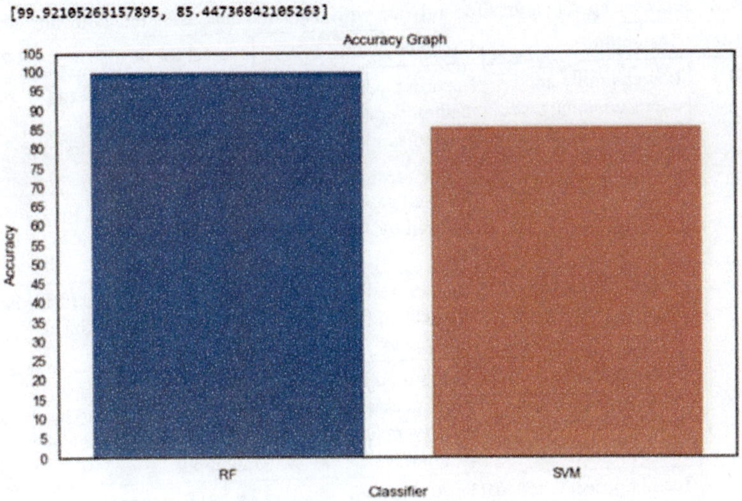

Fig. 5 Comparison between the two algorithms in terms of accuracy

Table 1 Test case specifications for software unit used to train the ML models

Test ID	Test input	Excepted result	Actual result	Remarks
T_01	Data upload in software unit	During this process, data has to be read from excel file	Data will be read from excel file	Pass
T_02	Data cleaning	This process has to read the uploaded data and clean it and store it back in a different excel sheet	The uploaded data will clean it and store it in a different excel sheet	Pass
T_03	Training model construction	During this process, training dataset has to read and process and create validation model	The training dataset will be read and process and create validation model	Pass
T_04	Validation model storing	The training model has to provide validation model which has to be saved	The training model will provided validation model which has to be saved	Pass
T_05	Testing process	During this process, test data has to read from the data which is pass to validation model and provide prediction as output	The test data will read from the data which is pass to validation model and provide prediction as output	Pass

Table 2 Test case specifications for software with real-time data using the hardware

Test ID	Test input	Excepted result	Actual result	Remarks
T_01	Power supply and environmental gases	Recorded values from the hardware measured by the gas sensors are detected	Gases are detected and values are recorded using the gas sensors in the hardware unit	Pass
T_02	Sensors in hardware	Detected gases are captured by Arduino Uno	Detected gases are captured by Arduino Uno	Pass
T_03	Arduino Uno data	Data is saved in the IOT cloud called ThinkSpeak	Data is saved in the IOT cloud called ThinkSpeak	Pass
T_04	IOT cloud data	Gases stored in ThinkSpeak are read by the software and used as raw input	Gases stored in ThinkSpeak are read by the software and used as raw input	Pass
T_05	Testing process	The AQI is then calculated using both random forest and SVM model	The AQI is then calculated using both random forest and SVM model	Pass

Fig. 6 AQI representation

AQI Category	AQI Range
Good	0 – 50
Satisfactory	51 – 100
Moderately polluted	101 – 200
Poor	201 – 300
Very poor	301 – 400
Severe	> 400

- Poor (201–300): Breathing discomfort to people on prolonged exposure.
- Very poor (301–400): Respiratory illness to people on prolonged exposure.
- Severe (>401): Severe respiratory effects even on healthy people.

Sensor data: The gas sensors capture readings at the interval of 5 s. These readings are reflected on the serial monitor.

The below screenshot (Fig. 7) represents the various levels of the gaseous contaminants and depicts the predicted AQI using RF (pre) and SVM (pre1).

The AQI index of the entered gases, calculated by random forest model is 2, indicating satisfactory level of the atmospheric air. In comparison with this, the support vector machine model having accuracy lesser than that of random forest model stated the same 2, indicating moderate air quality in the instant environment.

The central pollution control board has an AQI calculator that is used to calculate the accuracy of the result.

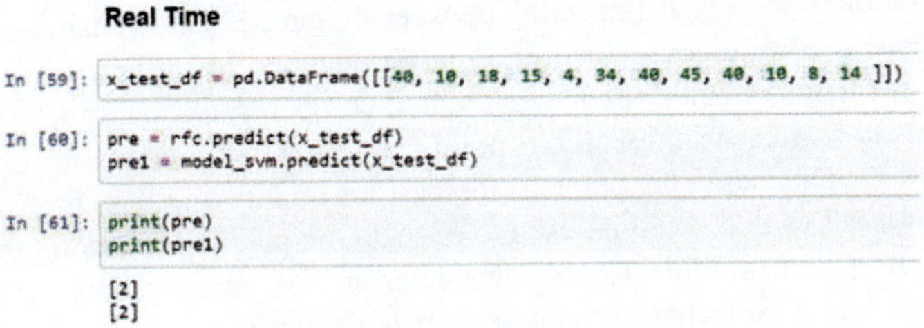

Fig. 7 Predicting AQI for the given data

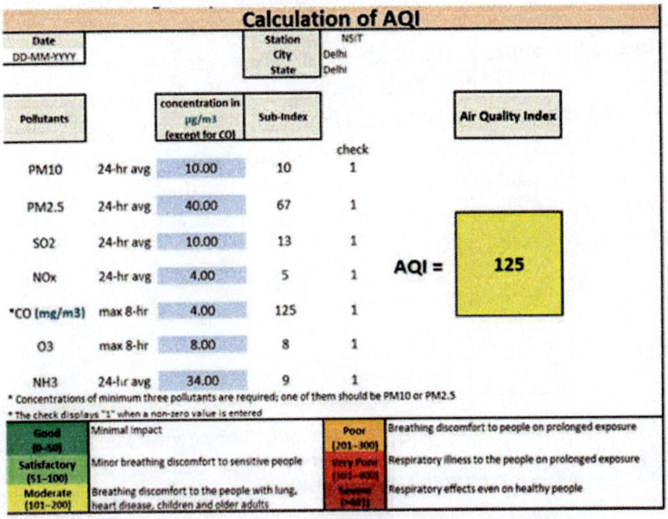

Fig. 8 AQI calculator given by CPCB used for testing the output obtained

Testing with the AQI calculator supplied by CPCB:

Thus, from the image above (Fig. 8), the AQI has been calculated to be 125 indicating that it lies in level 2 on our scale from 1 to 5, rightly demonstrated by the random forest model and support vector machine model.

4 Conclusion

The system designed comprises of three parts—the hardware component, IoT cloud storage (ThingSpeak), and software unit used to model and train the data. The hardware component collects the environmental data which is stored in the IoT cloud. From the cloud, data is read by the software and passes it to the validation model which predicts the result (an index between 0 and 5). Thus, enabling us to identify the air quality index, which is a metric intended to explain air quality safety impacts. It helps in health management and compares air quality levels at various locations or towns. Additionally, it points out potentially dangerous environments and their frequencies. This helps to forecast air quality and thus prepare pollution control measures that could be used to educate the public. It is useful to those who are suffering from multiple illnesses that deteriorate or are caused by poor air quality. Real-time data is collected from the sensors in the hardware component powered by Arduino Uno connected to the laptop. This data is sent to the ThingSpeak IoT cloud from where it is sent to be tested on the trained model using support vector machine and random forest. The AQI index of the mean of the last 20 readings (number can be changed) is taken and displayed on a scale from 0 to 5.

For future enhancements, the hardware can be made compact into a single device with the help of a 3D printer. Thus, making it easily portable to be placed in any location around the house or outside. Here, alarms can be set with alert messages when the AQI levels are very high, thus posing hazardous threats. Based on the detected values, we can switch on exhaust or air purifier, or it can instruct one to wear a mask. System can also provide warning messages in case of high levels of pollutants in the environment.

References

1. WHO: How Air Pollution is Destroying Our Health. https://www.who.int/airpollution/news-and-events/how-air-pollution-is-destroying-our-health. Last accessed 2020/05/01
2. USGS Nitrogen and Water: https://www.usgs.gov/special-topic/water-science-school/sci-ence/nitrogen-and-water?qt-science_center_objects=0#qt-science_center_objects. Last accessed 2020/05/01
3. WHO Ambient (Outdoor) Air Pollution: https://www.who.int/news-room/fact-sheets/de-tail/ambient-(outdoor)-air-quality-and-health. Last accessed 2020/05/01
4. Monia, Gupta, A., Sharma, S.: Predictive analysis of air pollution using collaborative filtering prediction algorithm. In: IEEE International Conference on Electrical, Computer and Communication Technologies (ICECCT) 2019, pp. 1–8, IEEE, Coimbatore (2019)
5. Xie, J.: Deep neural network for PM2.5 pollution forecasting based on manifold learning. In: International Conference on Sensing, Diagnostics, Prognostics, and Control (SDPC) 2017, pp. 236–240, Shanghai (2017)
6. Ayele, T., Mehta, R.: Air pollution monitoring and prediction using IoT. In: Second International Conference on Inventive Communication and Computational Technologies (ICICCT) 2018, 1741–1745 (2018)
7. Kök, İ., Şimşek, M.U., Özdemir, S.: A deep learning model for air quality prediction in smart cities. In: IEEE International Conference on Big Data (Big Data) 2017, pp. 1983–1990. IEEE, Boston (2017)
8. Soh, P., Chang, J., Huang, J.: Adaptive deep learning-based air quality prediction model using the most relevant spatial-temporal relations. IEEE Access **6**, 38186–38199 (2018)
9. Zeng, Y., Chang, Y.S., Fang, Y.H.: Data visualization for air quality analysis on Big data plat-form. In: International Conference on System Science and Engineering (ICSSE) 2019, pp. 313–317. IEEE, Dong Hoi (2019)
10. Khaefi, M.R., Pramestri, Z., Amin, I., Lee, J.G.: Nowcasting air quality by fusing insights from meteorological data, satellite imagery and social media images using deep learning. In:

IEEE/ACM International Conference on Advances in Social Networks Analysis and Mining (ASONAM) 2018, pp. 393–396. IEEE, Barcelona (2018)
11. Hu, Z., Bai, Z., Bian, K., Wang, T., Song, L.: Real-time fine-grained air quality sensing networks in smart city: design, implementation, and optimization. IEEE Internet of Things J. **6**(5), 7526–7542 (2019)
12. Ao, D., Cui, Z., Gu, D.: Hybrid model of air quality prediction using *K*-means clustering and deep neural network. In: Chinese Control Conference (CCC) 2019, pp. 8416–8421. IEEE, Guangzhou (2019)
13. Madaan, D., Dua, R., Mukherjee, P., Lall, B.: VayuAnukulani: adaptive memory networks for air pollution forecasting. In: IEEE SigPort (2019)
14. Rijal, N., Gutta, R.T., Cao, T., Lin, J., Bo, Q., Zhang, J.: Ensemble of deep neural networks for estimating particulate matter from images. In: IEEE 3rd International Conference on Image, Vision and Computing (ICIVC) 2018, pp. 733–738. IEEE, Chongqing (2018)
15. Qi, Z., Wang, T., Song, G., Hu, W., Li, X., Zhang, Z.: Deep air learning: interpolation, prediction, and feature analysis of fine-grained air quality. IEEE Trans. Knowl. Data Eng. **30**(12), 2285–2297 (2018)
16. Ghoneim, O.A., Doreswamy, Manjunatha, B.R.: Forecasting of ozone concentration in smart city using deep learning. In: International Conference on Advances in Computing, Communications and Informatics (ICACCI) 2017, pp. 1320–1326. IEEE, Udupi (2017)
17. Tao, Q., Liu, F., Li, Y., Sidorov, D.: Air pollution forecasting using a deep learning model based on 1D convnets and bidirectional GRU. IEEE Access **7**, 76690–76698 (2019)

Comparison of Various Classifiers for Indian Sign Language Recognition Using State of the Art Features

Pradip Patel and Narendra Patel

Abstract Indian sign language (ISL) is used by deaf people in India for their internal communication. Sign language recognition (SLR) system recognizes gestures of sign language and converts them into text and voice, thus enabling deaf people to communicate with others. In this paper, we have presented an automatic system for ISL hand gesture recognition. Modern features like Histogram of Oriented Gradient (HOG), Speeded Up Robust Features (SURF), Fourier Descriptors, Hu Moments, and Zernike Moments have been extracted from training images of our own created dataset. Using these features, most commonly used classifiers like Support Vector Machine (SVM), Artificial Neural Network (ANN), K-Nearest Neighbors (KNN), and Linear Discriminant Analysis (LDA) have been trained and evaluated. Comparative performance analysis of these classifiers along with experimental results is the main topic of discussion included in this paper. Moreover, a user-friendly GUI of the system allows user to select any combination of features and classifiers to perform gesture recognition from both stored file and live camera. During experiment, it is found that the system is able to recognize gestures with a recognition rate up to 98.70%.

Keywords Sign Language Recognition · Hand Segmentation · Feature Extraction · Classification

1 Introduction

Sign languages are the only fully accessible languages for deaf people where different gestures are used to convey meaning. In sign language, all the alphabets, numbers, and words of our language are assigned gestures. These gestures [1] are basically of

P. Patel (✉)
Gujarat Technological University, Ahmedabad, Gujarat, India
e-mail: pradippatel@ldce.ac.in

N. Patel
BVM Engineering College, Vallabh Vidyanagar, Gujarat, India
e-mail: nmpatel@bvmengineering.ac.in

H. Sharma et al. (eds.), *Congress on Intelligent Systems*, Advances in Intelligent Systems and Computing 1334, https://doi.org/10.1007/978-981-33-6981-8_55

two types: static gestures and dynamic gestures. Indian Sign Language (ISL) (see Fig. 1) is used by deaf people in India. It helps them to connect with other deaf persons and the larger deaf community. But it is difficult for them to exchange a few words with people who are not deaf, as they do not recognize sign language. Use of human interpreter as intermediately can solve this communication gap. But such types of interpreters are both costly and not easily available all the time. Instead, a computer-based system can be developed which can recognize gestures of sign language and convert them into text and audio. Development of such system is more complicated for ISL than other sign languages because it includes

- Both single-handed and double-handed gestures with complex hand shapes.
- Both static and dynamic hand gestures.
- Facial expressions, Head/Body postures.

From implementation point of view, the field of ISL recognition has many technical challenges. Due to complexity of ISL gestures, hand tracking and segmentation become tricky. Furthermore, to carry out research work in the field of ISL recognition, no standard dataset exists. Gesture recognition approaches [2] are mainly of two types—hardware based and vision based. In hardware-based approach, user is required to wear special hardware like data glove. Vision-based approach makes use of image processing [3] and computer vision techniques [4]. It is found more appropriate and realistic as compared to hardware-based approach.

This paper presents a vision-based system for ISL hand gesture recognition. With overall accuracy of 98.70%, it recognizes gestures of numbers (0–9) and alphabets (A–Z) both from stored file and live camera. Comparison of various classifiers for different features extraction techniques is also presented with experimental results. Section 2 of the paper describes literature survey. Section 3 explains various steps of proposed system such as preprocessing, feature extraction, and classification. It is followed by implementation results and conclusion in Sects. 4 and 5, respectively.

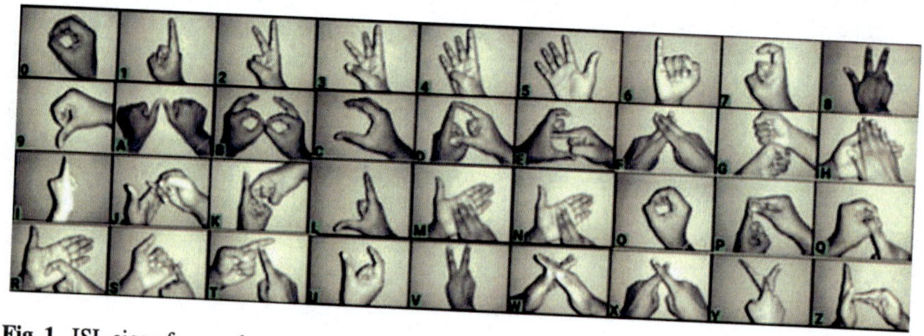

Fig. 1 ISL signs for numbers and alphabets

2 Literature Survey

Recently, with the improvement of science and technology, many researchers have developed methods for ISL recognition. System developed in [5] provided accuracy of 96.25%. It used Eigen vector as feature and Eigen value weighted Euclidean distance classifier for 24 different alphabets. A multimodal framework-based system is proposed in [6] that considered both facial expression and sign gesture with two sensors: Kinect and Leap motion. Hidden Markov Model classifier is used, and Independent Bayesian Classification Combination approach is used to advance recognition performance. Recognition rates achieved were 96.05% for single and 94.27% for double hand gestures. In [7], authors applied Euclidean distance transformation on the preprocessed binary image. On resultant image, row and column projection is applied. Both central moments and Hu Moments are used as features. System achieved recognition rates of 94.37% for ANN and 92.12% for SVM. System developed in [8] used invariant Krawtchouk moment-based features and provided 97.9% accuracy. The system proposed in [9] performed preprocessing and segmentation in HSV color space. Then, features like Hu Moments and motion trajectory were extracted and used to train SVM. The accuracy achieved was 97.5%. The application proposed in [10] used Microsoft Kinect to capture images and used Scale Invariant Feature Transform (SIFT) features. It achieved average accuracy rate of 90.68%. A system based on 2D FFT Fourier Descriptors for feature extraction was proposed in [1], where vector codebook is created using LBG and template matching is done using a simple Euclidean Distance method. The accuracy of the system was 92.91%. System developed in [2] used combination of Hu invariant moment and structural shape descriptors as features. For classification, a Multiclass SVM is used which achieved recognition rate of 96.23%. Authors developed a system in [11], where hand region is detected using skin segmentation with YCbCr and HSV color models. HOG features are then extracted from segmented images and are used to train SVM classifier. An automatic ISL finger spelling recognition is proposed in [12]. It also used skin color-based segmentation for hand region detection. A feed forward neural network is trained using distance transform-based shape features, and it achieved accuracy of 91.11%. Another different system, which first performs categorization of gestures into single-handed or double-handed, is proposed in [13]. With combination of HOG and SIFT features, KNN classifier achieved accuracy of 91%. Hence, ISL recognition is modern area of research. We have built our system with most common feature extraction and classification techniques. We have considered different feature extraction techniques such as HOG, SURF, Fourier Descriptors, Hu Moments, and Zernike Moments. Using these features, we have trained classifiers like SVM, ANN, KNN, and LDA. A user-friendly GUI of the system helps user to select any combination of features and classifiers. It also provides user an option to select stored file or live camera as an input.

3 Proposed System

The proposed system works in two phases: training and testing. For the duration of training phase, the classifier is trained by using the images of training dataset. As standard ISL dataset is not available, we have formed our own training dataset. It consists 220 images for each alphabet (A–Z) and number (0–9) resulting in total 7920 images. These images have been captured using external web camera. During training, these images are inputted to system along with their class labels. Various steps like preprocessing, feature extraction, and classifier training are then performed in that order. Once the training completes, testing takes place where an unknown test image is given as input to the system for classification. Testing also involves preprocessing, feature extraction, and sign recognition. Figure 2 shows entire flow of our system.

Fig. 2 Flow of proposed system

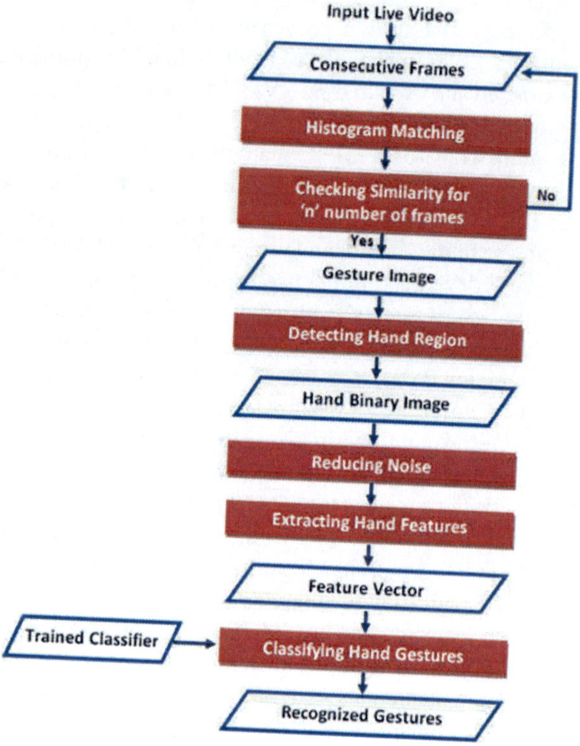

Fig. 3 Preprocessing steps (**a**) input image, (**b**) image in YCbCr color space, (**c**) image with removed face region, (**d**) hand detection using BLOB, (**e**) cropped image

3.1 Preprocessing

Preprocessing includes various steps necessary for hand segmentation. In our system, face detection is first performed in order to remove face region from input image. It results in an image with hand region as biggest skin color object. Skin color detection [14, 15] algorithm is then used to detect hand region. For that, input image is first converted from RGB to YCbCr color space using Eq. 1. Thresholding is then performed to detect skin color pixels as shown in Eq. 2. This results in binary image. Subsequently, hand region is detected by finding biggest binary linked object from the image. After that, image cropping is performed by finding bounding box of hand region. Eventually, resultant image is scaled to 110 by 110 pixels. Finally, morphological filters are applied to this binary image in order to remove noise and segmentation errors (see Fig. 3).

$$\begin{aligned} Y &= 0.299R + 0.587G + 0.114B \\ Cr &= 128 + 0.5R - 0.418G - 0.081B \\ Cb &= 128 - 0.168R - 0.331G + 0.5B \end{aligned} \tag{1}$$

$$75 < Cb < 135 \text{ and } 130 < Cr < 180 \text{ and } Y > 80 \tag{2}$$

3.2 Feature Extraction

Feature extraction is imperative step where important image features are computed. HOG, SURF, Fourier Descriptors, Hu Moments, and Zernike Moments are most common features for any image recognition system. HOG and SURF are gradient-based features, Fourier Descriptors are shape oriented features, and Hu Moments and Zernike Moments are region oriented features. These features not only describe the shape with high accuracy, but also reduce the data dimension to a great extent. Out of these features, HOG and SURF are very fast and efficient features. SURF, Fourier Descriptors, Hu Moments, and Zernike Moments also provide invariance

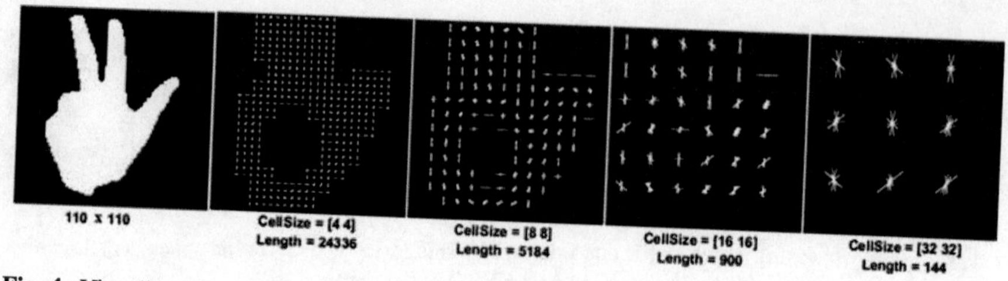

Fig. 4 Visualization of HOG features for different cell size

with respect to scale, translation, and rotation. So, we have preferred these features in our system.

Histogram of Oriented Gradients (HOGs)

HOGs [16] are commonly used features in many object recognition systems [11, 13]. For a given image, HOG features can be computed with following steps.

Step 1. Partition the image into cells and, for each cell, calculates histogram of the edge orientations.

Step 2. Normalize these local histograms for a group of cells called block.

Step 3. Calculate final feature vector by combining normalized histograms.

For an image of our dataset, visual representation along with corresponding length of HOG features for different cell size is shown in Fig. 4. In our system, we have used cell size of 32 × 32 yielding feature vector of length 144.

Speeded Up Robust Features (SURF)

SURF [17] features are based on orientation of the gradient in localized region. They can be computed as below.

Step 1. Generate integral image from the input image.

Step 2. Generate response map at each point using box filters [17]. Scale spaces are formed by using filters of different sizes (9 × 9, 15 × 15, 21 × 21, 27 × 27, ….).

Step 3. Detect interest points by performing non-maximum suppression in a 3 × 3 × 3 neighborhood over scales.

Step 4. Rotate the neighborhood sampling window to align with dominant direction of the feature.

Step 5. Divide the neighborhood into 16 subsquare which are further divided into 4 squares. The descriptor for each subsquare over its four quadrants is computed as $v = (\sum dx, \sum dy, \sum |dx|, \sum |dy|)$ resulting in final vector of size 64.

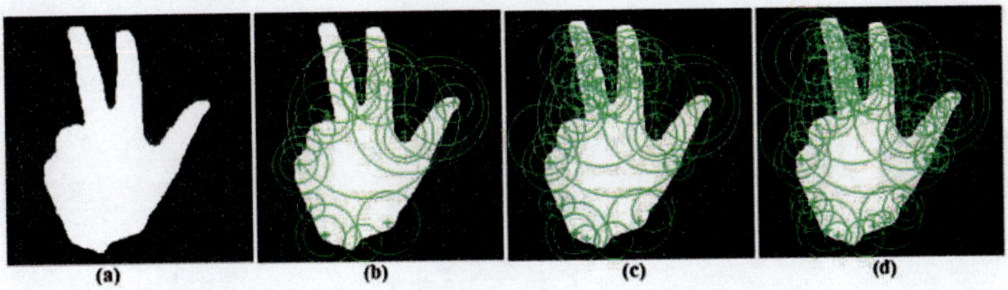

Fig. 5 SURF visualization (a) input image, (b) 25 key points, (c) 50 key points, and (d) 75 key points

Figure 5 shows visual representation of SURF features of image selected from our dataset.

Fourier Descriptors

Fourier descriptors [3] are widely used features to represent object in image. To calculate these features, coordinates of boundary points of hand region are represented as the sequence of complex numbers $s(k) = x(k) + jy(k)$ for $k = 0, 1, 2, \ldots K-1$. Fourier transform is then applied to this sequence.

$$a(u) = \frac{1}{K}\sum_{k=0}^{K-1} s(k)e^{-2\pi uk/K} \tag{3}$$

We have used only first 20 descriptor values in our system. These descriptors are made rotation invariant by keeping only magnitude values and ignoring phase information. They are made scale invariant by dividing all descriptors by DC component.

Hu Moments

Originally defined by M. Hu in [18], moments uniquely represent the shape of an object. The moment [3] of order $(p+q)$ of an image $f(x, y)$ is

$$m_{pq} = \sum_x \sum_y x^p y^q f(x, y) \tag{4}$$

where $p = 0, 1, 2, \ldots$ and $q = 0, 1, 2, \ldots$. The corresponding central moments of order $(p+q)$ is defined as

$$\mu_{pq} = \sum_x \sum_y (x-\bar{x})^p (y-\bar{y})^q f(x, y) \tag{5}$$

where

$$\bar{x} = \frac{m_{10}}{m_{00}}, \bar{y} = \frac{m_{01}}{m_{00}}$$

The normalized central moments of order $(p+q)$ are defined as

$$\eta_{pq} = \frac{\mu_{pq}}{\mu_{00}^{\gamma}} \quad (6)$$

where

$$\gamma = \frac{p+q}{2} + 1$$

for $(p+q) = 2, 3, \ldots$. A final set of seven invariant moments are

$$\begin{aligned}
\varphi_1 &= \eta_{20} + \eta_{02} \\
\varphi_2 &= (\eta_{20} - \eta_{02})^2 + 4\eta_{11}^2 \\
\varphi_3 &= (\eta_{30} - 3\eta_{12})^2 + (3\eta_{21} - \eta_{03})^2 \\
\varphi_4 &= (\eta_{30} + \eta_{12})^2 + (\eta_{21} + \eta_{03})^2 \\
\varphi_5 &= (\eta_{30} - 3\eta_{12})(\eta_{30} + \eta_{12})\left[(\eta_{30} + \eta_{12})^2 - 3(\eta_{21} + \eta_{03})^2\right] \\
&\quad + (3\eta_{21} - \eta_{03})(\eta_{21} + \eta_{03})\left[3(\eta_{30} + \eta_{12})^2 - (\eta_{21} + \eta_{03})^2\right] \\
\varphi_6 &= (\eta_{20} - \eta_{02})\left[(\eta_{30} + \eta_{12})^2 - (\eta_{21} + \eta_{03})^2\right] \\
&\quad + 4\eta_{11}(\eta_{30} + \eta_{12})(\eta_{21} + \eta_{03}) \\
\varphi_7 &= (3\eta_{21} - \eta_{03})(\eta_{30} + \eta_{12})\left[(\eta_{30} + \eta_{12})^2 - 3(\eta_{21} + \eta_{03})^2\right] \\
&\quad + (3\eta_{12} - \eta_{30})(\eta_{21} + \eta_{03})\left[3(\eta_{30} + \eta_{12})^2 - (\eta_{21} + \eta_{03})^2\right]
\end{aligned} \quad (7)$$

Zernike Moments

Zernike Moments [19] are image moment-based descriptors that represent shape of objects very accurately with relatively few data points. They are also independent to object translation, scale, and rotation. To compute these moments, the binary image is first mapped to the unit circle with its origin at the image's center (see Fig. 6). Only pixels inside the circle are considered for further computation.

Zernike moment [20] of order p and repetition q for function $f(x, y)$ is defined as

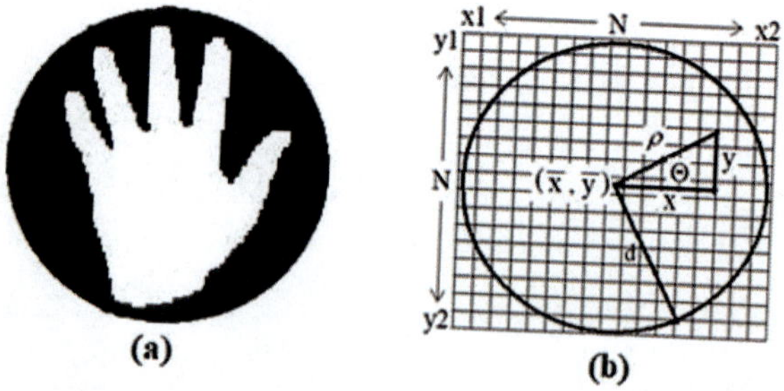

Fig. 6 Unit disk mapping process (**a**) computational range, (**b**) parameters of unit circle

$$Z_{pq} = \frac{p+1}{\pi} \sum_x \sum_y V^*_{pq}(x, y) f(x, y), x^2 + y^2 \leq 1 \tag{8}$$

$$V_{pq}(\rho, \Theta) = R_{pq}(\rho) e^{iq\Theta} \tag{9}$$

$$R_{pq}(\rho) = \sum_{\substack{k=|q| \\ |p-k|=\text{even}}}^{p} \frac{(-1)^{\frac{p-k}{2}} \frac{p+k}{2}!}{\frac{p-k}{2}! \frac{k-q}{2}! \frac{k+q}{2}!} \rho^k \tag{10}$$

Here, p is order, q is repetition, V^*_{pq} is complex conjugate of V_{pq}, R_{pq} is radial polynomial, and $0 \leq p \leq \infty, |q| \leq p, |p - q| =$ even. In the developed system, we have used Zernike moments of order up to 7 resulting into total 20 moments.

3.3 *Classification*

Vectors of features extracted during feature extraction are given as an input to train the classifier. Once trained, the classifier classifies the test image into one of possible classes. In our system, we have considered four different classifiers: Artificial Neural Network (ANN), K-Nearest Neighbors (KNN), Linear Discriminant Analysis (LDA), and Support Vector Machine (SVM).

Artificial Neural Network (ANN)
An ANN [7] classifier is based on the neural system of human brain. A neural network consists of one input layer, one output layer, and multiple hidden layers with each layer fully connected to the following layer. Back propagation algorithm is used to train a neural network where the pairs of training inputs and outputs are presented to the network. Inputs are given at the input layer, and the output of the network is obtained. Finally, the variation between real and estimated results is computed, and this inaccuracy value is propagated backwards. The algorithm tries to reduce the error until the neural network learns the training data.

K-Nearest Neighbors (KNN)
The KNN [13] is another classification method based on nearest training samples based on features. An element is classified as belonging to a class for which maximum elements are near around it. For classification, k elements, which are nearest to the test element, are considered. Euclidean is used as the default distance.

Linear Discriminant Analysis (LDA)
The LDA [6] makes predictions based on estimation of the probability that a new set of inputs belongs to each class. The class that has the highest probability becomes the output class. The LDA model uses Bayes theorem to estimate the probabilities.

Fig. 7 Gesture recognition from live camera

Support Vector Machine (SVM)
The SVM [2] is a popular classifier defined by hyper plane which separates two classes. It is originally developed as a binary classifier. When an input is presented to SVM, it predicts which of two possible classes forms the output. Multiclass Support Vector Machine (MSVM) is used to solve multiclass problem. It divides problem into several two-class problems that can be solved directly by using multiple SVMs.

4 Experimental Results

The proposed system is capable of recognizing gestures of 26 alphabets (A–Z) and 10 numbers (0–9) of ISL. With the help of developed GUI as shown in Fig. 7, the system recognizes gesture from both stored image and live camera. To measure the accuracy of our system, out of 220 images for each gesture, 160 images were used for training and 60 images were used for testing. Accuracy has been measured using cross validation with k-folds having $k = 10$. Figure 8 shows gesture-wise recognition rates for various features while Fig. 9 depicts recognition rates of different classifiers.

During experimentation, it is observed that, for all classifiers under consideration, SVM proved as best classifiers with comparison to others classifiers because it tries to find optimal margins that separate classes. It is also found that HOG and SURF provided better accuracy and speed because of their efficient procedure for localized edge orientation calculation. SURF, Fourier descriptors, Hu moments, and Zernike moments also provided rotation invariance. Average times required to compute various features for a given image of our dataset are as shown in Table 1.

5 Conclusion

A vision-based real-time hand gesture recognition system for Indian Sign Language is presented in this paper. The system works in two phases, training and testing. Each

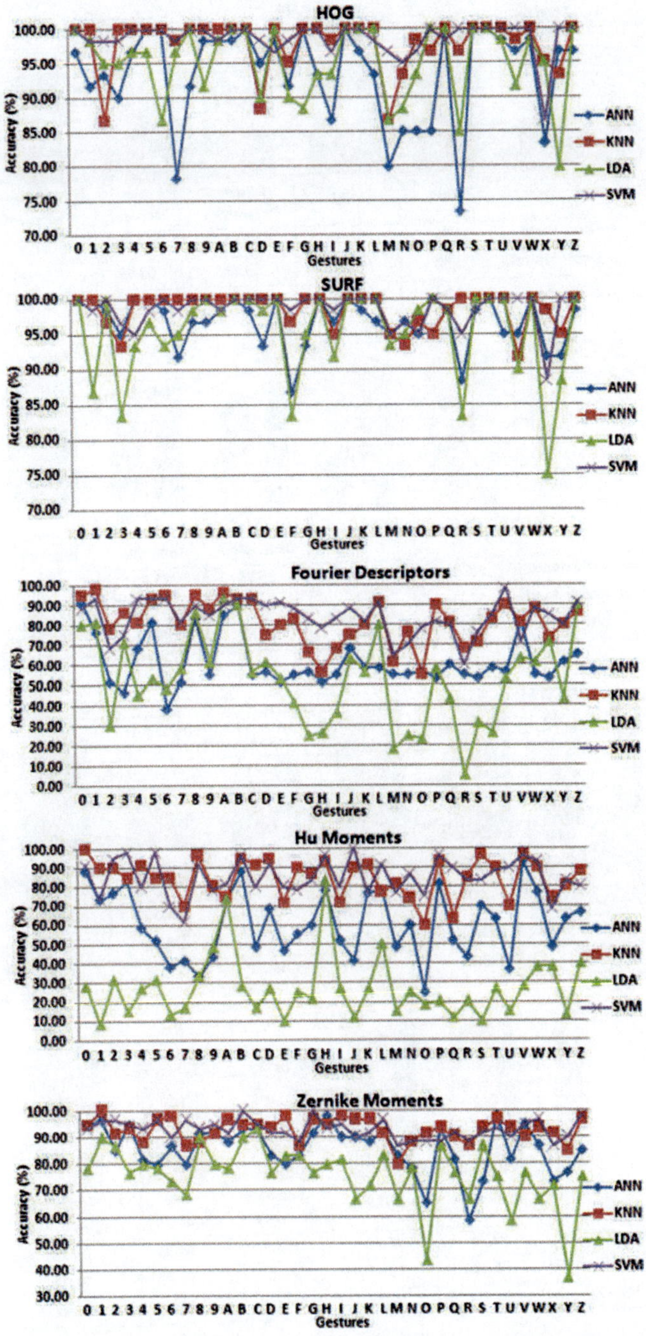

Fig. 8 Gesture-wise recognition rates for various features

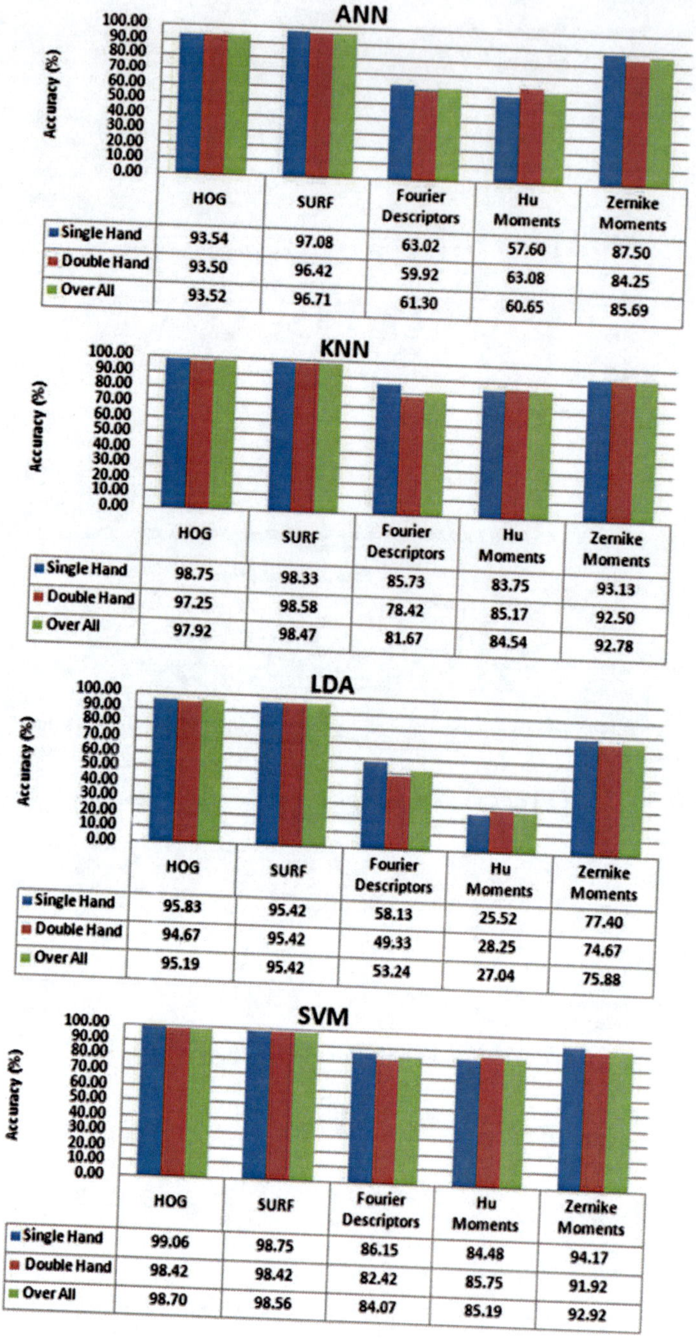

ANN

	HOG	SURF	Fourier Descriptors	Hu Moments	Zernike Moments
Single Hand	93.54	97.08	63.02	57.60	87.50
Double Hand	93.50	96.42	59.92	63.08	84.25
Over All	93.52	96.71	61.30	60.65	85.69

KNN

	HOG	SURF	Fourier Descriptors	Hu Moments	Zernike Moments
Single Hand	98.75	98.33	85.73	83.75	93.13
Double Hand	97.25	98.58	78.42	85.17	92.50
Over All	97.92	98.47	81.67	84.54	92.78

LDA

	HOG	SURF	Fourier Descriptors	Hu Moments	Zernike Moments
Single Hand	95.83	95.42	58.13	25.52	77.40
Double Hand	94.67	95.42	49.33	28.25	74.67
Over All	95.19	95.42	53.24	27.04	75.88

SVM

	HOG	SURF	Fourier Descriptors	Hu Moments	Zernike Moments
Single Hand	99.06	98.75	86.15	84.48	94.17
Double Hand	98.42	98.42	82.42	85.75	91.92
Over All	98.70	98.56	84.07	85.19	92.92

Fig. 9 Results obtained for classifiers using different feature extraction methods

Table 1 Computational time for different features

Sr. No.	Features	Computation time (seconds)
1	Histogram of Oriented Gradients	0.0015
2	Speeded Up Robust Features	0.0051
3	Fourier Descriptors	0.2251
4	Hu Moments	0.0017
5	Zernike Moments	0.5075

of these two phases includes various modules like preprocessing, feature extraction, and classification. Preprocessing consists of hand detection and noise removal. During feature extraction, dimension reduction of images is performed by extracting important image features. We have computed various features like HOG, SURF, Fourier Descriptors, Hu Moments, and Zernike Moments for images of our dataset. Using these features, we have trained four different classifiers like ANN, KNN, LDA, and SVM. From the experimental results, it is observed that the HOG and SURF provided better accuracy. Features such as SURF, Fourier Descriptors, Hu Moments, and Zernike Moments also provided invariance with respect to scale, translation, and rotation. Calculation of HOG features is very fast and efficient while that of Zernike is very slow. For our dataset, the system provided a recognition rate of 98.70% using HOG features and SVM classifier.

References

1. Badhe, P.C., Kulkarni, V.: Indian sign language translator using gesture recognition algorithm. In: IEEE International Conference on Computer Graphics, Vision and Information Security, pp. 195–200. IEEE, Bhubaneswar, India (2016)
2. Dixit, K., Jalal, A.S.: Automatic Indian sign language recognition system. In 3rd IEEE International Advance Computing Conference, pp. 883–887. IEEE, Ghaziabad, India (2013)
3. Gonzalez, R.C., Woods, R.E.: Digital Image Processing, 4th edn. Pearson Education (2018)
4. Forsyth, D.A., Ponce, J.: Computer Vision A Modern Approach, 2nd edn. Pearson Education (2015)
5. Singha, J., Das, K.: Recognition of Indian sign language in live video. Int. J. Comput. Appl. **70**(19), 17–22 (2013)
6. Kumar, M.: Conversion of sign language into text. Int. J. Appl. Eng. Res. **13**(9), 7154–7161 (2018)
7. Rokade, Y.I., Jadav, P.M.: Indian sign language recognition system. Int. J. Eng. Technol. **9**(3), 189–196 (2017)
8. Kaur, B., Joshi, G., Vig, R.: Indian sign language recognition using krawtchouk moment-based local features. The Imag. Sci. J. **65**(3), 171–179 (2017)
9. Raheja, J.L., Mishra, A., Chaudhary, A.: Indian sign language recognition using SVM. Pattern Recogn. Image Anal. **26**(2), 434–441 (2016)
10. Ansari, Z.A., Harit, G.: Nearest neighbour classification of Indian sign language gestures using kinect camera. Indian Acad. Sci. **41**(2), 161–182 (2016)

11. Chaudhary, D., Beevi, S.: Spotting and recognition of hand gesture for Indian sign language using skin segmentation with YCbCr snd HSV color models under different lighting conditions. Int. J. Innov. Adv. Comput. Sci. **6**(9), 426–435 (2017)
12. Adithya, Vinod, and Gopalakrishnan, U.: Artificial neural network based method for Indian sign language recognition. In: IEEE Conference on Information and Communication Technologies, pp.1080–1085. IEEE, Thuckalay, Tamil Nadu, India (2013).
13. Gupta, B., Shukla, P., Mittal, A.: K-nearest correlated neighbor classification for Indian sign language gesture recognition using feature fusion. In: International Conference on Computer Communication and Informatics, pp. 1–5. IEEE, Coimbatore, India (2016).
14. Kolkur, S., Kalbande, D., Shimpi, P., Bapat, C., Jatakia, J.: Human skin detection using RGB, HSV and YCbCr color models. Adv. Intell. Syst. Res. **137**, 324–332 (2017)
15. Yusuf, A., Mohamad, F., Sufyanu, Z.: Human face detection using skin color segmentation and watershed algorithm. Am. J. Artif. Intell. **1**(1), 29–35 (2017)
16. Dalal, N., Triggs, B.: Histograms of oriented gradients for human detection. In: IEEE Computer Society Conference on Computer Vision and Pattern Recognition, vol. 1, pp. 886–893. IEEE, San Diego, CA, USA (2005).
17. Bay. H, Tuytelaars T., Van Gool, L.: SURF: Speeded Up Robust Features. In: European Conference on Computer Vision, vol. 3951, pp 404–417. Springer, Heidelberg (2006).
18. Hu, M.: Visual pattern recognition by moment invariants. IRE transactions on Information Theory **8**(2), 179–187 (1962)
19. Zernike, F.: Beugungstheorie des Schneidenverfahrens und Seiner Verbesserten Form, der Phasenkontrastmethode. Physica **1**(8), 689–704 (1934)
20. Bhaskara Rao, P., Vara Prasad, D., Pavan Kumar, Ch.: Feature extraction using Zernike Moments. Int. J. Latest Trends Eng. Technol. **2**(2), 228–234 (2013)

Effectiveness of Software Metrics on Reliability for Safety Critical Real-Time Software

Shobha S. Prabhu and H. L. Shashirekha

Abstract Safety critical software is a key component of any critical system, and whenever there is a failure in this software, the system malfunctions with effect on safety of life or mission. Reliability is one of the quality factors and performance evaluator for this critical software. High reliability is expected of such software in its design, development and maintenance in order to increase the quality of the software for the system. Reliability metrics are derived for this kind of software during the planning phase, and enhancement of reliability metrics is achieved during the design and development phase. When the software is being designed, the reliability metrics are taken as the factors of foundation on which the software is built. In this paper, the software metrics which form the basis for proving their effectiveness on reliability for airborne engine control safety critical software are described. Development process based on these metrics not only influenced the reliability enhancement of the critical software but also improved the performance and efficiency of the embedded real-time system.

Keywords Safety critical software · Software reliability · Reliability metrics

1 Introduction

An embedded computer consisting of real-time software based on specific requirements with deterministic timing schedule constitutes the safety critical systems. Real-time systems are considered to be functioning precisely when the system output is logically correct along with meeting the strict timelines, regardless of the system load or adverse environmental conditions. Overall, these systems should behave in a predictable way irrespective of unpredictable external or internal stimulus. Embedded

S. S. Prabhu (✉)
Gas Turbine Research Establishment, Defense R&D Organisation, Bangalore, India

H. L. Shashirekha
Department of Computer Science, Mangalore University, Mangalore, India

H. Sharma et al. (eds.), *Congress on Intelligent Systems*, Advances in Intelligent Systems and Computing 1334, https://doi.org/10.1007/978-981-33-6981-8_56

software is developed for the functioning of these systems where the failures can critically affect the systems in terms of safety of life or mission. Examples of such software are present in the field of medical and aircraft systems, where defects directly cause loss of life/mission. This kind of software keeps the reliability as the key factor during various phases of software development life cycle such as planning, design, development and testing. Reliability, being an important quality factor of software, plays a vital role in the efficiency of these systems. Software reliability management talks about the process of reliability optimization through emphasis on software error prevention, fault detection/removal and use of measurements to maximise reliability irrespective of project constraints [1]. Metrics are a measurement of the level by which a system, component or process possesses a given characteristic/feature. Metrics help in getting a measurement of testability, maintainability, clarity, complexity, consistency, modularity and reliability. The goal of software metrics is to control and efficiently identify the essential parameters that affect software development [2]. In this paper, appropriate metrics are identified, studied and analysed to check their effectiveness on reliability for safety critical real-time software.

As the definition goes, software reliability is the probability of failure-free operation of software (in a system) over a specified time within a specific environment for a specified purpose. According to the above statement, reliability highlights the availability of the system with continued performance of the software with respect to functionality along with external conditions with which it is operated during the complete operational period. This shows that the reliability depends on various measurable quality factors, which could be improved during any/all the stages of software development life cycle to obtain a better performance, functionality and safety [3]. Application of quality attributes at all phases of the development life cycle highlighting error prevention, specifically in the early life cycle, is the foundation of building high reliability software. Metrics are needed at each development phase to measure applicable quality attributes [1], which reflect the characteristics of reliability. The safety critical software is expected to have proactive fault tolerance and correction of identified design errors through the stringent development process, while keeping strict vigil on the quality/reliability metrics.

The glimpse of this paper is: Section 2 explains the relevance of reliability and metrics considered for the safety critical software for the case study. Section 3 highlights the motivation obtained as an outcome of the survey carried out on the similar work. Section 4 presents the strategies followed in implementing the above concepts during the design and development of the software life cycle. Section 5 highlights the effectiveness of metrics on software reliability improvement and thereby on the software quality improvement. Section 6 concludes the methodologies with future scope for expansion of this work.

2 Current Work

Case under study is safety critical software developed for embedded real-time controller (SACSERC) system which provides engine control and monitoring of propulsion system over the complete flight envelope of the unmanned aerial vehicle [4]. This embedded software features the complete aero-engine control functionality along with efficient performance, safety and reliability. Unique features of this software include—execution of the expected functionality as per system requirements, hard real-time behaviour with scheduling methodologies for deterministic timing requirements, optimally well-planned software design strategies, complete traceability from system requirements to embedded object code, etc. Software of this system being safety critical in nature, is expected to have mean time to fail (MTTF) of 1 Lakh hours (continuously running for more than 10 years). This will lead to a failure once in 10^5 hours with a reliability score of more than 0.9. SACSERC being an embedded system, its reliability depends on the reliability of both software and hardware. As the software is being developed for the safety critical system, it is necessary to embed enormous safety and quality features in it. This enforces the software to implement safety and emergency features alongside normal functioning to take care of abnormal events with high reliability. Hence, airborne DO-178C process with Level A (highest level of safety) along with its in-built stringent guidelines is followed in achieving better reliability by minimising faults/failures [5]. The software reliability attributes are measured, so that they can be monitored and enhanced if required. Until now, there does not exist a single good method of measuring software reliability for software with varied scope and nature. Sufficient experimental studies are carried out on the SACSERC software to identify the metrics, prove their effectiveness on software reliability and enhancing the reliability by modifying the software through moderating the metrics is discussed in this paper.

Software failures are one of the most dominant causes of failures in today's critical systems. System reliability directly depends on the software reliability and in turn on failure-free functioning of the software. Failures are caused by faults which occur during different software life cycle phases, and it is required to eliminate these faults to build a reasonably reliable system. In reality, even though it is not possible to eliminate all the faults in the software, usage of sound software engineering process helps in producing better software in terms of reliability. The top-level metrics which are considered for the case study are mean time to fail (MTTF) and mean time to recover (MTTR) which are directly linked with the failure density and recovery time. Failure-related incidents and their recovery duration need to be minimised in order to increase the MTTF and decrease the MTTR. Identifying potential areas of problems and rectifying them at early phases prevents ripple effects at later stages and increases MTTF [4]. Hence, the metrics which affect the factors related to MTTF and MTTR are studied for their impact, and attempt has been made to moderate them. These metrics are identified through the measurement of structural complexity, interrelationship among the software components and transparency of software modules in the case study. Following are the metrics considered:

- ***Cyclomatic complexity***—A metrics which gives sneak peek into the design aspect of the software which has indirect influence on testing.
- ***Coupling***—A metrics which measures modularity as well as dependency factors of software modules.
- ***Percentage of comments***—A derived metrics which is a measure of clarity and dependent on lines of code (LOC) (complete coded lines along with comments excluding the blank lines).

Overall, the intention of this study is to enhance the reliability of the software by moderating these metrics which contribute to MTTF and MTTR, through restructuring the software architecture during software development life cycle.

3 Survey on Related Work

Even though research goes on continuously in the field of software reliability, till now there is no conclusion about a common methodology using which improvement can be done in reliability for all types of software. It is specific to each type of software and is the responsibility of the designers or developers to find a suitable methodology. In this context, the safety critical software which is developed for the airborne system demands reliability enhancement as a continuous effort to establish improvements in the quality. A number of literatures have been studied to derive suitable methodology for the work presented here.

To improve the quality and reliability of products, projects depend on software metrics for identification of critical areas where problems or failures may occur. Identification of these metrics which help in detection and correction of requirement faults are utmost important for prevention of errors at later stages of the life cycle is opined by Rosenberg [1]. Lockhart et al. [6] highlight the fact that consistent and concise system specification removes design errors prior to implementation which is crucial for developing reliable software systems for safety critical applications. These techniques also include static analysis for the removal of development errors before the release of the software in order to build better reliability into the critical embedded systems.

Most important and sought-after characteristic of software that is accurate and trustworthy is its reliability. Hence, the evaluation of software engineering metrics with deterministic quantitative model of failure patterns for reliability enhancement is proposed in [2]. Krasich [7] presents the methodology to improve reliability through early planning and assessment during the software development process. It highlights the fact that faults which have a high impact on reliability are discovered early in design, analysis and test, so that reliability improvement can be attempted with identification and correction of potential faults at every phase. Certification standard for avionics software, which provides guidance and data concerning verification-based software life cycle processes, is the essence given by Moy et al. [8]. Fang et al. [9] have brought out the necessity of achieving higher reliability by reducing faults through keeping eye on the metrics of software product, especially in the case of real-time safety critical software.

In a nut shell, the literature survey has revealed the impact of metrics and its relevance to reducing failures in the study of reliability of safety critical real-time software. In the context of airborne software, several areas were focused in terms of evaluating the improvement on reliability in order to reduce the MTTR and increase the MTTF. Keeping in view of the metrics at different software phases, removal of remotely occurring faults becomes prime important in increasing the reliability.

This paper highlights the software metrics for SACSERC, their effectiveness and the strategies adopted in order to enhance the reliability of the safety critical embedded software.

4 Methodology

The following paragraphs detail the methodology adopted for the above-mentioned strategies in the context of SACSERC for avionics application.

Working in tandem with DO-178C based software lifecycle processes, the quality and reliability are considered as basic non-functional characteristics in the case study. During the development of software, importance is given to reliable, traceable and error-free code, by which one can ensure better quality of software. It is necessary to measure and monitor the metrics, so that impact of the same on reliability could be studied. At the same time, these metrics provide the developer with certain details about the structure of software architecture, clarity and testability of modules, which are useful during maintenance period. During the development phase, it is also mandatory to expose the software to rigorous static and dynamic analyses, which in turn help in exposing the errors/faults (which otherwise may escape testing techniques) prior to releasing the object code. A combination of these analyses, alongside keeping a watch on the software metrics, is a better option while focusing on the quality attributes [10] such as reliability, maintainability, testability and efficiency.

4.1 Cyclomatic Complexity

It is generally accepted that more complex modules are more difficult to understand and have a higher probability of defects than less complex modules. Thus, complexity has a direct impact on overall quality specifically on maintainability [1]. Cyclomatic complexity is a metrics which is used to indicate the complexity of a program by directly measuring the number of linearly independent paths through a program's source code. It is a primary measure of the soundness of the module, which also is a strong indicator of testing effort. It is considered as a good testability metrics because it determines the number of test cases needed to test all paths in a software module (test coverage). The high complexity programs contain more errors, and detecting them is more difficult through testing [11]. Hence, developers who would like to simplify testing often re-write the code for reducing complexity. This also may affect the higher coupling factor. The developer has to work out a suitable and optimum software architecture for reducing complexity as well as coupling together to maintain good cohesion.

The SACSERC was developed as per the software design and coding standard specifically prepared for this software. The value of cyclomatic complexity stated in these standards was as low as 9 initially. During the software design phase, it was found that 9 was too less to accommodate few of the important and complex control function requirements. If the cyclomatic complexity number is increased to 15, it was logically difficult to handle too many test cases to complete the 100% test coverage during normal testing or regression testing. It is also observed that the maintenance becomes bit difficult with more complex module. There needed a trade-off between the complexity and all other dependant factors in order to have the increased reliability. The restructuring of software architecture was carried out to optimise the complexity, maintenance cost/time and testing/verification efforts.

Figure 1 depicts the effect of cyclomatic complexity on testability and maintainability metrics which in turn affect the reliability. Exercises are carried out by moderating different software architectures to finalise the cyclomatic complexity. It is observed from the figure that cyclomatic complexity of 9 (initial) has resulted in a higher testing effort and MTTR. To ease the situation, it was decided to go for cyclomatic complexity of 15. This again complicated the testing as well as MTTR as evident from Fig. 1. With several experiments, the cyclomatic complexity was finalised to 12, which gave the minimum testing effort (of 22 h) and least MTTR (of 120 min) during a demand of minor change in the software.

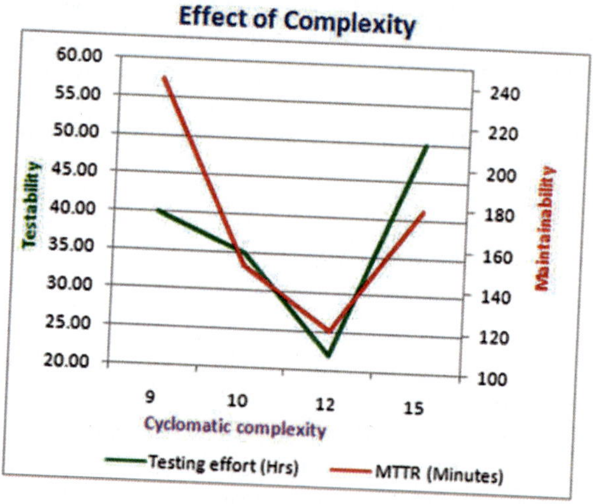

Fig. 1 Effect of complexity on reliability factors

4.2 Coupling

Coupling is the degree of interdependence between software modules or the strength of relationships between modules and represented as a measurement of modularity. Low coupling and high cohesion are often a sign of a well-structured software product with a good design and support the general goals of high readability and maintainability. In order to have the complete structural coverage, two types of couplings are discussed here: *Data Coupling and Control Coupling*. Both these coupling studies are intended to identify any code structure that was not exercised during the requirements-based testing [12]. In DO 178C process, data and control coupling objectives are to provide a measurement and assurance of the correctness of software components, interactions and dependencies between modules [13]. They can indicate issues during integration tests such as data loss across an interface, components adversely affecting each other, sub-modules not contributing to desired functionality and problems faced with global data against the intended/designed requirements.

Data coupling represents the dependence of a software module on data not exclusively under the control of that software component. Control coupling represents the manner or degree by which one software module influences the execution of another software component [12]. Extensive exercises have been carried out to decide on the coupling percentages, specifically keeping reliability in view during requirement-based testing and coverage analysis of the software. For complex software like SACSERC, deciding on the upper limit for coupling study is a major task as its impact is directly felt on the subsequent activities in the software life cycle such as review, analysis, testing and maintenance. It has been decided to have an upper limit of 20% for data coupling and 15% for control coupling to check if the data/modules are loosely or tightly coupled. These upper limits are finalised on the basis of low software maintenance, high cohesion among modules, low testing effort and low MTTR.

In SACSERC, there are 614 modules and 205 variables. The architecture of the software is designed in such a way that modules need to be loosely coupled with high cohesion of individual modules. Data or control coupling are categorised as given below:

(a) **Loosely Coupled Data (or Control)**: If by changing a specific variable (or module), the number of modules getting affected is less than 20% (or 15%) of the total number of modules, then the specific data (or control) is said to be loosely coupled.
(b) **Tightly Coupled Data (or Control)**: If by changing a specific variable (or module), the number of modules getting affected is more than 20% (or 15%) of the total number of modules, then the specific data (or control) is said to be tightly coupled.

Data coupling categories for all 205 variables for SACSERC software are worked out, and the result is shown in Fig. 2. From this, it can be noticed that 194 variables (data) (95%) are having coupling percentage of less than 10 and coupling percentage

Fig. 2 Effect of data coupling on reliability factors

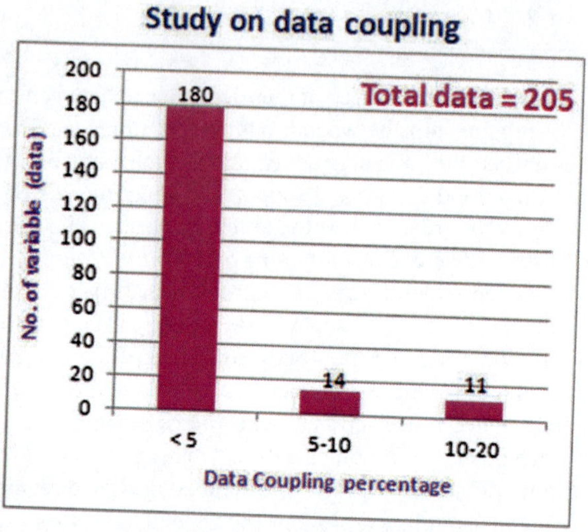

of 10–20 is observed in 11 variables (data) (5%). All variables are within the upper limit of data coupling percentage which is 20%. This shows that the dependency of modules on data is well within the limit and changes to the software can be handled with a low MTTR and low testing effort.

Control coupling categories for each of the 614 modules for this software are worked out, and the result is shown in Fig. 3. From this, it can be noticed that 589 modules (96%) are having coupling percentage of less than 10 and 25 modules (4%) are having coupling percentage of 10–15. All modules are within the upper limit of control coupling percentage which is 15%. This shows that the dependency of

Fig. 3 Effect of control coupling on reliability factors

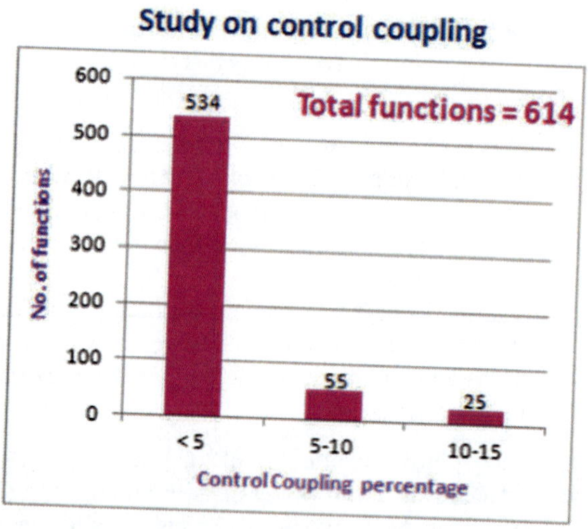

modules on other modules is well within the upper limit and changes to the software can be handled with a low MTTR and low testing effort.

4.3 Percentage of Comments

This is a measure of clarity which gives the details of each module. SACSERC is developed in C programming language as it is flexible, mature and easily portable. During the analysis of SACSERC, it was found that few metrics are dependent on lines of code (LOC). Analysis is carried out on those metrics which are keeping LOC as their prime factor and showing their efficacy on the software reliability indirectly. One such metrics is percentage of comments for a module. This is computed with the ratio of number of lines of comments to LOC of a module and converted into percentage. Even though there are many definitions of LOC [11], coded lines along with comments excluding the blank lines are considered as the LOC for this case study. Comments are written for declaration part as well as code part of the module separately to give transparency among development team members. Comments are useful in deriving the test cases at the time of verification and clarity during maintenance. The comments are bridging the gap between the developer of the module and other personnel involved in various stages of software development life cycle. The comments in a module play a vital role during structural coverage activity (intends to identify any code structure that was not exercised during the testing) and further till the active life of the software.

During the initial stages of implementation of SACSERC, minimum comments (5%) were introduced for all modules as it is assumed that detailed software design gives sufficient clarity. During the verification stage, confusion aroused during the testing and analysis phases due to insufficient and minimum comments. This affected the cost, time and development effort of the project. Hence, detailed and explanatory comments were introduced to eliminate any miscommunication among personnel involved in different phases of development. Further, lower (5%) and upper (40%) limits were introduced for the comments in order to control the optimum and necessary details only to check the completeness and correctness of implementation. Comments played a significant role during the static and dynamic analyses in eliminating hidden errors, in turn increasing MTTF. Detailed comments also helped in reduction of time during the maintenance phase which in turn reduced the MTTR. As code coverage (degree to which the source code has been tested) of 100% is mandatory for safety critical airborne software, optimum comments for a module helped in achieving structural coverage analysis in a shorter duration.

We have a total of 614 modules as seen from Sect. 4.2. Among them, 12% of modules (73) have below 5% of comments, 57% (350) have 5–20% of comments and 31% (191) have 20–40% of comments. To arrive at this final decision on the percentage of comments, several aspects were considered. Modules with a complex logic may require more comments, and short modules with simple logic may not need much description. Some modules are clearly explained in the design phase,

Fig. 4 Effect of clarity on reliability factors

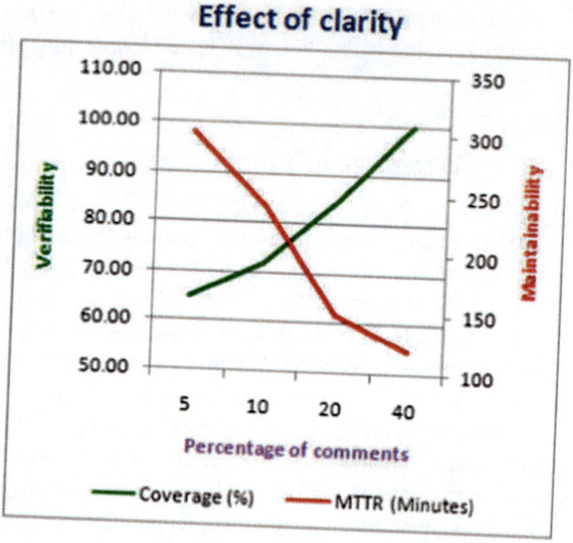

and hence, they may not need more description. All these aspects contributed for the study to decide on the amount of comments needed for each of the modules. Figure 4 depicts the effect of this percentage of comments on verification and maintenance metrics. It is also observed that as the percentage of comments optimised from 5 to a range of 5–40, coverage increased from 65 to 100%. Meanwhile the MTTR reduced from 300 to 120 min. It is derived that when the clarity is better, the testing becomes easy and it is possible to get the test coverage easily. Similarly, right percentage of comments help in achieving the software maintenance in a shorter time when there is complete clarity. This shows that the percentage of comments is having direct effect on verifiability as well as maintenance and hence contributed immensely in enhancing reliability of the said software.

5 Comparative Study

When the reliability considerations are not part of software development process, the metrics are computed for record of statistics rather than utilising them for software quality improvement. Hence, the usage of metrics is never highlighted. Since reliability is part of DO-178C process, which is followed in the current case study, measurements of critical features through metrics were made use for this study. The reliability enhancements were indirectly evaluated by measuring, monitoring and improving these metrics during the development life cycle of this critical software. The above paragraphs show the efforts made in exhibiting the effectiveness of software metrics on reliability for safety critical software. Metrics such as percentage of comments, cyclomatic complexity and coupling were studied and moderated to achieve the improvement in reliability (and also quality) through reduction of MTTR and increase in MTTF. The numbers which are moderated for each of these metrics may not be suitable for all types of software. In particular, even the safety critical software for different applications may not adapt to the same metric values. Normalisation of these metrics and their values is not uniform for different varieties of software.

6 Conclusions and Future Work

Reliability is an indicator of quality and performance of safety critical real-time software for an airborne system and is enhanced through some of the software metrics. In this paper, the metrics which are affecting the reliability are studied thoroughly and moderated (either increased or decreased) in order to enhance the reliability of the system significantly. Metrics such as percentage of comments has been considered as a strong clarity metrics which is useful in all phases of development life cycle. Cyclomatic complexity and coupling, being indicators of testability, complexity, modularity, and maintainability are also considered for improvement. Attempts were made to improve these metrics and show their effect on MTTR and MTTF of the software. This paper emphasises enabling improved safety and reliability into the software by means of study on few affecting metrics and observed their effectiveness.

As a future work, process-related metrics such as percentage of bidirectional traceability with object code, compliance to coding standards, absence of dead code, and existence of run time errors may be explored and analysed to bring the quality and reliability features of the airborne real-time embedded safety critical software to a improved/higher level. Feasibility study may be attempted to normalise the metrics for a particular type of software, if not for general embedded safety critical software.

Acknowledgements The authors are thankful to Director, Gas Turbine Research Establishment, Defense R&D Organisation, who permitted them to publish this paper. The authors thank all the personnel who are associated with this work.

References

1. Rosenberg, L., Hammer, T., Shaw, J.: Software metrics and reliability. In: IEEE International Symposium on Software Reliability Engineering (1998)
2. Kashyap, S., Tripathi, A., Sharma, K.: Analysis and ranking of software engineering metrics. In: 2nd International IEEE Conference on Computing for Sustainable Global Development (INDIACom), pp. 1654–1659. INSPEC Accession Number: 15109983 (2015).
3. Roger Pressman S.: Software Engineering a Practitioners Approach, 7th edition, The Mc-Graw Hill Publications (2011).

4. Prabhu, S.S., Kapil, H., Shashirekha, H.L.: Safety critical embedded software: significance and approach to reliability. In: 7th International Conference on Advances in Computing, Communication and Informatics, ICACCI-2018, pp. 449–455. https://doi.org/10.1109/ICACCI.2018.8554566, IEEE Xplore (2018).
5. Guidelines, Standard for: Software Considerations in Airborne Systems and Equipment Certification. RTCA DO-178B/C, Radio Technical Commission for Aeronautics (2011).
6. Lockhart, J., Purdy, C., Wilsey, P.: Formal methods for safety critical system specification. In: IEEE 57th International Midwest Symposium on Circuits and Systems (MWSCAS), pp. 201–204. https://doi.org/10.1109/MWSCAS.2014.6908387 (2014).
7. Krasich, M.: Modeling of SW reliability in early design with planning and measurement of its reliability growth. In: IEEE Conference on Reliability and Maintainability Symposium (RAMS). https://doi.org/10.1109/RAMS.2015.7105066, INSPEC Accession Number: 15112765 IEEE Xplore (2015).
8. Moy, Y., Ledinot, E., Delseny, H., Wiels, V., Monate, B.: Testing or formal verification: DO-178C alternatives and industrial experience. IEEE Softw. **30**, 50–57. https://doi.org/10.1109/MS.2013.43 (2013).
9. Fang, Q., Zhang, C., Ye, X., Shi, J., Zhang, X.: A new approach for developing safety critical software in automotive Industry. In: 5th IEEE International Conference on Software Engineering and Service Science (ICSESS), pp. 64–69. https://doi.org/10.1109/ICSESS.2014.6933515, INSPEC Accession Number: 14698700 (2014).
10. An article by Amir Ghahrai https://devqa.io/static-analysis-vs-dynamic-analysis-software-testing/. Last accessed on 8th January 2017.
11. Tashtoush, Y., Al-Maolegi, M., Arkok, B.: The correlation among software complexity metrics with case study. Int. J. Adv. Comput. Res. pp. 414–419. (ISSN (print): 2249-7277 ISSN (online): 2277-7970)Volume-4 Number-2 Issue-15 (2014).
12. Certification Authorities Software Team (CAST): Position Paper CAST-19 (Rev 2) Clarification of Structural Coverage Analyses of Data Coupling and Control Coupling. Federal Aviation Administration, USA (2004).
13. Maia, T., Souza, M.: A Practical Methodology for DO-178C Data and Control Coupling Objective Compliance. In: International Conference on Software Engineering Research and Practice, SERP'18, pp. 236–240. CSREA Press, ISBN: 1-60132-489-8 (2018).

Text Clustering Techniques for Voice of Customer Analysis

Zaheeruddin Ahmed, Sonu Mittal, and Harvir Singh

Abstract A comprehensive analysis of customers is very vital for any organization for customer enlightenment. However, conventional methods are not effective in analyzing customers concerns particularly when it is online. Data generated online is extremely complicated because of its unstructured behavior, and the data set in this format is often complex and very difficult to evaluate its overall model. Further, the application of conventional methods remains monotonous as many emerging techniques are proposed by the researchers, but the most effective approach will remain thought-provoking. In this paper, we introduce effective methods to analyze customer concerns with a computational approach that perfectly apply text clustering techniques. An effective clustering technique can be applied to a large set of unstructured data since online data often contain huge amount of noise, and we will discuss different text clustering techniques to evaluate and analyze the method of text extraction from customer statements mainly on Web and social media for customer analysis.

Keywords Text clustering · Customer analysis · Unstructured data

1 Introduction

Data analysis for voice of customer (VoC) is increasing the business value by bringing the perceptions of companies that need to improve their product and service practices. According to [1], VoC is the declaration or review made by customers on a product or service for their expectations, preferences or comments in any conversation. Therefore, a VoC analytics is a designed system of data collection, analysis, and action plan. In this process, organizations need to sketch the value of business

Z. Ahmed (✉)
Jaipur National University, Jaipur, India
e-mail: zaheeruddin@manipaldubai.com

S. Mittal · H. Singh
Engineering and Computer Sciences, Jaipur National University, Jaipur, India
e-mail: director.soet@jnujaipur.ac.in

H. Sharma et al. (eds.), *Congress on Intelligent Systems*, Advances in Intelligent Systems and Computing 1334, https://doi.org/10.1007/978-981-33-6981-8_57

with upright customer experience. A notion of customer feedback and reviews using text analytics will give business its insights to improve its value. This entire process requires to build a multichannel VoC analytics. Furthermore, the main objective of VoC analytics must be to gain better understanding of how customers distinguish products and services. Figure 1 shows steps involved in VoC analytics process.

The effective VoC analytics process creates value for business, but the analysis must produce accurate and meaningful insight [2]. First the process begins with identifying the questions, we want to answer, and this will return or gather data that we need to analyze. Once the data is gathered, then suitable analysis tools will be selected to draw conclusions and act. Some of the common data sources for VoC analysis could be online surveys or polls, social media comments or tweets, chat conversations, tickets or email reviews. All these sources are Web-based application from social media, polls, survey applications, or tweets which return unstructured form of data. The biggest challenge for any analysis is to deal with this form of data with complex data types.

In recent years, social media has become touchpoint for customers and business where both engage to exchange information. In this paper, we demonstrate that VoC analysis might be useful to perform effectively on data generated by customer interaction on polls. Customers will send feedback and reviews over poll application. The polls form a wealth of information for business to analyze. Considering the volume of unstructured data generated with the scale of it goes way beyond to analyze without effective mining techniques. This is where text analytic techniques can empower to build an effective VoC analytics that create value for any business.

Fig. 1 VoC process

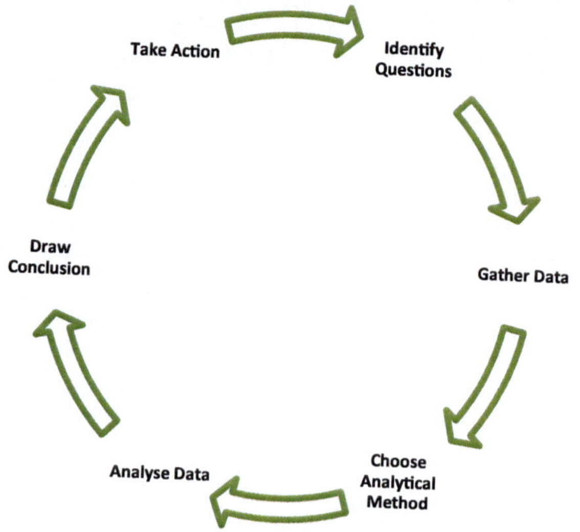

2 Clustering

2.1 *A Clustering Process*

The study of clustering is widely acknowledged in data mining because of its various applications, mainly in the field of data analysis. This process can be considered as the summarized model of data and interpreted in the procreative prototype. As per the definition, a cluster is a group of objects placed closely together, whereas clustering is a process of grouping a set of objects in a way that objects in the same group are more comparable to each other than in different groups. Clustering is one of the vital mechanisms of experimental data mining, and it organizes data into similar groups that shows internal structure of the data very useful in knowledge discovery [3].

2.2 *Clustering Application*

There are many applications in which clustering techniques are applied like data analysis, marketing research, and image processing but not restricted, and these applications address many problems that can be solved using different clustering methods [3]. Our research is to find the best technique to extract information from unstructured content using data mining.

Clustering is one such method for effectively analyzing meaningful data. Say for instance in data mining, clustering can be applied in classification and deviance analysis where a summary of precise data is helpful for understanding. Clustering can also be useful in the summarization of compatible user groups, and in this case, the parameters or ratings provided by the users can be used with collaborative filtering techniques for references. Another similar application of clustering analysis is in business customer segment which is like collaborative filtering produces a similar group based on arbitrary elements about the customers in the data. Dimensionality reduction method is one of the data summarization techniques which is used in many applications like linear discriminant analysis and machine learning, and this technique can be considered as clustering summarization [3–5].

Pattern recognition is one of the widely used common techniques for data analysis. In social media applications, clustering analysis can be used to perform trend detection that can be clustered using streaming data to find dynamic patterns of changes. Further in social networking analysis, clustering is applied to determine the groups in the primary network, this helps in understanding the structure of a community network. Another interesting area of clustering is biological data analysis which is the study of human genome for collecting human gene data which is sequential structured data that clustering can perform gathering of similar patterns [3–5].

In this paper, we will highlight text clustering for data analysis. In unstructured data, the purpose of similar segments has many applications. The analysis of such complex data can be multimodal which is one of the growing challenges in its classification and segmentation.

2.3 Clustering Areas

Data clustering areas are mainly of three categories: (1) technique, (2) data type, and (3) variations as shown in Fig. 2.

In technique category, various methods like density-based, probabilistic, distance-based, spectral, and dimensionality reduction-based techniques are used and may work in different scenarios. In addition to this different kind of data types, unstructured data has challenges and requires specialized techniques, in case of different applications, generating different data types with different attributes. For example, Web data on social media producing different formats of data needs data type technique of clustering process. If the data produces by the application which has separation between different kinds of properties like behavior, a clustering variation technique will be most suitable for data analysis. Further, checking the performance of clustering a through cluster validation is required with detailed insights.

Fig. 2 VoC proceubsection sample

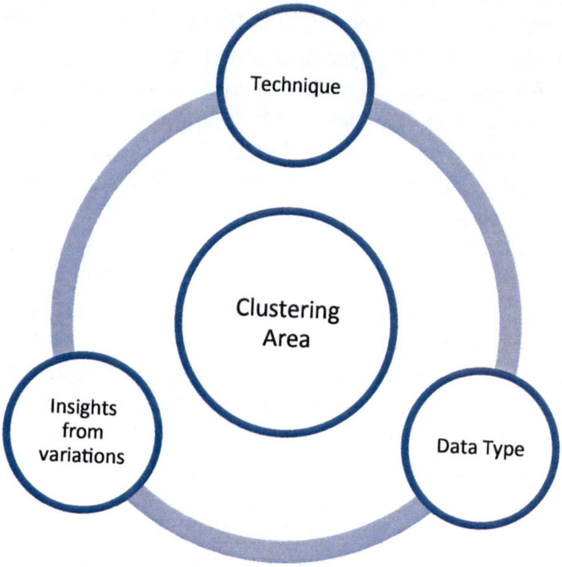

2.4 *Effect of Data Types in Clustering Analysis*

The data type will have a great impact on clustering analysis. Data now has become big data with different types, and it can be drawn from any type such as distinct, unstructured, or structured. First, we will discuss different types of data and the common clustering process. One of the most common types of data is real data set where data is characteristically distinct. This type of data sets leads to many challenges for clustering. In this case, the common clustering technique such as *k*-means are applied for analysis. Another form of data is text which is very widespread in Web and represented in a vector space format. The technique for clustering text data can also be challenging as text data has several properties that is high dimensional and sparse. In text data clustering, the common methods are scatter–gather, *K*-means, and agglomerative [3].

Time-series data is common in systems like sensor data, stock market data, or business forecasting data. Time-series data values are dependent on one another. Clustering in this type of data depend upon correlation-based online analysis or shape-based offline analysis. Some data on the Web creates discrete sequences like logs data or biological data. The contextual property of such data matches to the placement instead of time. There are numerous challenges in analyzing such data objects. Clustering analysis takes different approaches like suffix trees for sequence clustering but not restricted. Network data is one of the general forms of data representation. Many such data types can be represented as a graph, with its values on the edges. Thus, clustering graph data becomes another challenge [3–5].

2.5 *Cluster Methods*

The most common clustering methods are feature selection methods, probabilistic models, distance-based algorithms, density and grid-based methods, leveraging dimensionality reduction methods, matrix factorization and co-clustering, spectral methods, the high-dimensional scenario, scalable techniques, streaming algorithms, and the big data framework data [3]. In this research, we opted to apply selection method. In this method, the process will cluster set of objects based on similarities. Then, divide objects into clusters according to data similarity with which better clustering is achieved. Presumably assuming that between-clusters similarity is minimized, and the within-cluster similarity is maximized [6].

3 Text Clustering

Information retrieval is one of the growing areas of research that involves text analysis. Clustering is one technique that support in text analysis. Clustering of text is the application of analysis to documents based on structured or unstructured data. The process of clustering of text involves many techniques like machine learning or deep learning to characterize and understand meaningful data. In the process of text extraction or clustering, the set of text are extracted from any document to analyze the occurrence of words found in the document comparing other terms, and then the clusters of words can be identified [7, 8].

Text clustering algorithms is applied on group of words to break down unstructured data from the documents generated from resources like Web social media or blogs and turn it into a matrix model to tag relevant data (text clustering]. The process of text clustering can be used in variety of business needs and setting. The application of text clustering could also be for product development, customer support, customer segmentation, sentiment analysis, opining analysis, intension analysis, voice of employee, document analysis, voice of customers, financial industry, sentiment analysis, document structuring analysis, and corporate reputation. Another useful application of clustering text could be marketing copy for understanding the advertising performance, and this type of application creates unstructured data which is where text clustering can be very useful. IT can analyze the topics and words with the highest rates enabling us to create relevant copy of data [7, 8].

4 Related Work

In our literature review and related work conducted by other research, we will discuss some of the key areas of text analysis for VoC. In our research study, we found five related methods that the present research is conducted for VoC application: (1) Attention model of deep learning, (2) Tone base matrix based on online conversations, (3) User reviews extraction for sentiment analysis, (4) Measurement method, (5) Naïve Bayes method.

The authors in [9] describe attention method and highlights the critical issues related to customer reviews and proposed a solution that can retrieve perceptions from customer reviews significant in identifying key customer statements for rating products online. This technique according to authors can help in identifying key concerns responsible for poor product rating when applied with deep learning attention method. In another experiment by authors in [10], introduce tone-based metrics resulting from online customer conversations to quantify their concerns. This approach flawlessly incorporates specific tone analysis with features to enable multi-faceted conversation analysis.

The authors in [11] presented a methodology for justifying recommendations that relies on the information extracted from user reviews. The idea behind this approach is to apprehend the justification as a summary of the most relevant aspects, obtained by analyzing its reviews. Authors proposed an idea designed as framework including aspect extraction, sentiment analysis, and text summarization to gather reviews.

The study of authors in [12] suggested measurement method on semantic similarity being characterized in sentences learnt by semantic labeling. This method was assessed by comparing with the method of substring similarity for similarity measurement for existing documents. The results in this process identify the proposed method performed similar with the conventional method for the plagiarized documents which were rarely modified, whereas it had improved results for paraphrasing sentences which were changed in structure.

Authors in [13] proposed a naïve Bayes method to analyses biological material information. Accordingly, the proposed method finds the scientific literature of the feature is text, and then the naive Bayesian method is used to classify the same literature into different topic types. This process was assessed through the model test standard in data mining to verify the validity.

5 Experimental Analysis of VOC Over Poll

CollPoll, a Web and mobile-based application for campus management platform, is to address the growing complexity and competition. The application supports in campus polls for management of curriculum, timetable management, learning management system, placement management, campus workflow digitization, fees management, and infrastructure management [14]. In the process of VoC analysis at our university for hostel students concerns, we applied CollPoll application. We followed the following procedure for our experiment as shown in Fig. 3.

Gather: To gather data, we prepare questioner on the profiles created for CollPoll application. In this process, we have created poll for student concerns for various academic issues. The entire set of survey questions were upload in our CollPoll platform, see Fig. 4.

Process and analyze data: The next step was to configure and run analysis and then study what students' concerns are, what are they talking about, how they feel about our services and why they feel so that way. Here is where we applied clustering mechanism for text to retrieve student's intention. Various business intelligent tools like Excel and Tableau are used to create stories and dashboards to analyze students' issues.

Use of attention mechanism: Deep learning as part of artificial intelligence emulates the mechanisms of human brain in processing data and helps in creating patterns for decision making. This technique learns unsupervised way from unstructured or semi-structured data for instance multimedia text file or an image. One of the applications of attention mechanism in deep learning is to classify the key sentences within a review [9], and this mechanism is also known as attention model based on

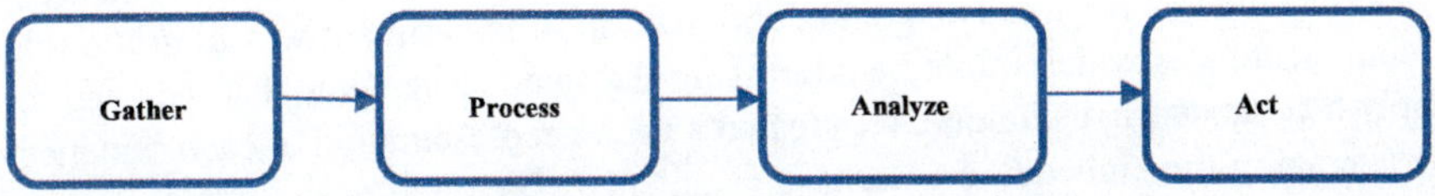

Fig. 3 VoC voice of customer analysis process

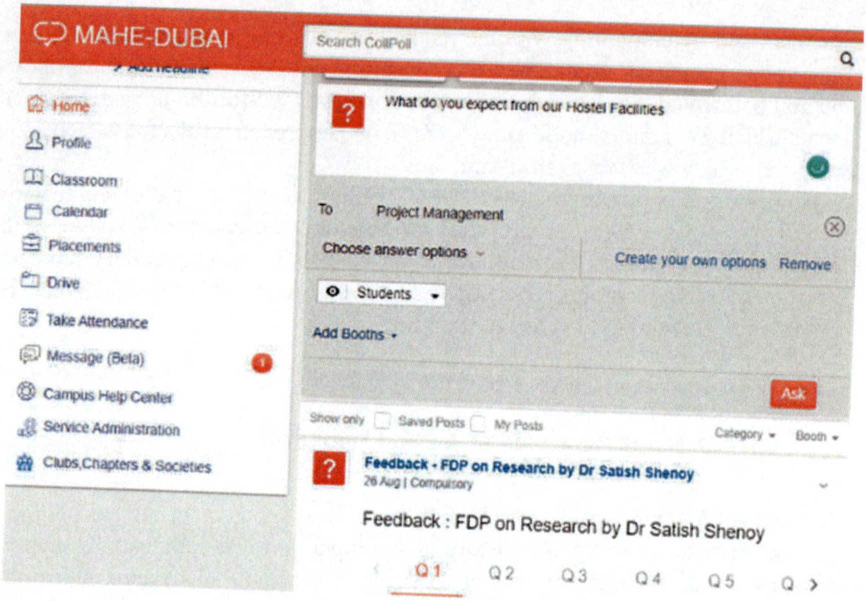

Fig. 4 Collpoll survey application

the visual attention by concentrating on certain parts of the text sentence whereas paying lesser attention to the remaining text.

The mechanism of deep learning attention model helps in improvising task of machine transformation by allowing the interpreter to choose the parts of a sentence to pay attention. In general, this mechanism uses deep learning techniques to identify parts of the input text which are significant in determining the outcome. In our experiment to hostel student reviews, this mechanism was used to identify key phrases within student concern analysis that has more influence in determining the IT service rating given by the hostel students.

One example of such key phrase identification in attention mechanism is shown in Fig. 5. The blue color sentence indicates that got attention.

6 Text Clustering After Attention Mechanism

Once the imperative sentences retrieved after the application of attention model, the next step was to perform text clustering on all the important retrieved sentences made by students review. The goal of this section of experiment was to group the similar sentences which might be addressing the same issue. Now that we need to apply text clustering technique, we created a vector representation of each sentence by averaging the embedding vectors in a sentence. Then, the most popular cluster

> Range is not enough. At night internet doesn't work sometimes The Wi-Fi isn't consistent, A lot of times the Wi-Fi simply stops working. Internet goes off almost an hour's in the night Sometimes the wi-fi just disconnects, mostly at night It sometimes doesn't work The internet only connects in few spots of the room The wi-fi has error every 3/4 day, Wi-fi goes off in between during nights, The internet gets really slow during the night, speed is very slow, Wi-Fi connectivity in the room keeps differing from place to place. And doesn't work at all in the corridors. The internet goes off sometimes by itself. Sometimes the net is slow, the wi-fi keeps disconnecting on my phone at night, the wi-fi keeps disconnecting on my phone at night, Sometimes the connection gets disconnected only from one device..., The internet gets disconnected automatically at night, and takes a while to reconnect. It's connected but no internet several times, sometimes there is internet connectivity issue at nighttime. Random disconnects at night, Speed in the night is very slow

Fig. 5 Attention sentences in blue color

algorithm *K*-means was used to cluster all the similar sentences. In this experiment, the optimal number of clusters was found using an approach called elbow method.

The K-means elbow is a method that implements the "elbow" technique to indicate the optimal number of clusters by appropriate model with a range of values. If the chart resembles a line arm, then the point of inflection on the curve is good indication. Then, the visualizer "elbow" will be annotated with a dashed line [15]. In our experiment, we applied this technique to perceive the appropriate graph.

In our experimental process of VoC, we applied process in which we first identified a sensitive sentence by using a technique of research called scoring method. This method was applied for each cluster of sentences gathered from students' feedback. We then extracted each word of a sentence in the cluster assigning a score based on appearance of each word that appears in the cluster. Then, for every phrase in that cluster, we calculated a sentence score by summing up the word scores and dividing it by number of words in the sentence. Then, we designated the sentence with the maximum score. This method was most suitable for choosing sensitive sentences for every cluster. The selected sentence reflected most of the sentences within a cluster. Then, a representative sentence was used to envision the impact of each cluster in the student reviews, and Fig. 6 shows the steps involved in the process for evaluation.

Act: Once the relevant data model is prepared after cluster analysis, we accomplish the results with powerful stories that demonstration how students interact with our services and how our institution should allocate resources. We need to identify and explain common topics, trends and patterns which were very challenging. Touch root causes and finds supporting data. We were supposed to understand the factors driving customer shake and affecting their allegiance. This makes us easy for recommendations. In the final concluding analysis, we will apply visualization techniques for bringing inferences of the data set.

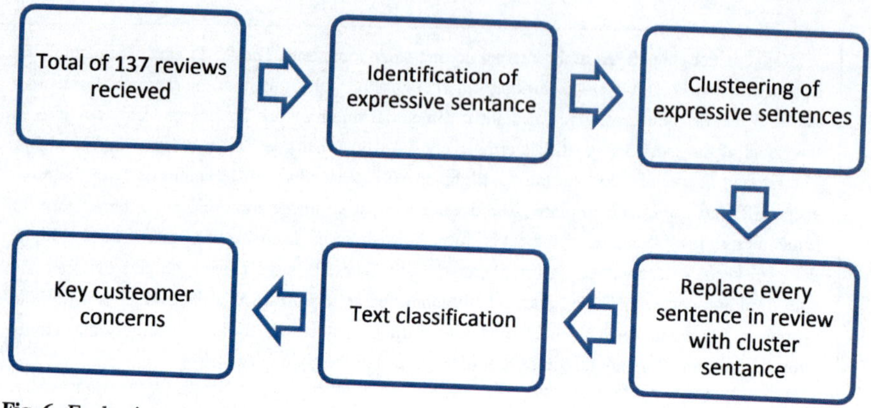

Fig. 6 Evaluation of students concerns

7 Challenges

Understanding unstructured data: One of the biggest challenges was unstructured data retrieved from the surveys. But we see recent techniques in clustering of text empower to build an effective VoC analytics that create value for any business. The overall experience was complicated exercise. We believe that for any VoC analysis to be successful it must be accurate. Another challenge was contradictory or inconsistent data received by the students. We begin our experiment analysis with the assumption that students in similar segments will have similar needs. It was found that students were consistently giving contradictory feedback, and it would be impractical to build an efficient process if the outcome from the process does not meet the needs of the customers. The experiment also finds VoC as educational challenge and again requires coordination between academics and operations to create a student-centric values in the organization.

8 Conclusion

In this research work, we have been able to identify the key elements in text clustering that supports in customer concerns from online polls. Our explanation also provides which of the customer issues are critical in determining the services expressed by the hostel students. In conclusion, we draw the interpretation of the results of this analysis as: students are having problem with Wi-Fi connectivity during night hours, and this problem has been common for most of the hostel students. With this analysis, we may now take the fourth section of our VoC process that is action. The best course of action could be increasing the bandwidth of Internet during night hours to uphold customer experience. This VoC analysis with text clustering techniques discussed in this paper if applied on unstructured data from online media can process customer reviews and produce constructive output which the analyst can relate to the business. As part of the future work, we are looking forward to applying on more complex data structures which involves multimedia data.

Acknowledgements The authors would like to acknowledge the valuable inputs given by the guide, professors and colleagues at Jaipur National University. Further, we are very grateful to the staff of IT department at MAHE Dubai for providing the data set. The support of all with inputs and discussions have been valuable in preparing this article paper.

References

1. International Six Sigma Institute, DMAIC Process for Voice of Customer Analysis https://www.sixsigma-institute.org/Six_Sigma_DMAIC_Process_Define_Phase_Capturing_Voice_Of_Customer_VOC.php. Last accessed 2016/11/21.
2. Lexalytics, Voice of Customer Analytics What it Is and How to Do It. https://www.lexalytics.com/lexablog/effective-voice-of-customer-voc-analytics. Last accessed 2020/05/1.
3. Aggarwal, C.C., Reddy, C.K. (eds.): Data Clustering Algorithms and Applications, CRC Press, Taylor & Francis Group (2014).
4. Everitt, B.S., Landau, S., Leese, M., Stahl, D.: Cluster Analysis, 5th edn. Wiley (2011).
5. Xu, R., Wunsch, II, D.C.: Clustering, IEEE Press, Wiley (2008)
6. Chormunge, S., Jena, S.: Correlation based feature selection with clustering for high dimensional data. J. Electr. Syst. Inf. Technol. 5(3), 542–549 (2018)
7. Inside BigData: What is Text Clustering, https://insidebigdata.com/2018/07/26/what-is-text-clustering. Last accessed 2020/06/09
8. MeaningCLoud: Text Clustering, https://www.meaningcloud.com/products/text-clustering. Last accessed 2020/06/12.
9. Tiwari, H., Sait, S., Rizvi, M.I.H., Damera-Venkata, N.: Identifying the relative importance of customer issues on product ratings through machine learning. In: DocEng'18: Proceedings of the ACM Symposium on Document Engineering (2018).
10. Liu, X., Xu, A., Sinha, V., Akkiraju, R.: Voice of customer: a tone-based analysis system for online user engagement. In: CHI EA'18 Extended Abstracts of the 2018 CHI Conference on Human Factors in Computing Systems (2018).
11. Musto, C., Rossiello, G., de Gemmis, M., Lops, P., Semeraro, G.: Combining text summarization and aspect-based sentiment analysis of users' reviews to justify recommendations. In: Proceedings of the 13th ACM Conference on Recommender Systems, pp. 383–387 (2018).
12. Lee, E., Lynn, H.M., Kim, H., Yeom, S., Kim, P.: Semantic analysis for paraphrase identification using semantic role labeling. In: SAC'19 Proceedings of the 34th ACM/SIGAPP Symposium on Applied Computing, pp. 2135–2138 (2019).
13. Zhang, H., Wei, H., Tang, Y., Pu, Q.: Research on classification of scientific and technological documents based on Naive Bayes. In: ICMLC'19 Proceedings of the 2019 11th International Conference on Machine Learning and Computing, pp. 327–331 (2019).
14. CollPoll: Go Beyond an ERP, https://www.collpoll.com/. Last accessed 2020/06/9.
15. YellowBrick: Elbow Method, https://www.scikit-yb.org/en/latest/api/cluster/elbow.html. Last accessed 2020/06/11.

Low Profile Wide Band Microstrip Antenna for 5G Communication

Pankaj Jha, Ram Lal Yadava, and Seema Nayak

Abstract In this paper, low profile microstrip antenna has been presented for 5G and wideband communication applications. The proposed antenna is designed with FR-4 substrate with rectangular radiator, partial ground, and with modified feed position. The proposed antenna obtained 10 dB impedance bandwidth for a wide range of frequency from 2.9 to 10 GHz. The gain of antenna is changing from 2.2 to 4.4 dB, and radiation efficiency is having maximum value of 90%. This antenna can be applicable for 5G communication and specific UWB application.

Keywords 5G · Microstrip antenna · Gain · Efficiency

1 Introduction

In 5G communication, with increasing communication demand, there has been a demand of higher data rate with bandwidth, keeping in mind the radio wave propagation and bandwidth; countries have opted for the range of 3 and 5 GHz frequencies for 5G services [1]. There are high and stable gain requirements with compact antenna for transmitting and receiving signals of 5G wireless network. Microstrip antenna cost is low, and its demand is high due to its low profile. Microstrip antenna can be designed according to demand. Due to its compact size and portable application, it is being used in WLAN [2, 3], Wi-Fi [4], and Bluetooth [5], 5G sub 6 GHz [6–15] and specific UWB application [15]. However, enhancing the gain with its reduced size in a microstrip antenna is a challenging problem, on which the researchers have been working for a long time.

P. Jha (✉)
Department of Electronics & Communication Engineering, IIMT College of Engineering, Greater Noida, UP, India

R. L. Yadava · S. Nayak
Department of Electronics & Communication Engineering, Galgotia College of Engineering & Technology, Greater Noida, UP, India

H. Sharma et al. (eds.), *Congress on Intelligent Systems*, Advances in Intelligent Systems and Computing 1334, https://doi.org/10.1007/978-981-33-6981-8_58

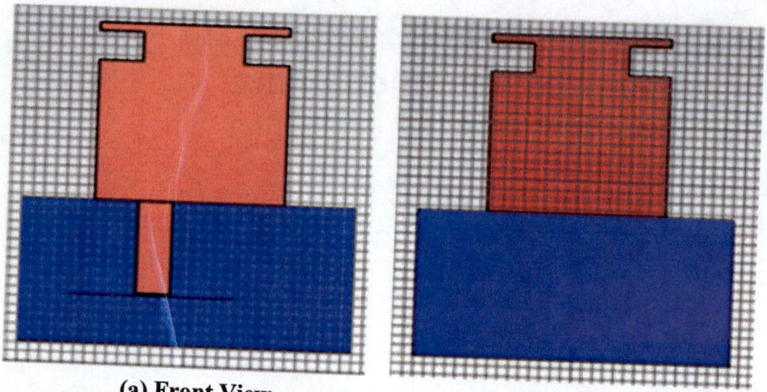

(a) Front View

(b) Back View

Fig. 1 Proposed microstrip antenna design

2 Antenna Design

The dimension of proposed antenna is designed with 17 mm × 18 mm × 1.6 mm rectangular radiator. Fr-4 substrate is used to design the antenna with dielectric constant 4.4. The substrate dimensions are used 30 × 40 × 1.6 mm. There are two slots used in the radiator to improve the impedance bandwidth of the antenna with dimensions of 4 mm × 3 mm. The feed line is used a transmission line with 10 mm length and 3 mm width. The 10 mm partial ground is used to achieve the wideband impedance bandwidth. The top view and back view of proposed antenna are given in Fig. 1, where blue part is ground plane and pink part is the top plane.

3 Simulation Results Analysis

The proposed antenna is designed with the CST 2018. The 10 dB simulated bandwidth of the proposed antenna is depicted in Fig. 2. It is clear from the figure that the proposed antenna is achieved the 10 dB impedance bandwidth from 2.9 GHz to 10 GHz. The resonant frequency is achieved from at 4 GHz, 7.1 GHz, and at 9 GHz frequencies.

The simulated value of gain is varying between 2.2 and 4.4 dB for operating bandwidth. The maximum gain is achieved at 5 and 8 GHz frequency. The simulation of the proposed antenna is shown in Fig. 3. The efficiency of the proposed antenna is achieved higher than 60%, but it maintains 80% up to 7 GHz as represented in Fig. 4. The omnidirectional radiation pattern of the proposed antenna is shown in Fig. 5, and the figure shows that it is linearly polarized with more than 20 dB difference between co-plane and cross-plane radiation pattern. In Table 1, the comparative analysis is done with recently published papers [6, 10, 11] and [12]. It is concluded that the

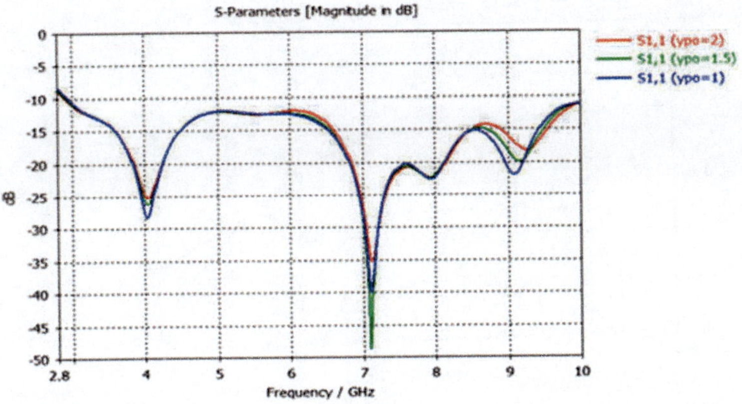

Fig. 2 Simulated $|S_{11}|$ in dB of the proposed antenna

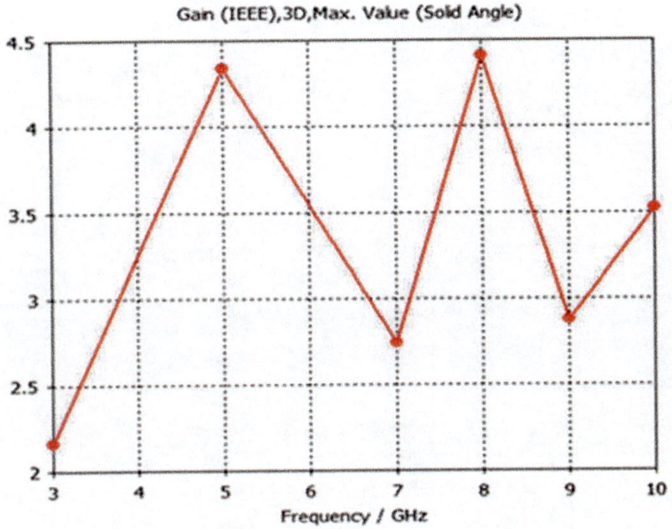

Fig. 3 Simulated gain of the proposed antenna

size of proposed antenna is miniaturized with Ref. [10, 11] and [12] with the higher bandwidth. The gain and efficiency are also better than the Ref. [10].

4 Conclusion

In this paper, microstrip antenna has been presented for 5G and specific UWB communication. The $|S_{11}|$ in dB of the proposed antenna is obtained for entire range

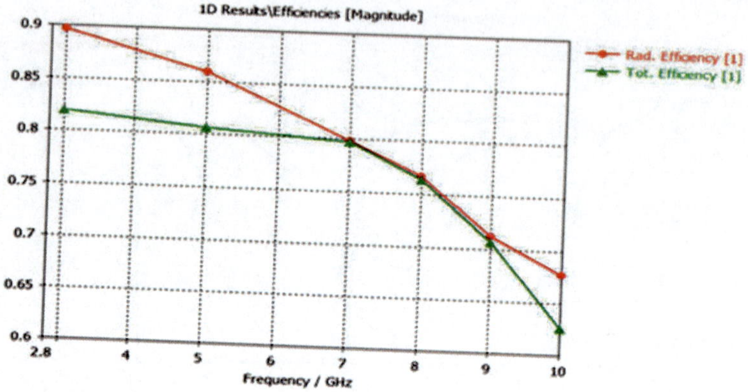

Fig. 4 Simulated efficiency of the proposed antenna

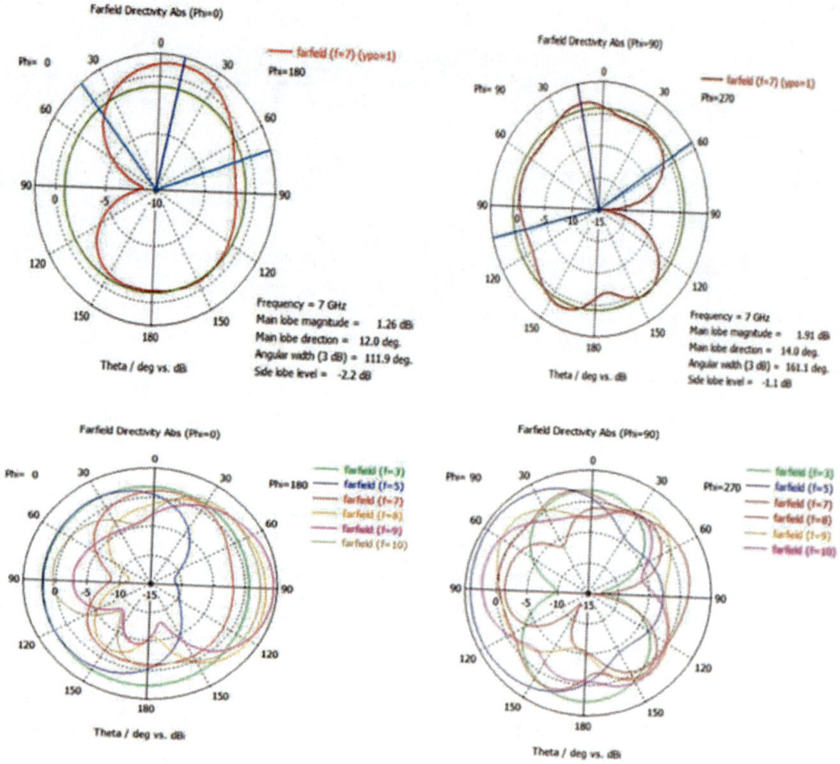

Fig. 5 Omnidirectional radiation pattern of the proposed antenna

Table. 1 Comparison table of the proposed antenna with recently published papers

References	Patch size in (mm)	BW (GHz)	Gain	Efficiency
[7]	17.30 × 10.4 × 1.6	**5.8**	5.6 dBi	NA
[11]	50 × 19.75 × 0.8	**2.5–4.8**	1.8–2.9 dB	60–90
[12]	47 × 19 × 0.8	**3.2–5.2**	2.1–4.1 dBi	NA
[13]	40 × 30 × 1.6	**3–5.64**	1.7–3.2 dB	74–90
Proposed work	30 × 40 × 1.6	**2.9–10**	2.2–4.4 dB	90–63

2.9–10 GHz. The maximum gain obtained from the proposed antenna is 2.2–4.4 dB. The efficiency is very good at sub 6 GHz band. The omnidirectional radiation pattern is obtained from the proposed antenna. So, the proposed microstrip antenna can be applicable for 5G, Wi-Fi, WLAN, and specific UWB communication system.

References

1. An, W., et al.: Low-profile and wideband microstrip antenna with stable gain for 5G wireless applications. IEEE Antennas Wirel. Propag. Lett. **17**(4), 621–624 (2018)
2. Ghosh, C.K., Parui, S.K.: Design, analysis and optimization of a slotted microstrip patch antenna array at frequency 5.25 GHz for WLAN-SDMA system. Int. J. Electr. Eng. Inf. **2**(2), 102–112 (2010).
3. Mutiara, A.B., Refianti, Rahmansyah: Design of microstrip antenna for wireless communication at 2.4 GHz (2012).
4. Tecpoyotl-Torres, M., Vera-Dimas, J.G., Vargas-Chable, P., Damián-Morales, J.A., Torres-Cisneros, M., Sánchez-Mondragón, J.: Antenna prototypes for indoor and outdoor Wi-Fi communication. In: 2009 III Conference of University of Guanajuato IEEE Students Chapter (IEEExPO), pp. 37–41. IEEE (2009).
5. Majumder, A.: Design of an H-shaped microstrip patch antenna for Bluetooth applications. Int. J. Innov. Appl. Stud. **3**(4), 987–994 (2013)
6. Balti, E., Johnson, B.K.: Sub-6 GHz microstrip antenna: design and radiation modeling. arXiv preprint arXiv:1901.05254 (2019).
7. Zhang, W., Weng, Z., Wang, L.: Design of a dual-band MIMO antenna for 5G smartphone application. In: 2018 International Workshop on Antenna Technology (iWAT). IEEE, 2018.
8. Alibakhshikenari, M., et al.: Wideband sub-6 GHz self-grounded bow-tie antenna with new feeding mechanism for 5G communication systems. In: 2019 13th European Conference on Antennas and Propagation (EuCAP). IEEE, 2019.
9. Ren, Z., Zhao, A.: Dual-band MIMO antenna with compact self-decoupled antenna Pairs for 5G mobile applications. IEEE Access 7:1 (2019).
10. Sekeljic, N., Yao, Z., Hsu, H.-H.: 5G broadband antenna for sub-6 GHz wireless applications. In: 2019 IEEE International Symposium on Antennas and Propagation and USNC-URSI Radio Science Meeting. IEEE, 2019.
11. Tang, X., et al.: Ultra-wideband patch antenna for sub-6 GHz 5G communications. In: 2019 International Workshop on Electromagnetics: Applications and Student Innovation Competition (iWEM). IEEE, 2019.
12. Jha, P., Singh, S., Yadava, R.: Wideband Sub-6 GHz Micro strip Antenna: Design & Fabrication. No. 2589. EasyChair, 2020.

13. Liu, C., Li, B., Li, Y.: A compact meandered dipole-like antenna for sub-6 GHz communication applications. In: 2018 International Applied Computational Electromagnetics Society Symposium-China (ACES). IEEE, 2018.
14. Zeng, J., Luk, K.-M.: Single-layered broadband magnetoelectric dipole antenna for new 5G application. IEEE Antennas Wirel. Propag. Lett. **18**(5), 911–915 (2019)
15. Verma, R.K., Kumar, A., Yadava, R L.: Compact multiband CPW fed sub 6 GHz frequency reconfigurable antenna for 5G and specific UWB applications. J. Commun. **15**(4), 345–349 (2020).

Analyzing the Impact of Software Requirements Measures on Reliability Through Fuzzy Logic

Syed Wajahat Abbas Rizvi

Abstract In today's scenario, the information technology is impacting every domain of our lives. The need of reliability as a quality attribute has been gaining importance. At the same time, it is a well-accepted fact that if the software requirements are handled carefully then the final software will have a high probability of reliability. With this spirit, researcher has identified a set of requirement stage measures and analyze their impact on the reliability of the software. This study has shown quantitively that the metrics based on requirements phase impact on the reliability of the final software. The paper also presents a comprehensive sensitivity analysis that highlights how various requirements phase measures influence the software reliability positively or negatively.

Keywords Early-stage requirements measures · Software Reliability · Fuzzy Inference System · Reliability Quantification

1 Introduction

In the current century, each and every part of society is relying software directly or indirectly. Every domain of human life has depended on the software applications developed for that specific domain. E-commerce applications are not used in the urban areas but the people living in villages are also using information technology applications [1]. At the same time reliability of these software has also become an important attribute [2]. It is the factor on which various industries are competing in the Industry for establishing them as the best company. Looking at the existing literature the software reliability can be defined as the probability of failure-free operations for a specified period of time in a specific environment [3].

A significant number of unfortunate events had already occurred because of the failure of software applications in different domains [4]. It could be noticed from the existing literature that 70–80% of all the faults in software are introduced during the

---

S. W. A. Rizvi (✉)
Department of Computer Science and Engineering, Amity University Uttar Pradesh, Lucknow, India
e-mail: swarizvi@lko.amity.edu

H. Sharma et al. (eds.), *Congress on Intelligent Systems*, Advances in Intelligent Systems and Computing 1334, https://doi.org/10.1007/978-981-33-6981-8_59

requirements phase [5]. And this is the stage of development life cycle that had not been given required focus while estimating the software reliability. Therefore, there is a need that the measure form requirements phase needs to be investigated for their influence that they have on the reliability of the finally delivered software.

The rest of the paper is organized as follows; Sect. 2 describes the state-of-art on how requirements level measures play a key role in ensuring reliability of the finally developed software. Section 3 presents the overview of some of reliability relevant software requirements metrics. Section 4 quantitatively analyzed the short-listed requirement level measures in the form of sensitivity analysis and finally the paper concludes in Sect. 5.

2 Related Work

Significant number of reliability-related research studies can be witnessed in the last three decades [6–15]. The researcher has already critically reviewed some of these studies in earlier papers [16–18]. In a research [19] authors have given a model for reliability estimation. The study has worked on five metrics that have been used for quantifying the reusability. In another study [20] three requirements-based measure has been used. In a research study [21], also some design level object-oriented metrics have been used. But these researches have neither quantitively justify nor analyzed the impact of requirement level metrices on software reliability. After going through the above, it is required to do some quantitative analysis that can analyze the impact of requirements measure on reliability.

2.1 *Research Gap*

It is being observed during the review of literature that requirements level measures are generally ignored while quantifying software reliability. As the errors introduced because of improper requirements analysis are having severe impact on the reliability of the finally delivered software. Therefore, consideration of requirements stage measures along with their impact on the software reliability is an attempt to fill this gap.

3 Software Requirements Measures

In the last thirty years, a significant number of software metrics have been proposed those are related to prediction or estimating the reliability of the software. These existing metrics can be noticed in the literature. Researchers has proposed these metrics at various stages of software development like requirements, design, and

Table 1 Requirements stage measures

S. No	Requirements metrics
1	RS (Requirements Stability)
2	RSDR (Regularity of Specification and Docu. Reviews)
3	RFD (Requirement Defect Density)
4	Scale of New Functionality Implemented
5	RC (Complexity of New Functionality)
6	QDI (Quality of Documentation Inspected)
7	PM (Process Maturity)
8	RIW (Review Inspection and Walkthrough)
9	RCR (Requirement Change Request)
10	ERT (Experience of Requirement Team)
11	RM (Requirements Management)
12	DSM (Development Staff Motivation)

coding. At the same time, a good number of reliability prediction models based on these software metrics have also been proposed in the last two decades. All these models predicting or estimating the reliability by incorporating a considerable number of software metrics. These metrics are categorized as traditional, object-oriented, and process metrics. However, looking at the existing literature it does not seem appropriate to consider all types of reliability relevant metrics in order to predict the corresponding software reliability.

It is well-accepted fact that the root cause of most of the reliability-related problems lies in the faulty software requirements. Therefore, the researcher has focused in this study only on the requirements metrics. After reviewing the state of the art very thoroughly the researcher has identified the above software requirements metrics as listed in Table 1. Out of the 12 metrics the researcher has shortlisted the four software requirements metrics as Requirements Stability (RS), Review Inspection and Walk-through (RIW), Complexity of New Functionality (RC), and Requirement Defect Density (RFD).

3.1 Requirements Stability

RS (Requirements Stability) is the software requirement metric that is inversely proportional to number of change requests coming from the client. It evaluates how stable the requirements are? It further assists to get more insight into the system's reliability and quality. The actual aim of this metric, as the researcher noticed from the existing literature, is to organize, control, and track changes to the originally specified requirements. This metric also provides an indication of the completeness, stability, and understanding of the requirements. As the frequency of proposed changes from the client side increases that will give rise to the complexity of the requirements specifications. Therefore, it could be easily inferred that higher the change request, higher the complexity in the proposed requirements that will act as a trigger to make the final software less reliable. Or in other words lesser the number of client's change request, lower the complexity in the proposed requirements that will act as a trigger to make the final software more reliable.

More Change Requests ⇒ More Complex Requirements ⇒ Less Reliable

Low Requirements Stability ⇒ High Complexity ⇒ Less Reliability

3.2 *Review, Inspections, and Walkthrough*

RIW (Review, Inspections, and Walkthrough) is an effective technique for uncovering requirements errors and improving software quality. If the number of RIWs will be higher, the probability of existing requirements defect will become low. That will result in higher reliability of the final product. Because each review and walkthrough will reduce the number of requirements defects and consequently the reliability will improve.

High RIW ⇒ Detect/Remove More Defects ⇒ SRS
Complexity will Reduce ⇒ Reliability Improves

3.3 *Requirements Fault Density*

The third shortlisted metric RFD (Requirements Fault Density) calculates the fraction of faulty or defective requirements specification in the SRS. Requirement fault density provides an indicator of the software quality of developing software during requirement analysis phase. If the fault density is high in SRS, then the reliability of the finally developed software will be low. On the other hand, if defect density is low in the SRS then definitely there will be higher probability of reliable software

High Fault Density ⇒ High Requirements Complexity ⇒ Low Reliability

3.4 *Complexity of New Functionality*

The RC (Complexity of New Functionality) notice or rate the complexity of the newly added functions in the SRS. Therefore, it is quite easy to correlate this requirement metric with the key factor of software reliability. It could be further described that the more complex requirements give rise to the higher probability of low reliability.

$$\text{High value of RC} \Rightarrow \text{Make the SRS Complex} \Rightarrow \text{Lower Reliability}$$

Therefore, in summarized form, it can be shown how the four requirements metrics are associated with reliability. RS α Reliability; RC α 1/Reliability; RFD α 1/Reliability; RIW α Reliability.

4 Analysis Phase

To analyze the impact of software requirements measures on software reliability the researcher has collected the relevant data on the four selected metrics from well-established and reputed software companies at Noida. After performing the implementation through fuzzy logic toolbox the researcher has analyzed the implementation outcomes in the form of sensitivity analysis as described in the following sub-sections.

4.1 *Sensitivity Analysis*

In this analysis, the impact of the different combinations of the four requirements level measure on the single variable that is requirement level reliability (RLR) has been analyzed. As the value of RLR depends on four requirements stage metrics RS, RIW, RC, and RFD.

In order to control the reliability at the requirements level more proactively it is better to know its impact. These figures are presenting the various combinations of input variables along with their impact on the output variable. In the following section, it will be discussed.

4.2 *Analysis of Requirements Measures*

The author has inferred the following observations while changing the values of requirements stage measures:

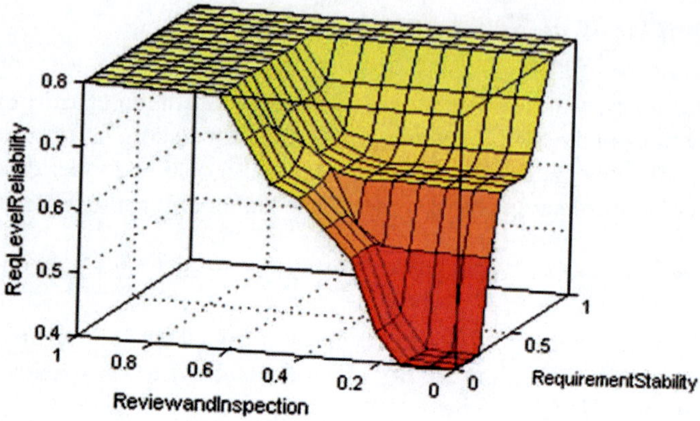

Fig. 1 Requirements stability versus requirements level reliability

(a) ***Individual Variation***

As the value of RS moves towards 0–1 the value of RLR also increases from 0 to 1. As the value of RS moves towards 1–0 the value of RLR also decreases towards 0.

As the value of RIW moves towards 0–1 the value of RLR also moves from 0 to 1.

As the value of RIW decreases from 1 to 0 the RLR also decreases in the same direction.

As the value of RC moves from 1 to 0 the value of RLR moves in opposite direction (0–1).

As the value of RC increases from 0 to 1 the value of RLR decreases towards 0.

As the value of RFD moves towards 1 the value of RLR moves in reverse direction (1–0).

As the value of RFD decreases, the value of RLR increases from 0 to 1 (Fig. 1).

(b) ***Combinational Variation***

After going through aforementioned empirical observations in the form of individual and combinational variations, it could be concluded that "Higher the value of RS the more reliable the requirements will be"; "Higher the value of RC the less reliable the requirements will be"; "Higher the value of RIW the more reliable the requirements will be"; "Higher the value of RFD the less reliable the requirements will be". Noticing the values of Table 2 and Fig. 3, it could be easily inferred that for any constant value of RS, RIW, and RC, the values of the metric RFD have a negative impact on the RLR. As the value of RFD increases, it forces the RLR to decrease. In other words, reliability of the software will decrease as the density of faults in the SRS increases (Fig. 2).

Table 2 Sensitivity of RLR with respect to RFD

RS = 0%		RS = 25%		RS = 50%		RS = 75%		RS = 100%	
RIW = 0%		RIW = 25%		RIW = 50%		RIW = 75%		RIW = 100%	
RC = 0%		RC = 25%		RC = 50%		RC = 75%		RC = 100%	
RFD	RLR	RFD	RLR	RFD	RLR	RFD	RLR	RFD	RLR
0.0	0.633	0.0	0.544	0.0	0.665	0.0	0.800	0.0	0.800
0.2	0.639	0.2	0.544	0.2	0.665	0.2	0.800	0.2	0.800
0.4	0.633	0.4	0.544	0.4	0.665	0.4	0.641	0.4	0.633
0.6	0.569	0.6	0.427	0.6	0.602	0.6	0.642	0.6	0.642
0.8	0.400	0.8	0.263	0.8	0.486	0.8	0.641	0.8	0.637
1.0	0.400	1.0	0.263	1.0	0.486	1.0	0.641	1.0	0.633

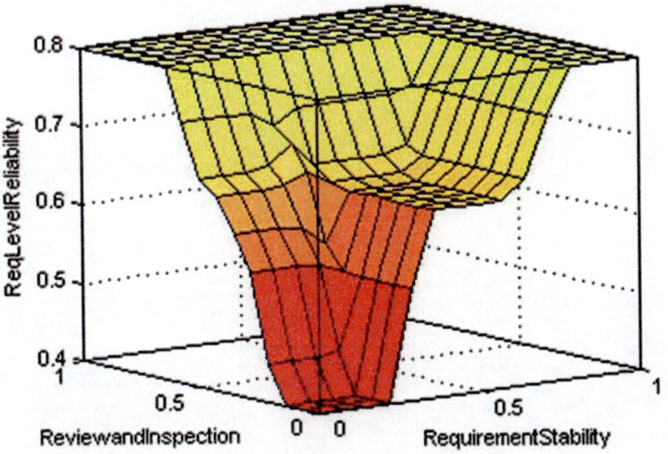

Fig. 2 Requirements stability and RIW versus requirements level reliability

Similarly noticing the values of Table 3 and Fig. 2, it could be easily inferred that for any constant value of RIW, RFD and RC, the values of RS has a positive impact on the RLR. As the value of RS increases, it forces the RLR to increase also. In other words, the developing software will be more reliable if the requirements have stability, i.e. frequency of change request is as low as possible.

Noticing the values of Table 4 and Fig. 4, it could be easily inferred that for any constant value of RS, RIW, and RFD, the metric RC has a negative impact on the RLR. When the value of RC increases, it forces the RLR to decrease. In other words, Reliability of the software will decrease as the functionalities get complicated in the SRS.

Noticing the values of Tables 5, 6 and Fig. 1, it could be easily inferred that for any constant value of RS, RFD, and RC, the metric RIW has a positive impact on the RLR. When the value of RIW increases, it forces the RLR to increase also. In

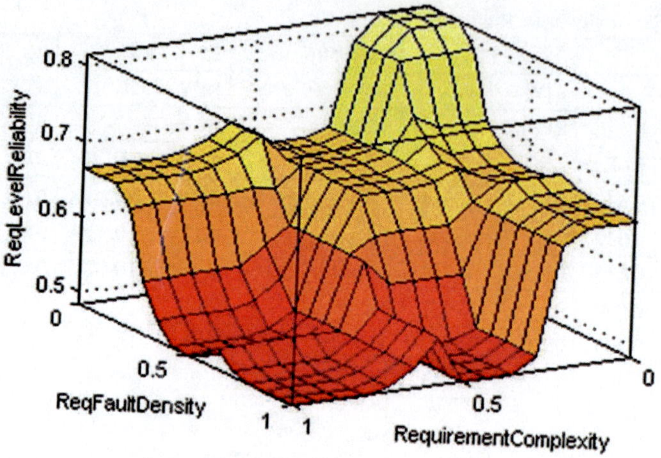

Fig. 3 Requirements fault density and requirements complexity versus requirements level reliability

Table 3 Sensitivity of RLR with respect to RS

RFD = 0%		RFD = 25%		RFD = 50%		RFD = 75%		RFD = 100%	
RIW = 0%		RIW = 25%		RIW = 50%		RIW = 75%		RIW = 100%	
RC = 0%		RC = 25%		RC = 50%		RC = 75%		RC = 100%	
RS	RLR	RS	RLR	RS	RLR	RS	RLR	RS	RLR
0.0	0.633	0.0	0.544	0.0	0.665	0.0	0.400	0.0	0.400
0.2	0.640	0.2	0.544	0.2	0.669	0.2	0.400	0.2	0.400
0.4	0.635	0.4	0.583	0.4	0.665	0.4	0.639	0.4	0.635
0.6	0.639	0.6	0.583	0.6	0.665	0.6	0.639	0.6	0.639
0.8	0.800	0.8	0.724	0.8	0.800	0.8	0.639	0.8	0.639
1.0	0.800	1.0	0.724	1.0	0.800	1.0	0.639	1.0	0.633

Table 4 Sensitivity of RLR with respect to RC

RS = 0%		RS = 25%		RS = 50%		RS = 75%		RS = 100%	
RIW = 0%		RIW = 25%		RIW = 50%		RIW = 75%		RIW = 100%	
RFD = 0%		RFD = 25%		RFD = 50%		RFD = 75%		RFD = 100%	
RC	RLR	RC	RLR	RC	RLR	RC	RLR	RC	RLR
0.0	0.633	0.0	0.551	0.0	0.665	0.0	0.800	0.0	0.800
0.2	0.642	0.2	0.551	0.2	0.676	0.2	0.800	0.2	0.800
0.4	0.633	0.4	0.551	0.4	0.665	0.4	0.641	0.4	0.633
0.6	0.400	0.6	0.265	0.6	0.534	0.6	0.645	0.6	0.645
0.8	0.400	0.8	0.271	0.8	0.486	0.8	0.641	0.8	0.637
1.0	0.400	1.0	0.271	1.0	0.486	1.0	0.641	1.0	0.633

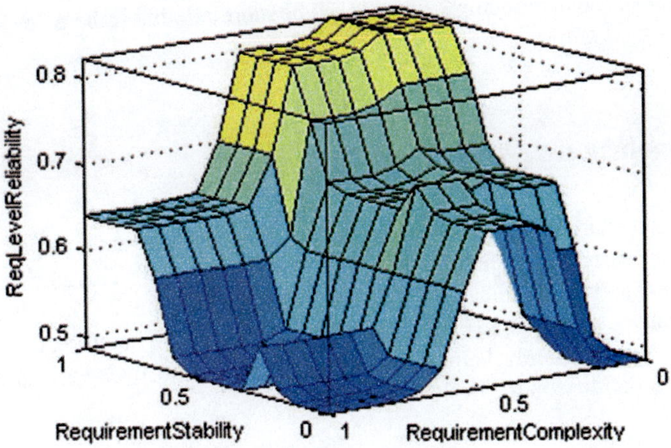

Fig. 4 Requirements stability versus requirements level reliability

Table 5 Sensitivity of RLR with respect to RIW

RS = 0%		RS = 25%		RS = 50%		RS = 75%		RS = 100%	
RFD = 0%		RFD = 25%		RFD = 50%		RFD = 75%		RFD = 100%	
RC = 0%		RC = 25%		RC = 50%		RC = 75%		RC = 100%	
RIW	RLR	RIW	RLR	RIW	RLR	RIW	RLR	RIW	RLR
0.0	0.633	0.0	0.544	0.0	0.639	0.0	0.400	0.0	0.400
0.2	0.639	0.2	0.544	0.2	0.639	0.2	0.400	0.2	0.400
0.4	0.633	0.4	0.544	0.4	0.639	0.4	0.641	0.4	0.633
0.6	0.800	0.6	0.700	0.6	0.800	0.6	0.642	0.6	0.642
0.8	0.800	0.8	0.700	0.8	0.800	0.8	0.641	0.8	0.636
1.0	0.800	1.0	0.700	1.0	0.800	1.0	0.641	1.0	0.633

Table 6 Sensitivity of RLR with respect to RS and RFD

RIW = 0%		RIW = 25%		RIW = 50%		RIW = 75%		RIW = 100%	
RC = 0%		RC = 25%		RC = 50%		RC = 75%		RC = 100%	
RS, RFD	RLR	RS, RFD	RLR	RS, RFD	RLR	RS, RFD	RLR	RS, RFD	RLR
0.0	0.633	0.0	0.544	0.0	0.486	0.0	0.639	0.0	0.633
0.2	0.640	0.2	0.544	0.2	0.492	0.2	0.640	0.2	0.640
0.4	0.400	0.4	0.566	0.4	0.665	0.4	0.639	0.4	0.635
0.6	0.400	0.6	0.556	0.6	0.602	0.6	0.642	0.6	0.642
0.8	0.639	0.8	0.685	0.8	0.639	0.8	0.639	0.8	0.639
1.0	0.633	1.0	0.685	1.0	0.639	1.0	0.639	1.0	0.633

other words, the developing software will be more reliable if the SRS has reviewed and inspected more.

5 Conclusion

Generally, requirements level measures are ignored and not get the required attention as far as the early stage reliability quantification is concerned. This study has not only focused exclusively but also analyzed using the fuzzy toolbox. The outcome of the analysis has been discussed in the form of critical findings. Therefore, these identified measures will definitely help the researchers to perform their early prediction of software reliability on the basis of these measures.

References

1. Yadav, D.K., Chaturvedi, S.K., Misra, R.B.: Early software defects prediction using fuzzy logic. Int. J. Perform. Eng. **8**(4), 399–408 (2012)
2. Rizvi, S.W.A., Khan, R.A.: Maintainability estimation model for object-oriented software in design phase (MEMOOD). J. Comput. **2**(4), 26–32 (2010)
3. Djenadic, S., Ignjatovic, D., Tanasijevic, M., Bugaric, U., Jankovic, I., Subaranovic, T.: Development of the availability concept by using fuzzy theory with AHP correction, a case study: bulldozers in the open-pit lignite mine. Energies **12**(21), 1–18 (2019)
4. Rizvi, S.W.A., Singh, V.K., Khan, R.A.: The state of the art in software reliability prediction: software metrics and fuzzy logic perspective. Adv. Intell. Syst. Comput. **433**, 629–637 (2016)
5. Tanasijevic, M., Jovancic, P., Ivezic, D., Bugaric, U., Djuric, R.: A fuzzy-based decision support model for effectiveness evaluation—a case study of examination of bulldozers. Int. J. Indus. Eng. Theory. **26**(6), 878–897 (2019)
6. Rizvi, S.W.A., Singh, V.K., Khan, R.A.: Fuzzy logic based software reliability quantification framework: early stage perspective (^{FL}SRQF). Elsevier Proc. Comput. Sci. **89**, 359–368 (2016)
7. Kabir, S., Papadopulos, Y.: A review of applications of fuzzy sets to safety and reliability engineering. Int. J. Approx. Reason. **100**, 29–55 (2018)
8. Rizvi, S.W.A., Singh, V.K., Khan, R.A.: Software reliability prediction using fuzzy inference system: early stage perspective. Int. J. Comput. Appl. **145**(10), 16–23 (2016)
9. Kraidi, L., Shah, R., Matipa, W., Borthwick, F.: Using stakeholders' judgement and fuzzy logic theory to analyze the risk influencing factors in oil and gas pipeline projects: case study in Iraq, stage II. Int. J. Crit. Infrastruct. Prot. **28**, 100337 (2020)
10. Gallab, M., Bouloiz, H., Alaoui, Y.L., Tkiouat, M.: Risk assessment of maintenance activities using fuzzy logic. Proc. Comput. Sci. **148**, 226–235 (2019)
11. Rizvi, S.W.A, Khan, R.A.: A critical review on software maintainability models. In: Proceedings of the Conference on Cutting Edge Computer and Electronics Technologies, pp. 144–148 (2009)
12. Chandima, R.M., Antosz, K.: Development of a Risk Matrix and extending the risk-based maintenance analysis with fuzzy logic. Proc. Eng. **182**, 602–610 (2017)
13. He, P., Li, B., Liu, X., Chen, J., Ma, Y.: An empirical study on software defect prediction with a simplified metric set. Inf. Softw. Technol. **59**, 170–190 (2015)
14. Jaderi, F., Ibrahim, Z.Z., Zahiri, R.Z.: Criticality analysis of petrochemical assets using risk based maintenance and the fuzzy inference system. Process Saf. Environ. Prot. **121**, 312–325 (2019)

15. Chang, Y., Wu, Y., Chen, G., Ye, J., Chen, B., Xu, L., Zhou, J., Yin, Z., Ren, K.: Comprehensive risk assessment of deepwater drilling riser using fuzzy Petri net model. Process Saf. Environ. Prot. **117**, 483–497 (2018)
16. Shooman, M.L.: Yes, software reliability can be measured and predicted. In: Proceedings of the 1987 Fall Joint Computer Conference on Exploring Technology, IEEE Computer Society, pp. 121–122 (1987)
17. Khatatneh, K., Mustafa, T.: Software reliability modeling using soft computing technique. Eur. J. Sci. Res. **26**(1), 154–160 (2009)
18. Rizvi, S.W.A., Khan, R.A.: Improving software requirements through formal methods. Int. J. Inf. Comput. Technol. **3**(11), 1217–1223 (2013)
19. Bhuyan, M.K., Mohapatra, D.P., Sethi, S.: A Survey of computational intelligence approaches for software reliability prediction. ACM SIGSOFT Softw. Eng. Notes **39**(2), 1–10 (2014)
20. Radjenovic, D., Hericko, M., Torkar, R., Zivkovic, A.: Software fault prediction metrics: a systematic literature review. Inf. Softw. Technol. **55**(8), 1397–1418 (2013)
21. Zhao, L., Zhang, J., Yang, J., Chu, Y.: Software reliability growth model based on fuzzy wavelet neural Network. In: 2nd International Conference on Future Computer and Communication (ICFCC). pp. 664–668. IEEE, Wuhan (2010)

Social Network Opinion Mining and Sentiment Analysis: Classification Approaches, Trends, Applications and Issues

Amit Pimpalkar and R. Jeberson Retna Raj

Abstract Sentiment Analysis (SA) or Opinion Mining (OM) probes the content stated about a topic of interest and classifies them as +ve, −ve or neutral hold on the person's opinion, emotion, judgment articulated in it. Nowadays people's opinions and reviews on blogs, forums, review websites, E-commerce sites are most vital information in the process decision-making for both organization and users. SA is an information revelation and knowledge discovery technique inherited from data mining. The manual investigation of such immense number of reviews is actually unthinkable so computerized frameworks are expected to do analysis and accordingly classify of information to take related decisions. In general, SA attempts to make some decisions on the base of an author standpoint that may be his or her judgment, mood or evaluation. In this study, we have described the essential concepts, various methods, trends, applications and key issues of OMSA in social networks. We tried to analyze SA literatures from various perspectives and presented those techniques and methods in a systematic manner. With brief descriptions of the algorithms, different SA taxonomical techniques classified along with their advantages and limitations. The critical opportunities that lie ahead for SA are defined and addressed on the basis of those sources. We attempted to present the recent applications and challenging issues in Sentiment Identification and presents evaluation metrics for SA.

Keywords Sentiment analysis · Opinion mining · Natural language processing · Social network analysis · Machine learning

1 Introduction

SA is a contribution of Natural Language Processing (NLP) which guarantees with the computational measures of belief, emotion, unbiased and impartiality inside the given text. The term sentiment includes feelings, conclusions, behavior, emotions,

A. Pimpalkar (✉) · R. Jeberson Retna Raj
School of Computing, Sathyabama Institute of Science and Technology, Chennai, India

R. Jeberson Retna Raj
e-mail: jebersonretnaraj.it@sathyabama.ac.in

H. Sharma et al. (eds.), *Congress on Intelligent Systems*, Advances in Intelligent Systems and Computing 1334, https://doi.org/10.1007/978-981-33-6981-8_60

and others. As the size of number of users and uses of social media are swiftly expanding, Social Network Analysis (SNA) has become a massive instrument for specialists and analysts in social computing [11]. OM includes reading evaluations, sentiments or the mindset of the writer from the written content. This uses the principles of NLP, facts mining and system getting to know to carry out the research [3]. SA is premeditated to be a classification problem in Computational Linguistics (CL). It involves NLP on many levels and inherits its challenges. There exists a wide of applications that could advantage from their consequence, which includes news analytics, advertising, query answering, knowledge bases and so forth. The real challenge in this area is to ameliorate the machine's capability to recognize and understand texts in customer opinions in the equivalent manner as human are capable of. Taking advantage from the number of opinions articulated on the internet in particular form social media is important for lots companies and establishments, regardless of whether its miles in phrases of product remarks, public satisfaction, or investor mood estimation. To deal with this issue, three distinctive key concerns are investigated [6].

The main motivation behind the SA is to uncover individual's persuasion on explicit topics, for instance, politics, movies, products and so forth. In this paper, our major contribution is an action on prior works done in SA. This work will help beginners to have a panoramic outlook of the whole field of research aiming at SA. The rest of the paper is arranged as follows: We discuss the related work including techniques used along with their domain and the performance of the given systems in Sect. 2. Section 3 introduces taxonomy for sentiment analysis models. Section 4 shows some of the applications in sentiment identification with open issues in Sect. 5, and we conclude with Sect. 6.

2 Literature Review

An emotion dictionary (SentiDict) which has several +ve and—the labels formulated by Estrada et al. [8]. The authors followed a four-step progression method to the learning models of DL with CNN and LSTM layers. Pathak et al. [24] and Rahman et al. [30] placed forth the categorization of traits to be taken into consideration for Deep Learning (DL) based SA and deflates the function of DL tactics for sentiment evaluation. In [40] investigated the predictive liveliness of historical information, sentiments to forecast monetary news sentiments. In [28] focused on the content of YouTube gamers' conversation on Facebook to discover widespread variations in terms in user-generated Facebook metrics and sentiments. To categorize tweets an Adaptive Neuro Fuzzy Inference System tweet recommended hybrid model was suggested by Jain et al. [17] using NB and SVM classifiers. The TextJSM system for the automatic generation of text-sentiment analysis was suggested for resolving predictive and descriptive problems [19]. Their studies confirmed that the TextJSM parallel performance offers high-quality sentiment analysis and parallelization.

To research stock trends primarily based on tweets and historical facts [21] projected a learning model and also compared it with existing conventional strategies NB Bernoulli, Random Forest (RF) and SVM which usually that specialize in one thing historic information or SA. In research [9] the micro-running a blog approach of the device gaining knowledge of method was proposed wherein the emoticons labelled to show the +ve, −ve or the neutral end result. NB, ME and SVM algorithm have been used to generalize the given records [32]. KNN, *K*-Means and LDA were compared with a manual prepared stop word lists on four different domains like politics, agriculture, economy, and entertainment data set to achieve better accuracy by Rani and Lobiyal [31] and Zhou [41]. Lexicon fusion and classification methods using all three possible Genetic Algorithms (GA) functions were illustrated [23] on Bing Liu's product review data sets. They suggested Hierarchical word scoring function as a specialized method for preparation of dynamic lexicons in aspect-based SA. Indra et al. [15] performed DF-IDF, *N*-grams clustering and Topic ranking is used to compare the two methods, document pivot and BN-grams, for detecting trending topics on Indonesian tweets. Performance of Learning Vector Quantization (LVQ) algorithm was observed by Stylianidis et al. [36].

With the intention to extract human opinion a specific concern by counting contents on social media [16] proposed the method which first builds a corpus based on a small set of +ve and—via hashtags associated with a given problem. They also classified the posts into a number of classes and estimation the sentiment credence. An exhaustive systematic evaluation, aims to talk about both technical elements of OMSA and non-technical factor within the shape of application areas have been offered by Shayaa et al. [35]. An emerged towards the interpretation, retrieval and evaluation of the social network was evolved by way of [5]. The statistics changed into filtered in line with applicable criteria and analyzed the usage of SA tools tailor-made particularly to the data source. Sports activity domains with Twitter data set the usage of WEKA sentiment ratings have been calculated. The approach becomes tested by means of making use of it to 2 instances, study questions, confirming beyond the findings on cultural and gender differences in SA. A three-section model of the SA for product assessment web sites, aiming at the detection of the maximum interesting components for producing better results proposed by Dragoni [7]. In [14] authors stated the strategies of textual content segmentation, Part Of Speech (POS) tagging and concept labeling.

In [38] authors developed two efficient function selection strategies, termed Maximum Discrimination (MD) and MDx2 methods, for text classification. Hogenboom [12] Advised that the polarity of natural language textual content should not be decided entirely based on what is said, however primarily based on how the message of a text is con-video as well. Wang and Dong [39] Compares illustration of the text inside the exploratory setting for SA and particularly focuses on the sensitivity of the evaluation to sentence period. Two representations compared were Bag-of-Words (BoW) and nine Dimensional Vectors (9Dim). In [34] the problem of reputation discovery through Frequent Named Entities and aggregation of sentiments with weight calculation algorithm addressed. So for to investigate the sentimental contextual text of such massive statistics the Senti_lexi set of rules was used with that it delivered the higher capability inside the SA of the contextual text of emoticons [20, 26, 27]. The authors proposed new undertaking decomposition method for Aspect Based Sentiment Analysis (ABSA), which includes three most important subtasks: aspect term extraction, aspect term polarity estimation and aspect term aggregation [25]. In [18] the authors described C-Feel-It, a gadget which could tap opinion content material in posts from Twitter. The aim was to construct a sentiment lexicon which is domain unbiased. They endorse a ML Based Sentiword Lexicon (MLBSL) on the Amazon records set which contains opinions from exclusive domains. This MLBSL yields an improvement over existing methods and automatic-built lexicons like SentiWordNet. Additionally, they offer a development in calculation method utilized in reviews SA [10].

Table 1 summaries the different works of literature cited for the study of SA of the last decade. After carefully reviewed each article, we summarize it according to the dataset and algorithm used by the various researchers along with their efficiencies as displayed below.

Table 1 Summary of various reviewed literatures with used dataset, algorithm and its accuracy

Refs. No.	Dataset used for evaluation	Algorithm and approach	Accuracy %
[8]	Education sentiTEXT, eduSERE	CNN, LSTM	93
[23]	Customer review, SentiWordNet	GA	91.8
[16]	Twitter, US Election event	NB	90.21
[34]	Twitter, Stanford NLP	Frequent named entities and aggregation of sentiments with weight calculation	90.0
[14]	Probase	Randomized approximation, weighted vote	89.0
[13]	Hotel public opinion datasets	SVM	86.75
[17]	Politics domain Twitter	NB and SVM	84.3
[28]	Video game, YouTube-Facebook	Grounded theory	82.3
[33]	Twitter, earthquake and independence	NB, SVM, RF	80.0
[36]	Movies, IMDb archive	LVQ	79.91
[40]	Stock Market	RNN with LSTM	79.01
[1]	StockTwits	SVM	76.65
[7]	Restaurant and laptop	LDA	65.12
[2]	Twitter	DT learner and *k*-NN	64.6
[41]	Election event, Twitter dataset	LDA	64.0
[21]	Stock Market, Twitter	NB, SVM	56.5

3 Taxonomy for Sentiment Analysis Models

For this study, we considered various research papers available between 2010 and April 2020 using the trustworthy database which includes Scopus, Science Direct, IEEE, ACM and some peer-reviewed refereed journal. The information from those researches is extracted and the crucial findings are presented into Fig. 1.

3.1 Levels of SA Classification

SA can take place at different levels like Document, Sentence or Aspect/Feature level. Those levels had been mentioned in this sub-section.

Document Level In this technique, sentiments are extracted from the whole review or a paragraph or a document and an entire opinion is classed primarily based on the sentiment of the opinion holder. The goal is to classify as +ve, −ve, or neutral. It could additionally be dealt with as a regression dilemma, as an example, conveying the score ranking inside the variety of 5 stars or 1 star for movie evaluations.

Sentence Level This degree of sentiment type targets to decide the sentiment from a single sentence. This technique typically entails two steps: [4]

- *Subjectivity classification into one of two classes: objective and subjective*
- *Sentiment classification of subjective sentences into two classes: +ve and −ve*

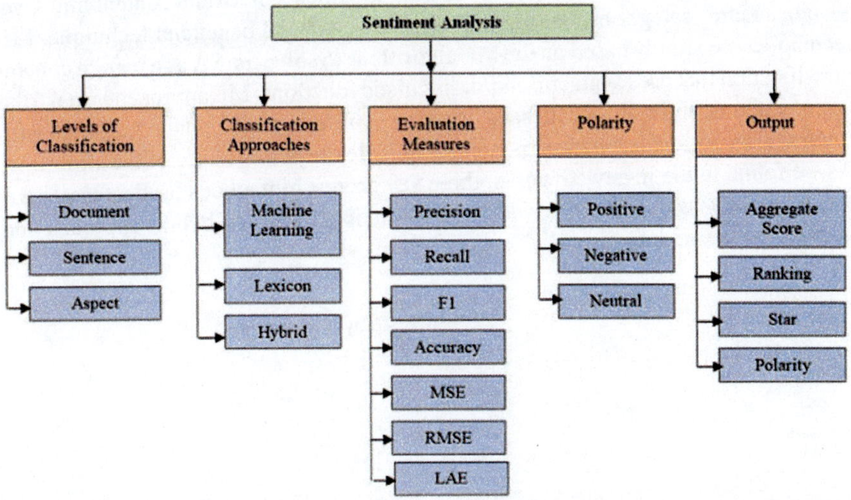

Fig. 1 Taxonomy for sentiment analysis models

As sentences are much shorter in comparison to document or a report, semantic and syntactic capabilities acquired through POS tagger, lexicons and parse trees used for classification at sentence level. Just like document level theories, sentence level also believes that each sentence holds sentiment of a single entity.

Aspect-Based (Feature) Sentiment Analysis (ABSA) The aim is to perceive and extract the item capabilities known as features that have been observed on by means of the opinion holder and determine whether the opinion is +ve, −ve, or neutral. Feature synonyms are aggregated and a feature-based outline of multiple critiques is produced. This stage assumes that a single entity is presented to document.

Locating sentiments for feature may be divided into 4 subtasks:

- *Textual Content Extraction*: It extracts sentences and phrases from the opinions.
- *Sentiment Classification*: It runs a classifier on every extracted sentence and phrase to determine if it's +ve, −ve, or neutral.
- *Aspect Component (Feature) Extractor*: It performs aspect term extraction for those that explicit a sentiment and determines the polarity for each of the components recognized.
- *Aspect Polarity Aggregation:* It is an organization of the emotions for aspects together and convey a very last review as +ve, −ve, or neutral.

3.2 *Classification Approaches*

Sentiment Classification paradigms may be alienated into ML, Lexicon Based and Hybrid approach. For the classification of the sentiments of textual content into +ve, −ve or neutral categories ML are the most powerful and beneficial technique. ML techniques are simply based on the ML algorithms to solve the SA as a textual content classification that uses syntactic and/or linguistic functions. ML approaches make the use of training and testing data sets. Training is used to prepare the algorithms with training data set and test dataset is used to validate the performance of the algorithm. As we found in the reviewed works, three key groups of methods used to assess SA are; machine learning approach, lexicon-based approach, and hybrid approach and they are presented into Fig. 2.

Machine Learning-Based Approach

There are ranges of ML algorithms used to classify sentiments which are classified as supervised ML algorithm like *k*-NN, SVM, ME, NB, DT, etc. and unsupervised ML algorithm such as *K*-means, NN, PCA, SVD, HMM, CNN, RNN, LSTM, etc. The supervised ML methods rely on the labelled training data. In textual content classification, it's so often hard to create these categorized labelled data, although it is much simpler to acquire the unlabelled documents. The unsupervised ML techniques triumph over these problems. Sometimes supervised and unsupervised procedures may be blended collectively to have better results. Supervised machine learning models require us to add a tag to texts manually before they can start forecasting themselves.

This means we'll first, have to identify a series of tags that we'll use later to label the documents by hand before the ML model can do it alone. The model is introduced by a "teacher" with labelled training data and their expected outputs, and the aim is to learn a rule of thumb connecting inputs into the final outputs. The model will be trained to label the texts with the expected tags and the amount of texts that we categorize would also impact the model conviction. Using an unsupervised machine learning approach, documents that include identical terms or sentences can

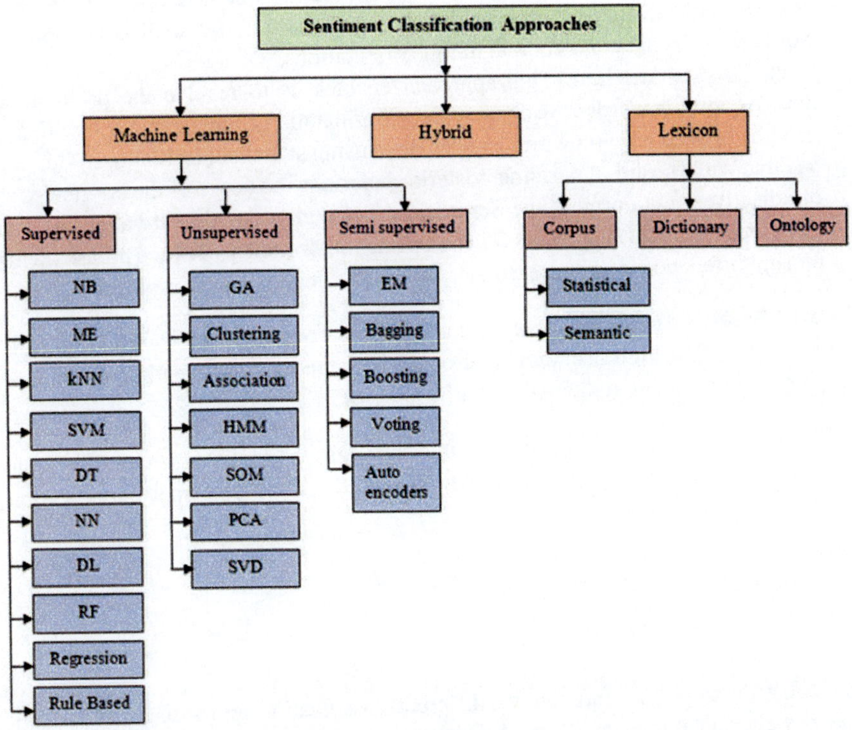

Fig. 2 Taxonomy for sentiment classification approach

be grouped together without any prior training by a classification model. The learning algorithm does not assign any labels, and it is left alone to find order in its data. Unsupervised learning algorithms enable one to handle quite complex computational tasks than supervised learning. While unsupervised learning may be more erratic relative to other methods of learning. Any algorithm in ML needs to learn from the results. Yet even with loads of worldwide information, such as texts, images, and many more, in reality, only a very small fraction data is labelled, heuristically or through a person. Semi-supervised learning is both supervised and unsupervised learning, where the teacher provides inadequate training in which of several of the missed goal outputs.

Lexicon-Based Approach This is based on a lexicon, a set of acknowledged and pre-accumulated sentiment terms. It is further classified as follows which uses statistical or semantically scheme to find sentiment polarity.

- **Dictionary-based approach**: A small set of sentiment words that are recognized as seeds with known +ve or −ve orientations are firstly manually collected. The set of rules then grows this set by way of looking inside the WordNet or some other on-line dictionary for their synonyms and antonyms. The recently found phrases are introduced to the seed list which follows the new iteration. The iterative method ends whilst no more new words can be determined to be added to the seed list. The major disadvantage of this approach is the lack of ability to discover opinion words with the area and context unique orientations.
- **Corpus-based method**: This approach enables us to resolve the problem of locating opinion words with context precise orientations. Identifying co-incidence patterns or seed opinion words may be done using statistical techniques. This can be done with the aid of deriving posterior polarities the usage of the co-incidence of adjectives in a corpus. The Semantic method gives sentiment values at once and relies on distinct standards for computing the similarity between phrases. This precept offers comparable sentiment values to semantically near phrases.

Hybrid-Based Approach The hybrid approaches combine each technique for the betterment of SA, high accuracy, and could be generous with sentiment lexicons playing a key position in the majority of strategies.

The works surveyed were classified into those groups as discussed earlier. Table 2 summarizes the strengths and weaknesses of the three principal classification approaches to reviewing the literature.

3.3 *Trends Classification*

To identify the trends of sentiment via different classification approaches and domain wise the classification which are elaborated earlier sub-section we have done the analysis of the existing literature which we have been considered for study in this

Table 2 Comparison of various machine learning methods of SA on their strength and weakness

Approach	Method	Strengths	Weakness
Supervised learning	Regression	Computing costs aren't high Simple to understand and execute	Under fitting
	K-NN	Implementation is simple Testing is performed in a simpler way	Noise sensitive Testing is exhaustive
	Bayesian belief network	It renders the processing easier Better speed and accuracy for larger data sets Framework can be effectively viewed Calculation fast and effective Not impaired by some unrelated characteristics	Has independent attributes In some cases where there is dependency among variables, it does not provide exact results
	Maximum entropy	Provides distribution appropriately Random variables are statistically independent	Needs additional human interaction, resources or metadata Not able to model the data that entail $p(x) = 1$ or 0
	Decision tree	Easy to implement No domain knowledge requirements Minimizes the ambiguity of complex decisions Interpretation is simple Can handle numerical and categorical data too	Space limitation Over fitting Classifier output is depending on dataset size If the form of dataset is numeric then a complex decision tree is created

(continued)

Table 2 (continued)

Approach	Method	Strengths	Weakness
	SVM	Higher accuracy Manage dynamic, nonlinear data points with ease Problems with alignment are not as frequent as other forms The dimensionality of the data collection is loosely reliant Produces precise and robust gradations It solves the multi-class problem by breaking it into two classes like one-against-one and one-against-all in pairs	High costs on computing Selecting the right kernel function is the main problem Specific kernel functions for each dataset give various effects It needs some time to equate the training cycle with other processes Mostly solve the binary class problem only Hard to understand subsequent pattern
	Neural network	Recognize complex interrelationships involving control variables factors effectively Capable of handling noisy data	Costly on computation Miniature accuracy Local Minima Over-fitting Black-box templates Processing is difficult to interpret High processing time is required
	Deep learning	High performance in many problems Larger dataset can process Covenant with deep architecture Easy to train algorithms Model needs minimal or no domain expertise Foster multi-task learning Breed pragmatic and labelled synthetic data Better on issues of pattern detection Cross-modal processing is possible Mostly supervised modes of learning	Data-hungry, thus not appropriate for low data scale Challenging to interpret High costs on computing Mode collapsing Vanishing gradient Typically a black box, yet tough to comprehend

(continued)

Table 2 (continued)

Approach	Method	Strengths	Weakness
	Ensemble	Overcome over fitting	High costs on computing Lent preparation
	Rule based	Generate more easy to understand models Carries out a greater description of emotions on the phrase level than the term level	Suitable for handling imbalanced data Output and accuracy depend entirely on the defining rules
	Generalization	Predictive Dictionary is not required Confirm strong classification accuracy More suitable for specific domains	Affected by linguistic variations problems In most instances, classifiers trained on the datasets of one domain do not function with other domains
Unsupervised learning	*K*-means	Quick and efficient Less challenging method	Needs pre-cluster amounts Poor behaviour of categorical features Do not discover non-convex shape in the cluster The effect differs with outlier of presence
	Lexicon Based	Labelled data and the learning experience aren't really necessary	Needs strong language tools that aren't always accessible Lacks context or domain based classification capability Sarcastic words Dual meaning words Linguistics rules Object attribute co-references

(continued)

Table 2 (continued)

Approach	Method	Strengths	Weakness
	Partitioning	Larger dataset can process Simple to understand and executes fast	Strong initializing responsiveness, noise and outliers
	Hierarchical	Capabilities of visualization Easy to implement No need to indicate specified number of clusters in advance	Poor visualization for larger dataset For certain instances, have the difficulty of cubic time and sluggish Slow and low accuracy Memory requirement is on higher side Once a decision is made regarding merge or split point, that can't be undone Performance is very poor in the presence of noise and outliers Neither scalable
	Density based	No need to indicate specified number of clusters in advance Accommodate arbitrary shaped clusters Handle complex, non-static data Acted well even there are outliers	Sample points are not treated for different densities Not so good for big datasets Parameter selection is very tricky Slow to process Performance depends upon the distance measure
Semi-supervised learning approach		Able to use unlabelled data in their training processes and boost labelling in systems where the data labelled is unavailable	Graph-based methods: distribute labels to unlabelled records, while the efficacy of the mechanism of mark dissemination depends on the estimation of variations between feature vectors Wrapper-based methods: the system operates even if the greatest trust forecasts are correct Theme-based method: depends on the trust level chosen and the number of topics inferred

paper. The results are given in Figs. 3 and 4. As per our literature study, we found that ML approaches increasing day by day as the computing power and huge availability of dataset along with ease of use of inbuilt libraries of emerging programming languages like Python, R, etc. It doesn't mean that the popularity of Lexicon Based and Hybrid approaches have gone down as these are giving better results for existing corpus and betterment of ML algorithms. According to our findings for last four years, we observed that most researchers are inclined to the stock market as the global indices of stock are going down like anything due to the pandemic Covid-19

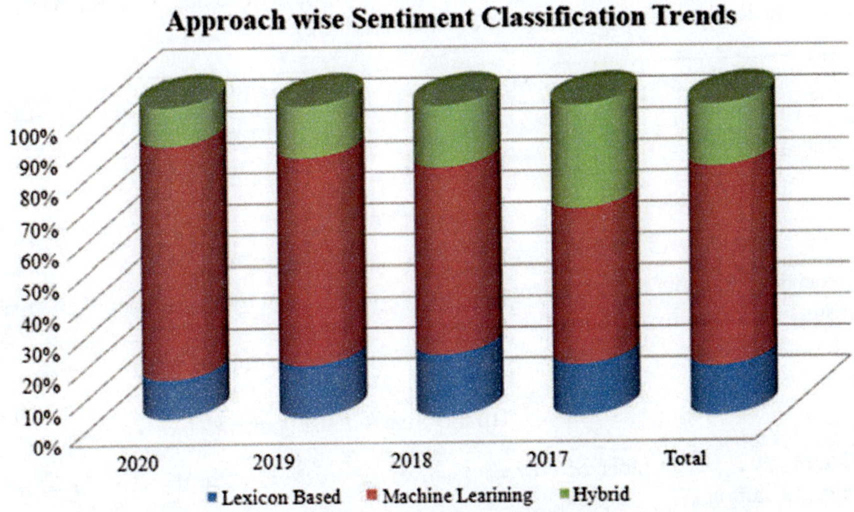

Fig. 3 Percentages of articles according to algorithm for sentiment classification

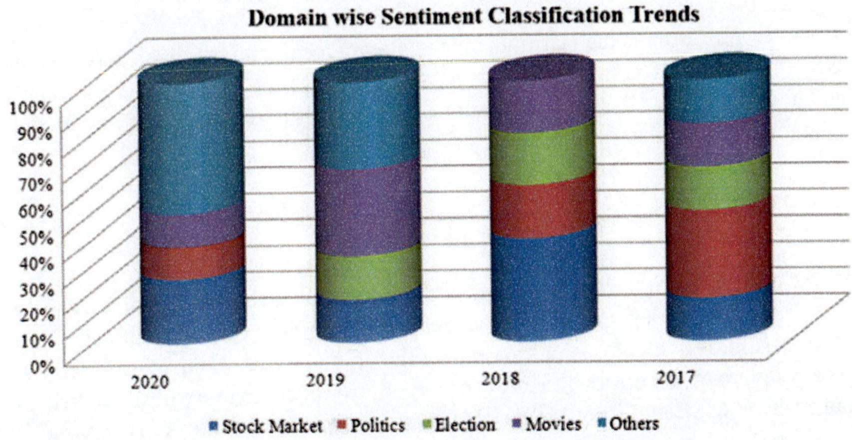

Fig. 4 Percentages of articles according to Domain for sentiment classification

outbreak. Apart from that election, politics, and movie reviews are the most favorable area of researchers in the last decade and they all go hand by hand.

3.4 Evaluation Metrics for SA

Evaluation metrics which are commonly used for SA have been discussed in this section. A confusion matrix is a tabular matrix which shows the performance of classification algorithms and the matrix can be illustrated as below on the basis of same various evaluation metrics of SA are based [22].

$n = 100$	Predicted: NO	Predicted: YES	
Actual: NO	TrueNegative = 40	FalsePositive = 10	50
Actual: YES	FalseNegative = 5	TruePositive = 45	50
	45	55	

Precision: It is the number of true positive results divided by the number of positive results predicted by the classifier.

$$\text{Precision} = \frac{\text{TruePositive}}{\text{TruePositive} + \text{FalsePositive}} \quad (1)$$

Recall: It is the number of correct positive results divided by the number of *all* relevant samples.

$$\text{Recall} = \frac{\text{TruePositive}}{\text{TruePositive} + \text{FalseNegative}} \quad (2)$$

F1 Score: $F1$ rating is the Harmonic mean among recall and precision and it varies from 0 to 1. It tells you preciseness of your classifier, as well as how robust it's far. The greater the $F1$ score, the better is the overall performance of any model. Mathematically, it may be expressed as:

$$F1 = \frac{2 * \text{Precision} * \text{Recall}}{\text{Precision} + \text{Recall}} \quad (3)$$

Mean Squared Error: Mean Squared Error (MSE) takes the average of the square of the difference among the predicted values and the original values specifically used for regression. It is much easier to compute the gradient with the help of MSE. As, we take square of the errors, the effect of larger error become more observable than smaller error. MSE and Root MSE (RMSE) are computed as follows.

$$\text{MSE} = \frac{1}{n}\sum_{i=1}^{n}\left(\hat{t}_i - t_i\right)^2 \tag{4}$$

$$\text{RMSE} = \sqrt{\frac{1}{n}\sum_{i=1}^{n}\left(\hat{t}_i - t_i\right)^2} \tag{5}$$

A lower value of MSE and RMSE signifies the improved performance of the model.

Least Absolute Error (LAE): It is commonly used evaluation measure to estimate the error of classification. Effectively, LAE describes the magnitude of the residuals. It is given as

$$\text{LAE} = \sum_{i=1}^{n}\left|\hat{t}_i - t_i\right| \tag{6}$$

where

n Number of test instances
$\hat{t}_i$ Vector of predicted sentiment values
t_i Vector of actual sentiment values

4 Application Area in Sentiment Analysis

SA is not limited to one application, but they facilitate enormous application in different vicinities to assist in the decision-making process [4, 42]. The application of SA is varying from business, marketing, social networking, politics, health, Academics, Research and Development of public action. The foremost applications areas of OMSA are as follows:

1. **Decision Making for Buying Service Product or Brand:** Purchasing of an item or selecting some services, taking the right choice is not an intense errand at this point. By this strategy, people can without trouble assess other's belief and experience about an item and also can easily assess the contending brands.
2. **Influence and Public Voice Analysis:** On the other side, the makers can accumulate the pundit's feelings about their product; brand or services and they can improve the high-caliber of their item. Strategic policymakers can use this hidden knowledge in creating new inhabitant responsive policies and promote the business.
3. **Link Analysis and Recommendation Systems:** Link analysis is primarily used to identify the connections between nodes. By characterizing the individuals' feelings into +ve or −ve, the framework can easily identify which should be getting recommended and which one should not get recommended.

4. **Trend Analysis and Topic Detection:** Supervising content available on social media is straightforwardly possible to detect trends. It also can able to identify the +ve, −ve or neutral content.

5 Open Issues in Sentiment Analysis

A prominent shortcoming of the SA via ML-based and lexicon-based approach at the document level is the reviews with a lot of noise which might be habitually assigned a neutral score. The reason for that is that the technique can be unsuccessful to detect any sentiment. The followings are the same usually diagnosed challenges in SA [4].

1. **Direct Versus Comparison Opinion**—Direct conclusion offers +ve or −ve input about the item legitimately though examination feeling contrasts the article and some other comparative articles [29].
2. **Domain Independence**—Whenever the person passes opinion about any topic, it may be possible that one feature set may give very excellent results in one domain; simultaneously it may perform extremely poor in some other area. "Go read the book" communicate +ve sentiment in case of a book-related reviews, whereas it becomes a—the sentiment for reviews of the movie [29].
3. **Named Entity Recognition**—What is the individual honestly speaking about, for example, are 300 Spartans is a film or set of Greeks or only a number?
4. **Parsing**—What is the subject and the object of the sentence, which one does the verb and/or adjective clearly refer to?
5. **Sarcasm**—If we don't know the first author's sentiments; we don't have a grasp of idea whether good means really good or bad [30, 37].
6. **Ambiguous or More Complex Situations**—"The hotel room is on the ground floor right by the reception". Is that +ve, −ve or its neutral?
7. **Linguistic Complications**—In current literature, there are still more challenges exist with conditional sentences in terms of language like in Chinese, Hindi, Arabic, etc.
8. **Scale**—What is the quantity of statistics input as a percentage of the entire universe of users? 10% of the Twitter or social networking sites give you a rough idea of what's going on, however, the outcomes are nowhere near the resolution in case you get to 50% of the reviews.
9. **Detection of Fake Reviews**—The internet may contain both good and garbage reviews given by the opinion holder. For compelling SA, these spam substances ought to be dispensed with before handling. This should be possible by recognizing duplicates, by distinguishing anomalies and by thinking about the notoriety of the commentator.

But, there are still a few demanding situations to overcome before SA can turn out to be a greater ideal mechanism. As an instance, human judgment remains a way accurate as a measure in SA. Automated systems can't discriminate sarcasm from the honest text, nor do they always properly analyze the precise context of the sentence.

Uses of acronyms like 'lol' or phrase abbreviations along with smiley's and symbols also additionally pose interpretation challenges. Furthermore, it's additionally not likely that an automatic machine could identify biased or fake opinions on a product or service.

6 Conclusion

The carried out systematic literature presents facts on studies on SA in social media. In this paper, diverse strategies for SA methods had been mentioned, including ML, ensemble methods, and dictionary (lexicon) based strategies. ML algorithms, like CNN, LSTM and GA outperformed the other methods and gives over 90% accuracy along with the NB, ME, and SVM, carried out the accuracy of about 80% while n-gram and bigram version were utilized. There are various techniques delivered through researches, nevertheless, the most common approach makes use of Lexicon based approach is SentiWordnet and TF-IDF even as for ML is NB and SVM. With the intention to enhance the quality and accuracy of the result, it is found that a combination of lexical and ML technique works best. The most famous social media website to extract data is Twitter as most of the cited paper here used Twitter as their main dataset and that is because of the richness of Twitter content, their availability along accessibility. Each day twitters users post more that millions of tweets on almost any subject matter with hashtags. This indicates that social media is turning into a treasured supply of information. The motive is to perform the responsibilities alongside finishing concern matter versions amongst units of files and examine the majority sentiments the use of precise algorithms and based totally mostly on to enhance the adaptation of the system.

The −ve aspect of an NLP-based method is that it'd probably carry out very poorly when used in the grammatically wrong textual content. As a whole lot of the sentiment behaviour, content to be had on the net ought to fall into that class, strategies to hit upon, and possibly correct terrible English would be necessary before to be used on a larger scale. However, it is also observed that less interest was shown by the researchers to other social media likes of Facebook, YouTube, WordPress, blogs, and many other social media platforms. The information on each social media is probably distinctive and it is significant to investigate other sources as mentioned earlier which might open to new expertise and conclusion. Further exploration is necessary to increase a generic version of the SA that may be useful for a unique type of statistics, which can investigate the different capabilities of social media platforms to achieve customers' opinions and getting a bigger perspective of SA utility. As a result, an exploration into the causes of those overall performance instabilities might be a fascinating course for destiny works. Further work is needed on additional recuperating both the accuracy of the classification and also improving the rate at which this huge amount of information must be processed.

References

1. Batra, R., Daudpota, S.M.: Integrating StockTwits with sentiment analysis for better prediction of stock price movement. In iCoMET, pp. 1–5 (2018)
2. Baydogan, C., Alatas, B.: Sentiment analysis using Konstanz information miner in social networks (2018), pp. 1–5
3. Bhatia, S., Sharma, M., Bhatia, K.K.: Internet of Things and Big Data Analytics Toward Next Generation Intelligence, Studies in Big Data, vol. 30 (2018)
4. Bing, L.: Sentiment analysis and subjectivity. In: Handbook of Natural Language Processing, 2nd edn. Synthesis Lectures on Human Language Technologies. Morgan and Claypool Publisher (2010)
5. Branz, L., Brockmann, P.: Sentiment analysis of twitter data: towards filtering, analyzing and interpreting social network data. In: DEBS, pp. 238–241. ACM (2018)
6. Daniel, J.S.O., Paulo, H.S.B., José, R.P., Daniely, A.B.: The application of the sentiment analysis technique in social media as a tool for social management practices at the governmental level. Braz. J. Publ. Adm. **53**(1), 235–251 (2017)
7. Dragoni, M.: Computational advertising in social networks: an opinion mining-based approach. In: Proceedings of ACM SAC Conference, pp. 1798–1804 (2018)
8. Estrada, M.L.B., Cabada, R.Z., Bustillos, R.O., Graff, M.: Opinion mining and emotion recognition applied to learning environments. Expert Syst. Appl. 113265 (2020)
9. Haddi, E.: Sentiment analysis: text pre-processing, reader views and cross domains. Dissertation, Brunel University London (2015)
10. Hamouda, A., Marei, M., Rohaim, M.: Building machine learning based senti-word lexicon for sentiment analysis. J. Adv. Inf. Technol. **2**(4), 199–203 (2011)
11. Hao, F., Park, D., Pei, Z.: When social computing meets soft computing: opportunities and insights. Hum. Centric Comput. Inf. Sci. **8**, 8 (2018)
12. Hogenboom, A.: Sentiment analysis of text guided by semantics and structure. Dissertation, Erasmus University Rotterdam (2015)
13. Hu, Y.: Text mining and data information analysis for network public opinion. Data Sci. J. **18**(7), 1–8 (2019)
14. Hua, W., Wang, Z., Wang, H, Zheng, K., Zhou, X.: Understand short texts by harvesting and analyzing semantic knowledge. IEEE Trans. Knowl. Data Eng. **29**(3), 499–512 (2017)
15. Indra, Winarko, E., Pulungan, R.: Trending topics detection of Indonesian tweets using BN-grams and Doc-p. J. King Saud Univ. CIS (2018)
16. Imane, E.A., Gahi, Y., Messoussi, R., Chaabi, Y., Todoskoff, A., Kobi, A.: A novel adaptable approach for sentiment analysis on big social data. J. Big Data **5**, 12 (2018)
17. Jain, D.K., Kumar, A., Sharma, V.: Tweet recommender model using adaptive neuro-fuzzy inference system. Futur. Gener. Comput. Syst. **5571**, 1–24 (2020)
18. Joshi, A., Balamurali, A.R., Bhattacharyya, P., Mohanty, R.: C-feel-it: a sentiment analyzer for micro-blogs. In: The 49th Annual Meeting of the Association for Computational Linguistics: Human Language Technologies, Portland, Oregon, USA, pp. 127–132 (2011)
19. Kotelnikov, E.V.: TextJSM: text sentiment analysis method. Autom. Documentation Math Linguist. **52**(1), 24–34 (2018)
20. Li, G., Liu, F.: A clustering-based approach on sentiment analysis. In: International Conference on Intelligent Systems and Knowledge Engineering, Hangzhou, pp. 331–337 (2010)
21. Li, M., Yang, C., Zhang, J., Puthal, D., Luo, Y., Li, J.: Stock market analysis using social networks. ACM (2018)
22. Mishra, A.: https://towardsdatascience.com/metrics-to-evaluate-your-machine-learning-algorithm-f10ba6e38234. Last accessed 2020/04/10
23. Mohammad, E.M., Mohammad, S.A., Hamidreza, K.: Aspect-based sentiment analysis using adaptive aspect-based lexicons. In: Expert Systems with Applications (2020)
24. Pathak, A.R., Agarwal, B., Pandey, M., Rautaray, S.: Application of deep learning approaches for sentiment analysis. In: Deep Learning-Based Approaches for Sentiment Analysis. Algorithms for Intelligent Systems, pp. 1–31 (2020)

25. Pavlopoulos, I.: Aspect based sentiment analysis. Dissertation, Department of Informatics, Athens University of Economics and Business (2014)
26. Pimpalkar, A.: Review of online product using rule based and fuzzy logic with smileys. Int. J. Comput. Technol. **1**(1), 39–44 (2014)
27. Pimpalkar, A.: A sentimental analysis of movie reviews involving fuzzy rule-based. Int. J. Artif. Intell. Knowl. Discov. **3**(2), 9–14 (2013)
28. Poecze, F., Ebster, C., Strauss, C.: Social media metrics and sentiment analysis to evaluate the effectiveness of social media posts. Procedia Comput. Sci. 660–666 (2018)
29. Pozzi, F.A., Fersini, E., Messina, E., Liu, B.: Challenges of sentiment analysis in social networks. In: Sentiment Analysis in Social Networks, pp. 1–11 (2017)
30. Rahman, L., Sarowar, G., Kamal, S.: Teenagers sentiment analysis from social network data. In: Social Networks Science: Design, Implementation, Security, and Challenges, pp. 3–23 (2018)
31. Rani, R., Lobiyal, D.K.: Performance evaluation of text-mining models with Hindi stopwords lists. J. King Saud Univ. CIS (2020)
32. Rashid, A., Asif, S., Butt, N.A., Ashraf, I.: Feature level opinion mining of educational student feedback data using sequential pattern mining and association rule mining. Int. J. Comput. Appl. **81**(10) (2013)
33. Ruz, G.A., Henriquez, P.A., Mascareno, A.: Sentiment analysis of Twitter data during critical events through Bayesian networks classifiers. Future Gener. CS (2020)
34. Seghouani, N.B., Bugiotti, F., Hewasinghage, M., Isaj, S., Quercini, G.: A frequent named entities-based approach for interpreting reputation in Twitter. Data Sci. Eng. **3**, 86–100 (2018)
35. Shayaa, S., Jaafar, N.I., Bahri, S., Sulaiman, A., Wai, P.S., Chung, Y.W., Piprani, A.Z., Ali, M., Garadi, A.: Sentiment analysis of big data: methods, applications, and open challenges 1–26 (2018)
36. Stylianidis, M., Galiotou, E., Sgouropoulou, C., Skourlas, C.: Opinion mining using an LVQ neural network. In: Proceedings of the 21st Pan-Hellenic Conference on Informatics, Article 61, pp. 1–5 (2017)
37. Sukthanker, R., Poria, S., Cambria, E., Thirunavukarasu, R.: Anaphora and coreference resolution: a review. Inf. Fusion **59**, 139–162 (2019)
38. Tang, B., Kay, S., He, H.: Toward optimal feature selection in Naïve Bayes for text categorization. IEEE Trans. Knowl. Data Eng. **28**(9), 2508–2521 (2016)
39. Wang, J., Dong, A.: A comparison of two text representations for sentiment analysis. In: International Conference on Computer Application and System Modelling, pp. 35–39 (2010)
40. Wataru, S., Irena, V., Hideaki, A.: Enhanced news sentiment analysis using deep learning methods. J. Comput. Soc. Sci. **2**, 33–46 (2019)
41. Zhou, X., Tao, X., Rahman, M.M., Zhang, J.: Coupling topic modelling in opinion mining for social media analysis. In: Proceedings of International Conference on Web Intelligence, pp. 533–540 (2017)
42. Zulfadzli, D., Haliyana, K.: Sentiment analysis in social media and its application: systematic literature review. Procedia Comput. Sci. **161**, 707–714 (2019). (Fifth Information Systems International Conference)

Dynamic Analysis of a Novel Modular Robot

Omkar D. Dixit and Vishal V. Dhende

Abstract A vast amount of research is being carried out in the field of modular robotics. The research is mainly focused on improving locomotion and docking capabilities. This paper presents dynamic analysis of a novel modular robot. The modular robot has nested configuration and thus can be assembled in various forms and can be adapted for various environments. Due to the availability of both unconventional and wheeled locomotion, dynamic analysis is carried out for each configuration separately. Euler–Lagrangian approach is used for analysis of unconventional locomotion. A MATLAB program is created to solve the equations of dynamics. While robot locomotion analysis is done to determine torque requirement for wheeled locomotion. Validation of analysis is done with the help of virtual robotic experimentation platform (VREP)—a robotic simulator.

Keywords Robot dynamics · Modular robot · Locomotion analysis

1 Introduction

Modular robot consists of multiple small units assembled together either manually or automatically which can transmit forces, moments, electrical power, and can communicate throughout the robot. Modular robots find their application in various fields like search and investigation in hazardous or constricted environments, in disaster affected area or even in education. Size of the modular robots is decreased drastically with the advent of smaller controllers, batteries, and other electronic components. Most of the modular robots are comprised of docking mechanisms, locomotion actuators, sensors, controllers, and batteries to be accommodated in compact space. This leads to increased mechanical complexity of these robots.

O. D. Dixit (✉) · V. V. Dhende
Department of Mechanical Engineering, Walchand College of Engineering, Sangli, MH 416415, India

V. V. Dhende
e-mail: vishal.dhende@walchandsangli.ac.in

H. Sharma et al. (eds.), *Congress on Intelligent Systems*, Advances in Intelligent Systems and Computing 1334, https://doi.org/10.1007/978-981-33-6981-8_61

Locomotion and autonomy are major challenges in modular robots [1]. Locomotion trend in modular robotics is currently focused toward enabling individual modules to move independently and to implement wheel-based locomotion [2]. Modular robots are classified as homogeneous and heterogeneous modular robots. Homogeneous modular robots have all the modules similar in every aspect, while heterogeneous modular robots have two or more types of modules with specific functions. Lyder et al. in [3], describe the design of such a heterogeneous modular robot named Odin. Whereas Spröwitz et al. in [4] present the hardware of Roombots robot. Ryland in [5], describe geometric considerations in design of iMobot. Both Roombots and iMobot belong to homogeneous modular robots category. Li et al. in [6] performed kinematic analysis of a novel self-reconfigurable modular robot using screw theory. A C++ based framework is presented by Meisster et al. in [7], which helps in generation of kinematic and dynamic equations for modular robots and also demonstrates the use of framework for two projects. Ning et al. in [8] present kinematic and dynamic analysis of modular multi-legged robot with various leg configurations. Baca et al. in [9] present design and simulation of a modular aerial system.

Due to the presence of only single type of module, docking and communication throughout the robot are simplified thus making homogeneous modular robots more popular than heterogeneous ones. But implementation of wheeled locomotion is difficult in homogeneous modular robots due to the constraint of single type of module. As a solution to this, a novel heterogeneous modular robot is presented here.

The proposed robot is to have three types of modules viz. controller module, locomotion module, and sensor module. Controller module is designed to accommodate microcontroller along with a battery and a servo motor connected to removable link. The link is used to connect a controller module with another controller module for serial configuration. Locomotion module will consist of a DC motor, a motor driver and Wheel, and derive power from controller module. The sensor module is designed to accommodate camera sensor with Wi-Fi for data transmission to user's mobile and ultrasonic sensor for obstacle avoidance. Sensor module will have dedicated power source as simultaneous operation of camera and Wi-Fi demands source to deliver large continuous current. Figure 1 shows the CAD models of all three types of modules. Some panels of modules are hidden to show details.

This work presents the dynamic analysis of aforementioned modular robot. Analysis is carried out in two parts. The organization of paper is as follows: Sect. 2 depicts the kinematics of the robot and trajectory planning for selected poses. Section 3 presents the dynamic analysis for both unconventional and wheeled locomotion also shows the validation results. And, Sect. 4 draws conclusion.

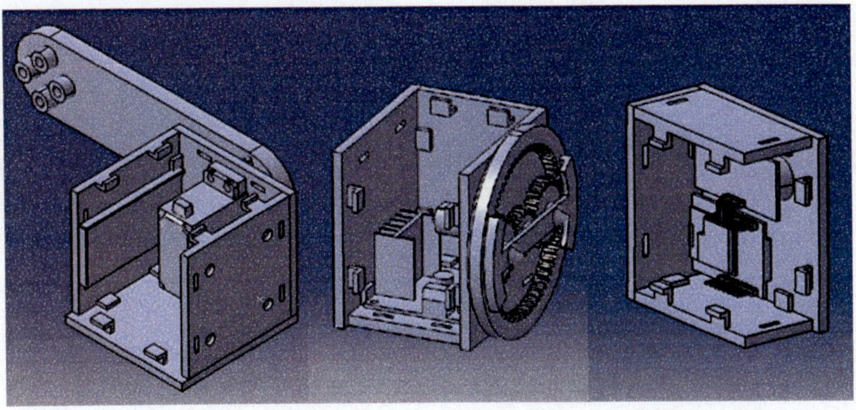

Fig. 1 Controller module, locomotion module, and sensor module (From left to right)

2 Kinematic Analysis and Trajectory Planning

Robot kinematics is the study of robotic mechanisms without the consideration of forces or torques responsible for the motion. Inverse kinematics computes the joint angles of robot on the basis of given final position and orientation. To represent pose of a body, a reference coordinate frame is attached to the body usually at its mass center, according to rules proposed by Denavit and Hartenberg in 1955 [10]. Trajectory planning focuses on implementing jerk-free motion in a robotic system. We need to find joint torque, and in particular cycle time, variation of this torque should be smooth. The expression of joint torque, which is given in next topic, has angular displacement, angular velocity and angular acceleration as variables. So, to ensure the smooth variation of joint torque, the angular velocity and angular acceleration should be continuous. In proposed robot, kinematic analysis and trajectory planning are done to calculate acceleration of links of robot. This acceleration is taken as input for dynamic analysis.

2.1 Poses Selected for Trajectory Planning

Figure 2 shows the assigned coordinate system to the proposed modular robot. It should be noted that, the axis systems are assigned at the farthest end of the joint, where the link connects to another module. This is done in order to get the reactive torque. Here, the cubical bodies represent the volume of the robot. Both the body and link are representative in nature and do not show the final topology of robot. Figure 2 also shows the Denavit and Hartenberg (DH) parameter table for said robot.

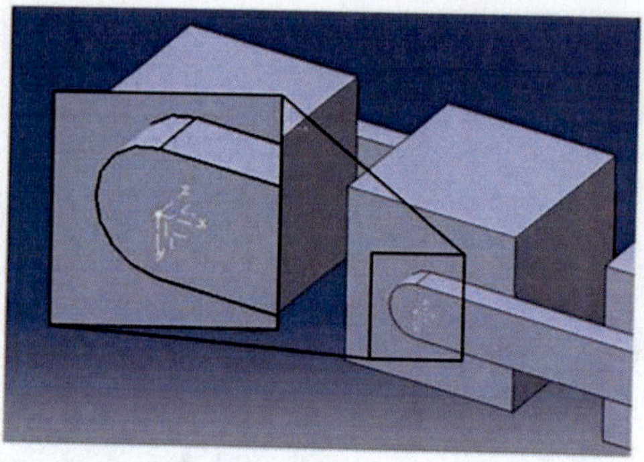

Frame	θ	*d*	α	*a*
1	θ_1	0	0	100
2	θ_2	0	0	100
3	θ_3	0	0	100
4	θ_4	0	0	100

DH parameters table

Fig. 2 Coordinate system assignment to joints and DH parameters table of robot

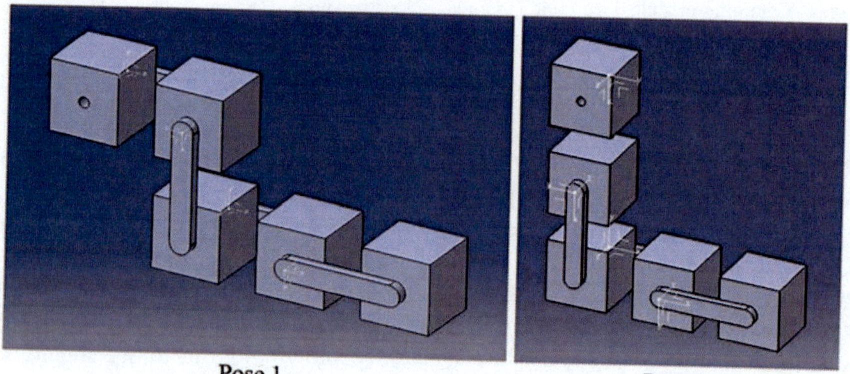

Pose 1 Pose 2

Fig. 3 Two of the selected poses for dynamic analysis

Kinematic analysis is only required for serial configuration of the robot. Few poses are selected for the analysis. Figure 3 shows two of these poses. Kinematic analysis is done by assuming the base link to be fixed. In reality, when robot moves, none of the links is fixed. But poses are selected such that very few links are rotated and other links are stationary so as to counteract the reactive force. With sufficient number of modules connected behind the rotating links base link can be assumed as fix with reasonable accuracy. The output of kinematic analysis is the angular displacement of each joint. These displacements are converted in joint-space coordinates from Cartesian coordinates and are used as inputs for trajectory planning.

Table 1 gives the variation of all the joint angles with respect to time. The time steps are selected on the basis of assumed maximum velocity of actuator. Pose 1 is achieved in three steps. Pose 2 achieved at the end of first time step and then the robot returns to its original pose. As can be seen, joint 3 and 4 are crucial for torque calculation, thus only variation of joint 3 and 4 is plotted.

Table 1 Change in joint angle (in radians) during given time for pose 1 and pose 2

Pose	Joint	Time steps (milliseconds)		
		0–800	801–1600	1601–2400
Pose 1	Joint 1	0–0	0–0	0–0
	Joint 2	0–0	0–0	0–0
	Joint 3	0–0	0–(−1.57)	(−1.57)–(−1.57)
	Joint 4	0–(−1.57)	(−1.57)–(−1.57)	(−1.57)–1.57
Pose 2	Joint 1	0–0	0–0	NA
	Joint 2	0–0	0–0	NA
	Joint 3	0–(−1.57)	(−1.57)–0	NA
	Joint 4	0–0	0–0	NA

2.2 *Trajectory Planning*

Trajectory planning is accomplished by applying a trajectory function to the angular displacement. There are several trajectory functions that can be used. One of them is the cubic polynomial function. Angular displacement is considered as cubic function of time. We have considered a simple case, where joint velocities at start and end are zero and joint angles are known from kinematics. This gives us the initial conditions for solving the equation. Time required for joint to completely change its pose is selected on the basis of requirement of task completion and speed of commonly available actuators. Following is the required trajectory planning equation.

$$\theta(t) = \theta_i + \frac{3(\theta_f - \theta_i)}{t_f^2} t^2 + \left(-\frac{2(\theta_f - \theta_i)}{t_f^3} t^3 \right) \tag{1}$$

This equation is used to fit a smooth curve between joint angle and time. A MATLAB program is created to plot the graph of various joint angles against time. Figure 4 shows the plot of joint angles against time for presented kinematic poses.

3 Dynamic Analysis

Robot dynamics is the study of relationship between various forces on robot and accelerations they produce. This section presents use of dynamic analysis applied to a novel modular robot to select actuators. The modular robot has both unconventional and wheeled locomotion systems. So, the dynamic analysis is carried out in two parts. The first part depicts use of Lagrangian approach for finding out torque requirement of unconventional locomotion. The second part depicts use of locomotion analysis to determine torque requirement of wheeled locomotion.

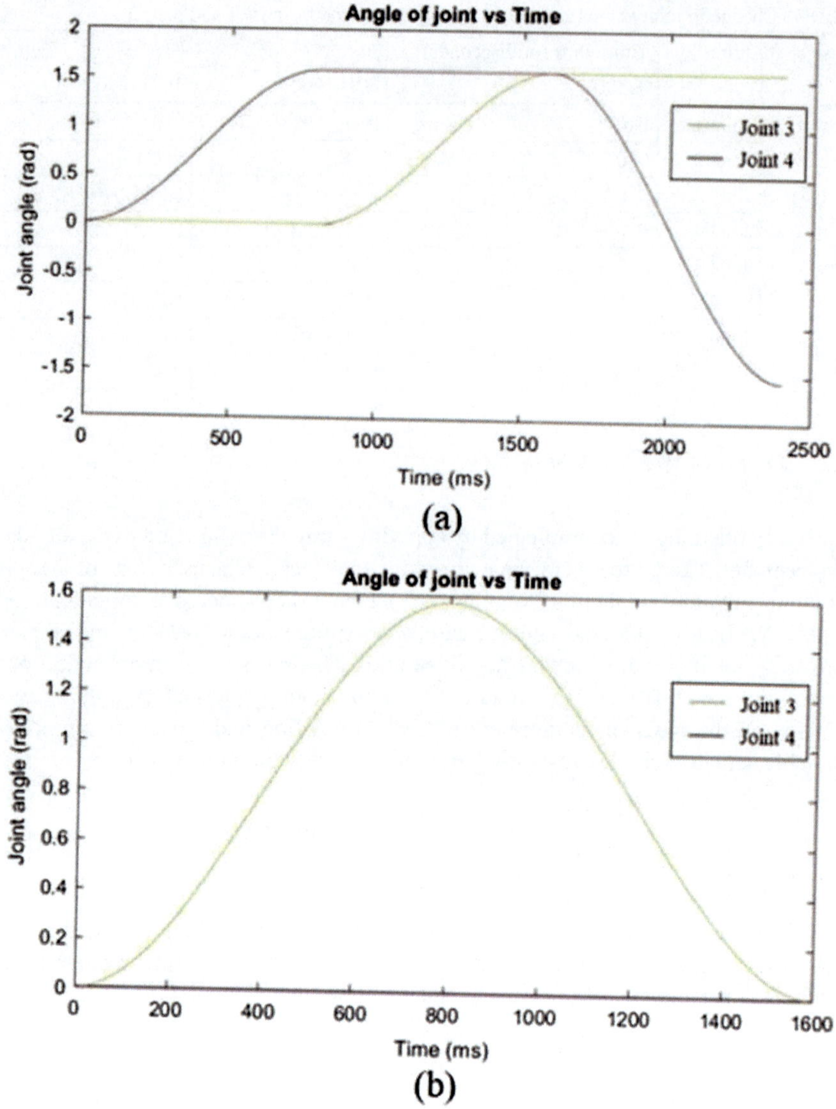

Fig. 4 Trajectory planning for **a** Pose 1 and **b** Pose 2

3.1 Dynamic Analysis for Unconventional Locomotion

Inverse dynamic analysis is helpful in determining the torque requirements of robotic joints from inertia properties of links and their acceleration. Lagrange-Euler approach is used to solve the inverse dynamics problem [11]. Joint torque consists of various components like inertia, centrifugal, Coriolis components. Euler–Lagrange equation is as follows.

$$\frac{\mathrm{d}}{\mathrm{d}t}\left(\frac{\partial L}{\partial \dot{q}_i}\right) - \frac{\partial L}{\partial q_i} = \tau_i \tag{2}$$

L in above equation represents Lagrangian and is the difference between kinetic and potential energy of the system. Where q_i and $\dot{q}_i$ represent angular displacement and velocity of ith joint for this particular case. And τ_i is the reaction torque at ith joint. Total kinetic and potential energy of n link serial manipulator are given by following equations.

$$\text{K.E.} = \frac{1}{2}\sum_{i=1}^{n}\sum_{a=1}^{i}\sum_{b=1}^{i}\left[Tr\left(U_{ia}J_i(U_{ib})^T\right)\dot{q}_a\dot{q}_b\right] \tag{3}$$

$$P = \sum_{i=1}^{n} -m_i\bar{g}\left({}^0_iT\,{}^i_i\bar{r}\right) \tag{4}$$

where $U_{ij} = \frac{\partial {}^0_iT}{\partial q_j}$, $U_{ijk} = \frac{\partial U_{ij}}{\partial q_k}$.

In above equation, ${}^0_iT$ is transformation matrix of link i with respect to origin, J_i is inertia matrix, m_i is mass of the ith link, $\bar{g}$ is gravity vector, and ${}^i_i\bar{r}$ is position of point i on ith link with respect to ith coordinate system. Therefore, the Lagrangian is given by,

$$L = \frac{1}{2}\sum_{i=1}^{n}\sum_{a=1}^{i}\sum_{b=1}^{i}\left[Tr\left(U_{ia}J_i(U_{ib})^T\right)\dot{q}_a\dot{q}_b\right] + \sum_{i=1}^{n} m_i\bar{g}\left({}^0_iT\,{}^i_i\bar{r}\right) \tag{5}$$

Putting this value of L in Eq. 1, and solving we get,

$$\tau_i = \sum_{c=1}^{n} D_{ic}\ddot{q}_c + \sum_{c=1}^{n}\sum_{d=1}^{n} h_{icd}\dot{q}_c\dot{q}_d + C_i \tag{6}$$

In Eq. 6, first term is the inertia term, second term is centrifugal and Coriolis term, and the last term is gravity term. It can be seen that acceleration and velocity of joints are required for torque calculation. Joint angles were determined in kinematic analysis from the position data of links. Angular velocity and angular acceleration were computed as a function of time in trajectory planning. Now, these angular velocity and acceleration can be substituted in Eq. 6 to get joint torque.

A MATLAB program is written to solve the equation. Various functions were created to compute all the terms in equation. As the weight of modules is less (around 350 gm), the gravity term is neglected. The program takes angular positions in radians of various links and time steps in milliseconds as input and computes joint torques accordingly for given time interval. The program also plots the variation of required torque against time. Inertia tensor of the module can be calculated

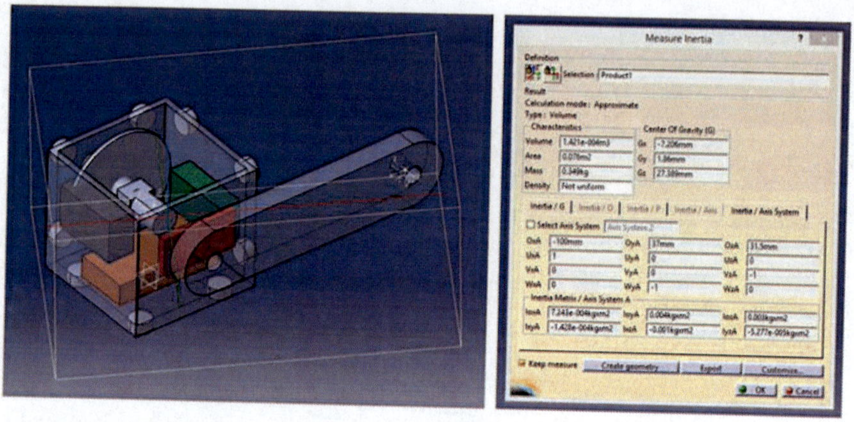

Fig. 5 CAD model of modular robot in system design phase and its inertia properties

by calculating individual inertia tensor of components, but the process is lengthy. CATIA can calculate inertia of a solid model. So, an approximate model of the robot module is created in CATIA with all the electronic and mechanical components in place. Materials are applied to model, such that weight of the components matches the actual weight. Inertia tensor and center of gravity are directly taken from CATIA. Figure 5 shows the CATIA model and inertia properties toolbox.

Figure 6 show the output of MATLAB program for dynamic analysis, when plot of joint angle vs time is given as input to the program for pose 1 and pose 2, respectively. It can be seen that, maximum torque for pose 1 is close to 0.5 N m and is acted upon joint 3. And maximum torque for pose 2 is close to 0.7 N m and is also acted upon joint 3.

3.2 *Wheeled Locomotion Analysis*

Robot locomotion analysis is done to calculate the torque and power requirement of the robot for locomotion purpose. Trafficability, Maneuverability, and Terrainability are three major performance indices [12]. Trafficability is concerned with robots ability to transverse on soft or hard terrain without the loss of traction. Maneuverability is concerned with navigation through environment. And terrainability is concerned with robots ability to negotiate terrain irregularities. Due to intended urban environment usage of robot and its small size, terrainability and maneuverability can be neglected. Trafficability considers sinking of wheels in ground, motion resistance and soil thrust and traction. For firm contact between ground and robot wheels, motion resistance is the only resistance considered in trafficability.

When a robot moves on paved surfaces or roads, it consumes energy to overcome the rolling resistance between tires and the ground, as well as gravitational and inertial forces. For speeds more than 60 kmph, aerodynamic forces also contribute in energy

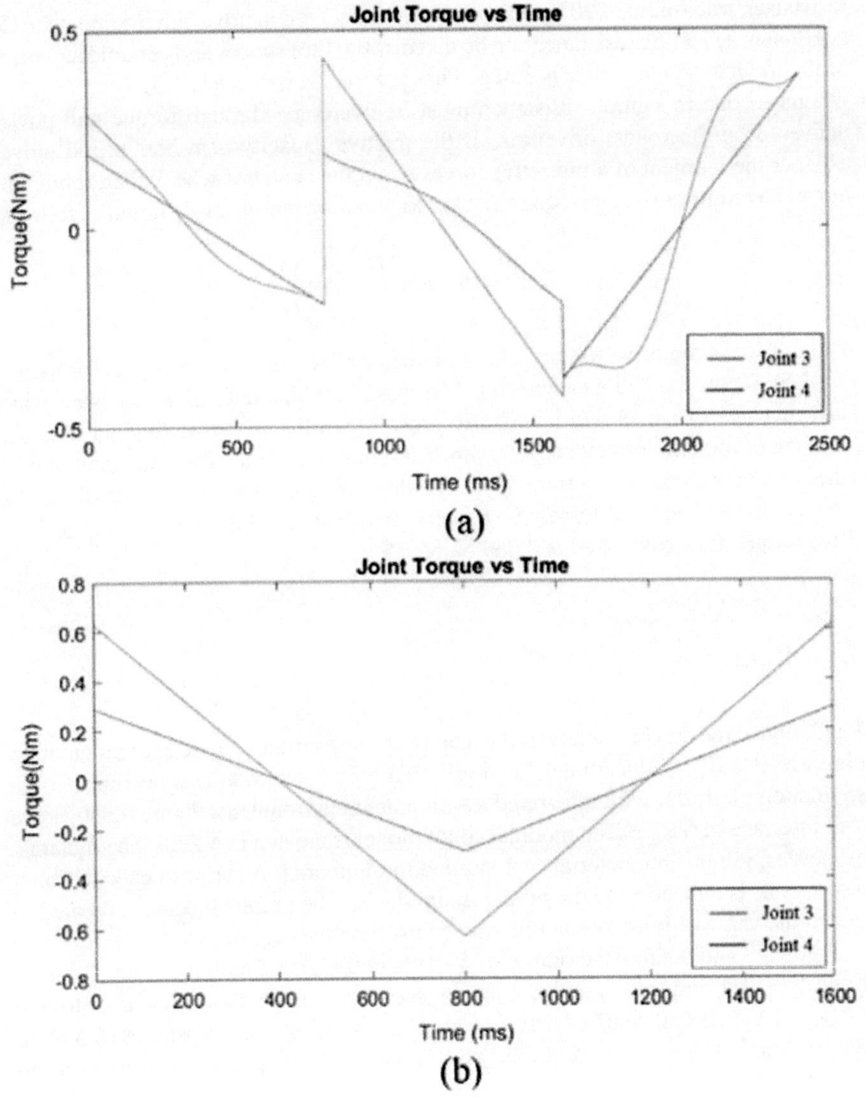

Fig. 6 Output of MATLAB program for **a** Pose 1 and **b** Pose 2

loss. Equation 7 can be used to calculate total motion resistance:

$$R_m = \mu W + W \sin\theta \tag{7}$$

where R_m indicates motion resistance, μ is the coefficient of friction between wheel and ground, W is the weight of the robot and θ is gradient of ground. In above equation first term represents rolling resistance and the second term represents gradient resistance. Maximum of 20° gradient can be considered in urban environments. Also coefficient of rolling resistance can be taken as 0.04 for rubber and ceramic and mass of the robot as given earlier is 350 g. Thus the total resistance is 1.31 N.

Losses due to motion resistance must be overcome through torque and power delivered by the robots drivetrain. If the tractive capacity of robots prime mover exceeds the moment of all resistive forces about the center of wheel, then robot will move. So minimum torque requirement of locomotion motor can be found as follows.

$$T_{\text{drive}} = R_m \times \left(\frac{D_w}{2} - \delta\right) \tag{8}$$

where T_{drive} is the drive torque, D_w is diameter of wheel, and δ is the deformation of wheel (0 for hard rubber wheels). The maximum diameter of wheel is selected such that size of wheel will be slightly greater than the size of individual module. The size of module is decided as 60 mm × 60 mm × 60 mm. Thus, the diameter of wheels is considered as 70 mm. Thus, the drive torque required is 0.046 N m. We have neglected frictional losses. So a factor of safety of 1.5 is considered. The final drive torque, thus calculated is 0.069 N m.

3.3 *Validation*

Validation of the dynamic analysis is carried out using virtual robotic experimentation platform (VREP) (Educational Version). VREP is a versatile and powerful robot simulation platform, with integrated development environment. Figure 7 shows the user interface of VREP. The modular robot model is created in VREP. The dynamic properties, joints, and materials are applied to the model. A script to execute robot movements is attached and the model is simulated. The graph objects are created to record the value of joint torque in N-m against time in seconds.

Figures 8 and 9 show the output of VREP simulation of modular robot for pose 1 and pose 2. From these graphs, it can be seen that, (a) Maximum value of torque for pose 1 occurs at the end of first time step and is approximately equal to 0.5 N m, and (b) Maximum value of torque for pose 2 occurs at the start of first time step and is approximately equal to 0.75 N m.

4 Conclusion

In this work, dynamic analysis of a novel modular robot by Euler–Lagrangian approach is presented. To derive and solve the dynamics equations, a MATLAB program was written whose main aim was to obtain the torque variation of robot joints. The output of the program is validated with the help of VREP robotic simulator. The maximum values of torque given by the program and simulator are found

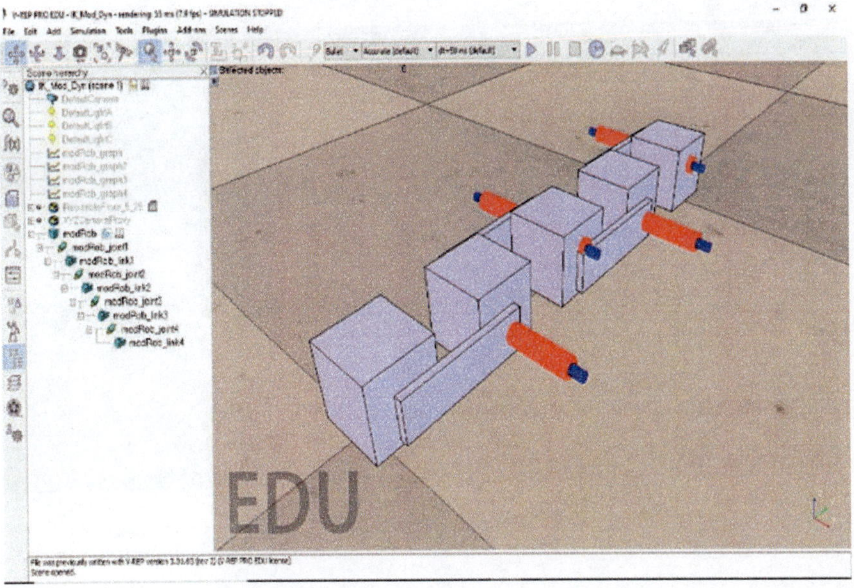

Fig. 7 VREP software UI with model of the proposed robot

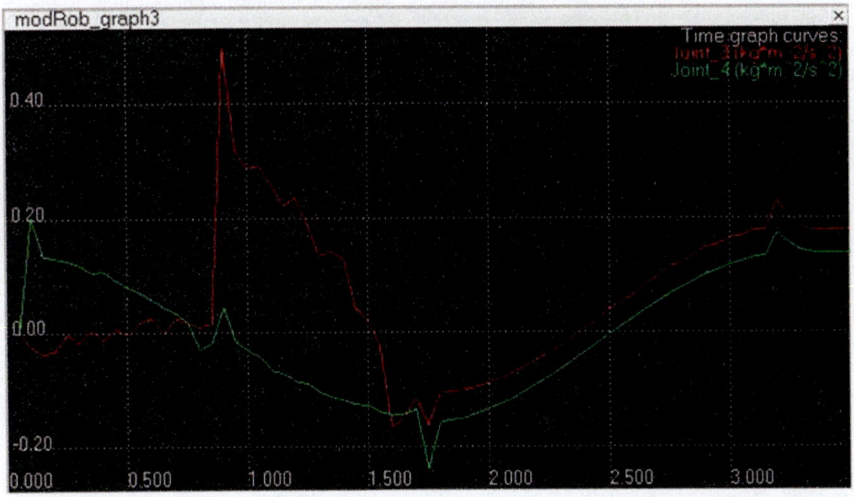

Fig. 8 Result of simulation in VREP for pose 1

to be very close with less than $\pm$ 0.1 N m variations. A small variation in graphs from program and simulator is observed, which is partly due to the neglected gravitational and frictional components in MATLAB program. But the trend of the graphs is almost similar. Locomotion analysis of the robot is also done for wheeled locomotion.

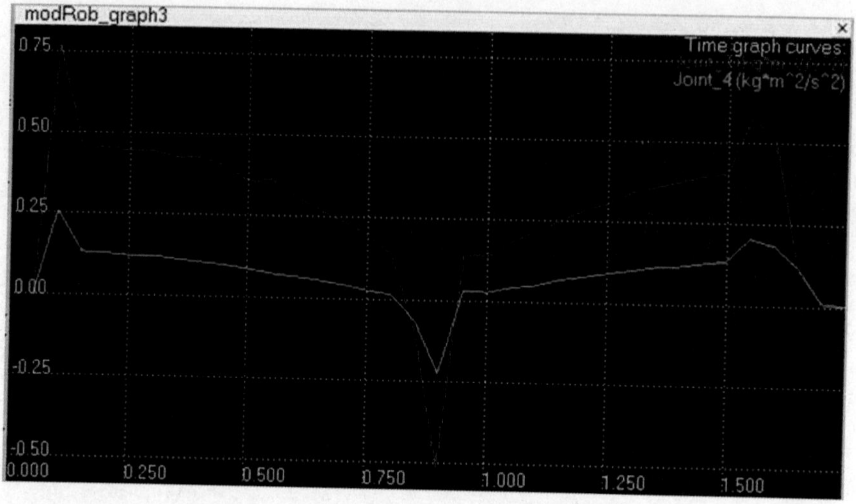

Fig. 9 Result of simulation in VREP for pose 2

References

1. Moubarak, P., Ben-Tzvi, P.: Modular and reconfigurable mobile robotics. Robot. Autonom. Syst. **60**, 1648–1663 (2012)
2. Brunete, A., Ranganath, A., Segovia, S., de Frutos, J.P., Hernando, M., Gambao, E.: Current trends in reconfigurable modular robots design. Int. J. Adv. Robot. Syst., 1–21 (2017)
3. Lyder, A., Garcia, R.F.M., Stoy, K.: Mechanical design of Odin, an extendable heterogeneous deformable modular robot. In: IEEE/RSJ International Conference on Intelligent Robots and Systems, pp. 883–888. IEEE, France (2008)
4. Spröwitz, A., Moeckel, R., Vespignani, M., Bonardi, S., Ijspeert, A.J.: Roombots: a hardware perspective on 3D self-reconfiguration and locomotion with a homogeneous modular robot. Robot. Autonom. Syst. **62**, 1016–1033 (2014)
5. Ryland, G.G.: Design of iMobot, an intelligent reconfigurable mobile robot with novel locomotion. In: IEEE International Conference on Robotics and Automation, pp. 60–65. IEEE, Alaska (2010)
6. Li, Y., Zhu, S., Wang, Z., Zhang, L., Ma, X., Cui, Z.: The kinematics analysis of a novel self-reconfigurable modular robot based on screw theory. In: 3rd International Conference on Mechanical, Industrial, and Manufacturing Engineering, pp. 23–30. China (2016)
7. Meister, E., Nosov, E., Levi, P.: Automatic onboard and online modelling of modular and self-reconfigurable robots. In: 6th IEEE Conference on Robotics, Automation and Mechatronics (RAM). IEEE, Philippines (2013)
8. Ning, M., Shao, L., Chen, F., Li, M., Zhang, C., Zhang, Q.: Modeling and analysis of a modular multilegged robot with improved fault tolerance and environmental adaptability. Math. Probl. Eng. (2019)
9. Baca, J., Jimenez, N., Winfield, K., Tay, S., Tijerina, B., Baierlipp, H., Cortez, J., Moreno, H.: Design and simulation analysis of a modular aerial system. In: Martínez, A., Moreno, A., Carrera, I.G., Campos, A., Baca, J. (eds.), Proceedings of the 2nd Latin American Congress on Automation and Robotics, Cali 2019, Lecture Notes in Networks and Systems, vol. 112, pp. 112–118. Springer, Colombia (2019)
10. Denavit, J., Hartenberg, R.S.: A kinematic notation for lower-pair mechanisms based on matrices. J. Appl. Mech. **22**, 215–221 (1955)

11. NPTEL Robotics course. https://nptel.ac.in/courses/112/105/112105249/. Last accessed 22 Jan 2020
12. Dimitrios, S.: Apostolopoulos: analytical configuration of wheeled robotic locomotion. Ph.D. Thesis, The Robotic Institute, Carnegie Mellon University, Pennsylvania (2001)

Design and Analysis of Mechanical Properties of Simple Cloud-Based Assembly Line Robots

Sivakumar Srrinivas and Ramachandran Hari Krishnan

Abstract Industrial robots have been extensively used in factory shop floors to do repetitive, tedious and dangerous operations such as assembly, painting, packing and welding. To further widen the functional capabilities, the existing pre-programmed robots are integrated with networking concepts to cultivate the networked robotics culture. This study provides an insight on design of a simple automated robotic assembly line, which serves the purpose of screwing bolts to the workpiece incoming through the assembly line. This finds its application in an assembly line of any production industry, where difficulty is faced by line workers in repeatedly joining parts using bolts and nuts. The assembly line is designed to consist of two robots—a 4-DoF SCARA type robot and a 5-DoF pick and place robot. The kinematic model and study for both the robots were performed, and 3D CAD models of both the robots were made and assembled. Analysis of various mechanical properties of both the robots such as stress, strain, displacement and buckling load are performed to ensure nominal operations. The assembly line is made to virtually show how the robots will work in tandem to achieve the intended purpose. Technological advancement of cloud computing can be integrated to the assembly line for improved performance parameters.

Keywords Industrial robot · Cloud robotics · Assembly line · Internet of things · SCARA robot

1 Introduction

The manufacturing industry has been evolving continuously for the past few decades. Starting from the introduction of the automated assembly line in the 1940s, implementation of NC and CNC machine tools, to the current advanced operation technologies, robots have served as much as the human workforce [1]. An industrial robot is defined as a common-utility machine, which can be re-programmed to carry

S. Srrinivas · R. Hari Krishnan (✉)
School of Mechanical Engineering, SASTRA Deemed to be University, Thanjavur, Tamil Nadu, India
e-mail: harikrishnan.r@mech.sastra.edu

H. Sharma et al. (eds.), *Congress on Intelligent Systems*, Advances in Intelligent Systems and Computing 1334, https://doi.org/10.1007/978-981-33-6981-8_62

out functionalities like learning, reasoning and judgement which are more human-like. Automobile industries implement a lot of these robotic technologies to improve the flexibility in their production. Robots are mainly used in welding, assembly, spray-coating and material handling operations [1].

Over the years, industrial production is being driven by world-wide competition and the need for continuous adaptation to the ever-changing customer requirements is the need of the hour. This can only be improved by dynamic improvements in the existing manufacturing facilities. Industry 4.0 is one such promising approach that is based on the factory integration in terms of manufacturing, business and company's value-chain aspects. Recent technological booms such as cloud computing, Internet of Things (IoT) [2] and Big data analysis are being extensively exploited in many recent developments in the field of networking and advanced computing. When robotic technologies and advanced networking concepts are integrated together as one system, the result is a cognitive solution called 'cloud robotics'. Home automation devices are small-scale examples of how cloud robotics work.

The concept of cloud robotics when integrated with factory setup as in a manufacturing industrial line has a number of advantages that can improve production rates, time to market for new products and more importantly achieving customer-oriented product development [3]. This also benefits the company, as huge resources spent on the currently used pre-programmed robots can be replaced with cloud technology that serves as an extended source of computational power, storage and memory.

2 Literature Review

Mahil and Al-Durra [4] had developed a mathematical model for a 2-DOF serial manipulator. A traditional approach of using D-H parameters and Lagrangian equations has been followed. In order to control this dynamic system, the mathematical model has been linearised. This resulted in the development of a full-state feedback controller and a full-order observer.

A 4-DOF robot arm was designed and implemented to work as a pick and place robotic arm which finds its usage in the assembly line of a manufacturing industry [5]. With the help of MATLAB Simulink, the forward and inverse kinematics of the manipulator was studied and simulated. Using a data acquisition (DAQ) card and MATLAB, a digital PID controller was designed and implemented to control the angular movement of the links by controlling the DC motors. Using a digital camera and image processing toolbox in MATLAB, a simple graphic user interface (GUI) application was developed to demonstrate the working of the robotic arm. The arm identifies certain objects present within its workspace based on their shape and colour, moves towards them, picks and places them at a certain position. The working prototype of the 4-DOF robot was made from scratch with light wood, so that it would be easier to handle it, but this led to some problems due to its weakness and instability in its motion.

A smart cloud robot [6] that is controlled using an Arduino UNO that communicates with a cloud data base using Raspberry Pi was developed by Prabha et al. The system has an autonomous robot which is employed to acquire environmental data such as temperature, humidity and Air Quality Index (AQI) along with Global Positioning System (GPS) coordinates. This information is then stored in the cloud and can be accessed through a Web browser whenever necessary.

Botta et al. [7] had focused their attention on the integration of cloud and IoT which is also known as the cloud–IoT paradigm. In many literature works, both cloud and IoT have been used separately. In order to bridge this research gap, many components of IoT like radio frequency identification (RFID) and wireless sensor networks (WSN) are integrated with cloud. This new cloud–IoT paradigm finds its application in various fields such as ambient-assisted living (AAL), electronic healthcare records (EHRs), smart cities, smart home and metering, video surveillance, smart logistics, smart mobility and automotive systems.

Yang et al. [8] had designed a platform that operates with intelligent robot systems, to improve the grasping efficiency of pick and place robots when they work in unstructured environments. All required experiments were carried out, and after analysing the data, it is concluded that the system had provided with better results.

Reddy et al. [9] had examined the design of a 6-DOF robotic arm, i.e., a pick and place arm in order to place a stack of sheet onto a shearing machine, thereby proposing an auto-feeding mechanism to improve the productivity effectively. The main objective of proposing such an autonomous mechanism is to reduce the risk attached with the manual stacking and de-stacking of sheets onto a machine. The mechanical design of the robot was made keeping in mind the functionality, reliability, motion range and speed, payload, reach and axes. The CAD design models of the links, the pneumatic cylinder and the joints were all made in CREO Parametric 3.0 and assembled using Autodesk Inventor 2017 software. The model was then exported to FEA to perform stress analysis using a dynamic simulation module. S–N curve was also plotted as a result of fatigue analysis.

Low cost and high precision robots are the order for the day in the modern industrial transformation. To achieve this, Noshahi et al. [10] had undertaken the design and fabrication of a low-cost 4-DoF pick and place manipulator. To tackle with problems related to precision and control, they selected SCARA configuration for the robot.

While most of the companies produce high load robotic arms, the development of light-weight and affordable robots is still an issue to be addressed. A 4-DoF robotic arm made of plastic with a gripper as an end-effector was developed for the purpose of sorting multiple materials [11]. Torque calculations were done, and appropriate servo motors were chosen for its operation. The model was designed in SolidWorks and then assembled. ANSYS software was used to analyse the deformation that the gripper may undergo under various conditions.

Many researchers have tried to depict the working mechanism of an industrial robot manipulator arm by studying robot kinematics and dynamics and simulating the link movements in software like MATLAB and Simulink. Also, recently the researchers are looking forward to propose novel ideas to interlink the existing robot technologies with the emerging networking technologies such as cloud and IoT. This study is also one such attempt of developing a simple robotic assembly line that can be made to work in a networked cloud-based system architecture, to improve major production constraints such as customer satisfaction, production time and productivity.

3 Detailed Design

The main objective of this work is to put forward an assembly line to serve the functionality of assembling parts together using fasteners and screws. To accomplish the operation, two robots are being planned to be used in the assembly line with first robot to place the screws at the required points on the workpiece and the second one to screw it down. All the aspects related to the designing of a robot from basic kinematic study to functional 3D model design for each of the robot in the assembly line have been performed. On finalising the prototype design, the models were analysed for their behaviour when acted upon mechanical load conditions.

3.1 *Kinematic Framework and Kinematic Study*

As the assembly line consists of two robots, forward kinematic study of the both the manipulators have been carried out. The configurations of the two robots are as follows:

(a) A 4-DoF SCARA type robot with the configuration—RRRP
(b) A 5-DoF pick and place robot arm with articulated robot configuration.

Forward Kinematic model of 4-DoF SCARA robot: Selective compliance assembly robotic arm (SCARA) is the most commonly deployed robot in assembly lines. As the name suggests, its primary function is to carry out assembly operations. In general, SCARA robots have four degrees of freedom, but some have reduced degrees of freedom. The usual configuration for a SCARA robot is RRPR, whereas to satisfy the requirements of the proposed assembly line, i.e., to drill down screws to the incoming workpiece, the robot arm has been designed to have configuration of RRRP. So, the robot would have three rotational joints and one linear movement.

Figure 1 shows the proposed 4-DoF SCARA type robot positioned in its home position, and Fig. 2 shows its frame assignments. From Fig. 2, it can be observed that three-axis frames are fixed at each joint and end-effector of the robot. Here [Z_b, X_b, Y_b] denote the base frame, and [Z_{ee}, X_{ee}, Y_{ee}] denote the end-effector frame.

The link and joint parameters also known as D-H parameters of the 4-DoF SCARA type robot are given in Table 1. Based on the D-H parameters in Table

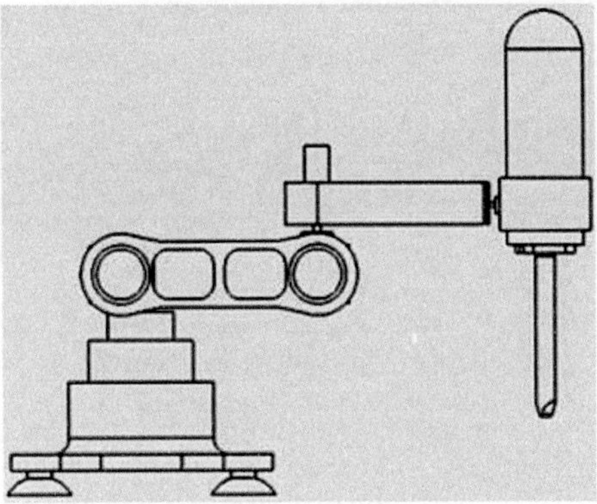

Fig. 1 4-DoF SCARA type robot positioned in its home position

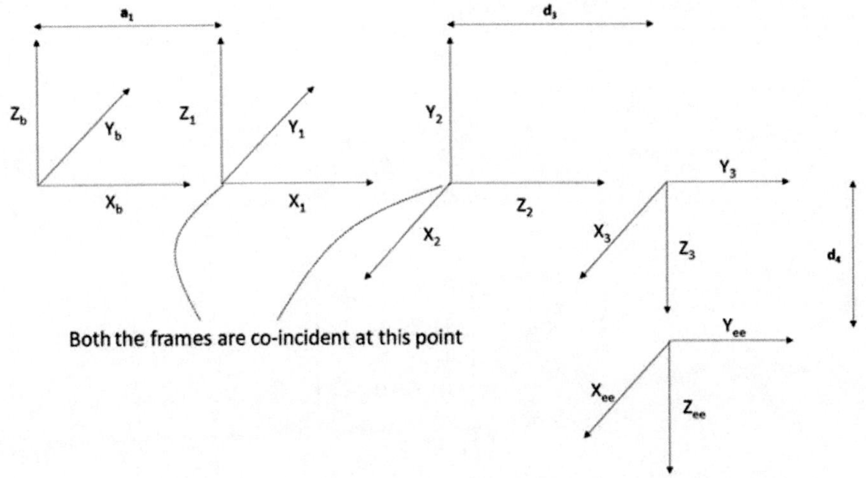

Fig. 2 Frame assignments for 4-DoF SCARA type robot

Table 1 D-H parameters of the 4-DoF SCARA type robot

Link	θ (°)	d (mm)	a (mm)	α (°)
1	θ_1	0	a_1	0
2	$\theta_{+2} + 90$	0	0	90
3	θ_3	d_3	0	90
4	0	d_4	0	0

1, D-H matrices corresponding to each coordinate frames are formulated. The four D-H matrices are ${}^0T_1$, ${}^1T_2$, ${}^2T_3$ and ${}^3T_4$.

The homogenous transformation matrix ${}^0T_4$ can be represented as product of all the four transformations. Hence

$$^0T_4 = ^0T_1 \cdot ^1T_2 \cdot ^2T_3 \cdot ^3T_4$$

$$^0T_4 = \begin{bmatrix} \cos\theta_1 & -\sin\theta_1 & 0 & a_1\cos\theta_1 \\ \sin\theta_1 & \cos\theta_1 & 0 & a_1\sin\theta_1 \\ 0 & 0 & 1 & 0 \\ 0 & 0 & 0 & 1 \end{bmatrix} \begin{bmatrix} -\sin\theta_2 & 0 & \cos\theta_2 & 0 \\ \cos\theta_2 & 0 & \sin\theta_2 & 0 \\ 0 & 1 & 0 & 0 \\ 0 & 0 & 0 & 1 \end{bmatrix}$$

$$\begin{bmatrix} \cos\theta_3 & 0 & \sin\theta_3 & 0 \\ \sin\theta_3 & 0 & -\cos\theta_3 & 0 \\ 0 & 1 & 0 & d_3 \\ 0 & 0 & 0 & 1 \end{bmatrix} \begin{bmatrix} 1 & 0 & 0 & 0 \\ 0 & 1 & 0 & 0 \\ 0 & 0 & 1 & d_4 \\ 0 & 0 & 0 & 1 \end{bmatrix}$$

$$^0T_4 = \begin{bmatrix} -\sin(\varphi)\cos\theta_3 & \cos(\varphi) & -\sin(\varphi)\sin\theta_3 & -d_4\sin(\varphi)\sin\theta_3 + d_3\cos(\varphi) + a_1\cos\theta_1 \\ \cos(\varphi)\cos\theta_3 & \sin(\varphi) & \cos(\varphi)\sin\theta_3 & d_4\cos(\varphi)\sin\theta_3 + d_3\sin(\varphi) + a_1\sin\theta_1 \\ \sin\theta_3 & 0 & -\cos\theta_3 & -d_4\cos\theta_3 \\ 0 & 0 & 0 & 1 \end{bmatrix} \quad (1)$$

where

$$\varphi = \theta_1 + \theta_2 \quad (2)$$

Therefore,

$$^RT_H = \begin{bmatrix} n_x & o_x & a_x & p_x \\ n_y & o_y & a_y & p_y \\ n_z & o_z & a_z & p_z \\ 0 & 0 & 0 & 1 \end{bmatrix} \quad (3)$$

$$^0T_4 = ^RT_H \quad (4)$$

From Eq. (3), the position coordinates of the manipulator end-effector is given by:

$$P_x = -d_4\sin(\varphi)\sin\theta_3 + d_3\cos(\varphi) + a_1\cos\theta_1 \quad (5)$$

$$P_y = d_4\cos(\varphi)\sin\theta_3 + d_3\sin(\varphi) + a_1\sin\theta_2 \quad (6)$$

$$P_z = -d_4\cos\theta_3 \quad (7)$$

The end-effector's orientation matrix is defined by the first three rows and three columns of the transformation matrix in Eq. (1).

Forward Kinematic model of 5-DoF pick and place robot: Pick and place robots are generally used for loading and un-loading purposes in manufacturing industries. In the proposed assembly line, a robot is planned to be used to pick the screws from a pallet and place them onto the incoming workpiece, wherever needed. The configuration for this robot is RRRRR which resembles an articulated arm configuration.

Figure 3 shows the proposed 5-DoF pick and place robot positioned in its home position, and Fig. 4 shows its frame assignments. From Fig. 4, it can be observed that three-axis frames are fixed at each joint and end-effector of the robot. Here [Z_b, X_b, Y_b] denote the base frame, and [Z_{ee}, X_{ee}, Y_{ee}] denote the end-effector frame.

D-H parameters of the 5-DoF pick and place robot are given in Table 2. Based on the D-H parameters in Table 2, D-H matrices corresponding to each coordinate frames are formulated. The five D-H matrices are ${}^0T_1$, ${}^1T_2$, ${}^2T_3$, ${}^3T_4$ and ${}^4T_5$.

The homogenous transformation matrix ${}^0T_5$ can be represented as product of all the four transformations. Hence

$$ {}^0T_5 = {}^0T_1 \cdot {}^1T_2 \cdot {}^2T_3 \cdot {}^3T_4 \cdot {}^4T_5 $$

$$ {}^0T_5 = \begin{bmatrix} \cos\theta_1 & -\sin\theta_1 & 0 & 0 \\ \sin\theta_1 & \cos\theta_1 & 0 & 0 \\ 0 & 0 & 1 & 0 \\ 0 & 0 & 0 & 1 \end{bmatrix} \begin{bmatrix} \cos\theta_2 & -\sin\theta_2 & 0 & 0 \\ \sin\theta_2 & \cos\theta_2 & 0 & 0 \\ 0 & 0 & 1 & 0 \\ 0 & 0 & 0 & 1 \end{bmatrix} $$

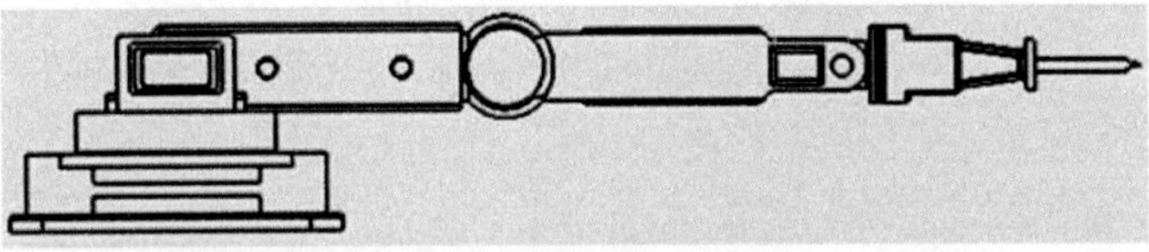

Fig. 3 DoF pick and place robot positioned in its home position

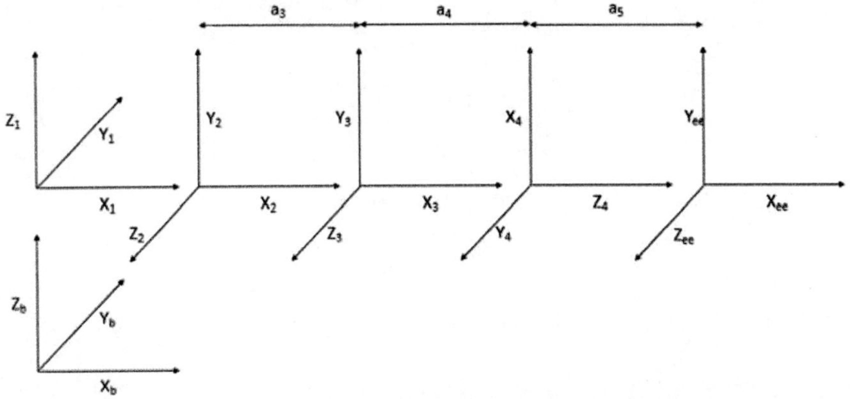

Fig. 4 Frame assignments for 5-DoF pick and place robot

Table 2 D-H parameters of the 5-DoF pick and place robot

Link	θ (°)	d (mm)	a (mm)	α (°)
1	θ_1	0	0	0
2	θ_2	0	0	0
3	θ_3	0	a_3	0
4	θ_4	0	a_4	90
5	θ_5	0	a_5	90

$$\begin{bmatrix} \cos\theta_3 & -\sin\theta_3 & 0 & a_3\cos\theta_3 \\ \sin\theta_3 & \cos\theta_3 & 0 & a_3\sin\theta_3 \\ 0 & 0 & 1 & 0 \\ 0 & 0 & 0 & 1 \end{bmatrix} \begin{bmatrix} \cos\theta_4 & 0 & \sin\theta_4 & a_4\cos\theta_4 \\ \sin\theta_4 & 0 & -\cos\theta_4 & a_4\sin\theta_4 \\ 0 & 1 & 0 & 0 \\ 0 & 0 & 0 & 1 \end{bmatrix}$$

$$\begin{bmatrix} \cos\theta_4 & 0 & \sin\theta_4 & a_4\cos\theta_4 \\ \sin\theta_4 & 0 & -\cos\theta_4 & a_4\sin\theta_4 \\ 0 & 1 & 0 & 0 \\ 0 & 0 & 0 & 1 \end{bmatrix}$$

$$^0T_5 = \begin{bmatrix} \cos\varphi\cos\theta_5 & \sin\varphi & \cos\varphi\sin\theta_5 & a_5\cos\varphi\cos\theta_5 + a_4\cos\varphi + a_3\cos\beta \\ \sin\varphi\cos\theta_5 & -\cos\varphi & \sin\varphi\cos\theta_5 & a_5\sin\varphi\sin\theta_5 + a_4\sin\varphi + a_3\sin\beta \\ \sin\theta_5 & 0 & -\cos\theta_5 & a_5\sin\theta_5 \\ 0 & 0 & 0 & 1 \end{bmatrix} \tag{8}$$

where

$$\varphi = \theta_1 + \theta_2 + \theta_3 + \theta_4 \text{ and } \beta = \theta_1 + \theta_2 + \theta_3 \tag{9}$$

Therefore,

$$^RT_H = \begin{bmatrix} n_x & o_x & a_x & p_x \\ n_y & o_y & a_y & p_y \\ n_z & o_z & a_z & p_z \\ 0 & 0 & 0 & 1 \end{bmatrix} \tag{10}$$

$$^0T_5 =^R T_H \tag{11}$$

From Eq. (10), the position coordinates of the manipulator end-effector is given by:

$$P_x = a_5\cos\varphi\cos\theta_5 + a_4\cos\varphi + a_3\cos\beta \tag{12}$$

$$P_y = a_5 \sin\varphi \sin\theta_5 + a_4 \sin\varphi + a_3 \sin\beta \tag{13}$$

$$P_z = a_5 \sin\theta_5 \tag{14}$$

The end-effector's orientation matrix is defined by the first three rows and three columns of the transformation matrix in Eq. (8).

3.2 *Design and Assembly of the Robots*

Design and assembly of links of 4-DoF SCARA type Robot: The 4-DoF SCARA type robot has four links and four joints in which the first three are rotational joints and the last one is a linear joint. Every link is attached to its preceding link through servo motors which are actuated to provide desired angular displacement for its corresponding links. The end-effector is an exception here, as it makes a linear displacement with respect to Link 3. It slides up and down through the respective link. All the links have been designed as required, and the final design assembly of the 4-DoF SCARA type Robot has been made in SolidWorks 2018, and the same is shown in Fig. 5.

The 4-DOF SCARA type robot has been designed in such a way that it can screw bolts placed on a plane parallel to the floor as well as a plane perpendicular to the floor. The working in both cases is shown in Fig. 6.

Design and assembly of links of 5-DoF Pick and place robot: The 5-DoF pick and place robot has five links and five revolute joints. Similar to the 4-DoF SCARA robot, every link is attached to its preceding link through servo motors which are actuated to provide desired angular displacement for its corresponding links. The final design assembly of the 5-DoF pick and place robot has been made in SolidWorks 2018 and is shown in Fig. 7.

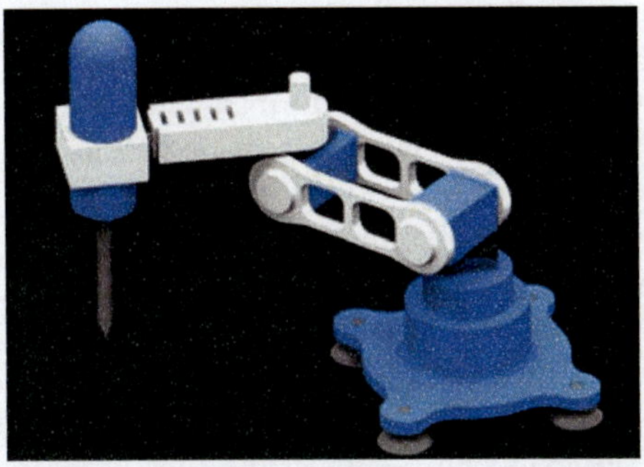

Fig. 5 CAD model of 4-DoF SCARA type robot

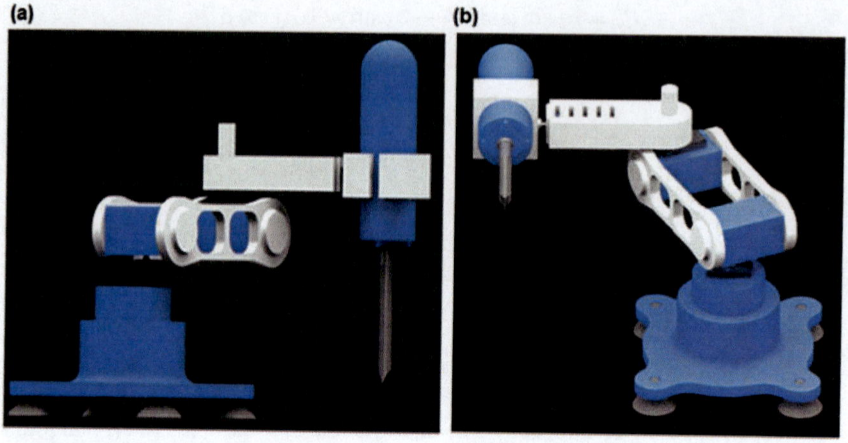

Fig. 6 **a** Screwing to a parallel plane; **b** Screwing to a perpendicular plane

Fig. 7 CAD model of 5-DoF pick and place robot

4 Simulations and Analysis

Static analysis emphasising on displacement, stress and strain analysis and buckling load analysis were performed on both the robot models making use of SolidWorks 2018. The loading conditions for the analysis of both the robots have been given in Tables 3 and 4.

The plots obtained after performing displacement, stress, strain and buckling load analysis on 4-DoF SCARA type robot are shown in Fig. 8. Similarly, the plots obtained after performing displacement, stress, strain and buckling load analysis on 5-DoF pick and place robot are given in Fig. 9.

From the displacement, stress and strain plots of both the robots, it can be inferred that the models are under the zone of least displacement, stress and strain, respectively. This shows that the models of the SCARA type robot and the pick and place robot work fine under the mechanical load conditions specified previously.

Table 3 Loading conditions for the analysis of 4-DoF SCARA type robot

Joint	Description	Magnitude	Direction
Floor-to-base	Fixture	NA	Upwards
Base-to-Link 1	Load–Torque	0.34 N cm	Counterclockwise
Link 1-to-Link 2	Load–Torque	0.34 N cm	Counterclockwise
Link 2-to-Link 3	Load–Torque	0.34 N cm	Counterclockwise
End-effector	Load–Force	2.94 N	Downwards
Gravity	Force	9.81 m/s^2	Downwards

Table 4 Loading conditions for the analysis of 5-DoF pick and place robot

Joint	Description	Magnitude	Direction
Floor-to-base	Fixture	NA	Upwards
Base-to-Link 1	Load–Torque	0.34 N cm	Counterclockwise
Link 1-to-Link 2	Load–Torque	0.34 N cm	Counterclockwise
Link 2-to-Link 3	Load–Torque	0.34 N cm	Counterclockwise
Link 3-to-Link 4	Load–Torque	0.34 N cm	Counterclockwise
Link 4-to-Link 5	Load–Torque	0.34 N cm	Counterclockwise
End-effector	Load–Force	2.94 N	Downwards
Gravity	Force	9.81 m/s^2	Downwards

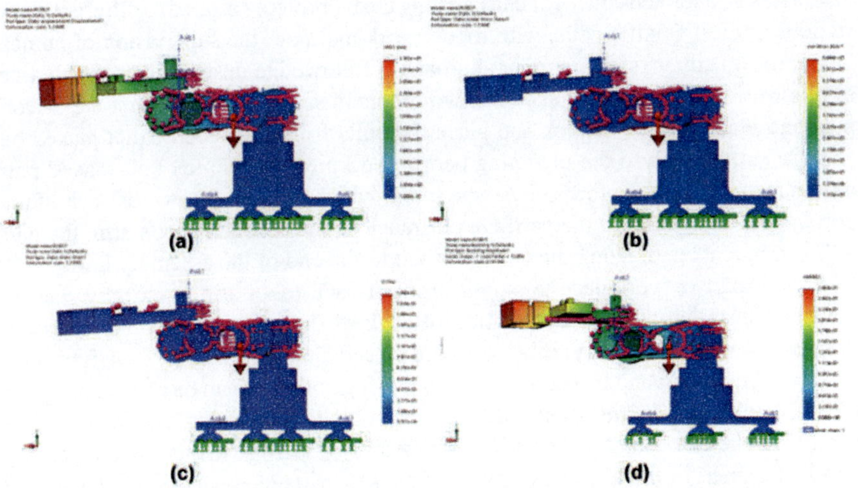

Fig. 8 **a** Displacement plot; **b** stress plot; **c** strain plot and **d** buckling load plot of SCARA type robot

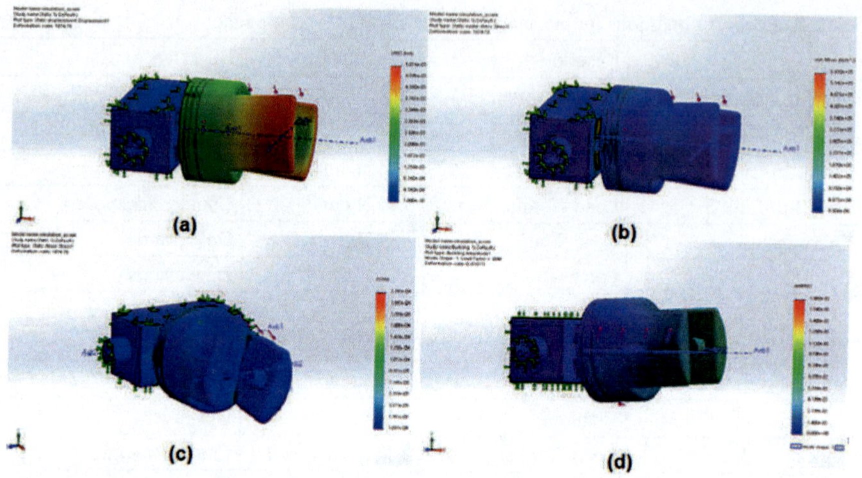

Fig. 9 **a** Displacement plot; **b** stress plot; **c** strain plot and **d** buckling load plot of pick and place robot

5 Discussion and Conclusions

Any assembly line or a production line in current manufacturing industries would have a motor-operated conveyor belt carrying the workpieces around to different work stations or even smaller cells, with robots working under the supervision of human labourers, on either side. A similar approach is followed to design the assembly line of this work. The idea is to place the two robots in such a way that, first the 5-DoF pick and place robot will pick and place the bolts from a wooden pallet placed on an adjacent table on to the incoming part or workpiece. Then, the bolt-loaded part moves towards the 4-DoF SCARA type robot, which will move towards each of the bolt and screw them into the part. On clearance of the second robot's arm, the part will continue its path along the conveyor belt to the end of the assembly line.

For the part or workpiece to stop in front of each robot arm, proximity sensors are to be installed at the very position of each of the robots. Assembly lines are designed to work either way; robot arms are placed along one side (left or right side) of the conveyor belt and inspection and monitoring activities can be carried out from the opposite side. The robots can also be placed on both the sides, and inspection if needed can be carried out at the end of the assembly line. Here in this work, assembly line is designed in the later way. Some of the key parts/elements in the design are (1) conveyor belt with side frame and rollers, and light indicators for direction of flow, (2) proximity sensors for each robot station and (3) platforms for the robots. All these, along with the robots are assembled in SolidWorks to provide an idea about the final production line, and the same is shown in Fig. 10.

Commercial industrial robots used in present day are manufactured by traditional techniques like forging, casting, sheet metal operations, etc. To cut down heavy manufacturing costs incurred in implementing these methods, latest technological

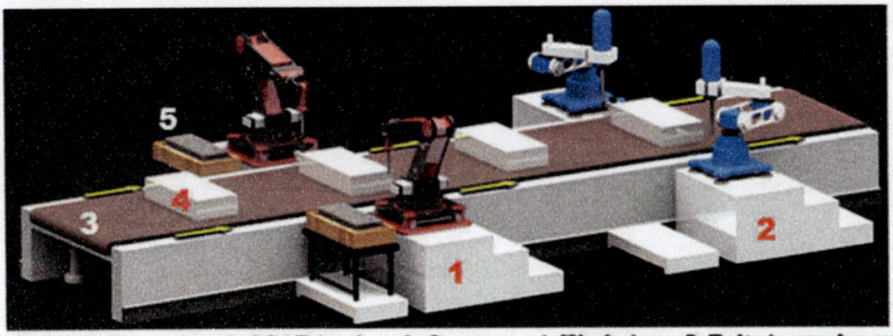

1. Pick and place robot, 2. SCARA robot, 3. Conveyor, 4. Workpiece, 5. Bolts in wooden pallet

Fig. 10 Proposed assembly line with 5-DoF pick and place robot and SCARA type robot

developments like additive manufacturing (AM) can be involved in fabrication of the robot links. Some of the commonly used AM techniques are stereolithography (SLA), polyjet, fused deposition modelling (FDM), 3D printing (3DP) and so on. Of these, FDM process is the recommended process for fabricating robot links, and the material predominantly used are polylactic acid (PLA) and acrylonitrile butadiene styrene (ABS).

In industry, to control the assembly line including the robots, programmable logic circuits (PLC) are used. For a smaller version, microcontrollers are most commonly preferred control blocks that run on previously coded and fed programs. To reduce the heavy storage problems dealt while using these pre-programmed robots, latest technological advancements like cloud computing, Internet of Things (IoT), software as a service (SAAS), etc., can be linked with existing robotics modules that gives rise to cloud robotics.

6 Conclusion

Cloud-controlled robots are employed in commercial industrial sectors to reduce the cost, time to market and production rates of new product development. Cloud robots serve as an extended source of computational power, storage and memory. Industries dealing with rapidly evolving products find it difficult to cope with existing traditional automated systems in order to meet customer expectations. By combining robots with cloud technologies, fast production of customised products can be achieved. In this work, a simple automated production line with a 4-DoF SCARA type robot and a 5-DoF pick and place robot has been designed. The two robots are intended to work together with an integrated interface to perform the basic assembly operation of screwing bolts to incoming workpieces. Forward kinematics analysis was performed on both the robots followed by detailed design of the links and assembly. The designed model of both the robots were simulated and analysed for stress, strain, deformation and buckling load under mechanical load conditions. The proposed assembly line has been designed in SolidWorks, and the need for cloud and IoT based automation has been discussed.

References

1. Golnazarian, W., Hall, E.L.: Intelligent industrial robots. In: IMTS Manufacturing Conference (2002)
2. Rojko, A.: Industry 4.0 concept: background and overview. Int. J. Interact. Mobile Technol. **11**(5), 77–90 (2017)
3. What is Cloud Robotics? https://roboearth.ethz.ch/cloud_robotics/index.html. Last accessed 5 June 2020
4. Mahil, S.M., Al-Durra, A.: Modeling analysis and simulation of 2-DOF robotic manipulator. In: IEEE 59th International Midwest Symposium on Circuits and Systems, pp. 1–4. Abu Dhabi (2016)
5. Ahmed, A.S.E., Abel-Lateef, M.K., et al.: Design, implementation and digital control of a robotic arm. B.Sc. Project Thesis, Faculty of Engineering, Computer & Systems Department, Minia University, Egypt (2009)
6. Prabha, S.S., Antony, A.J.P., et al.: Smart cloud robot using Raspberry Pi. In: 2014 International Conference on Recent Trends in Information Technology, pp. 1–5. Chennai (2014)
7. Botta, A., de Donato, W., et al.: Integration of cloud computing and internet of things: a survey. Futur. Gener. Comput. Syst. **56**, 684–700 (2016)
8. Yang, Q., Feng, B., et al.: Design and analysis of intelligent data acquisition system based on robotic arms. In: 3rd International Conference on Robotics and Automation Engineering, pp. 5–9. China (2018)
9. Reddy, G.R., Eranki, V.K.P.: Design and structural analysis of a robotic arm. Dissertation, Department of Mechanical Engineering, Blekinge Institute of Technology, Sweden (2017)
10. Noshahi, S.F., Farooq, A., et al.: Design and fabrication of an affordable SCARA 4-DOF robotic manipulator for pick and place object. In: 14th International Joint Symposium on Artificial Intelligence and Natural Language Processing, pp. 1–5. Thailand (2019)
11. Yi, K.Z., Razali, Z.B.: Design, analysis and fabrication of robotic arm for sorting of multi-materials. Project Report, School of Mechatronic Engineering, University of Malaysia Perlis (2013)

Impact of Dynamic Metrics on Maintainability of System Using Fuzzy Logic Approach

Manju and Pradeep Kumar Bhatia

Abstract In software development life cycle, software maintainability is the most important and crucial phase as it requires more effort as compared to other phases. To minimize the cost of software maintenance, it is necessary to predict software maintainability during the early phases of software development life cycle. Most of the metrics available in existing literature to calculate maintainability factor were static, i.e., based on design patterns. Therefore, in this paper, authors propose a fuzzy logic model to measure maintainability of software that takes static and dynamic metrics set as input and maintainability as output. Static metrics were collected using CodeMR tool, i.e., an eclipse plugin whereas dynamic event tracing was done by AspectJ, an implementation of aspect oriented programming on an eclipse platform. Further, the proposed fuzzy model was applied on 15 java classes to calculate maintainability at compile time and run time based on static and dynamic metrics. Lastly, proposed model was validated using analytical hierarchy processing (AHP) technique and it is concluded that proposed fuzzy model gives satisfactory result in case of dynamic metrics. Hence, it is concluded that dynamic metrics are better predictor of maintainability factor as compared to static metrics that helps the software industry to measure maintainability of software early in advance.

Keywords Fuzzy · AHP · Maintainability · AspectJ

1 Introduction

Software maintenance is the most crucial activity in the software development life cycle. It can consume around 70% of the cost of the entire life cycle [1]. Thus, a good maintainability prediction can reduce the cost of the software product. Therefore, evaluation of maintainability in the early phases of software development life cycle is very essential. Thus, if evaluation is done accurately, it can be useful in taking important decisions related to the software, efficiency of the maintenance process, comparing time effort, productivity, and costs among different projects. Metrics

Manju (✉) · P. K. Bhatia
Guru Jambheshwar University of Science &Technology, Hisar, India

H. Sharma et al. (eds.), *Congress on Intelligent Systems*, Advances in Intelligent Systems and Computing 1334, https://doi.org/10.1007/978-981-33-6981-8_63

[2] are measures to evaluate maintainability. There are many metrics available to calculate the maintainability factor of a software system. We choose four factors, namely complexity, coupling, cohesion, and size as input to our model to evaluate maintainability at compile time however to measure maintainability at runtime we choose run time values of these four factors. Software maintainability is a crucial factor to measure as most of the cost of software depends on this factor. However, we cannot predict the maintainability of a class by using crisp logic, i.e., yes or no. Therefore, for flexibility, we need fuzzy logic so that we can say maintainability of a class is low, medium, or high [3]. Hence, fuzzy logic provides us flexibility to predict the maintainability of a class in the form of low, medium, and high rather than crisp logic. Therefore, the main purpose of this paper is to evaluate software maintainability by using fuzzy-layered evaluation method and validate it by using analytic hierarchy process. AHP [4] is used since it helps to capture both subjective and objective evaluation measures. Static and dynamic metrics used in the current study are described in Table 1.

The rest of the paper is divided into four sections. Section 2 presents related work to the current study. Section 3 explains the proposed formula and proposed model to calculate maintainability. Section 4 presents experimental study and validation of proposed fuzzy model and lastly, Sect. 5 explains the conclusions and the future directions referenced to the current study.

Table 1 Static and dynamic metrics description

Static Metrics	Description
LCOM96a	number of attributes invoked by all the methods in a class at compile time [2]
WMC	Sum of the complexity of the methods of a class [5]
CBO	Count of the number of other classes to which it is coupled [5]
CLOC	The number of all nonempty, non-commented lines of the body of the class [6]
Dynamic Metrics	Description
DLCOM	$\frac{\left(\frac{1}{a}\sum_{i=1}^{m}\sum_{j=1}^{a}a_{ij}\right)-m}{1-m}+\frac{a}{m\sum_{i=1}^{m}\sum_{j=1}^{a}r_{ij}+m\sum_{i=1}^{m}\sum_{j=1}^{a}w_{ij}}+\frac{\sum_{i=1}^{m}\sum_{j=1}^{m}c_{ij}}{a*m}$ It is defined as combination of Runtime Lack of cohesion of methods (RLCOM), Runtime Attribute Access Rate (RAAR), and Runtime method method call (RMMC) where a is the total number of attributes accessed at runtime, m is the total number of times methods of a class executed at runtime, and a_{ij} is the count of method j executed at runtime access attribute *i*. r_{ij} is count of number of times method *i* reads attribute *j* at runtime, and w_{ij} is count of number of times method *i* writes attribute j at runtime, c_{ij} is a count of a method *i* calling another method *j* at runtime [7]
DWMC	Number of times methods of a class executed at runtime [8]
DCBO	Count of the number of other classes to which it is coupled at runtime [8]
DLOC	Total number of times lines of a class executed at runtime [8]

2 Related Work

Many empirical studies have been conducted to validate the strong relationship between design metrics and respective maintainability. These studies are useful as it helps in predicting the maintainability of the software in the early phases of the software project development. Jin and Liu [9], have taken Li and Henry, CK metrics, and size metrics as input and applied support vector machine (SVM) and clustering technique to find the maintainability of software. Al-Jamimi et al. [10] and Aljamaan et al. [11] also took the same set of metrics taken by Jin and Liu, 2010 as input. Al-jamimi proposed the fuzzy logic model and Aljamaan applied various machine learning algorithms, i.e., multilayer perceptron (MLP), radial basis function network (RBF), support vector machines (SVMs), and M5 model tree (M5P) algorithms to find the maintainability factor of software. Sharma and Chug [8] found the maintainability of software based on static and dynamic metrics by using multilayer perceptron (MLP), linear regression, SMOreg and Gaussian process classification algorithms and concluded that dynamic metrics are superior than static metrics to estimate maintainability factor of software. Kumar and Rath [12] proposed a neuro-fuzzy model to find maintainability factor based on same set of metrics taken by Jin and Liu, and concluded that the proposed neuro-fuzzy model is better than various existing models with least errors. Hence, from the literature study, authors concluded that most of the research is done in the area of static metrics and very little emphasis is given to dynamic metrics. Therefore, in this paper, authors tried to find the maintainability of software based on static and dynamic metrics with the help of fuzzy logic approach.

3 Proposed Work

3.1 Proposed Formula

A formula is proposed based on metrics described in Table 1. Our approach is to derive a formula to measure maintainability of a class based on following principles:

- Excessive complexity indicates class is highly complex and hard to maintain. Therefore, WMC and DWMC have a negative impact on class maintainability [13].
- High lack of cohesion indicates class subdivision is not accurate so decrease maintainability of software. Therefore, LCOM and DLCOM have a negative impact on class maintainability.
- High CBO means more complex design. Hence more time and effort required to maintain the software system. Therefore, CBO and DCBO have a negative impact on class maintainability.

- A very high count of class size might indicate that a class or method is trying to do much more work and should be split up. It also indicates that the class might be hard to maintain. Therefore, CLOC and DLOC have a negative impact on class maintainability.

From above-discussed principles, we can conclude that maintainability is inversely proportional to class complexity, coupling, cohesion, and sizes as shown below:

$$\mathrm{M_t} \propto 1/(\text{Lack of Cohesion} + \text{Coupling} + \text{Complexity} + \text{Size})$$

Therefore, maintainability of a class can be calculated by using the formula given below,

Maintainability of a class based on static metrics

$$(M_s) = 1/(w_1 * \ \mathrm{WMC} + w_2 * \mathrm{LCOM}(96a) + w_3 * \mathrm{CBO} + w_4 * \mathrm{CLOC}) \quad (1)$$

Maintainability of a class based on dynamic metrics

$$(M_D) = 1/(w_1 * \mathrm{DWMC} + w_2 * \mathrm{DLCOM} + w_3 * \mathrm{DCBO} + w_4 * \mathrm{DLOC}) \quad (2)$$

where w_1,w_2, w_3, and w_4 are the weights to be calculated by AHP technique described in Sect. 4.

3.2 *Proposed Fuzzy Logic Model*

Proposed fuzzy logic model, Mamdani's Fuzzy Inference System (MFIS) took complexity, cohesion, coupling, and size of a class as input and maintainability as output. We have taken three terms like low, medium, and high and four values as input to the system so according to the formula, 3^4 rules were developed [3]. Hence, a total of 81 rules were generated for proposed MFIS. After fuzzification process was completed, we took the fuzzy sets for output variable that required defuzzification. For defuzzification, the input was fuzzy set and output was a singleton value. The triangular membership function was used to define the values of membership function as low, medium, and high for each metric. The proposed model is shown in Fig. 1.

There were three triangular membership functions defined as low, medium, and high for factors taken as input. In this MFIS, four factors, i.e., complexity, cohesion, coupling, and size were taken as input and maintainability was taken as output. Therefore, 3^4 rules were generated. The proposed MFIS and the rule viewer of the proposed MFIS are shown in Fig. 2a, b, respectively.

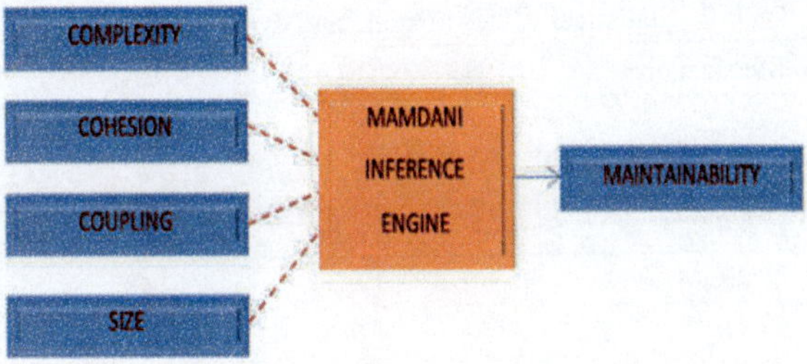

Fig. 1 Proposed fuzzy model

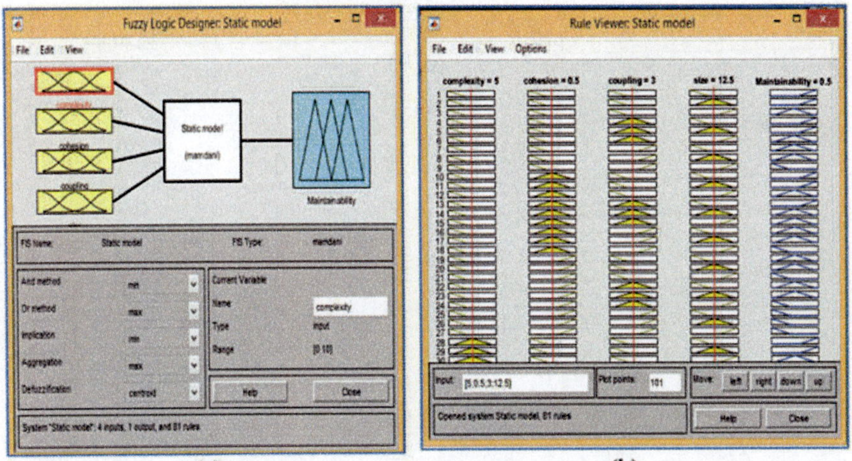

(a) (b)

Fig. 2 **a** Proposed Mamdani's fuzzy inference system. **b** Rule viewer of proposed model

4 Experimental Study

An experimental study was conducted on 15 classes coded in Java taken from Web [14]. Static metrics were extracted from CodeMR [6] tool, i.e., an eclipse plugin [15] and dynamic metrics were extracted by using AspectJ [16], an implementation of aspect oriented programming on an eclipse platform. The statistical values of static and dynamic metrics, computed with using MATLAB tool are shown in Table 2.

The proposed model was applied on 15 java classes and maintainability of classes was calculated. The results are shown in Table 3, whereas W denotes WMC, *D* denotes DWMC, L_s denotes LCOM, DL_D denotes DLCOM, C denotes CBO, CL denotes CLOC, DL denotes DLOC, F_S denotes maintainability calculated by the fuzzy model based on static metrics, and F_D denotes maintainability calculated by the fuzzy model based on dynamic metrics. Coupling between objects and dynamic coupling between objects metric values was same for static and dynamic metrics; therefore, only CBO is mentioned in Table 3.

Table 2 Statistical values of static and dynamic metrics

Metric name	MIN	MAX	MEAN	MEDIAN	STD.DEV
WMC	3	9	5.53	5	2.09
DWMC	2	24	8.53	6	6.47
LCOM	0	0.66	0.23	0.125	0.26
DLCOM	0	1	0.48	0.5	0.36
CBO	0	6	0.93	0	1.66
DCBO	0	6	0.93	0	1.66
CLOC	7	25	14.46	14	5.09
DLOC	4	50	18.13	10	14.04

Table 3 Static and dynamic maintainability evaluated by proposed MFIS

Class name	W	D	L_S	DL_D	C	CL	DL	F_S	F_D
Manager	3	2	0	0.5	1	7	4	0.68	0.85
Saving Account	5	12	0.25	0.23	0	16	26	0.55	0.16
IglooHouse	6	5	0	0.027	1	12	10	0.60	0.59
Car	9	24	0.583	0.653	0	19	38	0.39	0.14
Circle	4	3	0	1	0	9	7	0.84	0.50
Stack	7	16	0.125	0.275	0	20	22	0.49	0.16
Employee	5	6	0	0.314	0	20	0	0.54	0.15
Department	7	6	0.66	0.88	0	15	10	0.46	0.42
Notification Collection	4	5	0.5	0.73	3	14	22	0.50	0.24
TwoDShape	6	2	0.33	0.5	0	10	5	0.53	0.78
Cricket data	9	14	0.625	0.88	1	25	37	0.15	0.17
Square	8	4	0	1	0	14	7	0.50	0.48
ShoppingCart	3	4	0	0	2	14	10	0.55	0.60
CarFactory	4	6	0	0	6	15	10	0.50	0.41
Counter-Terrorist	3	9	0.5	0.32	0	7	14	0.60	0.27

4.1 Validation Using AHP

The proposed fuzzy model was validated by using standard analytic hierarchy process (AHP) technique given by Saaty [4]. Table 4 shows the calculated values of four factors, i.e., complexity, cohesion, coupling, and size. If the value of consistency index is less than 0.1, then the decision values of four factors were accepted; otherwise, it is rejected.

Consistency Index (CI) = $(\lambda_{max} - n)/(n - 1)$ where $n = 4$.

Table 4 AHP decision values

Factors	Complexity	Cohesion	Coupling	Size	Eigen vector (*w*)
Complexity	1	1/3	3	7	0.26
Cohesion	3	1	7	9	0.60
Coupling	1/3	1/7	1	3	0.10
Size	1/7	1/9	1/3	1	0.04
Total					1.000

$$= (4.07 - 4)/3 = 0.023$$

Consistency Ratio (CR) = 0.023/0.88 = 0.026 < 0.1, for $n = 4$.

Index of consistency = 0.88 [4]. Hence, decision values were accepted.

The decision values came out using AHP technique became corresponding weight values for these metrics, i.e., $w_1 = 0.26$, $w_2 = 0.60$, $w_3 = 0.10$, and $w_4 = 0.04$. Thus, by putting these weight values i.e., w_1, w_2, w_3, w_4in Eqs. (1) and (2) described earlier in Sect. 3.1, the value of maintainability was calculated for static and dynamic metrics. After that, values obtained from the proposed fuzzy model and AHP technique were compared and relative error (RE) and relative square error (RSE) was calculated as shown in Table 5. In Table 5, F_S and F_D denote maintainability calculated by the fuzzy model based on static and dynamic metrics, respectively. M_S and M_D denote

Table 5 Comparison of proposed fuzzy model and AHP calculated maintainability factor

Class name	F_S	M_S	RE	RSE	F_D	M_D	RE	RSE
Manager	0.68	0.86	0.16	0.0256	0.85	0.92	0.07	0.0049
SavingAccount	0.55	0.45	0.11	0.0121	0.16	0.23	0.007	0.0049
IglooHouseBuilder	0.60	0.47	0.13	0.0169	0.59	0.55	0.04	0.0016
Car	0.39	0.29	0.10	0.01	0.14	0.12	0.02	0.0004
Circle	0.84	0.71	0.13	0.0169	0.50	0.60	0.10	0.01
Stack	0.49	0.37	0.12	0.0144	0.16	0.19	0.03	0.0009
Employee	0.54	0.48	0.06	0.0036	0.15	0.18	0.01	0.001
Department	0.46	0.35	0.11	0.0121	0.42	0.40	0.02	0.0004
NotificationCollection	0.50	0.45	0.05	0.0025	0.24	0.34	0.10	0.01
TwoDShape	0.53	0.46	0.07	0.0049	0.78	0.78	0.00	00
Cricket data	0.15	0.26	0.11	0.0121	0.17	0.17	0.00	00
Square	0.50	0.38	0.12	0.0144	0.48	0.52	0.04	0.0016
ShoppingCartVisitor	0.55	0.65	0.10	0.01	0.60	0.61	0.01	0.001
CarFactory	0.50	0.45	0.05	0.0025	0.41	0.39	0.02	0.0004
Counter-Terrorist	0.60	0.74	0.14	0.0196	0.27	0.25	0.02	0.0004

maintainability calculated by AHP technique based on static and dynamic metrics, respectively.

4.2 Analysis Results

The maintainability calculated by the fuzzy model and AHP technique for static metrics was different than maintainability factor calculated by using dynamic metrics. Mean absolute error for compile time or static metrics is 0.104 and mean relative square error is 0.106 while in the case of dynamic metrics, MAE is 0.036 and RMSE [3] is 0.044, i.e., low as compared to static metrics. The results indicate that dynamic metrics are better predictor of maintainability of classes as compared to static metrics. Further, it is found that maintainability calculated by the fuzzy model using dynamic metrics and AHP technique using Eq. (2) was very close as compared to maintainability calculated by the fuzzy model using static metrics and corresponding values calculated by AHP technique by using Eq. (1) as shown in Fig. 3. Some observations drawn from the current study are described as follows:

- Complexity of a class increases at runtime in most of the cases except in the case where all the methods of a class did not execute at least once at run time.
- There were lots of variation between compile time and run time lack of cohesion metric values. In most of the cases, lack of cohesion metric value was 0 at compile time that indicates class is highly cohesive but at run time, values of these classes were not 0 and near to 1 that indicate class is not cohesive at runtime.

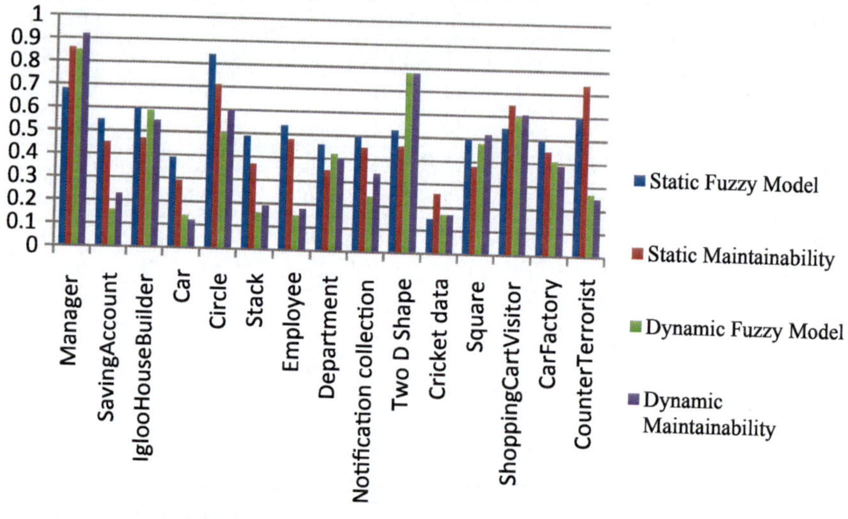

Fig. 3 Fuzzy maintainability versus AHP

Therefore, dynamic metrics predicts better as compared to static metrics to find maintainability of a class.

- Compile time and run time coupling metrics values were almost same.
- Size of a class at runtime was nearly equal to size of a class at runtime in most of the cases that indicates classes having good use of conditional statements in their methods have lesser line of code at runtime and methods of a class, that executes more than once at runtime have almost double line of code at runtime.

5 Conclusion

This paper proposed the MFIS to measure the maintainability of object-oriented software systems. The proposed model took four inputs, i.e., cohesion, coupling, complexity, size and one output, i.e., maintainability. We proposed a formula based on these inputs to evaluate the maintainability factor. A total of 81 rules were generated for the proposed MFIS. An experimental study was done on 15 java classes and maintainability was calculated by using static and dynamic metrics as input to proposed MFIS. Further, results were validated by using AHP technique and we observed that maintainability factor evaluated by proposed fuzzy model and maintainability measured by using AHP technique was almost same for all the 15 classes in the case of dynamic metrics. Therefore, it is concluded that dynamic metrics are better predictor of maintainability as compared to static metrics. This model is helpful for the software industry to predict the maintainability of object-oriented software systems in advance that would be very beneficial for developers too in reference to cost benefits. In the future, the model will be more refined by taking consideration of other object-oriented metrics and more number of projects. In this paper, we have taken only 15 java classes, so more work is required by taking large projects so that maintainability of a project can be calculated easily in advance.

Acknowledgements The author is grateful for support from the Guru Jambheshwar University of Science and Technology for providing lab facility to run Matlab tool and Prof. Pradeep kumar Bhatia for guiding me throughout during the study completion.

References

1. Pressman, R.S.: Software Engineering—A Practitioner's Approach, 7th edn. McGraw Hill (2005)
2. Sharma, T.: tusharma.in/technical/revisiting-lcom/. Last accessed 25 June 2020
3. Manju, Bhatia, PK.: Maintainability model of object oriented software based on fuzzy logic approach. Int. J. Eng. Sci. Invention (IJESI) **8**(5), 76–83 (2019)
4. Saaty, T.L.: How to make a decision: the analytic hierarchy process. Eur. J. Oper. Res. **48**, 9–26 (1990)
5. Chidamber, S.R., Kemerer, C.F.: A metrics suite for object oriented design. IEEE Trans. Software Eng. **20**(6), 476–493 (1994)

6. CodeMR guide (2020). https://www.codemr.co.uk/docs/codemr-intellij-userguide.pdf. Last accessed 25 June 2020
7. Manju, Bhatia, PK.: Measurement of dynamic cohesion using aspect oriented approach. Int. J. Res. Anal. Rev. (IJRAR) **6**(2), 438–432 (2019)
8. Sharma, H., Chug, A.: Dynamic metrics are superior than static metrics in maintainability prediction: An empirical case study. In: 4th International Conference on Reliability, Infocom Technologies and Optimization (ICRITO), IEEE, pp. 1–6 (2015). https://doi.org/10.1109/ICRITO.2015.7359354
9. Jin, C., Liu, J.A.: Applications of support vector machine and unsupervised learning for predicting maintainability using object oriented metrics. In: Second International Conference on Multimedia and Information Technology, pp. 24–27, Kaifeng, IEEE (2010)
10. Al-Jamimi, H., Ahmed, M.: Prediction of software maintainability using fuzzy logic. In: 3rd International Conference on Software Engineering and Service Science (ICSESS), pp. 702–705 (2012)
11. Aljamaan, H., Elish, MO., Ahmad, I.: An ensemble of computational intelligence models for software maintenance effort prediction. In: Advances in Computational Intelligence, pp. 592–603 (2013)
12. Kumar, L., Rath, S.K.: Software maintainability prediction using hybrid neural network and fuzzy logic approach with parallel computing concept. Int. J. Syst. Assur. Eng. Manage. **8**(1), 1487–1502. https://doi.org/10.1007/s13198-017-0618-4
13. Dubey, S.K., Rana, A.: A comprehensive assessment of object-oriented software systems using metrics approach. Int. J. Comput. Sci. Eng. **2**(8), 2726–2730 (2010)
14. Design patterns (2020). https://www.geeksforgeeks.org/chain-responsibility-design-pattern/. Last accessed 25 June 2020
15. Eclipse guide (2020). https://www.eclipse.org/aspectj/doc/next/progguide/printable.html. Last accessed 25 June 2020
16. AspectJ tutorial Homepage (2020). https://o7planning.org/en/10257/java-aspect-oriented-programming-tutorial-with-aspectj. Last accessed 25 June 2020

Author Index

H. Sharma et al. (eds.), *Congress on Intelligent Systems*, Advances in Intelligent Systems and Computing 1334, https://doi.org/10.1007/978-981-33-6981-8